Inequalities in Matrix Algebras

GRADUATE STUDIES
IN MATHEMATICS **251**

Inequalities in Matrix Algebras

Eric Carlen

AMERICAN
MATHEMATICAL
SOCIETY

Providence, Rhode Island

2020 *Mathematics Subject Classification.* Primary 47A63, 47C15, 46L60, 46L53.

Library of Congress Cataloging-in-Publication Data

Names: Carlen, Eric, 1957- author

Title: Inequalities in matrix algebras / Eric Anders Carlen.

Description: Providence, Rhode Island : American Mathematical Society, [2025] | Series: Graduate studies in mathematics, 1065-7339 ; 251 | Includes bibliographical references and index.

Identifiers: LCCN 2024061832 | ISBN 9781470479237 hardback | ISBN 9781470480264 paperback | ISBN 9781470480257 ebook

Subjects: LCSH: Matrix inequalities | AMS: Operator theory – General theory of linear operators – Operator inequalities | Operator theory – Individual linear operators as elements of algebraic systems – Operators in C^*- or von Neumann algebras | Functional analysis – Selfadjoint operator algebras (C^*-algebras, von Neumann (W^*-) algebras, etc.) – Applications of selfadjoint operator algebras to physics. | Functional analysis – Selfadjoint operator algebras (C^*-algebras, von Neumann (W^*-) algebras, etc.) – Noncommutative probability and statistics.

Classification: LCC QA188 .C367 2025 | DDC 512.9/434–dc23/eng/20250522

LC record available at https://lccn.loc.gov/2024061832

Graduate Studies in Mathematics ISSN: 1065-7339 (print); 2376-9203 (online)

DOI: https://doi.org/10.1090/gsm/251

This book was written in Lungavilla, Provincia di Pavia, and is dedicated to
Ester whose enthusiasm and encouragement have brought it to life.

Contents

Preface

Since Werner Heisenberg invented modern quantum mechanics in 1925, which Born and Jordan reformulated as "matrix mechanics" two months later, matrix algebras and matrix inequalities have been a fundamental part of the theory. The Heisenberg uncertainty principle is an inequality that expresses a fundamental limit on the accuracy to which position and momentum can be measured together, and it in turn is a consequence of the commutation relation linking these quantum observables, expressed as self-adjoint operators in a *matrix algebra.*

Schrödinger and Dirac soon provided other perspectives on quantum dynamics, Schrödinger in terms of what is now known as the Schrödinger equation. While the Schrödinger equation gives the evolution of the state of a quantum system, only rarely can it be solved for systems of physical interest. Moreover, due to the peculiar rules for performing measurements of quantum observables, even if one knows a system's exact state, one still does not know what result any particular measurement will yield. Partly for these reasons, inequalities are particularly important in quantum mechanics, especially in quantum statistical mechanics and quantum information theory. Often they provide the best available information on how solutions of the Schrödinger equation must behave and what can happen when a measurement is made.

The subject of this book has its roots in quantum mechanics, and quantum applications will be discussed, but the subject is mathematics: it is a book about a beautiful branch of mathematics that has grown up alongside quantum mechanics from the same beginnings about a century ago. To read this book, one need not know anything about quantum mechanics, and the mathematical prerequisites are minimal: good undergraduate courses in mathematical analysis, linear algebra, and elementary probability will suffice. All of the mathematics will be developed in full on these modest foundations. Along the way, ideas from quantum mechanics are introduced because while our subject is a branch of mathematics, it was shaped by the physics that grew up alongside it.

Our focus is on inequalities in algebras of operators on finite-dimensional Hilbert spaces, and thus in matrix algebras. On the one hand, for a large part of quantum information theory and quantum computing, this is the full realm of applications. In the finite-dimensional setting, the beautiful interplay between algebra and geometry that is the source of the inequalities to be studied can be brought into clear view without the obstruction of various technical difficulties: in finite dimensions every subspace of a Hilbert space is closed, there is no distinction between strong and weak topologies, etc. On the other hand, many of the results discussed here readily, or even trivially, extend to the infinite-dimensional case, and remarks and exercises show the way.

This is not a text on matrix analysis per se. Excellent texts by Bhatia [35] and by Horn and Johnson [115, 116], for example, cover the classic material in this subject in a thorough and clear manner. Nor is this a text on quantum information, quantum communication, or quantum computing per se. The books by Nielsen and Chuang [169], Wilde [232], and Kitaev, Shen, and Vyalyi [129] provide clear and comprehensive treatment of the mature parts of these topics. This is instead a text on a branch of mathematics that, nurtured by its applications, has in recent years matured into a well-articulated theory.

The material in Chapter 1 will be familiar to many readers, and most of it can be found in standard texts, but we include it for completeness and to establish notation. The system of notation used here is commonly used in mathematical physics and lies somewhere between what is generally used in physics and in pure mathematics. It may take some getting used to, but it bridges the literature produced by two communities that have shaped the subject.

Already in Chapter 2, the focus moves somewhat away from that of standard texts on matrix analysis. Multilinear algebra—tensor products of Hilbert spaces—and the notion of *entanglement* are fundamentally important in quantum mechanics.

Chapter 3 begins the serious work on proving inequalities and covers some fundamental ideas concerning monotonicity and convexity for functions of operators. This material can be found in various texts, but the aim here is to introduce the most important tools, from the beginning, but in relatively few pages.

Chapter 4 begins the treatment of matrix algebras proper, and it presents the fundamentals of the theory of von Neumann algebras on finite-dimensional Hilbert spaces and includes the structure theorem for them, and it begins the study of *conditional expectations* onto a sub-von Neumann algebra of another von Neumann algebra. This subject is closely connected with the quantum theory of measurement, which is introduced here, along with the version of Naimark's dilation theorem that is relevant for finite quantum systems. However, the main application of the results of this chapter will come after various species of positive maps are introduced in Chapter 5.

Chapter 5 begins the study of various types of positivity preserving maps from one von Neumann algebra to another, the most important being the class of *completely positive maps*. Completely positive maps that preserve traces are known as *quantum operations*, and as explained here following the work of Kraus, they describe the general change that can be made on the state of a quantum system through physical processes. The notion of complete positivity is due to Stinespring, as is the structure theorem for such maps. This is proved here along with some fundamental inequalities associated to completely positive maps. One of these is a so-called Schwarz inequality, originally due to Kadison, and extended by Choi. It says that if Φ is a completely positive unital map, then $\Phi(A^*A) \geq \Phi(A)^*\Phi(A)$ for all A in the von Neumann algebra $\mathcal{M}$ on which Φ is defined. As Choi showed, the set of A such that both A and A^* yield equality in the Schwarz inequality is an algebra. As Lindblad showed, analysis of the structure of this algebra leads to a fundamental theorem of quantum mechanics, the *no-cloning theorem*, which is proved and discussed in this chapter.

Chapters 6 and 7 delve more deeply into proving trace inequalities and entropy inequalities in particular. The centerpiece result here is the *strong subadditivity of the quantum entropy*, SSA, which had been conjectured by Lanford and Robinson in 1968, and was proved by Lieb and Ruskai in 1973, building on work, also from 1973, in which Lieb had solved the 1963 Wigner-Yanase-Dyson conjecture. The methods introduced in these breakthrough results had a profound impact on research in this subject, as will be made clear in Chapters 8, 9, and 10. A number of the inequalities proved in Chapters 6 and 7 can be proved by various other methods, especially the majorization methods that are discussed in Chapter 11, and this is the favored approach in most matrix analysis texts. Here instead, we rely on convex geometry methods and variational formulas in particular. These methods have proved to be very powerful for the kinds of problems that will arise in later chapters.

Chapter 8 is devoted to consequences and refinements of SSA. In particular, there is a full treatment of the determination of the cases of equality, a result of Hayden, Josza, Petz, and Winter. The proof makes essential use of the structure theory for von Neumann algebras on a finite-dimensional Hilbert space, and other ideas introduced in connection with the no-cloning theorem. This is another case in which operator algebraic methods are inextricably linked with the analysis of inequalities. Also in Chapter 8 we give a proof by Petz of SSA that is operator algebraic in character and which is fundamental for determining the cases of equality in SSA.

Chapter 9 is devoted to the quantitative study of *entanglement*, the property of quantum mechanics that leads to what Einstein called "spooky action at a distance". Although in 1935 Schrödinger identified it as the fundamental feature of quantum mechanics that separates it from classical mechanics, it was not until the work of Bell in the 1960s that efforts to quantify and measure its effects began to be made. The 2022 Nobel Prize in Physics was awarded to Aspect, Clauser, and Zellinger for their theoretical and experimental work to prove the reality of strange effects of entanglement to which Einstein had so strongly objected. One could write an entire book on this subject alone. Chapter 9 gathers a number of the most important results of a theoretical nature and is intended to pave the way to the study of the many open question in this area.

Chapter 10 treats a range of convexity, concavity, and monotonicity inequalities, and it discusses the relations between them. The chapter draws on recent advances in the use of variational methods for proving such inequalities.

Chapter 11 covers majorization methods, again in a self-contained manner. While this chapter presents some new proofs of results obtained earlier, it introduces a fundamental method due to Weyl for proving matrix inequalities. Weyl's method was used by Araki to extend the *Lieb-Thirring trace inequality* into a new range of exponents, producing what is now known as the *Araki-Lieb-Thirring inequality*. Though other methods give part of the range of exponents treated by Araki, no other method gives them all. Some recent multivariate extensions of the Araki-Lieb-Thirring inequality are proved in this chapter, using Weyl's method as one ingredient, the other being complex interpolation.

Chapter 12 provides an introduction to the Tomita-Takesaki theory of von Neumann algebras. The Tomita-Takesaki theory was developed to treat the decidedly infinite-dimensional Type III von Neumann algebras, and this theory provides the framework for transplanting results concerning matrix algebras into a truly general setting. Indeed, immediately after Lieb proved the fundamental *Lieb concavity theorem*, a fundamental trace inequality that resolved the Wigner-Yanase-Dyson conjecture, Araki developed a trace-free von Neumann algebra version (reducing exactly to Lieb's result in the matricial case) and, moreover, he even used Lieb's method of proof, adapted to the machinery of Tomita-Takesaki theory. Some years later, Petz realized that the methods of Tomita-Takesaki theory provides a powerful means of proving new inequalities in the finite-dimensional case. The final chapter of this book treats Tomita-Takesaki theory in our finite-dimensional setting and illustrates its application in this setting. The aim is to provide a clear picture of the content of the theory, and how it may be profitably applied in the finite-dimensional setting. It may also serve to provide a motivated introduction to the full theory for those who are inclined to tackle the technical difficulties inherent in dealing with Type III von Neumann algebras.

In writing this book, I have aimed for a clean and clear presentation of the theory and its many methods. A number of theorems are proved several ways to illustrate various methods. Fundamental results such as SSA are proved in various different ways in different chapters. It took some time for the new ideas that went into the proof of SSA—and the operator algebraic developments of the 1970s—to be put into a context encompassing them all. This began to be done in the second half of the 1970s by Ando, Choi, Uhlmann, Pusz, and Woronowicz, among others. Their work led to beautiful extensions of the breakthrough results of the early 1970s with beautifully simple proofs, and better yet, more closely connected with Kraus's ideas on quantum operations. These developments continued, and by now the context appears clear.

A distinguishing feature of the presentation in this book, even in the treatment of the more classical material, is that I have made maximal use of convex geometry and Legendre transforms, in particular as a tool for giving simple proofs of many deep inequalities. In some ways, it would have been natural to cover majorization methods early in the book, say, just before Chapter 4 on von Neumann algebras. I have not done so because this is a textbook designed to be useful for a one-semester course, and in such a course it is necessary to arrive at the material on completely positive maps in Chapter 5 before too late in the semester. However, Chapter 11 has been written so that it could be taught much earlier.

Appendix A treats convex geometry, and it is written so that it can be used as a self-contained primer on convex geometry, and in a course it might be used that way. Appendix B treats complex interpolation theory, supporting the applications made of this in Chapters 6 and 11.

Each chapter is followed by a set of exercises, over 150 in total. Some are simple warm-up exercises, but others develop applications of the theory that perhaps deserved treatment here, but for the fact that this would have added considerably to the length of the book. The exercises are an integral part of the text.

I am grateful to a number of colleagues for their comments and suggestions, particularly Nilanjana Datta, Rupert Frank, Elliott Lieb, Brian Kennedy, Suleyman Ulusoy, and Haonan Zhang. I would like to particularly thank two Rutgers graduate students, David Hererra and Trung Nghia Nguyen, who have read drafts carefully, corrected many errors, and provided many useful suggestions.

I was fortunate to have been introduced to this subject by Elliott Lieb, with whom I have worked for many years on many papers. I am deeply grateful for our many conversations, and these conversations have informed my perspective on the subject presented here.

Chapter 1

Hilbert space basics

1.1. Inner product spaces and Hilbert spaces

Let V be a vector space over the complex numbers $\mathbb{C}$. For now we make no assumption about the dimension of V. An **inner product** on V is a function on $V \times V$ that sends $(\mathbf{x}, \mathbf{y}) \in V \times V$ to a complex number $\langle \mathbf{x}, \mathbf{y} \rangle$ that is linear in $\mathbf{y}$ and conjugate linear in $\mathbf{x}$, and is such that $\langle \mathbf{x}, \mathbf{x} \rangle > 0$ for all nonzero $\mathbf{x} \in V$. An **inner product space** $(V, \langle \cdot, \cdot \rangle)$ is vector space V equipped with an inner product $\langle \cdot, \cdot \rangle$. Vectors $\mathbf{x}, \mathbf{y} \in V$ are **orthogonal** in case $\langle \mathbf{x}, \mathbf{y} \rangle = 0$. We shall also encounter real inner product spaces, in which case V is a vector space over $\mathbb{R}$, and $\langle \mathbf{x}, \mathbf{y} \rangle$ is a bilinear form on V with $\langle \mathbf{x}, \mathbf{x} \rangle > 0$ for all nonzero $\mathbf{x} \in V$. For the rest of this chapter, we focus on the complex case, and inner product spaces are complex unless they are explicitly indicated to be real.

Example 1.1. Let $\mathbb{C}^n$ be the vector space of all vectors $\mathbf{z} = (z_1, \ldots, z_n)$ with n complex entries, and the usual rules for addition and scalar multiplication. The **standard inner product** on $\mathbb{C}^n$ is given by $\langle \mathbf{z}, \mathbf{w} \rangle = \sum_{j=1}^{n} \overline{z_j} w_j$. Note that $\langle \mathbf{z}, \mathbf{z} \rangle = \sum_{j=1}^{n} |z_j|^2$ from which is easy to see that $\langle \mathbf{z}, \mathbf{z} \rangle > 0$ for all nonzero $\mathbf{z}$. It is easy to check that this is an inner product on $\mathbb{C}^n$.

Given an inner product space $(V, \langle \cdot, \cdot \rangle)$, define a function $\| \cdot \|$ on V with values in $[0, \infty)$ by

$$\|\mathbf{v}\| := \langle \mathbf{v}, \mathbf{v} \rangle^{1/2}.$$

This function is called the **norm associated to the inner product**. A **unit vector** is a vector $\mathbf{u}$ such that $\|\mathbf{u}\| = 1$. Since $\|\mathbf{x} + \mathbf{x}\|^2 = \|\mathbf{x}\|^2 + \|\mathbf{y}\|^2 + \langle \mathbf{x}, \mathbf{y} \rangle + \langle \mathbf{y}, \mathbf{x} \rangle$, $\Im(\langle \mathbf{x}, \mathbf{y} \rangle) = -\Im(\langle \mathbf{y}, \mathbf{x} \rangle)$. Replacing $\mathbf{y}$ by $i\mathbf{y}$, $\Re(\langle \mathbf{x}, \mathbf{y} \rangle) = \Re(\langle \mathbf{y}, \mathbf{x} \rangle)$. Hence for all $\mathbf{x}, \mathbf{y}$,

$$\overline{\langle \mathbf{x}, \mathbf{y} \rangle} = \langle \mathbf{y}, \mathbf{x} \rangle.$$

The fundamental inequality relating the norm and the inner product is the **Cauchy-Schwarz inequality**.

Theorem 1.2 (Cauchy-Schwarz Inequality). *Let* $(V, \langle \cdot, \cdot \rangle)$ *be an inner product space. For all* $\mathbf{x}, \mathbf{y} \in V$,

$$(1.1.1) \qquad |\langle \mathbf{x}, \mathbf{y} \rangle| \leq \|\mathbf{x}\|\|\mathbf{y}\|,$$

and when neither $\mathbf{x}$ *nor* $\mathbf{y}$ *is zero, there is equality in* (1.1.1) *if and only if* $\mathbf{x}$ *is a complex multiple of* $\mathbf{y}$, *or equivalently,* $\mathbf{y}$ *is a complex multiple of* $\mathbf{x}$.

Proof. We may suppose that both $\mathbf{x}$ and $\mathbf{y}$ are nonzero. Define $\mathbf{u} := \|\mathbf{x}\|^{-1}\mathbf{x}$ and $\mathbf{v} := e^{i\theta}\|\mathbf{y}\|^{-1}\mathbf{y}$, where $\theta \in [0, \pi]$ is such that $\langle \mathbf{u}, \mathbf{v} \rangle = |\langle \mathbf{u}, \mathbf{v} \rangle|$. Then $\|\mathbf{u} - \mathbf{v}\|^2 = \|\mathbf{u}\|^2 + \|\mathbf{v}\|^2 - 2|\langle \mathbf{u}, \mathbf{v} \rangle|$, so that

$$(1.1.2) \qquad |\langle \mathbf{u}, \mathbf{v} \rangle| = 1 - \frac{1}{2}\|\mathbf{u} - \mathbf{v}\|^2.$$

Multiplying both sides of (1.1.2) through by $\|\mathbf{x}\|\|\mathbf{y}\|$ yields

$$|\langle \mathbf{x}, \mathbf{y} \rangle| \leq \|\mathbf{x}\|\|\mathbf{y}\| - \frac{1}{2}\left\|\|\mathbf{y}\|\mathbf{x} - e^{i\theta}\|\mathbf{x}\|\mathbf{y}\right\|^2.$$

This proves (1.1.1), and there is equality if and only if $\|\mathbf{y}\|\mathbf{x} = e^{i\theta}\|\mathbf{x}\|\mathbf{y}$, In this case, when both $\mathbf{x}$ and $\mathbf{y}$ are nonzero, $\mathbf{x}$ and $\mathbf{y}$ are complex multiples of one another, and one readily checks that when $\mathbf{x}$ and $\mathbf{y}$ are complex multiples of one another, there is equality in (1.1.1). $\qquad\square$

Remark 1.3. If the inner product space $(V, \langle \cdot, \cdot \rangle)$ is real, the only change is that the condition for equality simplifies to $\|\mathbf{y}\|\mathbf{x} = \pm\|\mathbf{x}\|\mathbf{y}$, with essentially the same proof.

The norm on an inner product space satisfies **Minkowski's inequality**:

Theorem 1.4 (Minkowski's inequality for an inner product space). *Let* $(V, \langle \cdot, \cdot \rangle)$ *be an inner product space. For all* $\mathbf{x}, \mathbf{y} \in V$,

$$(1.1.3) \qquad \|\mathbf{x} + \mathbf{y}\| \leq \|\mathbf{x}\| + \|\mathbf{y}\|.$$

If both $\mathbf{x}$ *and* $\mathbf{y}$ *are nonzero, there is equality if and only if* $\mathbf{x} = c\mathbf{y}$ *for some* $c \geq 0$.

Proof. We may assume that $\mathbf{x} + \mathbf{y} \neq 0$. Define $\mathbf{u} := \|\mathbf{x} + \mathbf{y}\|^{-1}(\mathbf{x} + \mathbf{y})$, and note that $\|\mathbf{u}\| = 1$. Then

$$\|\mathbf{x} + \mathbf{y}\| = \|\mathbf{x} + \mathbf{y}\|^{-1}\langle \mathbf{x} + \mathbf{y}, \mathbf{x} + \mathbf{y} \rangle = \langle \mathbf{u}, \mathbf{x} \rangle + \langle \mathbf{u}, \mathbf{y} \rangle.$$

By the Cauchy-Schwarz inequality, $\langle \mathbf{u}, \mathbf{x} \rangle + \langle \mathbf{u}, \mathbf{y} \rangle \leq \|\mathbf{x}\| + \|\mathbf{y}\|$. The statement about cases of equality follows from the statement about cases of equality in the Cauchy-Schwarz inequality. $\qquad\square$

The norm $\| \cdot \|$ on an inner product space $(V, \langle \cdot, \cdot \rangle)$ determines a metric on V. We recall some relevant definitions:

Definition 1.5 (Metric spaces and completeness). A **metric** $d_{\mathcal{X}}$ on a set $\mathcal{X}$ is a function on $\mathcal{X} \times \mathcal{X}$ such that

   (i) $d_{\mathcal{X}}(x, y) \geq 0$ for all $x, y \in \mathcal{X}$ with equality if and only if $x = y$;

   (ii) $d_{\mathcal{X}}(x, y) = d_{\mathcal{X}}(y, x)$ for all $x, y \in \mathcal{X}$; and

   (iii) $d_{\mathcal{X}}(x, z) \leq d_{\mathcal{X}}(x, y) + d_{\mathcal{X}}(y, z)$ for all $x, y, z \in \mathcal{X}$.

A **metric space** $(\mathcal{X}, d_{\mathcal{X}})$ is a set $\mathcal{X}$ equipped with a metric $d_{\mathcal{X}}$ on $\mathcal{X}$. The inequality in (iii) is known as the **triangle inequality**.

A sequence $\{x_n\}_{n \in \mathbb{N}}$ in a metric space $(\mathcal{X}, d_{\mathcal{X}})$ is a **Cauchy sequence** if for all $\epsilon > 0$ there is an $N_\epsilon \in \mathbb{N}$ such that for all $m, n \geq N_\epsilon$, $d_{\mathcal{X}}(x_m, x_n) < \epsilon$. A metric space $(\mathcal{X}, d_{\mathcal{X}})$ is **complete** if, whenever $\{x_n\}_{n \in \mathbb{N}}$ is a Cauchy sequence in it, then there is an $x \in \mathcal{X}$ such that $\lim_{n \to \infty} d_{\mathcal{X}}(x_n, x) = 0$.

Now let $(V, \langle \cdot, \cdot \rangle)$ be an inner product space, and let $\| \cdot \|$ be the associated norm. Define the function $d(\mathbf{x}, \mathbf{y})$ on $V \times V$ by $d(\mathbf{x}, \mathbf{y}) := \|\mathbf{x} - \mathbf{y}\|$. Then evidently $d(\mathbf{x}, \mathbf{y}) \geq 0$ with equality if and only if $\mathbf{x} = \mathbf{y}$, and $d(\mathbf{x}, \mathbf{y}) = d(\mathbf{y}, \mathbf{x})$. As a consequence of (1.1.3) for all $\mathbf{x}, \mathbf{y}, \mathbf{z} \in V$,

$$\|\mathbf{x} - \mathbf{z}\| = \|(\mathbf{x} - \mathbf{y}) - (\mathbf{z} - \mathbf{y})\| \leq \|\mathbf{x} - \mathbf{y}\| + \|\mathbf{y} - \mathbf{z}\|,$$

which may be written as $d(\mathbf{x}, \mathbf{z}) \leq d(\mathbf{x}, \mathbf{y}) + d(\mathbf{y}, \mathbf{z})$, showing that the triangle inequality is satisfied. This shows that the function d is a **metric** on V, and $d(\mathbf{x}, \mathbf{y}) = \|\mathbf{x} - \mathbf{y}\|$ may be interpreted as the distance between $\mathbf{x}$ and $\mathbf{y}$ in the metric determined by the inner product.

Note that $\|\mathbf{x} + \mathbf{y}\| \leq \|\mathbf{x}\| + \|\mathbf{y}\|$ so that $\|\mathbf{y}\| \geq \|\mathbf{x} + \mathbf{y}\| - \|\mathbf{x}\|$. Similarly, $\|\mathbf{x}\| = \|(\mathbf{x} + \mathbf{y}) - \mathbf{y}\| \leq \|\mathbf{x} + \mathbf{y}\| + \|\mathbf{y}\|$, so that $\|\mathbf{y}\| \geq \|\mathbf{x}\| - \|\mathbf{x} + \mathbf{y}\|$. Altogether, $\|\mathbf{y}\| \geq |\|\mathbf{x} + \mathbf{y}\| - \|\mathbf{x}\||$. Therefore, $\lim_{\mathbf{y} \to 0} \|\mathbf{x} + \mathbf{y}\| = \|\mathbf{x}\|$ which shows that the norm is a continuous function on $\mathcal{H}$ with respect to the norm topology.

Definition 1.6 (Hilbert space). A **Hilbert space** is an inner product space $(V, \langle \cdot, \cdot \rangle)$ real or complex, that is a complete metric space for the metric determined by the inner product.

The completeness property of Hilbert spaces is essential in the proof of the following fundamental theorem, known as the *projection lemma*. Recall that a subset C of a vector space V is **convex** in case for all $\mathbf{x}, \mathbf{y} \in C$ and all $0 < t < 1$, $(1 - t)\mathbf{x} + t\mathbf{y} \in C$.

Lemma 1.7 (Projection lemma). *Let C be a nonempty closed convex subset of $\mathcal{H}$. Then there exists a unique $\mathbf{c}_0 \in C$ such that*

$$\|\mathbf{c}_0\| \leq \|\mathbf{c}\| \quad \text{for all} \quad \mathbf{c} \in C.$$

The proof is given in Appendix A; see Lemma A.2. If one is not already familiar with the proof, it would be good to read it at this time to appreciate the fundamental role of completeness. The projection lemma says that the norm function, which is continuous, has a minimum on any closed convex set—*even though the set need not be compact.* The first of many applications of it that we shall make is to the existence of orthogonal complements.

1.1.1. Orthogonality.

Definition 1.8. Let $\mathcal{H}$ be a Hilbert space. For any set $S \subset \mathcal{H}$, the **orthogonal complement** of S, $S^\perp$, is the subset of $\mathcal{H}$ given by

$$S^\perp := \{\mathbf{x} \in \mathcal{H} \ : \ \langle \mathbf{y}, \mathbf{x} \rangle = 0 \quad \text{for all} \quad \mathbf{y} \in S\}$$

$$(1.1.4) \qquad\qquad = \bigcap_{\mathbf{y} \in S} \{\mathbf{x} \in \mathcal{H} \ : \ \langle \mathbf{y}, \mathbf{x} \rangle = 0\}.$$

Since for every $\mathbf{y} \in \mathcal{H}$, the function $\mathbf{x} \mapsto \langle \mathbf{y}, \mathbf{x} \rangle$ is continuous, the set $\{\mathbf{x} \in \mathcal{H} \ : \ \langle \mathbf{y}, \mathbf{x} \rangle = 0\}$ is closed and, moreover, it is a subspace of $\mathcal{H}$ since it is the kernel of linear transformation. Hence for any $S \subset \mathcal{H}$, $S^\perp$ is the intersection closed subspaces of $\mathcal{H}$, and hence $S^\perp$ is always a closed subspace of $\mathcal{H}$.

Some useful containment properties may be deduced from (1.1.4): if $S_1 \subseteq S_2 \subseteq \mathcal{H}$, then $\{0\} \subseteq S_2^\perp \subseteq S_1^\perp$. However, if $S_2 = \overline{S_1}$, the closure of S_1, then $S_2^\perp = S_1^\perp$. This is because $\mathbf{x} \mapsto \langle \mathbf{y}, \mathbf{x} \rangle$ is continuous, so that it is equal to zero on all of S_1 if and only if it is equal to zero on all of $\overline{S_1} = S_2$.

Theorem 1.9. *Let $\mathcal{H}$ be a Hilbert space, and let $\mathcal{K}$ be a closed subspace of $\mathcal{H}$. Then $\mathcal{H} = \mathcal{K} \oplus \mathcal{K}^\perp$, meaning that every $\mathbf{x} \in \mathcal{H}$ can be written as $\mathbf{x} = \mathbf{v} + \mathbf{w}$ for a uniquely determined $\mathbf{v} \in \mathcal{K}$ and $\mathbf{w} \in \mathcal{K}^\perp$.*

Proof. Since $\mathcal{K}$ is closed and convex, so is $C := \mathbf{x} - \mathcal{K} = \{\mathbf{x} - \mathbf{v} \ : \ \mathbf{v} \in \mathcal{K}\}$. Hence Lemma 1.7 may be applied. Let $\mathbf{c}_0$ be the element of C of minimal norm. For all $\mathbf{v} \in \mathcal{K}$ and all $t \in \mathbb{R}$, $\mathbf{c}_0 + t\mathbf{v} \in C$, and hence $f(t) := \|\mathbf{c}_0 + t\mathbf{v}\|^2$ has a strict minimum at $t = 0$. Differentiating,

$$0 = f'(0) = \langle \mathbf{c}_0, \mathbf{v} \rangle + \langle \mathbf{v}, \mathbf{c}_0 \rangle = \langle \mathbf{c}_0, \mathbf{v} \rangle + \overline{\langle \mathbf{c}_0, \mathbf{v} \rangle} = 2\Re\langle \mathbf{c}_0, \mathbf{v} \rangle.$$

Hence $\Re\langle \mathbf{c}_0, \mathbf{v} \rangle = 0$ for all $\mathbf{v} \in \mathcal{K}$. Replacing $\mathbf{v}$ by $i\mathbf{v} \in \mathcal{K}$ yields $\Im\langle \mathbf{c}_0, \mathbf{v} \rangle = 0$ for all $\mathbf{v} \in \mathcal{K}$. Altogether, this proves that $\mathbf{c}_0 \in \mathcal{K}^\perp$. By the definition of C, $\mathbf{c}_0 = \mathbf{x} - \mathbf{v}$ for some $\mathbf{v} \in \mathcal{K}$. Thus, defining $\mathbf{w} := \mathbf{c}_0$, $\mathbf{x} = \mathbf{v} + \mathbf{w}$ with $\mathbf{v} \in \mathcal{K}$ and $\mathbf{w} \in \mathcal{K}^\perp$. This proves the existence of such a decomposition of $\mathbf{x}$. Now suppose $\mathbf{x} = \mathbf{v}' + \mathbf{w}'$ with $\mathbf{v}' \in \mathcal{K}$ and $\mathbf{w}' \in \mathcal{K}^{\perp}$. Then $\mathbf{v} - \mathbf{v}' = \mathbf{w}' - \mathbf{w}$. The left side belongs to $\mathcal{K}$ and the right side to $\mathcal{K}^{\perp}$, and hence both sides belong to $\mathcal{K} \cap \mathcal{K}^{\perp} = \{0\}$. This proves the uniqueness of the decomposition. $\square$

Corollary 1.10. *Let $\mathcal{V}$ be a subspace of a Hilbert space $\mathcal{H}$. Then $\mathcal{V}^{\perp\perp} = \overline{\mathcal{V}}$, the closure of $\mathcal{V}$.*

Proof. By Theorem 1.9, $\mathcal{H} = \overline{\mathcal{V}} \oplus \overline{\mathcal{V}}^{\perp}$. By the remarks following Definition 1.8, $\mathcal{V}^{\perp} = \overline{\mathcal{V}}^{\perp}$, and hence $\mathcal{H} = \overline{\mathcal{V}} \oplus \mathcal{V}^{\perp}$ and this means that $\mathcal{V}^{\perp\perp} = \overline{\mathcal{V}}$. $\square$

Definition 1.11. A set $\mathcal{U}$ of vectors in a Hilbert space $\mathcal{H}$ is **orthonormal** in case

(1) each $\mathbf{u} \in \mathcal{U}$ is a unit vector,

(2) every pair of distinct vectors in $\mathcal{U}$ is orthogonal.

If $\mathcal{K}$ is a subspace of $\mathcal{H}$, an orthonormal set $\mathcal{U} \subset \mathcal{K}$ is **maximal** in $\mathcal{K}$ in case no orthonormal subset of $\mathcal{K}$ strictly contains $\mathcal{U}$.

Let $\{\mathbf{u}_1, \dots, \mathbf{u}_n\} \subseteq \mathcal{U}$, $\mathcal{U}$ orthonormal. Then for any $z_1, \dots, z_n \in \mathbb{C}$,

$$(1.1.5) \qquad \left\| \sum_{j=1}^{n} z_j \mathbf{u}_j \right\|^2 = \sum_{i,j=1}^{n} \overline{z}_i z_j \langle \mathbf{u}_i, \mathbf{u}_j \rangle = \sum_{j=1}^{n} |z_j|^2.$$

Hence $\{\mathbf{u}_1, \dots, \mathbf{u}_n\}$ is linearly independent. Since $\{\mathbf{u}_1, \dots, \mathbf{u}_n\}$ is an arbitrary finite subset of $\mathcal{U}$, $\mathcal{U}$ is linearly independent.

For any orthonormal set $\{\mathbf{u}_1, \dots, \mathbf{u}_n\}$, $\mathrm{span}(\{\mathbf{u}_1, \dots, \mathbf{u}_n\})$ denotes its linear span. The elements $\mathbf{x}$ of $\mathrm{span}(\{\mathbf{u}_1, \dots, \mathbf{u}_n\})$ have the form $\mathbf{x} = \sum_{j=1}^{n} z_j \mathbf{u}_j$. Taking the inner product with $\mathbf{u}_k$ yields

$$(1.1.6) \qquad \langle \mathbf{u}_k, \mathbf{x} \rangle = \sum_{j=1}^{n} z_j \langle \mathbf{u}_k, \mathbf{u}_j \rangle = z_k.$$

Thus the vector of coefficients $(z_1, \dots, z_n)$, which is uniquely determined by the linear independence, may be obtained simply by taking inner products. Moreover, it then follows from (1.1.5) that for any $\mathbf{x} \in \mathrm{span}(\{\mathbf{u}_1, \dots, \mathbf{u}_n\})$,

$$\|\mathbf{x}\|^2 = \sum_{j=1}^{n} |\langle \mathbf{u}_j, \mathbf{x} \rangle|^2.$$

Lemma 1.12. *Let $\{\mathbf{u}_1, \dots, \mathbf{u}_n\}$ be an orthonormal subset of a Hilbert space $\mathcal{H}$. Then $\mathrm{span}(\{\mathbf{u}_1, \dots, \mathbf{u}_n\})$ is a closed subspace of $\mathcal{H}$.*

Proof. Define $T : \mathcal{H} \to \mathcal{H}$ by $T(\mathbf{x}) = \mathbf{x} - \sum_{j=1}^{n} \langle \mathbf{u}_j, \mathbf{x} \rangle \mathbf{u}_j$. Since $\mathbf{x} \mapsto \langle \mathbf{u}_j, \mathbf{x} \rangle$ is continuous, $\mathbf{x} \mapsto T(\mathbf{x})$ is continuous. Hence $\ker(T) = \{\mathbf{x} : T(\mathbf{x}) = 0\}$ is closed. However, $T(\mathbf{x}) = 0$ if and only if $\mathbf{x} = \sum_{j=1}^{n} \langle \mathbf{u}_j, \mathbf{x} \rangle \mathbf{u}_j$ and by (1.1.6), this is the case if and only if $\mathbf{x} \in \mathrm{span}(\{\mathbf{u}_1, \ldots, \mathbf{u}_n\})$. $\square$

Lemma 1.13. *Every nonzero finite-dimensional subspace $\mathcal{K}$ of a Hilbert space $\mathcal{H}$ contains an orthonormal set $\{\mathbf{u}_1, \ldots, \mathbf{u}_d\}$, where d is the dimension of $\mathcal{K}$ and $\mathcal{K} = \mathrm{span}(\{\mathbf{u}_1, \ldots, \mathbf{u}_d\})$.*

Proof. Choose any unit vector $\mathbf{u}_1 \in \mathcal{K}$. If $d = 1$, we are done. If $d > 1$, then by Lemma 1.12, $\mathrm{span}(\{\mathbf{u}_1\})$ is closed, so that by Theorem 1.9, $\mathcal{H} = \mathrm{span}(\{\mathbf{u}_1\}) \oplus (\mathrm{span}(\{\mathbf{u}_1\}))^\perp$. Therefore, $(\mathrm{span}(\{\mathbf{u}_1\}))^\perp \cap \mathcal{K}$ is a nontrivial subspace, and we may choose a unit vector $\mathbf{u}_2$ in this space. If $d = 2$ we are done. If not, for the same reasons, $(\mathrm{span}(\{\mathbf{u}_1, \mathbf{u}_2\}))^\perp \cap \mathcal{K}$ is a nontrivial subspace, and we may choose a unit vector $\mathbf{u}_3$ in this space. In d steps we arrive at the orthonormal set $\{\mathbf{u}_1, \ldots, \mathbf{u}_d\}$ in $\mathcal{K}$. If $\mathrm{span}(\{\mathbf{u}_1, \ldots, \mathbf{u}_d\}$ were a proper subspace of $\mathcal{K}$, we could continue the procedure one more step and produce a set of $d + 1$ orthonormal vectors in $\mathcal{K}$. Since orthonormal sets are linearly independent, and since $\mathcal{K}$ is d-dimensional, this is impossible. Hence $\mathcal{K} = \mathrm{span}(\{\mathbf{u}_1, \ldots, \mathbf{u}_d\})$. $\square$

Theorem 1.14. *Every finite-dimensional subspace $\mathcal{K}$ of any Hilbert space $\mathcal{H}$ is closed in $\mathcal{H}$.*

Proof. Let $\mathcal{K}$ be a d-dimensional subspace of $\mathcal{H}$. By Lemma 1.13, $\mathcal{K}$ contains an orthonormal set $\{\mathbf{u}_1, \ldots, \mathbf{u}_d\}$, and $\mathcal{K} = \mathrm{span}(\{\mathbf{u}_1, \ldots, \mathbf{u}_d\})$. By Lemma 1.12, $\mathrm{span}(\{\mathbf{u}_1, \ldots, \mathbf{u}_d\})$ is closed. $\square$

Theorem 1.15. *Let $\mathcal{H}$ be a Hilbert space of finite dimension n. Let $\mathcal{K}$ be any subspace of $\mathcal{H}$ of dimension d. Then the dimension of $\mathcal{K}^\perp$ is $n - d$.*

Proof. $\mathcal{K}$ is closed by Theorem 1.14, and hence by Theorem 1.9, $\mathcal{H} = \mathcal{K} \oplus \mathcal{K}^\perp$. By Lemma 1.13, there is a an orthonormal set $\{\mathbf{u}_1, \ldots, \mathbf{u}_d\}$ in $\mathcal{K}$, and $\mathrm{span}(\{\mathbf{u}_1, \ldots, \mathbf{u}_d\}) = \mathcal{K}$. Let m denote the dimension of $\mathcal{K}^\perp$. By Lemma 1.13, there is a complete orthonormal set $\{\mathbf{v}_1, \ldots, \mathbf{v}_m\}$ in $\mathcal{K}^\perp$, and $\mathrm{span}(\{\mathbf{v}_1, \ldots, \mathbf{v}_m\}) = \mathcal{K}^\perp$. Then $\{\mathbf{u}_1, \ldots, \mathbf{u}_d, \mathbf{v}_1, \ldots, \mathbf{v}_m\}$ is orthonormal in $\mathcal{H}$, and this set spans $\mathcal{K} \oplus \mathcal{K}^\perp = \mathcal{H}$ and is also linearly independent since it is orthonormal. Therefore, it is a basis of $\mathcal{H}$, and hence $n = d + m$. $\square$

Since orthonormal sets are linearly independent, a spanning orthonormal set in a finite-dimensional Hilbert space $\mathcal{H}$ is a vector space basis of $\mathcal{H}$, and is called an **orthonormal basis** for $\mathcal{H}$. This is the situation that concerns us in this book. However, a few words about how the infinite-dimensional case differs from the finite-dimensional case may be useful.

In an infinite-dimensional Hilbert space, there will exist infinite orthonormal sets, and their linear span (the space of all finite linear combinations) is an infinite-dimensional subspace of $\mathcal{H}$, that is generally not closed in $\mathcal{H}$. (Theorem 1.14 does not apply). However, if the orthonormal set is countable and maximal, its closure is all of $\mathcal{H}$.

1.2. Linear functionals and the Riesz representation theorem

Let $(\mathcal{V}, \langle \cdot, \cdot \rangle)$ be an inner product space. A **linear functional** is a linear transformation $\mathbf{f}$ from $\mathcal{V}$ to $\mathbb{C}$.

Theorem 1.16 (Riesz representation theorem for Hilbert space). *Let $\mathcal{H}$ be a Hilbert space and let $\mathbf{f}$ be a linear functional on $\mathcal{H}$. Then $\mathbf{f}$ is continuous if and only if $\ker(\mathbf{f}) = \{\mathbf{x} \in \mathcal{H} : \mathbf{f}(\mathbf{x}) = 0\}$ is closed in $\mathcal{H}$, and in this case there is a uniquely determined vector $\mathbf{v_f} \in \mathcal{H}$ such that for all $\mathbf{x} \in \mathcal{H}$,*

$$(1.2.1) \qquad \mathbf{f}(\mathbf{x}) = \langle \mathbf{v_f}, \mathbf{x} \rangle.$$

Proof. Suppose $\mathbf{f}$ is continuous. Let $\mathcal{K} = \ker(\mathbf{f})$. Since $\{0\}$ is closed in $\mathbb{C}$, and $\mathbf{f}$ is continuous, $\mathcal{K}$ is a closed subspace of $\mathcal{H}$. In summary, when $\mathbf{f}$ is continuous, $\mathcal{K} = \ker(\mathbf{f})$ is closed.

Next, suppose $\mathcal{K} = \ker(\mathbf{f})$ is closed. To avoid trivialities, assume that $\mathbf{f}$ is not identically zero. By Theorem 1.9, $\mathcal{H} = \mathcal{K} \oplus \mathcal{K}^\perp$. Since $\mathbf{f}$ is not identically zero on $\mathcal{K}^\perp$, and since $\mathbf{f}$ is linear, there exists $\mathbf{w} \in \mathcal{K}^\perp$ such that $\mathbf{f}(\mathbf{w}) = 1$. For any $\mathbf{y} \in \mathcal{K}^\perp$, $\mathbf{y} - \mathbf{f}(\mathbf{y})\mathbf{w} \in \ker(\mathbf{f}) = \mathcal{K}$, but $\mathbf{y} - \mathbf{f}(\mathbf{y})\mathbf{w}$ also belongs to $\mathcal{K}^\perp$, and hence is zero. This shows that for all $\mathbf{y} \in \mathcal{K}^\perp$, $\mathbf{y} = \mathbf{f}(\mathbf{y})\mathbf{w}$. Therefore, every $\mathbf{x} \in \mathcal{H}$ has a unique decomposition $\mathbf{x} = \mathbf{v} + z\mathbf{w}$ with $\mathbf{v} \in \ker(\mathbf{f}) = \mathcal{K}$, $z \in \mathbb{C}$, and $\mathbf{w} \in \mathcal{K}^\perp$ with $\mathbf{f}(\mathbf{w}) = 1$. Simple calculations yield

$$\mathbf{f}(\mathbf{x}) = \langle \|\mathbf{w}\|^{-2}\mathbf{w}, \mathbf{x} \rangle$$

for all $\mathbf{x} \in \mathcal{H}$. Hence (1.2.1) is satisfied with $\mathbf{v_f} = \|\mathbf{w}\|^{-2}\mathbf{w}$. Since the function $\mathbf{x} \mapsto \langle \mathbf{v_f}, \mathbf{x} \rangle$ is evidently continuous, this proves that $\mathbf{f}$ is continuous whenever $\ker(\mathbf{f})$ is closed, and then we have the representation (1.2.1). $\qquad \square$

Corollary 1.17. *A linear functional $\mathbf{f}$ on a Hilbert space $\mathcal{H}$ is continuous if and only if it is **bounded** in the sense that for some finite C,*

$$(1.2.2) \qquad |\mathbf{f}(\mathbf{x})| \leq C\|\mathbf{x}\| \quad \text{for all} \quad \mathbf{x} \in \mathcal{H}.$$

Proof. By Theorem 1.16, if $\mathbf{f}$ is continuous, then for some $\mathbf{v_f}$, $\mathbf{f}(\mathbf{x}) = \langle \mathbf{v_f}, \mathbf{x} \rangle$ for all $\mathbf{x} \in \mathcal{H}$. By the Cauchy-Schwarz inequality $|\mathbf{f}(\mathbf{x})| \leq \|\mathbf{v_f}\|\|\mathbf{x}\|$ so that (1.2.2) is satisfied with $C = \|\mathbf{v_f}\|$.

For the converse, assume (1.2.2). Then for any $\mathbf{x_0}, \mathbf{x} \in \mathcal{H}$,

$$|\mathbf{f}(\mathbf{x}) - \mathbf{f}(\mathbf{x_0})| = |\mathbf{f}(\mathbf{x} - \mathbf{x_0})| \leq C\|\mathbf{x} - \mathbf{x_0}\|.$$

Hence for all $\epsilon > 0$, whenever $\|\mathbf{x} - \mathbf{x_0}\| \leq \epsilon/C$, $|\mathbf{f}(\mathbf{x}) - \mathbf{f}(\mathbf{x_0})| \leq \epsilon$. Therefore, $\mathbf{f}$ is continuous. $\qquad \square$

1.3. Bounded linear transformations

Definition 1.18. Let $\mathcal{H}$ and $\mathcal{K}$ be Hilbert spaces. A linear transformation A from $\mathcal{H}$ to $\mathcal{K}$ is **bounded** in case for some finite constant C,

$$(1.3.1) \qquad \|A\mathbf{x}\| \leq C\|\mathbf{x}\| \quad \text{for all} \quad \mathbf{x} \in \mathcal{H},$$

and then we define

$$(1.3.2) \qquad \|A\| := \sup\{\|A\mathbf{u}\| \ : \ \mathbf{u} \in \mathcal{H}, \|\mathbf{u}\| = 1\}.$$

Lemma 1.19. *Let $\mathcal{H}$ and $\mathcal{K}$ be Hilbert spaces. A linear transformation A from $\mathcal{H}$ to $\mathcal{K}$ is bounded if and only if it is continuous.*

Proof. Suppose that A is continuous. Let $B_{\mathcal{K}}$ be the open unit ball in $\mathcal{K}$; that is $B_{\mathcal{K}} := \{\mathbf{y} \in \mathcal{K} \ : \ \|\mathbf{y}\| < 1\}$. Since A is continuous, the set $\{\mathbf{x} \in \mathcal{H} \ : \ A\mathbf{x} \in B_{\mathcal{K}}\}$ is open, and it contains zero. Therefore it contains some neighborhood of zero, meaning that for some $r > 0$ if $\|\mathbf{x}\| < r$, then $\|A\mathbf{x}\| < 1$. Since A is linear, it follows that (1.3.1) is satisfied with $C = 1/r$, so that A is bounded.

For the converse, suppose that (1.3.1) is satisfied. For any $\mathbf{x}_0, \mathbf{x} \in \mathcal{H}$,

$$\|A\mathbf{x} - A\mathbf{x}_0\| = \|A(\mathbf{x} - \mathbf{x}_0)\| \leq C\|\mathbf{x} - \mathbf{x}_0\|$$

so that for all $\epsilon > 0$, whenever $\|\mathbf{x} - \mathbf{x}_0\| \leq \epsilon/C$, $\|A\mathbf{x} - A\mathbf{x}_0\| \leq \epsilon$. Therefore, A is continuous. $\qquad\square$

Definition 1.20. Let $\mathcal{H}$ and $\mathcal{K}$ be two Hilbert spaces. Let $\mathcal{B}(\mathcal{H}, \mathcal{K})$ denote the set of all continuous linear transformations defined on $\mathcal{H}$ with values in $\mathcal{K}$. When $\mathcal{K} = \mathcal{H}$, we simply write $\mathcal{B}(\mathcal{H})$ in place of $\mathcal{B}(\mathcal{H}, \mathcal{H})$. The identity transformation in $\mathcal{B}(\mathcal{H})$ is denoted by $\mathbb{1}$.

Lemma 1.21. *Let $\mathcal{H}$ and $\mathcal{K}$ be two Hilbert spaces. The function $A \mapsto \|A\|$ in $\mathcal{B}(\mathcal{H}, \mathcal{K})$ defined in (1.3.2) has the following properties:*

(1) For all $A \in \mathcal{B}(\mathcal{H}, \mathcal{K})$, $\|A\| \geq 0$ with equality if and only if $A = 0$.

(2) For all $A \in \mathcal{B}(\mathcal{H}, \mathcal{K})$ and all $z \in \mathbb{C}$, $\|zA\| = |z|\|A\|$.

(3) For all $A, B \in \mathcal{B}(\mathcal{H}, \mathcal{K})$, $\|A + B\| \leq \|A\| + \|B\|$.

Proof. It is evident from the definition that $\|A\| \geq 0$. Moreover, if $A \neq 0$, there is some unit vector $\mathbf{u}$ such that $A\mathbf{u} \neq 0$, and then $\|A\| \geq \|A\mathbf{u}\| > 0$. This proves (1). Next, since for all $z \in \mathbb{C}$ and $\mathbf{y} \in \mathcal{K}$, $\|z\mathbf{y}\| = |z|\|\mathbf{y}\|$, for all unit vectors $\mathbf{u} \in \mathcal{H}$, $\|zA\mathbf{u}\| = |z|\|A\mathbf{u}\|$, and taking the supremum over all $\mathbf{u}$ yields (2). Next, since for all $\mathbf{y}_1, \mathbf{y}_2 \in \mathcal{K}$, $\|\mathbf{y}_1 + \mathbf{y}_2\| \leq \|\mathbf{y}_1\| + \|\mathbf{y}_2\|$, for all unit vectors $\mathbf{u} \in \mathcal{H}$, $\|(A + B)\mathbf{u}\| = \|A\mathbf{u} + B\mathbf{u}\| \leq \|A\mathbf{u}\| + \|B\mathbf{u}\|$, and

$$\sup\{\|A\mathbf{u}\| + \|B\mathbf{u}\| \ : \ \|\mathbf{u}\| = 1\}$$
$$\leq \sup\{\|A\mathbf{u}\| \ : \ \|\mathbf{u}\| = 1\} + \sup\{\|B\mathbf{u}\| \ : \ \|\mathbf{u}\| = 1\},$$

and this proves (3). $\qquad\square$

The three properties proved in Lemma 1.21 are exactly the properties the function $A \mapsto \|A\|$ must have to be a norm on the space, and to define a metric, or notion of distance, on the space. Hence we make the following definition:

Definition 1.22. The **operator norm** on $\mathcal{B}(\mathcal{H}, \mathcal{K})$ is the function $A \mapsto \|A\|$ in $\mathcal{B}(\mathcal{H}, \mathcal{K})$ defined in (1.3.2).

The corresponding distance function d on $\mathcal{B}(\mathcal{H}, \mathcal{K})$ is given by $d(A, B) = \|A - B\|$. By (1) of Lemma 1.21 $d(A, B) \geq 0$ with equality if and only if $A = B$, and by (2) of Lemma 1.21 $d(A, B) = d(B, A)$. To prove that the triangle inequality is satisfied, consider $A, B, C \in \mathcal{B}(\mathcal{H}, \mathcal{K})$. Then by (3) of Lemma 1.21,

$$d(A, C) = \|A - C\| = \|(A - B) + (B - C)\|$$
$$\leq \|A - B\| + \|B - C\| = d(A, B) + d(B, C).$$

The topology associated to this metric is the **operator norm topology**.

Theorem 1.23. *Equipped with operator norm topology, $\mathcal{B}(\mathcal{H}, \mathcal{K})$ is a complete metric space.*

Proof. Let $\{A_k\}_{k \in \mathbb{N}}$ be a Cauchy sequence. Let N_ϵ be such that for $n, m \geq N_\epsilon$, $\|A_m - A_n\| \leq \epsilon$. Fix any $\mathbf{x} \in \mathcal{H}$. Then for $m, n > N_\epsilon$,

$$\|A_m \mathbf{x} - A_n \mathbf{x}\| \leq \|A_m - A_n\| \|\mathbf{x}\| \leq \epsilon \|\mathbf{x}\|.$$

Therefore, $\{A_k \mathbf{x}\}_{k \in \mathbb{N}}$ is a Cauchy sequence in $\mathcal{K}$, and since $\mathcal{K}$ is complete, it has a limit, $A\mathbf{x}$, in $\mathcal{K}$. That is, for all $\mathbf{x} \in \mathcal{H}$, $A\mathbf{x} = \lim_{k \to \infty} A_k \mathbf{x}$ exists in $\mathcal{K}$. Since each A_k is linear, it is then evident that $\mathbf{x} \mapsto A\mathbf{x}$ is a linear transformation from $\mathcal{H}$ to $\mathcal{K}$. We now show that A is a bounded linear transformation. Since $\{A_k\}_{k \in \mathbb{N}}$ is a Cauchy sequence, there is an N such that $\|A_n - A_m\| \leq 1$ for all $m, n \geq N$. Let $\mathbf{u}$ be any unit vector in $\mathcal{H}$. Then $\|A\mathbf{u}\| \leq \|A_N \mathbf{u}\| + \|(A - A_N)\mathbf{u}\| \leq \|A_N\| + 1$. Therefore, $\|A\| \leq \|A_N\| + 1$. The norm convergence of A_k to A follows the same way. $\qquad\square$

1.3.1. The adjoint operation. Let $\mathcal{H}$ and $\mathcal{K}$ be two Hilbert spaces. For any $A \in \mathcal{B}(\mathcal{H}, \mathcal{K})$ and any $\mathbf{v} \in \mathcal{K}$, define a linear functional $\mathbf{f}$ on $\mathcal{H}$ by $\mathbf{f}(\mathbf{w}) = \langle \mathbf{v}, A\mathbf{w} \rangle$ for $\mathbf{w} \in \mathcal{H}$. By the Cauchy-Schwarz inequality and the definition of the operator norm, $|\mathbf{f}(\mathbf{w})| \leq (\|\mathbf{v}\| \|A\|) \|\mathbf{w}\|$ so that this linear functional is continuous. By the Riesz representation theorem there exists a unique vector in $\mathcal{H}$, denoted $A^* \mathbf{v}$, such that

$$(1.3.3) \qquad \langle \mathbf{v}, A\mathbf{w} \rangle = \langle A^* \mathbf{v}, \mathbf{w} \rangle$$

for all $\mathbf{w} \in \mathcal{H}$. Let $\mathbf{v}_1, \mathbf{v}_2 \in \mathcal{K}$ and let $z_1, z_2 \in \mathbb{C}$. Then

$$\langle A^*(z_1 \mathbf{v}_1 + z_2 \mathbf{v}_2), \mathbf{w} \rangle = \langle (z_1 \mathbf{v}_1 + z_2 \mathbf{v}_2), A\mathbf{w} \rangle$$
$$= \langle z_1 A^* \mathbf{v}_1 + z_2 A^* \mathbf{v}_2, \mathbf{w} \rangle.$$

This shows that A^* is a linear transformation from $\mathcal{K}$ to $\mathcal{H}$. Moreover, by (1.3.3) and the Schwarz inequality, taking $\mathbf{v}$ to be any unit vector in $\mathcal{K}$ such that $A^*\mathbf{v} \neq 0$, and taking $\mathbf{w} := \|A^*\mathbf{v}\|^{-1}A^*\mathbf{v}$, which is a unit vector,

$$\|A^*\mathbf{v}\| = \langle A^*\mathbf{v}, \mathbf{w}\rangle = \langle \mathbf{v}, A\mathbf{w}\rangle \leq \|\mathbf{v}\|\|A\|\|\mathbf{w}\| = \|A\|.$$

This proves that $\|A^*\| \leq \|A\|$ so that $A^* \in \mathcal{B}(\mathcal{K}, \mathcal{H})$. Applying the same arguments to A^*, we conclude that $\|A^{**}\| \leq \|A^*\|$. However, it is easy to see from the symmetry in (1.3.3) that $A^{**} = A$. Therefore, $\|A^*\| = \|A\|$. Finally, it is easy to see from (1.3.3) that the map $A \mapsto A^*$ is a conjugate linear map from $\mathcal{B}(\mathcal{H}, \mathcal{K})$ to $\mathcal{B}(\mathcal{K}, \mathcal{H})$.

Definition 1.24. Let $\mathcal{H}$ and $\mathcal{K}$ be two Hilbert spaces. For $A \in \mathcal{B}(\mathcal{H}, \mathcal{K})$, its **adjoint** is the map $A^* \in \mathcal{B}(\mathcal{K}, \mathcal{H})$ such that (1.3.3) is satisfied for all $\mathbf{v} \in \mathcal{K}$ and all $\mathbf{w} \in \mathcal{H}$. A linear transformation $A \in \mathcal{B}(\mathcal{H})$ is **self-adjoint** in case $A^* = A$, **skew adjoint** in case $A = -A^*$, and **normal** in case $AA^* = A^*A$. The set of self-adjoint operators on $\mathcal{H}$ is denoted by $\mathcal{B}^{\text{s.a.}}(\mathcal{H})$.

For $A \in \mathcal{B}(\mathcal{H}, \mathcal{K})$, the **range** of A is the set of $\mathbf{y} \in \mathcal{K}$ such that $\mathbf{y} = A\mathbf{x}$ for some $\mathbf{x} \in \mathcal{H}$. The **kernel** or **nullspace** of A is $\{\mathbf{x} \in \mathcal{H} \; : \; A\mathbf{x} = 0\}$. We always write $\text{ran}(A)$ to denote the range of A and $\ker(A)$ to denote the kernel of A. Both are easily seen to be subspaces, of $\mathcal{K}$ and $\mathcal{H}$ respectively. Since $\ker(A) = \{\mathbf{v} \in \mathcal{H} \; : \; A\mathbf{v} = 0\}$ and since $\{0\}$ is closed in $\mathcal{K}$ and A is continuous, $\ker(A)$ is always a closed subspace of $\mathcal{H}$, even when $\mathcal{H}$ is infinite dimensional. However, if $\mathcal{H}$ is infinite dimensional, it may be that $\text{ran}(A)$ is not closed.

Theorem 1.25 (Four subspaces theorem). *Let $\mathcal{H}$ and $\mathcal{K}$ be two Hilbert spaces. For all $A \in \mathcal{B}(\mathcal{H}, \mathcal{K})$,*

$$(1.3.4) \qquad \ker(A^*) = \text{ran}(A)^\perp \quad \textit{and} \quad \ker(A) = \text{ran}(A^*)^\perp.$$

Proof. Let $\mathbf{v} \in \ker(A^*)$. Then the right side of (1.3.3) is zero, so that $\langle \mathbf{v}, A\mathbf{w}\rangle = 0$ for all $\mathbf{w} \in \mathcal{H}$. Hence $\mathbf{v} \in \text{ran}(A)^\perp$. This shows $\ker(A^*) \subseteq \text{ran}(A)^\perp$. Next, let $\mathbf{v} \in \text{ran}(A)^\perp$. Then the left side of (1.3.3) is zero, so that $\langle A^*\mathbf{v}, \mathbf{w}\rangle = 0$ for all $\mathbf{w} \in \mathcal{H}$. Hence $\mathbf{v} \in \ker(A^*)$, and therefore $\text{ran}(A)^\perp \subseteq \ker(A^*)$. Altogether, we have $\ker(A^*) = \text{ran}(A)^\perp$. This proves the first equality in Theorem 1.25, and the second follows from the first upon replacing A by A^* since $A^{**} = A$. $\qquad \square$

Remark 1.26. Taking orthogonal complements in (1.3.4), and using Corollary 1.10 which says that $\mathcal{V}^{\perp\perp} = \overline{\mathcal{V}}$ yields

$$(1.3.5) \qquad \ker(A^*)^\perp = \overline{\text{ran}(A)} \quad \text{and} \quad \ker(A)^\perp = \overline{\text{ran}(A^*)}.$$

If $\mathcal{H}$ and $\mathcal{K}$ are finite dimensional, the ranges of A and A^* are closed by Theorem 1.14, and (1.3.5) simplifies to

$$(1.3.6) \qquad \ker(A^*)^\perp = \text{ran}(A) \quad \text{and} \quad \ker(A)^\perp = \text{ran}(A^*).$$

However, in the infinite-dimensional setting, (1.3.6) need not be true; in general one only has (1.3.5).

Definition 1.27. Let $\mathcal{H}$ and $\mathcal{K}$ be Hilbert spaces of finite dimensions m and n, respectively. For $K \in \mathcal{B}(\mathcal{H}, \mathcal{K})$, the **rank** of K, denoted $\mathrm{rank}(K)$, is the dimension of the range of K, and the **nullity** of K, denoted $\mathrm{nullity}(K)$, is the dimension of the kernel of K.

Theorem 1.28. *Let $\mathcal{H}$ and $\mathcal{K}$ be Hilbert spaces of finite dimensions m and n, respectively. Let $K \in \mathcal{B}(\mathcal{H}, \mathcal{K})$. Then*

$$(1.3.7) \qquad \mathrm{rank}(K) = \mathrm{rank}(KK^*) = \mathrm{rank}(K^*K) = \mathrm{rank}(K^*)$$

and

$$(1.3.8) \qquad \mathrm{rank}(K) + \mathrm{nullity}(K) = m.$$

Proof. By Theorem 1.25 and Remark 1.26, $\mathrm{ran}(K^*) = \ker(K)^\perp$. Then by Theorem 1.15,

$$(1.3.9) \qquad \mathrm{rank}(K^*) + \mathrm{nullity}(K) = m.$$

The same argument applied to K^*K yields $\mathrm{rank}(K^*K) + \mathrm{nullity}(K^*K) = m$.

However, since for all $\mathbf{x} \in \mathcal{H}$, $\langle \mathbf{x}, K^*K\mathbf{x} \rangle = \|K\mathbf{x}\|^2$, $K^*K\mathbf{x} = 0$ if and only if $K\mathbf{x} = 0$. That is $\ker(K^*K) = \ker(K)$. Therefore, (1.3.9) becomes $\dim(\mathrm{ran}(K^*K)) + \dim(\ker(K)) = m$, and then from (1.3.9) we conclude $\mathrm{rank}(K^*) = \mathrm{rank}(K^*K)$. Let $\{\mathbf{u}_1, \ldots, \mathbf{u}_r\}$ be an orthonormal basis for $\mathrm{ran}(K)$. Then $\{K^*\mathbf{u}_1, \ldots, K^*\mathbf{u}_r\}$ spans the range of K^*K which is also the range of K^*. Therefore, $\mathrm{rank}(K^*) \leq r = \mathrm{rank}(K)$. Altogether, $\mathrm{rank}(K^*K) = \mathrm{rank}(K^*) \leq \mathrm{rank}(K)$. The same reasoning with K and K^* interchanged leads to $\mathrm{rank}(KK^*) = \mathrm{rank}(K) \leq \mathrm{rank}(K^*)$. This proves (1.3.7), and then (1.3.8) follows from (1.3.7) and (1.3.9). $\qquad\qquad\square$

Remark 1.29. Since (1.3.8) does not refer to the adjoint of K or to the inner product on $\mathcal{H}$ in any way, it is simply a statement about K as a linear transformation on an m-dimensional Hilbert space. We can always equip any m-dimensional vector space with some inner product, making it a Hilbert space $\mathcal{H}$. Then any linear transformation from V into itself may be regarded as belonging to $\mathcal{B}(\mathcal{H})$. Then $K^* \in \mathcal{B}(\mathcal{H})$ will depend on the choice of the inner product, but by using any such inner product, one arrives at (1.3.8) which does not depend on the choice of inner product.

The statement (1.3.8), regarded as a statement about linear transformations from an m-dimensional vector space V into itself, is often called the **fundamental theorem of linear algebra**. In particular, it says that a linear transformation K from V into itself is one-to-one (that is, $\ker(K) = 0$) if and only if it maps V onto V (that is, $\mathrm{ran}(K) = V$), and hence if either of these conditions is known to be true, then both are true, and K is invertible.

1.3.2. Orthogonal projections. Let $\mathcal{H}$ be a Hilbert space of arbitrary dimension, and let $\mathcal{K}$ be a nontrivial closed subspace of $\mathcal{H}$. By Theorem 1.9, $\mathcal{H} = \mathcal{K} \oplus \mathcal{K}^\perp$. Define a linear transformation P on $\mathcal{H}$ by $P\mathbf{x} = \mathbf{v}$ where $\mathbf{x} = \mathbf{v} + \mathbf{w}$ with $\mathbf{v} \in \mathcal{K}$ and $\mathbf{w} \in \mathcal{K}^\perp$. Since $\mathbf{v}$ and $\mathbf{w}$ are orthogonal,

$$\|\mathbf{x}\|^2 = \|\mathbf{v} + \mathbf{w}\|^2 = \|\mathbf{v}\|^2 + \|\mathbf{w}\|^2 \geq \|\mathbf{v}\|^2 = \|P\mathbf{x}\|^2,$$

and there is equality if and only if $\mathbf{w} = 0$. Hence $\|P\| = 1$. For the trivial cases, $\mathcal{K} = 0$ and $\mathcal{K} = \mathcal{H}$ define $P = 0$ and $P = \mathbb{1}$, respectively.

Since P is linear and bounded, $P \in \mathcal{B}(\mathcal{H})$. The operator P is called the **orthogonal projection onto** $\mathcal{K}$. For any $\mathbf{x} \in \mathcal{H}$, $P\mathbf{x} \in \mathcal{K}$, so $P^2\mathbf{x} = P\mathbf{x}$. That is, $P^2 = P$. Moreover, P is self-adjoint. To see this, let $\mathbf{x}, \mathbf{y}$ be any two vectors in $\mathcal{H}$ with decompositions $\mathbf{x} = \mathbf{v} + \mathbf{w}$ and $\mathbf{y} = \mathbf{v}' + \mathbf{w}'$ in $\mathcal{K} \oplus \mathcal{K}^\perp$. Then

$$\langle \mathbf{x}, P\mathbf{y} \rangle = \langle \mathbf{v} + \mathbf{w}, \mathbf{v}' \rangle = \langle \mathbf{v}, \mathbf{v}' \rangle = \langle \mathbf{v}, \mathbf{v}' + \mathbf{w} \rangle = \langle P\mathbf{x}, \mathbf{y} \rangle.$$

In fact, the two properties $P = P^2$ and $P = P^*$ are an algebraic characterization of orthogonal projections.

Lemma 1.30. *Let $\mathcal{H}$ be a Hilbert space, and let $P \in \mathcal{B}(\mathcal{H})$ satisfy $P = P^*$ and $P^2 = P$. Then $\mathrm{ran}(P)$ is a closed subspace of $\mathcal{H}$, and P is the orthogonal projection onto $\mathrm{ran}(P)$.*

Proof. Let P be self-adjoint with $P^2 = P$. Define $Q \in \mathcal{B}(\mathcal{H})$ by $Q := \mathbb{1} - P$ so that $P + Q = \mathbb{1}$. We claim that $\mathrm{ran}(P) = \ker(Q)$.

Suppose that $\mathbf{x} \in \ker(Q)$. Then $\mathbf{x} = (P + Q)\mathbf{x} = P\mathbf{x}$, so that $\mathbf{x} \in \mathrm{ran}(P)$. Thus $\ker(Q) \subseteq \mathrm{ran}(P)$. Next suppose that $\mathbf{x} \in \mathrm{ran}(P)$. Since $P^2 = P$, $P\mathbf{x} = \mathbf{x}$. Therefore, $\mathbf{x} = (P + Q)\mathbf{x} = \mathbf{x} + Q\mathbf{x}$, and hence $Q\mathbf{x} = 0$. Thus $\mathrm{ran}(P) \subseteq \ker(Q)$, and altogether, $\mathrm{ran}(P) = \ker(Q)$.

Since Q is continuous, $\ker(Q) = \mathrm{ran}(P)$ is closed. By Theorem 1.25 and $P = P^*$, $\mathrm{ran}(P)^\perp = \ker(P)$, and then by Theorem 1.9, $\mathcal{H} = \mathrm{ran}(P) \oplus \ker(P)$. For any $\mathbf{x} \in \mathcal{H}$, let $\mathbf{x} = \mathbf{v} + \mathbf{w}$ be its decomposition in $\mathrm{ran}(P) \oplus \ker(P)$. Then $P\mathbf{x} = \mathbf{v}$, and hence P is the orthogonal projection onto $\mathrm{ran}(P)$. $\square$

Definition 1.31. Let $\mathcal{H}$ be a Hilbert space. An operator $P \in \mathcal{B}(\mathcal{H})$ is an **orthogonal projection** in case $P = P^*$ and $P^2 = P$.

Let $\mathcal{H}$ be a Hilbert space, and let $\mathcal{K}_1$ and $\mathcal{K}_2$ be two subspaces of $\mathcal{H}$. The subspaces are **mutually orthogonal** in case every $\mathbf{x} \in \mathcal{H}_1$ is orthogonal to every $\mathbf{y} \in \mathcal{K}_2$.

Let P be an orthogonal projection onto an r-dimensional subspace of $\mathcal{H}$, and let $\{\mathbf{u}_1, \ldots, \mathbf{u}_r\}$ be an orthonormal basis for the range of P. For every $\mathbf{x} \in \mathcal{H}$, $P\mathbf{x} \in \mathrm{ran}(P)$ and hence

$$(1.3.10) \qquad P\mathbf{x} = \sum_{j=1}^{r} \langle \mathbf{u}_j, P\mathbf{x} \rangle \mathbf{u}_j = \sum_{j=1}^{r} \langle P\mathbf{u}_j, \mathbf{x} \rangle \mathbf{u}_j = \sum_{j=1}^{r} \langle \mathbf{u}_j, \mathbf{x} \rangle \mathbf{u}_j.$$

There is another way to write (1.3.10), and many other formulas, that will prove to be very useful. This uses *part of* the **Dirac notation** system:

Definition 1.32. Let $\mathcal{H}$ and $\mathcal{K}$ be Hilbert spaces. For $\mathbf{x} \in \mathcal{K}$, $\mathbf{y} \in \mathcal{H}$, $|\mathbf{x}\rangle\langle\mathbf{y}|$ denotes the operator in $\mathcal{B}(\mathcal{H}, \mathcal{K})$ such that for all $\mathbf{z} \in \mathcal{H}$,

$$|\mathbf{x}\rangle\langle\mathbf{y}|\mathbf{z} := \langle\mathbf{y}, \mathbf{z}\rangle\mathbf{x}.$$

It is left as an easy exercise to check that $(|\mathbf{x}\rangle\langle\mathbf{y}|)^* = |\mathbf{y}\rangle\langle\mathbf{x}|$ and

$$(|\mathbf{x}\rangle\langle\mathbf{y}|)\,(|\mathbf{z}\rangle\langle\mathbf{w}|) = \langle\mathbf{y}, \mathbf{z}\rangle|\mathbf{x}\rangle\langle\mathbf{w}|.$$

In particular, for any unit vector $\mathbf{u}$, $|\mathbf{u}\rangle\langle\mathbf{u}|$ is self-adjoint and $(|\mathbf{u}\rangle\langle\mathbf{u}|)^2 = |\mathbf{u}\rangle\langle\mathbf{u}|$. Hence $|\mathbf{u}\rangle\langle\mathbf{u}|$ is a rank one orthogonal projection.

In the notation of Dirac, for $\mathbf{x}, \mathbf{y} \in \mathcal{H}$, $|\mathbf{x}\rangle\langle\mathbf{y}|$ is the **outer product** of $\mathbf{x}$ and $\mathbf{y}$, while $\langle\mathbf{x}|\mathbf{y}\rangle := \langle\mathbf{x}, \mathbf{y}\rangle$ is the **inner product** of $\mathbf{x}$ and $\mathbf{y}$. We shall continue to write $\langle\mathbf{x}, \mathbf{y}\rangle$ for the inner product, while using Dirac notation for the outer product. This is standard in the mathematical physics literature and is a reasonable compromise for our interdisciplinary subject.

Lemma 1.33. *Let $\mathcal{H}$ and $\mathcal{K}$ be Hilbert spaces, and let $\mathcal{H}$ have finite dimension n. Let $\{\mathbf{u}_1, \dots, \mathbf{u}_n\}$ be an orthonormal basis for $\mathcal{H}$. Then for all $A \in \mathcal{B}(\mathcal{H}, \mathcal{K})$,*

$$(1.3.11) \qquad A = \sum_{j=1}^{n} |A\mathbf{u}_j\rangle\langle\mathbf{u}_j|.$$

Proof. For every $\mathbf{x} \in \mathcal{H}$, $\mathbf{x} = \sum_{j=1}^{n}\langle\mathbf{u}_j, \mathbf{x}\rangle\mathbf{u}_j$ and hence

$$A\mathbf{x} = \sum_{j=1}^{n}\langle\mathbf{u}_j, \mathbf{x}\rangle A\mathbf{u}_j = \left(\sum_{j=1}^{n} |A\mathbf{u}_j\rangle\langle\mathbf{u}_j|\right)\mathbf{x}. \qquad \square$$

Returning to (1.3.10), let $\{\mathbf{u}_1, \dots, \mathbf{u}_n\}$ be an orthonormal basis of $\mathcal{H}$ such that $\{\mathbf{u}_1, \dots, \mathbf{u}_r\}$ is an orthonormal basis for the range of P. Then by Lemma 1.33,

$$(1.3.12) \qquad P = \sum_{j=1}^{r} |\mathbf{u}_j\rangle\langle\mathbf{u}_j|,$$

which expresses P as a sum of rank one projections.

1.3.3. Unitary operators.

Definition 1.34. Let $\mathcal{H}$ and $\mathcal{K}$ be Hilbert spaces. An operator $U \in \mathcal{B}(\mathcal{H}, \mathcal{K})$ is **unitary** in case $U^*U = \mathbb{1}_{\mathcal{H}}$ and $UU^* = \mathbb{1}_{\mathcal{K}}$. (Hence U is invertible, and U^* is its inverse, and $U^* \in \mathcal{B}(\mathcal{K}, \mathcal{H})$ is also unitary.)

Lemma 1.35. *$\mathcal{H}$ and $\mathcal{K}$ be Hilbert spaces. Suppose that $U \in \mathcal{B}(\mathcal{H}, \mathcal{K})$ is unitary. Then U maps $\mathcal{H}$ onto $\mathcal{K}$, and U is an isometry; that is, for all $\mathbf{x} \in \mathcal{H}$, $\|U\mathbf{x}\| = \|\mathbf{x}\|$.*

Proof. Since $UU^* = \mathbb{1}_{\mathcal{K}}$, for each $\mathbf{y} \in \mathcal{K}$, $\mathbf{y} = U(U^*\mathbf{y})$, so that U maps $\mathcal{H}$ onto $\mathcal{K}$. Since $U^*U = \mathbb{1}_{\mathcal{H}}$, for all $\mathbf{x} \in \mathcal{H}$,

$$\|\mathbf{x}\|^2 = \langle \mathbf{x}, U^*U\mathbf{x} \rangle = \langle U\mathbf{x}, U\mathbf{x} \rangle = \|U\mathbf{x}\|^2. \qquad \square$$

Lemma 1.36. *$\mathcal{H}$ and $\mathcal{K}$ be finite-dimensional Hilbert spaces. Then there exist unitary operators in $\mathcal{B}(\mathcal{H}, \mathcal{K})$ if and only if $\dim(\mathcal{H}) = \dim(\mathcal{K})$. If $\dim(\mathcal{H}) = \dim(\mathcal{K}) = n$, and $\{\mathbf{u}_1, \ldots, \mathbf{u}_n\}$ is any orthonormal basis for $\mathcal{H}$, then U is unitary if and only if $\{U\mathbf{u}_1, \ldots, U\mathbf{u}_n\}$ is orthonormal in $\mathcal{K}$.*

Proof. Suppose that $U \in \mathcal{B}(\mathcal{H}, \mathcal{K})$ is unitary, and that $\{\mathbf{u}_1, \ldots, \mathbf{u}_n\}$ is an orthonormal basis for $\mathcal{H}$. Then for each i, j

$$\langle U\mathbf{u}_i, U\mathbf{u}_j \rangle = \langle \mathbf{u}_i, U^*U\mathbf{u}_j \rangle \langle \mathbf{u}_i, \mathbf{u}_j \rangle = \delta_{i,j}.$$

Hence $\{U\mathbf{u}_1, \ldots, U\mathbf{u}_n\}$ is orthonormal in $\mathcal{K}$. Moreover, the dimension of $\mathcal{K}$ must be at least n, the dimension of $\mathcal{H}$. The same argument applied to the unitary U^* shows that the dimension of $\mathcal{H}$ is at least as large as the dimension of $\mathcal{K}$, and hence the dimensions are equal.

Let both $\mathcal{H}$ and $\mathcal{K}$ have dimension n. Let $\{\mathbf{u}_1, \ldots, \mathbf{u}_n\}$ be an orthonormal basis of $\mathcal{H}$. By Lemma 1.33 $U := \sum_{j=1}^{n} |U\mathbf{u}_j\rangle\langle\mathbf{u}_j|$. If $\{\mathbf{v}_1, \ldots, \mathbf{v}_n\} := \{U\mathbf{u}_1, \ldots, U\mathbf{u}_n\}$ is orthonormal, then $U = \sum_{j=1}^{n} |\mathbf{v}_j\rangle\langle\mathbf{u}_j|$, and

$$U^*U = \sum_{j=1}^{n} |\mathbf{u}_j\rangle\langle\mathbf{u}_j| = \mathbb{1}_{\mathcal{H}} \quad \text{and} \quad UU^* = \sum_{j=1}^{n} |\mathbf{v}_j\rangle\langle\mathbf{v}_j| = \mathbb{1}_{\mathcal{K}},$$

and hence U is unitary. $\qquad \square$

1.4. The trace and the Hilbert-Schmidt inner product

Lemma 1.37. *Let $\mathcal{H}$ be an n-dimensional Hilbert space. Let $\{\mathbf{u}_1, \ldots, \mathbf{u}_n\}$ and $\{\mathbf{v}_1, \ldots, \mathbf{v}_n\}$ be two orthonormal bases for $\mathcal{H}$. Let $A \in \mathcal{B}(\mathcal{H})$. Then*

$$\sum_{j=1}^{n} \langle \mathbf{u}_j, A\mathbf{u}_j \rangle = \sum_{k=1}^{n} \langle \mathbf{v}_k, A\mathbf{v}_k \rangle.$$

Proof. Since $\mathbb{1} = \sum_{k=1}^{n} |\mathbf{v}_k\rangle\langle\mathbf{v}_k|$,

$$A = \mathbb{1}A\mathbb{1} = \sum_{k,\ell=1}^{n} |\mathbf{v}_k\rangle\langle\mathbf{v}_k|A|\mathbf{v}_\ell\rangle\langle\mathbf{v}_\ell| = \sum_{k,\ell=1}^{n} \langle\mathbf{v}_k, A\mathbf{v}_\ell\rangle|\mathbf{v}_k\rangle\langle\mathbf{v}_\ell|.$$

Therefore, $\langle \mathbf{u}_j, A\mathbf{u}_j \rangle = \sum_{k,\ell=1}^{n} \langle \mathbf{v}_k, A\mathbf{v}_\ell \rangle \langle \mathbf{u}_j, \mathbf{v}_k \rangle \langle \mathbf{v}_\ell, \mathbf{u}_j \rangle$. Now summing on j, and using $\sum_{j=1}^{n} |\mathbf{u}_j\rangle\langle\mathbf{u}_j| = \mathbb{1}$,

$$\sum_{j=1}^{n} \langle \mathbf{u}_j, A\mathbf{u}_j \rangle = \sum_{k,\ell=1}^{n} \langle \mathbf{v}_k, A\mathbf{v}_\ell \rangle \left(\sum_{j=1}^{n} \langle \mathbf{v}_\ell, \mathbf{u}_j \rangle \langle \mathbf{u}_j, \mathbf{v}_k \rangle \right)$$

$$= \sum_{k,\ell=1}^{n} \langle \mathbf{v}_k, A\mathbf{v}_\ell \rangle \delta_{k,\ell} = \sum_{k=1}^{n} \langle \mathbf{v}_k, A\mathbf{v}_k \rangle. \qquad \square$$

By Lemma 1.37, the function $A \mapsto \sum_{j=1}^{n} \langle \mathbf{u}_j, A\mathbf{u}_j \rangle$ where $\{\mathbf{u}_1, \ldots, \mathbf{u}_n\}$ is an orthonormal basis of $\mathcal{H}$ does not depends on the choice of the orthonormal basis.

Definition 1.38. Let $\mathcal{H}$ be a finite-dimensional Hilbert space. The **trace** of A is the number $\mathrm{Tr}[A]$ given by

$$\mathrm{Tr}[A] := \sum_{j=1}^{n} \langle \mathbf{u}_j, A\mathbf{u}_j \rangle$$

where $\{\mathbf{u}_1, \ldots, \mathbf{u}_n\}$ is an orthonormal basis of $\mathcal{H}$.

On an infinite-dimensional Hilbert space, there are operators $A \in \mathcal{B}(\mathcal{H})$ and infinite orthonormal sets $\{\mathbf{u}_k\}_{k\in\mathbb{N}}$ such that the infinite sum $\sum_{k=1}^{\infty} \langle \mathbf{u}_k, A\mathbf{u}_k \rangle$ does not converge. For example, take $A = \mathbb{1}$.

The trace can be defined for infinite-dimensional Hilbert spaces $\mathcal{H}$, but not as a function on all of $\mathcal{B}(\mathcal{H})$. It is only defined on special set of operators $\mathcal{T}(\mathcal{H})$, called the **trace class operators**. Operators in this set are *approximately finite rank*, and hence approximately finite dimensional, in a very strong sense. See [**198**] for the extension to infinite dimensions.

Theorem 1.39. *Let $\mathcal{H}$ and $\mathcal{K}$ be finite-dimensional Hilbert spaces. Then for all $A \in \mathcal{B}(\mathcal{H}, \mathcal{K})$ and all $B \in \mathcal{B}(\mathcal{K}, \mathcal{H})$,*

$$(1.4.1) \qquad\qquad \mathrm{Tr}[AB] = \mathrm{Tr}[BA],$$

$$(1.4.2) \qquad\qquad \mathrm{Tr}[A^*A] \geq 0,$$

with equality if and only if $A = 0$, and for all $\mathbf{x} \in \mathcal{H}$ and $\mathbf{y} \in \mathcal{K}$,

$$(1.4.3) \qquad\qquad \mathrm{Tr}[|\mathbf{x}\rangle\langle\mathbf{y}|A] = \langle \mathbf{y}, A\mathbf{x} \rangle.$$

Proof. Let $\{\mathbf{u}_1, \dots, \mathbf{u}_n\}$ be an orthonormal basis of $\mathcal{H}$, and let $\{\mathbf{v}_1, \dots, \mathbf{v}_m\}$ be an orthonormal basis for $\mathcal{K}$. Then since $\sum_{k=1}^{m} |\mathbf{v}_k\rangle\langle\mathbf{v}_k| = \mathbb{1}_{\mathcal{K}}$, $BA = \sum_{k=1}^{m} B|\mathbf{v}_k\rangle\langle\mathbf{v}_k|A$,

$$\mathrm{Tr}[BA] = \sum_{j=1}^{n} \left\langle \mathbf{u}_j, \left(\sum_{k=1}^{m} B|\mathbf{v}_k\rangle\langle\mathbf{v}_k|A\right)\mathbf{u}_j \right\rangle = \sum_{j=1}^{n}\sum_{k=1}^{m} \langle\mathbf{u}_j, B\mathbf{v}_k\rangle\langle\mathbf{v}_k, A\mathbf{u}_j\rangle$$

$$= \sum_{k=1}^{m}\sum_{j=1}^{n} \langle\mathbf{v}_k, A\mathbf{u}_j\rangle\langle\mathbf{u}_j, B\mathbf{v}_k\rangle = \sum_{k=1}^{m} \left\langle \mathbf{v}_k, \left(\sum_{j=1}^{n} A|\mathbf{u}_j\rangle\langle\mathbf{u}_j|B\right)\mathbf{v}_k \right\rangle$$

$$= \mathrm{Tr}[AB].$$

This proves (1.4.1). Taking $B = A^*$, the first line of the above calculation gives

$$\mathrm{Tr}[A^*A] = \sum_{j=1}^{n}\sum_{k=1}^{m} |\langle\mathbf{v}_k, A\mathbf{u}_j\rangle|^2 = \sum_{j=1}^{n} \|A\mathbf{u}_j\|^2.$$

Therefore $\mathrm{Tr}[A^*A] \geq 0$, and $\mathrm{Tr}[A^*A] = 0$ if and only if $\|A\mathbf{u}_j\| = 0$ for $j = 1, \dots, n$, and hence if and only if $A = 0$.

To prove (1.4.3), let $\{\mathbf{u}_1, \dots, \mathbf{u}_n\}$ be an orthonormal basis for $\mathcal{H}$. Then

$$\mathrm{Tr}[|\mathbf{x}\rangle\langle\mathbf{y}|A] = \sum_{j=1}^{n} \langle\mathbf{u}_j, |\mathbf{x}\rangle\langle\mathbf{y}|A\mathbf{u}_j\rangle = \sum_{j=1}^{n} \langle\mathbf{y}, A\mathbf{u}_j\rangle\langle\mathbf{u}_j, \mathbf{x}\rangle = \langle\mathbf{y}, A\mathbf{x}\rangle. \qquad \square$$

Remark 1.40. Let $\mathcal{H}$ be a finite-dimensional Hilbert space, and let $A_1, \dots, A_m \in \mathcal{B}(\mathcal{H})$. Let π be a cyclic permutation of $\{1, \dots, m\}$. Then repeated application of (1.4.1) shows that

$$\mathrm{Tr}[A_1 A_2 \cdots A_m] = \mathrm{Tr}[A_{\pi(1)} A_{\pi(2)} \cdots A_{\pi(m)}].$$

The identity (1.4.1) is often referred to as **cyclicity of the trace**.

The map $(A, B) \mapsto \mathrm{Tr}[A^*B]$ on $\mathcal{B}(\mathcal{H}, \mathcal{K})$ is conjugate linear in the first variable, and linear in the second. By (1.4.2) it satisfies the positivity condition required of an inner product, and hence is an inner product.

Definition 1.41 (Hilbert-Schmidt inner product on $\mathcal{B}(\mathcal{H}, \mathcal{K})$). Let $\mathcal{H}$ and $\mathcal{K}$ be finite-dimensional Hilbert spaces. The **Hilbert-Schmidt inner product on** $\mathcal{B}(\mathcal{H}, \mathcal{K})$, $\langle\cdot, \cdot\rangle_{\mathrm{HS}}$, is defined by

$$\langle A, B\rangle_{\mathrm{HS}} := \mathrm{Tr}[A^*B].$$

For $A \in \mathcal{B}(\mathcal{H})$,

$$\|A\|_2 := \langle A, A\rangle_{\mathrm{HS}}^{1/2}$$

is the norm associated to the Hilbert-Schmidt inner product; it is called the **Hilbert-Schmidt norm** on $\mathcal{B}(\mathcal{H}, \mathcal{K})$.

If $\{\mathbf{u}_1, \dots, \mathbf{u}_n\}$ is any orthonormal basis of $\mathcal{H}$, then

$$\mathrm{Tr}[A^*B] = \sum_{j=1}^{n}\langle\mathbf{u}_j, A^*B\mathbf{u}_j\rangle = \sum_{j=1}^{n}\langle A\mathbf{u}_j, B\mathbf{u}_j\rangle.$$

Hence

$$(1.4.4) \qquad \langle A, B\rangle_{\mathrm{HS}} = \sum_{j=1}^{n}\langle A\mathbf{u}_j, B\mathbf{u}_j\rangle.$$

Lemma 1.42. *Let $\mathcal{H}$ and $\mathcal{K}$ be Hilbert spaces of dimension n and m, respectively. For all $A \in \mathcal{B}(\mathcal{H}, \mathcal{K})$, $\|A\| \leq \|A\|_2 \leq n^{1/2}\|A\|$.*

Proof. For any unit vector $\mathbf{u} \in \mathcal{H}$, let $\{\mathbf{u}_1, \dots, \mathbf{u}_n\}$ be an orthonormal basis for $\mathcal{H}$ such that $\mathbf{u}_1 = \mathbf{u}$. Then

$$\|A\mathbf{u}\|^2 = \langle\mathbf{u}, A^*A\mathbf{u}\rangle \leq \sum_{j=1}^{n}\langle\mathbf{u}_1, A^*A\mathbf{u}_1\rangle = \mathrm{Tr}[A^*A] = \|A\|_2^2.$$

and hence $\|A\| \leq \|A\|_2$ which is the first inequality. Next,

$$\|A\|_2^2 = \mathrm{Tr}[A^*A] = \sum_{j=1}^{n}\langle\mathbf{u}_j, A^*A\mathbf{u}_j\rangle = \sum_{j=1}^{n}\|A\mathbf{u}_j\|^2 \leq n\|A\|^2.$$

This proves the second inequality. $\qquad\qquad\square$

Theorem 1.43. *Let $\mathcal{H}$ and $\mathcal{K}$ be finite-dimensional Hilbert spaces. Then $\mathcal{B}(\mathcal{H}, \mathcal{K})$ equipped with the Hilbert-Schmidt inner product is a Hilbert space.*

Proof. By Lemma 1.42, a sequence $\{A_j\}_{j\in\mathbb{N}}$ is Cauchy in the operator norm if and only if it is Cauchy in the Hilbert-Schmidt norm, and also that $\lim_{j\to\infty}\|A_j - A\| = 0$ if and only if $\lim_{j\to\infty}\|A_j - A\|_2 = 0$. Therefore, since $\mathcal{B}(\mathcal{H}, \mathcal{K})$ is complete in the operator norm, it is complete in the Hilbert-Schmidt norm. $\qquad\square$

Lemma 1.44. *Let $\{\mathbf{u}_1, \dots, \mathbf{u}_n\}$ and $\{\mathbf{v}_1, \dots, \mathbf{v}_m\}$ be orthonormal bases of dimensional Hilbert spaces $\mathcal{H}$ and $\mathcal{K}$, respectively. Then*

$$(1.4.5) \qquad \{|\mathbf{v}_i\rangle\langle\mathbf{u}_j|\}_{1\leq i\leq m,\ 1\leq j\leq n}$$

is an orthonormal basis of $\mathcal{B}(\mathcal{H}, \mathcal{K})$ equipped with the Hilbert-Schmidt inner product, and for each $A \in \mathcal{B}(\mathcal{H}, \mathcal{K})$,

$$(1.4.6) \qquad A = \sum_{i,j}\langle\mathbf{v}_i, A\mathbf{u}_j\rangle|\mathbf{v}_i\rangle\langle\mathbf{u}_j|.$$

In particular, the dimension of $\mathcal{B}(\mathcal{H}, \mathcal{K})$ is mn.

Proof. For each j, $A\mathbf{u}_j = \sum_{i=1}^{m} \langle \mathbf{v}_i, A\mathbf{u}_j \rangle \mathbf{v}_i$, and substituting this into (1.3.11) yields (1.4.6). This shows that the set in (1.4.5) spans $\mathcal{B}(\mathcal{H}, \mathcal{K})$.

Next, by (1.4.3), $\mathrm{Tr}[(|\mathbf{v}_i\rangle\langle\mathbf{u}_j|)^* |\mathbf{v}_k\rangle\langle\mathbf{u}_\ell|] = \mathrm{Tr}[|\mathbf{u}_j\rangle\langle\mathbf{v}_i||\mathbf{v}_k\rangle\langle\mathbf{u}_\ell|] = \delta_{i,k}\delta_{j,\ell}$, and this shows that the set specified in (1.4.5) is orthonormal. This completes the proof that the set in (1.4.5) is an orthonormal basis for $\mathcal{B}(\mathcal{H}, \mathcal{K})$, and therefore, that the dimension of $\mathcal{B}(\mathcal{H}, \mathcal{K})$ is mn. $\square$

Definition 1.45. Let $\mathcal{H}$ and $\mathcal{K}$ be finite-dimensional Hilbert spaces. Let $\Phi : \mathcal{B}(\mathcal{H}) \to \mathcal{B}(\mathcal{K})$ be a linear transformation. Then $\Phi^\dagger : \mathcal{B}(\mathcal{K}) \to \mathcal{B}(\mathcal{H})$, the **Hermitian conjugate** of Φ, is the unique linear transformation such that for all $A \in \mathcal{B}(\mathcal{K})$ and $B \in \mathcal{B}(\mathcal{H})$,

$$(1.4.7) \qquad \langle A, \Phi(B) \rangle_{\mathrm{HS}} = \langle \Phi^\dagger(A), B \rangle_{\mathrm{HS}}.$$

Remark 1.46. We shall encounter other inner products on $\mathcal{B}(\mathcal{H})$, but throughout this book, the symbol $\dagger$ denotes the Hermitian conjugate with respect to the Hilbert-Schmidt inner product.

1.4.1. The matrix representation of operators. It is often convenient to represent operators from one finite-dimensional Hilbert space to another in terms of matrices. For example, this is convenient whenever determinants enter our considerations.

Definition 1.47. Let $\mathcal{H}$ and $\mathcal{K}$ be finite-dimensional Hilbert spaces. Let $\{\mathbf{u}_1, \dots, \mathbf{u}_n\}$ be an orthonormal basis of $\mathcal{H}$ and let $\{\mathbf{v}_1, \dots, \mathbf{v}_m\}$ be an orthonormal basis of $\mathcal{K}$. Let $M_{m,n}(\mathbb{C})$ denote the $m \times n$ matrices with entries in $\mathbb{C}$. For $A \in \mathcal{B}(\mathcal{H}, \mathcal{K})$, define $[A]$ to be the $m \times n$ matrix with i,j-th entry $[A]_{i,j}$ given by

$$[A]_{i,j} := \langle \mathbf{v}_i, A\mathbf{u}_j \rangle = \langle |\mathbf{v}_i\rangle\langle\mathbf{u}_j|, A \rangle_{\mathrm{HS}}.$$

Then $[A]$ is called the **matrix representation** of A for the orthonormal bases $\{\mathbf{u}_1, \dots, \mathbf{u}_n\}$ and $\{\mathbf{v}_1, \dots, \mathbf{v}_m\}$, and the map $A \mapsto [A]$ from $\mathcal{B}(\mathcal{H}, \mathcal{K})$ to $M_{m,n}(\mathbb{C})$ is called the **matrix representation map** for these orthonormal bases.

Formula (1.4.6) can now be written as

$$(1.4.8) \qquad A = \sum_{i,j} [A]_{i,j} |\mathbf{v}_i\rangle\langle\mathbf{u}_j|.$$

Since by Lemma 1.44, $\{|\mathbf{v}_i\rangle\langle\mathbf{u}_j|\}_{1 \le i \le m,\ 1 \le j \le n}$ is an orthonormal basis of $\mathcal{B}(\mathcal{H}, \mathcal{K})$, (1.4.8) is the expansion of A in this orthonormal basis, and the entries of the matrix are coefficients in this orthonormal expansion.

Now let us consider operators in $\mathcal{B}(\mathcal{H})$, using only the single orthonormal basis $\{\mathbf{u}_1, \dots, \mathbf{u}_n\}$ for both the domain and the range. Let $M_n(\mathbb{C})$ denote the $n \times n$ matrices with entries in $\mathbb{C}$. Then $A \in \mathcal{B}(\mathcal{H})$ has the matrix representation $[A] \in M_n(\mathbb{C})$ given by $[A]_{i,j} = \langle \mathbf{u}_i, A\mathbf{u}_j \rangle$.

Theorem 1.48. *Let $\mathcal{H}$ be a Hilbert space of finite dimension n, and let $\{\mathbf{u}_1, \ldots, \mathbf{u}_n\}$ be an orthonormal basis in $\mathcal{H}$. Let $\Phi : \mathcal{B}(\mathcal{H}) \to M_n(\mathbb{C})$ be the matrix representation for this orthonormal basis, that is*

$$(1.4.9) \qquad\qquad \Phi(A) = [A].$$

Then:

(1) Φ is linear, and for all $A, B \in \mathcal{B}(\mathcal{H})$, $\Phi(AB) = [A][B] = \Phi(A)\Phi(B)$, where the product on the right is matrix multiplication.

(2) For all $A \in \mathcal{B}(\mathcal{H})$, $\Phi(A^*) = [A^*] = [A]^*$, where $[A]^*$ denotes the Hermitian conjugate of $[A] \in M_n(\mathbb{C})$.

(3) $\Phi(\mathbb{1}) = \mathbb{1}$, where on the right, $\mathbb{1}$ denotes the identity matrix in $M_n(\mathbb{C})$.

(4) For all $A \in \mathcal{B}(\mathcal{H})$, the set of eigenvalues of $\Phi(A) = [A]$ is simply the set of eigenvalues of A, independent of the orthonormal basis of $\mathcal{H}$ used to form the matrix representation map.

(5) For all $A \in \mathcal{B}(\mathcal{H})$, $\mathrm{Tr}[A] = \sum_{j=1}^{n}[A]_{j,j} = \mathrm{Tr}[\Phi(A)]$ so that $\Phi : \mathcal{B}(\mathcal{H}) \to M_n(\mathbb{C})$ is a trace preserving map. Consequently, with $\mathcal{B}(\mathcal{H})$ and $M_n(\mathbb{C})$ both equipped with their Hilbert-Schmidt inner products, Φ is unitary.

Proof. Evidently, the map sending A to its matrix representative is linear.

To prove (1), note that since $\sum_{k=1}^{n} |\mathbf{u}_k\rangle\langle\mathbf{u}_k| = \mathbb{1}$, for all $A, B \in \mathcal{B}(\mathcal{H})$,

$$[AB]_{i,j} = \langle \mathbf{u}_i, AB\mathbf{u}_j \rangle = \left\langle \mathbf{u}_i, A\left(\sum_{k=1}^{n} |\mathbf{u}_k\rangle\langle\mathbf{u}_k|\right) B\mathbf{u}_j \right\rangle$$

$$= \sum_{k=1}^{n} \langle \mathbf{u}_i, A\mathbf{u}_k \rangle\langle \mathbf{u}_k, B\mathbf{u}_j \rangle = \sum_{k=1}^{n} [A]_{i,k}[B]_{k,j}.$$

To prove (2), note that for each i, j, $[A^*]_{i,j} = \langle \mathbf{u}_i, A^*\mathbf{u}_j \rangle = \langle A\mathbf{u}_i, \mathbf{u}_j \rangle = \overline{[A]_{j,i}}$. To prove (3), note that for each i, j, $[\mathbb{1}]_{i,j} = \langle \mathbf{u}_i, \mathbf{u}_j \rangle = \delta_{i,j}$.

To prove (4), for any $A \in \mathcal{B}(\mathcal{H})$ and any $\mathbf{x} \in \mathcal{H}$, define $x_j = \langle \mathbf{u}_j, \mathbf{x} \rangle$, $j = 1, \ldots, n$. Then $A\mathbf{x} = \lambda\mathbf{x}$ if and only if $\sum_{j=1}^{n}[A]_{i,j}x_j = \lambda x_i$ for each $i = 1, \ldots, n$. This provides a one-to-one correspondence between eigenvalues and eigenvectors of A and its matrix representative $[A]$. Hence, the eigenvalues of A are precisely the roots of the characteristic polynomial of the matrix representative of A, independent of the basis chosen to form the matrix representative.

To prove (5), by (1.4.9), $\mathrm{Tr}[\Phi(A)] = \sum_{j=1}^{n}[A]_{j,j} = \sum_{j=1}^{n}\langle \mathbf{u}_j, A\mathbf{u}_j \rangle = \mathrm{Tr}[A]$. Combining this with (1) and (2), $\mathrm{Tr}[\Phi(A)^*\Phi(B)] = \mathrm{Tr}[A^*B]$, and then by Lemma 1.36, Φ is unitary. $\qquad\square$

Let $\{\mathbf{v}_1, \ldots, \mathbf{v}_n\}$ be another orthonormal basis for $\mathcal{H}$. Let $\widehat{[A]}_{i,j}$ be the matrix representative of A with respect to this orthonormal basis. Then

$$(1.4.10) \qquad \widehat{[A]}_{i,j} = \langle \mathbf{v}_i, A\mathbf{v}_j \rangle = \sum_{k,\ell} \langle \mathbf{v}_i, \mathbf{u}_k \rangle \langle \mathbf{u}_k, A\mathbf{u}_\ell \rangle \langle \mathbf{u}_\ell, \mathbf{v}_j \rangle.$$

Define $U \in \mathcal{B}(\mathcal{H})$ by $U = \sum_{k=1}^n |\mathbf{u}_k\rangle\langle\mathbf{v}_k|$. One readily checks that $U^*U = UU^* = \mathbb{1}$, so that U is unitary. Then $[U]_{i,j} = \langle \mathbf{u}_i, U\mathbf{u}_j \rangle = \langle \mathbf{v}_i, \mathbf{u}_j \rangle$. Now (1.4.10) can be written as $\widehat{[A]} = [U]^*[A][U]$. Since U is unitary, $[U]^*$ is the inverse of $[U]$, so that $\widehat{[A]}$ is a similarity transform of $[A]$. Since the determinants of any two similar $n \times n$ matrices are equal, the determinant of $[A]$ is independent of the choice of basis. This justifies the following definition:

Definition 1.49. Let $\mathcal{H}$ be a finite-dimensional Hilbert space. Then for $A \in \mathcal{B}(\mathcal{H})$, the **determinant** of A, $\det(A)$, is defined to be the determinant of any matrix representative $[A]$ of A.

Theorem 1.48 says that the algebra of operators on an n-dimensional Hilbert space can be identified, through the arbitrary choice of an orthonormal basis, with $M_n(\mathbb{C})$, the algebra of $n \times n$ matrices. The matrix algebra picture is often useful conceptually and for practical computation.

1.5. The spectral theorem

Definition 1.50 (Spectrum). Let $\mathcal{H}$ be a Hilbert space of any dimension, and let $A \in \mathcal{B}(\mathcal{H})$. The **spectrum** of A, denoted $\sigma(A)$, is the set of $\lambda \in \mathbb{C}$ such that $\lambda\mathbb{1} - A$ is not invertible in $\mathcal{B}(\mathcal{H})$.

If $\mathcal{H}$ is finite dimensional, then by the fundamental theorem of linear algebra, $\lambda\mathbb{1} - A$ is not invertible if and only if $\ker(\lambda\mathbb{1} - A)$ is nonzero, meaning that there is a nonzero vector $\mathbf{v}$ such that $A\mathbf{v} = \lambda\mathbf{v}$. Therefore, for operators A on a finite-dimensional Hilbert space $\mathcal{H}$, the spectrum $\sigma(A)$ is the set of eigenvalues of A. For $\lambda \in \sigma(A)$, let $\mathcal{S}_\lambda(A) := \ker(\lambda\mathbb{1} - A)$ be the λ-**eigenspace** of A.

Lemma 1.51. *Let $A \in \mathcal{B}(\mathcal{H})$ be self-adjoint. Every eigenvalue $\lambda \in \sigma(A)$ is necessarily real. Moreover, if $\lambda, \mu \in \sigma(A)$ and $\lambda \neq \mu$, then the corresponding eigenspaces, $\mathcal{S}_\lambda(A)$ and $\mathcal{S}_\mu(A)$, are mutually orthogonal.*

Proof. If λ is an eigenvalue of A, there is a unit vector $\mathbf{u}$ such that $A\mathbf{u} = \lambda\mathbf{u}$. Then $\lambda = \lambda\langle\mathbf{u},\mathbf{u}\rangle = \langle\mathbf{u}, A\mathbf{u}\rangle = \langle A\mathbf{u},\mathbf{u}\rangle = \overline{\lambda}\langle\mathbf{u},\mathbf{u}\rangle = \overline{\lambda}$. Next, suppose λ and μ are two distinct eigenvalues of A. Let $\mathbf{x}$ and $\mathbf{y}$ be any two eigenvectors with $A\mathbf{x} = \lambda\mathbf{x}$ and $A\mathbf{y} = \mu\mathbf{y}$. Then $\mu\langle\mathbf{x},\mathbf{y}\rangle = \langle\mathbf{x}, A\mathbf{y}\rangle = \langle A\mathbf{x},\mathbf{y}\rangle = \lambda\langle\mathbf{x},\mathbf{y}\rangle$. Since $\lambda \neq \mu$, $\langle\mathbf{x},\mathbf{y}\rangle = 0$. $\qquad\square$

Let $\mathbf{u}, \mathbf{v}$ be two orthogonal unit vectors in a Hilbert space $\mathcal{H}$. Define $X \in \mathcal{B}(\mathcal{H})$ by $X := |\mathbf{u}\rangle\langle\mathbf{v}|$. Since for any $\mathbf{w}$ in $\mathcal{H}$, $X\mathbf{w}$ is a multiple of $\mathbf{u}$, and since

$X\mathbf{u} = 0$, $\sigma(X) = \{0\}$, and $\mathcal{S}_0(X)$ is spanned by $\mathbf{u}$. Note that X is not self-adjoint. *However, nonzero self-adjoint operators on a finite-dimensional Hilbert space always have at least one nonzero eigenvalue:*

Lemma 1.52. *Let $\mathcal{H}$ be a Hilbert space of finite dimension d, real or complex, and let $A \in \mathcal{B}(\mathcal{H})$ be self-adjoint. If $\langle \mathbf{x}, A\mathbf{x} \rangle = 0$ for all $\mathbf{x} \in \mathcal{H}$, then $A = 0$. If $A \neq 0$, then A has at least one nonzero eigenvalue λ, which is necessarily real.*

Proof. We give the proof in the complex case; the real case is almost the same, but simpler. Since A is self-adjoint, for all $\mathbf{y}, \mathbf{z} \in \mathcal{H}$,

$$\mathfrak{R}(\langle \mathbf{y}, A\mathbf{z} \rangle) = \frac{1}{2}(\langle \mathbf{z} + \mathbf{y}, A(\mathbf{z} + \mathbf{y}) \rangle - \langle \mathbf{z}, A\mathbf{z} \rangle - \langle \mathbf{y}, A\mathbf{y} \rangle),$$

Consequently, if $\langle \mathbf{x}, A\mathbf{x} \rangle = 0$ for all $\mathbf{x} \in \mathcal{H}$, then $\mathfrak{R}(\langle \mathbf{y}, A\mathbf{z} \rangle) = 0$ for all $\mathbf{y}, \mathbf{z} \in \mathcal{H}$. Replacing $\mathbf{y}$ by $i\mathbf{y}$, we conclude the same for the imaginary part, and hence if $\langle \mathbf{x}, A\mathbf{x} \rangle = 0$ for all $\mathbf{x} \in \mathcal{H}$, then $\langle \mathbf{y}, A\mathbf{z} \rangle = 0$ for all $\mathbf{y}, \mathbf{z} \in \mathcal{H}$. Taking $\mathbf{y} = A\mathbf{z}$ yields $\|A\mathbf{z}\| = 0$ for all $\mathbf{z}$, and hence $A = 0$.

Now suppose that $A \neq 0$. Let $S := \{\mathbf{u} \in \mathcal{H} \,:\, \|\mathbf{u}\| = 1\}$ denote the unit sphere in $\mathcal{H}$. This is a closed and bounded set in $\mathcal{H}$, and hence S is compact by the Heine-Borel theorem. Define the function $f(\mathbf{u}) := \langle \mathbf{u}, A\mathbf{u} \rangle$ on S. Since A is self-adjoint, $f(\mathbf{u})$ is real valued, and evidently f is continuous.

By Weierstrass's theorem, continuous real-valued functions on compact sets have minimizers and maximizers, and hence there exist vectors $\mathbf{u}_{\min}, \mathbf{u}_{\max} \in S$ such that

$$f(\mathbf{u}_{\min}) \leq f(\mathbf{u}) \leq f(\mathbf{u}_{\max}) \quad \text{for all} \quad \mathbf{u} \in S.$$

At least one of $f(\mathbf{u}_{\min}), f(\mathbf{u}_{\max})$ is nonzero, since otherwise $f(\mathbf{u}) = 0$ for all unit vectors $\mathbf{u}$. By the first part of the theorem, this would mean that $A = 0$ which is not the case. Replacing A by $-A$ if needed, we may assume that $f(\mathbf{u}_{\max}) \neq 0$.

We claim that $\mathbf{u}_{\max}$ is an eigenvector of A with eigenvalue $f(\mathbf{u}_{\max})$. To see this, let $\mathbf{v}$ be any unit vector in $\mathcal{H}$ that is orthogonal to $\mathbf{u}_{\max}$. For $t \in \mathbb{R}$, define $\mathbf{u}(t) = (1 + t^2)^{-1/2}(\mathbf{u}_{\max} + t\mathbf{v}) \in S$. Therefore, $f(\mathbf{u}(0)) \geq f(\mathbf{u}(t))$ for all t. Differentiating, one finds

$$0 = \frac{\mathrm{d}}{\mathrm{d}t}f(\mathbf{u}(t))\Big|_{t=0} = \langle \mathbf{u}_{\max}, A\mathbf{v} \rangle + \langle \mathbf{v}, A\mathbf{u}_{\max} \rangle = \langle A\mathbf{u}_{\max}, \mathbf{v} \rangle + \langle \mathbf{v}, A\mathbf{u}_{\max} \rangle,$$

where the last equality comes from the self-adjointness of A. That is, $\mathfrak{R}(\langle A\mathbf{u}_{\max}, \mathbf{v} \rangle) = 0$. Replacing $\mathbf{v}$ by $i\mathbf{v}$ yields $\mathfrak{I}(\langle A\mathbf{u}_{\max}, \mathbf{v} \rangle) = 0$, and hence $\langle A\mathbf{u}_{\max}, \mathbf{v} \rangle = 0$. Therefore, $A\mathbf{u}_{\max} \in \{\mathbf{u}_{\max}\}^{\perp\perp} = \mathrm{span}(\{\mathbf{u}_{\max}\})$, and hence $A\mathbf{u}_{\max} = \lambda\mathbf{u}_{\max}$ for some number λ. Taking the inner product with $\mathbf{u}_{\max}$,

$$\lambda = \langle \mathbf{u}_{\max}, \lambda\mathbf{u}_{\max} \rangle = \langle \mathbf{u}_{\max}, A\mathbf{u}_{\max} \rangle = f(\mathbf{u}_{\max}) \neq 0. \qquad \square$$

Lemma 1.53. *Let $A \in \mathcal{B}^{\mathrm{s.a.}}(\mathcal{H})$ be nonzero, and let $\mathcal{H}$ be a finite-dimensional Hilbert space. Let λ be a nonzero eigenvalue of A.*

Let P_λ be the orthogonal projection onto the λ-eigenspace. Define $B := A - \lambda P_\lambda$, which is also self-adjoint. The nonzero eigenvalues of B, if any, are precisely the nonzero eigenvalues of A with the exception of λ.

Moreover, if μ is a nonzero eigenvalue of B, $\mathcal{S}_\mu(B) = \mathcal{S}_\mu(A)$.

Proof. Since for all $\mathbf{x} \in \mathcal{H}$, $P_\lambda \mathbf{x}$ is in the λ-eigenspace, $A P_\lambda \mathbf{x} = \lambda P_\lambda \mathbf{x}$, and hence $A P_\lambda = \lambda P_\lambda$. Then since $P_\lambda + P_\lambda^\perp = \mathbb{1}$, $A = A P_\lambda + A P_\lambda^\perp = \lambda P_\lambda + A P_\lambda^\perp$. Therefore

$$B = A - \lambda P_\lambda = A P_\lambda^\perp.$$

Since $B = B^*$, $A P_\lambda^\perp = (A P_\lambda^\perp)^* = P_\lambda^\perp A$, and hence

$$B = A P_\lambda^\perp = P_\lambda^\perp A.$$

Suppose that $B \neq 0$. By Lemma 1.52, B has at least one nonzero eigenvalue μ. Let $\mathbf{u}$ be a nonzero vector such that $B\mathbf{u} = \mu\mathbf{u}$. Since $B = P_\lambda^\perp A$, $B\mathbf{u}$ is in the range of $P_\lambda^\perp$, and hence $P_\lambda^\perp \mathbf{u} = \mathbf{u}$. But then $A\mathbf{u} = A P_\lambda^\perp \mathbf{u} = B\mathbf{u} = \mu\mathbf{u}$. Therefore, $\mu \in \sigma(A)$, and $\mathcal{S}_\mu(B) \subseteq \mathcal{S}_\mu(A)$.

However, since $P_\lambda^\perp \mathbf{u} = \mathbf{u}$, $\mathbf{u} \in \mathcal{S}_\lambda(A)^\perp$, and hence, $\mu \neq \lambda$. If λ is the only nonzero eigenvalue of A, then B has no nonzero eigenvalues, and then by Lemma 1.52, $B = 0$ so that $A = \lambda P_\lambda$.

Conversely, suppose μ is a nonzero eigenvalue of A other than λ, and let $\mathbf{u}$ be a nonzero vector such that $A\mathbf{u} = \mu\mathbf{u}$. By Lemma 1.51, $\mathbf{u} \in \mathcal{S}_\lambda(A)^\perp$, and hence $P_\lambda^\perp \mathbf{u} = \mathbf{u}$. Therefore $B\mathbf{u} = A P_\lambda^\perp \mathbf{u} = A\mathbf{u} = \mu\mathbf{u}$, and hence $\mu \in \sigma(B)$, and then $\mathcal{S}_\mu(A) \subseteq \mathcal{S}_\mu(B)$. $\qquad\square$

The most important theorem about self-adjoint operators is the *spectral theorem*, and we are now ready to prove the finite-dimensional version.

Theorem 1.54 (Spectral theorem). *Let $\mathcal{H}$ be a finite-dimensional Hilbert space, and let $A \in \mathcal{B}(\mathcal{H})$ be self-adjoint. Then $\sigma(A) \subset \mathbb{R}$, and for each $\lambda \in \sigma(A)$, let P_λ be the orthogonal projection onto the eigenspace corresponding to the eigenvalue λ. Then*

$$(1.5.1) \qquad \mathbb{1} = \sum_{\lambda \in \sigma(A)} P_\lambda \qquad \text{and} \qquad A = \sum_{\lambda \in \sigma(A)} \lambda P_\lambda.$$

Proof. We proceed by induction on the the cardinality of $\sigma(A) \backslash \{0\}$, denoted $|\sigma(A) \backslash \{0\}|$. The case $A = 0$ is trivial, so suppose that $A \neq 0$. By Lemma 1.52, $\sigma(A)$ contains at least one nonzero eigenvalue λ, and hence $|\sigma(A) \backslash \{0\}| \geq 1$.

Define $B := A - \lambda P_\lambda$. By Lemma 1.53, if $|\sigma(A)\backslash\{0\}| = 1$, then $B = 0$ so that $A = \lambda P_\lambda$. If the λ-eigenspace of A is $\mathcal{H}$, then $A = \lambda \mathbb{1}$. Otherwise, since $\mathbb{1} = P_\lambda + P_\lambda^\perp$, $A = AP_\lambda + AP_\lambda^\perp = \lambda P_\lambda + AP_\lambda^\perp = A + AP_\lambda^\perp$, and hence $AP_\lambda^\perp = 0$. Thus $P_\lambda^\perp = P_0$, the orthogonal projection onto the 0-eigenspace of A, and $P_0 + P_\lambda = \mathbb{1}$. This proves (1.5.1) when $|\sigma(A)\backslash\{0\}| = 1$.

If $|\sigma(A)\backslash\{0\}| = m > 1$, then by Lemma 1.53 $|\sigma(B)\backslash\{0\}| = m - 1$. Making the inductive hypothesis that the theorem has been established whenever the cardinality of the nonzero spectrum is $m - 1$, we apply this to B to write

$$\sum_{\mu \in \sigma(B)} P_\mu = \mathbb{1} \quad \text{and} \quad \sum_{\mu \in \sigma(B)} \mu P_\mu = B.$$

Then since $A = B + \lambda P_\lambda$, it follows from Lemma 1.53 that

$$(1.5.2) \qquad A = \lambda P_\lambda + \sum_{\mu \in \sigma(B)} \mu P_\mu = \sum_{\mu \in \sigma(A)} \mu P_\mu.$$

By Lemma 1.51, $\{P_\mu \ : \ \mu \in \sigma(A)\backslash\{0\}\}$ is a mutually orthogonal set of projections. Hence $P := \sum_{\mu \in \sigma(A)\backslash\{0\}} P_\mu$ is an orthogonal projection and

$$AP = \left(\sum_{\mu \in \sigma(A)\backslash\{0\}} \mu P_\mu \right) P = \sum_{\mu \in \sigma(A)\backslash\{0\}} \mu P_\mu P = \sum_{\mu \in \sigma(A)\backslash\{0\}} \mu P_\mu = A.$$

It follows that $A(\mathbb{1} - P) = 0$, so that $P_0 := \mathbb{1} - P$ is the orthogonal projection onto the kernel of A which is the 0-eigenspace of A. Therefore, $\sum_{\mu \in \sigma(A)} P_\mu = \mathbb{1}$. Together with (1.5.2), this completes the proof of (1.5.1). $\qquad\square$

Corollary 1.55. *Let $\mathcal{H}$ be a Hilbert space of finite dimension d. For every self-adjoint $A \in \mathcal{B}(\mathcal{H})$, there exists an orthonormal basis $\{\mathbf{u}_1, \ldots, \mathbf{u}_d\}$ of $\mathcal{H}$ consisting of eigenvectors of A. That is, $A\mathbf{u}_j = \lambda_j \mathbf{u}_j$. Then in Dirac notation, $A = \sum_{j=1}^d \lambda_j |\mathbf{u}_j\rangle\langle\mathbf{u}_j|$.*

Proof. Write each P_λ appearing in (1.5.1) in the form (1.3.12), and collect all of the orthonormal vectors used onto the single set $\{\mathbf{u}_1, \ldots, \mathbf{u}_d\}$. $\qquad\square$

The spectral theorem is extremely useful for dealing with functions of self-adjoint operators. We begin with something more simple. One can always form linear combinations of positive powers of operators, and hence polynomials in operators, self-adjoint or not. Let $p(z)$ be a polynomial

$$p(z) = \sum_{j=0}^d c_j z^j,$$

where the coefficients $c_j, 0 \le j \le d$, belong to $\mathbb{C}$. For $A \in \mathcal{B}(\mathcal{H})$, define $A^0 := \mathbb{1}$ and $p(A) \in \mathcal{B}(\mathcal{H})$ by $p(A) := \sum_{j=0}^d c_j A^j$.

Lemma 1.56. *Let $\mathcal{H}$ be a finite-dimensional Hilbert space. Let $A \in \mathcal{B}(\mathcal{H})$ be self-adjoint, and let $A = \sum_{\lambda \in \sigma(A)} \lambda P_\lambda$ be its spectral decomposition. For each $\lambda \in \sigma(A)$, define the polynomial*

$$(1.5.3) \qquad\qquad p_\lambda(x) = \prod_{\lambda' \in \sigma(H) \setminus \{\lambda\}} \frac{\lambda' - x}{\lambda' - \lambda}.$$

Then p_λ has real coefficients and $P_\lambda = p_\lambda(A)$. That is, each P_λ is a polynomial in A, and the polynomial has real coefficients.

Proof. By the explicit formula (1.5.3) and Lemma 1.51, the coefficients of $p_\lambda(x)$ are real. Also by the explicit formula, if $\mathbf{x}$ is an eigenvector of A with eigenvalue $\lambda' \neq \lambda$, then $(\lambda' - A)\mathbf{x} = 0$, and hence $p_\lambda(A)\mathbf{x} = 0$. However, if $A\mathbf{x} = \lambda\mathbf{x}$, then $(\lambda' - A)\mathbf{x} = (\lambda' - \lambda)\mathbf{x}$, and hence $p_\lambda(A)\mathbf{x} = \mathbf{x}$. Therefore, by Corollary 1.55, $p_\lambda(A)$ and P_λ have the same action on an orthonormal basis of $\mathcal{H}$, so that $p_\lambda(A) = P_\lambda$. $\qquad\qquad\square$

There is a simple but significant extension of the spectral theorem. Any operator $A \in \mathcal{B}(\mathcal{H})$ can be decomposed as $A = H + iK$ where

$$H := \frac{1}{2}(A + A^*) \quad \text{and} \quad K := \frac{1}{2i}(A - A^*).$$

Note that H and K are self-adjoint. They are called the **real and imaginary parts** of A. Recall that an operator $A \in \mathcal{B}(\mathcal{H})$ is **normal** in case its real and imaginary parts commute, or, what is the same thing, if A and A^* commute. For example, all unitary operators are normal.

Definition 1.57. Let $\mathcal{H}$ be a finite-dimensional Hilbert space. The **spectral radius** of $A \in \mathcal{B}(\mathcal{H})$, $\rho(A)$, is defined by

$$\rho(A) = \max_{\lambda \in \sigma(A)} \{|\lambda|\}.$$

By definition, there is some $\lambda \in \sigma(A)$ such that $|\lambda| = \rho(A)$, and then there is a unit vector $\mathbf{u}$ such that $A\mathbf{u} = \lambda\mathbf{u}$. Then $\|A\mathbf{u}\| = \rho(A)\|\mathbf{u}\| = \rho(A)$, and this shows that $\rho(A) \leq \|A\|$. For normal operators, this inequality is saturated.

Theorem 1.58. *Let $\mathcal{H}$ be a finite-dimensional Hilbert space. Every normal operator $A \in \mathcal{B}(\mathcal{H})$ has a spectral decomposition of the form (1.5.6) that differs from the spectral decomposition (1.5.1) of a self-adjoint operator only in that the eigenvalues are complex in (1.5.6). Moreover, for every normal operator $A \in \mathcal{B}(\mathcal{H})$,*

$$(1.5.4) \qquad\qquad \|A\| = \rho(A) = \max\{|\lambda| \; : \; \lambda \in \sigma(A)\}.$$

Proof. Let $A \in \mathcal{B}(\mathcal{H})$ be normal, with $A = H + iK$ being its decomposition into real and imaginary parts. Let

$$(1.5.5) \qquad H = \sum_{\lambda \in \sigma(H)} \lambda P_\lambda \quad \text{and} \quad K = \sum_{\mu \in \sigma(K)} \mu Q_\mu$$

be the spectral decompositions of H and K, respectively.

Since each P_λ is a polynomial in H, and each Q_μ is a polynomial in K, and since H and K commute, each P_λ commutes with each Q_μ. Therefore, each $P_\lambda Q_\mu$ is an orthogonal projection—though $P_\lambda Q_\mu$ will be the orthogonal projection on to the trivial subspace $\{0\}$ if $\lambda + i\mu$ is not an eigenvalue of $A = H + iK$.

Since $\sum_{\mu \in \sigma(K)} Q_\mu = \mathbb{1}$, for each $\lambda \in \sigma(H)$, $\sum_{\mu \in \sigma(K)} P_\lambda Q_\mu = P_\lambda$. Likewise, for each $\mu \in \sigma(K)$, $\sum_{\lambda \in \sigma(H)} P_\lambda Q_\mu = Q_\mu$. From either of these identities we obtain $\mathbb{1} = \sum_{\lambda \in \sigma(H)} \sum_{\mu \in \sigma(K)} P_\lambda Q_\mu$, and using both identities in (1.5.5) yields $A = H + iK = \sum_{\lambda \in \sigma(H)} \sum_{\mu \in \sigma(K)} (\lambda + i\mu) P_\lambda Q_\mu$. It follows that

$$\sigma(A) = \{\lambda + i\mu \ : \ \lambda \in \sigma(H), \ \mu \in \sigma(K) \text{ and } P_\lambda Q_\mu \neq 0\}.$$

For each $\nu \in \sigma(A)$, define $R_\nu = \sum_{\{\lambda \in \sigma(H), \ \mu \in \sigma(K) \ : \ \lambda + i\mu = \nu\}} P_\lambda Q_\mu$. Then R_μ is a nonzero orthogonal projection, and

$$(1.5.6) \qquad \mathbb{1} = \sum_{\nu \in \sigma(A)} R_\nu \quad \text{and} \quad A = \sum_{\nu \in \sigma(A)} \nu R_\nu,$$

which proves the first part of the theorem. For the second part, (1.5.6) yields $A^*A = \sum_{\nu \in \sigma(A)} |\nu|^2 R_\nu$, so that for any unit vector $\mathbf{u}$,

$$\|A\mathbf{u}\|^2 = \langle \mathbf{u}, A^*A\mathbf{u} \rangle = \sum_{\nu \in \sigma(A)} |\nu|^2 \langle \mathbf{u}, R_\nu \mathbf{u} \rangle$$

$$\leq \max_{\lambda \in \sigma(A)} \{|\lambda|^2\} \sum_{\nu \in \sigma(A)} \langle \mathbf{u}, R_\nu \mathbf{u} \rangle = \max_{\lambda \in \sigma(A)} \{|\lambda|^2\}.$$

This proves that $\|A\| \leq \max_{\lambda \in \sigma(A)} \{|\lambda|\}$. If $\nu \in \sigma(A)$ satisfies $|\nu| = \max_{\lambda \in \sigma(A)} \{|\lambda|\}$, and $\mathbf{u}$ is a normalized eigenvector of A with eigenvalue ν, then $\|A\mathbf{u}\| = |\nu|$, so the upper bound is attained. This proves (1.5.4). $\qquad \square$

1.6. The functional calculus for normal operators

Definition 1.59. Let $\mathcal{H}$ be a finite-dimensional Hilbert space. Let $A \in \mathcal{B}(\mathcal{H})$ be normal with the spectral decomposition

$$(1.6.1) \qquad \mathbb{1} = \sum_{\lambda \in \sigma(A)} P_\lambda \quad \text{and} \quad A = \sum_{\lambda \in \sigma(A)} \lambda P_\lambda.$$

Let $\mathcal{C}(\sigma(A))$ denote the space of all complex valued functions on $\sigma(A)$. For $f \in \mathcal{C}(\sigma(A))$, define $f(A) \in \mathcal{B}(\mathcal{H})$ to be the normal operator given by

$$(1.6.2) \qquad f(A) = \sum_{\lambda \in \sigma(A)} f(\lambda) P_\lambda.$$

In fact, $\mathcal{C}(\sigma(A))$ is not only a set of functions, it is an algebra: given any $f, g \in \mathcal{C}(\sigma(A))$, and any $z, w \in \mathbb{C}$, define scalar multiplication, addition, and multiplication in $\mathcal{C}(\sigma(A))$ by

$$(1.6.3) \qquad (zf + wg)(\lambda) := zf(\lambda) + wg(\lambda) \quad \text{and} \quad fg(\lambda) := f(\lambda)g(\lambda)$$

for all $\lambda \in \sigma(A)$. Define a norm function $\|\|\cdot\|\|$ on $\mathcal{C}(\sigma(A))$ by

$$\|\|f\|\| = \max_{\lambda \in \sigma(A)} |f(\lambda)|.$$

Then one readily checks that the Minkowski inequality is satisfied; that is, for all $f, g \in \mathcal{C}(\sigma(A))$, $\|\|f + g\|\| \leq \|\|f\|\| + \|\|g\|\|$. Defining $d(f, g) := \|\|f - g\|\|$, one readily checks that this is a metric on $\mathcal{C}(\sigma(A))$, and that if $\{f_n\}_{n \in \mathbb{N}}$ is a sequence in $\mathcal{C}(\sigma(A))$, and $f \in \mathcal{C}(\sigma(A))$, then $\lim_{n \to \infty} \|\|f_n - f\|\| = 0$ if and only if $\lim_{n \to \infty} f_n(\lambda) = f(\lambda)$ for each λ. (On a finite set there is no difference between pointwise and uniform convergence.) Moreover, the norm $\|\|\cdot\|\|$ respects the multiplicative structure on the algebra $\mathcal{C}(\sigma(A))$ in that for all $f, g \in \mathcal{C}(\sigma(A))$, $\|\|fg\|\| \leq \|\|f\|\|\,\|\|g\|\|$ as one readily checks. Finally, it is an easy exercise to check that, as a consequence of the completeness of the complex numbers, $\mathcal{C}(\sigma(A))$ is complete in its norm topology. We have already observed that $\mathcal{B}(\mathcal{H})$ is complete in its norm topology. The map $f \mapsto f(A)$ respects the associated algebraic and metric structures on $\mathcal{C}(\sigma(A))$ and $\mathcal{B}(\mathcal{H})$:

Theorem 1.60 (Spectral calculus). *Let $\mathcal{H}$ be a finite-dimensional Hilbert space. Let $A \in \mathcal{B}(\mathcal{H})$ be normal with the spectral decomposition* (1.6.1). *Let Φ_A denote the map from $\mathcal{C}(\sigma(A))$ to $\mathcal{B}(\mathcal{H})$ given by*

$$(1.6.4) \qquad \Phi_A(f) := \sum_{\lambda \in \sigma(A)} f(\lambda)P_\lambda = f(A).$$

Then:

(1) *Φ_A is linear and for all $f, g \in \mathcal{C}(\sigma(A))$, $\Phi_A(fg) = \Phi_A(f)\Phi_A(g)$.*

(2) *Φ_A is isometric; that is, for all $f \in \mathcal{C}(\sigma(A))$, $\|\Phi_A(f)\| = \|\|f\|\|$.*

(3) *For any $f \in \mathcal{C}(\sigma(A))$ and any $g \in \mathcal{C}(\sigma(f(A)))$, so that $g \circ f \in \mathcal{C}(\sigma(A))$,*

$$(1.6.5) \qquad g(f(A)) = (g \circ f)(A).$$

(4) *If $p(z)$ is any polynomial in the real variable z, $p(z) = \sum_{j=1}^{d} c_j z^j$,*

$$(1.6.6) \qquad \Phi_A(p) = \sum_{j=1}^{d} c_j A^j.$$

Proof. The linearity of Φ_A is immediate from the definitions (1.6.3) and (1.6.4). To prove the second part of (1), note that for $\lambda \neq \mu$, $P_\lambda P_\mu = 0$, and hence

$$f(A)g(A) = \sum_{\lambda, \mu \in \sigma(A)} f(\lambda)g(\mu)P_\lambda P_\mu = \sum_{\lambda \in \sigma(A)} f(\lambda)g(\lambda)P_\lambda = fg(A).$$

Next, (2) follows from (1.5.4) so that $\|f(A)\| = \max_{\lambda \in \sigma(A)}\{|f(\lambda)|\} = \|\!|f|\!\|$. Two applications of the definition (1.6.2) yield (1.6.5), proving (3).

To prove (1.6.6), use (1) and induction to show that $A^j = \sum_{\lambda \in \sigma(A)} \lambda^j P_\lambda$ for all $j \in \mathbb{N}$. Now (1.6.6) follows from (1.6.2). $\qquad\square$

Remark 1.61. There is a version of Theorem 1.60 that is valid for normal operators on an infinite-dimensional Hilbert space. In this version, the formula (1.6.2) that defines Φ_A involves Lebesgue integrals instead of sums, and it is more technically involved. However, with this one difference, the rest of the theorem is the same. See [164] for a particularly elegant treatment.

Example 1.62. As a first example of the application of Theorem 1.60, suppose that $A \in \mathcal{B}(\mathcal{H})$ is self-adjoint and $\sigma(A) \subset (0, \infty)$. Then the logarithm function $f(x) = \log(x)$ is well-defined on $\sigma(A)$, and we use (1.6.2),

$$\log(A) = \sum_{\lambda \in \sigma(A)} \log(\lambda) P_\lambda.$$

Now take $g(x) = \exp(x)$ so that $g \circ f(x) = x$ for all $x > 0$. Let $H := \log(A)$. By (1.6.5), $\exp(H) = \exp(\log(A)) = A$. Moreover, define the degree N polynomial $p_N(z) := \sum_{k=1}^{N} z^k/k!$. Then $\lim_{N \to \infty} \|\!|\exp - p_N|\!\| = 0$, and hence by part (2) of Theorem 1.60,

$$\lim_{N \to \infty} \left\| \exp(H) - \sum_{k=0}^{N} \frac{1}{k!} H^k \right\| = 0.$$

Thus, the exponential function defined using the spectral calculus is the same as the one defined in terms of the convergent power series for the exponential.

Example 1.63. As a second example of the application of Theorem 1.60, suppose that $A \in \mathcal{B}(\mathcal{H})$ is self-adjoint and $\sigma(A) \subset [0, \infty)$. Then the square root function $f(x) = x^{1/2}$ is well-defined on $\sigma(A)$, and we may use (1.6.2) to define

$$A^{1/2} = \sum_{\lambda \in \sigma(A)} \lambda^{1/2} P_\lambda.$$

Then by (1) of Theorem 1.60, $(A^{1/2})^2 = A$, so that $A^{1/2}$ is indeed a square root of A. In fact, it is the unique square root of A with a natural positivity property. (See Exercise 16.) This brings us to the next definition.

Definition 1.64 (Positive operators). An operator $A \in \mathcal{B}(\mathcal{H})$ is **positive semi-definite** in case $\langle \mathbf{v}, A\mathbf{v} \rangle \geq 0$ for all $\mathbf{v} \in \mathcal{H}$, in which case we write $A \geq 0$, and it is **positive definite** in case $\langle \mathbf{v}, A\mathbf{v} \rangle > 0$ for all $\mathbf{v} \neq 0$ in $\mathcal{H}$ in which case we write $A > 0$. The set of positive semidefinite operators in $\mathcal{B}(\mathcal{H})$ is denoted by $\mathcal{B}^+(\mathcal{H})$, and the set of positive definite operators in $\mathcal{B}(\mathcal{H})$ is denoted by $\mathcal{B}^{++}(\mathcal{H})$.

Theorem 1.65. *For any Hilbert space $\mathcal{H}$, every $A \in \mathcal{B}^+(\mathcal{H})$ is self-adjoint. Consequently, for any $A \in \mathcal{B}(\mathcal{H})$, $A \in \mathcal{B}^+(\mathcal{H})$ if and only if $A = K^*K$ for some $K \in \mathcal{B}(\mathcal{H})$.*

Proof. Let $A \in \mathcal{B}^+(\mathcal{H})$. Then for all $\mathbf{v}$, $\langle A\mathbf{v}, \mathbf{v} \rangle = \overline{\langle \mathbf{v}, A\mathbf{v} \rangle} = \langle \mathbf{v}, A\mathbf{v} \rangle$. Therefore, for all $\mathbf{v}, \mathbf{w} \in \mathcal{H}$, $\langle \mathbf{v} + \mathbf{w}, A(\mathbf{v} + \mathbf{w}) \rangle = \langle A(\mathbf{v} + \mathbf{w}), \mathbf{v} + \mathbf{w} \rangle$, and after cancellation and rearrangement, $\langle \mathbf{v}, A\mathbf{w} \rangle - \langle A\mathbf{w}, \mathbf{v} \rangle = \langle A\mathbf{v}, \mathbf{w} \rangle - \langle \mathbf{w}, A\mathbf{v} \rangle$, and this is the same as $\mathfrak{I}(\langle \mathbf{v}, A\mathbf{w} \rangle) = \mathfrak{I}(\langle A\mathbf{v}, \mathbf{w} \rangle)$. Replacing $\mathbf{w}$ by $i\mathbf{w}$, we conclude that $\mathfrak{R}(\langle \mathbf{v}, A\mathbf{w} \rangle) = \mathfrak{R}(\langle A\mathbf{v}, \mathbf{w} \rangle)$ as well, and hence $\langle \mathbf{v}, A\mathbf{w} \rangle = \langle A\mathbf{v}, \mathbf{w} \rangle$ for all $\mathbf{v}, \mathbf{w} \in \mathcal{H}$.

For the second part, if $A = K^*K$, then for any $\mathbf{v} \in \mathcal{H}$, $\langle \mathbf{v}, A\mathbf{v} \rangle = \langle K\mathbf{v}, K\mathbf{v} \rangle \geq 0$, so that $A \in \mathcal{B}^+(\mathcal{H})$.

If $A \in \mathcal{B}^+(\mathcal{H})$, by the first part of this theorem, A is self adjoint. We continue the proof for $\mathcal{H}$ finite dimensional, but the argument can be adapted to the general case. By Theorem 1.54, A has the spectral decomposition (1.5.1). Since $\mathcal{H}$ is finite dimensional, $\sigma(A)$ consists of eigenvalues of A. If λ is an eigenvalue of A, there is a unit vector $\mathbf{u} \in \mathcal{H}$ such that $A\mathbf{u} = \lambda\mathbf{u}$, and then $\lambda = \langle \mathbf{u}, A\mathbf{u} \rangle \geq 0$. Hence for every $A \in \mathcal{B}^+(\mathcal{H})$, $\sigma(A) \subset [0, \infty)$, and we may define $K := A^{1/2}$, where $A^{1/2}$ is defined as in Example 1.63, and as shown there, $A = K^*K$. $\qquad\square$

Lemma 1.66. *Let $\mathcal{H}$ be a finite-dimensional Hilbert space. Every $A \in \mathcal{B}(\mathcal{H})$ can be written as a linear combination of four operators in $\mathcal{B}^+(\mathcal{H})$:*

$$(1.6.7) \qquad\qquad A = X_1 - X_2 + iX_3 - iX_4$$

with $X_1, X_2, X_3, X_4 \in \mathcal{B}^+(\mathcal{H})$.

Proof. Define $H := \frac{1}{2}(A + A^*)$ and $K := \frac{1}{2i}(A - A^*)$, so that $X = H + iK$ where H and K are self-adjoint. Now define $f_+(x) := \frac{1}{2}(x + |x|)$ and $f_-(x) := \frac{1}{2}(|x| - x)$, $x \in \mathbb{R}$, so that $x = f_+(x) - f_-(x)$, with $f_+(x), f_-(x) \geq 0$ for all x. Then defining

$$X_1 := f_+(H), \quad X_2 := f_-(H), \quad X_3 := f_+(K), \quad X_4 := f_-(K),$$

using Theorem 1.60, $X_1, X_2, X_3, X_4 \in \mathcal{B}^+(\mathcal{H})$, and (1.6.7) is satisfied. $\qquad\square$

Definition 1.67. Let $\mathcal{H}$ be a Hilbert space. For self-adjoint $H, K \in \mathcal{B}(\mathcal{H})$ we write $H \geq K$ in case $H - K \in \mathcal{B}^+(\mathcal{H})$ and $H > K$ in case $H - K \in \mathcal{B}^{++}(\mathcal{H})$.

One readily checks that for all self-adjoint H, K, L:

    (1) $H \geq H$.

    (2) If $H \geq K$ and $K \geq H$, then $H = K$.

    (3) If $H \geq K$ and $K \geq L$, then $H \geq L$.

Thus the relation specified in Definition 1.67 specifies a **partial order** on the self-adjoint elements of $\mathcal{B}(\mathcal{H})$.

1.7. Quantum mechanics: measurement and dynamics

The mathematics discussed in the first sections of this chapter is precisely the mathematics needed to formulate the basic rules of quantum mechanics.

In quantum mechanics, the state of a physical system is completely specified by a unit vector ψ in a Hilbert space $\mathcal{H}$, and the observables are represented by self-adjoint operators A on $\mathcal{H}$. When we refer to ψ as representing the state of the system, we mean that encoded into ψ is a complete description of what can be said in advance about about the result of any possible measurement that can be made of the system, before the measurement is made. Even knowing ψ, one cannot in general say exactly what the measurement will yield because quantum mechanics is inherently probabilistic, as we now explain. But as far as is known, and in the current orthodox view, ψ provides as complete a description of the state of the system as is possible.

When $\mathcal{H}$ is finite dimensional, the nature of self-adjoint operators is particularly simple: Let A be self-adjoint in $\mathcal{B}(\mathcal{H})$. Then by Theorem 1.54, there is a set $\{P_1, \dots, P_m\}$ of mutually orthogonal projections on $\mathcal{H}$ and real numbers $\{\lambda_1, \dots, \lambda_m\}$, the eigenvalues of A, such that

$$\sum_{j=1}^{m} P_j = \mathbb{1} \qquad \text{and} \qquad \sum_{j=1}^{m} \lambda_j P_j = A,$$

where $\mathbb{1}$ denotes the identity operator on $\mathcal{H}$. The range of the projection P_j is the eigenspace of A with eigenvalue λ_j.

According to the laws of quantum mechanics, when an experiment is performed to **measure** the value of the observable A, the value will always be one of the numbers $\{\lambda_1, \dots, \lambda_n\}$, and even if the state ψ of the system is known exactly, the result will be random: Let p_j denote the probability that the observation yields the the result λ_j. Since the outcome is one of the numbers $\{\lambda_1, \dots, \lambda_m\}$, we must have $\sum_{j=1}^{m} p_j = 1$.

By the laws of quantum mechanics,

$$(1.7.1) \qquad p_j = \langle \psi, P_j \psi \rangle = \|P_j \psi\|^2,$$

where the second equality uses $P_j = P_j^2$ and $P_j = P_j^*$. Since $\sum_{j=1}^{m} P_j = \mathbb{1}$, $\sum_{j=1}^{m} p_j = \langle \psi, \left(\sum_{j=1}^{m} P_j \right) \psi \rangle = \langle \psi, \psi \rangle = 1$. Hence the p_j's specified in (1.7.1) are indeed probabilities.

Now take the average over very many independent measurements of A, each time with the system prepared to be in the state ψ. By the *law of large numbers*, the result is very likely to be close to

$$\sum_{j=1}^{m} \lambda_j \langle \psi, P_j \psi \rangle = \langle \psi, A\psi \rangle.$$

This is called the **expected value of the observable A in the state ψ**.

A distinctive feature of quantum mechanics that the very act of measuring an observable changes the state of the system. Suppose that the system has the state ψ, and the observable A is measured. The value λ_j is found with probability p_j. Given that λ_j was the observed value, after the measurement, the state of the system is then no longer ψ, but is instead

(1.7.2) $$\phi := \|P_j\psi\|^{-1} P_j\psi.$$

Since $\sum_{j=1}^{m} P_j = \mathbb{1}$, any unit vector $\psi \in \mathcal{H}$ can be written as

(1.7.3) $$\psi = \sum_{j=1}^{m} P_j\psi,$$

which expresses ψ as a **superposition** of eigenfunction of A. During measurement of A, the state *collapses* onto a single eigenvector ϕ as in (1.7.2).

This random selection of a single component in the superposition (1.7.3) during a measurement of the observable A is one of the deep mysteries of quantum mechanics. Because of this, in quantum mechanics there are no passive measurements. The very act of performing the measurement changes the state of the system, with one exception, namely if ψ happens to be an eigenvector of A. In that case, the result of the observation is certain: if $A\psi = \lambda_j\psi$, the value λ_j will be observed, and the state of the system will remain ψ.

For any operator A on $\mathcal{H}$, and any $\mathbf{x}, \mathbf{y} \in \mathcal{H}$, $\mathrm{Tr}[|\mathbf{x}\rangle\langle\mathbf{y}|A] = \langle \mathbf{y}, A\mathbf{x} \rangle$ and therefore, (1.7.1) can be written as $p_j = \mathrm{Tr}[|\psi\rangle\langle\psi|P_j]$. Strictly speaking, it is the rank one projection $|\psi\rangle\langle\psi|$ and not the unit vector ψ itself that represents the state of the quantum system: multiplying ψ by a phase $e^{i\theta} \neq 1$ does not change $|\psi\rangle\langle\psi|$, and hence does not change the probabilities p_j, though it does change ψ. However, it is common practice to refer to ψ as **the state vector**. Also for this reason, the Hilbert space $\mathcal{H}$ is often called the **state space** of the system.

1.7.1. The wave function picture. In many quantum problems, there is a preferred set of observables consisting of a maximal commuting set of self-adjoint operators on the state space $\mathcal{H}$. It is then natural to use an orthonormal basis consisting of eigenvectors of the preferred observables. The eigenvectors are only determined up to phases; that is, up to multiplication by $e^{i\theta}$ for some $\theta \in [0, 2\pi)$. However, there is often even a natural choice of the phases, and hence a preferred orthonormal basis.

When an orthonormal basis $\{\mathbf{u}_1, \ldots, \mathbf{u}_m\}$ is specified for $\mathcal{H}$, we may regard $\mathcal{H}$ as a Hilbert space of square summable functions. Then we can express every ψ in terms of its coordinate vector $(\psi_1, \ldots, \psi_n) \in \mathbb{C}^n$ where $\psi_j := \langle \mathbf{u}_j, \psi \rangle$, $1 \leq j \leq n$. We may think of the coordinate vector as the complex valued function ψ defined on the set $\{1, \ldots, n\}$ by setting $\psi(j) := \psi_j$, $j = 1, \ldots, n$.

These functions ψ are often referred to as **wave functions**, and such a representation of $\mathcal{H}$ gives us the wave function picture. The additional structure present in such cases if often useful.

Definition 1.68 (Hilbert spaces of square summable functions). Let $\mathcal{X} = \{x_1, \ldots, x_n\}$ be some finite set. A **nondegenerate measure** μ on $\mathcal{X}$ is a function μ on $\mathcal{X}$ with values in $(0, \infty)$. For two complex valued functions ϕ, ψ on $\mathcal{X}$, define their $L^2(\mathcal{X}, \mu)$ **inner product** by

$$\langle \phi, \psi \rangle_{L^2(\mathcal{X}, \mu)} := \sum_{x \in \mathcal{X}} \overline{\phi(x)} \psi(x) \mu(x).$$

This evidently defines is a positive definite sesquilinear form on the complex functions on $\mathcal{X}$. Define $L^2(\mathcal{X}, \mu)$ to be the complex Hilbert space consisting of complex functions ψ on $\mathcal{X}$ equipped with the $L^2(\mathcal{X}, \mu)$ inner product.

The Hilbert space $L^2(\mathcal{X}, \mu)$ has a natural orthonormal basis $\{\phi_1, \ldots, \phi_n\}$, where for $1 \leq j \leq n$,

$$(1.7.4) \qquad \phi_j(x_k) := \begin{cases} 1 & k = j \\ 0 & k \neq j. \end{cases}$$

Now let $\mathcal{H}$ be a complex Hilbert space with dimension d. Choose any orthonormal basis $\{\mathbf{u}_1, \ldots, \mathbf{u}_d\}$ of $\mathcal{H}$. Define $\mathcal{X} := \{1, \ldots, d\}$, and define μ to be counting measure on $\mathcal{X}$. That is, for $E \subset \mathcal{X}$, $\mu(E)$ is the number of elements in E. For any vector $\mathbf{v} \in \mathcal{H}$, define a function $\psi_{\mathbf{v}}$ on $\mathcal{X}$ by

$$\psi_{\mathbf{v}}(j) := \langle \mathbf{u}_j, \mathbf{v} \rangle, \qquad 1 \leq j \leq d.$$

For any $\mathbf{v}, \mathbf{w} \in \mathcal{H}$, $\langle \psi_{\mathbf{v}}, \psi_{\mathbf{w}} \rangle_{L^2(\mu)} = \sum_{j=1}^{d} \overline{\langle \mathbf{u}_j, \mathbf{v} \rangle} \langle \mathbf{u}_j, \mathbf{w} \rangle = \langle \mathbf{v}, \mathbf{w} \rangle$.

Therefore, $\mathbf{v} \mapsto \psi_{\mathbf{v}}$ is an isometry of $\mathcal{H}$ into $L^2(\mathcal{X}, \mu)$, and since both Hilbert spaces have dimension d, this map U is a unitary transformation from $\mathcal{H}$ onto $L^2(\mathcal{X}, \mu)$. In fact, $U\mathbf{u}_j = \phi_j$ where ϕ_j is given by (1.7.4). The inverse transformation U^* is given by $U^*\psi = \sum_{j=1}^{n} \psi(j) \mathbf{u}_j$.

1.7.2. Quantum dynamics. In contrast with the measurement process, which is random, the time evolution of quantum states $\psi \in \mathcal{H}$ is deterministic, and is governed by the Schrödinger equation

$$(1.7.5) \qquad i\hbar \frac{\mathrm{d}}{\mathrm{d}t}\psi(t) = H\psi(t)$$

where $\hbar$ is the reduced Planck's constant, and H is a self-adjoint operator in $\mathcal{B}(\mathcal{H})$ called the **Hamiltonian**, which is the observable representing the energy of the system. We shall work in units in which $\hbar = 1$.

Define the unitary operator $U(t) := e^{-itH}$. Then $U(t)\psi_0$ is the solution to the equation (1.7.5) with initial data ψ_0. In principle, one can construct a device to exert arbitrary forces on the constituents of a quantum system, so that the Hamiltonian governing the time evolution of the system can be any self-adjoint operator H. Since every unitary operator U is normal, the spectral calculus for normal operators permits U to be written in the form $U = e^{it_0 H}$ for any $t_0 \neq 0$, and some self-adjoint H. Thus, every unitary $U \in \mathcal{B}(\mathcal{H})$ represents the time evolution of the states of a quantum system for a time t_0 as governed by some Hamiltonian H.

Exercises

(1) Let $\mathcal{H}$ be a Hilbert space. Show that for all $\mathbf{x}, \mathbf{y} \in \mathcal{H}$,

$$\left|\|\mathbf{x}\| - \|\mathbf{y}\|\right| \leq \|\mathbf{x} - \mathbf{y}\|.$$

Then show that if $\{\mathbf{x}_n\}_{n\in\mathbb{N}}$ is any Cauchy sequence in $\mathcal{H}$, then $\lim_{n\to\infty}\|\mathbf{x}_n\|$ exists.

(2) Let $\mathbf{x}, \mathbf{y} \in \mathcal{H}$, $\mathcal{H}$ a Hilbert space. Show that $(|\mathbf{x}\rangle\langle\mathbf{y}|)^* = |\mathbf{y}\rangle\langle\mathbf{x}|$.

(3) Let $\mathcal{H}$ be a finite-dimensional Hilbert space. Let $f : \mathcal{B}(\mathcal{H}) \to \mathbb{C}$ be any linear functional with the properties that

$$f(A^*A) \geq 0 \quad \text{for all} \quad A \in \mathcal{B}(\mathcal{H}) \quad \text{and} \quad f(\mathbb{1}) = 1.$$

 (a) Show that f is bounded and hence continuous with respect to the Hilbert-Schmidt norm.

 (b) Show that there exists an $X_f \in \mathcal{B}(\mathcal{H})$ such that for all $Y \in \mathcal{B}(\mathcal{H})$,

$$f(Y) = \langle X_f, Y\rangle_{\mathrm{HS}}.$$

 (c) Show that $X_f \in \mathcal{B}^+(\mathcal{H})$ (i.e., $\langle \mathbf{x}, X_f\mathbf{x}\rangle \geq 0$ for all $\mathbf{x} \in \mathcal{H}$), and show that $\mathrm{Tr}[X_f] = 1$ and that $\|f\| = \|X_f\|_2$.

 (d) Let $\mathcal{H}$ be a finite-dimensional Hilbert space, and let $X, Y \in \mathcal{B}(\mathcal{H})$ commute; that is $[X, Y] := XY - YX = 0$. Let $\lambda \in \sigma(X)$ and suppose that $\mathbf{v}$ is an eigenvector of X with eigenvalue λ.

  (i) Show that $X(Y\mathbf{v}) = \lambda Y\mathbf{v}$ and hence the eigenspaces of X are invariant under Y.

(ii) Suppose in addition that X and Y are self-adjoint. Show that there exists an orthonormal basis $\{\mathbf{u}_1, \ldots, \mathbf{u}_d\}$ of $\mathcal{H}$ such that for each j, $\mathbf{u}_j$ is an eigenvector of both X and Y.

(4) Let $\mathcal{H}$ be ℓ^2, the Hilbert space of all square summable functions f on $\mathbb{N}$ with the inner product $\langle f, g \rangle = \sum_{j=1}^{N} \overline{f(j)} g(j)$.

  (a) Show that $\mathcal{H}$ is a Hilbert space. (This is an infinite-dimensional Hilbert space whose analysis does require the theory of the Lebesgue integral).

  (b) Define $A \in \mathcal{B}(\mathcal{H})$ by $Af(j) = j^{-1}f(j)$. Show that A is self-adjoint and compute $\|A\|$.

  (c) Show that $f(j) := j^{-2/3}$ is in $\mathcal{H}$, but not in $\mathrm{ran}(A)$, and hence that $\mathrm{ran}(A)$ is not closed—something that cannot happen in the finite-dimensional setting.

(5) Let $X \in \mathcal{B}^+(\mathcal{H})$. Show that $\mathbf{v} \in \ker(X)$ if and only if $\langle \mathbf{v}, X\mathbf{v} \rangle = 0$.

(6) Let $X \in \mathcal{B}^+(\mathcal{H})$, and let P be an orthogonal projection in $\mathcal{H}$. Show that if $PXP = 0$, then $X = P^\perp X P^\perp$.

(7) Let $\mathcal{H}$ be a finite-dimensional Hilbert space. Show that for $X, Y \in \mathcal{B}(\mathcal{H})$, $X = Y$ if and only if $\mathrm{Tr}[AX] = \mathrm{Tr}[AY]$ for all $A \in \mathcal{B}^{++}(\mathcal{H})$.

(8) Let $\mathcal{H}$ be a Hilbert space and let $\{\mathbf{x}_1, \ldots, \mathbf{x}_n\}$, $\{\mathbf{y}_1, \ldots, \mathbf{y}_n\} \subset \mathcal{H}$. Define $X = \sum_{i,j=1}^{n} \langle \mathbf{y}_i, \mathbf{y}_j \rangle |\mathbf{x}_i\rangle\langle\mathbf{x}_j|$. Show that $X \in \mathcal{B}^+(\mathcal{H})$. Under what conditions is $X \in \mathcal{B}^{++}(\mathcal{H})$?

(9) Let $\mathcal{H}$ be a finite-dimensional Hilbert space, and let $X, Y \in \mathcal{B}(\mathcal{H})$ be self-adjoint. Suppose that X and $XY + YX$ are positive semidefinite and X is invertible. Show that then Y is positive semidefinite.

(10) Let $\mathcal{H}$ be a finite-dimensional Hilbert space. Let $\varphi : \mathcal{B}(\mathcal{H}) \to \mathbb{C}$ be a linear functional such that for all $X, Y \in \mathcal{B}(\mathcal{H})$, $\varphi(XY) = \varphi(YX)$. Show that there is a constant c such that for all X, $\varphi(X) = c\,\mathrm{Tr}[X]$.

(11) Let $A, B \in \mathcal{B}^{\mathrm{s.a.}}(\mathcal{H})$, $\mathcal{H}$ a finite-dimensional Hilbert space. Show that $\|A\| \leq \|A + iB\|$ and $\|B\| \leq \|A + iB\|$.

(12) Let $\mathcal{H} = \mathbb{C}^2$ with its standard inner product. The **Pauli matrices** σ_x, σ_y, and σ_z are defined by

$$\sigma_x := \begin{bmatrix} 0 & 1 \\ 1 & 0 \end{bmatrix}, \quad \sigma_y := \begin{bmatrix} 0 & -i \\ i & 0 \end{bmatrix}, \quad \text{and} \quad \sigma_z := \begin{bmatrix} 1 & 0 \\ 0 & -1 \end{bmatrix}.$$

  (a) For a vector $\mathbf{v} = (v_x, v_y, v_z) \in \mathbb{R}^3$, define $\mathbf{v} \cdot \sigma = v_x\sigma_x + v_y\sigma_y + v_z\sigma_z$. Show that if $A \in M_2(\mathbb{C})$ is a traceless and self-adjoint, then $A = \mathbf{v} \cdot \sigma$, where $v_x = \frac{1}{2}\mathrm{Tr}[A\sigma_x]$, $v_y = \frac{1}{2}\mathrm{Tr}[A\sigma_y]$ and $v_z = \frac{1}{2}\mathrm{Tr}[A\sigma_z]$. Show that $(\mathbf{v}\cdot\sigma)^2 = \|\mathbf{v}\|^2 \mathbb{1}$ and that the eigenvalues of $\mathbf{v} \cdot \sigma$ are $\pm\|\mathbf{v}\|$.

  (b) Use Lemma 1.56 to show that for $A = \mathbf{v} \cdot \sigma$, the two spectral projections of A, $P_{\|\mathbf{v}\|}$ and $P_{-\|\mathbf{v}\|}$ are

$$P_{\|\mathbf{v}\|} = \frac{\|\mathbf{v}\|\mathbb{1} + A}{2\|\mathbf{v}\|} \quad \text{and} \quad P_{-\|\mathbf{v}\|} = \frac{\|\mathbf{v}\|\mathbb{1} - A}{2\|\mathbf{v}\|}.$$

(c) Let $\psi = 2^{-1/2}\left(\begin{smallmatrix} 1 \\ 1 \end{smallmatrix}\right)$ be thought of as a quantum state. Suppose that starting from this state, successive measurements of $A = \mathbf{v} \cdot \sigma$ and then $B = \mathbf{w} \cdot \sigma$ are made. What is the probability that the successive observed values are $\|\mathbf{v}\|$ and then $\|\mathbf{w}\|$?

(13) Let $\mathcal{H}$ be a finite dimensional Hilbert space. Let $A, B \in \mathcal{B}^+(\mathcal{H})$. Then $\mathrm{Tr}[AB] \geq 0$. Give at least two proofs of this fundamental fact: first, using a special choice of an orthonormal basis with which to compute the trace, and second, using the cyclicity of the trace.

(14) Let $\mathcal{H}$ be a finite dimensional Hilbert space. Let $\mathcal{V}$ be a subspace of $\mathcal{H}$, and let P denote the orthogonal projection onto $\mathcal{V}$. Let $A \in \mathcal{B}(\mathcal{H})$ be self-adjoint. Show that $A\mathbf{v} \in \mathcal{V}$ for all $\mathbf{v} \in \mathcal{V}$ if and only if A commutes with P.

(15) Let $\mathcal{H}$ be a finite dimensional Hilbert space. Let $A, B \in \mathcal{B}(\mathcal{H})$ be self-adjoint with spectral projections $A = \sum_{\lambda \in \sigma(A)} \lambda P_\lambda$ and $B = \sum_{\mu \in \sigma(B)} \lambda Q_\mu$. Show that A commutes with B if and only if for each $\mu \in \sigma(B)$, A commutes with Q_μ, which is the case if and only for each $\lambda \in \sigma(A)$, B commutes with P_λ. Then go on to show that when A and B commute, for each $\lambda \in \sigma(A)$ and each $\mu \in \sigma(B)$, $R_{\lambda,\mu} := P_\lambda Q_\mu$ is an orthogonal projection, and

$$\sum_{\lambda \in \sigma(A), \mu \in \sigma(B)} R_{\lambda,\mu} = \mathbb{1} \,,$$

and moreover

$$A = \sum_{\lambda \in \sigma(A), \mu \in \sigma(B)} \lambda R_{\lambda,\mu} \quad \text{and} \quad B = \sum_{\lambda \in \sigma(A), \mu \in \sigma(B)} \mu R_{\lambda,\mu} \,.$$

(16) Let $\mathcal{H}$ be a finite dimensional Hilbert space. Show that for all $A \in \mathcal{B}^+(\mathcal{H})$, there exists a unique $B \in \mathcal{B}^+(\mathcal{H})$ such that $B^2 = A$.

(17) Let $\mathcal{H}$ be a finite dimensional Hilbert space, and $A, B \in \mathcal{B}^+(\mathcal{H})$. Show that

$$\begin{aligned} \mathrm{rank}(AB) &= \mathrm{rank}(A) - \dim(\mathrm{ran}(A) \cap \ker(B)) \\ &= \mathrm{rank}(B) - \dim(\mathrm{ran}(B) \cap \ker(A)) \,. \end{aligned}$$

Chapter 2

Tensor products of Hilbert spaces

2.1. Tensor products

The most interesting quantum systems have many components, for example the many atoms in a crystal. The Hilbert space for such a composite system is constructed out of other Hilbert spaces corresponding to the individual components of the full system. For example, consider a system consisting of two atoms in a crystal. Then since the positions of the atoms are fixed, the only relevant degrees of freedom may be finitely many internal states of each atom. Let $\mathcal{H}_1$ be the state space for the first atom, and let $\mathcal{H}_2$ be the state space for the second atom. The Hilbert space for the combined system of the two atoms is a Hilbert space $\mathcal{H}_1 \otimes \mathcal{H}_2$, called the *tensor product* of $\mathcal{H}_1$ and $\mathcal{H}_2$, which is constructed out of $\mathcal{H}_1$ and $\mathcal{H}_2$. For this reason, tensor products of Hilbert spaces will be prominent in our considerations, and we begin by recalling some basic facts. The tensor product construction is particularly transparent in the wave function picture introduced in Section 1.7.1, which is always available once orthonormal bases are selected. Therefore, we first discuss tensor products in the wave function picture.

Let $\mathcal{X}_1$ and $\mathcal{X}_2$ be finite sets of cardinality m and n, respectively. Let μ_1 and μ_2 be nondegenerate measures on $\mathcal{X}_1$ and $\mathcal{X}_2$, respectively. That is, for $j = 1, 2$, $\mu_j(x) > 0$ for all $x \in \mathcal{X}_j$, and for any $E \subset \mathcal{X}_j$, the measure $\mu_j(E)$ of E is

$$\mu_j(E) = \sum_{x \in E} \mu_j(x).$$

For $j = 1, 2$, let $\mathcal{H}_j$ be the Hilbert space $\mathcal{H}_j := L^2(\mathcal{X}_j, \mu_j)$ as in Definition 1.68. Define the **tensor product of the measures** μ_1 and μ_2 to be the measure $\mu_1 \otimes \mu_2$ on $\mathcal{X}_1 \times \mathcal{X}_2$ given by

$$\mu_1 \otimes \mu_2(x_1, x_2) = \mu_1(x_1)\mu_2(x_2).$$

The **tensor product of** $\mathcal{H}_1$ **and** $\mathcal{H}_2$, $\mathcal{H}_1 \otimes \mathcal{H}_2$, is simply $L^2(\mathcal{X}_1 \times \mathcal{X}_2, \mu_1 \otimes \mu_2)$; that is, the space of complex valued functions on $\mathcal{X}_1 \times \mathcal{X}_2$ equipped with the inner product

$$\langle \phi, \psi \rangle_{\mathcal{H}_1 \otimes \mathcal{H}_2} = \sum_{(x_1, x_2) \in \mathcal{X}_1 \times \mathcal{X}_2} \overline{\phi(x_1, x_2)} \psi(x_1, x_2) \mu_1(x_1)\mu_2(x_2).$$

If we think of the elements ψ of $L^2(\mathcal{X}_1 \times \mathcal{X}_2, \mu_1 \otimes \mu_2)$ as quantum mechanical *wave functions* for a finite quantum system, the x_1 and x_2 variables can be thought of as corresponding to different components of a system, for example, the two atoms in a crystal, as discussed above.

The extension to more components is obvious; simply consider functions with more variables, each corresponding to a different subsystem:

$$L^2(\mathcal{X}_1, \mu_1) \otimes \cdots \otimes L^2(\mathcal{X}_N, \mu_N) = L^2(\mathcal{X}_1 \times \cdots \times \mathcal{X}_N, \mu_1 \otimes \cdots \otimes \mu_N).$$

Definition 2.1 (Tensor product). Given $\phi \in \mathcal{H}_1 = L^2(\mathcal{X}_1, \mu_1)$ and $\psi \in \mathcal{H}_2 = L^2(\mathcal{X}_2, \mu_2)$, define $\phi \otimes \psi \in \mathcal{H}_1 \otimes \mathcal{H}_2 = L^2(\mathcal{X}_1 \times \mathcal{X}_2, \mu_1 \otimes \mu_2)$ by

$$\phi \otimes \psi(x_1, x_2) = \phi(x_1)\psi(x_2).$$

The bilinear map

$$(\phi, \psi) \mapsto \phi \otimes \psi$$

from $\mathcal{H}_1 \times \mathcal{H}_2$ to $\mathcal{H}_1 \otimes \mathcal{H}_2$ is called the **tensor product** on $\mathcal{H}_1 \times \mathcal{H}_2$. The tensor product is extended as above to an arbitrary finite number of factors.

Lemma 2.2. *Let* $\mathcal{H}_1 = L^2(\mathcal{X}_1, \mu_1)$ *and* $\mathcal{H}_2 = L^2(\mathcal{X}_2, \mu_2)$. *Then for* $\phi_1, \phi_2 \in \mathcal{H}_1$ *and* $\psi_1, \psi_2 \in \mathcal{H}_2$,

$$(2.1.1) \qquad \langle \phi_1 \otimes \psi_1, \phi_2 \otimes \psi_2 \rangle_{\mathcal{H}_1 \otimes \mathcal{H}_2} = \langle \phi_1, \phi_2 \rangle_{\mathcal{H}_1} \langle \psi_1, \psi_2 \rangle_{\mathcal{H}_2}.$$

If $\{\phi_1, \ldots, \phi_m\}$ *and* $\{\psi_1, \ldots, \psi_n\}$ *are orthonormal bases of* $\mathcal{H}_1$ *and f* $\mathcal{H}_2$, *respectively, then*

$$\{\phi_i \otimes \psi_j\}_{1 \leq i \leq m, \ 1 \leq j \leq n}$$

is an orthonormal basis of $\mathcal{H}_1 \otimes \mathcal{H}_2$.

Proof. By the definitions,

$$\langle \phi_1 \otimes \psi_1, \phi_2 \otimes \psi_2 \rangle_{\mathcal{H}_1 \otimes \mathcal{H}_2}$$

$$= \sum_{(x_1, x_2)} \overline{\phi_1(x_1)\psi_1(x_2)} \phi_2(x_1)\psi_2(x_2)\mu_1(x_1)\mu_2(x_2)$$

$$= \langle \phi_1, \phi_2 \rangle_{\mathcal{H}_1} \langle \psi_1, \psi_2 \rangle_{\mathcal{H}_2}.$$

This proves (2.1.1), and the final statement is then an immediate consequence since the dimension of $L^2(\mathcal{X}_1 \times \mathcal{X}_2, \mu_1 \otimes \mu_2)$ is the product of the dimensions of $L^2(\mathcal{X}_1, \mu_1)$ and $L^2(\mathcal{X}_2, \mu_2)$. $\qquad\square$

To pass from the L^2 setting to the general setting of finite-dimensional Hilbert spaces, use the fact that every finite-dimensional Hilbert space can be identified with an L^2 space as explained in Section 1.7.1.

In particular, for any two finite-dimensional Hilbert spaces $\mathcal{H}_1$ and $\mathcal{H}_2$, this construction yields the spaces $L^2(\mathcal{X}_1, \mu_1)$ and $L^2(\mathcal{X}_2, \mu_2)$ and unitaries U_j from $\mathcal{H}_j$ to $L^2(\mathcal{X}_j, \mu_j)$, $j = 1, 2$.

Then for $\mathbf{x} \in \mathcal{H}_1$ and $\mathbf{y} \in \mathcal{H}_2$, we use the L^2 tensor product on $L^2(\mathcal{X}_1, \mu_1)$ and $L^2(\mathcal{X}_2, \mu_2)$ to define

$$\mathbf{x} \otimes \mathbf{y} = U_1\mathbf{x} \otimes U_2\mathbf{y} \in L^2(\mathcal{X}_1, \mu_1) \otimes L^2(\mathcal{X}_2, \mu_2).$$

Having identified $\mathcal{H}_1$ and $L^2(\mathcal{X}_1, \mu_1)$ with $\mathcal{H}_2$ with $L^2(\mathcal{X}_2, \mu_2)$, it is natural to refer to $L^2(\mathcal{X}_1, \mu_1) \otimes L^2(\mathcal{X}_2, \mu_2)$ as $\mathcal{H}_1 \otimes \mathcal{H}_2$, and then evidently we have a bilinear map

$$(\mathbf{x}, \mathbf{y}) \mapsto \mathbf{x} \otimes \mathbf{y}$$

from $\mathcal{H}_1 \times \mathcal{H}_2$ to $\mathcal{H}_1 \otimes \mathcal{H}_2$. Moreover, it follows immediately from the definitions that for all $\mathbf{x}_1, \mathbf{x}_2 \in \mathcal{H}_1$ and all $\mathbf{y}_1, \mathbf{y}_2 \in \mathcal{H}_2$,

$$\langle \mathbf{x}_1 \otimes \mathbf{y}_1, \mathbf{x}_2 \otimes \mathbf{y}_2 \rangle_{\mathcal{H}_1 \otimes \mathcal{H}_2} = \langle \mathbf{x}_1, \mathbf{x}_2 \rangle_{\mathcal{H}_1} \langle \mathbf{y}_1, \mathbf{y}_2 \rangle_{\mathcal{H}_2}.$$

Of course, this construction depends on a choice of orthonormal bases for $\mathcal{H}_1$ and $\mathcal{H}_2$, but only in a trivial way:

Let $\{\mathbf{u}_1, \dots, \mathbf{u}_n\}$ and $\{\mathbf{v}_1, \dots, \mathbf{v}_m\}$ be orthonormal bases of $\mathcal{H}_1$ and $\mathcal{H}_2$. Then $\{\mathbf{u}_i \otimes \mathbf{v}_j\}_{1 \le i \le n,\ 1 \le j \le m}$ is an orthonormal basis of $\mathcal{H}_1 \otimes \mathcal{H}_2$, and the general element ψ of $\mathcal{H}_1 \otimes \mathcal{H}_2$ has the expansion

$$(2.1.2) \qquad \psi = \sum_{i=1}^{n} \sum_{j=1}^{m} \psi(i, j) \mathbf{u}_i \otimes \mathbf{v}_j.$$

The function $\psi(i, j)$ on $\mathcal{X}_1 \times \mathcal{X}_2$ is the wave function representing ψ in $L^2(\mathcal{X}_1 \times \mathcal{X}_2, \mu_1 \otimes \mu_2)$ for the bases $\{\mathbf{u}_1, \dots, \mathbf{u}_m\}$ and $\{\mathbf{v}_1, \dots, \mathbf{v}_n\}$. Indeed,

$$\|\psi\|^2 = \sum_{i=1}^{n} \sum_{j=1}^{m} |\psi(i, j)|^2.$$

Now let $\{\tilde{\mathbf{u}}_1, \dots, \tilde{\mathbf{u}}_n\}$ and $\{\tilde{\mathbf{v}}_1, \dots, \tilde{\mathbf{v}}_m\}$ be another pair of orthonormal bases of $\mathcal{H}_1$ and $\mathcal{H}_2$. Then

$$\mathbf{u}_i = \sum_{k=1}^{n} \langle \tilde{\mathbf{u}}_k, \mathbf{u}_i \rangle \tilde{\mathbf{u}}_k \quad \text{and} \quad \mathbf{v}_j = \sum_{\ell=1}^{m} \langle \tilde{\mathbf{v}}_\ell, \mathbf{v}_j \rangle \tilde{\mathbf{v}}_\ell.$$

Define matrices $U \in M_n(\mathbb{C})$ and $V \in M_m(\mathbb{C})$ by $U_{k,i} := \langle \tilde{\mathbf{u}}_k, \mathbf{u}_i \rangle$ and $V_{\ell,j} := \langle \tilde{\mathbf{v}}_\ell, \mathbf{v}_j \rangle$. Note that U and V are unitary matrices. Then

$$(2.1.3) \qquad \mathbf{u}_i \otimes \mathbf{v}_j = \sum_{k=1}^{n} \sum_{\ell=1}^{m} U_{k,i} V_{\ell,j} \tilde{\mathbf{u}}_k \otimes \tilde{\mathbf{v}}_\ell.$$

Substituting (2.1.3) into (2.1.2)

$$(2.1.4) \quad \psi = \sum_{k=1}^{n} \sum_{\ell=1}^{m} \widetilde{\psi}(k,\ell) \tilde{\mathbf{u}}_k \otimes \tilde{\mathbf{v}}_\ell, \quad \text{where} \quad \widetilde{\psi}(k,\ell) = \sum_{i=1}^{n} \sum_{j=1}^{m} U_{k,i} V_{\ell,j} \psi(i,j).$$

The function $\widetilde{\psi}(k,\ell)$ on $\mathcal{X}_1 \times \mathcal{X}_2$ is the wave function representing ψ in $L^2(\mathcal{X}_1 \times \mathcal{X}_2, \mu_1 \otimes \mu_2)$ for the bases $\{\tilde{\mathbf{u}}_1, \ldots, \tilde{\mathbf{u}}_n\}$ and $\{\tilde{\mathbf{v}}_1, \ldots, \tilde{\mathbf{v}}_m\}$.

While the wave function of any element ψ of $\mathcal{H}_1 \otimes \mathcal{H}_2$ depends on the choice of bases, the dependence is only through the individual *change of basis* unitary matrices as expressed in (2.1.4), and

$$\sum_{k=1}^{n} \sum_{\ell=1}^{m} \widetilde{\psi}(k,\ell) \tilde{\mathbf{u}}_k \otimes \tilde{\mathbf{v}}_\ell = \psi = \sum_{i=1}^{n} \sum_{j=1}^{m} \psi(i,j) \mathbf{u}_i \otimes \mathbf{v}_j.$$

2.1.1. Tensor products according to Murray and von Neumann. The theory we have described so far is how everyone thought about tensor products for the first dozen years of modern quantum mechanics, and is how many people think about them today. However, because of the fundamental importance of tensor products, it is desirable to have a construction of the tensor product of Hilbert spaces that is independent of any choice of bases. This appears in the literature for the first time in 1938, in a paper of Murray and von Neumann [**159**, Chapter II]. It is a robust approach, adapted to the consideration of infinite tensor products. However, we start with tensor products of two Hilbert spaces, and only discuss finite products.

Let $\mathcal{H}_1$ and $\mathcal{H}_2$ be two finite-dimensional Hilbert spaces. A **conjugate bilinear functional on** $\mathcal{H}_1 \times \mathcal{H}_2$ is a function $\mathbf{f}$ on $\mathcal{H}_1 \times \mathcal{H}_2$ with values in $\mathbb{C}$ such that for all $\mathbf{x}_1, \mathbf{x}_2 \in \mathcal{H}_1$, all $\mathbf{y}_1, \mathbf{y}_2 \in \mathcal{H}_2$, all $z, w \in \mathbb{C}$,

$$\mathbf{f}(z\mathbf{x}_1 + \mathbf{x}_2, \mathbf{y}_1) = \bar{z}\mathbf{f}(\mathbf{x}_1, \mathbf{y}_1) + \mathbf{f}(\mathbf{x}_2, \mathbf{y}_1) \quad \text{and}$$

$$\mathbf{f}(\mathbf{x}_1, w\mathbf{y}_1 + \mathbf{y}_2) = \overline{w}\mathbf{f}(\mathbf{x}_1, \mathbf{y}_1) + \mathbf{f}(\mathbf{x}_1, \mathbf{y}_2).$$

It is an easy exercise to verify that linear combinations of conjugate bilinear functionals are conjugate bilinear functionals, so the set of conjugate bilinear functionals forms a vector space.

Definition 2.3. As a vector space, the tensor product $\mathcal{H}_1 \otimes \mathcal{H}_2$ of two Hilbert spaces $\mathcal{H}_1$ and $\mathcal{H}_2$ is the vector space of all conjugate bilinear functionals on $\mathcal{H}_1 \times \mathcal{H}_2$.

Lemma 2.4. *Let $\mathcal{H}_1, \mathcal{H}_2$ be finite-dimensional Hilbert spaces. Let $\mathbf{f}, \mathbf{g} \in \mathcal{H}_1 \otimes \mathcal{H}_2$. Let $\{\mathbf{u}_1, \dots, \mathbf{u}_n\}$ and $\{\mathbf{v}_1, \dots, \mathbf{v}_m\}$ be orthonormal bases of $\mathcal{H}_1$ and $\mathcal{H}_2$, respectively. Define*

$$\langle \mathbf{f}, \mathbf{g} \rangle = \sum_{i=1}^{n} \sum_{j=1}^{m} \overline{\mathbf{f}(\mathbf{u}_i, \mathbf{v}_j)} \mathbf{g}(\mathbf{u}_i, \mathbf{v}_j).$$

Then $\langle \mathbf{f}, \mathbf{f} \rangle = 0$ if and only if $\mathbf{f} = 0$. Moreover, if $\{\tilde{\mathbf{u}}_1, \dots, \tilde{\mathbf{u}}_n\}$ and $\{\tilde{\mathbf{v}}_1, \dots, \tilde{\mathbf{v}}_m\}$ are any other pair of orthonormal bases of $\mathcal{H}_1$ and $\mathcal{H}_2$, then for all $\mathbf{f}, \mathbf{g} \in \mathcal{H}_1 \otimes \mathcal{H}_2$,

$$(2.1.5) \qquad \sum_{i=1}^{n} \sum_{j=1}^{m} \overline{\mathbf{f}(\mathbf{u}_i, \mathbf{v}_j)} \mathbf{g}(\mathbf{u}_i, \mathbf{v}_j) = \sum_{i=1}^{n} \sum_{j=1}^{m} \overline{\mathbf{f}(\tilde{\mathbf{u}}_i, \tilde{\mathbf{v}}_j)} \mathbf{g}(\tilde{\mathbf{u}}_i, \tilde{\mathbf{v}}_j).$$

Proof. If $\langle \mathbf{f}, \mathbf{f} \rangle = 0$, then $\sum_{i=1}^{n} \sum_{j=1}^{m} |\mathbf{f}(\mathbf{u}_i, \mathbf{v}_j)|^2 = 0$, and hence $\mathbf{f}(\mathbf{u}_i, \mathbf{v}_j) = 0$ for all i, j. For $\mathbf{x} \in \mathcal{H}_1$ and $\mathbf{y} \in \mathcal{H}_2$, expand $\mathbf{x} = \sum_{i=1}^{n} x_i \mathbf{u}_i$ and $\mathbf{y} = \sum_{j=1}^{m} y_j \mathbf{v}_j$. Then $\mathbf{f}(\mathbf{x}, \mathbf{y}) = \sum_{i=1}^{n} \sum_{j=1}^{m} \overline{x_i y_j} \mathbf{f}(\mathbf{u}_i, \mathbf{v}_j) = 0$ for all $\mathbf{x}, \mathbf{y}$ so that $\mathbf{f} = 0$.

Next, expand

$$(2.1.6) \qquad \mathbf{u}_i = \sum_{k=1}^{n} \langle \tilde{\mathbf{u}}_k, \mathbf{u}_i \rangle \tilde{\mathbf{u}}_k \quad \text{and} \quad \mathbf{v}_j = \sum_{\ell=1}^{m} \langle \tilde{\mathbf{v}}_\ell, \mathbf{v}_j \rangle \tilde{\mathbf{v}}_\ell.$$

Define $U \in M_n(\mathbb{C})$ and $V \in M_m(\mathbb{C})$ by $U_{k,i} := \langle \tilde{\mathbf{u}}_k, \mathbf{u}_i \rangle$ and $V_{\ell,j} := \langle \tilde{\mathbf{v}}_\ell, \mathbf{v}_j \rangle$. Note that U and V are unitary. Substituting (2.1.6) into $\mathbf{f}(\mathbf{u}_i, \mathbf{v}_j)$,

$$\mathbf{f}(\mathbf{u}_i, \mathbf{v}_j) = \sum_{k=1}^{n} \sum_{\ell=1}^{m} \overline{U_{k,i} V_{\ell,j}} \mathbf{f}(\tilde{\mathbf{u}}_k, \tilde{\mathbf{v}}_\ell),$$

and likewise for $\mathbf{g}(\mathbf{u}_i, \mathbf{v}_j)$. Therefore,

$$(2.1.7) \quad \sum_{i=1}^{n} \sum_{j=1}^{m} \overline{\mathbf{f}(\mathbf{u}_i, \mathbf{v}_j)} \mathbf{g}(\mathbf{u}_i, \mathbf{v}_j)$$

$$= \sum_{k=1}^{n} \sum_{\ell=1}^{m} \sum_{k'=1}^{n} \sum_{\ell'=1}^{m} \left(\sum_{i=1}^{n} \sum_{j=1}^{m} U_{k,i} V_{\ell,j} \overline{U_{k',i} V_{\ell',j}} \right) \overline{\mathbf{f}(\tilde{\mathbf{u}}_k, \tilde{\mathbf{v}}_\ell)} \mathbf{g}(\tilde{\mathbf{u}}_{k'}, \tilde{\mathbf{v}}_{\ell'}).$$

However, since U and V are unitary,

$$\sum_{i=1}^{n} \sum_{j=1}^{m} U_{k,i} V_{\ell,j} \overline{U_{k',i} V_{\ell',j}} = \sum_{i=1}^{n} \sum_{j=1}^{m} U_{k,i} U_{i,k'}^* V_{\ell,j} V_{j,\ell'}^* = \delta_{k,k'} \delta_{\ell,\ell'}.$$

Substituting this into (2.1.7) yields (2.1.5). $\qquad\square$

This shows that $(\mathcal{H}_1 \otimes \mathcal{H}_2, \langle \cdot, \cdot \rangle)$ is an inner product space. We now construct an orthonormal basis.

Definition 2.5. For $\mathbf{u} \in \mathcal{H}_1$ and $\mathbf{v} \in \mathcal{H}_2$, define $\mathbf{u} \otimes \mathbf{v} \in \mathcal{H}_1 \otimes \mathcal{H}_2$ to be the conjugate bilinear functional such that for all $\mathbf{x} \in \mathcal{H}_1$ and all $\mathbf{y} \in \mathcal{H}_2$,

$$(\mathbf{u} \otimes \mathbf{v})(\mathbf{x}, \mathbf{y}) = \langle \mathbf{x}, \mathbf{u} \rangle \langle \mathbf{y}, \mathbf{v} \rangle.$$

It is easy to check that $\mathbf{u} \otimes \mathbf{v}$ is in fact a conjugate bilinear functional, and hence a vector in $\mathcal{H}_1 \otimes \mathcal{H}_2$.

Lemma 2.6. *Let* $\{\mathbf{u}_1, \dots, \mathbf{u}_n\}$ *and* $\{\mathbf{v}_1, \dots, \mathbf{v}_m\}$ *be orthonormal bases of* $\mathcal{H}_1$ *and* $\mathcal{H}_2$, *respectively. Then:*

(1) $\mathrm{span}(\{\mathbf{u}_i \otimes \mathbf{v}_j \ : \ 1 \leq i \leq n, \ 1 \leq j \leq m\}) = \mathcal{H}_1 \otimes \mathcal{H}_2$;

(2) *for all* $1 \leq i, k \leq n$, $1 \leq j, \ell \leq m$, $\langle \mathbf{u}_i \otimes \mathbf{v}_j, \mathbf{u}_k \otimes \mathbf{v}_\ell \rangle = \delta_{i,k} \delta_{j,\ell}$.

Consequently, $\{\mathbf{u}_i \otimes \mathbf{v}_j \ : \ 1 \leq i \leq n, \ 1 \leq j \leq m\}$ *is an orthonormal basis of* $\mathcal{H}_1 \otimes \mathcal{H}_2$, *and the dimension of* $\mathcal{H}_1 \otimes \mathcal{H}_2$ *is* mn.

Proof. For any $\mathbf{x} \in \mathcal{H}_1$ and any $\mathbf{y} \in \mathcal{H}_2$, expand $\mathbf{x} = \sum_{i=1}^n x_i \mathbf{u}_i$ and $\mathbf{y} = \sum_{j=1}^m y_j \mathbf{v}_j$. For any $\mathbf{f} \in \mathcal{H}_1 \otimes \mathcal{H}_2$, define $f_{i,j} := \mathbf{f}(\mathbf{u}_i, \mathbf{v}_j)$. Then

$$\mathbf{f}(\mathbf{x}, \mathbf{y}) = \sum_{i=1}^n \sum_{j=1}^m \mathbf{f}(\mathbf{u}_i, \mathbf{v}_j) \overline{x_i y_j} = \left(\sum_{i=1}^n \sum_{j=1}^m f_{i,j} \mathbf{u}_i \otimes \mathbf{v}_j \right)(\mathbf{x}, \mathbf{y}).$$

This shows that $\mathbf{f}$ is a linear combination of the vectors $\mathbf{u}_i \otimes \mathbf{v}_j$, and hence (1) is proved.

Next, by the definition of the inner product,

$$\langle \mathbf{u}_i \otimes \mathbf{v}_j, \mathbf{u}_k \otimes \mathbf{v}_\ell \rangle = \sum_{r=1}^n \sum_{s=1}^m \overline{\mathbf{u}_i \otimes \mathbf{v}_j(\mathbf{u}_r, \mathbf{v}_s)} \mathbf{u}_k \otimes \mathbf{v}_\ell(\mathbf{u}_r, \mathbf{v}_s)$$

$$= \sum_{r=1}^n \sum_{s=1}^m \overline{\langle \mathbf{u}_r, \mathbf{u}_i \rangle \langle \mathbf{v}_s, \mathbf{v}_j \rangle} \langle \mathbf{u}_r, \mathbf{u}_k \rangle \langle \mathbf{v}_s, \mathbf{v}_\ell \rangle$$

$$= \sum_{r=1}^n \sum_{s=1}^m \langle \mathbf{u}_i, \mathbf{u}_r \rangle \langle \mathbf{u}_r, \mathbf{u}_k \rangle \langle \mathbf{v}_j, \mathbf{v}_s \rangle \langle \mathbf{v}_s, \mathbf{v}_\ell \rangle$$

$$= \delta_{i,k} \delta_{j,\ell}.$$

This proves (2). $\qquad\qquad\qquad\qquad\qquad\qquad\qquad\qquad\qquad\qquad\qquad\qquad\quad\square$

Using Lemma 2.6, it is an easy exercise to show that for all $\mathbf{x}, \mathbf{z} \in \mathcal{H}_1$ and all $\mathbf{y}, \mathbf{w} \in \mathcal{H}_2$,

$$\langle \mathbf{x} \otimes \mathbf{y}, \mathbf{z} \otimes \mathbf{w} \rangle = \langle \mathbf{x}, \mathbf{z} \rangle \langle \mathbf{y}, \mathbf{w} \rangle.$$

Finally, $(\mathcal{H}_1 \otimes \mathcal{H}_2, \langle \cdot, \cdot \rangle)$ is a Hilbert space. The only matter that has not been addressed is the completeness, but what we have explained so far identifies $\mathcal{H}_1 \otimes \mathcal{H}_2$ with $\mathbb{C}^{mn}$, which is complete.

To form the tensor product of a finite set of finite-dimensional Hilbert spaces, $\mathcal{H}_1,\dots,\mathcal{H}_r$, proceed in the same way, starting with the introduction of the vector space of all r-conjugate linear functionals on $\mathcal{H}_1 \times \cdots \times \mathcal{H}_r$, which then is the tensor product $\mathcal{H}_1 \otimes \cdots \otimes \mathcal{H}_r$ as a vector space.

2.1.2. Tensor products of operators. Let $\mathcal{H}_1, \mathcal{H}_2$ and $\mathcal{K}_1, \mathcal{K}_2$ be finite-dimensional Hilbert spaces. By Theorem 1.43, $\mathcal{B}(\mathcal{H}_1,\mathcal{K}_1)$ and $\mathcal{B}(\mathcal{H}_2,\mathcal{K}_2)$, when equipped with their respective Hilbert-Schmidt inner products, are Hilbert spaces. We can then form their tensor product $\mathcal{B}(\mathcal{H}_1,\mathcal{K}_1) \otimes \mathcal{B}(\mathcal{H}_2,\mathcal{K}_2)$, just as we can with any pair of Hilbert spaces.

Let $\{\mathbf{u}_1,\dots,\mathbf{u}_m\}$ and $\{\mathbf{v}_1,\dots,\mathbf{v}_n\}$ be orthonormal bases of $\mathcal{H}_1$ and $\mathcal{H}_2$, respectively. Let $\{\tilde{\mathbf{u}}_1,\dots,\tilde{\mathbf{u}}_p\}$ and $\{\tilde{\mathbf{v}}_1,\dots,\tilde{\mathbf{v}}_q\}$ be orthonormal bases of $\mathcal{K}_1$ and $\mathcal{K}_2$, respectively. By Lemma 1.44

$$\{|\tilde{\mathbf{u}}_i\rangle\langle\mathbf{u}_j|\}_{1\le i\le p,\ 1\le j\le m} \quad \text{and} \quad \{|\tilde{\mathbf{v}}_k\rangle\langle\mathbf{v}_\ell|\}_{1\le k\le q,\ 1\le \ell\le n}$$

are orthonormal bases of $\mathcal{B}(\mathcal{H}_1,\mathcal{K}_1)$ and $\mathcal{B}(\mathcal{H}_2,\mathcal{K}_2)$, respectively. Hence

$$\{|\tilde{\mathbf{u}}_i\rangle\langle\mathbf{u}_j| \otimes |\tilde{\mathbf{v}}_k\rangle\langle\mathbf{v}_\ell|\}_{1\le i\le p,\ 1\le j\le m,\ 1\le k\le q,\ 1\le\ell\le n}$$

is an orthonormal basis for $\mathcal{B}(\mathcal{H}_1,\mathcal{K}_1) \otimes \mathcal{B}(\mathcal{H}_2,\mathcal{K}_2)$.

There is another Hilbert space involving tensor products that is associated to the pairs $\mathcal{H}_1, \mathcal{H}_2$ and $\mathcal{K}_1, \mathcal{K}_2$, namely $\mathcal{B}(\mathcal{H}_1 \otimes \mathcal{H}_2, \mathcal{K}_1 \otimes \mathcal{K}_2)$. Using the same orthonormal bases for $\mathcal{H}_1, \mathcal{H}_2$ and $\mathcal{K}_1, \mathcal{K}_2$,

$$\{|\tilde{\mathbf{u}}_i \otimes \tilde{\mathbf{v}}_k\rangle\langle\mathbf{u}_j \otimes \mathbf{v}_\ell|\}_{1\le i\le p,\ 1\le j\le m,\ 1\le k\le q,\ 1\le\ell\le n}$$

is an orthonormal basis for $\mathcal{B}(\mathcal{H}_1 \otimes \mathcal{H}_2, \mathcal{K}_1 \otimes \mathcal{K}_2)$.

Evidently, $\mathcal{B}(\mathcal{H}_1,\mathcal{K}_1) \otimes \mathcal{B}(\mathcal{H}_2,\mathcal{K}_2)$ and $\mathcal{B}(\mathcal{H}_1 \otimes \mathcal{H}_2, \mathcal{K}_1 \otimes \mathcal{K}_2)$ have the same dimension, namely $mnpq$. We define a unitary transformation $U : \mathcal{B}(\mathcal{H}_1,\mathcal{K}_1) \otimes \mathcal{B}(\mathcal{H}_2,\mathcal{K}_2) \to \mathcal{B}(\mathcal{H}_1 \otimes \mathcal{H}_2, \mathcal{K}_1 \otimes \mathcal{K}_2)$ by

$$(2.1.8) \qquad U(|\tilde{\mathbf{u}}_i\rangle\langle\mathbf{u}_j| \otimes |\tilde{\mathbf{v}}_k\rangle\langle\mathbf{v}_\ell|) = |\tilde{\mathbf{u}}_i \otimes \tilde{\mathbf{v}}_k\rangle\langle\mathbf{u}_j \otimes \mathbf{v}_\ell|$$

for all $1 \le i \le p$, $1 \le j \le m$, $1 \le k \le q$, $1 \le \ell \le n$.

Lemma 2.7. *Let $\mathcal{H}_1, \mathcal{H}_2$ and $\mathcal{K}_1, \mathcal{K}_2$ be finite-dimensional Hilbert spaces, and let $U : \mathcal{B}(\mathcal{H}_1,\mathcal{K}_1) \otimes \mathcal{B}(\mathcal{H}_2,\mathcal{K}_2) \to \mathcal{B}(\mathcal{H}_1 \otimes \mathcal{H}_2, \mathcal{K}_1 \otimes \mathcal{K}_2)$ be the unitary transformation defined by (2.1.8). For $A \in \mathcal{B}(\mathcal{H}_1,\mathcal{K}_1)$, $B \in \mathcal{B}(\mathcal{H}_2,\mathcal{K}_2)$, $\mathbf{x} \in \mathcal{H}_1$ and $\mathbf{y} \in \mathcal{H}_2$,*

$$(2.1.9) \qquad U(A \otimes B)\mathbf{x} \otimes \mathbf{y} = A\mathbf{x} \otimes B\mathbf{y}.$$

Proof. Expanding in the bases introduced above,

$$A = \sum_{i,j}\langle\tilde{\mathbf{u}}_i, A\mathbf{u}_j\rangle|\tilde{\mathbf{u}}_i\rangle\langle\mathbf{u}_j| \quad \text{and} \quad B = \sum_{k,\ell}\langle\tilde{\mathbf{v}}_k, B\mathbf{v}_\ell\rangle|\tilde{\mathbf{v}}_k\rangle\langle\mathbf{v}_\ell|.$$

Therefore, $U(A \otimes B) = \sum_{i,j,k,\ell} \langle \tilde{\mathbf{u}}_i, A\mathbf{u}_j \rangle \langle \tilde{\mathbf{v}}_k, B\mathbf{v}_\ell \rangle |\tilde{\mathbf{u}}_i \otimes \tilde{\mathbf{v}}_k \rangle \langle \mathbf{u}_j \otimes \mathbf{v}_\ell|$. Hence,

$$U(A \otimes B)\mathbf{x} \otimes \mathbf{y} = \sum_{i,j,k,\ell} \langle \tilde{\mathbf{u}}_i, A\mathbf{u}_j \rangle \langle \tilde{\mathbf{v}}_k, B\mathbf{v}_\ell \rangle \langle \mathbf{u}_j, \mathbf{x} \rangle \tilde{\mathbf{u}}_i \otimes \langle \mathbf{v}_\ell, \mathbf{y} \rangle \tilde{\mathbf{v}}_k$$

$$= \sum_{i,k} \langle \tilde{\mathbf{u}}_i, A\mathbf{x} \rangle \langle \tilde{\mathbf{v}}_k, B\mathbf{y} \rangle \tilde{\mathbf{u}}_i \otimes \tilde{\mathbf{v}}_k = A\mathbf{x} \otimes B\mathbf{y}. \qquad \square$$

While a choice of orthonormal bases was used to define U, (2.1.9) shows that U does not depend on the choice of bases. It will be convenient to suppress the unitary U and simply identify $A \otimes B$ in $\mathcal{B}(\mathcal{H}_1, \mathcal{K}_1) \otimes \mathcal{B}(\mathcal{H}_2, \mathcal{K}_2)$ with the operator on $\mathcal{B}(\mathcal{H}_1 \otimes \mathcal{H}_2, \mathcal{K}_1 \otimes \mathcal{K}_2)$ specified in (2.1.9).

Definition 2.8. Let $\mathcal{H}_1, \mathcal{H}_2$ and $\mathcal{K}_1, \mathcal{K}_2$ be finite-dimensional Hilbert spaces. Then for $A \in \mathcal{B}(\mathcal{H}_1, \mathcal{K}_1)$, $B \in \mathcal{B}(\mathcal{H}_2, \mathcal{K}_2)$, $A \otimes B \in \mathcal{B}(\mathcal{H}_1 \otimes \mathcal{H}_2, \mathcal{K}_1 \otimes \mathcal{K}_2)$ is given by

$$(2.1.10) \qquad A \otimes B(\mathbf{x} \otimes \mathbf{y}) = A\mathbf{x} \otimes B\mathbf{y}$$

for all $\mathbf{x} \in \mathcal{H}_1$ and all $\mathbf{y} \in \mathcal{H}_2$. It is left as an easy exercise to check that $(A \otimes B)^* = A^* \otimes B^*$.

Example 2.9. Let $\mathbf{x}, \mathbf{y} \in \mathcal{H}$ and $\mathbf{z}, \mathbf{w} \in \mathcal{K}$. Then

$$(2.1.11) \qquad |\mathbf{x}\rangle\langle\mathbf{y}| \otimes |\mathbf{z}\rangle\langle\mathbf{w}| = |\mathbf{x} \otimes \mathbf{z}\rangle\langle\mathbf{y} \otimes \mathbf{w}|,$$

as one sees by applying both sides to $\mathbf{a} \otimes \mathbf{b}$ for $\mathbf{a} \in \mathcal{H}$ and $\mathbf{b} \in \mathcal{K}$.

The special case of (2.1.10), in which one of A or B is the identity, arises when we consider the relation between observables for a component of a quantum system in relation to observables on the whole system.

2.2. Partial traces

If $\mathcal{H}$ and $\mathcal{K}$ are the Hilbert spaces of two components of a quantum system, say two atoms in a crystal, then observables for the first atom—self-adjoint operators on $\mathcal{H}$—can also be considered as observables on the larger system consisting of the two atoms, i.e., as self-adjoint operators on $\mathcal{H} \otimes \mathcal{K}$.

In mathematical terms, there are natural embeddings $\Phi_{\mathcal{H}}$ of $\mathcal{B}(\mathcal{H})$ into $\mathcal{B}(\mathcal{H} \otimes \mathcal{K})$ and $\Phi_{\mathcal{K}}$ of $\mathcal{B}(\mathcal{K})$ into $\mathcal{B}(\mathcal{H} \otimes \mathcal{K})$ defined by

$$(2.2.1) \qquad \Phi_{\mathcal{H}}(A) := A \otimes \mathbb{1} \quad \text{and} \quad \Phi_{\mathcal{K}}(B) := \mathbb{1} \otimes B$$

for all $A \in \mathcal{B}(\mathcal{H})$ and $B \in \mathcal{B}(\mathcal{K})$.

Consider $\Phi_{\mathcal{H}}^{\dagger} : \mathcal{B}(\mathcal{H} \otimes \mathcal{K}) \to \mathcal{B}(\mathcal{H})$, the adjoint of $\Phi_{\mathcal{H}}$ with respect to the Hilbert-Schmidt inner product. By the definition (1.4.7), for $X \in \mathcal{B}(\mathcal{H} \otimes \mathcal{K})$ and $A \in \mathcal{B}(\mathcal{H})$,

$$(2.2.2) \qquad \mathrm{Tr}[\Phi_{\mathcal{H}}^{\dagger}(X)^* A] = \mathrm{Tr}[X^* \Phi_{\mathcal{H}}(A)] = \mathrm{Tr}[X^*(A \otimes \mathbb{1})],$$

where the first trace is taken over $\mathcal{H}$ and the last two traces are taken over $\mathcal{H} \otimes \mathcal{K}$. Since $\mathrm{Tr}[\Phi_{\mathcal{H}}^{\dagger}(X)^* A] = \mathrm{Tr}[X^*(A \otimes \mathbb{1})]$ is equivalent to $\mathrm{Tr}[A^* \Phi_{\mathcal{H}}^{\dagger}(X)] = \mathrm{Tr}[(A^* \otimes \mathbb{1})X]$, (2.2.2) is satisfied for all $A \in \mathcal{B}(\mathcal{H})$ if and only if for all $A \in \mathcal{B}(\mathcal{H})$,

$$\mathrm{Tr}[A\Phi_{\mathcal{H}}^{\dagger}(X)] = \mathrm{Tr}[(A \otimes \mathbb{1})X].$$

Definition 2.10 (Partial trace). Let $\mathcal{H}$ and $\mathcal{K}$ be finite-dimensional Hilbert spaces. For $X \in \mathcal{B}(\mathcal{H} \otimes \mathcal{K})$ the **partial trace of X over** $\mathcal{K}$, denoted $\mathrm{Tr}_{\mathcal{K}}[X]$, is $\Phi_{\mathcal{H}}^{\dagger}[X]$, where $\Phi_{\mathcal{H}}$ is defined in (2.2.1), so that $\mathrm{Tr}_{\mathcal{K}}[X]$ is the unique operator in $\mathcal{B}(\mathcal{H})$ such that

$$(2.2.3) \qquad \mathrm{Tr}[A \, \mathrm{Tr}_{\mathcal{K}}[X]] = \mathrm{Tr}[(A \otimes \mathbb{1})X]$$

for all $A \in \mathcal{B}(\mathcal{H})$. Interchanging the roles of $\mathcal{H}$ and $\mathcal{K}$ yields the partial trace of A over $\mathcal{H}$, denoted $\mathrm{Tr}_{\mathcal{H}}[A]$.

Lemma 2.11. *Let $\{\mathbf{v}_1, \dots, \mathbf{v}_m\}$ be any orthonormal basis for $\mathcal{K}$, and let $X \in \mathcal{B}(\mathcal{H} \otimes \mathcal{K})$. Then for all $\mathbf{x}, \mathbf{y} \in \mathcal{H}$,*

$$(2.2.4) \qquad \langle \mathbf{x}, \mathrm{Tr}_{\mathcal{K}}[X]\mathbf{y} \rangle = \sum_{j=1}^{m} \langle \mathbf{x} \otimes \mathbf{v}_j, X\mathbf{y} \otimes \mathbf{v}_j \rangle,$$

Proof. Define $A := |\mathbf{y}\rangle\langle\mathbf{x}|$ so that $\langle \mathbf{x}, \mathrm{Tr}_{\mathcal{K}}[X]\mathbf{y} \rangle = \mathrm{Tr}[(A \otimes \mathbb{1})X]$. Then by Example 2.9,

$$A \otimes \mathbb{1} = |\mathbf{y}\rangle\langle\mathbf{x}| \otimes \left(\sum_{j=1}^{m} |\mathbf{v}_j\rangle\langle\mathbf{v}_j| \right) = \sum_{j=1}^{m} |\mathbf{y} \otimes \mathbf{v}_j\rangle\langle\mathbf{x} \otimes \mathbf{v}_j|,$$

and hence $\mathrm{Tr}[(A \otimes \mathbb{1})X] = \sum_{j=1}^{m} \langle \mathbf{x} \otimes \mathbf{v}_j, X\mathbf{y} \otimes \mathbf{v}_j \rangle$. Now (2.2.4) follows from (2.2.3) $\qquad \square$

Example 2.12. Let $X = A \otimes B \in \mathcal{B}(\mathcal{H} \otimes \mathcal{K})$. By (2.2.4), for all $\mathbf{x}, \mathbf{y} \in \mathcal{H}$,

$$\langle \mathbf{x}, \mathrm{Tr}_{\mathcal{K}}[A \otimes B]\mathbf{y} \rangle = \sum_{j=1}^{m} \langle \mathbf{x} \otimes \mathbf{v}_j, A \otimes B\mathbf{y} \otimes \mathbf{v}_j \rangle$$

$$= \sum_{j=1}^{m} \langle \mathbf{x} \otimes \mathbf{v}_j, A\mathbf{y} \otimes B\mathbf{v}_j \rangle = \langle \mathbf{x}, A\mathbf{y} \rangle \sum_{j=1}^{m} \langle \mathbf{v}_j, B\mathbf{v}_j \rangle$$

$$= \langle \mathbf{x}, A\mathbf{y} \rangle \, \mathrm{Tr}[B].$$

Since $\mathbf{x}$ and $\mathbf{y}$ are arbitrary, $\mathrm{Tr}_{\mathcal{K}}[A \otimes B] = \mathrm{Tr}[B]A$ and in the same way, we find $\mathrm{Tr}_{\mathcal{H}}[A \otimes B] = \mathrm{Tr}[A]B$.

To get explicit formulas for the partial traces, choose orthonormal bases $\{\mathbf{u}_1,\ldots,\mathbf{u}_n\}$ and $\{\mathbf{v}_1,\ldots,\mathbf{v}_m\}$ for $\mathcal{H}$ and $\mathcal{K}$, respectively. Taking $\mathbf{x} = \mathbf{u}_k$ and $\mathbf{y} = \mathbf{u}_\ell$ in (2.2.4) yields

$$\langle \mathbf{u}_k, \mathrm{Tr}_{\mathcal{K}}[X]\mathbf{u}_\ell \rangle = \sum_{j=1}^{m} \langle \mathbf{u}_k \otimes \mathbf{v}_j, X\mathbf{u}_\ell \otimes \mathbf{v}_j \rangle.$$

Then since for any $A \in \mathcal{B}(\mathcal{H})$, $A = \sum_{k,\ell=1}^{n} \langle \mathbf{u}_k, A\mathbf{u}_\ell \rangle |\mathbf{u}_k\rangle\langle\mathbf{u}_\ell|$,

$$\mathrm{Tr}_{\mathcal{K}}[X] = \sum_{k,\ell=1}^{n} \left(\sum_{j=1}^{m} \langle \mathbf{u}_k \otimes \mathbf{v}_j, X\mathbf{u}_\ell \otimes \mathbf{v}_j \rangle \right) |\mathbf{u}_k\rangle\langle\mathbf{u}_\ell|$$

$$(2.2.5) \qquad = \sum_{k,\ell=1}^{n} \left(\sum_{j=1}^{m} [X]_{(k,j),(\ell,j)} \right) |\mathbf{u}_k\rangle\langle\mathbf{u}_\ell|,$$

where the $[X]_{(k,i),(\ell,j)}$ are the entries of the matrix representative of X for the basis $\{\mathbf{u}_\ell \otimes \mathbf{v}_j\}$.

Since $\mathrm{Tr}[X] = \sum_{i=1}^{n}\sum_{j=1}^{m}[X]_{(i,j),(i,j)}$ is the *full* trace of X, by (2.2.5), the matrix representative of the partial trace is obtained by equating only the second pair of indices and summing over those, and hence the name *partial trace*.

Lemma 2.13. *Let $\mathcal{H}$ and $\mathcal{K}$ be finite-dimensional Hilbert spaces. Then for all* $X \in \mathcal{B}(\mathcal{H} \otimes \mathcal{K})$,

$$\mathrm{Tr}[\mathrm{Tr}_{\mathcal{K}}[X]] = \mathrm{Tr}[X],$$

where the first trace is taken on $\mathcal{H}$, and the second on $\mathcal{H} \otimes \mathcal{K}$. Moreover,

$$X \in \mathcal{B}^+(\mathcal{H} \otimes \mathcal{K}) \quad \Rightarrow \quad \mathrm{Tr}_{\mathcal{K}}[X] \in \mathcal{B}^+(\mathcal{H}).$$

Proof. The first part is an immediate consequence of (2.2.5). For the second, note that by Lemma 2.11, for all $\mathbf{x} \in \mathcal{H}$,

$$\langle \mathbf{x}, \mathrm{Tr}_{\mathcal{K}}[X]\mathbf{x} \rangle = \sum_{j=1}^{m} \langle \mathbf{x} \otimes \mathbf{v}_j, X\mathbf{x} \otimes \mathbf{v}_j \rangle,$$

and if $X \in \mathcal{B}^+(\mathcal{H} \otimes \mathcal{K})$, then each summand on the right is nonnegative. $\square$

Obviously, we can interchange the roles of $\mathcal{H}$ and $\mathcal{K}$ defining the partial trace of $X \in \mathcal{B}(\mathcal{H} \otimes \mathcal{K})$ over $\mathcal{H}$ to be the element of $\mathcal{B}(\mathcal{K})$ given by

$$(2.2.6) \qquad \langle \mathbf{x}, \mathrm{Tr}_{\mathcal{H}}[X]\mathbf{y} \rangle = \sum_{i=1}^{n} \langle \mathbf{u}_i \otimes \mathbf{x}, X(\mathbf{u}_i \otimes \mathbf{y}) \rangle,$$

where $\mathbf{x}, \mathbf{y}$ are any vectors in $\mathcal{K}$. Alternatively, using the same orthonormal bases introduced above,

$$\mathrm{Tr}_{\mathcal{H}}[X] = \sum_{\ell,j=1}^{m} \left(\sum_{i=1}^{n} [X]_{(i,\ell),(i,j)} \right) |\mathbf{v}_\ell\rangle\langle\mathbf{v}_j|.$$

2.3. Measurements on multicomponent systems and entanglement

In a multicomponent quantum system with Hilbert space $\mathcal{H} \otimes \mathcal{K}$, it is often the case that only observables on one of $\mathcal{H}$ or $\mathcal{K}$ are available to a particular observer. For instance, suppose we have two experimentalists, Hal and Katy, and an atom with Hilbert space $\mathcal{H}$ is in Hal's laboratory, while an atom with Hilbert space $\mathcal{K}$ is in Katy's laboratory. The two laboratories are very far apart, and Hal can only access observables on $\mathcal{H}$, while Katy can only access observables on $\mathcal{K}$.

However, if the two atoms were once close together and interacting, their state may be represented by a general unit vector ψ in the system state space $\mathcal{H} \otimes \mathcal{K}$.

Suppose that Hal decides to measure an observable represented by a self-adjoint operator A on $\mathcal{H}$. Let $A = \sum_{\lambda \in \sigma(A)} \lambda P_\lambda$ be its spectral decomposition so that $A \otimes \mathbb{1} = \sum_{\lambda \in \sigma(A)} \lambda(P_\lambda \otimes \mathbb{1})$ is the spectral decomposition of $A \otimes \mathbb{1}$. Define

$$(2.3.1) \qquad \rho_{\mathcal{H}} := \mathrm{Tr}_{\mathcal{K}}[|\psi\rangle\langle\psi|] \in \mathcal{B}(\mathcal{H}).$$

By Lemma 2.13, $\rho_{\mathcal{H}} \in \mathcal{B}^+(\mathcal{H})$ and $\mathrm{Tr}[\rho_{\mathcal{H}}] = 1$.

Let p_λ denote the probability that when a measurement of A, or rather $A \otimes \mathbb{1}$, is made, the observed value is λ. Then

$$(2.3.2) \qquad p_\lambda = \langle \psi, P_\lambda \otimes \mathbb{1}\,\psi \rangle = \mathrm{Tr}[|\psi\rangle\langle\psi|P_\lambda \otimes \mathbb{1}] = \mathrm{Tr}[\rho_{\mathcal{H}}P_\lambda],$$

where the final equality comes from the definition (2.2.3).

After the measurement, if λ is the observed value, in which case $p_\lambda > 0$, the state of the system in $\mathcal{H} \otimes \mathcal{K}$ is

$$\phi_\lambda := \frac{1}{p_\lambda^{1/2}}(P_\lambda \otimes \mathbb{1})\psi.$$

It is left as an exercise to show that

$$(2.3.3) \qquad \mathrm{Tr}_{\mathcal{K}}[|\phi_\lambda\rangle\langle\phi_\lambda|] = \frac{1}{p_\lambda}P_\lambda \rho_{\mathcal{H}} P_\lambda.$$

We may regard this as representing the new state of the system, as far as observations of observables on $\mathcal{H}$ are concerned, after the measurement of A in state $\rho_{\mathcal{H}}$ has resulted in the value λ.

Effectively, as far as observations on the first component are concerned, the state of the system is given by the operator $\rho_{\mathcal{H}} \in \mathcal{B}(\mathcal{H})$.

Definition 2.14 (Density matrices and pure states). Let $\mathcal{H}$ be a finite-dimensional Hilbert space. A **density matrix** on $\mathcal{H}$ is a positive semidefinite operator $\rho \in \mathcal{B}(\mathcal{H})$ such that $\mathrm{Tr}[\rho] = 1$. The set of all density matrices on $\mathcal{H}$ is denoted $\mathfrak{S}_{\mathcal{H}}$, and is also referred to as the set of **states** of quantum systems on $\mathcal{H}$. A state $\rho \in \mathfrak{S}_{\mathcal{H}}$ is a **pure state** in case $\rho = |\psi\rangle\langle\psi|$ for some unit vector $\psi \in \mathcal{H}$.

Definition 2.15 (Support of a density matrix). For $\rho \in \mathfrak{S}_{\mathcal{H}}$, let $\rho = \sum_{\lambda \in \sigma(\rho)} \lambda P_\lambda$ be its spectral decomposition. The orthogonal projection

$$P_\rho := \sum_{\lambda \in \sigma(\rho),\, \lambda > 0} P_\lambda,$$

which is the projection onto $\mathrm{ran}(\rho) = \mathrm{ker}(\rho)^\perp$, is the **support** of ρ, and the subspace $\mathrm{ran}(\rho)$ is called the **supporting space** of ρ, denoted $\mathrm{supp}(\rho)$.

The discussion at the beginning of this section gives us the rules concerning measurements of a quantum system on $\mathcal{H}$ whose state is given by a density matrix $\rho \in \mathfrak{S}_{\mathcal{H}}$. If A is a self-adjoint operator on $\mathcal{H}$ with spectral decomposition $A = \sum_{\lambda \in \sigma(A)} \lambda P_\lambda$, we have from (2.3.2) that the probability that a measurement of A in the state ρ results in the value λ is

$$p_\lambda = \mathrm{Tr}[\rho P_\lambda].$$

If the result of the measurement of A is λ, then $p_\lambda > 0$ and after the measurement the state of the system is given by the density matrix,

$$\frac{1}{p_\lambda} P_\lambda \rho P_\lambda.$$

There are two density matrices naturally associated to any $\psi \in \mathcal{H} \otimes \mathcal{K}$, namely $\rho_{\mathcal{H}}$ as specified in (2.3.1) and $\rho_{\mathcal{K}} := \mathrm{Tr}_{\mathcal{H}}[|\psi\rangle\langle\psi|] \in \mathcal{B}(\mathcal{K})$. These two density matrices are closely related. The following discussion is based on a 1935 paper of Schrödinger [**193**] in which he introduced the concept of *entanglement*.

Consider a system of two atoms that have finitely many internal degrees of freedom, say certain *spins* or angular momentum degrees of freedom. Suppose that the Hilbert spaces for these degrees of freedom for the two atoms are $\mathcal{H}$ and $\mathcal{K}$, two finite-dimensional Hilbert spaces. The Hilbert space for the bipartite system is then $\mathcal{H} \otimes \mathcal{K}$.

Suppose the atoms are brought together and an observation of some observable L on $\mathcal{H} \otimes \mathcal{K}$ is made. Suppose the eigenspaces of L are one dimensional. Then, if the value $\lambda \in \sigma(L)$ is found, after the measurement the state is given by $\psi \in \mathcal{H} \otimes \mathcal{K}$, where ψ is an eigenvector of L with eigenvalue λ. We can contemplate any observable L on $\mathcal{H} \otimes \mathcal{K}$ with any outcome λ, and so ψ can be any unit vector in $\mathcal{H} \otimes \mathcal{K}$.

For example, consider two atoms with an intrinsic *spin* angular momentum. We might prepare the system by measuring a component of the total angular momentum, say finding zero, so that the two individual angular momenta cancel. The angular momenta of the two atoms are the *internal degrees of freedom* with which we are concerned.

Now suppose the atoms are moved far apart without disturbing the internal degrees of freedom; these are still described by ψ, but now one atom is in Hal's laboratory and the other is in Katy's.

Let A be self-adjoint on $\mathcal{H}$, and let $\{\mathbf{u}_1, \dots, \mathbf{u}_n\}$ be an orthonormal basis of $\mathcal{H}$ consisting of eigenvectors of A: $A\mathbf{u}_j = \lambda_j \mathbf{u}_j$ for $j = 1, \dots, n$. Following Schrödinger, we suppose for now that the eigenspaces are all one dimensional. Let $\{\mathbf{v}_1, \dots, \mathbf{v}_m\}$ be any orthonormal basis for $\mathcal{K}$. Then expanding ψ in the corresponding orthonormal basis of $\mathcal{H} \otimes \mathcal{K}$, $\psi = \sum_{i=1}^{n} \sum_{j=1}^{m} \psi_{i,j} \mathbf{u}_i \otimes \mathbf{v}_j$ yields

$$(2.3.4) \qquad \psi = \sum_{i=1}^{n} \mathbf{u}_i \otimes \mathbf{y}_i, \quad \text{where} \quad \mathbf{y}_i := \sum_{j=1}^{m} \psi_{i,j} \mathbf{v}_j.$$

The spectrum of the observable $A \otimes \mathbb{1}$ is of course simply $\{\lambda_1, \dots, \lambda_n\}$, and each (nonzero) $\mathbf{u}_i \otimes \mathbf{y}_i$ is an eigenvector with eigenvalue λ_i. When Hal performs his measurement of the observable $A \otimes \mathbb{1}$ on the whole system in state ψ, the rules say that he finds the value λ_i with probability $\|\mathbf{u}_i \otimes \mathbf{y}_i\|^2 = \|\mathbf{y}_i\|^2$, and that after the measurement, given that λ_i is the result (so that $\|\mathbf{y}_i\|^2 \neq 0$), the state of the system is

$$(2.3.5) \qquad \phi = \|\mathbf{y}_i\|^{-1} \mathbf{u}_i \otimes \mathbf{y}_i.$$

Now let B be self-adjoint on $\mathcal{K}$, and suppose that Katy decides to measure $\mathbb{1} \otimes B$ for her atom. Let $B = \sum_{\mu \in \sigma(B)} \mu Q_\mu$ be the spectral decomposition of B. After Hal's experiment, so that the state is ϕ given by (2.3.5), the probability that Katy finds the value μ is

$$p_{\mu,i} = \|\mathbf{y}_i\|^{-2} \langle \mathbf{y}_i, Q_\mu \mathbf{y}_i \rangle = \|\mathbf{y}_i\|^{-2} \|Q_\mu \mathbf{y}_i\|^2.$$

However, if Hal had not done his experiment, the state would still be ψ, and then the probability that Katy finds the value μ would be

$$p_\mu = \sum_{i=1}^{n} \langle \mathbf{y}_i, Q_\mu \mathbf{y}_i \rangle = \sum_{i=1}^{n} \|Q_\mu \mathbf{y}_i\|^2.$$

Even though there is no interaction between the two atoms, which could be lightyears apart, what Hal did in his laboratory has influenced what goes on in Katy's laboratory, and has done so instantly. Although Hal cannot exert any forces on Katy's atom, he has managed to, as Schrödinger put it, **steer** it into another state. Schrödinger described this relation between two noninteracting systems as **entanglement**. He said [**193**] of entanglement that "I would call that not *one*, but rather *the* characteristic trait of quantum mechanics, the one that enforces its entire departure from classical lines of thought."

Schrödinger's thought experiment has actually been carried out. The 2022 Nobel prize in physics went to three experimentalists who did this, and found that the results exactly matched what the rules predict.

To get a clearer understanding of entanglement, Schrödinger asked when in an expansion of the form (2.3.4) the vectors $\{\mathbf{y}_1, \dots, \mathbf{y}_n\}$ will be mutually orthogonal. (Recall that they are not assumed to be normalized, and some may even be zero). He proved [**193**] the following:

Theorem 2.16 (Schrödinger, 1935). *Let $\mathcal{H}$ and $\mathcal{K}$ be Hilbert spaces of dimension m and n, respectively. Let ψ be any unit vector in $\mathcal{H} \otimes \mathcal{K}$. Then there exists an integer $0 < r \leq \min\{m, n\}$ and $\{\lambda_1, \dots, \lambda_r\} \subset (0, 1]$ with $\sum_{j=1}^{r} \lambda_j = 1$, and orthonormal sets $\{\mathbf{u}_1, \dots, \mathbf{u}_r\}$ in $\mathcal{H}$ and $\{\mathbf{v}_1, \dots, \mathbf{v}_r\}$ in $\mathcal{K}$ such that*

$$(2.3.6) \qquad \psi = \sum_{i=1}^{r} \lambda_j^{1/2} \mathbf{u}_j \otimes \mathbf{v}_j,$$

$$(2.3.7) \qquad \mathrm{Tr}_{\mathcal{K}}[|\psi\rangle\langle\psi|] = \sum_{i=1}^{r} \lambda_i |\mathbf{u}_i\rangle\langle\mathbf{u}_i| \quad and \quad \mathrm{Tr}_{\mathcal{H}}[|\psi\rangle\langle\psi|] = \sum_{i=1}^{r} \lambda_i |\mathbf{v}_i\rangle\langle\mathbf{v}_i|.$$

Proof. Define $\rho_{\mathcal{H}} := \mathrm{Tr}_{\mathcal{K}}[|\psi\rangle\langle\psi|]$. Let $\{\mathbf{u}_1, \dots, \mathbf{u}_m\}$ be an orthonormal basis of $\mathcal{H}$ consisting of eigenvectors of $\rho_{\mathcal{H}}$; that is, $\rho_{\mathcal{H}} \mathbf{u}_j = \lambda_j \mathbf{u}_j$ for all j. We may suppose the indexing is such that if the rank of $\rho_{\mathcal{H}}$ is r, then $\lambda_j > 0$ for $1 \leq j \leq r$. Then $\sum_{j=1}^{r} \lambda_j = 1$, and

$$(2.3.8) \qquad \rho_{\mathcal{H}} = \sum_{j=1}^{r} \lambda_j |\mathbf{u}_j\rangle\langle\mathbf{u}_j|.$$

Let $\{\mathbf{w}_1, \dots, \mathbf{w}_n\}$ be any orthonormal basis of $\mathcal{K}$, and write

$$\psi = \sum_{i=1}^{m} \sum_{\ell=1}^{n} \psi_{i,\ell} \mathbf{u}_i \otimes \mathbf{w}_\ell = \sum_{i=1}^{m} \mathbf{u}_i \otimes \mathbf{y}_i,$$

where

$$(2.3.9) \qquad \psi_{i,\ell} = \langle \mathbf{u}_i \otimes \mathbf{w}_\ell, \psi \rangle \quad and \quad \mathbf{y}_i = \sum_{\ell=1}^{m} \psi_{i,\ell} \mathbf{w}_\ell.$$

We now show that $\mathbf{y}_1, \dots, \mathbf{y}_n$ are mutually orthogonal. For $1 \leq i, j \leq n$,

$$\langle \mathbf{y}_j, \mathbf{y}_i \rangle = \sum_{\ell=1}^{n} \overline{\psi_{i,\ell}} \psi_{j,\ell} = \overline{\langle \mathbf{u}_j \otimes \mathbf{w}_\ell, \psi \rangle} \langle \mathbf{u}_i \otimes \mathbf{w}_\ell, \psi \rangle$$

$$= \sum_{\ell=1}^{m} \langle \mathbf{u}_i \otimes \mathbf{w}_\ell, |\psi\rangle\langle\psi| \mathbf{u}_j \otimes \mathbf{w}_\ell \rangle = \langle \mathbf{u}_i, \rho_{\mathcal{H}} \mathbf{u}_j \rangle = \lambda_i \delta_{i,j},$$

where the first two equalities come from (2.3.9) and the third is the formula for the partial trace from Lemma 2.11.

Hence the vectors in $\mathbf{y}_1, \dots, \mathbf{y}_n$ are mutually orthogonal, and $\mathbf{y}_j = 0$ for $j > r$. Defining $\mathbf{v}_j = \lambda_j^{-1/2}\mathbf{y}_j$ for $1 \le j \le r$, $\{\mathbf{v}_1, \dots, \mathbf{v}_r\}$ is orthonormal and (2.3.9) becomes (2.3.6). Then (2.3.8) is the first part of (2.3.7), and the second part follows by symmetry in $\mathcal{H}$ and $\mathcal{K}$, though it also follows readily from (2.3.6) and (2.2.6). $\qquad\square$

Corollary 2.17. *Let $\mathcal{H}$ and $\mathcal{K}$ be two finite-dimensional Hilbert spaces. Let ψ be any unit vector in $\mathcal{H} \otimes \mathcal{K}$. Then $\rho_{\mathcal{H}} = \mathrm{Tr}_{\mathcal{K}}[|\psi\rangle\langle\psi|]$ and $\rho_{\mathcal{K}} = \mathrm{Tr}_{\mathcal{H}}[|\psi\rangle\langle\psi|]$ have the same nonzero eigenvalues with the same multiplicities.*

Proof. This is an immediate consequence of (2.3.7). $\qquad\square$

Remark 2.18. The decomposition (2.3.6) of ψ is often called the **Schmidt decomposition** in reference to work of Schmidt [**191**] on integral equations. He did not explicitly discuss tensor products.

Theorem 2.16 says that for any $\psi \in \mathcal{H} \otimes \mathcal{K}$, r and the numbers $\lambda_1, \dots, \lambda_r$ are uniquely determined by ψ—they are the rank and nonzero eigenvalues of $\mathrm{Tr}_{\mathcal{K}}[|\psi\rangle\langle\psi|]$. The number r is called the **Schmidt rank** of ψ, and $\lambda_1, \dots, \lambda_r$ are called the **Schmidt numbers**.

Definition 2.19. Let $\mathcal{H}$ and $\mathcal{K}$ be two finite-dimensional Hilbert spaces. A unit vector $\psi \in \mathcal{H} \otimes \mathcal{K}$ is **entangled** if the Schmidt rank of ψ is greater than 1, and is **separable** in case the Schmidt rank equals 1, in which case $\psi = \mathbf{u} \otimes \mathbf{v}$ is a simple product.

Now let us return to Hal and Katy. Suppose they produce a pair of atoms whose internal degrees of freedom are such that the state space for each atom is two dimensional. Thus we may identify $\mathcal{H}$ and $\mathcal{K}$ with $\mathbb{C}^2$ equipped with its standard inner product. Let $\mathbf{u}_1 = (1, 0)$ and $\mathbf{u}_2 = (0, 1)$. Suppose the state of the two particles is

$$\psi = \frac{1}{\sqrt{2}}(\mathbf{u}_1 \otimes \mathbf{u}_2 - \mathbf{u}_2 \otimes \mathbf{u}_1).$$

They each carry their atom off to their distant laboratories without disturbing the internal degrees of freedom. Eventually, they will be very far apart. Hal wishes to be able to send a simple message to Katy at some specified point in the future—such as "I am well" or "I am not well". Hal tells Katy, that at the specified point in the future, he will measure one of two observables, A or B, in his lab. After this measurement, the state of the atom in Katy's lab will be instantly changed. Hal says that to send the message, "I am well", he will measure A, and to send the message "I am not well", he will measure B. At the specified time, when Katy knows Hal will have performed his measurement, Katy is to observe her atom, and deduce what choice of measurement Hal made, and hence to receive his instantaneous message—transmitted faster than light can travel. Is this possible?

To investigate this mathematically, note first that ψ is entangled; the Schmidt rank is 2. Let the observables Hal will measure correspond to the self-adjoint matrices

$$A = \begin{bmatrix} 0 & 1 \\ 1 & 0 \end{bmatrix} \quad \text{and} \quad B = \begin{bmatrix} 1 & 0 \\ 0 & -1 \end{bmatrix}.$$

One easily computes that $\sigma(A) = \sigma(B) = \{-1, 1\}$ and that their spectral decompositions $A = P_1 - P_{-1}$ and $B = Q_1 - Q_{-1}$ are given by

$$P_1 = \frac{1}{2}\begin{bmatrix} 1 & 1 \\ 1 & 1 \end{bmatrix}, \quad P_{-1} = \frac{1}{2}\begin{bmatrix} 1 & -1 \\ -1 & 1 \end{bmatrix}, \quad Q_1 = \begin{bmatrix} 1 & 0 \\ 0 & 0 \end{bmatrix}, \quad Q_{-1} = \begin{bmatrix} 0 & 0 \\ 0 & 1 \end{bmatrix}.$$

If Hal decides to measure A, and finds the value 1, after his measurement the state of the system is

$$\frac{1}{\|(P_1 \otimes \mathbb{1})\psi\|}(P_1 \otimes \mathbb{1})\psi = \frac{1}{2}(\mathbf{u}_1 + \mathbf{u}_2) \otimes (\mathbf{u}_2 - \mathbf{u}_1).$$

Note that this state is separable; Hal's measurement has destroyed the entanglement. Computing the final states for each of the other three possibilities, we conclude the following.

If Hal measures A, after his measurement the state of the system is either $\frac{1}{2}(\mathbf{u}_1 + \mathbf{u}_2) \otimes (\mathbf{u}_2 - \mathbf{u}_1)$ or $\frac{1}{2}(\mathbf{u}_1 - \mathbf{u}_2) \otimes (\mathbf{u}_1 + \mathbf{u}_1)$, corresponding to the observed values 1 and -1. However, if Hal decides to measure B, after his measurement the state of the system is either $\mathbf{u}_1 \otimes \mathbf{u}_2$ or $\mathbf{u}_2 \otimes \mathbf{u}_1$, corresponding to the observed values 1 and -1. Therefore, the state ϕ of the atom in Katy's lab will now be either $\frac{1}{\sqrt{2}}(\mathbf{u}_1 + \mathbf{u}_2)$ or $\frac{1}{\sqrt{2}}(\mathbf{u}_1 - \mathbf{u}_2)$ if Hal measured A, or $\mathbf{u}_2$ or $\mathbf{u}_1$ if Hal measured B.

If Katy can do a measurement to decide which of these four states she now has, she will have received Hal's message. Suppose that Katy can *clone* many copies of the state ϕ. She can then repeatedly measure B in this state ϕ. If the state is $\mathbf{u}_2$ or $\mathbf{u}_1$, she will always find the value 1, or always find the value -1. However, if the state is either, $\frac{1}{\sqrt{2}}(\mathbf{u}_1 + \mathbf{u}_2)$ or $\frac{1}{\sqrt{2}}(\mathbf{u}_1 - \mathbf{u}_2)$, she will find the values 1 and -1 with equal probability each time she measures. If she finds a mix of outcomes, sometimes 1, sometimes -1, then she knows Hal measured A, and she has received the message "I am well". If she has made 20 measurements and the results are either all always either 1 or all always -1, then she is quite sure Hal measured B since if $\phi = \frac{1}{\sqrt{2}}(\mathbf{u}_1 + \mathbf{u}_2)$ or $\phi = \frac{1}{\sqrt{2}}(\mathbf{u}_1 - \mathbf{u}_2)$, the probability of this outcome is 2^{-20}; that is, less than one in a million. She has received, with high certainty, the message "I am not well", and as long as she can clone further copies of ϕ, she can continue to raise her certainty.

However, this scheme for communication depends on Katy's ability to clone the state ϕ. She does not learn anything from a single measurement. Moreover, if Hal and Katy shared 20 identical copies of ψ, and Hal performed a measurement of A 20 times, he would produce a random sequence of 20 values 1 and -1. Even if Katy now also measures A she would see a random sequence, just as she would if she had chosen to measure B. Without cloning, the message has arrived but cannot be read.

The set of states $|\phi\rangle\langle\phi|$ that Katy would need to be able to clone is non-commutative. The *no cloning theorem*, to be proved in Chapter 5, says that in fact, there is no device that can perfectly clone a noncommutative set of density matrices. In some sense Hal's message has arrived in Katy's lab, but she has no way of reading it. This scheme for faster-than-light communication was actually put forward by a physicist Nick Herbert in 1982 [**101**], and he recognized that its implementation required a cloning device. This proposal focused the attention of a number of researchers on the cloning issue, and several no-cloning theorems resulted [**73, 235**] asserting that no device could clone *all* quantum states. The example at hand does not require all states to be cloned, but later and more general results [**24, 150**] rule out cloning a noncommutative set of states. We shall return to cloning issue in Chapter 5. The nonlocal effect of measurement of entangled states in quantum mechanics bothered Einstein, and many others, enormously, already in the 1930s. It is surprising that only much later did physicists embrace entanglement as something deeply interesting and useful, and begin to work out the mathematical and physical theory of entanglement in detail.

2.4. Purification

We have seen that if $\mathcal{H}$ and $\mathcal{K}$ are finite-dimensional Hilbert spaces, and ψ is any unit vector in $\mathcal{H}\otimes\mathcal{K}$ representing the state of a bipartite system, the density matrix $\rho_{\mathcal{H}} = \mathrm{Tr}_{\mathcal{K}}[|\psi\rangle\langle\psi|]$ can be regarded as specifying the state of the system as far as observations only of observables of the form $A \otimes \mathbb{1}$ are concerned. It is an important fact that *every* density matrix ρ on $\mathcal{H}$ arises in this way by taking the partial trace of a pure state.

Definition 2.20 (Purification). Let $\mathcal{H}$ be a finite-dimensional Hilbert space. Let $\rho \in \mathfrak{S}_{\mathcal{H}}$. Then a pure state of the form $|\psi\rangle\langle\psi|$ on $\mathcal{H} \otimes \mathcal{K}$ for some other finite-dimensional Hilbert space $\mathcal{K}$ such that $\rho = \mathrm{Tr}_K[|\psi\rangle\langle\psi|]$ is called a **purification** of ρ. (One often refers to ψ itself as the purification.)

Theorem 2.21 (Purification theorem). *Let $\mathcal{H}$ be a finite-dimensional Hilbert space, and let $\rho \in \mathfrak{S}_{\mathcal{H}}$ have rank r and spectral decomposition $\rho = \sum_{i=1}^{r} \lambda_i |\mathbf{u}_i\rangle\langle\mathbf{u}_i|$. Let $\mathcal{K}$ be another Hilbert space of dimension at least r. Then ψ is a purification of ρ if and only if for some orthonormal set $\{\mathbf{v}_1,\ldots,\mathbf{v}_r\}$ in $\mathcal{K}$, $\psi = \sum_{i=1}^{r} \lambda_i^{1/2}\mathbf{u}_i \otimes \mathbf{v}_i$.*

Proof. This is a direct consequence of Corollary 2.17 to Schrödinger's theorem, Theorem 2.16. $\square$

There is a useful construction of a particular type of purification of ρ out of $\rho^{1/2}$ that will be of frequent use.

Theorem 2.22. *Let $\mathcal{H}$ be a finite-dimensional Hilbert space, and let $\widehat{\mathcal{H}}$ denote $\mathcal{B}(\mathcal{H})$ equipped with the Hilbert-Schmidt inner product. Let $\{\mathbf{u}_1, \ldots, \mathbf{u}_n\}$ be an orthonormal basis for $\mathcal{H}$, and define*

$$(2.4.1) \qquad \Psi := \sum_{j=1}^{n} \mathbf{u}_j \otimes \mathbf{u}_j.$$

Then the map $\mathcal{U} : \widehat{\mathcal{H}} \to \mathcal{H} \otimes \mathcal{H}$ given by

$$(2.4.2) \qquad \mathcal{U}(X) = (X \otimes \mathbb{1})\Psi = \sum_{j=1}^{n} X\mathbf{u}_j \otimes \mathbf{u}_j$$

is unitary from $\widehat{\mathcal{H}}$ to $\mathcal{H} \otimes \mathcal{H}$, and moreover, for all $X, Y \in \mathcal{B}(\mathcal{H})$,

$$(2.4.3) \qquad \mathrm{Tr}_2[|\mathcal{U}(X)\rangle\langle\mathcal{U}(Y)|] = XY^*.$$

In particular, for any $\rho \in \mathfrak{S}_{\mathcal{H}}$, $\psi := \mathcal{U}(\rho^{1/2}) = (\rho^{1/2} \otimes \mathbb{1})\Psi$ is a purification of ρ, and ψ can also be written as

$$(2.4.4) \qquad \psi = \sum_{i,j=1}^{n} \langle \mathbf{u}_i, \rho^{1/2}\mathbf{u}_j\rangle \mathbf{u}_i \otimes \mathbf{u}_j.$$

Proof. By (2.1.11), $|\mathcal{U}(X)\rangle\langle\mathcal{U}(Y)| = \sum_{i,j=1}^{n} |X\mathbf{u}_i\rangle\langle Y\mathbf{u}_j| \otimes |\mathbf{u}_i\rangle\langle\mathbf{u}_j|$, and therefore, since $\mathrm{Tr}[\mathbf{u}_i\rangle\langle\mathbf{u}_j|] = \delta_{i,j}$,

$$\mathrm{Tr}_2[|\mathcal{U}(X)\rangle\langle\mathcal{U}(Y)|] = \sum_{i}^{n} |X\mathbf{u}_i\rangle\langle Y\mathbf{u}_i| = X\left(\sum_{i}^{n} |\mathbf{u}_i\rangle\langle\mathbf{u}_i|\right)Y^* = X\mathbb{1}Y^*.$$

This proves (2.4.3), and now it follows that

$$\langle \mathcal{U}(Y), \mathcal{U}(X)\rangle_{\mathcal{H}\otimes\mathcal{H}} = \mathrm{Tr}[|\mathcal{U}(X)\rangle\langle\mathcal{U}(Y)|]$$

$$= \mathrm{Tr}[\mathrm{Tr}_2[|\mathcal{U}(X)\rangle\langle\mathcal{U}(Y)|]] = \mathrm{Tr}[XY^*] = \langle Y, X\rangle_{\widehat{H}}.$$

This shows that $\mathcal{U}$ is an isometry, and since $\widehat{\mathcal{H}}$ and $\mathcal{H} \otimes \mathcal{H}$ have the same dimension, it is unitary. Taking $X = Y = \rho^{1/2}$ in (2.4.3) proves the $\psi := \mathcal{U}(\rho^{1/2})$ is a purification of ρ, and then substituting $\rho^{1/2}\mathbf{u}_j = \sum_{i=1}^{n} \langle \mathbf{u}_i, \rho^{1/2}\mathbf{u}_j\rangle\mathbf{u}_i$ into yields (2.4.4). $\square$

2.4.1. Ensembles of quantum states. There is another way density matrices arise. Suppose the state of the system is one of a finite collection of pure states $|\psi_j\rangle\langle\psi_j|$, $j = 1,\ldots,s$. Suppose we do not know which of these states is actually the state of the system, but we have probabilities p_j, $j = 1,\ldots,s$ such that the actual state is $|\psi_j\rangle\langle\psi_j|$ with probability p_j.

Definition 2.23. An **ensemble of quantum pure states** on a Hilbert space $\mathcal{H}$ is a finite set
$$\{p_j,\ |\psi_j\rangle\langle\psi_j|\}_{1\le j\le s}$$
of pure states $|\psi_j\rangle\langle\psi_j|$, $j = 1,\ldots,s$, together with probabilities p_j, $j = 1,\ldots,s$, $\sum_{j=1}^s p_j = 1$.

Associated to any ensemble is the density matrix

$$(2.4.5) \qquad \rho := \sum_{j=1}^s p_j |\psi_j\rangle\langle\psi_j|.$$

Let $\rho \in \mathfrak{S}_{\mathcal{H}}$, and let $\{p_j,\ |\psi_j\rangle\langle\psi_j|\}_{1\le j\le s}$ be an ensemble such that (2.4.5) is satisfied. Then (2.4.5) is an **ensemble decomposition** of ρ. The set of all ensembles providing decompositions of ρ is denoted by $\mathrm{Ens}(\rho)$.

Let $\mathcal{H}$ be a finite-dimensional Hilbert space. Then any given $\rho \in \mathfrak{S}_{\mathcal{H}}$ arises as the density matrix of infinitely many ensembles of states in $\mathcal{H}$. One such ensemble comes from the spectral decomposition of ρ: If ρ has rank r, and $\rho := \sum_{j=1}^r \lambda_j |\mathbf{u}_j\rangle\langle\mathbf{u}_j|$ is a spectral decomposition of ρ, $\sum_{j=1}^r \lambda_j = 1$. Therefore, $\{\lambda_j,\ |\mathbf{u}_j\rangle\langle\mathbf{u}_j|\}_{1\le j\le r}$ is an ensemble whose density matrix is ρ.

A theorem of Hughston, Jozsa, and Wooters [**117**] gives a complete and explicit description of $\mathrm{Ens}(\rho)$. The proof uses purification and Schrödinger's theorem.

Theorem 2.24 (Hughston–Jozsa–Wooters theorem). *Let $\mathcal{H}$ be a finite-dimensional Hilbert space, and let $\rho \in \mathfrak{S}_{\mathcal{H}}$ have rank r. Let $\rho := \sum_{j=1}^r \lambda_j |\mathbf{u}_j\rangle\langle\mathbf{u}_j|$ be a spectral decomposition of ρ where $\lambda_j > 0$ for $j = 1,\ldots,r$.*

Then an ensemble $\{p_j,\ |\phi_j\rangle\langle\phi_j|\}_{1\le j\le s}$ has density matrix ρ if and only if there is an $s \times r$ matrix U whose r columns are orthonormal (and hence $s \ge r$) such that for all $j = 1,\ldots,s$,

$$(2.4.6) \qquad p_j = \sum_{i=1}^r |U_{j,i}|^2 \lambda_i \quad \text{and} \quad \phi_j = p_j^{-1/2} \sum_{i=1}^r \lambda_i^{1/2} U_{j,i} \mathbf{u}_i.$$

Proof. Let $\{p_j, |\phi_j\rangle\langle\phi_j|\}_{1\le j\le s} \in \mathrm{Ens}(\rho)$. Let $\mathcal{K}$ be an s-dimensional Hilbert space. Let $\{\mathbf{w}_1,\ldots,\mathbf{w}_s\}$ be orthonormal in $\mathcal{K}$ and define $\psi \in \mathcal{H} \otimes \mathcal{K}$ by

$$(2.4.7) \qquad \psi := \sum_{\ell=1}^s p_\ell^{1/2} \phi_\ell \otimes \mathbf{w}_\ell.$$

By Lemma 2.11 and (2.4.7), for any $\mathbf{x}, \mathbf{y} \in \mathcal{H}$, since $\langle \mathbf{w}_j, \mathbf{w}_\ell \rangle = \delta_{j,\ell}$,

$$\langle \mathbf{x}, \mathrm{Tr}_{\mathcal{K}}[|\psi\rangle\langle\psi|]\mathbf{y}\rangle = \sum_{j=1}^{s} \langle \mathbf{x} \otimes \mathbf{w}_j, |\psi\rangle\langle\psi|\mathbf{y} \otimes \mathbf{w}_j\rangle$$

$$= \sum_{j=1}^{s} p_j\langle \mathbf{x}, |\phi_j\rangle\langle\phi_j|\mathbf{y}\rangle.$$

Therefore, $\mathrm{Tr}_{\mathcal{K}}[|\psi\rangle\langle\psi|] = \sum_{j=1}^{s} p_j|\phi_j\rangle\langle\phi_j| = \rho$ so that ψ is a purification of ρ. By Schrödinger's theorem, there is an orthonormal set $\{\mathbf{v}_1, \ldots, \mathbf{v}_r\}$ in $\mathcal{K}$ so that

$$(2.4.8) \qquad \psi := \sum_{i=1}^{r} \lambda_i^{1/2}\mathbf{u}_i \otimes \mathbf{v}_i.$$

Since $\{\mathbf{w}_1, \ldots, \mathbf{w}_s\}$ is an orthonormal basis for $\mathcal{K}$, $\mathbf{v}_i = \sum_{j=1}^{s}\langle \mathbf{w}_j, \mathbf{v}_i\rangle \mathbf{w}_j$, and substituting this into (2.4.8) yields

$$\psi := \sum_{i=1}^{r} \lambda_i^{1/2}\mathbf{u}_i \otimes \left(\sum_{j=1}^{s}\langle \mathbf{w}_j, \mathbf{v}_i\rangle \mathbf{w}_j \right)$$

$$(2.4.9) \qquad = \sum_{j=1}^{s} \left(\sum_{i=1}^{r} \lambda_i^{1/2}\langle \mathbf{w}_j, \mathbf{v}_i\rangle \mathbf{u}_i \right) \otimes \mathbf{w}_j.$$

Comparing (2.4.7) and (2.4.9) and recalling that $\{\mathbf{w}_1, \ldots, \mathbf{w}_s\}$ is orthonormal yields

$$(2.4.10) \qquad p_j^{1/2}\phi_j = \sum_{i=1}^{r} \lambda_i^{1/2}\langle \mathbf{w}_j, \mathbf{v}_i\rangle \mathbf{u}_i$$

for $j = 1, \ldots, s$. Taking the norm on both sides of (2.4.10),

$$(2.4.11) \qquad p_j = \sum_{i=1}^{r} \lambda_i|\langle \mathbf{w}_j, \mathbf{v}_i\rangle|^2.$$

Define $U \in M_{s,r}(\mathbb{C})$ by $U_{k,\ell} = \langle \mathbf{w}_k, \mathbf{v}_\ell\rangle$. Then since

$$\sum_{k=1}^{s}\langle \mathbf{v}_j, \mathbf{w}_k\rangle\langle \mathbf{w}_k, \mathbf{v}_l\rangle = \langle \mathbf{v}_j, \mathbf{v}_\ell\rangle = \delta_{j,\ell},$$

the r columns of U are orthonormal in $\mathbb{C}^s$. Written in terms of U, (2.4.10) and (2.4.11) become (2.4.6).

For the converse, let p_j and ϕ_j, $j = 1, \ldots, s$ be given by (2.4.6) in terms of a spectral decomposition of ρ and $U \in M_{s,r}(\mathbb{C})$ with orthonormal columns.

Then each p_j is nonnegative, and

$$\sum_{j=1}^{s} p_j = \sum_{i=1}^{r}\sum_{j=1}^{s} \lambda_i |U_{j,i}|^2 = \sum_{i=1}^{r} \lambda_i = 1.$$

Also ϕ_j is a unit vector for $j = 1,\ldots,s$, and hence $\{|\phi_j\rangle\langle\phi_j|\}_{1\leq j\leq s}$ together with $\{p_j\}_{1\leq j\leq s}$ is a quantum ensemble. To see that it has density matrix ρ, compute using (2.4.6):

$$\sum_{j=1}^{s} p_j |\phi_j\rangle\langle\phi_j| = \sum_{j=1}^{s} |p_j^{1/2}\phi_j\rangle\langle p_j^{1/2}\phi_j|$$

$$= \sum_{i,k=1}^{r} \left(\sum_{j=1}^{s} \overline{U_{j,i}} U_{j,k} \lambda_i^{1/2}\lambda_k^{1/2} |\mathbf{u}_i\rangle\langle\mathbf{u}_k|\right)$$

$$= \sum_{i,k=1}^{r} \left(\delta_{i,k}\lambda_i^{1/2}\lambda_k^{1/2} |\mathbf{u}_i\rangle\langle\mathbf{u}_k|\right) = \rho. \qquad \square$$

2.5. The singular value decomposition

The following theorem is due to Beltrami [**28**] and Jordan [**121, 122**]. Schmidt [**191**] later developed a version for integral equations. An interesting account of the history can be found in [**202**]. It is essentially another formulation of Schrödinger's theorem [**193**] about tensor products (which came later), but it will be useful to prove it directly by adapting Schrödinger's proof to the operator setting.

Theorem 2.25 (Singular value decomposition). *Let $\mathcal{H}$ and $\mathcal{K}$ be Hilbert spaces of dimension m and n, respectively. Let $K \in \mathcal{B}(\mathcal{H},\mathcal{K})$, $K \neq 0$, and let $r > 0$ denote the rank of K. There are orthonormal sets $\{\mathbf{u}_1,\ldots,\mathbf{u}_r\} \subset \mathcal{H}$ and $\{\mathbf{v}_1,\ldots,\mathbf{v}_r\} \subset \mathcal{K}$ and positive numbers $\sigma_1,\ldots,\sigma_r$ such that for $j = 1,\ldots,r$, $K^*K\mathbf{u}_j = \sigma_j^2\mathbf{u}_j$ and*

$$(2.5.1) \qquad K = \sum_{j=1}^{r} \sigma_j |\mathbf{v}_j\rangle\langle\mathbf{u}_j|.$$

Proof. Let $\{\mathbf{u}_1,\ldots,\mathbf{u}_n\}$ be any orthonormal basis of $\mathcal{H}$, Then defining $\mathbf{y}_k := K\mathbf{u}_k$, $k = 1,\ldots,n$, we have

$$(2.5.2) \qquad K = \sum_{k=1}^{n} |\mathbf{y}_k\rangle\langle\mathbf{u}_k|.$$

Then for $1 \leq i, j \leq n$,

$$(2.5.3) \qquad \langle\mathbf{y}_i,\mathbf{y}_j\rangle = \langle\mathbf{u}_i, K^*K\mathbf{u}_j\rangle.$$

If we choose $\{\mathbf{u}_1, \ldots, \mathbf{u}_n\}$ to consist of eigenvectors of K^*K, so that $K^*K\mathbf{u}_j = \lambda_j \mathbf{u}_j$, $j = 1, \ldots, n$, then each $\lambda_j \geq 0$. Relabel as needed to arrange that $\lambda_j > 0$ for $j = 1, \ldots, r$, and $\lambda_j = 0$ for any other values of j. Then (2.5.3) becomes

$$(2.5.4) \qquad\qquad \langle \mathbf{y}_i, \mathbf{y}_j \rangle = \lambda_j \delta_{i,j}.$$

For $j = 1, \ldots, r$, define $\sigma_j = \lambda_j^{1/2}$, and define $\mathbf{v}_j = \sigma_j^{-1} \mathbf{y}_j$. By (2.5.4), $\{\mathbf{v}_1, \ldots, \mathbf{v}_r\}$ is orthonormal, and then (2.5.2) becomes (2.5.1). $\qquad\qquad\square$

Remark 2.26. The decomposition of K given by (2.5.1) is one way of writing the **singular value decomposition** (SVD) of K. The positive numbers in the set $\{\sigma_1, \ldots, \sigma_r\}$ are called the **singular values** of K.

2.5.1. The polar decomposition. The **polar decomposition** is another important operator decomposition that is closely related to the SVD.

Let $K \in \mathcal{B}(\mathcal{K}, \mathcal{H})$, $\mathcal{H}$ and $\mathcal{K}$ finite-dimensional Hilbert spaces. Then $K^*K \in \mathcal{B}^+(\mathcal{K})$, and we may use the spectral theorem to define $(K^*K)^{1/2}$.

Definition 2.27. For any $K \in \mathcal{B}(\mathcal{K}, \mathcal{H})$, $|K| \in \mathcal{B}^+(\mathcal{K})$, the **absolute value** of K, is defined by

$$|K| := (K^*K)^{1/2}.$$

Evidently, if $\ker(K) = 0$, K^*K and hence also $|K|$ are strictly positive, so that $|K|$ is invertible. We then define $U := K|K|^{-1}$, and have

$$U^*U = |K|^{-1}K^*K|K|^{-1} = (K^*K)^{-1/2}(K^*K)(K^*K)^{-1/2} = \mathbb{1},$$

so that U is unitary. Hence we have $K = U|K|$, which is the **polar decomposition** of K as a product of a unitary and a positive matrix. When $\ker(K) \neq 0$, we still have the polar decomposition, but now U will be a partial isometry instead of a unitary:

Definition 2.28 (Partial isometry). Let $\mathcal{H}$ and $\mathcal{K}$ be Hilbert spaces. An operator $U \in \mathcal{B}(\mathcal{H}, \mathcal{K})$ is a **partial isometry** in case U^*U is an orthogonal projection in $\mathcal{H}$. If U is a partial isometry, then the **final space** of U is $\operatorname{ran}(U) \subseteq \mathcal{K}$, and the **initial space** of U is $\ker(U)^\perp \subseteq \mathcal{H}$.

Remark 2.29. Suppose that U is a partial isometry, so that U^*U is an orthogonal projection. Then

$$\operatorname{ran}(U^*U) = \ker(U^*U)^\perp = \ker(U)^\perp,$$

so that U^*U is the orthogonal projection onto the initial space of U.

Observe that a unitary $U \in \mathcal{B}(\mathcal{H}, \mathcal{K})$ is a very special type of partial isometry since $U^*U = \mathbb{1}_{\mathcal{H}}$ and $UU^* = \mathbb{1}_{\mathcal{K}}$, and the identity is a very special type of orthogonal projection. Also, every orthogonal projection $P \in \mathcal{B}(H)$ is a partial isometry since $P = P^*$ and $P^2 = P$.

Lemma 2.30. *Let $\mathcal{H}$ and $\mathcal{K}$ be finite-dimensional Hilbert spaces. An operator $U \in \mathcal{B}(\mathcal{H}, \mathcal{K})$ is a partial isometry if and only if there exist orthonormal sets $\{\mathbf{v}_1, \ldots, \mathbf{v}_r\}$ in $\mathcal{K}$ and $\{\mathbf{u}_1, \ldots, \mathbf{u}_r\}$ in $\mathcal{H}$ such that*

$$(2.5.5) \qquad U = \sum_{j=1}^{r} |\mathbf{v}_j\rangle\langle\mathbf{u}_j|.$$

In this case, the initial space of U is $\mathrm{span}(\{\mathbf{u}_1, \ldots, \mathbf{u}_r\})$, *and the final space of U is* $\mathrm{span}(\{\mathbf{v}_1, \ldots, \mathbf{v}_r\})$.

Proof. Let $\{\mathbf{u}_1, \ldots, \mathbf{u}_r\}$ be an orthonormal basis for the initial space of U. Define $\mathbf{v}_j := U\mathbf{u}_j$, $j = 1, \ldots, r$. Then

$$(2.5.6) \qquad \langle \mathbf{v}_i, \mathbf{v}_j \rangle = \langle U\mathbf{u}_i, U\mathbf{u}_j \rangle = \langle \mathbf{u}_i, U^*U\mathbf{u}_j \rangle.$$

By Remark 2.29, U^*U is the orthogonal projection onto the initial space of U, and hence $U^*U\mathbf{v}_j = \mathbf{v}_j$ for each j. Then from (2.5.6), $\langle \mathbf{u}_j, \mathbf{u}_j \rangle = \langle \mathbf{v}_i, \mathbf{v}_j \rangle = \delta_{i,j}$. Therefore, $\{\mathbf{v}_1, \ldots, \mathbf{v}_r\}$ is orthonormal.

Next, completing $\{\mathbf{u}_1, \ldots, \mathbf{u}_r\}$ to an orthonormal basis $\{\mathbf{u}_1, \ldots, \mathbf{u}_n\}$ of $\mathcal{H}$, since $U\mathbf{u}_j = 0$ for $j > r$,

$$U = \sum_{j=1}^{n} |U\mathbf{u}_j\rangle\langle\mathbf{u}_j| = \sum_{j=1}^{r} |U\mathbf{u}_j\rangle\langle\mathbf{u}_j| = \sum_{j=1}^{r} |\mathbf{v}_j\rangle\langle\mathbf{u}_j|.$$

For the converse, since $\{\mathbf{u}_1, \ldots, \mathbf{u}_r\}$ and $\{\mathbf{v}_1, \ldots, \mathbf{v}_r\}$ are orthonormal,

$$U^*U = \left(\sum_{k=1}^{r} |\mathbf{u}_k\rangle\langle\mathbf{v}_k| \right)\left(\sum_{j=1}^{r} |\mathbf{v}_j\rangle\langle\mathbf{u}_j| \right) = \sum_{j=1}^{r} |\mathbf{u}_j\rangle\langle\mathbf{u}_j|,$$

which is the orthogonal projection onto the initial space of U, and likewise $UU^* = \sum_{j=1}^{r} |\mathbf{v}_j\rangle\langle\mathbf{v}_j|$ is the projector onto the final space of U. $\square$

A number of conclusions follow easily from Lemma 2.30 and its proof. First, if $U \in \mathcal{B}(\mathcal{H}, \mathcal{K})$ is a partial isometry, then $U^* \in \mathcal{B}(\mathcal{K}, \mathcal{H})$ is also a partial isometry. To see this, note that by Lemma 2.30, any partial isometry $U \in \mathcal{B}(\mathcal{H}, \mathcal{K})$ has the form (2.5.5). Then

$$UU^* = \left(\sum_{j=1}^{r} |\mathbf{u}_j\rangle\langle\mathbf{v}_j| \right)\left(\sum_{k=1}^{r} |\mathbf{v}_k\rangle\langle\mathbf{u}_j| \right) = \sum_{j=1}^{r} |\mathbf{u}_j\rangle\langle\mathbf{u}_j|,$$

and this is the orthogonal projection onto the final space of U. Therefore, UU^* is the orthogonal projection onto the final space of U, and U^*U is the orthogonal projection onto the initial space of U.

Theorem 2.31 (Unique polar factorization). *Let $\mathcal{H}$ and $\mathcal{K}$ be finite-dimensional Hilbert spaces. Then for $K \in \mathcal{B}(\mathcal{H}, \mathcal{K})$, there is a unique partial isometry $U \in \mathcal{B}(\mathcal{H}, \mathcal{K})$ such that $K = U|K|$ such that the initial space of U is $\operatorname{ran}(K^*) = \ker(K)^{\perp}$.*

Proof. Let $r \geq 0$ denote the rank of K. By Theorem 2.25 there are orthonormal sets $\{\mathbf{u}_1, \ldots, \mathbf{u}_r\} \subset \mathcal{H}$ and $\{\mathbf{v}_1, \ldots, \mathbf{v}_r\} \subset \mathcal{K}$ and positive numbers $\sigma_1, \ldots, \sigma_r$ such that $K = \sum_{j=1}^{r} \sigma_j |\mathbf{v}_j\rangle\langle\mathbf{u}_j|$. Then $|K| = \sum_{j=1}^{r} \sigma_j |\mathbf{u}_j\rangle\langle\mathbf{u}_j|$. Define $U := \sum_{j=1}^{r} |\mathbf{v}_j\rangle\langle\mathbf{u}_j|$. Then $K = U|K|$, and U is a partial isometry by Lemma 2.30. Since $U^*U := \sum_{j=1}^{r} |\mathbf{u}_j\rangle\langle\mathbf{u}_j|$, $\ker(U)^{\perp} = \ker(K)^{\perp}$. This proves the existence of U.

To prove the uniqueness, suppose that $V \in \mathcal{B}(\mathcal{H}, \mathcal{K})$ is a partial isometry such that $K = V|K|$ and $\ker(V)^{\perp} = \ker(K)^{\perp}$. For any $\mathbf{x} \in \mathcal{H}$,

$$V(|K|\mathbf{x}) = K\mathbf{x} = U(|K|\mathbf{x}).$$

Since $\operatorname{ran}(|K|) = \operatorname{ran}(K^*K) = \operatorname{ran}(K^*) = \ker(K)^{\perp}$, V and U agree on $\ker(K)^{\perp}$. Now let $\{\mathbf{v}_1, \ldots, \mathbf{v}_r\}$ be an orthonormal basis for $\ker(K)^{\perp}$, and extend it to an orthonormal basis $\{\mathbf{v}_1, \ldots, \mathbf{v}_n\}$. Since both U and V have $\ker(K)^{\perp}$ as their initial spaces, $U\mathbf{v}_j = V\mathbf{v}_j = 0$ for $j > r$. Therefore,

$$U = \sum_{j=1}^{n} |U\mathbf{v}_j\rangle\langle\mathbf{v}_j| = \sum_{j=1}^{r} |U\mathbf{v}_j\rangle\langle\mathbf{v}_j| = \sum_{j=1}^{r} |V\mathbf{v}_j\rangle\langle\mathbf{v}_j| = \sum_{j=1}^{n} |V\mathbf{v}_j\rangle\langle\mathbf{v}_j| = V. \quad \square$$

An important application of Lemma 2.30 yields a useful description of the relation between different purifications of the same density matrix ρ.

Theorem 2.32. *Let $\mathcal{H}$ be a finite-dimensional Hilbert space, and let $\rho \in \mathfrak{S}_{\mathcal{H}}$. Let $\mathcal{K}$ and $\widetilde{\mathcal{K}}$ be two Hilbert spaces. Let $\psi \in \mathcal{H} \otimes \mathcal{K}$ be a purification of ρ. Then $\widetilde{\psi} \in \mathcal{H} \otimes \widetilde{\mathcal{K}}$ is a purification of ρ if and only if there is a partial isometry $U \in \mathcal{B}(\mathcal{K}, \widetilde{\mathcal{K}})$ such that*

$$(2.5.7) \qquad \widetilde{\psi} = (\mathbb{1} \otimes U)\psi \quad and \quad (\mathbb{1} \otimes U^*U)\psi = \psi.$$

Proof. First, suppose that $\widetilde{\psi} \in \mathcal{H} \otimes \widetilde{\mathcal{K}}$ is a purification of ρ. Let $\rho = \sum_{k=1}^{r} \lambda_k |\mathbf{u}_k\rangle\langle\mathbf{u}_k|$ be a spectral decomposition of ρ, where r is the rank of ρ. By Theorem 2.21 there are orthonormal sets $\{\mathbf{v}_1, \ldots, \mathbf{v}_r\}$ in $\mathcal{K}$ and $\{\widetilde{\mathbf{v}}_1, \ldots, \widetilde{\mathbf{v}}_r\}$ in $\widetilde{\mathcal{K}}$ such that

$$\psi = \sum_{k=1}^{r} \lambda_k^{1/2} \mathbf{u}_k \otimes \mathbf{v}_k \quad \text{and} \quad \widetilde{\psi} = \sum_{k=1}^{r} \lambda_k^{1/2} \mathbf{u}_k \otimes \widetilde{\mathbf{v}}_k.$$

Define $U := \sum_{k=1}^{r} |\widetilde{\mathbf{v}}_k\rangle\langle\mathbf{v}_k|$. By Lemma 2.30, U is a partial isometry, and it is easy to check that (2.5.7) is satisfied.

Next, suppose that (2.5.7) is satisfied. Let r be the rank of U^*U, and let $\{\mathbf{v}_1,\ldots,\mathbf{v}_r\}$ be an orthonormal basis of $\mathrm{ran}(U^*U)$. Then there are uniquely determined vectors $\{\mathbf{w}_1,\ldots,\mathbf{w}_r\} \subset \mathcal{H}$ such that $\psi = \sum_{j=1}^{r} \mathbf{w}_j \otimes \mathbf{v}_j$. Define $\{\tilde{\mathbf{v}}_1,\ldots,\tilde{\mathbf{v}}_r\}$ in $\widetilde{\mathcal{K}}$ by $\tilde{\mathbf{v}}_j := U\mathbf{v}_j$, $j = 1,\ldots,r$. Then $\{\tilde{\mathbf{v}}_1,\ldots,\tilde{\mathbf{v}}_r\}$ is orthonormal since $\langle \tilde{\mathbf{v}}_i, \tilde{\mathbf{v}}_j \rangle = \langle \mathbf{v}_i, U^*U\mathbf{v}_j \rangle = \langle \mathbf{v}_i, \mathbf{v}_j \rangle = \delta_{ij}$. Therefore, $\tilde{\psi} = \sum_{j=1}^{r} \mathbf{w}_j \otimes \tilde{\mathbf{v}}_j$, where $\{\tilde{\mathbf{v}}_1,\ldots,\tilde{\mathbf{v}}_r\}$ is orthonormal, and hence $\mathrm{Tr}_{\widetilde{\mathcal{K}}}[|\tilde{\psi}\rangle\langle\tilde{\psi}|] = \sum_{j=1}^{r} |\mathbf{w}_j\rangle\langle\mathbf{w}_j| = \mathrm{Tr}_{\mathcal{K}}[|\psi\rangle\langle\psi|] = \rho$. $\square$

The polar factorization also yields a concrete description of all purifications $\rho \in \mathfrak{S}_{\mathcal{H}}$ in $\mathcal{H} \otimes \mathcal{H}$.

Theorem 2.33. *Let $\mathcal{H}$ be a finite-dimensional Hilbert space and let $\{\mathbf{u}_1,\ldots,\mathbf{u}_n\}$ be an orthonormal basis of it. Let $\rho \in \mathfrak{S}_{\mathcal{H}}$. Then $\psi \in \mathcal{H} \otimes \mathcal{H}$ is a purification of ρ if and only if*

$$\psi = \sum_{j=1}^{n} \rho^{1/2} U\mathbf{u}_j \otimes \mathbf{u}_j,$$

where U is a partial isometry whose final space is $\mathrm{ran}(\rho)$.

Proof. By Theorem 2.22, every ψ has the form $\psi = \sum_{j=1}^{n} X\mathbf{u}_j \otimes \mathbf{u}_j$ for some $X \in \mathcal{B}(\mathcal{H})$, and by (2.4.3), $\mathrm{Tr}_2[|\psi\rangle\langle\psi|] = XX^*$. Therefore, it suffices to show that $XX^* = \rho$ if and only if $X = \rho^{1/2}U$, where U is a partial isometry whose final space is $\mathrm{ran}(\rho)$. But $XX^* = \rho$ if and only if $|X^*| = \rho^{1/2}$, and then if $X^* = U^*\rho^{1/2}$ is the polar factorization of X^*, the initial space of U^* is $\mathrm{ran}(\rho)$, and $X = \rho^{1/2}U$. Conversely, if $X = \rho^{1/2}U$ and the final space of U is $\mathrm{ran}(\rho)$, $XX^* = \rho$. $\square$

2.5.2. The Moore-Penrose generalized inverse. Let $A \in M_{m,n}(\mathbb{C})$. Then for $\mathbf{b} \in \mathbb{C}^m$, there may be no solutions $\mathbf{x}$ to the equation $A\mathbf{x} = \mathbf{b}$, but there are always **least squares solutions**. That is, the set

$$(2.5.8) \qquad \{\mathbf{x} \ : \ \|A\mathbf{x} - \mathbf{b}\|^2 = \min_{\mathbf{y}\in\mathbb{C}^n}\{\|A\mathbf{y} - \mathbf{b}\|^2\}\}$$

is never empty. To see this, observe that $\{A\mathbf{y}-\mathbf{b} \ : \ \mathbf{y} \in \mathbb{C}^n\}$ is closed affine (and hence convex) set in $\mathbb{C}^m$. By the projection lemma, Lemma A.2, it contains a unique element $\mathbf{b}_0$ of minimal norm. By the construction of $\mathbf{b}_0$, the solution set of $A\mathbf{y} = \mathbf{b} + \mathbf{b}_0$ is not empty, and this is the set specified in (2.5.8). While the solution set may not be unique, it too is a closed affine subset, this time of $\mathbb{C}^n$, and hence there is a unique element $\mathbf{x}_0$ of this set that has minimal norm. We define $\mathbf{x}_0$ to be the **least length, least squares solution** of $A\mathbf{x} = \mathbf{b}$.

Theorem 2.34. *Let $A \in M_{m,n}(\mathbb{C})$ have the singular value decomposition $A = \sum_{j=1}^{r} \sigma_j |\mathbf{v}_j\rangle\langle\mathbf{u}_j|$ where r is the rank of A. Define $A^+ \in M_{n,m}(\mathbb{C})$,*

$$(2.5.9) \qquad A^+ = \sum_{j=1}^{r} \sigma_j^{-1} |\mathbf{u}_j\rangle\langle\mathbf{v}_j|.$$

Let $P_{\mathrm{ran}(A)}$ and $P_{\mathrm{ker}(A)}$, respectively, denote the orthogonal projections onto the range and kernel of A. Then

$$(2.5.10) \qquad \mathrm{ran}(A^+) = \ker(A)^\perp \quad and \quad \ker(A^+) = \mathrm{ran}(A)^\perp,$$

and

$$(2.5.11) \qquad AA^+ = P_{\mathrm{ran}(A)} \quad and \quad A^+A = P^\perp_{\ker(A)}.$$

Moreover, for all $\mathbf{b} \in \mathbb{C}^m$, the least length, least squares solution $\mathbf{x}_0$ to $A\mathbf{x} = \mathbf{b}$ is given by $\mathbf{x}_0 = A^+\mathbf{b}$.

Proof. By the definition of A^+, $\{\mathbf{v}_1,\ldots,\mathbf{v}_r\}$ and $\{\mathbf{u}_1,\ldots,\mathbf{u}_r\}$ are orthonormal bases for $\mathrm{ran}(A)$ and $\ker(A)^\perp$, and (2.5.10) is an immediate consequence of this. Moreover,

$$AA^+ = \sum_{j=1}^{r} |\mathbf{v}_j\rangle\langle\mathbf{v}_j| \quad and \quad A^+A = \sum_{j=1}^{r} |\mathbf{u}_j\rangle\langle\mathbf{u}_j|.$$

This proves (2.5.11). Therefore, for any $\mathbf{y} \in \mathbb{C}^n$, $\mathbf{b} \in \mathbb{C}^m$,

$$\|A\mathbf{y} - \mathbf{b}\|^2 = \|A\mathbf{y} - P_{\mathrm{ran}(A)}\mathbf{b}\|^2 + \|P^\perp_{\mathrm{ran}(A)}\mathbf{b}\|^2 \geq \|P^\perp_{\mathrm{ran}(A)}\mathbf{b}\|^2$$

with equality if $A\mathbf{y} = P_{\mathrm{ran}(A)}\mathbf{b}$. Since $P_{\mathrm{ran}(A)}\mathbf{b} \in \mathrm{ran}(A)$, such $\mathbf{y}$ exist. This shows that the minimum in (2.5.8) is achieved for any $\mathbf{y}$ such that $A\mathbf{y} = P_{\mathrm{ran}(A)}\mathbf{b}$. By the first equation in (2.5.11), $A^+\mathbf{b}$ has this property.

It remains to show that if $A\mathbf{y} = P_{\mathrm{ran}(A)}\mathbf{b}$, then $\|\mathbf{y}\| \geq \|A^+\mathbf{b}\|$ with equality only in case $\mathbf{y} = A^+\mathbf{b}$. For any such $\mathbf{y}$, $\mathbf{y} = (\mathbf{y} - A^+\mathbf{b}) + A^+\mathbf{b}$, and evidently $\mathbf{y} - A^+\mathbf{b} \in \ker(A)$ while by (2.5.10), $A^+\mathbf{b} \in \ker(A)^\perp$. Therefore, $\|\mathbf{y}\|^2 = \|\mathbf{y} - A^+\mathbf{b}\|^2 + \|A^+\mathbf{b}\|^2$. $\qquad\square$

The uniqueness of the least length, least squares solution of $A\mathbf{x} = \mathbf{b}$ then implies that A^+ does not depend on the choice of bases in (2.5.9), and is in fact uniquely determined by A. The $n \times m$ matrix A^+ is called the Moore-Penrose **generalized inverse** of A [155, 174].

Note that for $X \in M_n^+(\mathbb{C})$, X^+ is obtained from X by inverting all of the strictly positive eigenvalues of X in the spectral decomposition of X, and by leaving the null space as it is.

2.6. The block matrix picture for $\mathcal{B}(\mathcal{H} \otimes \mathcal{K})$

Let $\{\mathbf{u}_1,\ldots,\mathbf{u}_m\}$ and $\{\mathbf{v}_1,\ldots,\mathbf{v}_n\}$ be orthonormal bases for two Hilbert spaces $\mathcal{H}$ and $\mathcal{K}$.

We have seen how to use these bases to construct a unitary map from $\mathcal{H} \otimes \mathcal{K}$ onto $\mathcal{B}(\mathcal{K}, \mathcal{H})$, two Hilbert spaces of the same dimension mn.

There is another mn-dimensional Hilbert space associated to $\mathcal{H}$: Define $\bigoplus_n \mathcal{H}$ to be the direct sum of n copies of $\mathcal{H}$. That is, the elements of $\bigoplus_n \mathcal{H}$ are n-tuples $(\mathbf{x}_1, \dots, \mathbf{x}_n)$ of vectors in $\mathcal{H}$. We equip $\bigoplus_n \mathcal{H}$ with the obvious vector space structure and the inner product

$$\langle (\mathbf{x}_1, \dots, \mathbf{x}_n), (\mathbf{y}_1, \dots, \mathbf{y}_n) \rangle_{\bigoplus_n \mathcal{H}} = \sum_{j=1}^{n} \langle \mathbf{x}_j, \mathbf{y}_j \rangle_{\mathcal{H}},$$

which makes $\bigoplus_n \mathcal{H}$ a Hilbert space of dimension mn. We now show that $\mathcal{B}(\bigoplus_n \mathcal{H})$ can be identified with the space of $n \times n$ matrices with entries in $\mathcal{B}(\mathcal{H})$.

For $j = 1, \dots, n$, define $V_j : \bigoplus_n \mathcal{H} \to \mathcal{H}$ to be the linear transformation given by

$$(2.6.1) \qquad V_j(\mathbf{x}_1, \dots, \mathbf{x}_n) = \mathbf{x}_j.$$

It is easy to check that $V_j^* : \mathcal{H} \to \bigoplus_n \mathcal{H}$ is given by $V_j^* \mathbf{x} = (\mathbf{y}_1, \dots, \mathbf{y}_n)$ where $\mathbf{y}_j = \mathbf{x}$ and $\mathbf{y}_k = 0$ for $k \neq j$. Let P_j denote the orthogonal projection in $\bigoplus_n \mathcal{H}$ onto the range of V_j^*; that is, onto the subspace of $\bigoplus_n \mathcal{H}$ consisting of vectors whose only nonzero entry is in the jth place. Then

$$P_j = V_j^* V_j.$$

Lemma 2.35. *For any $A \in \mathcal{B}(\bigoplus_n \mathcal{H})$, for each $1 \leq i, j \leq n$, define*

$$(2.6.2) \qquad [A]_{i,j} := V_i A V_j^* \in \mathcal{B}(\mathcal{H}).$$

Then for all $(\mathbf{x}_1, \dots, \mathbf{x}_n) \in \bigoplus_n \mathcal{H}$,

$$(2.6.3) \qquad A(\mathbf{x}_1, \dots, \mathbf{x}_n) = \left(\sum_{j=1}^{n} [A]_{1,j} \mathbf{x}_j, \dots, \sum_{j=1}^{n} [A]_{n,j} \mathbf{x}_j \right).$$

Therefore, we may identify A with the $n \times n$ block matrix with entries in $\mathcal{B}(\mathcal{H})$ whose (i, j)-th entry is $[A]_{i,j} = V_i A V_j^*$.

Proof. For any $A \in \mathcal{B}\left(\bigoplus_n \mathcal{H} \right)$, by (2.6.2) and $\sum_{j=1}^{n} P_j = \mathbb{1}$,

$$A = \sum_{i,j=1}^{n} P_i A P_j = \sum_{i,j=1}^{n} V_i^* (V_i A V_j^*) V_j = \sum_{i,j=1}^{n} V_i^* [A]_{i,j} V_j.$$

Then (2.6.3) follows from (2.6.1). $\qquad\square$

We can interpret (2.6.3) as saying that A is represented by the **block matrix**

$$\begin{bmatrix} [A]_{1,1} & \cdots & [A]_{1,n} \\ \vdots & \ddots & \vdots \\ [A]_{n,1} & \cdots & [A]_{n,n} \end{bmatrix}.$$

2.6.1. Tensor products and block matrices. An orthonormal basis $\{\mathbf{v}_1, \dots, \mathbf{v}_n\}$ of $\mathcal{K}$ induces a useful unitary transformation $\mathcal{V} : \mathcal{H} \otimes \mathcal{K} \to \bigoplus_n \mathcal{H}$. As noted already in connection with Schrödinger's theorem, every $\psi \in \mathcal{H} \otimes \mathcal{K}$ has a uniquely determined expansion $\psi = \sum_{j=1}^n \psi_j \otimes \mathbf{v}_j$.

In terms of this expansion, define $\mathcal{V} : \mathcal{H} \otimes \mathcal{K} \to \bigoplus_n \mathcal{H}$ by

$$(2.6.4) \qquad\qquad \mathcal{V}\psi = (\psi_1, \dots, \psi_n).$$

Then $\|\mathcal{V}\psi\|^2_{\oplus_n \mathcal{H}} = \sum_{j=1}^n \|\psi_j\|^2_{\mathcal{H}} = \sum_{j=1}^n \|\psi_j \otimes \mathbf{v}_j\|^2_{\mathcal{H} \otimes \mathcal{K}} = \|\psi\|^2_{\mathcal{H} \otimes \mathcal{K}}$. This shows that $\mathcal{V}$ is an isometry. Then since $\dim(\mathcal{H} \otimes \mathcal{K}) = \dim(\bigoplus_n \mathcal{H})$, $\mathcal{V}$ is unitary.

If $A \in \mathcal{B}(\mathcal{H} \otimes \mathcal{K})$, then $\mathcal{V}A\mathcal{V}^* \in \mathcal{B}(\bigoplus_n \mathcal{H})$, and this allows us to transfer the block matrix picture for $\bigoplus_n \mathcal{H}$ to $\mathcal{B}(\mathcal{H} \otimes \mathcal{K})$.

For $1 \leq j \leq n$, define the linear transformation $W_j : \mathcal{H} \otimes \mathcal{K} \to \mathcal{H}$ by

$$(2.6.5) \qquad\qquad W_j := V_j \mathcal{V},$$

so that for ψ given by (2.6.4), $W_j \psi = \psi_j$. The adjoint W_j^* of W_j is given by $W_j^* = \mathcal{V}^* V_j^*$ so that for all $\mathbf{x} \in \mathcal{H}$,

$$(2.6.6) \qquad\qquad W_j^* \mathbf{x} = \mathbf{x} \otimes \mathbf{v}_j.$$

We can now rewrite (2.6.4) as $\mathcal{V}\psi = (W_1\psi, \dots, W_n\psi)$ and then

$$\mathcal{V}^*(\mathbf{x}_1, \dots, \mathbf{x}_n) = \sum_{j=1}^n \mathbf{x}_j \otimes \mathbf{v}_j = \sum_{j=1}^n W_j^* \mathbf{x}_j.$$

Theorem 2.36. *Let* $\mathcal{V} : \mathcal{H} \otimes \mathcal{K} \to \bigoplus_n \mathcal{H}$ *be the unitary defined in* (2.6.4). *For* $A \in \mathcal{B}(\mathcal{H} \otimes \mathcal{K})$, $\mathcal{V}A\mathcal{V}^* \in \mathcal{B}(\bigoplus_n \mathcal{H})$, *considered as an* $n \times n$ *block matrix with entries* $[\mathcal{V}A\mathcal{V}^*]_{i,j} \in \mathcal{B}(\mathcal{H})$, *is given by*

$$(2.6.7) \qquad\qquad [\mathcal{V}A\mathcal{V}^*]_{i,j} = W_i A W_j^*.$$

Likewise, for $B \in \mathcal{B}(\bigoplus_n \mathcal{H})$ *considered as an* $n \times n$ *block matrix whose* (i, j)-*th entry is* $[B]_{i,j} \in \mathcal{B}(\mathcal{H})$,

$$\mathcal{V}^* B \mathcal{V} = \sum_{i,j=1}^n [B]_{i,j} \otimes |\mathbf{v}_i\rangle\langle\mathbf{v}_j|.$$

Proof. By Lemma 2.35, and then (2.6.5), $[\mathcal{V}A\mathcal{V}^*]_{i,j} = V_i \mathcal{V}A\mathcal{V}^* V_j^* = W_i A W_j^*$, and this proves (2.6.7).

Next, any $\psi \in \mathcal{H} \otimes \mathcal{K}$ is of the form $\psi = \sum_{k=1}^{n} \psi_k \otimes \mathbf{v}_k$, and hence

$$\left(\sum_{i,j=1}^{n} [B]_{i,j} \otimes |\mathbf{v}_i\rangle\langle\mathbf{v}_j| \right) \psi = \sum_{i=1}^{n} \left(\sum_{j=1}^{n} [B]_{i,j} \psi_j \right) \otimes \mathbf{v}_i$$

$$= \sum_{i=1}^{n} W_i^* \left(\sum_{j=1}^{n} [B]_{i,j} W_j \psi \right)$$

$$= \sum_{i,j=1}^{n} W_i^* V_i B V_j^* W_j \psi = V^* B V \psi. \qquad \square$$

There is another useful way to think about representing operators on $\mathcal{H} \otimes \mathcal{K}$ as block matrices. Instead of starting from a unitary identification of the underlying Hilbert spaces $\mathcal{H} \otimes \mathcal{K}$ and $\bigoplus_n \mathcal{H}$, one can start from Lemma 2.6 and Example 2.9, which say that

$$\{|\mathbf{u}_i\rangle\langle\mathbf{u}_j| \otimes |\mathbf{v}_k\rangle\langle\mathbf{v}_\ell|\}_{\{1\leq i,j\leq m,\ 1\leq k,\ell\leq n\}}$$

is an orthonormal basis for $\mathcal{B}(\mathcal{H} \otimes \mathcal{K})$, equipped with the Hilbert-Schmidt inner product. Therefore, every $X \in \mathcal{B}(\mathcal{H} \otimes \mathcal{K})$ has an expansion

$$X = \sum_{i,k=1}^{m} \sum_{j,\ell=1}^{n} X_{(i,j),(k,\ell)} |\mathbf{u}_i\rangle\langle\mathbf{u}_j| \otimes |\mathbf{v}_k\rangle\langle\mathbf{v}_\ell|$$

with uniquely determined coefficients. Define the operators $X_{k,\ell} \in \mathcal{B}(\mathcal{H})$, $1 \leq j,\ell \leq m$ by $X_{k,\ell} := \sum_{i,k=1}^{m} X_{(i,j),(k,\ell)} |\mathbf{u}_i\rangle\langle\mathbf{u}_j|$. Then

$$(2.6.8) \qquad X = \sum_{k,\ell=1}^{n} X_{k,\ell} \otimes |\mathbf{v}_k\rangle\langle\mathbf{v}_\ell|.$$

This identifies X with the block matrix

$$(2.6.9) \qquad \begin{bmatrix} X_{1,1} & \cdots & X_{1,n} \\ \vdots & \ddots & \vdots \\ X_{n,1} & \cdots & X_{n,n} \end{bmatrix}.$$

Moreover, for all $\mathbf{x}, \mathbf{y} \in \mathcal{H}$,

$$\langle \mathbf{x}, X_{k,\ell} \mathbf{y}\rangle = \langle \mathbf{x} \otimes \mathbf{v}_k, X\mathbf{y} \otimes \mathbf{v}_\ell\rangle = \langle W_j^* \mathbf{x}, X W_j^* \mathbf{y}\rangle,$$

where the last equality is due to by (2.6.6). Therefore, by Theorem 2.36, this block matrix description of X coincides with our previous block matrix description.

Remark 2.37. The block matrix picture provides a useful perspective on the partial trace: By Example 2.12 and (2.6.8),

$$\mathrm{Tr}_{\mathcal{K}}[X] = \sum_{k,\ell=1}^{n} X_{k,\ell}\, \mathrm{Tr}[|\mathbf{v}_k\rangle\langle\mathbf{v}_\ell|] = \sum_{k=1}^{n} X_{k,k},$$

hence $\mathrm{Tr}_{\mathcal{K}}[X]$ is the *block trace* of the block matrix in (2.6.9). Also by Example 2.12 and (2.6.8),

$$\mathrm{Tr}_{\mathcal{K}}[X] = \sum_{k,\ell=1}^{n} \mathrm{Tr}[X_{k,\ell}]|\mathbf{v}_k\rangle\langle\mathbf{v}_\ell|.$$

Example 2.38 (Block matrix form of $X \otimes \mathbb{1}$). Let $X \in \mathcal{B}(\mathcal{H})$. Since $\mathbb{1} = \sum_{j=1}^{n} |\mathbf{v}_j\rangle\langle\mathbf{v}_j|$, $X \otimes \mathbb{1} = \sum_{j=1}^{n} X \otimes |\mathbf{v}_j\rangle\langle\mathbf{v}_j|$. Then by (2.6.8), $X \otimes \mathbb{1} \in \mathcal{B}(\mathcal{H} \otimes \mathcal{K})$ has the block matrix representation

$$\begin{bmatrix} X & & \\ & \ddots & \\ & & X \end{bmatrix}.$$

Example 2.39 (Block matrix form of $\mathbb{1} \otimes X$). Let $X \in \mathcal{B}(\mathcal{K})$. Define $X_{k,\ell} := \langle\mathbf{v}_k, X\mathbf{v}_\ell\rangle$ so that $X = \sum_{k,\ell=1}^{n} X_{k,\ell}|\mathbf{v}_k\rangle\langle\mathbf{v}_\ell|$, and hence

$$\mathbb{1} \otimes X = \sum_{k,\ell=1}^{n} X_{k,\ell}\mathbb{1} \otimes |\mathbf{v}_k\rangle\langle\mathbf{v}_\ell|.$$

Then by (2.6.8), $\mathbb{1} \otimes X \in \mathcal{B}(\mathcal{H} \otimes \mathcal{K})$ has the block matrix representation

$$\begin{bmatrix} X_{1,1}\mathbb{1}_{\mathcal{H}} & \cdots & X_{1,n}\mathbb{1}_{\mathcal{H}} \\ \vdots & \ddots & \vdots \\ X_{n,1}\mathbb{1}_{\mathcal{H}} & \cdots & X_{n,n}\mathbb{1}_{\mathcal{H}} \end{bmatrix}.$$

Conversely, every block matrix of this form is unitarily equivalent, as described in Theorem 2.36 to the operator $\mathbb{1} \otimes X$, where $X = \sum_{i,j=1}^{n} X_{i,j}|\mathbf{v}_i\rangle\langle\mathbf{v}_j|$.

2.7. Tensor powers of operators

Let $\mathcal{H}, \mathcal{K}$ be finite-dimensional Hilbert spaces. For $A \in \mathcal{B}(\mathcal{H})$ and $B \in \mathcal{B}(\mathcal{K})$, $A \otimes B \in \mathcal{B}(\mathcal{H} \otimes \mathcal{K})$ was defined in (2.1.10):

$$(2.7.1) \qquad A \otimes B(\mathbf{u} \otimes \mathbf{v}) = A\mathbf{u} \otimes B\mathbf{v} \qquad \text{for all} \qquad \mathbf{u} \in \mathcal{H}, \mathbf{v} \in \mathcal{K}.$$

If A and B are self-adjoint and $\{\mathbf{u}_1,\ldots,\mathbf{u}_m\}$ and $\{\mathbf{v}_1,\ldots,\mathbf{v}_n\}$ are orthonormal bases of $\mathcal{H}$ and $\mathcal{K}$, respectively, such that $A\mathbf{u}_i = \lambda_i\mathbf{u}_i$ for $1 \leq i \leq m$ and $B\mathbf{v}_j = \mu_j\mathbf{v}_j$ for $1 \leq j \leq n$, then

$$(2.7.2) \quad A \otimes B(\mathbf{u}_i \otimes \mathbf{v}_j) = \lambda_i\mu_j\mathbf{u}_i \otimes \mathbf{v}_j \qquad \text{for all} \qquad 1 \leq i \leq m, \ 1 \leq j \leq n.$$

In particular, if $A, B \geq 0$ then $\{\mathbf{u}_i \otimes \mathbf{v}_j\}_{1\leq i\leq m, \ 1\leq j\leq n}$ is an orthonormal basis of $\mathcal{H} \otimes \mathcal{K}$ consisting of eigenvectors of $A \otimes B$, and all of the eigenvalues are nonnegative. Thus:

Lemma 2.40. *Let $\mathcal{H}, \mathcal{K}$ be finite-dimensional Hilbert spaces. For all $A \in \mathcal{B}^+(\mathcal{H})$ and $B \in \mathcal{B}^+(\mathcal{K})$, $A \otimes B \in \mathcal{B}^+(\mathcal{H} \otimes \mathcal{K})$.*

For general $A \in \mathcal{B}(\mathcal{H})$ and $B \in \mathcal{B}(\mathcal{K})$, consider the SVDs of A and B: Let $\{\mathbf{u}_1,\ldots,\mathbf{u}_m\}$ and $\{\mathbf{v}_1,\ldots,\mathbf{v}_m\}$ be orthonormal bases of $\mathcal{H}$ and let $\{\sigma_1,\ldots,\sigma_m\}$ be a set of m nonnegative numbers arranged in decreasing order, such that $A = \sum_{i=1}^{m} \sigma_i |\mathbf{v}_i\rangle\langle\mathbf{u}_i|$. Let $\{\tilde{\mathbf{u}}_1,\ldots,\tilde{\mathbf{u}}_n\}$ and $\{\tilde{\mathbf{v}}_1,\ldots,\tilde{\mathbf{v}}_n\}$ be orthonormal bases of $\mathcal{K}$ and let $\{\tilde{\sigma}_1,\ldots,\tilde{\sigma}_n\}$ be a set of n nonnegative numbers arranged in decreasing order, such that $B = \sum_{j=1}^{n} \tilde{\sigma}_j |\tilde{\mathbf{v}}_j\rangle\langle\tilde{\mathbf{u}}_j|$. Then it is evident from (2.7.1) that for all $1 \le i \le m$, $1 \le j \le n$, $A \otimes B(\mathbf{u}_i \otimes \tilde{\mathbf{u}}_j) = \sigma_i \tilde{\sigma}_j \mathbf{v}_i \otimes \tilde{\mathbf{v}}_j$.

Thus, an SVD of $A \otimes B$ is given by

$$A \otimes B = \sum_{i,j} \sigma_i \tilde{\sigma}_j |\mathbf{v}_i \otimes \tilde{\mathbf{v}}_j\rangle\langle\mathbf{u}_i \otimes \tilde{\mathbf{u}}_j|.$$

Just as we can form the tensor product of an arbitrary finite number n of Hilbert spaces $\{\mathcal{H}_1,\ldots,\mathcal{H}_n\}$, $\bigotimes_{j=1}^{n} \mathcal{H}_j$, given an arbitrary set $\{A_1,\ldots,A_n\}$ with $A_j \in \mathcal{B}(\mathcal{H}_j)$, $j = 1,\ldots,n$, we can form $\bigotimes_{j=1}^{n} A_j \in \mathcal{B}(\bigotimes_{j=1}^{n} \mathcal{H}_j)$. There is no essential difference from the case $n = 2$ that we have discussed, and hence we freely use the construction going forward. The special case in which all of the Hilbert spaces and operators are the same is of particular interest. The following combinatoric lemma will be useful.

Lemma 2.41. *The number of vectors $(m_1,\ldots,m_d)$ whose entries are nonnegative integers and subject to the constraint that $\sum_{j=1}^{d} m_j = n$ is*

$$\binom{n+d-1}{d-1}.$$

Proof. Choose an arbitrary subset of $\{1,\ldots,n+d-1\}$ consisting of $d-1$ elements. For example, if $n = 5$ and $d = 3$, and the subset of $\{1,2,3,4,5,6,7\}$ chosen is $\{2,3\}$, consider the diagram

$$\bullet \quad \star \quad \star \quad \bullet \quad \bullet \quad \bullet \quad \bullet$$

in which the symbols in places corresponding to the chosen subset are stars and the rest are bullets. The stars divide the bullets into three sets of cardinalities $m_1 = 1$, $m_2 = 0$, and $m_3 = 4$. In general, there are $\binom{n+d-1}{d-1}$ ways to place the $d-1$ dividing stars. $\qquad\square$

Definition 2.42 (Tensor powers). We write $\mathcal{H}^{\otimes n}$ for the n-fold tensor product of $\mathcal{H}$ with itself, and for $A \in \mathcal{B}(\mathcal{H})$ we write $A^{\otimes n}$ to denote the operator

$$(2.7.3) \qquad A^{\otimes n}(\mathbf{v}_1 \otimes \cdots \otimes \mathbf{v}_n) := A\mathbf{v}_1 \otimes \cdots \otimes A\mathbf{v}_n.$$

When $\mathcal{H}$ is finite dimensional, for all $A \in \mathcal{B}(\mathcal{H})$, the spectrum of $A^{\otimes n}$ is necessarily very degenerate.

Lemma 2.43. *Let $\mathcal{H}$ be a d-dimensional Hilbert space with $d \geq 2$, and let $A \in \mathcal{B}(\mathcal{H})$ be self-adjoint. Let $n \geq 2$. Then while $\dim(\mathcal{H}^{\otimes n}) = d^n$, the cardinality of the spectrum of $A^{\otimes n}$, $|\sigma(A^{\otimes n})|$, is only polynomially large in n. Specifically,*

$$|\sigma(A^{\otimes n})| \leq \binom{n+d-1}{d-1} \leq \frac{(n+d-1)^{d-1}}{(d-1)!}.$$

Proof. Since $\dim(\mathcal{H}) = d$, the cardinality of the spectrum of A, $\sigma(A)$, is at most d, which is the case when A has d distinct eigenvalues $\{\lambda_1, \ldots, \lambda_d\}$. Then by the generalization of (2.7.2) to n factors, the spectrum of $\bigotimes^n A$ consists of the numbers of the form

$$\prod_{j=1}^{d} \lambda_j^{m_j} \qquad \text{where} \qquad \sum_{j=1}^{d} m_j = n.$$

Thus, the cardinality of $\sigma(\bigotimes^n A)$ is at most equal to the number of vectors $(m_1, \ldots, m_d)$ whose entries are nonnegative integers subject to the constraint that $\sum_{j=1}^{d} m_j = n$. Now apply Lemma 2.41. $\qquad\square$

2.8. Symmetric and antisymmetric tensor products

Let $\mathcal{H}$ denote a d-dimensional Hilbert space, and let $\bigotimes^n \mathcal{H}$ denote the tensor product of n copies of H. There is a natural unitary action of $\mathcal{S}_n$, the symmetric group on n letters, on $\bigotimes^n \mathcal{H}$. It is convenient to begin by selecting an orthonormal basis $\{\mathbf{u}_1, \ldots, \mathbf{u}_d\}$ of $\mathcal{H}$ so that we may identify $\mathcal{H}$ with complex valued functions on the set $X = \{1, \ldots, d\}$ equipped with counting measure, and then $\bigotimes^n \mathcal{H}$ is identified with the set of complex functions $\psi(x_1, \ldots, x_n)$ on $\{1, \ldots, d\}^n$ equipped with counting measure. For $\pi \in \mathcal{S}_n$ define the unitary operator $V_\pi \in \mathcal{B}(\bigotimes^n \mathcal{H})$ by

$$V_\pi \psi(x_1, \ldots, x_n) = \psi(x_{\pi^{-1}(1)}, \ldots, x_{\pi^{-1}(n)}).$$

Lemma 2.44. *Let $\pi \in \mathcal{S}_n$ and let $\{\psi_1, \ldots, \psi_n\}$ be any set of n vectors in $\mathcal{H}$. Then*

$$V_\pi(\psi_1 \otimes \cdots \otimes \psi_n) = \psi_{\pi(1)} \otimes \cdots \otimes \psi_{\pi(n)}.$$

Proof. In wave function terms, $\psi_1 \otimes \cdots \otimes \psi_n(x_1, \ldots, x_n) = \prod_{j=1}^{n} \psi_j(x_j)$, and hence

$$V_\pi(\psi_1 \otimes \cdots \otimes \psi_n)(x_1, \ldots, x_n) = \prod_{j=1}^{n} \psi_j(x_{\pi^{-1}(j)}) = \prod_{j=1}^{n} \psi_{\pi(j)}(x_j). \qquad\square$$

Going forward, we need a few facts about the symmetric group $\mathcal{S}_n$. Recall that each $\pi \in \mathcal{S}_n$ can be written as the product of either an even or an odd number of pair swaps, but never both. In the first case, π is said to be **even**, and in the second case, π is said to be **odd**. The map $\pi \mapsto \mathrm{sgn}(\pi) \in \{-1, 1\}$ is defined by $\mathrm{sgn}(\pi) = 1$ in case π is even and $\mathrm{sgn}(\pi) = -1$ in case π is odd. It is then easy to see that for all $\pi, \pi' \in \mathcal{S}_n$,

$$\mathrm{sgn}(\pi'\pi) = \mathrm{sgn}(\pi')\,\mathrm{sgn}(\pi),$$

and consequently

$$(2.8.1) \qquad\qquad \mathrm{sgn}(\pi^{-1}) = \mathrm{sgn}(\pi).$$

For all $A \in M_n(\mathbb{C})$, the determinant of A, $\det(A)$ is given by

$$(2.8.2) \qquad\qquad \det(A) = \sum_{\pi \in \mathcal{S}_n} \mathrm{sgn}(\pi) \prod_{j=1}^{n} A_{\pi(j),j}.$$

Proofs of all of these facts can be found in any thorough treatment of the determinant.

We now define an operator P_{FER} on $\bigotimes^n \mathcal{H}$ by

$$P_{\mathrm{FER}} := \frac{1}{n!} \sum_{\pi \in \mathcal{S}_n} \mathrm{sgn}(\pi) V_\pi.$$

Then it is easy to check using (2.8.1) that $P_{\mathrm{FER}}^* = P_{\mathrm{FER}}$ and $P_{\mathrm{FER}}^2 = P_{\mathrm{FER}}$, so that P_{FER} is an orthogonal projection in $\mathcal{B}(\bigotimes^n \mathcal{H})$. Even more simply, we define an operator P_{BOS} on $\bigotimes^n \mathcal{H}$ by

$$(2.8.3) \qquad\qquad P_{\mathrm{BOS}} := \frac{1}{n!} \sum_{\pi \in \mathcal{S}_n} V_\pi.$$

Again, it is easy to check that P_{BOS} is an orthogonal projection in $\mathcal{B}(\bigotimes^n \mathcal{H})$.

Definition 2.45. The subspace of $\bigotimes^n \mathcal{H}$ consisting of $\psi \in \bigotimes^n \mathcal{H}$ such that

$$V_\pi \psi = \mathrm{sgn}(\pi)\psi$$

for all $\pi \in \mathcal{S}_n$ is called the **fermionic subspace** of $\mathcal{H}^{\otimes n}$ and is denoted by $\bigwedge^n \mathcal{H}$.

The subspace of $\bigotimes^n \mathcal{H}$ consisting of $\psi \in \bigotimes^n \mathcal{H}$ such that

$$V_\pi \psi = \psi$$

for all $\pi \in \mathcal{S}_n$ is called the **bosonic subspace** of $\mathcal{H}^{\otimes n}$ and is denoted by $\bigvee^n \mathcal{H}$.

Lemma 2.46. *The range of P_{FER} is $\bigwedge^n \mathcal{H}$ and the range of P_{BOS} is $\bigvee^n \mathcal{H}$.*

Proof. This is left as a simple exercise. $\qquad\qquad\qquad\qquad\qquad\qquad \square$

Remark 2.47 (Fermions and bosons). The subspaces $\bigwedge^n \mathcal{H}$ and $\bigvee^n \mathcal{H}$ of $\bigotimes^n \mathcal{H}$ are important both physically and mathematically. It is an observed fact that physical particles are always of one of two types: **fermions** or **bosons**.

For a system of n identical fermions whose individual state space is $\mathcal{H}$, the n-particle state space is not all of $\bigotimes^n \mathcal{H}$, but only $\bigwedge^n \mathcal{H}$. Likewise, for a system of n identical bosons whose individual state space is $\mathcal{H}$, the n-particle state space is $\bigvee^n \mathcal{H}$.

2.8.1. Properties of $\bigwedge^n \mathcal{H}$ and $\mathcal{B}(\bigwedge^n \mathcal{H})$. For $\mathbf{v}_1, \dots, \mathbf{v}_n \in \mathcal{H}$, define the **wedge product**

$$\mathbf{v}_1 \wedge \cdots \wedge \mathbf{v}_n = \frac{1}{\sqrt{n!}} \sum_{\pi \in \mathcal{S}_n} \mathrm{sgn}(\pi) \mathbf{v}_{\pi(1)} \otimes \cdots \otimes \mathbf{v}_{\pi(n)}.$$

Evidently,

$$\mathbf{v}_1 \wedge \cdots \wedge \mathbf{v}_n = \sqrt{n!} P_{\mathrm{FER}}(\mathbf{v}_1 \otimes \cdots \otimes \mathbf{v}_n),$$

and hence $\mathbf{v}_1 \wedge \cdots \wedge \mathbf{v}_n \in \bigwedge^n \mathcal{H}$. Inner products of wedge products can be expressed in terms of determinants:

Lemma 2.48. *For all* $\{\mathbf{v}_1, \dots, \mathbf{v}_n, \mathbf{w}_1 \dots, \mathbf{w}_n\} \subset \mathcal{H}$,

$$(2.8.4) \qquad \langle \mathbf{v}_1 \wedge \cdots \wedge \mathbf{v}_n, \mathbf{w}_1 \wedge \cdots \wedge \mathbf{w}_n \rangle_{\otimes^n \mathcal{H}} = \det([\langle \mathbf{v}_i, \mathbf{w}_j \rangle]),$$

where $[\langle \mathbf{v}_i, \mathbf{w}_j \rangle]$ *is the* $n \times n$ *matrix whose* (i,j)*-th entry is* $\langle \mathbf{v}_i, \mathbf{w}_j \rangle$. *For any orthonormal basis* $\{\mathbf{u}_1, \dots, \mathbf{u}_d\}$ *of* $\mathcal{H}$, *and any* $(j_1, \dots, j_n) \in \{1, \dots, d\}^n$,

$$\mathbf{u}_{j_1} \wedge \cdots \wedge \mathbf{u}_{j_n} = 0 \quad \text{if} \quad j_\ell = j_m \quad \text{for any} \quad 1 \le \ell < m \le d,$$

and otherwise, $\mathbf{u}_{j_1} \wedge \cdots \wedge \mathbf{u}_{j_n}$ *is a unit vector in* $\bigwedge^n \mathcal{H}$. *Moreover,*

$$\{\mathbf{u}_{j_1} \wedge \cdots \wedge \mathbf{u}_{j_n} \ : \ 1 \le j_1 < \cdots < j_n \le d\}$$

is an orthonormal basis for $\bigwedge^n \mathcal{H}$, *which therefore has dimension* $\binom{d}{n}$ *for all* $n \le d$ *and* 0 *for* $n > d$; *that is, for* $n > d$, $\bigwedge^n \mathcal{H}$ *is the zero subspace of* $\bigotimes^n \mathcal{H}$.

Proof. By the definitions,

$$\langle \mathbf{v}_1 \wedge \cdots \wedge \mathbf{v}_n, \mathbf{w}_1 \wedge \cdots \wedge \mathbf{w}_n \rangle_{\otimes^n \mathcal{H}}$$

$$= \frac{1}{n!} \sum_{\pi, \pi' \in \mathcal{S}_n} \mathrm{sgn}(\pi) \mathrm{sgn}(\pi') \langle \mathbf{v}_{\pi(1)} \otimes \cdots \otimes \mathbf{v}_{\pi(n)}, \mathbf{w}_{\pi'(1)} \otimes \cdots \otimes \mathbf{w}_{\pi'(n)} \rangle.$$

Then since

$$\langle \mathbf{v}_{\pi(1)} \otimes \cdots \otimes \mathbf{v}_{\pi(n)}, \mathbf{w}_{\pi'(1)} \otimes \cdots \otimes \mathbf{w}_{\pi'(n)} \rangle$$

$$= \prod_{j=1}^{n} \langle \mathbf{v}_{\pi(j)}, \mathbf{w}_{\pi'(j)} \rangle = \prod_{j=1}^{n} \langle \mathbf{v}_j, \mathbf{w}_{\pi'(\pi^{-1}j)} \rangle,$$

and since for each $\pi' \in \mathcal{S}_n$, $\pi \mapsto \pi' \circ \pi^{-1}$ is a one-to-one map of $\mathcal{S}_n$ onto itself, and since $\mathrm{sgn}(\pi)\,\mathrm{sgn}(\pi') = \mathrm{sgn}(\pi' \circ \pi^{-1})$,

$$\langle \mathbf{v}_1 \wedge \cdots \wedge \mathbf{v}_n, \mathbf{w}_1 \wedge \cdots \wedge \mathbf{w}_n \rangle_{\otimes^n \mathcal{H}} = \sum_{\pi \in \mathcal{S}_n} \prod_{j=1}^{n} \mathrm{sgn}(\pi) \langle \mathbf{v}_j, \mathbf{w}_{\pi(j)} \rangle.$$

Now (2.8.4) follows from (2.8.2).

Since the determinant of a matrix vanishes if any two rows or columns are identical, the rest follows readily from (2.8.4). $\qquad\square$

Definition 2.49. For $A \in \mathcal{B}(\mathcal{H})$, define $\bigwedge^n A \in \mathcal{B}(\bigwedge^n \mathcal{H})$ by

$$(2.8.5) \qquad \left(\bigwedge^n A\right)(\mathbf{v}_1 \wedge \cdots \wedge \mathbf{v}_n) = A\mathbf{v}_1 \wedge \cdots \wedge A\mathbf{v}_n$$

for all $\mathbf{v}_1, \ldots, \mathbf{v}_n \in \mathcal{H}$.

Lemma 2.50. *Suppose that $A \in \mathcal{B}(\mathcal{H})$ is self-adjoint. Let $\{\mathbf{u}_1, \ldots, \mathbf{u}_d\}$ be an orthonormal basis of $\mathcal{H}$ consisting of eigenvectors of A so that for each $j = 1, \ldots, n$, $A\mathbf{u}_j = \lambda_j \mathbf{u}_j$. Then*

$$\bigwedge^n A = \sum_{1 \le j_1 < \cdots < j_n \le d} \left(\prod_{k=1}^{n} \lambda_{j_k}\right) |\mathbf{u}_{j_1} \wedge \cdots \wedge \mathbf{u}_{j_n}\rangle\langle \mathbf{u}_{j_1} \wedge \cdots \wedge \mathbf{u}_{j_n}|$$

gives the spectral decomposition of $\bigwedge^n A$, and in particular, if $A \ge 0$ and the eigenvalues of A are arranged in decreasing order; i.e., $\lambda_j \le \lambda_i$ for $j > i$, then for $n \le d$, the largest eigenvalue of $\wedge^n A$ is $\prod_{k=1}^{n} \lambda_k$.

Proof. It follows from (2.8.5) that for all $1 \le n \le d$ and all $1 \le j_1 < \cdots < j_n \le d$,

$$\bigwedge^n A(\mathbf{u}_{j_1} \wedge \cdots \wedge \mathbf{u}_{j_n}) = \left(\prod_{k=1}^{n} \lambda_{j_k}\right) \mathbf{u}_{j_1} \wedge \cdots \wedge \mathbf{u}_{j_n}.$$

Then by Lemma 2.48, this gives us an orthonormal basis of eigenvectors of $\bigwedge^n A$, and hence the spectral decomposition of $\bigwedge^n A$. $\qquad\square$

2.8.2. Properties of $\bigvee^n \mathcal{H}$ and $\mathcal{B}(\bigvee^n \mathcal{H})$. Let $\{\mathbf{u}_1, \ldots, \mathbf{u}_d\}$ be an orthonormal basis of $\mathcal{H}$. Then

$$\{\mathbf{u}_{j_1} \otimes \cdots \otimes \mathbf{u}_{j_n} \ : \ (j_1, \ldots, j_n) \in \{1, \ldots, d\}^n\}$$

is an orthonormal basis of $\mathcal{H}^{\otimes n}$. Consequently, applying P_{BOS} to each vector in this orthonormal basis yields a spanning set for $\bigvee^n \mathcal{H}$.

Let $S(n, d)$ denote the set of vectors $(m_1, \ldots, m_d)$ with nonnegative integer entries such that $\sum_{j=1}^{d} m_j = n$, which is a sort of discrete simplex that by Lemma 2.41 consists of $\binom{n+d-1}{d-1}$ points. Define a function $m : \{1, \ldots, d\}^n \to S(n, d)$ by

$$m(j_1, \ldots, j_n) = (m_1(j_1, \ldots, j_n), \ldots, m_d(j_1, \ldots, j_n)),$$

where $m_\ell(j_1, \ldots, j_n)$ is the cardinality of the set $\{k : j_k = \ell\}$.

For $(j_1, \ldots, j_n) \in \{1, \ldots, d\}^n$, write $(m_1, \ldots, m_d) := m(j_1, \ldots, j_n)$ for brevity. Then $\sum_{\pi \in S_n} V_\pi(\mathbf{u}_{j_1} \otimes \cdots \otimes \mathbf{u}_{j_n})$ is the sum of $\left(\frac{n!}{m_1! m_2! \cdots m_d!}\right)$ orthogonal vectors, each with length $m_1! \, m_2! \cdots m_d!$, coming from the repeated terms in the sum. Therefore,

$$\left\| \sum_{\pi \in S_n} V_\pi(\mathbf{u}_{j_1} \otimes \cdots \otimes \mathbf{u}_{j_n}) \right\| = m_1! \, m_2! \cdots m_d! \left(\frac{n!}{m_1! \, m_2! \cdots m_d!}\right)^{1/2}.$$

In particular,

$$\left\| P_{\mathrm{BOS}}(\mathbf{u}_{j_1} \otimes \cdots \otimes \mathbf{u}_{j_n}) \right\| = \left(\frac{m_1! \, m_2! \cdots m_d!}{n!}\right)^{1/2}.$$

Finally, $P_{\mathrm{BOS}}(\mathbf{u}_{j_1} \otimes \cdots \otimes \mathbf{u}_{j_n}) = P_{\mathrm{BOS}}(\mathbf{u}_{\tilde{j}_1} \otimes \cdots \otimes \mathbf{u}_{\tilde{j}_n})$ if and only if $m(j_1, \ldots, j_n) = m(\tilde{j}_1, \ldots, \tilde{j}_n)$, and that otherwise, these vectors are orthogonal.

Therefore, for each $(m_1, \ldots, m_d) \in S(n, d)$, define

$$\mathbf{u}_{(m_1, \ldots, m_d)} := \left(\frac{m_1! \, m_2! \cdots m_d!}{n!}\right)^{1/2} P_{\mathrm{BOS}}(\mathbf{u}_{j_1} \otimes \cdots \otimes \mathbf{u}_{j_n}),$$

where for $k = 1, \ldots, n$, $j_k = \ell$ for $\sum_{i=0}^{\ell-1} m_i < k \leq \sum_{i=0}^{\ell} m_i$.

Then $\{\mathbf{u}_{(m_1, \ldots, m_d)}, (m_1, \ldots, m_d) \in S(n, d)\}$ is an orthonormal basis for $\bigvee^n \mathcal{H}$, and by Lemma 2.41,

$$\dim\left(\bigvee^n \mathcal{H}\right) = \binom{n+d-1}{d-1}.$$

Let $\mathcal{U}(\mathcal{H})$ denote the group of unitaries on $\mathcal{H}$. Given $U \in \mathcal{U}(\mathcal{H})$, define $U^{\otimes n} \in \mathcal{B}(\mathcal{H}^{\otimes n})$ as in (2.7.3). The map $U \mapsto U^{\otimes n}$ is a representation of $\mathcal{U}$ on $\mathcal{H}^{\otimes n}$. Since $U^{\otimes n}$ commutes with each V_π, $\pi \in S_n$, $\bigvee^n \mathcal{H}$ is invariant under this representation. In fact, as we now show, $U \mapsto U^{\otimes n}$ is an irreducible representation of $\mathcal{U}$ on $\bigvee^n \mathcal{H}$, meaning that no nontrivial subspaces of $\bigvee^n \mathcal{H}$ are invariant under the action of $\mathcal{U}(\mathcal{H})$.

Lemma 2.51. *For all nonzero $\mathbf{v} \in \mathcal{H}$,*

$$(2.8.6) \qquad \mathrm{span}(\{U^{\otimes n}\mathbf{v}^{\otimes n} : U \in \mathcal{U}(\mathcal{H})\}) = \bigvee^n \mathcal{H}.$$

Proof. Define the polynomial $p(t_1, \ldots, t_d)$ on $\mathbb{R}^d$ with values in $\bigvee^n \mathcal{H}$ by

$$p(t_1, \ldots, t_d) := \left(\sum_{j=1}^{d} t_j \mathbf{u}_j \right)^{\otimes n} \in \bigvee^n \mathcal{H}.$$

Evidently, for all $(m_1, \ldots, m_d) \in S(n, d)$, the coefficient of $\prod_{j=1}^{d} t^{m_j}$ in $p(t_1, \ldots, t_d)$ is a nonzero multiple of $\mathbf{u}_{(m_1, \ldots, m_d)}$. Therefore, $\mathbf{u}_{(m_1, \ldots, m_d)}$ is a certain partial derivative of $p(t_1, \ldots, t_d)$, and all partial derivatives of $p(t_1, \ldots, t_d)$ belong to $\mathrm{span}(\{p(t_1, \ldots, t_d) \ : \ (t_1, \ldots, t_d) \in \mathbb{R}^d\})$, and hence this subspace of $\bigvee^n \mathcal{H}$ contains an orthonormal basis of $\bigvee^n \mathcal{H}$. Therefore

$$(2.8.7) \qquad \bigvee^n \mathcal{H} = \mathrm{span}(\{p(t_1, \ldots, t_d) \ : \ (t_1, \ldots, t_d) \in \mathbb{R}^d\}).$$

Given a unit vector $\mathbf{v} \in \mathcal{H}$, let $U \in \mathcal{U}$ be such that

$$\left(\sum_{j=1}^{d} t_j^2 \right)^{1/2} U \mathbf{v} = \left(\sum_{j=1}^{d} t_j \mathbf{u}_j \right).$$

Then,

$$p(t_1, \ldots, t_d) = \left(\sum_{j=1}^{d} t_j^2 \right)^{n/2} U^{\otimes n} \mathbf{v}^{\otimes n},$$

and hence

$$(2.8.8) \quad \mathrm{span}(\{p(t_1, \ldots, t_d) \ : \ (t_1, \ldots, t_d) \in \mathbb{R}^d\})$$

$$\subseteq \mathrm{span}(\{U^{\otimes n} \mathbf{v}^{\otimes n} \ : \ U \in \mathcal{U}(\mathcal{H})\}) \subseteq \bigvee^n \mathcal{H}.$$

Together, (2.8.7) and (2.8.8) prove (2.8.6). $\qquad\qquad\square$

Lemma 2.52. *Let $\{\mathbf{u}_1, \ldots, \mathbf{u}_d\}$ be any orthonormal basis of $\mathcal{H}$. Let ψ be any nonzero vector in $\bigvee^n \mathcal{H}$. Then*

$$(2.8.9) \qquad\qquad \mathbf{u}_1^{\otimes n} \in \mathrm{span}(\{U^{\otimes n} \psi \ : \ U \in \mathcal{U}(\mathcal{H})\}).$$

Proof. Let $\mathcal{V}$ denote the subspace on the right in (2.8.9). Since $\mathcal{V}$ is unchanged if we replace ψ by $V^{\otimes n} \psi$ for any $V \in \mathcal{U}(\mathcal{H})$, because of Lemma 2.51, we may assume that

$$\alpha_{(1, \ldots, 1)} := \langle \mathbf{u}_1^{\otimes n}, \psi \rangle \neq 0.$$

For $j = 2, \ldots, d$, define unitaries $U_j(\theta) \in \mathcal{U}(\mathcal{H})$ by

$$U_j(\theta) \mathbf{u}_k = \begin{cases} e^{i\theta} \mathbf{u}_j & k = j \\ \mathbf{u}_k & k \neq j, \end{cases}$$

and hence for all $(m_1, \ldots, m_d) \in S(n, d)$,

$$\frac{1}{2\pi} \int_0^{2\pi} U_j(\theta) \mathbf{u}_{(m_1, \ldots, m_d)} = \begin{cases} \mathbf{u}_{(m_1, \ldots, m_d)} & m_j = 0 \\ 0 & m_j \neq 0. \end{cases}$$

Then $\alpha_{(1, \ldots, 1)} \mathbf{u}_1^{\otimes n} = (2\pi)^{1-d} \int_{[0, 2\pi]^{d-1}} U_2(\theta_2) \cdots U_d(\theta_d) \psi \, d\theta_2 \cdots d\theta_d.$ $\square$

Theorem 2.53. *For all unit vectors $\psi \in \bigvee^n \mathcal{H}$,*

$$(2.8.10) \qquad \operatorname{span}\left(\{U^{\otimes n}\psi \ : \ U \in \mathcal{U}(\mathcal{H})\}\right) = \bigvee^n \mathcal{H}.$$

and with dU denoting normalized Haar measure on $\mathcal{U}(\mathcal{H})$,

$$(2.8.11) \qquad \int_{\mathcal{U}(\mathcal{H})} |U^{\otimes n}\psi\rangle\langle U^{\otimes n}\psi| \, dU = \binom{n+d-1}{d-1}^{-1} P_{\mathrm{BOS}}.$$

Proof. Let A denote the operator on the left side of (2.8.11). Then $A \geq 0$ and for all $V \in \mathcal{U}$, A commutes with $V^{\otimes n}$. Let ϕ be any eigenvector of A, so that $A\phi = \lambda\phi$. Then

$$A\left(V^{\otimes n}\phi\right) = V^{\otimes n}\left(A\phi\right) = \lambda V^{\otimes n}\phi.$$

Therefore, the eigenspace of A with eigenvalue λ is invariant under the action of $\mathcal{U}(\mathcal{H})$ on $\bigvee^n \mathcal{H}$. By Lemma 2.52, $\mathbf{u}_1^{\otimes n}$ belongs to this eigenspace, and since λ is arbitrary, $\mathbf{u}_1^{\otimes n}$ belongs to every eigenspace. Since eigenspaces of distinct eigenvalues are orthogonal, A can have only one eigenspace, and hence A is a multiple of the identity on $\bigvee^n \mathcal{H}$. Since $\mathrm{Tr}[A] = 1$, the multiple is the inverse of the dimension of $\bigvee^n \mathcal{H}$, which is $\binom{n+d-1}{d-1}^{-1}$.

Finally, regarding A as an operator on $\mathcal{H}^{\otimes n}$, it is evidently zero on the orthogonal complement of $\bigvee^n \mathcal{H}$, and this proves (2.8.11), and then (2.8.10) follows directly. $\square$

Exercises

(1) Let $X, Y \in \mathcal{B}(\mathcal{H} \otimes \mathcal{K})$ where $\mathcal{H}$ and $\mathcal{K}$ are finite-dimensional Hilbert spaces. Show that it need not be the case that $\mathrm{Tr}_{\mathcal{K}}[XY] = \mathrm{Tr}_{\mathcal{K}}[YX]$, but that for all $Z \in \mathcal{B}(\mathcal{H})$, $\mathrm{Tr}_{\mathcal{K}}[X(Z \otimes \mathbb{1})] = \mathrm{Tr}_{\mathcal{K}}[(Z \otimes \mathbb{1})X]$. (This is the partial cyclicity property of the partial trace).

(2) Prove formula (2.3.3).

(3) Let $\{\mathbf{u}_1,\ldots,\mathbf{u}_m\}$ and $\{\mathbf{v}_1,\ldots,\mathbf{v}_n\}$ be orthonormal bases for Hilbert spaces $\mathcal{H}$ and $\mathcal{K}$, respectively. Then $\{\mathbf{u}_i \otimes \mathbf{v}_j\}_{1\leq i\leq m,\, 1\leq j\leq n}$ is an orthonormal basis of $\mathcal{H} \otimes \mathcal{K}$ and $\{|\mathbf{u}_i\rangle\langle\mathbf{v}_j|\}_{1\leq i\leq m,\, 1\leq j\leq n}$ is an orthonormal basis for $\mathcal{B}(\mathcal{K},\mathcal{H})$. Define a unitary transformation $\mathcal{U} : \mathcal{H} \otimes \mathcal{K} \to \mathcal{B}(\mathcal{K},\mathcal{H})$ by $\mathcal{U}\mathbf{u}_i \otimes \mathbf{v}_j = |\mathbf{u}_i\rangle\langle\mathbf{v}_j|$ for $1 \leq i \leq m$ and $1 \leq j \leq n$. Then for $\psi \in \mathcal{H} \otimes \mathcal{K}$, define $K_\psi \in \mathcal{B}(\mathcal{K},\mathcal{H})$ by

$$K_\psi := \mathcal{U}\psi = \sum_{i=1}^{m}\sum_{j=1}^{n}\langle \mathbf{u}_i \otimes \mathbf{v}_j, \psi\rangle\, |\mathbf{u}_i\rangle\langle\mathbf{v}_j|.$$

Show that $\mathrm{Tr}_{\mathcal{K}}[|\psi\rangle\langle\phi|] = K_\psi K_\phi^*$ and that the matrix representative of $\mathrm{Tr}_{\mathcal{H}}[|\psi\rangle\langle\phi|]$ with respect to $\{\mathbf{v}_1,\ldots,\mathbf{v}_n\}$ is the transpose of the matrix representative of $K_\phi^* K_\psi$ with respect to this basis.

(4) Let $\mathcal{H}$ be a finite-dimensional Hilbert space, and let $A \in \mathcal{B}(\mathcal{H})$. Show that
$$\|A\| \leq 1 \text{ if and only if } \begin{bmatrix} \mathbb{1} & A \\ A^* & \mathbb{1} \end{bmatrix} \in \mathcal{B}^+(\mathcal{H} \oplus \mathcal{H}).$$

(5) Let $\mathcal{H}$ be a finite-dimensional Hilbert space and let $\{\mathbf{u}_1,\ldots,\mathbf{u}_n\}$ be an orthonormal basis for it. For $X \in \mathcal{B}(\mathcal{H})$, define the transpose of X, X^T with respect to $\{\mathbf{u}_1,\ldots,\mathbf{u}_n\}$ by $X^T := \sum_{i,j=1}^{n}\langle\mathbf{u}_j, X\mathbf{u}_i\rangle|\mathbf{u}_i\rangle\langle\mathbf{u}_j|$.

  (a) Show that if $X \in \mathcal{B}^{\mathrm{s.a.}}(\mathcal{H})$, then $X^T \in \mathcal{B}^{\mathrm{s.a.}}(\mathcal{H})$, and that for $X, Y \in \mathcal{B}(\mathcal{H})$, $(XY)^T = Y^T X^T$.

  (b) Show that if $P \in \mathcal{B}(\mathcal{H})$ is an orthogonal projection, then so is P^T. Also, show that if $U \in \mathcal{B}(\mathcal{H})$ is a partial isometry, then so is U^T

  (c) Let $\mathcal{H}$ be a finite-dimensional Hilbert space and let $\{\mathbf{u}_1,\ldots,\mathbf{u}_n\}$ be an orthonormal basis for it. Define $\Psi \in \mathcal{H} \otimes \mathcal{H}$ as in (2.4.1).

  (i) Show that for all $X \in \mathcal{B}(\mathcal{H})$, with X^T defined as in the previous exercise,
$$(X \otimes \mathbb{1})\Psi = (\mathbb{1} \otimes X^T)\Psi.$$

  (ii) Let $\mathcal{U} : \mathcal{B}(\mathcal{H}) \to \mathcal{H} \otimes \mathcal{H}$ be defined in (2.4.1) and (2.4.2) so that $\mathcal{U}(X) = (X \otimes \mathbb{1})\Psi$. Let $K \in \mathcal{B}(\mathcal{H})$ and $A, B \in \mathcal{B}^+(\mathcal{H})$. Show that
$$\langle\mathcal{U}(K), A \otimes B\mathcal{U}(K)\rangle = \mathrm{Tr}[K^* AKB^T].$$

(6) Let $\mathcal{H}$ be a finite-dimensional Hilbert space, and let $U \in \mathcal{B}(\mathcal{H})$ be a partial isometry. Show that there exists a unitary $V \in \mathcal{B}(\mathcal{H})$ such that $UU^*V = U$.

(7) Let $\mathcal{H}$ be a finite-dimensional Hilbert space. Let $U \in \mathcal{B}(\mathcal{H})$ be a partial isometry. Show that there is another partial isometry $\hat{U} \in \mathcal{B}(\mathcal{H})$ such that $U + \hat{U}$ is unitary.

(8) Let $\mathcal{H}$ be a finite-dimensional Hilbert space and let $K \in \mathcal{B}(\mathcal{H})$ be a contraction, so that $K^*K \leq \mathbb{1}$. Let $K = V|K|$ be the polar decomposition of A. Define $U \in \mathcal{B}(\mathcal{H} \oplus \mathcal{H})$ by

$$U := \begin{bmatrix} K & V(\mathbb{1} - |K|^2)^{1/2} \\ -V(\mathbb{1} - |K|^2)^{1/2} & K \end{bmatrix}.$$

Define $W : \mathcal{H} \oplus \mathcal{H} \to \mathcal{H}$ by $W(\mathbf{x},\mathbf{y}) = \mathbf{x}$. Show that U is unitary, W is a partial isometry, and that $V = WUW^*$.

(9) Prove Lemma 2.46.

Chapter 3

Monotonicity and convexity for operators

3.1. Monotonicity and convexity on the real line

Many inequalities to be discussed here are proved using arguments involving monotonicity or convexity—or both. We first recall the basic definitions for real valued functions f defined on a interval $I \subseteq \mathbb{R}$. The function f is **monotone increasing** in case for all $t > s$ in I, $f(t) \geq f(s)$. The function f is **strictly monotone increasing** in case for $t > s$ in I, $f(t) > f(s)$. The function f is (strictly) monotone decreasing in case $-f$ is (strictly) monotone increasing.

A real valued function f defined on a connected set I of $\mathbb{R}$ is **convex** in case for all $s, t \in I$ and all $0 \leq \lambda \leq 1$,

$$(3.1.1) \qquad f((1 - \lambda)s + \lambda t) \leq (1 - \lambda)f(s) + \lambda f(t),$$

and is **strictly convex** in case strict inequality holds in (3.1.1) whenever $s \neq t$ and $0 < \lambda < 1$. The function f is (strictly) concave in case $-f$ is (strictly) convex.

If f is a real valued convex function defined on a proper subset I of $\mathbb{R}$, we may extend the domain of definition of f to all of $\mathbb{R}$ by setting $f(t) = \infty$ for all $t \notin I$. Then the inequality (3.1.1) is still satisfied for all $s, t \in \mathbb{R}$. Therefore, without loss of generality, we may consider all convex functions f of one real variable to be defined on $\mathbb{R}$ with values in the partially extended real line $(-\infty, \infty]$. Such functions are strictly convex in case there is strict inequality in (3.1.1) for all distinct $s, t \in \mathbb{R}$ for which $f(s), f(t) < \infty$.

The definition (3.1.1) has a consequence known as **Jensen's inequality**.

Lemma 3.1. *Let f be convex on $I \subseteq \mathbb{R}$. Let $\{x_1,\ldots,x_k\} \subset I$. Let $\{\lambda_1,\ldots,\lambda_k\} \subset (0,1)$ with $\sum_{j=1}^{k} \lambda_j = 1$. Then*

$$(3.1.2) \qquad f\left(\sum_{j=1}^{k} \lambda_j x_j\right) \leq \sum_{j=1}^{k} \lambda_j f(x_j).$$

Moreover, if f is strictly convex on I, then there is equality in (3.1.2) if and only if $x_i = x_j$ for all $i, j = 1,\ldots,k$.

Proof. Define $t := x_k$, $\lambda := \lambda_k$, $s := (1-\lambda)^{-1} \sum_{j=1}^{k-1} \lambda_j x_j$. Then

$$f\left(\sum_{j=1}^{k} \lambda_j x_j\right) = f((1-\lambda)s + \lambda t).$$

By (3.1.1),

$$(3.1.3) \qquad f\left(\sum_{j=1}^{k} \lambda_j x_j\right) \leq \lambda_k f(x_k) + (1-\lambda_k)f\left(\sum_{j=1}^{k-1} \frac{\lambda_j}{1-\lambda_k} x_j\right).$$

For $1 \leq j \leq k-1$, define $\mu_j := \frac{\lambda_j}{1-\lambda_k} \in (0,1)$. Then $\sum_{j=1}^{k-1} \mu_j = 1$ and making the inductive hypothesis that (3.1.2) is valid for $k - 1$,

$$f\left(\sum_{j=1}^{k-1} \frac{\lambda_j}{1-\lambda_k} x_j\right) \leq \sum_{j=1}^{k-1} \frac{\lambda_j}{1-\lambda_k} f(x_j).$$

Combining this with (3.1.3) yields (3.1.2). The case $k = 2$ is definition (3.1.1) itself.

If there is equality in (3.1.2), there must be equality in (3.1.3), and then if f is strictly convex on I, $x_k = \sum_{j=1}^{k-1} \frac{\lambda_j}{1-\lambda_k} x_j$. Thus, x_k is an average of $\{x_1,\ldots,x_{k-1}\}$, and now a simple induction argument shows that if there is equality in (3.1.2), then the x_j's are all the same. $\qquad\qquad\square$

The following elementary theorems are basic to the subject.

Lemma 3.2. *Let X be self-adjoint in $\mathcal{B}(\mathcal{H})$, $\mathcal{H}$ a finite-dimensional Hilbert space. Let $f : \mathbb{R} \to (-\infty, \infty]$ be a convex function that is finite on $\sigma(X)$. Then for any unit vector $\mathbf{v} \in \mathcal{H}$.*

$$f(\langle \mathbf{v}, X\mathbf{v}\rangle) \leq \langle \mathbf{v}, f(X)\mathbf{v}\rangle.$$

There is equality when $\mathbf{v}$ is an eigenvector of X, and only in this case when f is strictly convex on an interval containing $\sigma(X)$.

Proof. Let $X = \sum_{\lambda \in \sigma(X)} \lambda P_\lambda$ be a spectral decomposition of X. Then $f(X) = \sum_{\lambda \in \sigma(X)} f(\lambda)P_\lambda$. Since $\sum_{\lambda \in \sigma(X)}\langle \mathbf{v}, P_\lambda \mathbf{v}\rangle = 1$ and f is convex, application of Jensen's inequality yields

$$\langle \mathbf{v}, f(X)\mathbf{v}\rangle = \sum_{\lambda \in \sigma(X)} \langle \mathbf{v}, P_\lambda \mathbf{v}\rangle f(\lambda)$$

$$\geq f\left(\sum_{\lambda \in \sigma(X)} \langle \mathbf{v}, P_\lambda \mathbf{v}\rangle \lambda\right) = f(\langle \mathbf{v}, X\mathbf{v}\rangle).$$

If f is strictly convex, the inequality is strict unless $\langle \mathbf{v}, P_\lambda \mathbf{v}\rangle$ is nonzero for only one $\lambda \in \sigma(X)$. In this case, $P_\lambda \mathbf{v} = \mathbf{v}$, and $\mathbf{v}$ is an eigenvector of X with eigenvalue λ. $\qquad\square$

We shall encounter convex functions $f : \mathbb{R} \to (-\infty, \infty]$ and operators $X \in \mathcal{B}(\mathcal{H})$ for which $f(X)$ is not defined as an element of $\mathcal{B}(\mathcal{H})$ because $f(s) = \infty$ for some $s \in \sigma(X)$. However, taking the trace of $f(X) = \sum_{\lambda \in \sigma(X)} f(\lambda)P_\lambda$ yields

$$\mathrm{Tr}[f(X)] = \sum_{\lambda \in \sigma(X)} f(\lambda)\,\mathrm{Tr}[P_\lambda],$$

which does make perfect sense: $\mathrm{Tr}[f(X)] = \infty$ if $f(\lambda) = \infty$ for some $\lambda \in \sigma(X)$. We therefore define $\mathrm{Tr}[f(X)] := \infty$ in case f is not finite on $\sigma(X)$.

Lemma 3.2 leads to a basic convexity theorem:

Theorem 3.3. *Let $\mathcal{B}^{\mathrm{s.a.}}(\mathcal{H})$ denote the self-adjoint elements of $\mathcal{B}(\mathcal{H})$, $\mathcal{H}$ a finite-dimensional Hilbert space. For every convex function $f : \mathbb{R} \to (-\infty, \infty]$, the function $X \mapsto \mathrm{Tr}[f(X)]$ from $\mathcal{B}^{\mathrm{s.a.}}(\mathcal{H})$ to $(-\infty, \infty]$ is convex. That is, for all $X, Y \in \mathcal{B}^{\mathrm{s.a.}}(\mathcal{H})$ and all convex f,*

$$\mathrm{Tr}[f(aX + (1-a)Y)] \leq a\,\mathrm{Tr}[f(X)] + (1-a)\,\mathrm{Tr}[f(Y)]$$

for all $a \in (0, 1)$. If f is strictly convex, then so is $X \mapsto \mathrm{Tr}[f(X)]$.

Proof. Let $\{\mathbf{u}_1, \ldots, \mathbf{u}_n\}$ be an orthonormal basis of $\mathcal{H}$ consisting of eigenvectors of $aX + (1-a)Y$. Then for each j,

$$(3.1.4) \qquad \langle \mathbf{u}_j, f(aX + (1-a)Y)\mathbf{u}_j\rangle = f\left(\langle \mathbf{u}_j, (aX + (1-a)Y)\mathbf{u}_j\rangle\right)$$

so that

$$\mathrm{Tr}[f(aX + (1-a)Y)] \leq \sum_{j=1}^{n}\left(af(\langle \mathbf{u}_j, X\mathbf{u}_j\rangle) + (1-a)f(\langle \mathbf{u}_j, Y\mathbf{u}_j\rangle)\right)$$

$$\leq \sum_{j=1}^{n}\left(a\langle \mathbf{u}_j, f(X)\mathbf{u}_j\rangle + (1-a)\langle \mathbf{u}_j, f(Y)\mathbf{u}_j\rangle\right)$$

$$(3.1.5) \qquad = a\,\mathrm{Tr}[f(X)] + (1-a)\,\mathrm{Tr}[f(Y)],$$

where the first inequality is due to the convexity of f, and the second is due to Lemma 3.2.

Now suppose that f is strictly convex, and $\mathrm{Tr}[f(X)], \mathrm{Tr}[f(Y)] < \infty$. Then for $0 < a < 1$, the first inequality in (3.1.5), due to the convexity of f, is strict unless for each $1 \leq j \leq n$,

$$(3.1.6) \qquad \langle \mathbf{u}_j, X\mathbf{u}_j\rangle = \langle \mathbf{u}_j, Y\mathbf{u}_j\rangle.$$

By the conditions for equality in Lemma 3.2, equality in the second inequality in (3.1.5) further requires that each $\mathbf{u}_j$ is an eigenvector of both X and Y. This, together with (3.1.6) implies that $X = Y$. Thus there is equality in (3.1.4) if and only if $X = Y$, and hence $X \mapsto \mathrm{Tr}[f(X)]$ is strictly convex. $\square$

Theorem 3.4. *Let* $Y \geq X \in \mathcal{B}^{\mathrm{s.a.}}(\mathcal{H})$, $\mathcal{H}$ *a finite-dimensional Hilbert space. Let* f *be a real valued function that is monotone increasing on* $\sigma(X) \cup \sigma(Y)$. *Then*

$$(3.1.7) \qquad \mathrm{Tr}[f(Y)] \geq \mathrm{Tr}[f(X)],$$

and if f *is strictly monotone increasing on* $\sigma(X) \cup \sigma(Y)$, *then there is equality in* (3.1.7) *if and only if* $Y = X$.

Proof. Since $\sigma(X) \cup \sigma(Y)$ is a finite set, f is the restriction to $\sigma(X) \cup \sigma(Y)$ of a continuously differentiable monotone increasing function defined on some open bounded interval I. We choose such an extension of f, which we shall also call f. If the original f is strictly monotone increasing, we may assume the extension has a strictly positive derivative f'.

Let $p(x)$ be a polynomial in $x \in \mathbb{R}$. Let $A \geq 0$, and let B be self-adjoint. Then $p(B + sA)$ is a continuously differentiable function of the real variable s. To see this, consider the case in which $p(x) = x^n$. Then by a telescoping sum calculation $(B + sA)^n - B^n = s\sum_{j=0}^{n-1}(B + sA)^{n-1-j}AB^j$, one finds

$$\frac{\mathrm{d}}{\mathrm{d}s}(B + sA)^n\bigg|_{s=0} = \sum_{j=0}^{n-1} B^{n-1-j}AB^j.$$

Taking the trace, the effects of noncommutativity vanish due to cyclicity of the trace, and we can can express the result in terms of p':

$$(3.1.8) \qquad \frac{\mathrm{d}}{\mathrm{d}s}\, \mathrm{Tr}[p(B+sA)]\Big|_{s=0} = \mathrm{Tr}[p'(B)A].$$

Now take $t \in (0,1)$, $A = Y - X \geq 0$, and $B = X + tA$. Then (3.1.8) becomes

$$\frac{\mathrm{d}}{\mathrm{d}s}\, \mathrm{Tr}[p(X+(t+s)A)]\Big|_{s=0} = \mathrm{Tr}[p'(X+tA)A].$$

Since $\mathrm{Tr}[p(X+tA)]$ and $\mathrm{Tr}[p'(X+tA)A]$ are continuous functions of t,

$$\mathrm{Tr}[p(Y)] - \mathrm{Tr}[p(X)] = \int_0^1 \mathrm{Tr}[p'(X+tA)A]\mathrm{d}t.$$

Approximating f by polynomials whose derivatives converge to f' uniformly on I, we deduce

$$\mathrm{Tr}[f(Y)] - \mathrm{Tr}[f(X)] = \int_0^1 \mathrm{Tr}[f'(X+tA)A]\mathrm{d}t.$$

Since f is increasing, $f'(x) \geq 0$ for all $x \in I$, and then by the spectral theorem, $f'(X+tA) \geq 0$. Likewise, if $f' > 0$ on I, then $f'(X+tA) > 0$. Therefore,

$$\mathrm{Tr}[f'(X+tA)A] = \mathrm{Tr}[A^{1/2}f'(X+tA)A^{1/2}] \geq 0,$$

and it is strictly positive unless $A = 0$ whenever $f'(X+tA) > 0$. $\qquad\square$

3.2. Operator monotonicity

Definition 3.5. Let I be an interval in $\mathbb{R}$. A function $f : I \to \mathbb{R}$ is **operator monotone increasing** in case for all $n \in \mathbb{N}$ and all self-adjoint $A, B \in M_n(\mathbb{C})$ with $\sigma(A), \sigma(B) \subset I$, $A \geq B$ implies $f(A) \geq f(B)$. The function f is **operator monotone decreasing** if $-f$ is operator monotone increasing.

Considering multiples of the identity, one sees that if f is monotone in the operator sense, then it must be monotone in the usual sense as a function from $\mathbb{R}$ to $\mathbb{R}$. The converse is not true.

Example 3.6 (Nonmonotonicity of the square). The function $f(x) - x^2$ is monotone on $[0, \infty)$ in the usual sense, but for self-adjoint $A, B \in M_n^+(\mathbb{C})$ and $t > 0$,

$$(A + tB)^2 = A^2 + t(AB + BA) + t^2 B^2.$$

For any choice of A and B such that $AB + BA$ has even one strictly negative eigenvalue, $(A + tB)^2 \geq A^2$ will fail for all sufficiently small t. It is easy to find such A and B in $M_n^+(\mathbb{C})$. For any nonzero self-adjoint A, let $\mathbf{u}$ be a unit vector that is not an eigenvector of A. Let $B := |\mathbf{u}\rangle\langle\mathbf{u}|$, and define $\mathbf{v} = A\mathbf{u}$. Then

$$AB + BA = |\mathbf{v}\rangle\langle\mathbf{u}| + |\mathbf{u}\rangle\langle\mathbf{v}|.$$

Since $\mathbf{v}$ is not a multiple of $\mathbf{u}$, this operator has one strictly negative eigenvalue. Thus, the square function is not operator monotone on $[0, \infty)$. $\square$

The inverse function $f(x) = x^{-1}$ on $(0, \infty)$ is a fundamental example of operator monotonicity.

Theorem 3.7. *Let $f(x) = x^{-1}$ on $(0, \infty)$. Then f is operator monotone decreasing on $\mathcal{B}^{++}(\mathcal{H})$.*

First proof of Theorem 3.7. For $A, B > 0$, define

$$X := (A + B)^{-1/2}A(A + B)^{-1/2},$$

$$(3.2.1) \qquad Y := (A + B)^{-1/2}B(A + B)^{-1/2}.$$

Then $X + Y = \mathbb{1}$. Therefore, if $\{\mathbf{u}_1, \ldots, \mathbf{u}_n\}$ is an orthonormal basis of $\mathcal{H}$ consisting of eigenvectors X, so that $X\mathbf{u}_j = \lambda_j\mathbf{u}_j$ for each j,

$$X = \sum_{j=1}^{n} \lambda_j |\mathbf{u}_j\rangle\langle\mathbf{u}_j| \quad \text{and} \quad Y = \sum_{j=1}^{n}(1 - \lambda_j)|\mathbf{u}_j\rangle\langle\mathbf{u}_j|$$

and

$$X^{-1} = \sum_{j=1}^{n} \lambda_j^{-1}|\mathbf{u}_j\rangle\langle\mathbf{u}_j| \quad \text{and} \quad Y^{-1} = \sum_{j=1}^{n}(1 - \lambda_j)^{-1}|\mathbf{u}_j\rangle\langle\mathbf{u}_j|.$$

Therefore, $X \geq Y$ if and only if $X^{-1} \leq Y^{-1}$. Then since

$$A - B = (A + B)^{1/2}(X - Y)(A + B)^{1/2}$$

and

$$A^{-1} - B^{-1} = (A + B)^{-1/2}(X^{-1} - Y^{-1})(A + B)^{-1/2},$$

$A \geq B$ if and only if $A^{-1} \leq B^{-1}$. $\square$

A second proof of Theorem 3.7 uses multivariable calculus.

Definition 3.8. A function $f : M_n(\mathbb{C}) \to M_n(\mathbb{C})$ is **differentiable at $X \in M_n(\mathbb{C})$** in case there is a linear transformation $\Lambda : M_n(\mathbb{C}) \to M_n(\mathbb{C})$ such that for all $Y \in M_n(\mathbb{C})$ and all $t \in \mathbb{R}$,

$$(3.2.2) \qquad \lim_{t \to 0} \frac{1}{t}\|f(X + tY) - f(X) - t\Lambda(Y)\| = 0.$$

For any f there is at most one $\Lambda : M_n(\mathbb{C}) \to M_n(\mathbb{C})$ such that Definition 3.8 is satisfied for all $Y \in M_n(\mathbb{C})$. When the limit in (3.2.2) exists, the **derivative of f at X** is the linear transformation $D_{f,X} : M_n(\mathbb{C}) \to M_n(\mathbb{C})$ defined by:

Definition 3.9.

$$D_{f,X}(Y) = \lim_{t \to 0} \frac{1}{t}\big(f(X + tY) - f(X)\big).$$

Second proof of Theorem 3.7. Simple algebra proves the following identity for $A > 0$ and H self-adjoint and such that $A + H > 0$:

$$(3.2.3) \qquad \frac{1}{A + H} - \frac{1}{A} = -\frac{1}{A + H}H\frac{1}{A} = -\frac{1}{A}H\frac{1}{A + H},$$

known as the **resolvent identity**.

For $A > 0$ and any $H \in M_n^{\mathrm{s.a.}}(\mathbb{C})$, for all t sufficiently small, $A + tH > 0$. Therefore, we may replace H by tH in (3.2.3) to obtain

$$\frac{1}{A + tH} - \frac{1}{A} = -t\frac{1}{A + tH}H\frac{1}{A} = -t\frac{1}{A}H\frac{1}{A + tH}.$$

Therefore, with $f(x) = x^{-1}$ on $(0, \infty)$,

$$(3.2.4) \qquad D_{f,A}(H) = \frac{\mathrm{d}}{\mathrm{d}t}\frac{1}{A + tH}\bigg|_{t=0} = -\frac{1}{A}H\frac{1}{A}.$$

It follows that if $A, H > 0$, $\frac{\mathrm{d}}{\mathrm{d}t}\frac{1}{A+tH} \le 0$. Now use the fundamental theorem of calculus. $\qquad\square$

3.2.1. Monotonicity of the logarithm. The example provided by Theorem 3.7 leads to many others:

Theorem 3.10. *The function $f(x) = \log(x)$ on $(0, \infty)$ is operator monotone increasing. That is for $X \in M_n^{++}(\mathbb{C})$,*

$$\log(X) = \int_0^\infty \left(\frac{1}{\lambda + 1} - \frac{1}{\lambda + X}\right)\mathrm{d}\lambda,$$

and for $Y \ge X$, $\log(Y) \ge \log(X)$.

First proof of Theorem 3.10. The logarithm has the integral representation

$$\log(x) = \int_0^\infty \left(\frac{1}{\lambda + 1} - \frac{1}{\lambda + x}\right)\mathrm{d}\lambda.$$

Since for each $\lambda \ge 0$, $X \mapsto (\lambda + X)^{-1}$ is operator monotone decreasing on $M_n^{++}(\mathbb{C})$ by Theorem 3.7, it follows that $f(x) = \log x$ is operator monotone increasing. $\qquad\square$

Second proof of Theorem 3.10. Let $X > 0$, and let H be self-adjoint. For all t sufficiently close to 0, $X + tH > 0$. By (3.2.3),

$$\log(X + tH) - \log(X) = \int_0^\infty \left(\frac{1}{\lambda + X} - \frac{1}{\lambda + (X + tH)} \right) d\lambda$$

$$= t \int_0^\infty \left(\frac{1}{\lambda + X} H \frac{1}{\lambda + (X + tH)} \right) d\lambda.$$

Therefore, the function $X \mapsto \log(X)$ is continuously differentiable on $M_n^{++}(\mathbb{C})$, and for all $H \in M_n^{\text{s.a.}}(\mathbb{C})$,

$$(3.2.5) \qquad D_{\log,X}(H) = \frac{d}{dt} \log(X + tH) \Big|_{t=0} = \int_0^\infty \left(\frac{1}{\lambda + X} H \frac{1}{\lambda + X} \right) d\lambda.$$

For $H \geq 0$, $D_{\log,X}(H) \in M_n^+(\mathbb{C})$, and then by the fundamental theorem of calculus, $\log(X + H) - \log(X) = \int_0^1 \left(\frac{d}{dt} \log(X + tH) \right) dt \geq 0$. $\qquad\square$

Let $\mathcal{H}_n$ denote the Hilbert space consisting of $M_n(\mathbb{C})$ equipped with the Hilbert-Schmidt inner product. The right-hand side of (3.2.5) makes sense even if H is not self-adjoint, and we define $D_{\log,X} \in \mathcal{B}(\mathcal{H}_n)$ by this integral formula.

It is easy and useful to write down an orthonormal basis of $\mathcal{H}_n$ consisting of eigenvectors of $D_{\log,X}$: Let $\{\mathbf{u}_1, \ldots, \mathbf{u}_n\}$ be an orthonormal basis of $\mathbb{C}^n$ consisting of eigenvectors of X, so that for $j = 1, \ldots, n$, $X\mathbf{u}_j = \lambda_j \mathbf{u}_j$ and $\lambda_j > 0$. Then $\{|\mathbf{u}_i\rangle\langle\mathbf{u}_j| : 1 \leq i, j \leq n\}$ is an orthonormal basis of $\mathcal{H}_n$ and for each i, j,

$$(3.2.6) \qquad D_{\log,X}(|\mathbf{u}_i\rangle\langle\mathbf{u}_j|) = \left(\int_0^\infty \frac{1}{\lambda + \lambda_i} \frac{1}{\lambda + \lambda_j} d\lambda \right) |\mathbf{u}_i\rangle\langle\mathbf{u}_j|.$$

Hence for each i, j, $|\mathbf{u}_i\rangle\langle\mathbf{u}_j|$ is an eigenvector of $D_{\log,X}$, and the eigenvalue is given by the integral appearing in (3.2.6). Evidently all of these eigenvalues are strictly positive, and hence $D_{\log,X}$ is invertible.

Lemma 3.11. *Let $X \in M_n^{++}(\mathbb{C})$ have n distinct eigenvalues, and let $\{\mathbf{u}_1, \ldots, \mathbf{u}_n\}$ be an orthonormal basis of $\mathbb{C}^n$ consisting of eigenvectors of X. Let $\mathbf{v}$ be any unit vector such that for $j = 1, \ldots, n$, $\langle \mathbf{u}_j, \mathbf{v} \rangle \neq 0$. Then $D_{\log,X}(|\mathbf{v}\rangle\langle\mathbf{v}|) > 0$.*

Proof. By (3.2.6), $D_{\log,X}(|\mathbf{v}\rangle\langle\mathbf{v}|) = \sum_{i,j} \left(\int_0^\infty \frac{\langle\mathbf{u}_i,\mathbf{v}\rangle}{\lambda+\lambda_j} \frac{\langle\mathbf{v},\mathbf{u}_j\rangle}{\lambda+\lambda_i} d\lambda \right) |\mathbf{u}_i\rangle\langle\mathbf{u}_j|$. Hence for any $\mathbf{x} := \sum_{j=1}^n x_j \mathbf{u}_j \in \mathbb{C}^n$,

$$\langle \mathbf{x}, D_{\log,X}(|\mathbf{v}\rangle\langle\mathbf{v}|)\mathbf{x} \rangle = \int_0^\infty \left| \sum_{j=1}^n \frac{x_j \langle \mathbf{u}_j, \mathbf{v} \rangle}{\lambda + \lambda_j} \right|^2 d\lambda.$$

Since the functions $g_j(\lambda) := (\lambda + \lambda_j)^{-1}$ are linearly independent, and because the $\langle \mathbf{u}_j, \mathbf{v} \rangle$ are all nonzero, the integral is strictly positive unless $\mathbf{x} = 0$. Thus, $D_{\log,X}(|\mathbf{v}\rangle\langle\mathbf{v}|) > 0$. $\qquad\square$

3.2.2. Nonmonotonicity of the exponential function. Let $\mathcal{H}$ be a Hilbert space, and let $X \in \mathcal{B}(\mathcal{H})$. The **exponential** of X, e^X, is defined by the power series

$$\tag{3.2.7} e^X = \sum_{m=0}^{\infty} \frac{1}{m!} X^m,$$

and it is easy to check that this sum converges in the operator norm. It follows that the exponential function $X \mapsto e^X$ is analytic, and, in particular, continuously differentiable. As a consequence, the exponential function is Lipschitz continuous on a bounded subset of $\mathcal{B}(\mathcal{H})$. The following explicit bound on the modulus of continuity is useful:

Lemma 3.12. *For all* $X, Y \in \mathcal{B}(\mathcal{H})$, $\|e^X - e^Y\| \le e^{\|X\| + \|Y\|} \|X - Y\|$.

Proof. Apply the **telescoping sum formula**:

$$\tag{3.2.8} X^m - Y^m = \sum_{k=1}^{m} X^{m-k}(X - Y)Y^{k-1},$$

which is valid for all positive integers m. Then

$$\|X^m - Y^m\| \le m(\|X\| + \|Y\|)^{m-1}\|X - Y\|,$$

and $\|e^X - e^Y\| \le \sum_{m=1}^{\infty} \frac{1}{m!}\|X^m - Y^m\|$. $\qquad\square$

Directly from the definition (3.2.7),

$$\tag{3.2.9} \frac{\mathrm{d}}{\mathrm{d}t}e^{tA} = Ae^{tA} = e^{tA}A.$$

However, what is much more useful is a formula for the derivative in t of e^{A+tB} for any A and B in $\mathcal{B}(\mathcal{H})$.

Lemma 3.13 (Duhammel's formula). *For all* $A, B \in \mathcal{B}(\mathcal{H})$,

$$\tag{3.2.10} e^{A+B} - e^A = \int_0^1 e^{s(A+B)}Be^{(1-s)A}\,\mathrm{d}s.$$

Consequently,

$$D_{\exp,A}(B) = \frac{\mathrm{d}}{\mathrm{d}t}e^{A+tB}\Big|_{t=0} = \int_0^1 e^{sA}Be^{(1-s)A}\,\mathrm{d}s.$$

Proof. Define $X(s) := e^{s(A+B)}e^{(1-s)A}$. Then by (3.2.9),

$$\frac{\mathrm{d}}{\mathrm{d}s}X(s) = e^{s(A+B)}Be^{(1-s)A}.$$

Now (3.2.10) follows from the fundamental theorem of calculus.

Next, replace B by tB in (3.2.10), and divide both sides by t. Then take the limit $t \to 0$ using Lemma 3.12 to show that $e^{s(A+tB)}$ converges to e^{sA} uniformly in $s \in [0,1]$ as $t \to 0$. $\qquad\square$

Since the logarithm function is the inverse of the exponential function, the inverse function theorem says that

$$D_{\log,e^A}(D_{\exp,A}(K)) = K$$

for all self-adjoint K. That is, the linear transformations $D_{\log,e^A}$ and $D_{\exp,A}$ are inverse to one another. In fact, one can easily check that if $\{\mathbf{u}_1,\ldots,\mathbf{u}_n\}$ is an orthonormal basis consisting of eigenvectors of A, $A\mathbf{u}_j = \lambda_j\mathbf{u}_j$, then

$$(3.2.11) \qquad D_{\exp,A}(|\mathbf{u}_i\rangle\langle\mathbf{u}_j|) = \left(\int_0^1 e^{s\lambda_i}e^{(1-s)\lambda_j}ds\right)|\mathbf{u}_i\rangle\langle\mathbf{u}_j|.$$

It is a simple calculus exercise to check that in fact

$$\left(\int_0^1 e^{s\lambda_i}e^{(1-s)\lambda_j}ds\right)^{-1} = \int_0^\infty \frac{1}{\lambda + e^{\lambda_i}}\frac{1}{\lambda + e^{\lambda_j}}d\lambda.$$

Definition 3.14. The **logarithmic mean** of $x, y > 0$, $\Lambda(x,y)$, is defined by

$$\Lambda(x,y) = \int_0^1 x^{1-s}y^s ds = \begin{cases} \dfrac{y-x}{\log y - \log x} & x \neq y \\ x & x = y \end{cases}.$$

Note that $\Lambda(x,y) = \Lambda(y,x)$. Since for each s, $x^{1-s}y^s$ is a weighted geometric mean of x and y, $\Lambda(x,y)$ is an average of weighted geometric means of x and y. By the arithmetic-geometric mean inequality and then the weighted version applied twice more,

$$x^{1/2}y^{1/2} = (x^{1-s}y^s)^{1/2}(x^s y^{1-s})^{1/2} \leq \frac{1}{2}\left(x^{1-s}y^s + x^s y^{1-s}\right) \leq \frac{x+y}{2}.$$

Thus, $\Lambda(x,y)$ lies between the geometric and arithmetic means of x and y:

$$(3.2.12) \qquad x^{1/2}y^{1/2} \leq \Lambda(x,y) \leq \frac{x+y}{2}.$$

We can now rewrite (3.2.6) and (3.2.11) as

$$D_{\log,X}(|\mathbf{u}_i\rangle\langle\mathbf{u}_j|) = \frac{1}{\Lambda(\lambda_i,\lambda_j)}|\mathbf{u}_i\rangle\langle\mathbf{u}_j|$$

and

$$D_{\exp,\log(X)}(|\mathbf{u}_i\rangle\langle\mathbf{u}_j|) = \Lambda(\lambda_i,\lambda_j)|\mathbf{u}_i\rangle\langle\mathbf{u}_j|.$$

Theorem 3.15. *The exponential function $f(x) = e^x$ is not operator monotone increasing. In fact, for each $H \in M_n^{\mathrm{s.a.}}(\mathbb{C})$ with n distinct eigenvalues, there exists an $A > 0$ and a unit vector $\mathbf{u}$ such that $\langle \mathbf{u}, e^{H+A}\mathbf{u}\rangle < \langle \mathbf{u}, e^H \mathbf{u}\rangle$. Nonetheless, for all $H \in M_n^{\mathrm{s.a.}}(\mathbb{C})$ and $A \in M_n^+(\mathbb{C})$,*

$$\mathrm{Tr}[e^{H+A}] \geq \mathrm{Tr}[e^H].$$

Proof. There exists a unit vector $\mathbf{v}$ such that $D_{\log, e^H}(|\mathbf{v}\rangle\langle\mathbf{v}|) > 0$ by Lemma 3.11. Now let $\mathbf{u} \in \mathbb{C}^n$ be any unit vector that is orthogonal to $\mathbf{v}$. For all $\epsilon > 0$ sufficiently small,

$$A := D_{\log, e^H}(|\mathbf{v}\rangle\langle\mathbf{v}|) - \epsilon D_{\log, e^H}(|\mathbf{u}\rangle\langle\mathbf{u}|) > 0.$$

Then

$$\lim_{t \to 0} \frac{1}{t}\left(e^{H+tA} - e^H\right) = D_{\log, e^H}^{-1}(A) = |\mathbf{v}\rangle\langle\mathbf{v}| - \epsilon|\mathbf{u}\rangle\langle\mathbf{u}|,$$

and hence $\left.\frac{\mathrm{d}}{\mathrm{d}t}\langle \mathbf{u}, e^{H+tA}\mathbf{u}\rangle\right|_{t=0} < 0$. The final statement is a consequence of the monotonicity of the exponential function and Theorem 3.4. $\qquad\square$

In general e^{A+B} is not equal to $e^A e^B$. The Lie product formula (also known at the Lie-Trotter product formula in the infinite-dimensional setting [211]) is a useful substitute:

Lemma 3.16 (Lie product formula). *For all $A, B \in M_n(\mathbb{C})$,*

$$e^{A+B} = \lim_{m \to \infty} (e^{A/m} e^{B/m})^m.$$

Proof. Apply the telescoping sum formula (3.2.8) with $X = e^{(A+B)/m}$ and $Y = e^{A/m} e^{B/m}$ to obtain

$$(3.2.13) \quad e^{A+B} - (e^{A/m} e^{B/m})^m$$

$$= \sum_{k=1}^{m} (e^{(A+B)/m})^{m-k} [e^{(A+B)/m} - e^{A/m} e^{B/m}](e^{A/m} e^{B/m})^{k-1}.$$

Because for all $X, Y \in M_n(\mathbb{C})$, $\|XY\| \leq \|X\|\|Y\|$ and $\|X + Y\| \leq \|X\| + \|Y\|$, it is easy to see that $\|e^X\| \leq e^{\|X\|}$. Therefore, $\|(e^{(A+B)/m})^{m-k}\| \leq e^{\frac{m-k}{m}(\|A\|+\|B\|)}$ and $\|(e^{A/m} e^{B/m})^{k-1}\| \leq e^{\frac{k-1}{m}(\|A\|+\|B\|)}$. Altogether,

$$\|(e^{(A+B)/m})^{m-k}[e^{(A+B)/m} - e^{A/m} e^{B/m}](e^{A/m} e^{B/m})^{k-1}\|$$

$$\leq \|e^{(A+B)/m}\|^{m-1}\|e^{(A+B)/m} - e^{A/m} e^{B/m}\|.$$

It is left to the reader to show (see Exercise 12) that there is a constant C depending only on $\|A\|$ and $\|B\|$ such that $\|e^{(A+B)/m} - e^{A/m}e^{B/m}\| \le Cm^{-2}$. Therefore,

$$(3.2.14) \quad \|(e^{(A+B)/m})^{m-k}[e^{(A+B)/m} - e^{A/m}e^{B/m}](e^{A/m}e^{B/m})^{k-1}\|$$

$$\le e^{\|A\|+\|B\|}Cm^{-2}.$$

Then (3.2.13) and (3.2.14) yield $\|e^{A+B} - (e^{A/m}e^{B/m})^m\| \le e^{\|A\|+\|B\|}Cm^{-1}$. $\square$

3.2.3. Power functions. Other important examples that follow from Theorem 3.7 concern the power functions $f(x) = x^r$ on $(0, \infty)$. The treatment of these functions closely parallels the treatment of the logarithm and exponential functions, and hence we shall be brief.

Lemma 3.17. *For all $0 < r < 1$ and all $x > 0$*

$$(3.2.15) \qquad x^{r-1} = \frac{\sin(\pi r)}{\pi} \int_0^\infty \lambda^{r-1}\frac{1}{\lambda + x}d\lambda,$$

$$(3.2.16) \quad x^r = \frac{\sin(\pi r)}{\pi} \int_0^\infty \lambda^r \left(\frac{1}{\lambda} - \frac{1}{\lambda + x}\right)d\lambda = \frac{\sin(\pi r)}{\pi} \int_0^\infty \lambda^{r-1}\frac{x}{\lambda + x}d\lambda,$$

and

$$(3.2.17) \qquad x^{r+1} = \frac{\sin(\pi r)}{\pi} \int_0^\infty \lambda^r \left(\frac{x}{\lambda} - 1 + \frac{\lambda}{\lambda + x}\right)d\lambda.$$

Proof. The integral on the right in (3.2.15) is clearly convergent. Making the change of variables $\lambda \to x\lambda$,

$$\int_0^\infty \lambda^{r-1}\frac{1}{\lambda + x}d\lambda = x^{r-1}\int_0^\infty \lambda^{r-1}\frac{1}{\lambda + 1}d\lambda.$$

For the proof of (3.2.15) it then remains to show that

$$\int_0^\infty \lambda^{r-1}\frac{1}{\lambda + 1}d\lambda = \frac{\pi}{\sin(\pi r)},$$

which can be done using contour integrals. However, in our applications, such as the following theorem, we never need the exact value of this constant.

Then since $x^r = xx^{r-1}$ and since $\frac{x}{\lambda+x} = \lambda\left(\frac{1}{\lambda} - \frac{1}{\lambda+x}\right)$, (3.2.16) follows directly from (3.2.15).

Then since $x^{r+1} = xx^r$ and since $x\left(\frac{1}{\lambda} - \frac{1}{\lambda+x}\right) = \frac{x}{\lambda} - 1 + \frac{\lambda}{\lambda+x}$, (3.2.17) follows directly from (3.2.16). $\square$

By the spectral theorem, for all $A \in M_n^{++}(\mathbb{C})$ and all $0 < r < 1$,

$$(3.2.18) \qquad A^{r-1} = \frac{\sin(\pi r)}{\pi} \int_0^\infty \lambda^{r-1}\frac{1}{\lambda + A}d\lambda,$$

$$(3.2.19) \quad A^r = \frac{\sin(\pi r)}{\pi} \int_0^\infty \lambda^{r-1} \frac{A}{\lambda + A}\, d\lambda = \frac{\sin(\pi r)}{\pi} \int_0^\infty \lambda^r \left(\frac{1}{\lambda} \mathbb{1} - \frac{1}{\lambda + A} \right) d\lambda,$$

and

$$A^{r+1} = \frac{\sin(\pi r)}{\pi} \int_0^\infty \lambda^r \left(\frac{1}{\lambda} A - \mathbb{1} + \frac{\lambda}{\lambda + A} \right) d\lambda.$$

Formulas (3.2.18) and (3.2.19) express powers of A in terms of $(\lambda + A)^{-1}$, which is monotone decreasing in A, and it follows immediately that $x \mapsto x^p$ is operator monotone decreasing for $-1 < p < 0$ and operator monotone increasing for $0 < p < 1$. The third formula expresses A^p, $1 < p < 2$, in terms of both A, which is monotone increasing, and $(\lambda + A)^{-1}$, which is monotone decreasing, yielding nothing as far as monotonicity is concerned. However, this formula will be useful when we discuss operator convexity.

Theorem 3.18. *For $r \neq 0$, let $f_r(x) = x^r$ on $(0, \infty)$. Then for $0 < r \le 1$, f_r is operator monotone increasing and for $-1 < r < 0$, f_r is operator monotone decreasing while for all other values of r, f_r is neither operator monotone increasing or decreasing.*

Proof. It remains to prove the final statement. It is left as a simple exercise to modify the proof of Lemma 3.11 to show that when $X \in M_n^{++}(\mathbb{C})$ has n distinct eigenvalues, for each $-1 < r < 0$ there exist vectors $\mathbf{v}$ such that $D_{f_r,X}(|\mathbf{v}\rangle\langle\mathbf{v}|)$ is strictly negative, and for each $0 < r < 1$ there exist vectors $\mathbf{v}$ such that $D_{f_r,X}(|\mathbf{v}\rangle\langle\mathbf{v}|)$ is strictly positive.

It is then simple to modify the proof of Theorem 3.15 to show that for each $-1 < r < 0$ there exists $A > 0$ such that $D_{f_{1/r},X}(A)$ is not negative semidefinite, and therefore $f_{1/r}$ cannot be operator monotone decreasing. Considering multiples of the identity, it is clear that it cannot be operator monotone increasing either. This takes care of f_p for $-\infty < p < -1$.

Likewise, the same reasoning shows that for each $0 < r < 1$ there exists $A > 0$ such that $D_{f_{1/r},X}(A)$ is not positive semidefinite, and therefore $f_{1/r}$ cannot be operator monotone increasing. Considering multiples of the identity, it is clear that it cannot be operator monotone decreasing either. This takes care of f_p for $1 < p < \infty$. $\qquad\square$

It turns out that *all* operator monotone increasing functions f on $(0, \infty)$ have an integral representation analogous to (3.2.16). This is a theorem of Löwner [**151**]. In practice, when we make an operator monotonicity argument in a specific case, we will have an explicit integral representation such as we have given for the logarithm $\log(x)$ or the power functions x^r, $-1 < r \le 0$ and $0 < r < 1$. However it is very satisfying to know that nothing is being left out when we work with integral representations. Hansen [95] has given an elegant treatment of Löwner's theorem, and Simon's book [199] contains many different proofs of it from a variety of perspectives.

3.3. Operator convexity

Definition 3.19. A real valued function f on a connected subset I of $\mathbb{R}$ is an **operator convex function** in case for all finite-dimensional Hilbert spaces $\mathcal{H}$, and all $A, B \in \mathcal{B}^{\text{s.a.}}(\mathcal{H})$ with $\sigma(A), \sigma(B) \subset I$, and all $0 < \lambda < 1$,

$$(3.3.1) \qquad f((1 - \lambda)A + \lambda B) \leq (1 - \lambda)f(A) + \lambda f(B),$$

and f is said to be **operator concave** if $-f$ is operator convex.

The induction used to prove Lemma 3.1 adapts directly to the operator setting so that if f is operator convex, and for $j = 1, \ldots, n$, $X_j \in \mathcal{B}^{\text{s.a.}}(\mathcal{H})$, $\lambda_j \geq 0$ and $\sum_{j=1}^{n} \lambda_j = 1$,

$$(3.3.2) \qquad f\left(\sum_{j=1}^{n} \lambda_j X_j\right) \leq \sum_{j=1}^{n} \lambda_j f(X_j).$$

Theorem 3.20 (Convexity of the inverse function). *The inverse function* $f(x) = x^{-1}$ *on* $(0, \infty)$ *is operator convex.*

First proof of Theorem 3.20. Let $A, B \in \mathcal{B}^{++}(\mathcal{H})$. As in the first proof of Theorem 3.7, define X and Y in terms of A and B by (3.2.1) so that X and Y commute and are positive. Since X and Y can be simultaneously diagonalized, it is evident that for all $0 < \lambda < 1$,

$$((1 - \lambda)X + \lambda Y)^{-1} \leq (1 - \lambda)X^{-1} + \lambda Y^{-1}.$$

Multiplying through on both sides by $(A + B)^{-1/2}$, we obtain

$$(3.3.3) \qquad ((1 - \lambda)A + \lambda B)^{-1} \leq (1 - \lambda)A^{-1} + \lambda B^{-1},$$

proving that f is operator convex. $\qquad\qquad\qquad\qquad\qquad\qquad \square$

Second proof of Theorem 3.20. Let $X \in \mathcal{B}^{++}(\mathcal{H})$ and $H \in \mathcal{B}^{\text{s.a.}}(\mathcal{H})$. For all t sufficiently close to 0, $X + tH \in \mathcal{B}^{++}(\mathcal{H})$, applying (3.2.4) twice yields

$$\frac{\mathrm{d}^2}{\mathrm{d}t^2} \frac{1}{X + tH} = 2\frac{1}{X + tH}H\frac{1}{X + tH}H\frac{1}{X + tH}$$

$$= 2\left|\left(\frac{1}{X + tH}\right)^{1/2} H \frac{1}{X + tH}\right|^2 \geq 0.$$

Hence for any unit vector $\mathbf{u}$, the function $g(t) := \langle \mathbf{u}, (X + tH)^{-1}\mathbf{u}\rangle$ is convex on the interval of t sufficiently small that $X + tH > 0$. Taking $X = (1-\lambda)A + \lambda B$, $H = B - A$, this interval includes $[-\lambda, 1-\lambda]$, and hence $g(0) \le (1-\lambda)g(-\lambda) + \lambda g(1-\lambda)$, and since $\mathbf{u}$ is any unit vector, this is the same as (3.3.3). $\qquad\square$

Corollary 3.21. *Let $f_r(x) = x^r$, $x > 0$, $r \in \mathbb{R}$. Then for $-1 \le r \le 0$ and $1 \le r \le 2$, f_r is operator convex, while for $0 \le r \le 1$, f_r is operator concave. Furthermore, the function $f(x) = \log(x)$ is operator concave on $(0, \infty)$.*

Proof. For f_r, the assertions for $r = 0, 1$ are trivial. For $r = 2$, note that for all $X, Y \in \mathcal{B}^{\text{s.a.}}(\mathcal{H})$,

$$\frac{X^2 + Y^2}{2} - \left(\frac{X + Y}{2}\right)^2 = \left(\frac{X - Y}{2}\right)^2 \ge 0,$$

which proves midpoint convexity. Then a simple iteration proves (3.3.1) for f_2 at all dyadic rational $0 < \lambda < 1$, and then the general case follows by continuity. For the other values of r and for the logarithm function, the assertions follow from Theorem 3.20 and the integral representation we have given. Alternatively, for $r = 2$, note that $f_2 = \lim_{r\uparrow 2} f_r$, and limits of convex functions are convex. $\qquad\square$

For early work on operator convexity, see [**29**]. In particular, in this paper it is proved that if f is operator convex on $(0, \infty)$, then $f'(x)$ is operator monotone on $(0, \infty)$. It then follows from Theorem 3.18 that $f(x) = x^p$ is not operator convex for any $p > 2$.

3.4. Pinching and the operator Jensen inequality

Theorem 3.22. *Let $\{P_1, \ldots, P_m\}$ be a set of mutually orthogonal projections on a Hilbert space $\mathcal{H}$ such that $\sum_{j=1}^{m} P_j = \mathbb{1}$. Define $\mathcal{P} : \mathcal{B}(\mathcal{H}) \to \mathcal{B}(\mathcal{H})$ by*

$$(3.4.1) \qquad\qquad \mathcal{P}(X) = \sum_{j=1}^{m} P_j X P_j.$$

Let f be an operator convex function on $(0, \infty)$. Then for all $X \in \mathcal{B}^+(\mathcal{H})$,

$$(3.4.2) \qquad\qquad f(\mathcal{P}(X)) \le \mathcal{P}(f(X))$$

and

$$(3.4.3) \qquad\qquad \mathcal{P}(X) \ge \frac{1}{m}X.$$

Proof. Let $\omega := e^{i2\pi/m}$. For $1 \le k \le m$, define $U_k := \sum_{j=1}^{m} \omega^{kj} P_j$. Then U_k is unitary and

$$\frac{1}{m}\sum_{k=1}^{m} U_k^* X U_k = \sum_{k,j,\ell=1}^{m} \omega^{-kj}\omega^{k\ell} P_j X P_\ell = \sum_{j,\ell=1}^{m} \delta_{j,\ell} P_j X P_\ell = \mathcal{P}(X).$$

Therefore,

$$f(\mathcal{P}(X)) = f\left(\frac{1}{m}\sum_{k=1}^{m} U_k^* X U_k\right) \le \frac{1}{m}\sum_{k=1}^{m} U_k^* f(X) U_k = \mathcal{P}(f(X))$$

since $f(U_k^* X U) = U_k^* f(X) U_k$ due to the unitarity of U_k. Finally, since $U_m = \mathbb{1}$, the initial calculation in the proof yields

$$\mathcal{P}(X) = \frac{1}{m}\sum_{k=1}^{m} U_k^* X U_k \ge \frac{1}{m}X. \qquad \Box$$

The map $\mathcal{P}$ defined in (3.4.1) was called a **pinching operation** by Davis [**70, 71**] because if we define $\mathcal{H}_j$ to be the range of P_j, so that $\mathcal{H} = \bigoplus_{j=1}^{k} \mathcal{H}_j$, $\mathcal{M}_{\mathcal{P}}(X)$ will be block diagonal in the corresponding block matrix representation of operators on $\bigoplus_{j=1}^{k} \mathcal{H}_j$, and the block diagonal entries of $\mathcal{P}(X)$ are the block diagonal entries of X—the others get *pinched out*. The inequality (3.4.3) is known as the **pinching inequality**. Unlike (3.4.2) it does not refer to operator convexity, and is more elementary. Indeed, it may be proved using only the Cauchy-Schwarz inequality, though twice and in two ways. For $X \ge 0$, let $X = \sum_{\lambda \in \sigma(X)} \lambda P_\lambda$ be its spectral decomposition. Since $\sum_{\lambda \in \sigma(X)} \lambda P_\lambda = \mathbb{1}$, for any $\psi \in \mathcal{H}$,

$$\begin{aligned}
\langle \psi, X\psi \rangle &= \sum_{\lambda,\mu \in \sigma(X)} \langle P_\lambda \psi, X P_\mu \psi \rangle = \sum_{\lambda,\mu \in \sigma(X)} \langle X^{\frac{1}{2}} P_\lambda \psi, X^{\frac{1}{2}} P_\mu \psi \rangle \\
&\le \sum_{\lambda,\mu \in \sigma(X)} \langle \psi, P_\lambda X P_\lambda \psi \rangle^{\frac{1}{2}} \langle \psi, P_\mu X P_\mu \psi \rangle^{\frac{1}{2}} \\
&\le \left(\sum_{\lambda,\mu \in \sigma(X)}\right)^{\frac{1}{2}} \left(\sum_{\lambda,\mu \in \sigma(X)} \langle \psi, P_\lambda X P_\lambda \psi \rangle \langle \psi, P_\mu X P_\mu \psi \rangle\right)^{\frac{1}{2}} \\
&= |\sigma(X)| \langle \psi, \mathcal{P}(X)\psi \rangle,
\end{aligned}$$

where $|\sigma(X)|$ is the cardinality of the spectrum of X. This gives a second, more elementary proof of (3.4.3). In the rest of this section, we focus on its deeper companion, (3.4.2).

The following theorem is due to Davis and Sherman [**69, 70**].

Theorem 3.23 (Sherman-Davis inequality). *For all operator convex functions* f, *and all orthogonal projections* $P \in \mathcal{B}(\mathcal{H})$,

$$(3.4.4) \qquad\qquad Pf(PXP)P \le Pf(X)P$$

for all $X \in \mathcal{B}^+(\mathcal{H})$.

Proof. Define $P_1 := P$ and let $P_2 := \mathbb{1} - P$. Define $\mathcal{P}$ by $\mathcal{P}(X) = P_1 X P_1 + P_2 X P_2$. Then because

$$P_1 f(P_1 X P_1) P_1 = P_1 f(P_1 X P_1 + P_2 X P_2) P_1 = P_1 f(\mathcal{P}(X)) P_1,$$

we have from (3.4.2) of Theorem 3.22 that

$$P f(PXP) P = P_1 f(\mathcal{P}(X)) P_1 \le P_1 \mathcal{P}(f(X)) P_1 = P f(X) P,$$

and we obtain (3.4.4). $\qquad\square$

Remark 3.24. If $f(0) = 0$, $P f(PXP) P = f(PXP)$, and (3.4.4) may be shortened to

$$f(PXP) \le P f(X) P.$$

Frequently in applications, $f(s) = s^{-1}$, where (3.4.4) must be used as is.

The next inequality is a variant of Theorem 3.23 due to Davis [**71**, Second Corollary to Theorem 7]. It is convenient to prove it using matrix representations, and hence we state it in matrix terms.

Theorem 3.25 (Operator Jensen inequality). *Let $V_1, \ldots, V_k \in M_{m,n}(\mathbb{C})$ satisfy*

$$(3.4.5) \qquad \sum_{j=1}^{k} V_j^* V_j = \mathbb{1}.$$

Then for any operator convex function f, and any $X_1, \ldots, X_k \in M_m^{\mathrm{s.a.}}(\mathbb{C})$,

$$(3.4.6) \qquad f\left(\sum_{j=1}^{k} V_j^* X_j V_j \right) \le \sum_{j=1}^{k} V_j^* f(X_j) V_j.$$

Proof. Without loss of generality, we may suppose that $km \ge n$ since we may increase k by including additional operators $V_j = 0$ which does not affect either (3.4.5) or (3.4.6).

Let $\{\mathbf{e}_1, \ldots, \mathbf{e}_n\}$ be the standard orthonormal basis for $\mathbb{C}^n$. Then for $1 \le i, j \le n$, $\sum_{\ell=1}^{k} \langle V_\ell \mathbf{e}_i, V_\ell \mathbf{e}_j \rangle = \sum_{\ell=1}^{k} \langle \mathbf{e}_i, V_\ell^* V_\ell \mathbf{e}_j \rangle = \langle \mathbf{e}_i, \mathbf{e}_j \rangle = \delta_{i,j}$.

Therefore, the $km \times n$ matrix written in block form as $\begin{bmatrix} V_1 \\ \vdots \\ V_k \end{bmatrix}$ has orthonormal columns. Extend this set of n orthonormal vectors in $\mathbb{C}^{km}$ to an orthonormal basis of $\mathbb{C}^{km}$, and putting these vectors in as the columns of an $km \times km$ matrix yields a unitary matrix $\mathcal{U}$.

Next, let $\mathcal{X}$ be the $km \times km$ matrix, again viewed as an $k \times k$ block matrix, which has X_j for its jth diagonal block, and zero for all off-diagonal blocks. Finally, let $\mathcal{P}$ be the $km \times km$ orthogonal projection with $\mathbb{1}_{n \times n}$ in the upper left block, and zero elsewhere.

The first n columns of $\mathcal{X}\mathcal{U}$ form the $km \times n$ block matrix $\begin{bmatrix} X_1V_1 \\ \vdots \\ X_kV_k \end{bmatrix}$. There-fore, $\mathcal{U}^*\mathcal{X}\mathcal{U}$ has $\sum_{j=1}^{k} V_j^* X_j V_j$ as its upper left $n \times n$ block.

Note that $f(\mathcal{X})$ has $f(X_j)$ in the jth diagonal block and zero elsewhere, and by the spectral theorem $f(\mathcal{U}^*\mathcal{X}\mathcal{U}) = \mathcal{U}^* f(\mathcal{X})\mathcal{U}$. By the same calculation that we have just made, $f(\mathcal{U}^*\mathcal{X}\mathcal{U}) = \mathcal{U}^* f(\mathcal{X})\mathcal{U}$ has $\sum_{j=1}^{k} V_j^* f(X_j)V_j$ as its upper left $n \times n$ block. Therefore, the inequality to be proved, (3.4.6), is equivalent to $\mathcal{P}f(\mathcal{P}\mathcal{U}^*\mathcal{X}\mathcal{U}\mathcal{P})\mathcal{P} \leq \mathcal{P}f(\mathcal{U}^*\mathcal{X}\mathcal{U})\mathcal{P}$, and this is true by (3.4.4) of the Sherman-Davis theorem. $\qquad\square$

Remark 3.26. Theorem 3.25 is a far-reaching generalization of (3.3.2). Take $\{\lambda_1,\dots,\lambda_k\} \subset (0,1)$ with $\sum_{j=1}^{k} \lambda_j = 1$, and define $V_j := \lambda_j^{1/2}\mathbb{1}$. Then (3.4.6) reduces to (3.3.2). It is remarkable that no additional restriction is required to take care of the *noncommutative weights* in (3.4.6).

In fact, while (3.3.2) is trivial for $k = 1$, (3.4.6) is not. Suppose that $m < n$, and V is a partial isometry whose initial space is $\mathbb{C}^n$; that is, such that $\ker(V) = 0$. Then $V^*V = \mathbb{1}$. It follows that for any $X \in M_m^{\text{s.a.}}(\mathbb{C})$ and any operator convex f,

$$(3.4.7) \qquad\qquad f(V^*XV) \leq V^* f(X)V.$$

3.5. The Golden-Thompson inequality

Theorem 3.27 (Golden-Thompson inequality). *Let $\mathcal{H}$ be a d-dimensional Hilbert space, and let $H, K \in \mathcal{B}(\mathcal{H})$ be self-adjoint. Then*

$$\mathrm{Tr}[e^{H+K}] \leq \mathrm{Tr}[e^H e^K].$$

Remark 3.28. By the cyclicity of the trace $\mathrm{Tr}[e^H e^K] = \mathrm{Tr}[e^{H/2}e^K e^{H/2}] = \mathrm{Tr}[e^{K/2}e^H e^{K/2}]$ so that while $e^H e^K$ is not in general positive or self-adjoint, it does have a positive trace.

The following proof is taken from [205] and is quite different from the original proofs of Golden [92] and Thompson [207]. It provides a first hint of how results on tensor products will be used to prove matrix inequalities.

We preface the proof with some remarks about logarithms of tensor powers. For $X \in \mathcal{B}(\mathcal{H})$, define $(X)_j \in \mathcal{B}(\mathcal{H}^{\otimes n})$ by

$$(X)_j(\mathbf{v}_1 \otimes \cdots \otimes \mathbf{v}_n) = \mathbf{w}_1 \otimes \cdots \otimes \mathbf{w}_n,$$

where $\mathbf{w}_j = X\mathbf{v}_j$ and $\mathbf{w}_k = \mathbf{v}_k$ for all other k. For $j \neq \ell$, $(X)_j$ and $(X)_\ell$ commute. This gives a decomposition of $X^{\otimes n}$ into the product of n commuting operators, $\prod_{j=1}^{n}(X)_j$. Moreover, when $X \in \mathcal{B}^{++}(\mathcal{H})$, each $(X)_j \in \mathcal{B}^{++}(\mathcal{H}^{\otimes n})$. Hence for $X \in \mathcal{B}^{++}(\mathcal{H})$

$$\log(X^{\otimes n}) = \sum_{j=1}^{n}(\log X)_j \qquad \text{since} \qquad X^{\otimes n} = \prod_{j=1}^{n}(e^{\log X})_j = e^{\sum_{j=1}^{n}(\log X)_j}.$$

Then for $A, B \in \mathcal{B}^{++}(\mathcal{H})$, $\log(A^{\otimes n}) + \log(B^{\otimes n}) = \sum_{j=1}^{n}(\log A + \log B)_j$ and hence

$$(3.5.1) \qquad e^{\log(A^{\otimes n})+\log(B^{\otimes n})} = (e^{\log A + \log B})^{\otimes n}.$$

Proof of Theorem 3.27. Define $A = e^H$ and $B = e^K$. The inequality to be proved is $\mathrm{Tr}[e^{\log A + \log B}] \leq \mathrm{Tr}[AB]$. Let $n \in \mathbb{N}$. Then

$$(3.5.2) \qquad \mathrm{Tr}[A^{\otimes n}B^{\otimes n}] = \mathrm{Tr}[(AB)^{\otimes n}] = \mathrm{Tr}[AB]^n,$$

and by (3.5.1),

$$(3.5.3) \qquad \mathrm{Tr}[e^{(\log A^{\otimes n})+(\log B^{\otimes n})}] = \mathrm{Tr}[e^{\log A + \log B}]^n.$$

Let $B^{\otimes n} = \sum_{j=1}^{m(n)} \lambda_j P_j$ be the spectral decomposition of $B^{\otimes n}$. We will use the fact that $m(n)$ grows only polynomially in n by Lemma 2.43. Since $B^{\otimes n} = \sum_{j=1}^{m(n)} P_j B^{\otimes n} P_j$, cyclicity of the trace yields

$$(3.5.4) \qquad \mathrm{Tr}[A^{\otimes n}B^{\otimes n}] = \sum_{j=1}^{m(n)} \mathrm{Tr}[P_j A^{\otimes n} P_j B^{\otimes n}] = \mathrm{Tr}[\mathcal{P}(A^{\otimes n})B^{\otimes n}],$$

where $\mathcal{P}$ is the pinching operation associated to the set of spectral projections of $B^{\otimes n}$ as defined in (3.4.1). Since $\mathcal{P}(A^{\otimes n})$ and $B^{\otimes n}$ are commuting positive operators, $\mathcal{P}(A^{\otimes n})B^{\otimes n} = e^{\log(\mathcal{P}(A^{\otimes n}))+\log(B^{\otimes n})}$. By the pinching inequality (3.4.3), $\mathcal{P}(A^{\otimes n}) \geq \frac{1}{m(n)}A^{\otimes n}$, and then by Theorem 3.10 asserting the monotonicity of the logarithm,

$$(3.5.5) \qquad \log(\mathcal{P}(A^{\otimes n})) \geq \log(A^{\otimes n}) - \log m(n).$$

Theorem 3.4 implies that $f(X) = \mathrm{Tr}[e^X]$ is a monotone increasing function on the set of self-adjoint operators in $\mathcal{B}(\mathcal{H})$, and applying this using (3.5.5) yields $\mathrm{Tr}[e^{\log(\mathcal{P}(A^{\otimes n}))+\log(B^{\otimes n})}] \geq \frac{1}{m(n)}\mathrm{Tr}[e^{\log(A^{\otimes n})+\log(B^{\otimes n})}]$. Combining this with (3.5.4) yields $tr[A^{\otimes n}B^{\otimes n}] \geq \frac{1}{m(n)}\mathrm{Tr}[e^{\log(A^{\otimes n})+\log(B^{\otimes n})}]$. By (3.5.2) and (3.5.3) this is equivalent to

$$\mathrm{Tr}[AB] \geq \left(\frac{1}{m(n)}\right)^{1/n} \mathrm{Tr}[e^{\log A + \log B}]$$

for all n. Finally, $\lim_{n\to\infty} m(n)^{-1/n} = 1$ by Lemma 2.43. $\qquad \square$

Exercises

(1) Prove that for all $X \in M_n(\mathbb{C})$ and all $0 < \lambda < 1$,

$$((1 - \lambda)X + \lambda Y)^*((1 - \lambda)X + \lambda Y) \leq (1 - \lambda)X^*X + \lambda Y^*Y.$$

(2) Show that the function f on $\mathbb{R}$ defined by $f(x) := \max\{0, x\}$ is not operator monotone.

(3) Show that the function f on $\mathbb{R}$ defined by $f(x) := |x|$ is not operator convex.

(4) Let $f : (0, \infty) \to (0, \infty)$. Show that if f is operator concave, then $1/f$ is operator convex.

(5) Let $f : (0, \infty) \to (0, \infty)$. For $X, Y \in \mathcal{B}^{++}(\mathcal{H})$ and $\lambda \in (0, 1)$, use the identity $\lambda Y = \lambda X + (1 - \lambda)\left(\frac{\lambda}{1-\lambda}(Y - X)\right)$ to show that if f is operator concave and $Y > X$, then $f(\lambda Y) \geq \lambda f(X)$. Then show that every operator concave function $f : (0, \infty) \to (0, \infty)$ is operator monotone.

(6) Show that $f(x) = x \log x$ is operator convex on $(0, \infty)$.

(7) Let $\mathbf{u}, \mathbf{v}$ be unit vectors in $\mathbb{C}^m$ with $\langle \mathbf{u}, \mathbf{v} \rangle = \alpha > 0$. Define $P := |\mathbf{u}\rangle\langle\mathbf{u}|$ and $Q := |\mathbf{v}\rangle\langle\mathbf{v}|$. Let $p \in \mathbb{N}$, $p > 2$. Show that $\frac{d^2}{dt^2}(P + tQ)^p\big|_{t=0}$ has a strictly negative eigenvalue for all sufficiently small α.

(8) Let f be operator convex such that $f(0) \leq 0$. Let V be a contraction in $\mathcal{B}(\mathcal{H}, \mathcal{K})$. Show for all $X \in \mathcal{B}^{\text{s.a.}}(\mathcal{K})$, $f(V^*XV) \leq V^*f(X)V$.

(9) Let $V_1, \ldots, V_k \in \mathcal{B}(\mathcal{H})$ satisfy $\sum_{j=1}^{k} V_j^*V_j = \mathbb{1}$. Show that for any convex function f (not necessarily operator convex) on $\mathbb{R}_+$, and any $X_1, \ldots, X_k \in \mathcal{B}^+(\mathcal{H})$, $\text{Tr}\left[f\left(\sum_{j=1}^{k} V_j^*X_jV_j\right)\right] \leq \sum_{j=1}^{k} \text{Tr}[V_j^*f(X_j)V_j]$.

(10) Let f be a convex function (not necessarily operator convex) on $\mathbb{R}_+$ such that $f(0) \leq 0$. Let V be a contraction in $\mathcal{B}(\mathcal{H})$. Show that for all $X \in \mathcal{B}^+(\mathcal{H})$, $\text{Tr}[f(V^*XV)] \leq \text{Tr}[V^*f(X)V]$.

(11) Show that for $X \in M_n(\mathbb{C})$ and $t \in [0, 1]$, $\|e^{tX} - \mathbb{1}\| \leq t\|X\|e^{\|X\|}$.

(12) Show that for $A, B \in M_n(\mathbb{C})$ and $m \in \mathbb{N}$, there are unit vectors $\mathbf{u}, \mathbf{v}$ such that $\|e^{(A+B)/m} - e^{A/m}e^{B/m}\| = \langle \mathbf{v}, (e^{(A+B)/m} - e^{A/m}e^{B/m})\rangle$. Then define the function $f(t) := \langle \mathbf{v}, (e^{t(A+B)} - e^{tA}e^{tB})\rangle$. Compute an upper bound on $f'(t)$ valid for $t \in (0, 1/m)$. Find a constant C depending only on $\|A\|$ and $\|B\|$ such that $\|e^{(A+B)/m} - e^{A/m}e^{B/m}\| \leq Cm^{-2}$, thus completing the proof of Lemma 3.16.

(13) The Lie product formula, Theorem 3.16, extends to an arbitrary finite set $\{A_1, \ldots, A_m\} \subset M_n(\mathbb{C})$. Prove that

$$\lim_{r \to 0} \left(e^{rA_1} \cdots e^{rA_m}\right)^{1/r} = e^{\sum_{j=1}^{m} A_j}.$$

(14) Show that for $0 < r < 1$, $x^{r+1} = \int_0^1 \left(\frac{x^2}{t+(1-t)x}\right) d\rho_r(t)dt$, where $\rho_r(t) := \frac{\sin(\pi r)}{\pi} \frac{1}{t^{1-r}(1-t)^r}$ is a probability density. This variation on Lemma 3.17 is sometimes useful.

(15) For unit vectors $\mathbf{u}, \mathbf{v} \in \mathbb{C}^2$, let $P := |\mathbf{u}\rangle\langle\mathbf{u}|$ and $Q := |\mathbf{v}\rangle\langle\mathbf{v}|$. Suppose that $\langle\mathbf{u}, \mathbf{v}\rangle = \alpha > 0$. Show that for all $n \in \mathbb{N}, n > 2$,

$$A := \frac{\mathrm{d}^2}{\mathrm{d}t^2}(P + tQ)^n\Big|_{t=0} = (PQ + QP) + \mathcal{O}(\alpha^2).$$

Then show that all $m > 2$, for all α sufficiently small (depending on m), $A \notin M_2^+(\mathbb{C})$, and hence $f(x) = x^n$ is not operator convex.

von Neumann algebras on finite-dimensional Hilbert spaces

4.1. Operator algebras

We shall study operators A on a Hilbert space $\mathcal{H}$ not only in isolation, but also as members of subalgebras $\mathcal{M}$ of $\mathcal{B}(\mathcal{H})$. While we are mainly concerned with the case in which $\mathcal{H}$ is finite dimensional, many of the results to be discussed are also valid in a far more general setting. Therefore, we begin our discussion in an infinite-dimensional setting so that when we do specialize to finite-dimensional problems, it will be clear where and how the assumption of finite dimensionality enters our considerations.

Definition 4.1. A vector subspace $\mathcal{M}$ of $\mathcal{B}(\mathcal{H})$ is a $*$-**algebra** in $\mathcal{B}(\mathcal{H})$ if for all $A, B \in \mathcal{M}$, $AB \in \mathcal{M}$ and $A^* \in \mathcal{M}$. A $*$-subalgebra $\mathcal{M}$ of $\mathcal{B}(\mathcal{H})$ is **unital** if it contains the multiplicative identity $\mathbb{1} \in \mathcal{B}(\mathcal{H})$. Note that $\mathcal{B}(\mathcal{H})$ itself is a unital $*$-algebra.

The operator norm respects the structure of $*$-subalgebras: It follows immediately from (1.3.2) that for all $A, B \in \mathcal{B}(\mathcal{H})$,

$$\|AB\| \le \|A\|\|B\| \quad \text{and} \quad \|A^*\| = \|A\|.$$

Definition 4.2. A **von Neumann algebra** on a finite-dimensional Hilbert space $\mathcal{H}$ is a $*$-subalgebra $\mathcal{M}$ of $\mathcal{B}(\mathcal{H})$ such that it contains a multiplicative identity E. That is, there exists $E \in \mathcal{M}$ such the

$$(4.1.1) \qquad EA = A = AE \quad \text{for all} \quad A \in \mathcal{M}.$$

Remark 4.3. The definition of a von Neumann algebra on a general Hilbert space $\mathcal{H}$ has one additional requirement on $\mathcal{M}$, namely that $\mathcal{M}$ be closed in $\mathcal{B}(\mathcal{H})$ in a topology known as the *weak operator topology on $\mathcal{B}(\mathcal{H})$*. The weak operator topology on $\mathcal{B}(\mathcal{H})$ is the weakest topology such that for each $\mathbf{x}, \mathbf{y} \in \mathcal{H}$, the linear functional $A \mapsto \langle \mathbf{x}, A\mathbf{y} \rangle$ is continuous. When $\mathcal{H}$ is finite dimensional, this is simply the metric topology defined by the operator norm.

When $\mathcal{H}$ is infinite dimensional, the weak operator topology is a strictly weaker topology than the norm topology. Von Neumann showed that closure in this topology ensures that a $*$-subalgebra of $\mathcal{B}(\mathcal{H})$ contains all of the spectral projections of its self-adjoint elements, which need not be the case for norm closed $*$-subalgebras when $\mathcal{H}$ is infinite dimensional. The importance of spectral projections in quantum mechanics motivated von Neumann's choice of the topology for closure.

When $\mathcal{H}$ is finite dimensional, we shall be able to prove this and other consequences of closure in the weak operator topology directly, without mentioning this topology. The rest of this section is devoted to results of this sort.

Remark 4.4. Taking the adjoint throughout (4.1.1) and then replacing A^* by A yields $E^*A = A = AE^*$. Then taking $A = E$ shows that $E = E^*$ and $E = E^2$. Thus, E is an orthogonal projection. Let $\mathcal{K}$ be the range of E. Then the restriction of E to $\mathcal{K}$ is the identity on $\mathcal{K}$; that is $E|_{\mathcal{K}} = \mathbb{1}_{\mathcal{K}}$. Moreover, since for each $X \in \mathcal{M}$, $X = EXE$, we may regard $\mathcal{M}$ as a von Neumann algebra of operators on $\mathcal{K}$, and then the identity is simply $\mathbb{1}_{\mathcal{K}}$. **For this reason, we shall simply denote the multiplicative identity in a von Neumann algebra $\mathcal{M}$ on a Hilbert space $\mathcal{H}$ by $\mathbb{1}$.** It may not be the identity on $\mathcal{H}$, but it is the identity on a subspace $\mathcal{K}$ of $\mathcal{H}$ that *carries $\mathcal{M}$*.

Theorem 4.5. *Let $\mathcal{H}$ be a finite-dimensional Hilbert space, and let $A \in \mathcal{B}(\mathcal{H})$ be self-adjoint. Let $A = \sum_{\lambda \in \sigma(A)} \lambda P_\lambda$ be the spectral decomposition of A provided by the spectral theorem. Then for each $\lambda \in \sigma(A)$, P_λ belongs to every von Neumann subalgebra $\mathcal{M}$ of $\mathcal{B}(\mathcal{H})$ that contains A.*

Moreover, for every von Neumann algebra $\mathcal{M}$ on $\mathcal{H}$:

   (i) *$\mathcal{M}$ is the linear span of the orthogonal projections contained in $\mathcal{M}$.*

   (ii) *$\mathcal{M}$ is the linear span of the unitaries contained in $\mathcal{M}$.*

Proof. By Lemma 1.56, each P_λ can be written as a polynomial in A, and hence $P_\lambda \in \mathcal{M}$.

To prove (i) note that since $\mathcal{M}$ is a $*$-algebra, every $A \in \mathcal{M}$ satisfies $A = H + iK$, where H and K are self-adjoint and belong to $\mathcal{M}$. The decompositions of H and K provided by the spectral theorem express them as belonging to the real linear span of a set of orthogonal projections that belong to $\mathcal{M}$.

With (i) proved, to prove (ii) it suffices to show that every orthogonal projection $P \in \mathcal{M}$ is a linear combination of two unitaries in $\mathcal{M}$. Since $P \in \mathcal{M}$, $P^\perp \in \mathcal{M}$, where $P^\perp = \mathbb{1} - P$ is the complementary projection. Define $U_\pm = P \pm P^\perp$. Then U_+ and U_- are unitary and belong to $\mathcal{M}$, and $P = \frac{1}{2}(U_+ + U_-)$. $\qquad\square$

Theorem 4.6. *Let $\mathcal{H}$ be a finite-dimensional Hilbert space and let $\mathcal{M}$ be a von Neumann subalgebra of $\mathcal{B}(\mathcal{H})$. For $A \in \mathcal{M}$, let $A = U|A|$ be the (unique) polar decomposition of A in which the initial space of U is the range of $|A|$. Then U and $|A|$ both belong to $\mathcal{M}$.*

Proof. Evidently, A^*A belongs to $\mathcal{M}$. Let $A^*A = \sum_{\lambda \in \sigma(A^*A)} \lambda P_\lambda$ be the spectral decomposition of A^*A. By Theorem 4.5, each spectral projection P_λ belongs to $\mathcal{M}$, and hence so does $|A| = (A^*A)^{1/2} = \sum_{\lambda \in \sigma(A^*A)} \lambda^{1/2} P_\lambda$. Define

$$U := \sum_{\lambda \in \sigma(A^*A),\ \lambda > 0} \lambda^{-1/2} A P_\lambda,$$

which also belongs to $\mathcal{M}$. Note that

$$U^*U = \sum_{\lambda,\mu \in \sigma(A^*A),\ \lambda > 0} \mu^{-1/2} \lambda^{-1/2} P_\mu |A|^2 P_\lambda = \sum_{\lambda \in \sigma(A^*A),\ \lambda > 0} P_\lambda$$

is the orthogonal projection onto $\mathrm{ran}(|A|) = \ker(A)^\perp$. Hence U is a partial isometry, and

$$U|A| = \sum_{\lambda,\mu \in \sigma(A^*A),\ \lambda > 0} \left(\lambda^{-1/2} A P_\lambda\right)\left(\mu^{1/2} P_\mu\right) = A \sum_{\lambda \in \sigma(A^*A),\ \lambda > 0} P_\lambda = A,$$

since $P_\lambda P_\mu = \delta_{\lambda,\mu} P_\lambda$. $\qquad\square$

Next, consider the SVD. Let $\mathcal{H}$ and $\mathcal{K}$ be finite-dimensional Hilbert spaces.

Theorem 4.7. *Let $\mathcal{H}$ and $\mathcal{K}$ be finite-dimensional Hilbert spaces. For $A \in \mathcal{B}(\mathcal{H}, \mathcal{K})$, let P denote the orthogonal projection onto $\ker(A)^\perp$. Then there are partial isometries $\{U_\sigma : \sigma \in \sigma(|A|)\backslash\{0\}\} \subset \mathcal{B}(\mathcal{H}, \mathcal{K})$ such that*

$$(4.1.2) \qquad P = \sum_{\sigma \in \sigma(|A|)\backslash\{0\}} U_\sigma^* U_\sigma \quad and \quad A = \sum_{\sigma \in \sigma(|A|)\backslash\{0\}} \sigma U_\sigma.$$

Remark 4.8. By Theorem 2.25, the numbers in $\sigma(|A|)\backslash\{0\}$ are the singular values of A. Choosing orthonormal bases $\{\mathbf{v}_1, \dots, \mathbf{v}_d\}$ and $\{\mathbf{u}_1, \dots, \mathbf{u}_d\}$ for the initial and final spaces of each U_σ, $U_\sigma = \sum_{j=1}^{d} |\mathbf{u}_j\rangle\langle\mathbf{v}_j|$. Substituting this into (4.1.2) yields the SVD in the form presented in Theorem 2.25.

Proof of Theorem 4.7. Let $A = U|A|$ be the polar decomposition of A. Let $|A| = \sum_{\sigma \in \sigma(|A|)} \sigma P_\sigma$ be the spectral decomposition of $|A| \in \mathcal{B}(\mathcal{H})$. Then U is a partial isometry whose initial space is $\mathrm{ran}(|A|)$. Since $\sum_{\sigma \in \sigma(|A|)\backslash\{0\}} P_\sigma$ is the orthogonal projection onto $\mathrm{ran}\,(|A|)$, which is U^*U, for each $\sigma \neq 0$, $U^*UP_\sigma = P_\sigma$, and hence

$$U_\sigma^*U_\sigma = P_\sigma U^*UP_\sigma = P_\sigma,$$

which proves the first part of (4.1.2). Next, by the definition of U_σ,

$$\sum_{\sigma \in \sigma(|A|)\setminus\{0\}} \sigma U_\sigma = \sum_{\sigma \in \sigma(|A|)\setminus\{0\}} \sigma UP_\sigma = U|A| = A. \qquad \square$$

Corollary 4.9. *Let $\mathcal{H}$ be a finite-dimensional Hilbert space. For all $A \in \mathcal{B}(\mathcal{H})$, let $A = \sum_{\sigma \in \sigma(|A|)\setminus\{0\}} \sigma U_\sigma$ be the SVD of A in the form provided by Theorem 4.7. Then each U_σ belongs to every von Neumann subalgebra $\mathcal{M}$ of $\mathcal{B}(\mathcal{H})$ that contains A.*

Proof. Since $U \in \mathcal{M}$ by Theorem 4.6 and $P_\sigma \in \mathcal{M}$ by Theorem 4.5, it follows that $U_\sigma = UP_\sigma \in \mathcal{M}$. $\qquad \square$

Theorem 4.10. *Let $\mathcal{H}$ be a finite-dimensional Hilbert space. If $A \in \mathcal{B}(\mathcal{H})$ is invertible, its inverse belongs to every von Neumann subalgebra $\mathcal{M}$ of $\mathcal{B}(\mathcal{H})$ that contains A.*

Proof. Let $A = U|A|$ be the polar decomposition of A. Since A is invertible, U is invertible and therefore unitary. By Theorem 4.6, $U \in \mathcal{M}$, and since $\mathcal{M}$ is a $*$-algebra, so is $U^* = U^{-1}$.

Let $\sum_{\lambda \in \sigma(|A|)} \lambda P_\lambda$ be the spectral decomposition of $|A|$. Since A is invertible, each $\lambda > 0$, and hence $|A|^{-1} = \sum_{\lambda \in \sigma(|A|)} \lambda^{-1} P_\lambda \in \mathcal{M}$. By Theorem 4.5 $A^{-1} = |A|^{-1}U^* \in \mathcal{M}$. $\qquad \square$

Theorem 4.10 has the following consequence. Let $A \in \mathcal{B}(\mathcal{H})$ and let $\mathcal{M}$ be any von Neumann subalgebra of $\mathcal{B}(\mathcal{H})$ containing A. One could define the spectrum of A in $\mathcal{M}$ to be the set of $\lambda \in \mathbb{C}$ for which $A - \lambda\mathbb{1}$ is not invertible in $\mathcal{M}$. But by Theorem 4.10, this would always be the spectrum of A in $\mathcal{B}(\mathcal{H})$ as we have already defined it.

4.2. Projections in von Neumann algebras

Theorem 4.11. *Let $\mathcal{M}$ be a commutative von Neumann algebra on a finite-dimensional Hilbert space $\mathcal{H}$. There exists a set $\{P_1, \ldots, P_n\}$ of mutually orthogonal projections in $\mathcal{M}$ such that $\mathcal{M} = \mathrm{span}(\{P_1, \ldots, P_n\})$.*

Proof. For self-adjoint $H \in \mathcal{M}$, the cardinality of the spectrum is at most the dimension of $\mathcal{H}$. Choose a self-adjoint $H \in \mathcal{M}$ such that $\sigma(H)$ has maximal cardinality. Let

$$H = \sum_{\lambda \in \sigma(H)} \lambda P_\lambda$$

be the spectral decomposition of H. We claim that $\{P_\lambda \; : \; \lambda \in \sigma(H)\}$ spans $\mathcal{M}$. If not, since every von Neumann algebra is the span of the orthogonal projections it contains, there is a nonzero orthogonal projection $Q \in \mathcal{M}$ that does not belong to the span of $\{P_\lambda \; : \; \lambda \in \sigma(H)\}$.

Since $\sum_{\lambda \in \sigma(H)} P_\lambda = \mathbb{1}$, $Q = \sum_{\lambda \in \sigma(H)} QP_\lambda$. Since $\mathcal{M}$ is commutative, QP_λ is an orthogonal projection and $0 \leq QP_\lambda \leq P_\lambda$. If for each λ, either $QP_\lambda = 0$ or else $QP_\lambda = P_\lambda$, then $Q = \sum_\lambda QP_\lambda$ would be in the span of $\{P_\lambda \; : \; \lambda \in \sigma(H)\}$. Hence there is at least one $\mu \in \sigma(H)$ such that $QP_\mu \neq 0$ and $QP_\mu \neq P_\mu$.

Since $Q^\perp = \mathbb{1} - Q$, $Q^\perp$ also belongs to $\mathcal{M}$, and $P_\mu = QP_\mu + Q^\perp P_\mu$. Hence QP_μ and $Q^\perp P_\mu$ are both nonzero orthogonal projections (because of the commutativity) belonging to $\mathcal{M}$. Let $t \in \mathbb{R} \backslash \sigma(H)$ and define

$$\widehat{H} = \sum_{\lambda \in \sigma(H) \backslash \{\mu\}} \lambda P_\lambda + \mu QP_\mu + tQ^\perp P_\mu.$$

Then $\sigma(\widehat{H}) = \sigma(H) \cup \{t\}$, and this contradicts the maximum cardinality of $\sigma(H)$. Hence $\{P_\lambda \; : \; \lambda \in \sigma(H)\}$ spans $\mathcal{M}$. $\qquad\square$

By Theorem 4.11, there is a linear isomorphism between a commutative von Neumann algebra $\mathcal{M}$ on a finite-dimensional Hilbert space, and the space of complex valued functions on some finite set $\mathcal{X}$. Indeed, by Theorem 4.11, there is a set $\{P_1, \ldots, P_n\}$ of mutually orthogonal projections in $\mathcal{M}$ such each $X \in \mathcal{M}$ has the form

$$X = \sum_{j=1}^{n} z_j P_j$$

for some complex numbers $\{z_1, \ldots, z_n\}$. Let $\mathcal{X} := \{1, \ldots, n\}$. Let $\mathcal{C}(\mathcal{X})$ denote the space of complex valued functions f on $\mathcal{X}$. Define a map from $\mathcal{C}(\mathcal{X})$ $\mathcal{M}$ by $f \mapsto \sum_{j=1}^{n} f(j)P_j$. It is readily checked that this map is a linear isomorphism between $\mathcal{C}(\mathcal{X})$ and $\mathcal{M}$.

In fact, for any finite set $\mathcal{X}$, $\mathcal{C}(\mathcal{X})$ is not only a vector space, but an algebra with multiplication defined by $fg(x) = f(x)g(x)$ for all $f, g \in \mathcal{C}(\mathcal{X})$ and all $x \in \mathcal{X}$. The constant function 1 is a multiplicative identity. There is also a natural conjugation defined in $\mathcal{C}(\mathcal{X})$: for $f \in \mathcal{C}(\mathcal{X})$, define $f^* \in \mathcal{C}(\mathcal{X})$ by $f^*(x) = \overline{f(x)}$. Thus, $\mathcal{C}(\mathcal{X})$ has all of the characteristics of a von Neumann algebra, except of course that is does not consist of operators on a Hilbert space. The following construction closes this gap.

Let $\mathcal{H}$ denote the Hilbert space $L^2(\mathcal{X}, \mu)$, as in Definition 1.68, where μ is counting measure on $\mathcal{X}$. Identify $f \in \mathcal{C}(\mathcal{X})$ with the **multiplication operator** M_f on $\mathcal{H}$ by

$$(4.2.1) \qquad\qquad M_f \psi(x) = f(x)\psi(x)$$

for all $f \in \mathcal{C}(\mathcal{X})$, $\psi \in \mathcal{H}$ and $x \in \mathcal{X}$. It is easily checked that the map $f \mapsto M_f$ is one-to-one from $\mathcal{C}(\mathcal{X})$ into $\mathcal{B}(\mathcal{H})$, and moreover this map is linear, and has the properties that $M_{fg} = M_f M_g$ and $M_{f*} = M_f^*$ for all $f, g \in \mathcal{C}(\mathcal{X})$. We may thus identify $\mathcal{C}(\mathcal{X})$ with its image in $\mathcal{B}(\mathcal{H})$ under this map, and this identifies $\mathcal{C}(\mathcal{X})$ with a commutative von Neumann algebra.

Definition 4.12. For any finte set $\mathcal{X}$, let μ be counting measure on $\mathcal{X}$. Then the identification of $f \in \mathcal{C}(\mathcal{X})$ with the multiplication operator M_f on $L^2(\mathcal{X}, \mu)$, as in (4.2.1), makes $\mathcal{C}(\mathcal{X})$ into a commutative von Neumann algebra. This is called the **standard von Neumann algebra structure** on $\mathcal{C}(\mathcal{X})$.

As explained above, every commutative von Neumann algebra on a finite-dimensional Hilbert space may be viewed, for some finite set $\mathcal{X}$, as $\mathcal{C}(\mathcal{X})$ with its standard von Neumann algebra structure. We now turn to the study of projections in the noncommutative setting.

Definition 4.13. Let P and Q be two projections in $\mathcal{B}(\mathcal{H})$. Then $P \vee Q$ is defined to be the projection onto $\mathrm{span}(\mathrm{ran}(P) \cup \mathrm{ran}(Q))$, and $P \wedge Q$ is defined to be the projection onto $\mathrm{ran}(P) \cap \mathrm{ran}(Q)$.

Note that in the usual partial order of self-adjoint operators, $P \vee Q \geq P, Q \geq P \wedge Q$. Let $V := \mathrm{ran}(P)$ and $W := \mathrm{ran}(Q)$. Then since

$$(\mathrm{span}(V \cup W))^{\perp} = (V \cup W)^{\perp} = V^{\perp} \cap W^{\perp},$$

and since $P, Q \geq 0$, $\ker(P + Q) = \ker(P) \cap \ker(Q)$,

$$\mathrm{ran}(P + Q) = (V^{\perp} \cap W^{\perp})^{\perp} = \mathrm{span}(V \cup W).$$

This shows that $P \vee Q$ is the orthogonal projection onto $\mathrm{ran}(P + Q)$.

Lemma 4.14. *Let $\mathcal{M}$ be a von Neumann algebra on a finite-dimensional Hilbert space $\mathcal{H}$. Let P and Q be any two projections in $\mathcal{M}$. Then $P \vee Q \in \mathcal{M}$ and $P \wedge Q \in \mathcal{M}$.*

Proof. Since $P \vee Q$ is the orthogonal projection onto $\mathrm{ran}(P + Q)$, and since all spectral projections of $P + Q$ belong to $\mathcal{M}$, so does $P \vee Q$. Likewise, $P \wedge Q$ is the spectral projection corresponding to the eigenvalue 2 of $P + Q$, and hence $P \wedge Q \in \mathcal{M}$. $\square$

Definition 4.15. Let $\mathcal{M}$ be a von Neumann algebra on a finite-dimensional Hilbert space $\mathcal{H}$. Let $\{P_j : j \in \mathcal{J}\}$ be an arbitrary set of orthogonal of projections in $\mathcal{M}$ indexed by some set $\mathcal{J}$. Then

$$\bigvee_{j \in \mathcal{J}} P_j$$

is defined to be the projection onto the span of $\bigcup_{j \in \mathcal{J}} \mathrm{ran}(P_j)$.

The following is another important closure theorem for projections in a von Neumann algebra.

Theorem 4.16. *Let $\mathcal{M}$ be a von Neumann algebra on a finite-dimensional Hilbert space $\mathcal{H}$. Let $\{P_j \ : \ j \in \mathcal{J}\}$ be an arbitrary set of orthogonal projections in $\mathcal{M}$. Then the projection $\bigvee_{j\in\mathcal{J}} P_j$ belongs to $\mathcal{M}$.*

Proof. By Lemma 4.15 and an obvious induction argument, for each finite set $\{j_1,\dots,j_m\} \subset \mathcal{J}$,

$$\bigvee_{j\in\{j_1,\dots,j_m\}} P_j \in \mathcal{M}.$$

Now we make use of the finite dimensionality of $\mathcal{H}$: Pick $\{j_1,\dots,j_m\} \subset \mathcal{J}$ so that the dimension of the range of $\widehat{P} := \bigvee_{j\in\{j_1,\dots,j_m\}} P_j$ is maximal for all finite subsets of $\mathcal{J}$.

For all $j \in \mathcal{J}$, $P_j \leq \widehat{P}$, since if not, we could adjoin j to $\{j_1,\dots,j_m\}$ and obtain a strictly larger range. Hence $\bigvee_{j\in\mathcal{J}} P_j = \widehat{P} \in \mathcal{M}$. $\qquad\square$

4.3. Commutants

Definition 4.17. Let $\mathcal{H}$ be a Hilbert space, and let $\mathcal{S} \subset \mathcal{B}(\mathcal{H})$. The **commutant** $\mathcal{S}'$ of $\mathcal{S}$ is the subset of $\mathcal{B}(\mathcal{H})$ given by

$$\mathcal{S}' = \{A \in \mathcal{B}(\mathcal{H}) \ : \ AB - BA = 0 \quad \text{for all} \quad B \in \mathcal{S}\}.$$

Lemma 4.18. *Let $\mathcal{H}$ be a Hilbert space. Let $\mathcal{S} \subseteq \mathcal{B}(\mathcal{H})$ be such that $A^* \in \mathcal{S}$ whenever $A \in \mathcal{S}$. Then $\mathcal{S}'$ is a von Neumann algebra.*

Proof of Lemma 4.18. It is evident that for all $\mathcal{S} \subset \mathcal{B}(\mathcal{H})$, $\mathbb{1} \in \mathcal{S}'$. The fact that $\mathcal{S}'$ is an algebra is equally evident, and the closure of $\mathcal{S}'$ under the involution when $\mathcal{S}$ is closed under the involution follows from the fact that $(AB-BA)^* = B^*A^* - A^*B^*$. In the finite-dimensional case, closure in the weak operator topology is automatic, but in general it is easy: $B \in \mathcal{S}'$ if and only if for all $A \in \mathcal{S}$ and all $\mathbf{x}, \mathbf{y} \in \mathcal{H}$, $\langle \mathbf{x}, AB\mathbf{y}\rangle = \langle \mathbf{x}, BA\mathbf{y}\rangle$, which is the same as

$$\langle A^*\mathbf{x}, B\mathbf{y}\rangle = \langle \mathbf{x}, BA\mathbf{y}\rangle.$$

Therefore, $\mathcal{S}'$ is closed in any topology in which the functions $B \mapsto \langle \mathbf{w}, B\mathbf{z}\rangle$ are continuous for each $\mathbf{w}, \mathbf{z} \in \mathcal{H}$, and the weakest of these is, by definition, the weak operator topology. Hence $\mathcal{S}'$ is closed in this topology. $\qquad\square$

In particular, if $\mathcal{M}$ is a von Neumann algebra on $\mathcal{H}$, then so is the commutant of $\mathcal{M}$, $\mathcal{M}'$. The double commutant $\mathcal{M}''$ is then also a von Neumann algebra on $\mathcal{H}$, but it is nothing new: according to the celebrated von Neumann double commutant theorem [223], $\mathcal{M}''$ is $\mathcal{M}$ itself. We treat the finite-dimensional case, avoiding the topological considerations that are necessary in infinite dimensions.

Theorem 4.19 (von Neumann double commutant theorem). *Let $\mathcal{M}$ be a von Neumann on a finite-dimensional Hilbert space $\mathcal{H}$. Then*

$$\mathcal{M}'' = \mathcal{M}.$$

Proof. We first show that for any $B \in \mathcal{M}''$ and any $\mathbf{v} \in \mathcal{H}$, there exists an $A \in \mathcal{M}$ such that

$$A\mathbf{v} = B\mathbf{v}.$$

To see this, define V to be the subspace of $\mathcal{H}$ given by

$$V = \{A\mathbf{v} \ : \ A \in \mathcal{M}\}.$$

Let P be the orthogonal projection onto V. By construction, V is invariant under the action of $\mathcal{M}$, and hence $PAP = AP$ for all $A \in \mathcal{M}$. Taking Hermitian conjugates, $PA^*P = PA^*$ for all $A \in \mathcal{M}$. Since $\mathcal{M}$ is a $*$-algebra, this implies $PA = AP$ for all $A \in \mathcal{M}$. That is, $P \in \mathcal{M}'$.

Consequently, for any $B \in \mathcal{M}''$, $BP = PB$, and so V is invariant under the action of $\mathcal{M}''$. Then since $\mathbf{v} \in V$, $B\mathbf{v} \in V$, and hence, by the definition of V, $B\mathbf{v} = A\mathbf{v}$ for some $A \in \mathcal{M}$.

We then apply what has just been proved to the $*$-subalgebra $\widehat{\mathcal{M}}$ of $\mathcal{B}(\bigoplus_n \mathcal{H})$ consisting of diagonal block matrices of the form

$$\widehat{A} := \begin{bmatrix} A & & & \\ & A & & \\ & & \ddots & \\ & & & A \end{bmatrix}, \quad A \in \mathcal{M}.$$

(The block matrix $\widehat{A}$ is sometimes called an **amplification** of A.)

One readily checks that $(\widehat{\mathcal{M}})'$ consists of $n \times n$ block matrices with entries in $\mathcal{M}'$, and then that $(\widehat{\mathcal{M}})''$ consists of diagonal block matrices of the form

$$\widehat{B} := \begin{bmatrix} B & & & \\ & B & & \\ & & \ddots & \\ & & & B \end{bmatrix}, \quad B \in \mathcal{M}''.$$

Let $\{\mathbf{v}_1, \dots, \mathbf{v}_n\}$ be any basis of $\mathcal{H}$, and let $\mathbf{v} = (\mathbf{v}_1, \dots, \mathbf{v}_n) \in \bigoplus_n \mathcal{H}$, where $\bigoplus_n \mathcal{H}$ denotes the direct sum of n copies of $\mathcal{H}$. Then

$$\widehat{A}\mathbf{v} = \widehat{B}\mathbf{v} \quad \Rightarrow \quad A\mathbf{v}_j = B\mathbf{v}_j \quad j = 1, \dots, n.$$

Since $\{\mathbf{v}_1, \dots, \mathbf{v}_n\}$ is a basis of $\mathcal{H}$, this means $B = A \in \mathcal{M}$. Since B was an arbitrary element of $\mathcal{M}''$, this shows that $\mathcal{M}'' \subset \mathcal{M}$. Since $\mathcal{M} \subset \mathcal{M}''$ is automatic, this proves that $\mathcal{M} = \mathcal{M}''$. $\quad\square$

Remark 4.20. This proof is adapted from a proof of Nelson [**165**], which can be consulted for the extension to the general case—easy for those familiar with the *strong operator topology*.

Example 4.21. Let $\mathcal{H} = \mathcal{H}_1 \otimes \mathcal{H}_2$ be the tensor product of two finite-dimensional Hilbert spaces. Let $\mathcal{M}$ be the subalgebra of $\mathcal{B}(\mathcal{H})$ consisting of operators of the form $\mathbb{1}_{\mathcal{H}_1} \otimes A$, $A \in \mathcal{B}(\mathcal{H}_2)$. Evidently $\mathcal{M}$ is a von Neumann subalgebra of $\mathcal{B}(\mathcal{H}_1 \otimes \mathcal{H}_2)$. It is easy to see that every operator of the form $B \otimes \mathbb{1}_{\mathcal{H}_2}$, $B \in \mathcal{B}(\mathcal{H}_1)$, belongs to $\mathcal{M}'$. We now show that in fact,

$$(4.3.1) \qquad \mathcal{M}' = \{B \otimes \mathbb{1}_{\mathcal{H}_2} \; : \; B \in \mathcal{B}(\mathcal{H}_1)\}.$$

To see this, it is convenient to use the block matrix representation of operators on $\mathcal{H}_1 \otimes \mathcal{H}_2$. Let $\{\mathbf{v}_1, \ldots, \mathbf{v}_n\}$ be an orthonormal basis of $\mathcal{H}_2$. As in Example 2.39, the corresponding block matrix representation of $\mathbb{1}_{\mathcal{H}_1} \otimes A$ is

$$\begin{bmatrix} a_{1,1}\mathbb{1}_{\mathcal{H}_1} & \cdots & a_{1,n}\mathbb{1}_{\mathcal{H}_1} \\ \vdots & \ddots & \vdots \\ a_{n,1}\mathbb{1}_{\mathcal{H}_1} & \cdots & a_{n,n}\mathbb{1}_{\mathcal{H}_1} \end{bmatrix},$$

where $a_{i,j} = \langle \mathbf{v}_i, A\mathbf{v}_j \rangle$. Let $C_{i,j} \in \mathcal{B}(\mathcal{H}_1)$ denote the (i, j)-th entry in the block matrix representing C. Then $(\mathbb{1}_{\mathcal{H}_1} \otimes A)C = C(\mathbb{1}_{\mathcal{H}_1} \otimes A)$ yields

$$\sum_{k=1}^{n} a_{i,k} C_{k,j} = \sum_{k=1}^{n} C_{i,k} a_{k,j}$$

for all choices of A and all i, j. Choose $1 \leq r, s \leq n$, and A so that $a_{i,j} = \delta_{i,r}\delta_{j,s}$. Then $\delta_{i,r} C_{s,j} = C_{i,r}\delta_{j,s}$. Taking $i \neq r$ and $j = s$, yields $C_{i,r} = 0$, hence all off-diagonal blocks of C are zero. Next, taking $s = j$ and $i = r$ yields $C_{r,r} = C_{s,s}$, and hence the diagonal entries of C are all the same. Thus C has the form

$$C = \begin{bmatrix} B & & \\ & \ddots & \\ & & B \end{bmatrix}$$

for some $B \in \mathcal{B}(\mathcal{H}_1)$ and therefore as in Example 2.38, $C = B \otimes \mathbb{1}_{\mathcal{H}_2}$, and this proves (4.3.1).

4.4. The structure theorem

Definition 4.22. Let $\mathcal{H}$ be a Hilbert space, and let $\mathcal{M}$ be a von Neumann algebra on $\mathcal{H}$. The **center** of $\mathcal{M}$ is $\mathcal{M} \cap \mathcal{M}'$. $\mathcal{M}$ is a **factor** in case $\mathcal{M}$ has a trivial center; that is, the center is spanned by $\mathbb{1}$.

Example 4.23. If $\mathcal{H}$ is a finite-dimensional Hilbert space and $\{\mathbf{u}_1, \ldots, \mathbf{u}_n\}$ is an orthonormal basis for $\mathcal{H}$, define $E_{i,j} = |\mathbf{u}_i\rangle\langle\mathbf{u}_j|$ for $1 \leq i, j \leq n$. It is easy to check that if $A \in \mathcal{B}(\mathcal{H})$ commutes with each $E_{i,j}$, then $A = \lambda\mathbb{1}$ for some $\lambda \in \mathbb{C}$. Thus, $\mathcal{B}(\mathcal{H})$ is a factor.

For a second example, let $\mathcal{H} = \mathcal{H}_1 \otimes \mathcal{H}_2$ where $\mathcal{H}_1$ and $\mathcal{H}_2$ are finite-dimensional Hilbert spaces. Let $\mathcal{M}$ denote the subalgebra of $\mathcal{B}(\mathcal{H})$ consisting of operators of the form $A \otimes \mathbb{1}_{\mathcal{H}_2}$ where $A \in \mathcal{B}(\mathcal{H}_1)$. Anything in $\mathcal{M} \cap \mathcal{M}'$ has the form $B \otimes \mathbb{1}_{\mathcal{H}_2}$, $B \in \mathcal{B}(\mathcal{H}_1)$ and $BA = AB$ for all $A \in \mathcal{B}(\mathcal{H}_1)$. By the first paragraph, for some $\lambda \in \mathbb{C}$, $B = \lambda\mathbb{1}_{\mathcal{H}_1}$, so that $B \otimes \mathbb{1}_{\mathcal{H}_2} = \lambda\mathbb{1}_{\mathcal{H}}$. Hence $\mathcal{M}$ is a factor.

The same reasoning shows that the subalgebra of $\mathcal{B}(\mathcal{H})$, consisting of operators of the form $\mathbb{1}_{\mathcal{H}_1} \otimes A$ where $A \in \mathcal{B}(\mathcal{H}_2)$, is also a factor. We shall prove in this section that there are no other examples.

It is useful that factors are so simple, as claimed in Example 4.23, since every von Neumann algebra $\mathcal{M}$ on a finite-dimensional Hilbert space can be decomposed into a direct sum of factors. In the process of proving this, a remarkable symmetry between $\mathcal{M}$ and $\mathcal{M}'$ will emerge. The starting point is an application of the double commutant theorem.

Lemma 4.24. *Let $\mathcal{H}$ be a finite-dimensional Hilbert space, and let $\mathcal{M}$ be a von Neumann subalgebra of $\mathcal{B}(\mathcal{H})$. Then $\mathcal{M}$ and $\mathcal{M}'$ have the same center.*

Proof. The center of $\mathcal{M}$ is, by definition, $\mathcal{M} \cap \mathcal{M}'$. The center of $\mathcal{M}'$ is $\mathcal{M}' \cap \mathcal{M}''$. Since $\mathcal{M}'' = \mathcal{M}$ by Theorem 4.19, the two centers coincide. $\qquad\square$

The center $\mathcal{Z}$ of $\mathcal{M}$ is evidently an abelian von Neumann algebra, and therefore, by Theorem 4.11, $\mathcal{Z}$ is spanned by a finite collection $\{P_1, \ldots, P_n\}$ of mutually orthogonal projections.

For each $1 \le j \le n$, let $\mathcal{K}_j$ denote the range of P_j. Then

$$(4.4.1) \qquad\qquad \mathcal{H} = \bigoplus_{j=1}^{n} \mathcal{K}_j.$$

For all $A \in \mathcal{M}$, and all j, $AP_j = AP_j^2 = P_j AP_j$. Define

$$\mathcal{M}_j := \mathcal{M}P_j = \{P_j X P_j \; : \; X \in \mathcal{M}\},$$

viewed as a subalgebra of $\mathcal{B}(\mathcal{K}_j)$, and then since $\sum_{j=1}^{n} P_j = \mathbb{1}_{\mathcal{H}}$,

$$(4.4.2) \qquad\qquad \mathcal{M} = \bigoplus_{j=1}^{n} \mathcal{M}_j.$$

Since P_j is the identity on $\mathcal{K}_j$, $\mathcal{M}_j$ is unital, and hence is a von Neumann algebra. Let Q be a projection in the center of $\mathcal{M}_j$, which necessarily belongs to $\mathcal{Z}$. Then Q belongs to the span of $\{P_1, \ldots, P_n\}$, and $P_j Q P_j = Q$, and hence $Q = P_j$. Thus, the only projection in the center of $\mathcal{M}_j$ is P_j, the identity in $\mathcal{M}_j$, and therefore $\mathcal{M}_j$ is a factor. Thus, every von Neumann algebra on a finite-dimensional Hilbert space is a direct sum of factors.

This decomposition of $\mathcal{M}$ as a direct sum of factors was produced using only the center of $\mathcal{M}$, and since $\mathcal{M}$ and $\mathcal{M}'$ have the same center $\mathcal{Z}$, the same set $\{P_1, \ldots, P_n\}$ of mutually orthogonal projections also gives us a decomposition of $\mathcal{M}'$ as a direct sum of factors. Decompose $\mathcal{H}$ as a direct sum exactly as in (4.4.1). Then define $(\mathcal{M}')_j := P_j \mathcal{M}' P_j$. This is a von Neumann subalgebra of $\mathcal{B}(\mathcal{K}_j)$, and it is a factor for the same reason $\mathcal{M}_j$ is a factor. It is also easy to verify that as such it is the commutant of $\mathcal{M}_j$, and hence we simply write $\mathcal{M}'_j$ in place of $(\mathcal{M}')_j$. Therefore, we also have the decomposition

$$(4.4.3) \qquad \mathcal{M}' = \bigoplus_{j=1}^{n} \mathcal{M}'_j$$

as a direct sum of factors acting on the same direct summands $\mathcal{K}_j$ in the direct sum (4.4.1).

We now prepare to prove the claim made at the end of Example 4.23 concerning the simple structure of factors.

Definition 4.25 (Minimal projection). Let $\mathcal{M}$ be a von Neumann algebra. A projection $P \in \mathcal{M}$ such that $0 < P < \mathbb{1}$ is **minimal** in case whenever $Q \in \mathcal{M}$ is a projection with $Q \le P$, either $Q = P$ or $Q = 0$.

Every factor on a finite-dimensional Hilbert space contains a minimal projection—any projection whose range has nonzero minimal dimension. This need not be true for factors on infinite-dimensional Hilbert spaces. Looking at (4.4.2), one might think *summand* would be better terminology than *factor*, but the following theorem justifies the terminology:

Theorem 4.26. *Let $\mathcal{M}$ be a factor on a finite-dimensional Hilbert space $\mathcal{H}$. Then there exist finite-dimensional Hilbert spaces $\mathcal{H}_1$ and $\mathcal{H}_2$ and a unitary $U : \mathcal{H} \to \mathcal{H}_1 \otimes \mathcal{H}_2$ such that*

$$U\mathcal{M}U^* = \mathbb{1}_{\mathcal{H}_1} \otimes \mathcal{B}(\mathcal{H}_2).$$

Lemma 4.27. *Let $\mathcal{M}$ be a von Neumann algebra on a Hilbert space $\mathcal{H}$. Let $P, Q \in \mathcal{M}$ be nonzero orthogonal projections, and suppose that $\mathcal{M}$ is a factor. Then there exists a nonzero partial isometry $U \in \mathcal{M}$ such that*

$$(4.4.4) \qquad UU^* \le P \quad and \quad U^*U \le Q.$$

Proof. For each unitary $V \in \mathcal{M}$, V^*PV is an orthogonal projection in $\mathcal{M}$. Define

$$\widehat{P} := \bigvee_{V \text{ unitary in } \mathcal{M}} V^*PV.$$

Then $\widehat{P}$ commutes with every unitary in $\mathcal{M}$, and hence it belongs to $\mathcal{M}'$, as well as to $\mathcal{M}$ by Theorem 4.16. Hence it belongs to the center of $\mathcal{M}$, but $\mathcal{M}$ is a factor, so $\widehat{P} = \mathbb{1}$. Therefore, $\widehat{P}Q = Q \ne 0$. Hence there is some unitary $V \in \mathcal{M}$ such that $V^*PVQ \ne 0$, or equivalently $X := PVQ \ne 0$.

Let $X = U|X|$ be the polar factorization of X. By Theorem 4.6, $U \in \mathcal{M}$. The final space of U is $\mathrm{ran}(X) \subseteq \mathrm{ran}(P)$, and the initial space of U is $\ker(X)^\perp \subseteq \ker(Q)^\perp = \mathrm{ran}(Q)$. Therefore, (4.4.4) is satisfied. $\qquad\square$

Proof of Theorem 4.26. The proof will make use of the block matrix representation of operators on a tensor product of Hilbert spaces as explained in Section 2.6.1. We shall first derive a block matrix representation of the operators in the factor $\mathcal{M}$, and then use the results of Section 2.6.1 to write this in tensor product form.

For the block matrix representation, we seek a decomposition of $\mathcal{H}$ of the form

$$(4.4.5) \qquad \mathcal{H} = \bigoplus_{j=1}^{m} \mathcal{K}_j$$

for some m, where the Hilbert spaces $\mathcal{K}_j$ all have the same dimension, and moreover, a set $\{U_1, \dots, U_m\}$ of partial isometries such that for each j, the initial space of U_j is $\mathcal{K}_j$, and the final space is $\mathcal{K}_1$. Note that in this case,

$$(4.4.6) \qquad U_i U_j^* = \delta_{i,j} U_1 U_1^*.$$

To produce such a decomposition, let P_1 be a minimal projection in $\mathcal{M}$. By Lemma 4.27, there is a nonzero partial isometry $U \in \mathcal{M}$ such that $UU^* \leq P_1$ and $U^*U \leq P_1^\perp$. Since P_1 is minimal, $UU^* = P_1$. Define $P_2 := U^*U$.

Define $\mathcal{K}_j = P_j\mathcal{H}$ for $j = 1, 2$. Define $U_1 := P_1$ and $U_2 := U$, with U as above. Since UU^* and U^*U have the same rank, $\mathcal{K}_1$ and $\mathcal{K}_2$ have the same dimension. For $j = 1, 2$, U_j is an isometry from $\mathcal{K}_j$ onto $\mathcal{K}_1$, and if $\mathcal{H} = \mathcal{K}_1 \oplus \mathcal{K}_2$, we have the decomposition (4.4.5) of $\mathcal{H}$ that we seek.

If not, repeat the argument made above with $P_1^\perp$ replaced by $(P_1 + P_2)^\perp$, thus producing a projection P_3 in $\mathcal{M}$ with $P_3 P_j = 0$ for $j = 1, 2$, and U_3, a partial isometry in $\mathcal{M}$ that maps $\mathcal{K}_3$, the range of P_3, onto $\mathcal{K}_1$. Evidently, after some finite number $m - 1$ of such steps, $\mathcal{H}$ is exhausted, and we have produced a set $\{P_1, \dots, P_m\}$ of mutually orthogonal projections in $\mathcal{M}$ and a set $\{U_1, \dots, U_n\}$ of partial isometries in $\mathcal{M}$ where the initial space of U_j is $\mathcal{K}_j$, and the final space of U_j is $\mathcal{K}_1$. Equivalently, the initial space of U_j^* is $\mathcal{K}_1$, and the final space of U_j^* is $\mathcal{K}_j$.

We claim that for any $A \in \mathcal{M}$, there are numbers $\lambda_{i,j} \in \mathbb{C}$, $1 \leq i, j \leq m$, such that

$$(4.4.7) \qquad A = \sum_{i,j=1}^{m} \lambda_{i,j} U_i^* U_j.$$

To see this observe that $A = \sum_{i,j=1}^{m} P_i A P_j = \sum_{i,j=1}^{m} U_i^*(U_i A U_j^*)U_j.$

Since for each j, the initial space of U_j^* is $\mathcal{K}_1$, $U_j^* = U_j^* P_1$. Therefore, $U_i A U_j^* = P_1 U_i A U_j^* P_1 \in P_1 \mathcal{M} P_1$. The von Neumann algebra $P_1 \mathcal{M} P_1$ is spanned by the projections it contains, and since P_1 is minimal, $P_1 \mathcal{M} P_1 = \mathbb{C} P_1$. Hence for some $\lambda_{i,j} \in \mathbb{C}$, $U_i^* A U_j = \lambda_{i,j} P_1$, yielding (4.4.7).

Define a linear transformation $\mathcal{U}$ from $\bigoplus_m \mathcal{K}_1$ to $\mathcal{H}$ by

$$\mathcal{U}(\mathbf{v}_1, \ldots, \mathbf{v}_m) = \sum_{j=1}^m U_j^* \mathbf{v}_j.$$

It is evident that this map is unitary. Moreover, for any $A \in \mathcal{M}$, using (4.4.7) and then (4.4.6),

$$A\mathcal{U}(\mathbf{v}_1, \ldots, \mathbf{v}_m) = \sum_{i,j,\ell=1}^m \lambda_{i,j} U_i^* U_j U_\ell^* \mathbf{v}_\ell$$

$$= \sum_{i,j=1}^m \lambda_{i,j} U_i^* \mathbf{v}_j = \mathcal{U}\left(\sum_{j=1}^m \lambda_{1,j} \mathbf{v}_j, \ldots, \sum_{j=1}^m \lambda_{m,j} \mathbf{v}_j \right).$$

That is, $\mathcal{U}^* A \mathcal{U} \in \mathcal{B}(\bigoplus_m \mathcal{K}_1)$ is represented by the $m \times m$ block matrix whose (i,j)-th entry is $\lambda_{i,j} \mathbb{1} \in \mathcal{B}(\mathcal{K}_1)$. Define $\mathcal{H}_1 = \mathbb{C}^m$ with its standard inner product and orthonormal basis so that $\mathcal{B}(\mathcal{H}_1)$ is identified with $M_m(\mathbb{C})$. Define $\Lambda \in M_m(\mathbb{C})$ to be the $m \times m$ matrix whose (i,j)-th entry is $\lambda_{i,j}$. Define $\mathcal{H}_2 = \mathcal{K}_1$. By Theorem 2.36 and Example 2.39, $\mathcal{U}^* A \mathcal{U}$ is the block matrix representation of $\mathbb{1} \otimes \Lambda$. $\square$

We now have the following structure theorem:

Theorem 4.28. *Let $\mathcal{M}$ be a von Neumann algebra on a finite-dimensional Hilbert space $\mathcal{H}$, and let $\mathcal{Z}$ be the center of $\mathcal{M}$. Then there is a finite set of mutually orthogonal projections $\{P_1, \ldots, P_m\}$ whose span is $\mathcal{Z}$. Define $\mathcal{H}_j$ to be the range of P_j. Then for each j, there are Hilbert spaces $\mathcal{H}_j^{(\ell)}$ and $\mathcal{H}_j^{(r)}$ and a unitary $U_j : \mathcal{H}_j \to \mathcal{H}_j^{(\ell)} \otimes \mathcal{H}_j^{(r)}$ such that $\mathcal{M}$ consists of the elements of $\mathcal{B}(\mathcal{H})$ of the form*

$$A = \bigoplus_{j=1}^m U_j^* (\mathbb{1}_{\mathcal{H}_j^{(\ell)}} \otimes A_j) U_j \qquad \text{where each} \qquad A_j \in \mathcal{B}(\mathcal{H}_j^{(r)}).$$

The commutant $\mathcal{M}'$ of $\mathcal{M}$ is the von Neumann subalgebra of $\mathcal{B}(\mathcal{H})$ consisting of operators B of the form

$$B = \bigoplus_{j=1}^m U_j^* (B_j \otimes \mathbb{1}_{\mathcal{H}_j^{(r)}}) U_j \qquad \text{where each} \qquad B_j \in \mathcal{B}(\mathcal{H}_j^{(\ell)}).$$

Proof. We have seen that the common center $\mathcal{Z}$ of $\mathcal{M}$ and $\mathcal{M}'$ is the span of a set $\{P_1, \ldots, P_m\}$ of mutually orthogonal projections, and these induce a direct sum decomposition (4.4.1) of $\mathcal{H}$, and direct sum decompositions (4.4.2) and (4.4.3) of $\mathcal{M}$ and $\mathcal{M}'$ as a direct sum of factors. By Theorem 4.26, each factor $\mathcal{M}_j$ in the decomposition of $\mathcal{M}$ is unitarily equivalent to a factor of the form $\mathbb{1}_{\mathcal{H}_j^{(\ell)}} \otimes \mathcal{B}(\mathcal{H}_j^{(r)})$. Then by Example 4.21, the commutant of $\mathbb{1}_{\mathcal{H}_j^{(\ell)}} \otimes \mathcal{B}(\mathcal{H}_j^{(r)})$ is $\mathcal{B}(\mathcal{H}_j^{(\ell)}) \otimes \mathbb{1}_{\mathcal{H}_j^{(r)}}$. $\qquad\square$

4.5. Weyl-Heisenberg groups on finite-dimensional Hilbert spaces

Let $\mathcal{M}$ be a von Neumann algebra on a finite-dimensional Hilbert space $\mathcal{H}$. Theorem 4.5 says that $\mathcal{M}$ is the span of the unitaries belonging to $\mathcal{M}$. The main theorem of this section is a much stronger and more useful statement: $\mathcal{M}$ is the span of a *finite group* of unitaries in $\mathcal{M}$. Using the structure theorem, Theorem 4.28, this is easy to prove once we have proved it for the case in which $\mathcal{M}$ is $\mathcal{B}(\mathcal{H})$ itself.

Definition 4.29. Let $\mathcal{H}$ be a Hilbert space of dimension d. Let $\{\mathbf{u}_1, \ldots, \mathbf{u}_d\}$ be any orthonormal basis of $\mathcal{H}$. For $j = 1, \ldots, d$, define the orthogonal projection $P_j = |\mathbf{u}_j\rangle\langle\mathbf{u}_j|$. Define three unitary groups as follows.

First, for $0 \leq k \leq d - 1$, define

$$U_k := \sum_{j=1}^{d} e^{2\pi i jk/d} P_j.$$

Then $U_0 = \mathbb{1}$. For any nonnegative integer p, let $(p)_{\mathrm{mod}\ d}$ denote the remainder when p is divided by d. That is, if $p = nd + r$ with n a nonnegative integer and $0 \leq r < d$, then $(p)_{\mathrm{mod}\ d} = r$. Using this notation, for each j, and each $0 \leq k, \ell \leq d - 1$, one readily checks that

$$U_k U_\ell = U_{(k+\ell)_{\mathrm{mod}\ d}},$$

and from this it is clear that $\{U_0, \ldots, U_{d-1}\}$ is a group of unitary operators.

Second, let V be the unitary *shift* operator on $\mathcal{H}$ such that $V\mathbf{u}_j = \mathbf{u}_{j-1}$ for $2 \leq j \leq d$ and $V\mathbf{u}_1 = \mathbf{u}_d$. For $0 \leq k \leq d - 1$, define

$$V_k := V^k.$$

Simple computations show that $V_k V_\ell = V_{(k+\ell)_{\mathrm{mod}\ d}}$. Hence $\{V_0, \ldots, V_{d-1}\}$ is a group of unitary operators.

Third, for $0 \leq m \leq d - 1$, define

$$Z_m := e^{-2\pi i m/d} \mathbb{1}.$$

It is obvious that $\{Z_0, \ldots, Z_{d-1}\}$ is a group of unitary operators.

Lemma 4.30. *For all $X \in \mathcal{B}(\mathcal{H})$,*

$$\frac{1}{d}\sum_{k=0}^{d-1} U_k^* X U_k = \sum_{j=1}^{d} P_j X P_j.$$

Proof.

$$\frac{1}{d}\sum_{k=0}^{d-1} U_k^* X U_k = \frac{1}{d}\sum_{k=0}^{d-1}\sum_{j,\ell=1}^{d} e^{-2\pi i k j/d} e^{2\pi i k \ell/d} P_j X P_\ell$$

$$= \sum_{j,\ell=1}^{d} \frac{1}{d}\sum_{k=0}^{d-1} e^{2\pi i k(\ell-j)/d} P_j X P_\ell$$

$$= \sum_{j,\ell=1}^{d} \delta_{j,\ell} P_j X P_\ell = \sum_{j=1}^{d} P_j X P_j. \qquad \square$$

Lemma 4.31. *For $0 \le k, \ell \le d-1$,*

$$U_k V_\ell = V_\ell U_k Z_{(k\ell)_{\mathrm{mod}\ d}}.$$

Consequently,

$$\mathcal{G} := \{U_k V_\ell Z_m \ : \ 0 \le k, \ell, m \le d-1\}$$

is a finite group of unitaries in $\mathcal{B}(\mathcal{H})$. Moreover, if $X \in \mathcal{B}(\mathcal{H})$ commutes with each $G \in \mathcal{G}$, then $X = \frac{1}{d}\operatorname{Tr}[X]\mathbb{1}$. In particular, X is a multiple of $\mathbb{1}$.

Proof. It is easy to check that or each j, $V^* P_j V = P_{(j+1)_{\mathrm{mod}\ d}}$, and then simple computations show that for each $0 \le k \le d-1$, $V^* U_k V = U_k Z_k$, or what is the same, for each $0 \le k \le d-1$, $U_k V = V U_k Z_k$. Iterating, we have the commutation relations

$$(4.5.1) \qquad U_k V_\ell = V_\ell U_k Z_{(k\ell)_{\mathrm{mod}\ d}}.$$

These commutation relations, together with the fact that each Z_m is a multiple of the identity, show that for each $0 \le k_1, k_2, \ell_1, \ell_2, m_1, m_2 \le d-1$,

$$(U_{k_1} V_{\ell_1} Z_{m_1})(U_{k_2} V_{\ell_2} Z_{m_2}) \in \mathcal{G} \quad \text{and} \quad (U_{k_1} V_{\ell_1} Z_{m_1})^{-1} = Z_{m_1}^* V_{\ell_1}^* U_{k_1}^* \in \mathcal{G}.$$

Thus $\mathcal{G}$ is a group.

Now suppose $X \in \mathcal{B}(\mathcal{H})$ commutes with each element of $\mathcal{G}$. In particular, $U_k^* X U_k = X$ for each $k = 0, \dots, d-1$, and then by Lemma 4.30,

$$X = \frac{1}{d}\sum_{k=0}^{d-1} U_k^* X U_k = \sum_{j=1}^{d} P_j X P_j = \sum_{j=1}^{d} \langle \mathbf{u}_j, X\mathbf{u}_j\rangle |\mathbf{u}_j\rangle\langle \mathbf{u}_j|.$$

This shows that the matrix representative of X in the basis $\{\mathbf{u}_1, \dots, \mathbf{u}_d\}$ is diagonal.

Also, for each ℓ, $V_\ell^* X V_\ell = X V_\ell^* V_\ell = X$, and since $\mathbf{u}_{d-\ell} = V_\ell \mathbf{u}_d$,

$$\langle \mathbf{u}_{d-\ell}, X\mathbf{u}_{d-\ell}\rangle = \langle V_\ell \mathbf{u}_d, X V_\ell \mathbf{u}_d\rangle = \langle \mathbf{u}_d, X\mathbf{u}_d\rangle.$$

Hence all the diagonal entries of the matrix representative of X in the basis $\{\mathbf{u}_1, \ldots, \mathbf{u}_d\}$ are the same. Therefore, X is a multiple of the identity; that is, $X = z\mathbb{1}$ for some z. But then $\mathrm{Tr}[X] = dz$, and hence $z = d^{-1}\mathrm{Tr}[X]$. $\qquad\square$

Remark 4.32. The commutation relations (4.5.1) are a discrete analogue of the Heisenberg canonical commutation relations $[Q, P] = i\hbar\mathbb{1}$ written in Weyl's unitary form,

$$e^{itQ}e^{isP} = e^{isP}e^{itQ}e^{-ist/\hbar}.$$

This discrete version was discussed by Weyl [**228**, Chapter 4, Section 14].

Definition 4.33. Let $\mathcal{H}$ be a Hilbert space of dimension d, and let $\{\mathbf{u}_1, \ldots, \mathbf{u}_d\}$ be any orthonormal basis in $\mathcal{H}$. Let $\mathcal{G}$ be the group constructed in Lemma 4.31. Then $\mathcal{G}$ is the **Weyl-Heisenberg group** corresponding to $\{\mathbf{u}_1, \ldots, \mathbf{u}_d\}$.

Lemma 4.34. *Let $\mathcal{H}$ be a finite-dimensional Hilbert space, and let $\mathcal{G}$ be any finite group of unitaries in $\mathcal{B}(\mathcal{H})$. Then the span of $\mathcal{G}$ is a von Neumann subalgebra $\mathcal{M}$ of $\mathcal{B}(\mathcal{H})$. Furthermore, when the commutant $\mathcal{G}'$ consists of multiples of the identity, $\mathcal{M} = \mathcal{B}(\mathcal{H})$. In particular, if $\mathcal{G}$ is the Weyl-Heisenberg group for an orthonormal basis $\{\mathbf{u}_1, \ldots, \mathbf{u}_d\}$ of $\mathcal{H}$, then $\mathrm{span}(\mathcal{G}) = \mathcal{B}(\mathcal{H})$.*

Proof. Any multiplicative group of unitaries on some Hilbert space $\mathcal{H}$ is closed under the $*$ operator, since for unitaries the inverse is the adjoint, and of course it is closed under multiplication. Hence the span $\mathcal{M}$ of this group is a $*$-subalgebra of $\mathcal{B}(\mathcal{H})$ containing the identity. When $\mathcal{H}$ is finite dimensional, the algebra is automatically closed, and hence $\mathcal{M}$ is a von Neumann subalgebra of $\mathcal{B}(\mathcal{H})$. Since $\mathcal{M}$ is the span of $\mathcal{G}$, $\mathcal{M}' = \mathcal{G}'$, and hence $\mathcal{M}'$ consists of multiples of the identity when $\mathcal{G}'$ does. In this case, $\mathcal{M}'' = \mathcal{B}(\mathcal{H})$, and then by the von Neumann double commutant theorem, $\mathcal{M} = \mathcal{B}(\mathcal{H})$. Lemma 4.31 says that when $\mathcal{G}$ is a Weyl-Heisenberg group, $\mathcal{G}'$ consists of multiples of the identity, and hence the final statement is proved. $\qquad\square$

The final statement of Lemma 4.34 is due to Schwinger [**194**]. His quite different proof does not use the double commutant theorem.

Theorem 4.35. *Let $\mathcal{M}$ be a von Neumann algebra on a finite-dimensional Hilbert space $\mathcal{H}$. Then there is a finite group $\mathcal{G}_{\mathcal{M}}$ of unitaries in $\mathcal{M}$ whose linear span is all of $\mathcal{M}$.*

Proof. Let $\mathcal{M}$ be a factor of the form $\mathcal{B}(\mathcal{H}^{(\ell)}) \otimes \mathbb{1}$ in $\mathcal{B}(\mathcal{H}^{(\ell)} \otimes \mathcal{H}^{(r)})$. Choose any orthonormal basis of $\mathcal{H}^{(\ell)}$, and let $\mathcal{G}$ be the corresponding Weyl-Heisenberg group in $\mathcal{B}(\mathcal{H}^{(\ell)})$. By Lemma 4.34, the linear span of $\mathcal{G}$ in $\mathcal{B}(\mathcal{H}^{(\ell)})$ is all of $\mathcal{B}(\mathcal{H}^{(\ell)})$. Hence the linear span of

$$\{U \otimes \mathbb{1} \ : \ U \in \mathcal{G}\}$$

is all of $\mathcal{M}$.

This proves the theorem when $\mathcal{M}$ is a factor. By Theorem 4.28, the general von Neumann algebra on $\mathcal{H}$ is a finite direct sum of factors, and one need only put the pieces together in the obvious way. $\qquad\square$

4.6. Conditional expectations in von Neumann algebras

Definition 4.36 (Conditional expectation). Let $\mathcal{M}$ be a von Neumann algebra on a finite-dimensional Hilbert space, and let $\mathcal{N}$ be a von Neumann subalgebra of $\mathcal{M}$. Equipping $\mathcal{M}$ with the Hilbert-Schmidt inner product, it is a Hilbert space, and we define the **conditional expectation** $\mathcal{E}_{\mathcal{N}}$ of $\mathcal{M}$ onto $\mathcal{N}$ to be the orthogonal projection of $\mathcal{M}$ onto $\mathcal{N}$. In other words, for $A \in \mathcal{M}$, $\mathcal{E}_{\mathcal{N}}(A)$ is the unique element of $\mathcal{N}$ such that

$$(4.6.1) \qquad \mathrm{Tr}[B\mathcal{E}_{\mathcal{N}}(A)] = \mathrm{Tr}[BA]$$

for all $B \in \mathcal{N}$.

Example 4.37. Let $\mathcal{M} = M_n(\mathbb{C})$ and let $\mathcal{N}$ be the subalgebra of $M_n(\mathbb{C})$ consisting of diagonal matrices. For $1 \le j \le n$, let $D_j \in \mathcal{N}$ have 1 in the jth diagonal entry, and 0 in every other entry. Evidently, $\{D_1, \dots, D_n\}$ is an orthonormal basis for $\mathcal{N}$, and hence for any $A \in \mathcal{M}$,

$$\mathcal{E}_{\mathcal{N}}(A) = \sum_{j=1}^{n} \langle D_j, A \rangle D_j = \mathrm{diag}(A),$$

where $\mathrm{diag}(A)$ denotes the diagonal part of A; that is

$$[\mathrm{diag}(A)]_{i,j} = \delta_{i,j} A_{i,j}.$$

In some quantum mechanical applications, the commutative subalgebra $\mathcal{N}$ can be viewed as the *classical part* of $\mathcal{M} = M_n(\mathbb{C})$, and then $\mathcal{E}_{\mathcal{N}}$ provides a useful means for comparing quantum with classical, as we shall see.

Example 4.38 (Partial trace and conditional expectation). Let $\mathcal{H}_1$ and $\mathcal{H}_2$ be two finite-dimensional Hilbert spaces. Let $\mathcal{M} = \mathcal{B}(\mathcal{H}_1 \otimes \mathcal{H}_2)$, and let $\mathcal{N}$ be the subalgebra consisting of operators of the form $X \otimes \mathbb{1}$, $X \in \mathcal{B}(\mathcal{H}_1)$. The conditional expectation $\mathcal{E}_{\mathcal{N}}$ is related to the partial trace by the following very useful formula:

$$(4.6.2) \qquad \mathcal{E}_{\mathcal{N}}(A) = \frac{1}{d_2} \mathrm{Tr}_2[A] \otimes \mathbb{1} = \mathrm{Tr}_2[A] \otimes \tau,$$

where $d_2 := \dim(\mathcal{H}_2)$ and $\tau := d_2^{-1}\mathbb{1}$ is the uniform density matrix on $\mathcal{H}_2$.

To see this, let $\{X_1, \dots, X_N\}$ be any orthonormal basis of $\mathcal{B}(\mathcal{H}_1)$. Then

$$\{d_2^{-1/2} X_1 \otimes \mathbb{1}, \dots, d_2^{-1/2} X_N \otimes \mathbb{1}\}$$

is an orthonormal basis for $\mathcal{N}$. Consequently, for any $A \in \mathcal{M} = \mathcal{B}(\mathcal{H}_1 \otimes \mathcal{H}_2)$,

$$(4.6.3) \qquad \mathcal{E}_{\mathcal{N}}(A) = \frac{1}{d_2} \sum_{j=1}^{N} \langle X_j \otimes \mathbb{1}, A \rangle X_j \otimes \mathbb{1}.$$

However, by definition,

$$\begin{aligned}
\langle X_j \otimes \mathbb{1}, A \rangle &= \mathrm{Tr}_{\mathcal{H}_1 \otimes \mathcal{H}_2}[(X_j^* \otimes \mathbb{1})A] \\
&= \mathrm{Tr}_{\mathcal{H}_1}[X_j^* \, \mathrm{Tr}_2[A]] = \langle X_j, \mathrm{Tr}_2[A] \rangle.
\end{aligned}$$

Using this in (4.6.3) yields

$$\mathcal{E}_{\mathcal{N}}(A) = \frac{1}{d_2} \left(\sum_{j=1}^{N} \langle X_j, \mathrm{Tr}_2[A] \rangle X_j \right) \otimes \mathbb{1} = \frac{1}{d_2} \mathrm{Tr}_2[A] \otimes \mathbb{1}.$$

Theorem 4.39 (Properties of the conditional expectation). *Let $\mathcal{M}$ be a von Neumann algebra on a finite-dimensional Hilbert space, and let $\mathcal{N}$ be a von Neumann subalgebra of $\mathcal{M}$. The conditional expectation $\mathcal{E}_{\mathcal{N}}$ of $\mathcal{M}$ onto $\mathcal{N}$ has the following properties:*

(1) *For all $A \in \mathcal{M}$ and all $X, Y \in \mathcal{N}$, $\mathcal{E}_{\mathcal{N}}(XAY) = X \mathcal{E}_{\mathcal{N}}(A) Y$.*

(2) *For all $A \in \mathcal{M}$, $\mathrm{Tr}[\mathcal{E}_{\mathcal{N}}(A)] = \mathrm{Tr}[A]$.*

Furthermore, $\mathcal{E}_{\mathcal{N}}$ is the unique linear operator from $\mathcal{M}$ to $\mathcal{N}$ with these two properties.

Proof. By definition, for all $A \in \mathcal{M}$, $A = \mathcal{E}_{\mathcal{N}}(A) + (A - \mathcal{E}_{\mathcal{N}}(A))$ is the unique decomposition of A into a sum of elements in $\mathcal{N}$ and $\mathcal{N}^{\perp}$. Then for $X, Y \in \mathcal{N}$,

$$XAY = X \mathcal{E}_{\mathcal{N}}(A) Y + X(A - \mathcal{E}_{\mathcal{N}}(A)) Y,$$

and since $X \mathcal{E}_{\mathcal{N}}(A) Y \in \mathcal{N}$, property (1) will be proved when we have shown that $X(A - \mathcal{E}_{\mathcal{N}}(A)) Y \in \mathcal{N}^{\perp}$: Let $Z \in \mathcal{N}$. Then

$$\begin{aligned}
\langle Z, X(A - \mathcal{E}_{\mathcal{N}}(A)) Y \rangle &= \mathrm{Tr}[Z^* X(A - \mathcal{E}_{\mathcal{N}}(A)) Y] \\
&= \mathrm{Tr}[(X^* Z Y^*)^* (A - \mathcal{E}_{\mathcal{N}}(A))] \\
&= \langle X^* Z Y^*, A - \mathcal{E}_{\mathcal{N}}(A) \rangle = 0,
\end{aligned}$$

since $X^* Z Y^* \in \mathcal{N}$ and $A - \mathcal{E}_{\mathcal{N}}(A) \in \mathcal{N}^{\perp}$. Next, property (2) follows from (4.6.1) since $\mathbb{1} \in \mathcal{N}$.

To prove the final statement, let $\Phi : \mathcal{M} \to \mathcal{N}$ be linear, and suppose that Φ has the properties (1) and (2) of $\mathcal{E}_{\mathcal{N}}$. Let $A \in \mathcal{M}$.

$$(4.6.4) \qquad A = \Phi(A) + (A - \Phi(A)).$$

Then for all $Z \in \mathcal{N}$, by (1) and then (2),

$$\mathrm{Tr}[Z(A - \Phi(A))] = \mathrm{Tr}[ZA] - \mathrm{Tr}[\Phi(ZA)] = 0.$$

Thus, $A - \Phi(A) \in \mathcal{N}^{\perp}$, and then (4.6.4) is the orthogonal decomposition of A into its components in $\mathcal{N}$ and $\mathcal{N}^{\perp}$. Therefore, Φ is the orthogonal projection onto $\mathcal{N}$, and hence $\Phi = \mathcal{E}_{\mathcal{N}}$. $\square$

The following theorem represents conditional expectations as averages over unitary conjugations, and it is the basis of many convexity inequalities.

Theorem 4.40. *Let $\mathcal{N}$ be any von Neumann subalagebra of $\mathcal{B}(\mathcal{H})$, let $\mathcal{H}$ be a finite-dimensional Hilbert space. Then there is a finite group $\mathcal{G}$ of unitaries in $\mathcal{B}(\mathcal{H})$ such that for all $A \in \mathcal{B}(\mathcal{H})$,*

$$\mathcal{E}_{\mathcal{N}}(A) := \frac{1}{|\mathcal{G}|} \sum_{G \in \mathcal{G}} G^{*}AG,$$

where $|\mathcal{G}|$ denotes the cardinality of $\mathcal{G}$.

Proof. By Theorem 4.35 there is a finite group $\mathcal{G}$ of unitaries in $\mathcal{N}'$ such that $\mathrm{span}(\mathcal{G}) = \mathcal{N}'$. Define $\Phi : \mathcal{B}(\mathcal{H}) \to \mathcal{B}(\mathcal{H})$ by

$$\Phi(A) := \frac{1}{|\mathcal{G}|} \sum_{G \in \mathcal{G}} G^{*}AG,$$

for all $A \in \mathcal{B}(\mathcal{H})$. We claim that the range of Φ is contained in $\mathcal{N}$. To see this, observe that by the double commutant theorem,

$$\mathcal{N} = \mathcal{N}'' = \mathrm{span}(\mathcal{G})' = \mathcal{G}'.$$

Since $\mathcal{G}$ is a group, for any $G_0 \in \mathcal{G}$ the map $G \mapsto GG_0^{*}$ simply permutes the elements of $\mathcal{G}$, and hence

$$G_0 \left(\sum_{G \in \mathcal{G}} G^{*}AG \right) = \sum_{G \in \mathcal{G}} (GG_0^{*})^{*}A(GG_0^{*})G_0 = \left(\sum_{G \in \mathcal{G}} G^{*}AG \right) G_0.$$

Therefore, $\sum_{G \in \mathcal{G}} G^{*}AG \in \mathcal{G}' = \mathcal{N}$, which proves the claim.

Next, for all $X, Y \in \mathcal{N}$, $\Phi(XAY) = X\Phi(A)Y$ since each $G \in \mathcal{N}'$, and by cyclicity of the trace, and the unitarity of each G,

$$\mathrm{Tr}[\Phi(A)] = \frac{1}{|\mathcal{G}|} \sum_{G \in \mathcal{G}} \mathrm{Tr}[G^{*}AG] - \frac{1}{|\mathcal{G}|} \sum_{G \in \mathcal{G}} \mathrm{Tr}[GG^{*}A] - \mathrm{Tr}[A].$$

Therefore, Φ has the two properties that by Theorem 4.39 characterize $\mathcal{E}_{\mathcal{N}}$, and hence $\Phi = \mathcal{E}_{\mathcal{N}}$. $\square$

4.7. Quantum state measurement and Naimark's theorem

Let $\mathcal{H}$ be a finite-dimensional Hilbert space, and let $\rho \in \mathfrak{S}_{\mathcal{H}}$ represent the state of a quantum system. What kind of measurement should we make to gather information on ρ?

For example, a basic problem of quantum communication theory can be formulated as follows: Bob has a "transmitter" that emits quantum states chosen from an "alphabet" $\{\rho_1,\ldots,\rho_n\}$. He chooses one and transmits it to Alice. Each time she receives a state, she makes an observation of some fixed observable A, and based on the outcome, she makes a guess as to which state was sent. What A should Alice choose to eliminate as much guesswork as possible?

Let $A = \sum_{\lambda \in \sigma(A)} \lambda P_\lambda$ be the spectral decomposition of A. For each $j = 1,\ldots,n$ define a classical probability density p_{ρ_j} on $\sigma(A)$ by

$$p_{\rho_j}(\lambda) := \mathrm{Tr}[\rho_j P_\lambda].$$

For quantum communication purposes, it would be highly desirable to choose, if possible, an observable A such that for each $\lambda \in \sigma(A)$, $p_{\rho_j}(\lambda) \approx 1$ for some unique value of j, and $p_{\rho_k}(\lambda) \approx 0$ for all others. Then when λ is observed, one could be reasonably certain that ρ_j was transmitted.

Example 4.41. Let ψ and ϕ be two unit vectors in a two-dimensional Hilbert space $\mathcal{H}$. Let $\rho = |\psi\rangle\langle\psi|$ and $\sigma = |\phi\rangle\langle\phi|$ be the two states we shall try to distinguish, knowing that the state we have received is one of these two. In a two-dimensional Hilbert space, every self-adjoint operator A has the form $a_1|\mathbf{u}_1\rangle\langle\mathbf{u}_1| + a_2|\mathbf{u}_2\rangle\langle\mathbf{u}_2|$ for some orthonormal basis $\{\mathbf{u}_1,\mathbf{u}_2\}$. Since $|\mathbf{u}_1\rangle\langle\mathbf{u}_1| + |\mathbf{u}_2\rangle\langle\mathbf{u}_2| = \mathbb{1}$, all information contained in a measurement of A is contained in a measurement of $P = |\mathbf{u}_1\rangle\langle\mathbf{u}_1|$.

After a measurement of P the result is either 1 or 0, and the probability of finding 1 is $\mathrm{Tr}[\rho P] = |\langle\mathbf{u}_1,\psi\rangle|^2$, if the state is ρ and $\mathrm{Tr}[\sigma P] = |\langle\mathbf{u}_1,\phi\rangle|^2$ if the state is σ. If ψ and ϕ are orthogonal, we can choose $\mathbf{u}_1$ so that these two probabilities are 0 and 1 (or 1 and 0), respectively, and then the measurement tells us with certainty which state was sent.

However, suppose they are not orthogonal. We can change ψ by a phase without affecting ρ, so there is no loss of generality in assuming that

$$\langle\psi,\phi\rangle = |\langle\psi,\phi\rangle| =: a, \quad 0 < a < 1,$$

and we do so. If we choose $\mathbf{u}_1$ to be orthogonal to ψ, then

$$\mathrm{Tr}[\rho P] = 0 \quad \text{and} \quad \mathrm{Tr}[\sigma P] = 1 - a^2.$$

Therefore, if the measured value is 1, the state cannot have been ρ and, hence, must have been σ. But

$$\mathrm{Tr}[\rho(\mathbb{1} - P)] = 1 \quad \text{and} \quad \mathrm{Tr}[\sigma(\mathbb{1} - P)] = a^2,$$

and hence if the measured value is 0, the result is ambiguous—both ρ and σ could have been sent, though if a is small, σ is unlikely to have been sent.

4.7.1. Measurements with ancilla. We now introduce a more elaborate type of measurement—measurements with **ancilla**. In particular, this type of measurement will enable us to distinguish between two nonorthogonal pure states ρ and σ with certainty, though not always. We seek an observable with three possible values corresponding to three outcomes: (1) the state is ρ, (2) the state is σ, and (3) inconclusive.

No self-adjoint operator on a two-dimensional Hilbert space has a spectrum consisting of three distinct eigenvalues, so we must go to a larger Hilbert space; we must bring in **ancillary degrees of freedom**. We consider a second Hilbert space $\mathcal{K}$ of some finite dimension d, and we fix a pure state $\omega = |\eta\rangle\langle\eta|$ on $\mathcal{K}$. We then perform a measurement using the state that we were sent, which is either $\rho \otimes \omega$ or $\sigma \otimes \omega$. The Hilbert space $\mathcal{H} \otimes \mathcal{K}$ is $2d$ dimensional, so even if $d = 2$, there will be self-adjoint operators with four distinct eigenvalues, which is more than the three we seek, though it will be convenient to choose a slightly larger ancillary space $\mathcal{K}$.

We seek a set $\{P_1, P_2, P_3\}$ of mutually orthogonal projections on $\mathcal{H} \otimes \mathcal{K}$ such that $P_1 + P_2 + P_3 = \mathbb{1}$ and such that

(4.7.1) $$\operatorname{Tr}[(\rho \otimes \omega)P_1] = 0 \quad \text{and} \quad \operatorname{Tr}[(\sigma \otimes \omega)P_2] = 0.$$

Now measure $H = P_1 + 2P_2 + 3P_3$. If the result is the eigenvalue 1, then with certainty σ was sent, and if the result is the eigenvalue 2, then with certainty ρ was sent, while if the result is the eigenvalue 3, the measurement was inconclusive.

If we can find such $\{P_1, P_2, P_3\}$, we have a better detection system than the once discussed in Example 4.41 in that now there is some probability of detecting *both* states ρ and σ unambiguously, and not only one or the other.

To see that such a set $\{P_1, P_2, P_3\}$ does exist, let us assume this for now and see what properties these projections must have as a result of (4.7.1). That done, the construction will be easy. Define

$$E_j = \operatorname{Tr}_{\mathcal{K}}[(\mathbb{1} \otimes \omega)P_j] = \operatorname{Tr}_{\mathcal{K}}[(\mathbb{1} \otimes \omega)P_j(\mathbb{1} \otimes \omega)], \quad j = 1, 2, 3.$$

To see the second equality, which shows that $E_j \geq 0$, use $\omega^2 = \omega$ and hence $(\mathbb{1} \otimes \omega)^2 = (\mathbb{1} \otimes \omega)$, together with the fact that for any $A \in \mathcal{B}(\mathcal{H})$, $(A \otimes \mathbb{1})(\mathbb{1} \otimes \omega) = (\mathbb{1} \otimes \omega)(A \otimes \mathbb{1})$. Then by the definition of the partial trace, and cyclicity of the full trace,

$$\begin{aligned}
\operatorname{Tr}[AE_j] &= \operatorname{Tr}[(A \otimes \mathbb{1})(\mathbb{1} \otimes \omega)P_j] = \operatorname{Tr}[(A \otimes \mathbb{1})(\mathbb{1} \otimes \omega)^2 P_j] \\
&= \operatorname{Tr}[(A \otimes \mathbb{1})(\mathbb{1} \otimes \omega)P_j(\mathbb{1} \otimes \omega)] \\
&= \operatorname{Tr}[A \operatorname{Tr}_{\mathcal{K}}[(\mathbb{1} \otimes \omega)P_j(\mathbb{1} \otimes \omega)]].
\end{aligned}$$

Therefore, $E_j \geq 0$ for each $j = 1, 2, 3$. From either fomula for E_j, the fact that $P_1 + P_2 + P_3 = \mathbb{1}$ yields $E_1 + E_2 + E_3 = \text{Tr}_{\mathcal{K}}[\mathbb{1} \otimes \omega] = \mathbb{1}$. Moreover, for any $A \in \mathcal{B}(\mathcal{H})$, $\text{Tr}[AE_j] = \text{Tr}[A \otimes \omega P_j]$. In particular, taking $A = \rho$ and $A = \sigma$,

$$\text{Tr}[\rho E_1] = \text{Tr}[\rho \otimes \omega P_1] = 0 \quad \text{and} \quad \text{Tr}[\sigma E_2] = \text{Tr}[\sigma \otimes \omega P_2] = 0.$$

Since $\text{Tr}[\rho E_1] = 0$, $E_1 \psi = 0$, and hence $E_1 = c_1 |\mathbf{f}_1\rangle\langle\mathbf{f}_1|$ for some $c_1 \geq 0$ and some unit vector $\mathbf{f}_1 \in \mathcal{H}$ that is orthogonal to ψ. Likewise, E_2 must have the form $E_2 = c_2 |\mathbf{f}_2\rangle\langle\mathbf{f}_2|$ for some $c_2 \geq 0$ and some unit vector $\mathbf{f}_2 \in \mathcal{H}$ that is orthogonal to ϕ. This determines $\mathbf{f}_1$ and $\mathbf{f}_2$ up to phases, and we choose $\mathbf{f}_1 = (\overline{\psi_2}, -\overline{\psi_1})$ where $\psi = (\psi_1, \psi_2)$ and choose $\mathbf{f}_2 = (\overline{\phi_2}, -\overline{\phi_1})$ where $\phi = (\phi_1, \phi_2)$. With this choice of phases, $\langle \mathbf{f}_1, \mathbf{f}_2 \rangle = \langle \psi, \phi \rangle \geq 0$, and moreover, $\mathbf{f}_1 + \mathbf{f}_2$ and $\mathbf{f}_1 - \mathbf{f}_2$ are orthogonal.

We want to choose c_1 and c_2 as large as possible, and the constraint is that $E_1 + E_2 + E_3 = \mathbb{1}$. Let us take $c_1 = c_2 = c$ to make our measurement symmetric in ρ and σ. Define $F := E_1 + E_2 = c(|\mathbf{f}_1\rangle\langle\mathbf{f}_1| + |\mathbf{f}_2\rangle\langle\mathbf{f}_2|)$. Then

$$F(\mathbf{f}_1 + \mathbf{f}_2) = c(1 + \langle \psi, \phi \rangle)(\mathbf{f}_1 + \mathbf{f}_2) \quad \text{and} \quad F(\mathbf{f}_1 - \mathbf{f}_2) = c(1 - \langle \psi, \phi \rangle)(\mathbf{f}_1 - \mathbf{f}_2).$$

Hence, the eigenvalues of F are $c(1 \pm \langle \psi, \phi \rangle)$. Define

$$\mathbf{v}_1 := \|\mathbf{f}_1 + \mathbf{f}_2\|^{-1}(\mathbf{f}_1 + \mathbf{f}_2) \quad \text{and} \quad \mathbf{v}_2 := \|\mathbf{f}_1 - \mathbf{f}_2\|^{-1}(\mathbf{f}_1 - \mathbf{f}_2),$$

so that $\{\mathbf{v}_1, \mathbf{v}_2\}$ is an orthognormal basis consisting of eigenvectors of F. Then $E_1 + E_2 = F$ has the spectral decomposition

$$E_1 + E_2 = c(1 + \langle \psi, \phi \rangle)|\mathbf{v}_1\rangle\langle\mathbf{v}_1| + c(1 - \langle \psi, \phi \rangle)|\mathbf{v}_2\rangle\langle\mathbf{v}_2|.$$

Since $E_1 + E_2 \leq \mathbb{1}$, the largest value we can choose for c is $(1 + \langle \psi, \phi \rangle)^{-1}$. Then

$$E_1 + E_2 = |\mathbf{v}_1\rangle\langle\mathbf{v}_1| + \frac{1 - \langle \psi, \phi \rangle}{1 + \langle \psi, \phi \rangle}|\mathbf{v}_2\rangle\langle\mathbf{v}_2|.$$

Finally,

$$(4.7.2) \qquad E_3 = \mathbb{1} - (E_1 + E_2) = \frac{2\langle \psi, \phi \rangle}{1 + \langle \psi, \phi \rangle}|\mathbf{v}_2\rangle\langle\mathbf{v}_2|$$

and

$$(4.7.3) \qquad E_1 = (1 + \langle \psi, \phi \rangle)^{-1}|\mathbf{f}_1\rangle\langle\mathbf{f}_1| \quad \text{and} \quad E_2 = (1 + \langle \psi, \phi \rangle)^{-1}|\mathbf{f}_2\rangle\langle\mathbf{f}_2|.$$

In this case, all three of E_1, E_2, and E_3 are rank one operators.

We can now construct the desired projections P_1, P_2, and P_3 in $\mathcal{H} \otimes \mathcal{K}$. Choose orthonormal bases $\{\mathbf{u}_1, \mathbf{u}_2\}$ and $\{\mathbf{e}_1, \mathbf{e}_2, \mathbf{e}_3\}$ of $\mathcal{H}$ and $\mathcal{K}$ where we have taken $\mathcal{K}$ to be three dimensional. Define an isometry V from $\mathcal{H}$ into $\mathcal{H} \otimes \mathcal{K}$ by $V\mathbf{x} = \sum_{j=1}^{3} E_j^{1/2}\mathbf{x} \otimes \mathbf{e}_j$. To see that this is an isometry, compute

$$(4.7.4) \qquad \langle V\mathbf{x}, V\mathbf{x} \rangle = \sum_{j=1}^{3}\langle E_j^{1/2}\mathbf{x}, E_j^{1/2}\mathbf{x} \rangle = \sum_{j=1}^{3}\langle \mathbf{x}, E_j\mathbf{x} \rangle = \langle \mathbf{x}, \mathbf{x} \rangle.$$

Define $\widehat{V} : \mathcal{H} \to \mathcal{H} \otimes \mathcal{K}$ by $\widehat{V}\mathbf{x} = \mathbf{x} \otimes \mathbf{e}_1$, which is evidently an isometry. Since V and $\widehat{V}$ have the same initial space, there is a (nonunique) unitary $\mathcal{U} \in \mathcal{B}(\mathcal{H} \otimes \mathcal{K})$ such that $V = \mathcal{U}\widehat{V}$. Finally, define a set of mutually orthogonal projections $\{P_1, P_2, P_3\}$ on $\mathcal{H} \otimes \mathcal{K}$ by

$$(4.7.5) \qquad P_j = \mathcal{U}^*(\mathbb{1} \otimes |\mathbf{e}_j\rangle\langle\mathbf{e}_j|)\mathcal{U}, \quad j = 1, 2, 3.$$

Note that $P_1 + P_2 + P_3 = \mathbb{1}$. Finally, we will take the ancillary state ω to be $|\mathbf{e}_1\rangle\langle\mathbf{e}_1|$. Then for any $\mathbf{x} \in \mathcal{H}$,

$$\begin{aligned}
\mathrm{Tr}[(|\mathbf{x}\rangle\langle\mathbf{x}| \otimes \omega)P_j] &= \langle \mathbf{x} \otimes \mathbf{e}_1, P_j \mathbf{x} \otimes \mathbf{e}_1 \rangle \\
&= \langle \mathcal{U}\mathbf{x} \otimes \mathbf{e}_1, \mathbb{1} \otimes |\mathbf{e}_j\rangle\langle\mathbf{e}_j|\mathcal{U}\mathbf{x} \otimes \mathbf{e}_1 \rangle \\
&= \sum_{k,\ell=1}^{3} \langle E_k^{1/2}\mathbf{x} \otimes \mathbf{e}_k, (\mathbb{1} \otimes |\mathbf{e}_j\rangle\langle\mathbf{e}_j|)E_\ell^{1/2}\mathbf{x} \otimes \mathbf{e}_\ell \rangle \\
&= \langle E_j^{1/2}\mathbf{x}, E_j^{1/2}\mathbf{x} \rangle = \mathrm{Tr}[|\mathbf{x}\rangle\langle\mathbf{x}|E_j].
\end{aligned}$$

By linearity, for all $\hat{\rho} \in \mathfrak{S}_{\mathcal{H}}$, and all $j = 1, 2, 3$, $\mathrm{Tr}[(\hat{\rho} \otimes \omega)P_j] = \mathrm{Tr}[\hat{\rho}E_j]$.

Example 4.42 (Unambiguous state discrimination). We return to the problem examined in Example 4.41, but now use the measurement with the ancillary state ω on the ancillary space $\mathcal{K}$ described just above. Let $H = P_1 + 2P_2 + 3P_3$ where projections P_j are given by (4.7.5). Taking the ancillary state $\omega = |\mathbf{e}_1\rangle\langle\mathbf{e}_1|$, measure the observable H in the state $\hat{\rho} \otimes \omega$, where $\hat{\rho}$ is the unknown state that we have received. The probabilities of observing the values 1, 2, or 3, respectively, are then

$$\mathrm{Tr}[\hat{\rho}E_1], \quad \mathrm{Tr}[\hat{\rho}E_2], \quad \text{and} \quad \mathrm{Tr}[\hat{\rho}E_3],$$

where the operators E_j are given by (4.7.2) and (4.7.3).

Observe that $|\langle\mathbf{f}_1, \phi\rangle| = \langle\mathbf{f}_2, \psi\rangle = \sqrt{1 - \langle\psi, \phi\rangle^2}$. Therefore,

$$\mathrm{Tr}[\rho E_1] = 0, \quad \mathrm{Tr}[\rho E_2] = 1 - \langle\psi, \phi\rangle, \quad \text{and} \quad \mathrm{Tr}[\rho E_3] = \langle\psi, \phi\rangle,$$

while

$$\mathrm{Tr}[\sigma E_1] = 1 - \langle\psi, \phi\rangle, \quad \mathrm{Tr}[\sigma E_2] = 0, \quad \text{and} \quad \mathrm{Tr}[\sigma E_3] = \langle\psi, \phi\rangle.$$

No matter which state is sent, the probability of an ambiguous result is $|\langle\psi, \phi\rangle|$, and otherwise the state is unambiguously determined. This procedure was developed by Ivanovic [**119**], Dieks [**74**], and Peres [**175**]. They also proved that this procedure is optimal; i.e., that no other procedure gives a lower probability of an ambiguous result independent of the transmitted state. Without bringing ancilla into the procedure, one cannot obtain a nonzero probability of unambiguous detection for both ρ and σ.

Hopefully these examples convey something of the utility of making measurements with ancillary variables and states. We now develop the mathematics of such measurements.

Definition 4.43 (Positive operator valued measure). Let $\mathcal{H}$ be a finite-dimensional Hilbert space. Let $\mathcal{X} = \{x_1, \ldots, x_m\}$ be any nonempty finite set. A **positive operator valued measure (POVM)** is a function $\mathcal{E} : \mathcal{X} \to \mathcal{B}^+(\mathcal{H})$ such that $\sum_{x \in \mathcal{X}} \mathcal{E}(x) = \mathbb{1}$. It will often be convenient to write E_x for $\mathcal{E}(x)$. Associated to every POVM $\mathcal{E}$ is a function also denoted $\mathcal{E}$ from $\mathfrak{S}_{\mathcal{H}}$ to probability densities p on $\mathcal{X}$ given by $p(x) = \mathrm{Tr}[\rho E_x]$.

A POVM $\mathcal{E}$ on $\mathcal{X}$ is a **projection valued measure (PVM)** on $\mathcal{X}$ in case for each $x \in \mathcal{X}$, $\mathcal{E}(x)$, is an orthogonal projection, in which case it is convenient to write P_x for $\mathcal{E}(x)$. Because $\sum_{x \in \mathcal{X}} P_x = \mathbb{1}$, $\{P_x : x \in \mathcal{X}\}$ is mutually orthogonal.

The POVM is **complete** in case each of the operators E_x is rank one.

There is a PVM associated to every self-adjoint operator A. Namely, let $A = \sum_{\lambda \in \sigma(A)} \lambda P_\lambda$ be the spectral decomposition of A. Take $\mathcal{X} := \sigma(A)$ and define $\mathcal{E} : \mathcal{X} \to \mathcal{B}^+(\mathcal{H})$ by $\mathcal{E}(\lambda) = P_\lambda$. Then the function from $\mathfrak{S}_{\mathcal{H}}$ to probability densities on $\sigma(A)$, also denoted $\mathcal{E}$, is given by $\mathcal{E}(\rho)(\lambda) = \mathrm{Tr}[\rho P_\lambda]$.

We have seen in the discussion leading up to Example 4.42 that a POVM $\mathcal{E} : \mathcal{X} \to \mathcal{B}^+(\mathcal{H})$ may arise from a PVM $\widehat{\mathcal{E}} : \mathcal{X} \to \mathcal{B}^+(\mathcal{H} \otimes \mathcal{K})$ through the formula

$$\mathcal{E}(x) = \mathrm{Tr}_{\mathcal{K}}[(\mathbb{1} \otimes \omega)\widehat{\mathcal{E}}(x)],$$

where ω is some ancillary density matrix on the ancillary space $\mathcal{K}$. Such a POVM can be understood as arising from a standard PVM on an enlarged system, including ancilla. In this case, the cardinality of $\mathcal{X}$ is limited only by the dimension of $\mathcal{H} \otimes \mathcal{K}$, instead of by the dimension of $\mathcal{H}$ as it is for a PVM. But since the ancilla space is arbitrary, one can arrange for $\mathcal{X}$ to be arbitrarily large.

A theorem of Naimark [162] says, roughly speaking, that every POVM comes from a PVM on a larger system including ancilla. The proof of Naimark's theorem in this finite-dimensional setting is closely patterned on the construction of the PVM used in Example 4.42.

Theorem 4.44 (Naimark's theorem). *Let $\mathcal{X}$ be a finite set, and let $\mathcal{H}$ be a finite-dimensional Hilbert space. Let $\mathcal{E} : \mathcal{X} \to \mathcal{B}^+(\mathcal{H})$ be a POVM. Then there exists a finite-dimensional Hilbert space $\mathcal{K}$, a PVM $\widehat{\mathcal{E}} : \mathcal{X} \to \mathcal{B}^+(\mathcal{H} \otimes \mathcal{K})$, and an isometry $V : \mathcal{H} \to \mathcal{H} \otimes \mathcal{K}$ such that for each x,*

$$(4.7.6) \qquad \mathcal{E}(x) = V^* \widehat{\mathcal{E}}(x) V.$$

Moreover, given any pure state $\omega = |\phi\rangle\langle\phi| \in \mathfrak{S}_{\mathcal{K}}$, the isometry can be taken to have the form

$$(4.7.7) \qquad V\mathbf{x} = \mathcal{U}(\mathbf{x} \otimes \phi)$$

for some unitary $\mathcal{U}$ on $\mathcal{H} \otimes \mathcal{K}$. Then with the dilation chosen in this form, for all $x \in \mathcal{X}$ and $\rho \in \mathfrak{S}_{\mathcal{H}}$,

$$(4.7.8) \qquad \mathrm{Tr}[\rho \mathcal{E}(x)] = \mathrm{Tr}[(\rho \otimes \omega)\widehat{\mathcal{E}}(x)].$$

Proof. Let $\mathcal{X}$ have cardinality m. Let $\mathcal{K}$ be an m-dimensional Hilbert space. Let $\{\mathbf{e}_x\}_{x \in \mathcal{X}}$ be an orthonormal basis of $\mathcal{K}$. Define a map $\mathcal{V} : \mathcal{H} \to \mathcal{H} \otimes \mathcal{K}$ by

$$(4.7.9) \qquad \mathcal{V}\mathbf{v} := \sum_{x \in \mathcal{X}} \mathcal{E}(x)^{1/2}\mathbf{v} \otimes \mathbf{e}_x.$$

The computation (4.7.4) shows that $\mathcal{V}$ is an isometry. Define

$$(4.7.10) \qquad \widehat{\mathcal{E}}(x) = \mathbb{1} \otimes |\mathbf{e}_x\rangle\langle\mathbf{e}_x|.$$

Then $\widehat{\mathcal{E}}$ is a PVM on $\mathcal{X}$, and for any $\mathbf{v}, \mathbf{w} \in \mathcal{H}$,

$$\langle \mathbf{v}, \mathcal{V}^*\widehat{\mathcal{E}}(x)\mathcal{V}\mathbf{w} \rangle = \langle \mathcal{V}\mathbf{v}, \widehat{\mathcal{E}}(x)\mathcal{V}\mathbf{w} \rangle$$

$$= \sum_{x',x'' \in \mathcal{X}} \langle E_{x'}^{1/2}\mathbf{v} \otimes \mathbf{e}_{x'}(\mathbb{1} \otimes |\mathbf{e}_x\rangle\langle\mathbf{e}_x|)E_{x''}^{1/2}\mathbf{w} \otimes \mathbf{e}_{x''} \rangle$$

$$= \langle \mathcal{E}(x)^{1/2}\mathbf{v}, \mathcal{E}(x)^{1/2}\mathbf{w} \rangle = \langle \mathbf{v}, \mathcal{E}(x)\mathbf{w} \rangle.$$

Since $\mathbf{v}$ and $\mathbf{w}$ are arbitrary, this proves (4.7.6).

Define the map $\mathcal{W} : \mathcal{H} \to \mathcal{H} \otimes \mathcal{K}$ by $\mathcal{W}\mathbf{x} := \mathbf{x} \otimes \phi$, and note that this too is an isometry. Since $\mathcal{V}$ and $\mathcal{W}$ have the same initial space, there is a (nonunique) unitary $\mathcal{U}$ such that $\mathcal{V} = \mathcal{U}\mathcal{W}$. For $x \in \mathcal{X}$, define the orthogonal projection $\mathcal{F}(x)$ by

$$\mathcal{F}(x) := \mathcal{U}^*\widehat{\mathcal{E}}(x)\mathcal{U}.$$

Then $\mathcal{F}$ is also a PVM on $\mathcal{X}$ with values in $\mathcal{B}(\mathcal{H} \otimes \mathcal{K})$, By the definitions, $\mathcal{W}^*\mathcal{F}(x)\mathcal{W} = \mathcal{V}^*\widehat{\mathcal{E}}(x)\mathcal{V}$, so that this is an alternate dilation of $\mathcal{E}$. Therefore, from (4.7.6),

$$\mathrm{Tr}[\rho \mathcal{E}(x)] = \mathrm{Tr}[\rho \mathcal{V}^*\widehat{\mathcal{E}}(x)\mathcal{V}] = \mathrm{Tr}[\rho \mathcal{W}^*\mathcal{F}(x)\mathcal{W}] = \mathrm{Tr}[\mathcal{W}\rho \mathcal{W}^*\mathcal{F}(x)].$$

By the definition of $\mathcal{W}$, $\mathcal{W}\rho\mathcal{W}^* = \rho \otimes |\phi\rangle\langle\phi|$. Therefore,

$$\mathrm{Tr}[\rho \mathcal{E}(x)] = \mathrm{Tr}[(\rho \otimes |\phi\rangle\langle\phi|)\mathcal{F}(x)].$$

Choosing $\mathcal{F}$ for the PVM $\widehat{\mathcal{E}}$, and letting ω denote $|\phi\rangle\langle\phi|$, this becomes (4.7.8). $\qquad\square$

Naimark dilations in which the isometry $\mathcal{V}$ has the form (4.7.7) have a particularly natural physical interpretation. The pure state ω is the state of the ancilla that we adjoin to the system on $\mathcal{H}$ whose state is ρ. The combined system is then evolved by subjecting it to some sort of forces. This evolution is given by a unitary $\mathcal{U}$ that, in principle, could be found by solving Schrödinger's equation for the time-dependent Hamiltonian describing the forces being applied. This results in the state $\mathcal{U}(\rho \otimes \omega)\mathcal{U}^*$. Now one makes a projective measurement in $\mathcal{H} \otimes \mathcal{K}$ using the PVM $\widehat{\mathcal{E}}$. The probability p_x of finding the value x is then given by $p_x = \mathrm{Tr}[\mathcal{U}(\rho \otimes \omega)\mathcal{U}^*\widehat{\mathcal{E}}(x)]$, and after the measurement, the state of the system is, by the usual rules,

$$(4.7.11) \qquad \frac{1}{p_x}\widehat{\mathcal{E}}(x)\mathcal{U}(\rho \otimes \omega)\mathcal{U}^*\widehat{\mathcal{E}}(x).$$

If $\rho = \psi$ is a pure state, then by (4.7.7),

$$\mathcal{U}(\rho \otimes \omega)\mathcal{U} = |\mathcal{U}\psi \otimes \phi\rangle\langle\mathcal{U}\psi \otimes \phi| = |V\psi\rangle\langle V\psi|.$$

Therefore, using (4.7.9) and (4.7.10), (4.7.11) becomes

$$\frac{1}{p_x}|E_x^{1/2}\psi\rangle\langle E_x^{1/2}\psi| \otimes |\mathbf{e}_x\rangle\langle\mathbf{e}_x|.$$

Tracing out the component on $\mathcal{K}$, the final state on $\mathcal{H}$ is the pure state $\frac{1}{p_x}|E_x^{1/2}\psi\rangle\langle E_x^{1/2}\psi|$. In particular, like a projective measurement, this type of measurement with ancilla takes pure states to pure states. In terms of state vectors, the transformation when the result is x is given by

$$\psi \mapsto \frac{1}{\sqrt{p_x}}\mathcal{E}(x)^{1/2}\psi.$$

In the definition of V in (4.7.9), we did not have to use the positive square root of $\mathcal{E}(x)$. We could replace $\mathcal{E}(x)^{1/2}$ by any other operator $A(x)$ such that $A^*(x)A(x) = \mathcal{E}(x)$. Then V is still an isometry, and the whole construction goes through from there, except now $\mathcal{U}$ is a different unitary. Different choices correspond to different laboratory procedures, and hence different final states, but the probabilities are the same.

4.7.2. Quantum tomography. Let $\mathcal{H}$ be a Hilbert space of dimension d. The span of $\mathfrak{S}_{\mathcal{H}}$ is $\mathcal{B}^{\mathrm{s.a.}}(\mathcal{H})$, which is a real Hilbert space of dimension d^2. Let $\mathcal{E}$ be a POVM on a finite set $\mathcal{X}$ with values $E_x \in \mathcal{B}^+(\mathcal{H})$. Recall that $\mathcal{E}$ determines the map, also denoted $\mathcal{E}$, from $\mathfrak{S}_{\mathcal{H}} \to \mathcal{P}(\mathcal{X})$ given by $\mathcal{E}(\rho)(x) = \mathrm{Tr}[E_x\rho]$. By linearity, this map extends to a linear map from $\mathcal{B}^{\mathrm{s.a.}}(\mathcal{H})$ to $\mathcal{F}(\mathcal{X})$, the space of real valued functions on $\mathcal{X}$.

We seek POVMs such that the map $\mathcal{E} : \mathcal{B}^{\mathrm{s.a.}}(\mathcal{H}) \to \mathcal{F}(\mathcal{X})$ is invertible, since in that case there is a measurement process by which the state ρ can be measured, provided we have many copies of ρ, and this is called **quantum tomography**

When there is an invertible linear map $\mathcal{E} : \mathcal{B}^{\mathrm{s.a.}}(\mathcal{H}) \to \mathcal{F}(X)$, these spaces must have the same dimension, and hence the cardinality of $\mathcal{X}$ must be d^2. Even without invertibility, just knowing that the kernel of this linear transformation is zero implies that for any two distinct $\rho, \sigma \in \mathfrak{S}_{\mathcal{H}}$, $\mathcal{E}(\rho) \neq \mathcal{E}(\sigma)$, but by the fundamental theorem of linear algebra, the map is invertible if and only if the kernel is zero and the dimensions of the domain and range are the same. If the map is one-to-one, the range must have dimension at least d^2.

Definition 4.45. Let $\mathcal{H}$ be a Hilbert space of finite dimension d, and let $\mathcal{X}$ be a set of cardinality at least d^2. A POVM on $\mathcal{X}$ with values in $\mathcal{B}^+(\mathcal{H})$ is **informationally complete** in case the map $\mathcal{E} : \mathfrak{S}_{\mathcal{H}} \to \mathcal{P}(\mathcal{X})$ is one-to-one, and it is **minimally informationally complete** in case this map is invertible.

Informationally complete POVMs were introduced and investigated by Prugovečki [**182**], and minimally informationally complete POVM's were first constructed and used by Caves, Fuchs, and Schack [**56**].

Theorem 4.46. *Let $\mathcal{H}$ be a Hilbert space of finite dimension d. Then there exist informationally complete POVM with values in $\mathcal{B}^+(\mathcal{H})$.*

Proof. By the spectral theorem, every $A \in \mathcal{B}^{\text{s.a.}}(\mathcal{H})$ is a linear combination of rank one projectors $|\mathbf{u}\rangle\langle\mathbf{u}|$, where each $\mathbf{u}$ is an eigenvector of A. Hence the real linear span of $\{|\mathbf{u}\rangle\langle\mathbf{u}| \ : \ \mathbf{u} \in \mathcal{H}, \ \|\mathbf{u}\| = 1\}$ is all of $\mathcal{B}^{\text{s.a.}}(\mathcal{H})$. Therefore, we may choose bases for $\mathcal{B}^{\text{s.a.}}(\mathcal{H})$ from this set—any linearly independent set of d^2 elements will be a basis. Let $\mathcal{X} := \{1, \ldots, d^2\}$, and let $\{\mathbf{u}_1, \ldots, \mathbf{u}_{d^2}\}$ be such that $\{|\mathbf{u}_x\rangle\langle\mathbf{u}_x|\}_{x \in \mathcal{X}}$ is linearly independent. Let $A \in \mathcal{B}^{\text{s.a.}}(\mathcal{H})$. If $\text{Tr}[|\mathbf{u}_x\rangle\langle\mathbf{u}_x|A] = 0$ for all A, then A is orthogonal to a spanning set in $\mathcal{B}^{\text{s.a.}}(\mathcal{H})$, and therefore $A = 0$.

Define $G := \sum_{x \in \mathcal{X}} |\mathbf{u}_x\rangle\langle\mathbf{u}_x|$. By what has just been noted, $\langle\mathbf{v}, G\mathbf{v}\rangle = \text{Tr}[|\mathbf{v}\rangle\langle\mathbf{v}|G] > 0$ for all nonzero $\mathbf{v}$, and hence G is invertible. Finally, defining

$$E_x = |G^{-1/2}\mathbf{u}_x\rangle\langle G^{-1/2}\mathbf{u}_x| = G^{-1/2}|\mathbf{u}_x\rangle\langle\mathbf{u}_x|G^{-1/2},$$

and then $\mathcal{E} : \mathcal{X} \to \mathcal{B}^+(\mathcal{H})$ by $\mathcal{E}(x) = E_x$, $\mathcal{E}$ is a POVM on $\mathcal{X}$ with values in $\mathcal{B}^+(\mathcal{H})$.

Now consider the map from $\mathcal{B}^{\text{s.a.}}(\mathcal{H})$ to $\mathcal{F}(\mathcal{X})$, also denoted by $\mathcal{E}$, and defined by $\mathcal{E}(A)(x) = \text{Tr}[E_x A] := f_A(x)$. If A is in the nullspace of this linear map, then for all x, $0 = \text{Tr}[E_x A] = \text{Tr}[G^{-1/2}AG^{-1/2}|\mathbf{u}_x\rangle\langle\mathbf{u}_x|]$, and hence $G^{-1/2}AG^{-1/2} = 0$, and then $A = 0$. Thus the map is one-to-one, and since the domain and range have the same dimension, this linear map is invertible. $\quad\square$

4.7.3. The pinching operation associated to a PVM. Let $\mathcal{X} = \{x_1, \ldots, x_m\}$ be a finite set, and let $\mathcal{E}$ be a PVM on $\mathcal{X}$ with values in $\mathcal{B}^+(\mathcal{H})$. Denote $\mathcal{E}(x_j)$ by P_j so that $\{P_1, \ldots, P_m\}$ is a set of mutually orthogonal projections such that $\sum_{j=1}^m P_j = \mathbb{1}$. Define $\mathcal{M}$ to be the span of $\{P_1, \ldots, P_m\}$. This is evidently a commutative von Neumann algebra.

Let $X \in \mathcal{M}'$. Then in particular, $X \in \{P_1, \ldots, P_m\}'$ and

$$X = \sum_{i,j=1}^m P_i X P_j = \sum_{i,j=1}^m X P_i P_j = \sum_{i,j=1}^m \delta_{i,j} X P_i P_j = \sum_{j=1}^m P_j X P_j.$$

Conversely, $X = \sum_{j=1}^{m} P_j X P_j$, then $P_i X = X P_i = P_i X P_i$ for all i, and so $X \in \{P_1, \ldots, P_m\}'$. This shows that the von Neumann subalgebra $\mathcal{M}' = \{P_1, \ldots, P_m\}'$ of $\mathcal{B}(\mathcal{H})$ is given by $\mathcal{M}' = \{X \in \mathcal{B}(\mathcal{H}) \ : \ X = \sum_{j=1}^{m} P_j X P_j\}$.

We have already encountered the map $\mathcal{P}$ defined by defined by $\mathcal{P}(X) = \sum_{j=1}^{m} P_j X P_j$ as the pinching operation defined in (3.4.1) of Theorem 3.22.

If $\rho \in \mathfrak{S}_{\mathcal{H}}$ and we perform a measurement corresponding to a PVM associated to the set $\{P_1, \ldots, P_m\}$ of mutually orthogonal projections, then after the measurement, the state belongs to

$$\{p_j^{-1} P_j \rho P_j \ : \ p_j := \mathrm{Tr}[\rho P_j] > 0\}.$$

In particular, the post measurement state belongs to $\mathcal{M}' = \{P_1, \ldots, P_m\}'$, and it is easy to see that every $X \in \mathcal{M}'$ is in the span of the post measurement state. The pinching operation $\mathcal{P}$ is sometimes called a **nonselective measurement**. If we fail to record the result of the observation, all we can say is that with probability p_j the state of the system is now $p_j^{-1} P_j \rho P_j$, so that *on average*, it is $\mathcal{P}(\rho)$.

We claim that $\mathcal{P} = \mathcal{E}_{\mathcal{M}'}$. By Theorem 4.39, it suffices to check that $\mathcal{P}$ has properties (i) and (ii) listed there. Let $A \in \mathcal{B}(\mathcal{H})$ and $X, Y \in \mathcal{M}'$. Then since $\mathcal{M}' = \{P_1, \ldots, P_m\}'$,

$$\mathcal{P}(XAY) = \sum_{j=1}^{m} P_j X A Y P_j = \sum_{j=1}^{m} X P_j A P_j Y = X\mathcal{P}(A)Y$$

and

$$\mathrm{Tr}[\mathcal{P}(A)] = \sum_{j=1}^{m} \mathrm{Tr}[P_j A P_j] = \sum_{j=1}^{m} \mathrm{Tr}[A P_j] = \mathrm{Tr}[A\mathbb{1}] = \mathrm{Tr}[A].$$

Therefore, by Theorem 4.39, $\mathcal{P} = \mathcal{E}_{\mathcal{M}'}$.

4.8. Representations of von Neumann algebras

Definition 4.47. Let $\mathcal{H}$ and $\mathcal{K}$ be two Hilbert spaces. Let $\mathcal{M}$ and $\mathcal{N}$ be unital $*$-subalgebras of $\mathcal{B}(\mathcal{H})$ and $\mathcal{B}(\mathcal{K})$, respectively. A linear map $\Phi : \mathcal{M} \to \mathcal{N}$ is a $*$-**homomorphism** in case for all $A_1, A_2 \in \mathcal{M}$,

$$(4.8.1) \qquad \Phi(A_1 A_2) = \Phi(A_1)\Phi(A_2),$$

$$\Phi(A)^* = \Phi(A^*).$$

A $*$-**isomorphism** is a $*$-homomorphism Φ that is invertible. A **representation** of $\mathcal{M}$ is a $*$-homomorphism π of $\mathcal{M}$ into $\mathcal{B}(\widehat{\mathcal{H}})$ for some Hilbert space $\widehat{\mathcal{H}}$. (The symbol π is traditional for representations.) A representation π is **faithful** if it is one-to-one.

It is evident that if $\Phi : \mathcal{M} \to \mathcal{N}$ is a $*$-homomorphism, then the range $\widetilde{\mathcal{N}}$ of Φ is a von Neumann subalgebra of $\mathcal{N}$, and moreover, for all $A \in \mathcal{M}$, $\Phi(\mathbb{1})\Phi(A) = \Phi(A) = \Phi(A)\Phi(\mathbb{1})$. Thus, $\Phi(\mathbb{1})$ is automatically the multiplicative identity in $\widetilde{\mathcal{N}}$.

Example 4.48. Let $\mathcal{H}$ be a finite-dimensional Hilbert space. Let $A \in \mathcal{B}(\mathcal{H})$ be normal with the spectral decomposition

$$\mathbb{1} = \sum_{\lambda \in \sigma(A)} P_\lambda \quad \text{and} \quad A = \sum_{\lambda \in \sigma(A)} \lambda P_\lambda.$$

As in Definition 1.59, let $\mathcal{C}(\sigma(A))$ denote the space of all complex valued functions on $\sigma(A)$. Equip $\mathcal{C}(\sigma(A))$ with its standard von Neumann algebra structure, as in Definition 4.12.

Define $\pi : \mathcal{C}(\sigma(A)) \to \mathcal{B}(\mathcal{H})$ by $\Phi(f) = f(A)$ using the spectral calculus as in (1.6.2). By Theorem 1.60, π is a faithful representation of the commutative von Neumann algebra $\mathcal{C}(\sigma(A))$ on $\mathcal{B}(\mathcal{H})$.

Example 4.49. Let $\mathcal{H}$ be a finite-dimensional Hilbert space and let $\mathcal{K}$ be a nontrivial subspace of $\mathcal{H}$. Let $P \in \mathcal{B}(\mathcal{H})$ be the orthogonal projection onto $\mathcal{K}$. Define $U \in \mathcal{B}(\mathcal{H}, \mathcal{K})$ by

$$(4.8.2) \qquad U\mathbf{x} = P\mathbf{x} \quad \text{for all} \quad \mathbf{x} \in \mathcal{H}.$$

Then U is a partial isometry. Define the linear map $\pi : \mathcal{B}(\mathcal{K}) \to \mathcal{B}(\mathcal{H})$ by

$$\pi(X) = U^*XU.$$

Note that $\pi(X^*) = \pi(X)^*$, and $\pi(\mathbb{1}_\mathcal{K}) = U^*U = P$, which is the multiplicative identity in the image of π. Since $UU^* = \mathbb{1}_\mathcal{K}$, $\pi(XY) = \pi(X)\pi(Y)$ for all $X, Y \in \mathcal{B}(\mathcal{K})$, and $U\pi(X)U^* = X$ for all $X \in \mathcal{B}(\mathcal{K})$, so that π is one-to-one. Thus, π is a representation of $\mathcal{B}(\mathcal{K})$ on $\mathcal{H}$.

Definition 4.50. Let $\mathcal{H}$ be a finite-dimensional Hilbert space, and let $\mathcal{K}$ be a nontrivial subspace of $\mathcal{H}$. The representation π of $\mathcal{B}(\mathcal{K})$ on $\mathcal{H}$ defined in Example 4.49 is the **canonical representation of** $\mathcal{B}(\mathcal{K})$ **on** $\mathcal{H}$.

The canonical representation π of $\mathcal{B}(\mathcal{K})$ on $\mathcal{H}$ comes up frequently in what follows. Because the partial isometry U defined in (4.8.2) is essentially the orthogonal projection P from $\mathcal{H}$ onto $\mathcal{K}$ considered as an element of $\mathcal{B}(\mathcal{H}, \mathcal{K})$ instead of $\mathcal{B}(\mathcal{H})$, it is standard practice to write $P\mathcal{B}(\mathcal{H})P$ for the image of $\mathcal{B}(\mathcal{K})$ under π.

The definition of a faithful representation Φ of a von Neumann algebra $\mathcal{M}$ only refers to algebraic properties of Φ, and yet the metric properties encoded into the norm are fundamentally important. This is not an oversight.

Theorem 4.51. *Let $\mathcal{M}$ be a von Neumann subalgebra of $\mathcal{B}(\mathcal{H})$, where $\mathcal{H}$ is a finite-dimensional Hilbert space. Let π be a faithful representation of $\mathcal{M}$ on a Hilbert space $\mathcal{K}$. Then π is an isometry; $\|\pi(X)\| = \|X\|$ for all $X \in \mathcal{M}$.*

Proof. Let $P := \pi(\mathbb{1})$ be the orthogonal projection in $\mathcal{K}$ that is the multiplicative identity in $\pi(\mathcal{M})$. Since $P\pi(X)P = \pi(X)$ for all $X \in \mathcal{M}$, we may regard $\pi(\mathcal{M})$ as a von Neumann algebra on the range of P, or, what is the same thing, we may assume that $\pi(\mathbb{1}_{\mathcal{H}}) = \mathbb{1}_{\mathcal{K}}$ after discarding an extraneous subspace of the original $\mathcal{K}$.

Then for all $X \in \mathcal{M}$, X is invertible in $\mathcal{M}$ if and only if $\pi(X)$ is invertible in $\mathcal{B}(\mathcal{K})$. Indeed, suppose that X is invertible. By (4.8.1) $\pi(X^{-1})$ is the inverse of $\pi(X)$ in $\mathcal{B}(\mathcal{K})$. However, if X is not invertible, there exists a nonzero $Y \in \mathcal{M}$ such that $XY = 0$. Then $\pi(X)\pi(Y) = 0$. Since π is faithful, $\pi(Y) \neq 0$, and hence $\pi(X)$ is not invertible in $\mathcal{B}(\mathcal{K})$.

It now follows, again using (4.8.1), that for $\lambda \in \mathbb{C}$, $(\lambda - X)$ is invertible in $\mathcal{M}$ if and only if $(\lambda - \pi(X))$ is invertible in $\mathcal{B}(\mathcal{K})$. In particular $\sigma(X) = \sigma(\pi(X))$. Since the norm of X is the spectral radius of $|X| = (X^*X)^{1/2}$, and likewise for $\pi(X)$, it follows that X and $\pi(X)$ have the same norm. $\qquad\square$

Any von Neumann algebra $\mathcal{M}$ is already an algebra of operators on some Hilbert space $\mathcal{H}$, and hence it may seem to be a matter of scant interest to represent $\mathcal{M}$ on another Hilbert space $\mathcal{K}$. However, it turns out that even representations of $M_m(\mathbb{C})$ will be very useful. The following definition provides the building blocks of a natural representation of $M_m(\mathbb{C})$.

Definition 4.52. Fix a positive integer m. Let $\mathcal{H}_m$ denote $M_m(\mathbb{C})$ equipped with the Hilbert-Schmidt inner product. For any $Z \in M_m(\mathbb{C})$, define the operators $L_Z, R_Z \in \mathcal{B}(\mathcal{H}_m)$ by $L_Z A = ZA$ and $R_Z A = AZ$, respectively. These are called the **operator of left multiplication by** Z, and the **operator of right multiplication by** Z, respectively.

Let $\mathcal{M}$ denote $M_m(\mathbb{C})$. Now define $\pi : \mathcal{M} \to \mathcal{B}(\mathcal{H}_m)$ by $\pi(X) = L_X$. Then π is obviously linear. Also, for $X, Y \in \mathcal{M}$, $L_Y L_X = L_{YX}$ and $L_{\mathbb{1}} = \mathbb{1}$. Finally, for $A, B, X \in \mathcal{M}$,

$$\langle A, L_X B\rangle_{\mathcal{H}_m} = \mathrm{Tr}[A^*XB] = \mathrm{Tr}[(X^*A)^*B] = \langle L_{X^*}A, B\rangle_{\mathcal{H}_m},$$

and hence $\pi(X^*) = \pi(X)^*$. This proves that π is a representation of $\mathcal{M}$ on the Hilbert space $\mathcal{H}_m$. Representations preserve positivity: Recall that $X \in M_m^+(\mathbb{C})$ if and only if $X = K^*K$ for some $K \in M_m(\mathbb{C})$. Then since π is a $*$-homomorphism, $\pi(X) = \pi(K^*K) = \pi(K)^*\pi(K) \geq 0$. Thus the representation of a positive matrix is positive. In fact, $X \in M_m^{++}(\mathbb{C})$ if and only if for some $\epsilon > 0$, $X = \epsilon\mathbb{1} + K^*K$ for some $K \in M_m(\mathbb{C})$. Then

$$\pi(X) = \pi(K^*K + \epsilon\mathbb{1}) = \pi(K)^*\pi(K) + \epsilon\mathbb{1} > 0.$$

Thus, the representation of a strictly positive matrix is strictly positive.

Define $\pi' : \mathcal{M} \to \mathcal{B}(\mathcal{H}_m)$ by $\pi'(Y) := R_Y$. The map π' is not quite a representation: as one can easily check, it is linear, $(R_Y)^* = R_{Y^*}$, and $R_{\mathbb{1}} = \mathbb{1}$.

However, $R_{XY} = R_Y R_X$—the order is reversed. Nonetheless, π' still has the important property that $\pi'(Y) \geq 0$ whenever $Y \geq 0$. To see this, note that $Y \geq 0$ if and only if $Y = L^*L$ for some $L \in M_m(\mathbb{C})$. Then

$$\pi'(Y) = \pi'(L^*L) = \pi'(L)\pi'(L)^* \geq 0.$$

Arguing as above, we see that $\pi'(Y) > 0$ whenever $Y > 0$.

Next observe that for $X, Y, A \in M_m(\mathbb{C})$, $L_X R_Y A = XAY = R_Y L_X A$. That is, L_X and R_Y commute. In fact, more is true:

Theorem 4.53. *Let $\pi(\mathcal{M})$ denote the image of π, $\{L_Z \ : \ Z \in \mathcal{M}\}$. Likewise, let $\pi'(\mathcal{M})$ denote the image of π', $\{R_Z \ : \ Z \in \mathcal{M}\}$. Then*

$$(\pi(\mathcal{M}))' = \pi'(\mathcal{M}) \quad and \quad (\pi'(\mathcal{M}))' = \pi(\mathcal{M}).$$

Proof. Let Φ be an operator on $\mathcal{H}_m$ that commutes with L_X for all $X \in \mathcal{M}$. Then for all $X, Y \in \mathcal{M}$,

$$X\Phi(Y) = L_X\Phi(Y) = \Phi(L_X Y) = \Phi(XY).$$

In particular, taking $Y = \mathbb{1}$, $X\Phi(\mathbb{1}) = \Phi(X)$. Therefore, defining $Z := \Phi(\mathbb{1})$, $\Phi(X) = R_Z X$ for all $X \in \mathcal{M}$. Hence every $\Phi \in (\pi(\mathcal{M}))'$ has the form R_Z for some $Z \in \mathcal{M}$, and as we have already noted, for all $Z \in \mathcal{M}$, $R_z \in (\pi(\mathcal{M}))'$. Hence $(\pi(\mathcal{M}))' = \pi'(\mathcal{M})$. The proof that $(\pi'(\mathcal{M}))' = \pi(\mathcal{M})$ is essentially the same. $\qquad\square$

Now suppose that $X, Y \in M_m^+(\mathbb{C})$. Since L_X and R_Y are commuting elements of $\mathcal{B}^+(\mathcal{H}_m)$, there is an orthonormal basis of $\mathcal{H}_m$ consisting of simultaneous eigenvectors of L_X and R_Y. In fact, one can easily construct such a basis.

Let $\{\mathbf{u}_1, \ldots, \mathbf{u}_m\}$ be an orthonormal basis of $\mathbb{C}^m$ consisting of eigenvectors of Y; $Y\mathbf{u}_j = \lambda_j\mathbf{u}_j$, $i = 1, \ldots, m$. Likewise, let $\{\mathbf{v}_1, \ldots, \mathbf{v}_m\}$ be an orthonormal basis of $\mathbb{C}^m$ consisting of eigenvectors of X; $X\mathbf{v}_i = \mu_i\mathbf{v}_i$. Then define $E_{i,j} = |\mathbf{v}_i\rangle\langle\mathbf{u}_j|$, $1 \leq i, j \leq m$, so that $\{E_{i,j} \ : \ 1 \leq i, j \leq m\}$ is an orthonormal set of m^2 vectors in $\mathcal{B}(\mathcal{H})$, and hence an orthonormal basis of $\mathcal{B}(\mathcal{H})$. Evidently,

$$L_X E_{i,j} = \mu_i E_{i,j} \qquad and \qquad R_Y E_{i,j} = \lambda_j E_{i,j}.$$

Hence, the spectrum of L_X is the same as the spectrum of X, but the multiplicities differ by a factor of m. Likewise, the spectrum of R_Y is the same as the spectrum of Y, but the multiplicities differ by a factor of m.

For $H \in M_m^{\text{s.a.}}(\mathbb{C})$, the spectral decompositions of H and $\pi(H)$ are related by

$$H = \sum_{j=1}^{m} \lambda_j|\mathbf{u}_j\rangle\langle\mathbf{u}_j| \quad and \quad \pi(H) = \sum_{i,j=1}^{m} \lambda_i|E_{i,j}\rangle\langle E_{i,j}|,$$

and hence for any complex valued function f on $\sigma(H)$,

$$\pi(f(H)) = f(\pi(H)).$$

Finally, as a direct consequence of the definitions, for all $X \in \mathcal{M}$,

$$(4.8.3) \qquad \operatorname{Tr}[X] = \langle \mathbb{1}, \pi(X)\mathbb{1}\rangle_{\mathcal{H}_m}.$$

On account of (4.8.3), this representation π allows one to investigate trace inequalities in $M_m(\mathbb{C})$ in terms of expectations in pure states, which turns out to be quite useful. This construction yields a representation π of $\mathcal{M} = M_m(\mathbb{C})$, on a Hilbert space $\mathcal{H}$ that is constructed from $\mathcal{M}$ itself and a positive linear functional on it, in this case the trace. It is a special case of the **Gelfand-Naimark-Segal (GNS) construction**, [90, 195] that plays a fundamental role in the final chapter. In the meantime, we shall get a lot of mileage out of this special case.

We summarize some conclusions in a theorem for later reference.

Theorem 4.54. *For any $X, Y \in M_m(\mathbb{C})$, R_X and L_Y commute. If $X, Y \in M_m^+(\mathbb{C})$, then $L_X, R_Y \in \mathcal{B}^+(\mathcal{H}_m)$, so that there is an orthonormal basis of $\mathcal{H}_m$ consisting of simultaneous eigenvectors of R_X and L_Y. Moreover, if $X, Y \in M_m^{++}(\mathbb{C})$, then $L_X, R_Y \in \mathcal{B}^{++}(\mathcal{H}_m)$.*

Exercises

(1) Let $\mathcal{M}$ be a von Neumann algebra on a finite-dimensional Hilbert space $\mathcal{H}$. Suppose that P is a nontrivial projection in $\mathcal{M}$. Define $\mathcal{N} := \{PXP : X \in \mathcal{M}\}$. Show that $\mathcal{N}$ is a von Neumann subalgebra of $\mathcal{B}(\mathcal{H})$ in which the multiplicative identity is P.

(2) (von Neumann) Let P and Q be two orthogonal projections in a finite-dimensional Hilbert spaces $\mathcal{H}$. Show that $P \wedge Q = \lim_{n \to \infty}(PQ)^n$.

(3) Let $\mathcal{H}$ be a finite-dimensional Hilbert space, and let $A \in \mathcal{B}(\mathcal{H})$ be normal. Let $A = \sum_{\lambda \in \sigma(A)} \lambda P_\lambda$ be the spectral decomposition of A. Let $\mathcal{M}$ be the von Neumann algebra consisting of all polynomials in A. Show that

$$\mathcal{M}' = \left\{ X \in \mathcal{B}(\mathcal{H}) : X = \sum_{\lambda \in \sigma(A)} P_\lambda X P_\lambda \right\}.$$

(4) Let $\mathcal{H}$ and $\mathcal{K}$ be finite-dimensional Hilbert spaces. Let $\mathcal{M}$ be the von Neumann algebra consisting of operators of the form $A \otimes \mathbb{1}$ in $\mathcal{B}(\mathcal{H} \otimes \mathcal{K})$. Describe the set of all minimal projections in $\mathcal{M}$.

(5) Let $\mathcal{H}$ be a finite-dimensional Hilbert space. Let $A \in \mathcal{B}(\mathcal{H})$ be normal. Let $\mathcal{M}$ be the von Neumann algebra consisting of all polynomials in A. Describe the set of all minimal projections in $\mathcal{M}$.

(6) Let $\mathcal{H}$ be a finite-dimensional Hilbert space.
   (a) Let P and Q be orthogonal projections in $\mathcal{B}(\mathcal{H})$. Show that there exists a partial isometry U such that $UU^* = P$ and $U^*U = Q$ if and only if $\mathrm{rank}(P) = \mathrm{rank}(Q)$.
   (b) Let $\mathcal{M}$ be a von Neumann subalgebra of $\mathcal{B}(\mathcal{H})$. Let P and Q be orthogonal projections in $\mathcal{M}$. Show that there need not be any partial isometry $U \in \mathcal{M}$ such that $UU^* = P$ and $U^*U = Q$ even if $\mathrm{rank}(P) = \mathrm{rank}(Q)$.

(7) Let P and Q be orthogonal projections on a finite-dimensional Hilbert space $\mathcal{H}$. Show that $P \vee Q = (P^\perp \wedge Q^\perp)^\perp$.

(8) Let $\mathcal{M}$ be a finite von Neumann subalgebra of $\mathcal{B}(\mathcal{H})$, where $\mathcal{H}$ is a finite-dimensional Hilbert space. Let $P \in \mathcal{M}$ be a projection, and let $\mathcal{K}$ be the range of P. Let $P\mathcal{M}P$ and $\mathcal{M}'P$ be considered as operator algebras on $\mathcal{K}$. Show that they are commutants of one another.

(9) Let $\mathcal{H}$ be a finite-dimensional Hilbert space. Let $\mathcal{M}$ be a factor in $\mathcal{B}(\mathcal{H})$. Show that if $X \in \mathcal{M}$, $Y \in \mathcal{M}'$ and $XY = 0$, then either $X = 0$ or $Y = 0$. Then show that the map $X \otimes Y \mapsto XY$ from $\mathcal{M} \otimes \mathcal{M}' \to \mathcal{B}(\mathcal{H})$ is a linear isomorphism.

(10) Let $\mathcal{M}$ be a von Neumann algebra. Let $P \in \mathcal{M}$ be a projection such that $PXP^\perp = 0$ for all $X \in \mathcal{M}$. Show $P \in \mathcal{M}'$.

(11) Let $\{\mathbf{u}_1, \ldots, \mathbf{u}_d\}$ be an orthonormal basis for a d-dimensional Hilbert space $\mathcal{H}$. Let U and V be the unitaries that generate the Weyl-Heisenberg group.
   (a) Show that for all $0 \le p, q \le d - 1$, $\mathrm{Tr}[U^p V^q] = \delta_{q,0}\delta_{p,0}d$.
   (b) Show that $\{d^{-1/2}U^p V^q : 0 \le q, p \le d - 1\}$ is an orthonormal basis for $\mathcal{B}(\mathcal{H})$ equipped with the Hilbert-Schmidt inner product.
   (c) Let μ denote the uniform measure on $\mathcal{X} := \{0, \ldots, d-1\}^2$ with total mass d, so that for $0 \le i, j \le d - 1$, $\mu(i, j) = d^{-1}$. Define a map $\mathcal{F} : \mathcal{B}(\mathcal{H}) \to L^2(\mathcal{X}, \mu)$ by $\mathcal{F}(X)(q, p) = \langle d^{-1/2}U^p V^q, X\rangle_{\mathrm{HS}}$. Show that $\mathcal{F}$ is unitary.
   (d) For any unit vector $\phi_0 \in \mathcal{H}$, define a map $\mathcal{G} : \mathcal{H} \to L^2(\mathcal{X}, \mu)$ by $\mathcal{G}(\psi)(q, p) = \mathcal{F}(|\psi\rangle\langle\phi_0|)$. Show that $\mathcal{G}$ is unitary.

(12) Let $\{\mathbf{u}_1, \ldots, \mathbf{u}_d\}$ be an orthonormal basis for a d-dimensional Hilbert space $\mathcal{H}$. Let U and V be the unitaries that generate the Weyl-Heisenberg group. Show that U and V have the same spectrum, namely $\sigma(U) = \sigma(V) = \{1, \omega, \ldots, \omega^{d-1}\}$, where $\omega = e^{i2\pi/d}$. Let $\{\mathbf{v}_1, \ldots, \mathbf{v}_d\}$ be any orthonormal basis of eigenvectors of V. Show that for all $1 \le i, j \le d$, $|\langle\mathbf{u}_i, \mathbf{v}_j\rangle|^2 = d^{-1}$.

(13) Let $\mathcal{M}$ be a von Neumann algebra on a finite-dimensional Hilbert space $\mathcal{H}$. Let $\mathcal{N}$ be a von Neumann subalgebra of $\mathcal{M}$. Let $\mathcal{E}_\mathcal{N}$ be the conditional expectation onto $\mathcal{N}$. Show that if $f : (0, \infty) \to \mathbb{R}$ is operator convex, $X \mapsto f(\mathcal{E}_\mathcal{N}(X))$ is convex on the positive elements of $\mathcal{M}$. Also show that if $f : (0, \infty) \to \mathbb{R}$ is convex, $X \mapsto \mathrm{Tr}[f(\mathcal{E}_\mathcal{N}(X))]$ is convex on the positive elements of $\mathcal{M}$.

(14) Let $\mathcal{H}, \mathcal{K}$ be finite-dimensional Hilbert spaces, and let $\Phi : \mathcal{B}(\mathcal{H}) \to \mathcal{B}(\mathcal{K})$ be a unital linear transformation such that $\|\Phi(X)\| \le \|X\|$ for all normal $X \in \mathcal{B}(\mathcal{H})$. Show that Φ is Hermitian; that is, that $\Phi(X)$ is self-adjoint whenever X is self-adjoint.

(15) Let $\mathcal{M}$ be a von Neumann algebra on a finite-dimensional Hilbert space $\mathcal{H}$. Let $\mathcal{N}$ be a von Neumann subalgebra of $\mathcal{M}$. Let $\Phi : \mathcal{M} \to \mathcal{N}$ be a linear map such that $\Phi(Y) = Y$ for all $Y \in \mathcal{N}$ and such that $\|\Phi(X)\| \leq \|X\|$ for all $X \in \mathcal{M}$. Show that Φ is a positive map.

Chapter 5

Positive linear maps and quantum operations

5.1. Positive linear maps

Definition 5.1 (Positive and Hermitian linear maps). A linear map $\Phi :$ $M_n(\mathbb{C}) \to M_m(\mathbb{C})$ is **positive** in case for all $X \geq 0$, $\Phi(X) \geq 0$. A linear map $\Phi : M_n(\mathbb{C}) \to M_m(\mathbb{C})$ is **Hermitian** in case for all $X \in M_n(\mathbb{C})$, $\Phi(X^*) = \Phi(X)^*$.

Lemma 5.2. *Every positive linear map* $\Phi : M_n(\mathbb{C}) \to M_m(\mathbb{C})$ *is Hermitian. Moreover, for all $A \in M_n(\mathbb{C})$,*

$$(5.1.1) \qquad \mathrm{ran}(\Phi(A)) \subseteq \mathrm{ran}(\Phi(\mathbb{1})).$$

Proof. By Lemma 1.66 every $A \in M_n(\mathbb{C})$ can be written as a linear combination of four elements of $M_n^+(\mathbb{C})$:

$$A = X_1 - X_2 + iX_3 - iX_4$$

with $X_j \in M_n^+(\mathbb{C})$ for $j = 1, 2, 3, 4$. Then

$$(5.1.2) \qquad \Phi(A) = \Phi(X_1) - \Phi(X_2) + i\Phi(X_3) - i\Phi(X_4).$$

By Theorem 1.65 positive operators are self-adjoint: $\Phi(X_j)^* = \Phi(X_j)$ for each j, and hence $\Phi(A)^* = \Phi(A^*)$. Moreover, since for each j, $0 \leq X_j \leq \|A\|\mathbb{1}$, $0 \leq \Phi(X_j) \leq \|A\|\Phi(\mathbb{1})$, which implies that $\ker(\Phi(\mathbb{1})) \subseteq \ker(\Phi(X_j))$, from which it follows that $\mathrm{ran}(\Phi(X_j)) \subseteq \mathrm{ran}(\Phi(\mathbb{1}))$. Therefore, (5.1.1) follows from (5.1.2). $\square$

Throughout this chapter, for $X, Y \in M_n(\mathbb{C})$, $\langle X, Y \rangle$ denotes the Hilbert-Schmidt inner product of X and Y, and we regard $M_n(\mathbb{C})$ as a Hilbert space with this inner product. For any linear map, $\Phi : M_n(\mathbb{C}) \to M_m(\mathbb{C})$ $\Phi^\dagger$ denotes the

Hilbert space adjoint of Φ when both the domain and range are equipped with the Hilbert-Schmidt inner product. That is, for all $X \in M_m(\mathbb{C})$ and $Y \in M_n(\mathbb{C})$

$$\langle \Phi^\dagger(X), Y \rangle = \langle X, \Phi(Y) \rangle.$$

Later we shall encounter other inner products, and hence other adjoints. However, the notation $\Phi^\dagger$ always denotes the canonical adjoint defined above.

Lemma 5.3. *A linear map* $\Phi : M_n(\mathbb{C}) \to M_m(\mathbb{C})$ *is positive if and only if*

$$\langle X, \Phi(Y) \rangle \geq 0 \quad \text{for all} \quad X \in M_m^+(\mathbb{C}), \ Y \in M_n^+(\mathbb{C}).$$

Consequently, Φ is positive if and only if $\Phi^\dagger$ is positive.

Proof. By definition, Φ is positive if and only if for all $Y \in M_n^+(\mathbb{C})$ and all $\mathbf{v} \in \mathbb{C}^m$,

$$0 \leq \langle \mathbf{v}, \Phi(Y)\mathbf{v} \rangle_{\mathbb{C}^n} = \langle |\mathbf{v}\rangle\langle\mathbf{v}|, \Phi(Y) \rangle,$$

and by the spectral theorem, every $X \in M_m^+(\mathbb{C})$ is a positive linear combination of operators of the form $|\mathbf{v}\rangle\langle\mathbf{v}|$. $\qquad\square$

Definition 5.4 (Unital and quasi unital maps). A linear map $\Phi : M_n(\mathbb{C}) \to M_m(\mathbb{C})$ is **unital** in case $\Phi(\mathbb{1}) = \mathbb{1}$, and it is **quasi unital** if and only if $\Phi(\mathbb{1})$ is an orthogonal projection.

Example 5.5. For $m > n$, define $\Psi : M_n(\mathbb{C}) \to M_m(\mathbb{C})$ by $\Psi(A) := \begin{bmatrix} A & 0 \\ 0 & 0 \end{bmatrix}$, where the right-hand side is the $m \times m$ matrix with A as its upper-left $n \times n$ block, and all other entries being zero. Then Ψ is quasi unital.

Lemma 5.6. *Let* $\Phi : M_n(\mathbb{C}) \to M_m(\mathbb{C})$ *be a positive quasi unital linear map. Then for all $A \in M_n(\mathbb{C})$,*

$$\Phi(A) = \Phi(\mathbb{1})\Phi(A) = \Phi(A)\Phi(\mathbb{1}).$$

Proof. When $\Phi(\mathbb{1})$ is an orthogonal projection, it is the orthogonal projection onto $\mathrm{ran}(\Phi(\mathbb{1}))$. Then $\Phi(A) = \Phi(\mathbb{1})\Phi(A)$ by Lemma 5.2. Then $\Phi(A) = \Phi(A)\Phi(\mathbb{1})$ for all $A \in M_n(\mathbb{C})$ follows by taking adjoints. $\qquad\square$

Definition 5.7 (Trace preserving and quasi trace preserving maps). A linear map $\Phi : M_n(\mathbb{C}) \to M_m(\mathbb{C})$ is **trace preserving** in case $\mathrm{Tr}[\Phi(X)] = \mathrm{Tr}[X]$ for all $X \in M_n(\mathbb{C})$, and it is **quasi trace preserving** if and only if there exists an orthogonal projection $P \in M_n(\mathbb{C})$ such that

$$\mathrm{Tr}[PX] = \mathrm{Tr}[\Phi(X)] \quad \text{for all} \quad X \in M_n(\mathbb{C}).$$

Theorem 5.8. $\Phi : M_n(\mathbb{C}) \to M_m(\mathbb{C})$ *is quasi unital if and only if the adjoint map* $\Phi^\dagger : M_m(\mathbb{C}) \to M_n(\mathbb{C})$ *is quasi trace preserving, In this case, the unique projection such that* $\mathrm{Tr}[PX] = \mathrm{Tr}[\Phi^\dagger(X)]$ *for all X is $P = \Phi(\mathbb{1})$. In particular,* $\Phi : M_n(\mathbb{C}) \to M_m(\mathbb{C})$ *is unital if and only if* $\Phi^\dagger : M_m(\mathbb{C}) \to M_n(\mathbb{C})$ *is trace preserving.*

Proof. Let $X \in M_m(\mathbb{C})$, and let P be an orthogonal projection on $M_m(\mathbb{C})$. Then $\text{Tr}[PX] = \langle P, X \rangle$ and $\text{Tr}[\Phi^\dagger(X)] = \langle \mathbb{1}, \Phi^\dagger(X) \rangle = \langle \Phi(\mathbb{1}), X \rangle$.

Therefore $\text{Tr}[PX] = \text{Tr}[\Phi^\dagger(X)]$ if and only if $\langle \Phi(\mathbb{1}), X \rangle = \langle P, X \rangle$. $\square$

5.2. k-positivity and complete positivity

There are some refined notions of positivity that are important both to the mathematical theory and to its quantum mechanical applications.

Definition 5.9 (k-positivity and complete positivity). Let $\Phi : M_n(\mathbb{C}) \to M_m(\mathbb{C})$ be a linear transformation. Φ is 2-**positive** in case the block matrix

$$\begin{bmatrix} \Phi(A) & \Phi(B) \\ \Phi(C) & \Phi(D) \end{bmatrix} \geq 0 \qquad \text{whenever} \qquad \begin{bmatrix} A & B \\ C & D \end{bmatrix} \geq 0.$$

For each integer $k > 2$, the condition of k-**positivity** is defined in the analogous manner, and Φ is **completely positive (CP)** if it is k-positive for all k.

For a treatment of complete positivity in a general infinite-dimensional context, see the book of Paulsen [**173**], and for more in the matrix setting, see the book of Bhatia [**36**].

Remark 5.10. Evidently, any positive linear combination of k-positive maps from $M_n(\mathbb{C})$ to $M_m(\mathbb{C})$ is k-positive. Consequently, any positive linear combination of CP maps from $M_n(\mathbb{C})$ to $M_m(\mathbb{C})$ is CP.

The following example shows that even the class of 2-positive maps is strictly contained within the class of positive maps.

Example 5.11. The canonical example is that in which $\Phi(A) = A^T$, the transpose of A, for all $A \in M_n(\mathbb{C})$. Then taking

$$A := \begin{bmatrix} 1 & 0 \\ 0 & 0 \end{bmatrix}, \quad B := \begin{bmatrix} 0 & 1 \\ 0 & 0 \end{bmatrix}, \quad C := \begin{bmatrix} 0 & 0 \\ 1 & 0 \end{bmatrix}, \quad D := \begin{bmatrix} 0 & 0 \\ 0 & 1 \end{bmatrix},$$

the block matrix $\begin{bmatrix} A & B \\ C & D \end{bmatrix}$ is positive semidefinite, while

$$\begin{bmatrix} \Phi(A) & \Phi(B) \\ \Phi(C) & \Phi(D) \end{bmatrix} = \begin{bmatrix} 1 & 0 & 0 & 0 \\ 0 & 0 & 1 & 0 \\ 0 & 1 & 0 & 0 \\ 0 & 0 & 0 & 1 \end{bmatrix},$$

and evidently -1 is an eigenvalue of this last matrix. Thus Φ is positive but not 2-positive.

However, many of the maps we have already encountered in this book turn out to be CP.

Example 5.12. Let $V \in M_{m,n}(\mathbb{C})$. Then the map $\Phi : X \mapsto V^*XV$ from $M_n(\mathbb{C})$ to $M_m(\mathbb{C})$ is evidently positive. Moreover, there is the factorization

$$\begin{bmatrix} \Phi(A) & \Phi(B) \\ \Phi(C) & \Phi(D) \end{bmatrix} = \begin{bmatrix} V^* & 0 \\ 0 & V^* \end{bmatrix} \begin{bmatrix} A & B \\ C & D \end{bmatrix} \begin{bmatrix} V & 0 \\ 0 & V \end{bmatrix}$$

from which it follows that Φ is 2-positive. The same factorization argument shows Φ is k-positive for all k, and hence is CP.

Then by Remark 5.10 for any set $\{V_1, \ldots, V_\ell\} \subset M_{m,n}(\mathbb{C})$,

$$\Phi(X) = \sum_{j=1}^{\ell} V_j^* X V_j$$

is CP. In fact, in some sense, this example is all there is: By a theorem of Kraus [133] and Choi [60, 62], to be proved later in this chapter, every CP map Φ from $M_n(\mathbb{C})$ to $M_m(\mathbb{C})$ has this form, and in fact, one can always find such a representation with $\ell \leq mn$. Such a representation is called a **Kraus representation**.

Example 5.13. By Example 5.12 and by Theorem 4.40, which expresses conditional expectations as averages over unitary conjugations, if $\mathcal{M}$ is any von Neumann algebra on a finite-dimensional Hilbert space and $\mathcal{N}$ is any von Neumann subalgebra of it, $\mathcal{E}_{\mathcal{N}}$ is CP.

Example 5.14 (∗-homomorphisms are CP). A linear map $\Phi : M_n(\mathbb{C}) \to M_m(\mathbb{C})$ is a ∗-homomorphism in case for all $X, Y \in M_n(\mathbb{C})$, $\Phi(XY) = \Phi(X)\Phi(Y)$ and $\Phi(X^*) = \Phi(X)^*$. A ∗-homomorphism is always CP. Let $\Phi : M_n(\mathbb{C}) \to M_m(\mathbb{C})$ be a ∗-homomorphism. Consider a positive semidefinite block matrix $\begin{bmatrix} X & K \\ K^* & Y \end{bmatrix}$ with entries in $M_n(\mathbb{C})$. This has a factorization

$$\begin{bmatrix} X & K \\ K^* & Y \end{bmatrix} = \begin{bmatrix} A & L \\ L^* & B \end{bmatrix}^* \begin{bmatrix} A & L \\ L^* & B \end{bmatrix},$$

and by the ∗-homomorphism property of Φ,

$$\begin{bmatrix} \Phi(X) & \Phi(K) \\ \Phi(K)^* & \Phi(Y) \end{bmatrix} = \begin{bmatrix} \Phi(A) & \Phi(L) \\ \Phi(L)^* & \Phi(B) \end{bmatrix}^* \begin{bmatrix} \Phi(A) & \Phi(L) \\ \Phi(L)^* & \Phi(B) \end{bmatrix} \geq 0.$$

The proof of k-positivity is analogous. Thus, Φ is CP.

As an example, let $\mathbf{u}$ be any unit vector in $\mathbb{C}^\ell$, and define $\Phi : M_n(\mathbb{C}) \to M_n(\mathbb{C}) \otimes M_\ell(\mathbb{C})$ by $\Phi(X) = X \otimes |\mathbf{u}\rangle\langle\mathbf{u}|$. It is evident that Φ is a ∗-homomorphism. Now let ω be a density matrix in $M_\ell(\mathbb{C})$, and write it in the form $\omega = \sum_{j=1}^{r} \lambda_j |\mathbf{u}_j\rangle\langle\mathbf{u}_j|$ where $\{\mathbf{u}_1, \ldots, \mathbf{u}_r\}$ is an orthonormal set of eigenvectors of ρ. Then the map

$$(5.2.1) \hspace{4cm} \Psi(X) = X \otimes \omega$$

can be written as a convex combination of ∗-homomorphisms, and then by Remark 5.10, Ψ is CP.

5.2.1. k-positivity and tensor products. Let $\mathcal{H}_n$ denote the Hilbert space consisting of $M_n(\mathbb{C})$ equipped with the Hilbert-Schmidt inner product. Let $\{e_1, \dots, e_k\}$ denote the standard orthonormal basis of $\mathbb{C}^k$. Define $E_{i,j} := |e_i\rangle\langle e_j|$ so that $\{E_{i,j} \ : \ 1 \leq i, j \leq k\}$ is an orthonormal basis of $\mathcal{H}_k$. Then any element X of $\mathcal{H}_n \otimes \mathcal{H}_k$ has the expansion

$$(5.2.2) \qquad X = \sum_{i,j=1}^{k} X_{i,j} \otimes E_{i,j},$$

where each $X_{i,j}$ belongs to $\mathcal{H}_n$. We may then identify X with the block matrix

$$\begin{bmatrix} X_{1,1} & \cdots & X_{1,k} \\ \vdots & \ddots & \vdots \\ X_{k,1} & \cdots & X_{k,k} \end{bmatrix}.$$

If Φ is a linear map from $\mathcal{H}_n$ to $\mathcal{H}_n$, the map $\Phi \otimes \mathbb{1}$, where $\mathbb{1}$ denotes the identity on $\mathcal{H}_k$, is given by

$$\Phi \otimes \mathbb{1}(X) = \sum_{i,j=1}^{k} \Phi(X_{i,j}) \otimes E_{i,j},$$

where we have used (5.2.2), and hence $\Phi \otimes \mathbb{1}(X)$ has the block matrix representation

$$\begin{bmatrix} \Phi(X_{1,1}) & \cdots & \Phi(X_{1,k}) \\ \vdots & \ddots & \vdots \\ \Phi(X_{k,1}) & \cdots & \Phi(X_{k,k}) \end{bmatrix}.$$

Therefore, an equivalent phrasing of the definition of k-positivity is that

$$\Phi \text{ is } k\text{-positive} \quad \Longleftrightarrow \quad \Phi \otimes \mathbb{1}_k \text{ is positive.}$$

In many treatments of the subject, this formulation is taken as the basic definition of k-positivity. It is in any case a useful perspective, as the proof of the next lemma shows.

It is left as an exercise to show that $(\Phi \otimes \mathbb{1}_k)^\dagger = \Phi^\dagger \otimes \mathbb{1}_k$. As a consquence of this and Lemma 5.3, we have:

Lemma 5.15. *Let Φ be a linear map from $M_n(\mathbb{C})$ to $M_m(\mathbb{C})$. Φ is k-positive if and only if $\Phi^\dagger$ is k-positive, and consequently, Φ is CP if and only if $\Phi^\dagger$ is CP.*

Proof. By Lemma 5.3, $(\Phi \otimes \mathbb{1})^\dagger$ is positive if and only if $\Phi \otimes \mathbb{1}$ is positive, Then since $(\Phi \otimes \mathbb{1})^\dagger = \Phi^\dagger \otimes \mathbb{1}$, Φ is positive if and only if $\Phi^\dagger$ is positive. $\qquad\square$

Some physically important CP maps arise from a tensor product structure.

Example 5.16. Let $\Psi_k : M_n(\mathbb{C}) \to M_n(\mathbb{C}) \otimes M_k(\mathbb{C})$ be defined by

$$(5.2.3) \qquad\qquad \Psi_k(A) = A \otimes \mathbb{1}.$$

Identifying $\mathbb{C}^n \otimes \mathbb{C}^k$ and $\bigoplus_k \mathbb{C}^n$ we may also think of Ψ_k as a map from $M_n(\mathbb{C})$ to $M_{nk}(\mathbb{C})$, and clearly Ψ_k is a $*$-homomorpishm, and hence is CP. Example 5.12 provides another way to see this since it is easy to write down a Kraus representation for Ψ_k starting from (5.2.3).

Let $\{\mathbf{u}_1, \ldots, \mathbf{u}_k\}$ be an orthonormal basis for $\mathbb{C}^k$. Every element $\mathbf{x}$ of $\mathbb{C}^n \otimes \mathbb{C}^k$ has a unique expression of the form

$$(5.2.4) \qquad\qquad \mathbf{x} = \sum_{j=1}^{k} \mathbf{x}_j \otimes \mathbf{u}_j,$$

where each $\mathbf{x}_j \in \mathbb{C}^n$. Using (5.2.4), define $V_j : \mathbb{C}^n \otimes \mathbb{C}^k \to \mathbb{C}^n$ by $V_j \mathbf{x} = \mathbf{x}_j$, $j = 1, \ldots, k$. Then simple computations show that V_j^* is given by $V_j^* \mathbf{v} = \mathbf{v} \otimes \mathbf{u}_j$. Then for $X \in M_n(\mathbb{C})$, and $\mathbf{x} \in \mathbb{C}^n \otimes \mathbb{C}^k$ given by (5.2.4)

$$\sum_{k=1}^{k} V_j^* X V_j \mathbf{x} = \sum_{j=1}^{k} V_k^*(X\mathbf{x}_j) = \sum_{j=1}^{k} (X\mathbf{x}_j) \otimes \mathbf{u}_k = (X \otimes \mathbb{1})\mathbf{x}.$$

This shows that

$$(5.2.5) \qquad\qquad \Psi_k(X) = \sum_{j=1}^{k} V_j^* X V_j$$

is a Kraus representation of Ψ_k, and hence Ψ_k is CP.

By Lemma 5.15, its adjoint, $\Psi_k^\dagger$, is CP and since Ψ_k is unital, $\Psi_k^\dagger$ is trace preserving. However, it is easy to see from (5.2.5) that for $Y \in M_n(\mathbb{C}) \otimes M_k(\mathbb{C})$,

$$\Psi_k^\dagger(Y) = \sum_{j=1}^{k} V_j Y V_j^*,$$

which is a Kraus representation of $\Psi_k^\dagger$, displaying it as CP.

From the definitions of $\{V_1, \ldots, V_k\}$, for any $\mathbf{v}, \mathbf{v}' \in \mathbb{C}^n$,

$$\left\langle \mathbf{v}', \sum_{j=1}^{k} V_j Y V_j^* \mathbf{v} \right\rangle = \sum_{j=1}^{k} \left\langle V_j^* \mathbf{v}', Y V_j^* \mathbf{v} \right\rangle$$

$$= \sum_{j=1}^{k} \langle \mathbf{v}' \otimes \mathbf{u}_j, Y(\mathbf{v} \otimes \mathbf{u}_j) \rangle = \langle \mathbf{v}', \mathrm{Tr}_2[Y]\mathbf{v} \rangle.$$

This identifies $\Psi_k^\dagger$ as the partial trace Tr_2, and provides a Kraus representation for the partial trace. In particular, the partial trace is CP.

It is useful to discuss this in the block matrix language. Let $\widehat{Y} \in M_{nk}(\mathbb{C})$ be given in the block matrix form

$$\widehat{Y} = \begin{bmatrix} Y_{1,1} & \cdots & Y_{1,k} \\ \vdots & \ddots & \vdots \\ Y_{k,1} & \cdots & Y_{k,k} \end{bmatrix}.$$

Since for $X \in M_n(\mathbb{C})$, $\langle \Psi_k(X), \widehat{Y} \rangle = \sum_{j=1}^{k} \mathrm{Tr}[X^* Y_{j,j}] = \left\langle X, \sum_{j=1}^{k} Y_{j,j} \right\rangle$,

$$\Psi_k^{\dagger}(Y) = \sum_{j=1}^{k} Y_{j,j},$$

which gives Tr_2 in its block matrix form. It is left as an exercise to show that with $\widehat{Y}$ as above,

$$\mathrm{Tr}_1[\widehat{Y}] = \begin{bmatrix} \mathrm{Tr}[Y_{1,1}] & \cdots & \mathrm{Tr}[Y_{1,k}] \\ \vdots & \ddots & \vdots \\ \mathrm{Tr}[Y_{k,1}] & \cdots & \mathrm{Tr}[Y_{k,k}] \end{bmatrix}.$$

5.3. Quantum operations

Let $\rho \in \mathfrak{S}_{\mathcal{H}}$, where $\mathcal{H}$ is a finite-dimensional Hilbert space. What is the most general operation we could perform on ρ? We might "attach" an ancillary state $\omega \in \mathfrak{S}_{\mathcal{K}}$ for some other finite-dimensional Hilbert space $\mathcal{K}$, resulting in $\rho \otimes \omega \in \mathfrak{S}_{\mathcal{H} \otimes \mathcal{K}}$. Then we may subject this state to the unitary evolution governed by some Hamiltonian H on $\mathcal{H} \otimes \mathcal{K}$, resulting in $\mathcal{U}(\rho \otimes \omega)\mathcal{U}^*$. Finally, one may restrict this state to $\mathcal{B}(\mathcal{H})$, taking the partial trace. The result is $\mathrm{Tr}_{\mathcal{K}}[\mathcal{U}(\rho \otimes)\omega\mathcal{U}^*]$. Somewhat more generally, if $\widetilde{\mathcal{H}} \otimes \widetilde{\mathcal{K}}$ is another tensor product factorization of $\mathcal{H} \otimes \mathcal{K}$, we may restrict to $\mathcal{B}(\widetilde{\mathcal{H}})$, producing $\mathrm{Tr}_{\widetilde{\mathcal{K}}}[\mathcal{U}(\rho \otimes \omega)\mathcal{U}^*]$.

Altogether, define the map

$$(5.3.1) \qquad \Phi(\rho) := \mathrm{Tr}_{\widetilde{\mathcal{K}}}[\mathcal{U}(\rho \otimes \omega)\mathcal{U}^*].$$

This is a trace preserving CP map because it is the composition of three trace preserving CP maps. In fact, we have seen in (5.2.1) of Example 5.14 that the map $X \mapsto X \otimes \omega$ is CP, and evidently it is trace preserving. Likewise, the map $Y \mapsto \mathcal{U}Y\mathcal{U}^*$ is CP by either Example 5.12 or Example 5.14, and evidently it is trace preserving. Finally, the partial trace is CP by Example 5.16, and again, it is evidently trace preserving. One can now compose more such operations, and since the composition of CP trace preserving maps is CP and trace preserving, every transformation of a quantum state that can be physically implemented corresponds to a CP trace preserving map.

There is a converse to this statement that will be proved below: Every CP trace preserving map Φ has a decomposition of the form (5.3.1). Anticipating this result, we make the following definition.

Definition 5.17 (Quantum operation). A **quantum operation** is a linear map $\Phi : \mathcal{B}(\mathcal{H}) \to \mathcal{B}(\widetilde{\mathcal{H}})$ that is CP and trace preserving. Another name for the same thing, widely used in quantum information theory, is **quantum channel**.

The notion of complete positivity is due to Stinespring [203]. One of his results is a structure theorem, known as Stinespring's factorization theorem, or as Stinespring's dilation theorem. As we shall see, this provides the factorization of any CP map of the form (5.2.1). It is for this reason that CP maps Φ play a fundamental role in quantum information theory and the quantum theory of open systems [68, 133].

5.4. Choi's theorem and Kraus representations

In this section the following **notation** is used:

- For $m, n \in \mathbb{N}$, $\mathcal{H}_{mn}$ denotes the space of $m \times n$ complex matrices equipped with the Hilbert-Schmidt inner product. When $n = m$, we use the shorter notation, $\mathcal{H}_n$.
- For $m, n \in \mathbb{N}$, $\widehat{\mathcal{H}}_{mn}$ will denote the Hilbert space consisting of $\mathcal{B}(\mathcal{H}_n, \mathcal{H}_m)$ equipped with the Hilbert-Schmidt inner product. When $n = m$, we use the shorter notation $\widehat{\mathcal{H}}_n$.

For computational purposes, it is convenient to express the Hilbert-Schmidt inner product on $\widehat{\mathcal{H}}_{nm}$ in the form (1.4.4). That is, for any orthonormal basis $\{F_{i,j}\}_{1\leq i,j\leq n}$ of $\mathcal{H}_n$, and $\Phi, \Psi \in \widehat{\mathcal{H}}_{n,m}$,

$$(5.4.1) \qquad \langle \Psi, \Phi \rangle_{\widehat{\mathcal{H}}_{n,m}} = \sum_{i,j=1}^{n} \langle \Psi(F_{i,j}), \Phi(F_{i,j}) \rangle_{\mathcal{H}_m}.$$

Next, given orthonormal bases $\{F_{i,j}\}_{1\leq i,j\leq m}$ and $\{G_{k,\ell}\}_{1\leq k,\ell\leq n}$ $\mathcal{H}_{mn}$, it follows from Lemma 1.44 that

$$\{|F_{i,j}\rangle\langle G_{k,\ell}|\}_{1\leq i,j\leq m,\, 1\leq k,\ell\leq n}$$

is an orthonormal basis for $\widehat{\mathcal{H}}_{mn}$. Given two vectors $\mathbf{u}$ and $\mathbf{v}$ in an inner product space, $|\mathbf{u}\rangle\langle\mathbf{v}|$ is sometimes called the outer product of $\mathbf{u}$ and $\mathbf{v}$. Since the vectors in $\mathcal{H}_{mn}$ are operators (or matrices), we also have the operator product at our disposal, and using this, it is possible to construct other orthonormal bases that are particularly useful for investigating complete positivity.

Definition 5.18. For $F \in \mathcal{H}_{mn}$ and $G \in \mathcal{H}_{nm}$, define $F \# G \in \widehat{\mathcal{H}}_{n,m}$ by

$$F\#G(X) := FXG$$

for all $X \in \mathcal{H}_n$.

Note that $G^* \in \mathcal{H}_{mn}$, so that $|F\rangle\langle G^*| \in \widehat{\mathcal{H}}_{n,m}$, as is $F\#G$, but the two are quite different: $|F\rangle\langle G^*|$ is always a rank one operator, but in general $F\#G$ is not. However, there is an important similarity: if $\{F_{i,j}\}_{1\leq i,j\leq n}$ and $\{G_{k,\ell}\}_{1\leq k,\ell\leq m}$ are orthonormal bases for for $\mathcal{H}_n$ and $\mathcal{H}_m$ respectively, then

$$\{F_{i,j}\#G_{k,\ell}\}_{1\leq i,j\leq n,\, 1\leq k,\ell\leq m}$$

is an orthonormal basis for $\widehat{\mathcal{H}}_{n,m}$. The next lemma [93] is the key to this.

Lemma 5.19. *For any* $F_1, F_2 \in M_{mn}(\mathbb{C})$ *and any* $G_1, G_2 \in M_{nm}(\mathbb{C})$,

$$\langle F_1\#G_1, F_2\#G_2\rangle_{\widehat{\mathcal{H}}_{nm}} = \langle F_1, F_2\rangle_{\mathcal{H}_{mn}}\langle G_1, G_2\rangle_{\mathcal{H}_{nm}}.$$

Proof. Let $\{\mathbf{u}_1,\ldots,\mathbf{u}_n\}$ be any orthonormal basis of $\mathbb{C}^n$. For $1 \leq i, j \leq n$, define $E_{i,j} := |\mathbf{u}_i\rangle\langle \mathbf{u}_j|$. Then $\{E_{i,j} \;:\; 1 \leq i, j \leq n\}$ is an orthonormal basis of $\mathcal{H}_n$, and by (5.4.1)

$$\langle F_1\#G_1, F_2\#G_2\rangle_{\widehat{\mathcal{H}}_{nm}} = \sum_{i,j=1}^{n} \mathrm{Tr}_{\mathcal{H}_m}[(F_1 E_{i,j} G_1)^*(F_2 E_{i,j} G_2)]$$

$$= \sum_{i,j=1}^{n} \mathrm{Tr}_{\mathcal{H}_m}[G_1^* E_{j,i} F_1^* F_2 E_{i,j} G_2]$$

$$= \sum_{i,j=1}^{n} \langle \mathbf{u}_i, F_1^* F_2 \mathbf{u}_i\rangle\langle \mathbf{u}_j, G_2 G_1^* \mathbf{u}_j\rangle$$

$$= \mathrm{Tr}_{\mathcal{H}_n}[F_1^* F_2]\, \mathrm{Tr}_{\mathcal{H}_m}[G_1^* G_2]. \qquad \square$$

It is convenient to index orthonormal bases of $\mathcal{H}_{mn}$ by ordered pairs $(i, j) \in \{1,\ldots,m\}\times\{1,\ldots,n\} =: \mathcal{J}_{mn}$. We use lower case Greek letters to denote elements of the index set. For $\alpha = (i, j) \in \mathcal{J}_{m,n}$, $\alpha' := (j, i) \in \mathcal{J}_{nm}$.

Theorem 5.20. *If* $\{F_\alpha\}_{\alpha\in\mathcal{J}_{nm}}$ *is any orthonormal basis of* $\mathcal{H}_{nm}$, *then* $\{F_\alpha^*\#F_\beta\}_{\alpha,\beta\in\mathcal{J}_{nm}}$ *is an orthonormal basis of* $\widehat{\mathcal{H}}_{nm}$, *so that every* $\Phi \in \widehat{\mathcal{H}}_{n,m}$ *has an expansion*

(5.4.2)
$$\Phi = \sum_{\alpha,\beta\in\mathcal{J}_{n,m}} (c_\Phi)_{\alpha,\beta} F_\alpha^*\#F_\beta,$$

with uniquely determined coefficients, namely

(5.4.3)
$$(c_\Phi)_{\alpha,\beta} = \langle F_\alpha^*\#F_\beta), \Phi\rangle_{\widehat{\mathcal{H}}_{nm}}.$$

Proof. If $\{F_\alpha\}_{\alpha\in\mathcal{J}_{mn}}$ and $\{G_\beta\}_{\beta\in\mathcal{J}_{nm}}$ are orthonormal bases in $\mathcal{H}_{mn}$ and $\mathcal{H}_{nm}$, respectively, then by Lemma 5.19 $\{F_\alpha\#G_\beta\}_{\alpha\in\mathcal{J}_{m,n},\, \beta\in\mathcal{J}_{nm}}$ is orthonormal in $\widehat{\mathcal{H}}_{nm}$, and has cardinality $n^2 m^2$. Therefore, $\{F_\alpha\#G_\beta\}_{\alpha\in\mathcal{J}_{m,n},\, \beta\in\mathcal{J}_{nm}}$ is an orthonormal basis in $\widehat{\mathcal{H}}_{nm}$.

Next, for any orthonormal basis $\{F_\alpha\}_{\alpha\in\mathcal{I}_{nm}}$ of $\mathcal{H}_{nm}$, $\{F_\alpha^*\}_{\alpha\in\mathcal{I}_{nm}}$ is an orthonormal basis of $\mathcal{H}_{mn}$, and hence $\{F_\alpha^*\#F_\beta\}_{\alpha,\beta\in\mathcal{I}_{nm}}$ is an orthonormal basis of $\widehat{\mathcal{H}}_{nm}$, from which the rest is immediate. $\qquad\qquad\qquad\qquad\qquad\qquad\qquad\qquad\square$

Definition 5.21 (Characteristic matrix). Given $\Phi\in\widehat{\mathcal{H}}_{nm}$, and an orthonormal basis $\{F_\alpha\}_{\alpha\in\mathcal{I}_{nm}}$ of $\mathcal{H}_{nm}$, its **characteristic matrix** for this orthonormal basis is the $nm\times nm$ matrix c_Φ whose (α,β)-th entry is $(c_\Phi)_{\alpha,\beta}$ as specified in (5.4.3).

Theorem 5.22 (Choi's theorem). *A linear operator Φ in $\mathcal{B}(\mathcal{H}_n,\mathcal{H}_m)$ is CP if and only if, for any orthonormal basis $\{F_\alpha\}_{\alpha\in\mathcal{I}_{nm}}$ of $\mathcal{H}_{nm}$, the corresponding characteristic matrix c_Φ is positive semidefinite.*

Moreover, when Φ is CP, it can be written in the form

$$\Phi(X) = \sum_{j=1}^{p} V_j^* X V_j,$$

where $p \le mn$, and the $n\times m$ matrices V_j can be computed in terms of the eigenvalues and eigenvectors of c_Φ and the basis $\{F_\alpha\}_{\alpha\in\mathcal{I}_{nm}}$.

Finally, Φ in $\mathcal{B}(\mathcal{H}_n,\mathcal{H}_m)$ is CP if and only if it is $\min\{m,n\}$-positive.

Proof. Let $\{\mathbf{u}_1,\dots,\mathbf{u}_n\}$ be any orthonormal basis of $\mathbb{C}^n$. For $1\le i,j\le n$, define $E_{i,j} := |\mathbf{u}_i\rangle\langle\mathbf{u}_j|$. Then $\{E_{i,j} \ : \ 1\le i,j\le n\}$ is an orthonormal basis of $\mathcal{H}_n$. The right side of (5.4.3) can be computed using this basis in (5.4.1) to compute the inner product:

$$(c_\Phi)_{\alpha,\beta} = \langle F_\alpha^*\#F_\beta, \Phi\rangle_{\widehat{H}_{nm}} = \sum_{i,j=1}^{n} \mathrm{Tr}[E_{i,j}^* F_\alpha \Phi(E_{i,j}) F_\beta^*].$$

For any nm complex numbers z_α, $\alpha\in\mathcal{I}_{n,m}$, define $G := \sum_{\alpha\in\mathcal{I}_{n,m}} z_\alpha F_\alpha^*$. Then

$$(5.4.4)\qquad \sum_{\alpha,\beta}\overline{z}_\alpha (c_\Phi)_{\alpha,\beta} z_\beta = \sum_{i,j=1}^{n} \mathrm{Tr}[E_{j,i} G^* \Phi(E_{i,j}) G].$$

Let P denote the $n\times n$ block matrix whose (i,j)-th entry is $\frac{1}{n}E_{i,j}$. Then

$$(P^2)_{i,j} = \frac{1}{n^2}\sum_{k=1}^{n} E_{i,k}E_{k,j} = \frac{1}{n}E_{i,j} = P_{i,j},$$

and P is evidently self-adjoint since $E_{i,j}^* = E_{j,i}$. Thus P is an orthogonal projection, and in particular, P is positive semidefinite. This proves that the $n\times n$ block matrix whose (i,j)-th entry is $E_{i,j}$ is positive semidefinite.

Therefore, if Φ is CP or even n-positive, the $n\times n$ block matrix whose (i,j)-th entry is $\Phi(E_{i,j})$ is positive semidefinite by the definition of n-positivity. It trivially follows that the $n\times n$ block matrix Q whose (i,j)-th entry is $G^*\Phi(E_{i,j})G$ is positive semidefinite. The right side of (5.4.4) is then the trace of the product of the positive semidefinite $nm \times nm$ matrices P and Q, and as such it is positive. Thus, whenever Φ is n-positive, c_Φ is positive semidefinite.

On the other hand, suppose that c_Φ is positive semidefinite. Let Λ be a diagonal matrix whose diagonal entries, λ_γ, are the eigenvalues of c_Φ, and let U be a unitary such that $c_\Phi = U^*\Lambda U$. Then by (5.4.2), for any $X \in \mathcal{H}_n$,

$$\Phi(X) = \sum_{\alpha,\beta,\gamma \in \mathcal{I}_{nm}} U^*_{\alpha,\gamma} \lambda_\gamma U_{\gamma,\beta} F^*_\alpha X F_\beta = \sum_{\gamma \in \mathcal{I}_{nm}} V^*_\gamma X V_\gamma,$$

where

$$(5.4.5) \qquad\qquad V_\gamma := \sqrt{\lambda_\gamma} \sum_{\alpha \in \mathcal{I}_{nm}} U_{\gamma,\alpha} F_\alpha.$$

This shows that whenever c_Φ is positive semidefinite, Φ is CP. Moreover in this Kraus representation, the Kraus operators are in one-to-one correspondence with the nonzero eigenvalues (repeated according to their multiplicity) of C_Φ. There can be no more than mn of these. As shown in the first part, when Φ is n-positive, c_Φ is positive semidefinite, and then by the second part, Φ is CP. Finally, since $\Phi^\dagger$ is CP if and only if Φ is CP, and since $\Phi^\dagger : \mathcal{H}_m \to \mathcal{H}_n$, what we have just proved shows that $\Phi^\dagger$ is CP if and only if it is m-positive, which is the case if and only if Φ is m-positive. $\qquad\square$

Remark 5.23. The formula (5.4.5) shows how to compute a Kraus representation of a CP map $\Phi : \mathcal{H}_n \to \mathcal{H}_m$ through the diagonalization of c_Φ.

Example 5.24. Let $\Phi : M_2(\mathbb{C}) \to M_2(\mathbb{C})$ be given by $\Phi(X) = \frac{1}{2}\mathrm{Tr}[X]\mathbb{1}$, and note that Φ is trace preserving and positive. To check whether it is CP, and find a Kraus representation if it is, we compute c_Φ. Let $\mathbf{e}_1 = (0,1)$ and $\mathbf{e}_2 = (0,1)$ so that $\{\mathbf{e}_1, \mathbf{e}_2\}$ is the standard orthonormal basis of $\mathbb{C}^2$, and define $F_\alpha := |\mathbf{e}_{\alpha_1}\rangle\langle\mathbf{e}_{\alpha_2}|$. Then $(c_\Phi)_{\alpha,\beta} = \langle F^*_\alpha \# F_\beta, \Phi\rangle_{\hat{H}_{nm}}$ and we shall also use the $|\mathbf{e}_i\rangle\langle\mathbf{e}_j|$ basis to compute the trace in the Hilbert-Schmidt inner product. We find

$$(c_\Phi)_{\alpha,\beta} = \langle F^*_\alpha \# F_\beta, \Phi\rangle_{\hat{H}_{nm}} = \sum_{i,j=1}^{2} \mathrm{Tr}[(F^*_\alpha E_{i,j} F_\beta)^* \Phi(E_{i,j})]$$

$$= \sum_{i,j=1}^{2} \mathrm{Tr}[E_{j,i} F_\alpha \Phi(E_{i,j}) F^*_\beta] = \sum_{i,j=1}^{2} \mathrm{Tr}\left[E_{j,i} F_\alpha \left(\tfrac{1}{2}\delta_{i,j}\mathbb{1}\right) F^*_\beta\right]$$

$$= \frac{1}{2}\sum_{j=1}^{2} \langle\mathbf{e}_j, F_\alpha F^*_\beta \mathbf{e}_j\rangle = \tfrac{1}{2}\langle F_\beta, F_\alpha\rangle_{\mathcal{H}_2} = \tfrac{1}{2}\delta_{\alpha,\beta}.$$

Therefore, $2c_\Phi$ is simply the identity map, which is positive, and hence Φ is CP. Moreover, c_Φ is already diagonal, and all of the eigenvalues are $\frac{1}{2}$, and hence the Kraus operators are simply $\left\{\frac{1}{\sqrt{2}}F_{i,j}\right\}_{1\le i,j\le 2}$. Indeed, one readily check that

$$\frac{1}{2}\sum_{i,j=1}^{2}F_{i,j}^{*}XF_{i,j}=\frac{1}{2}\sum_{i,j=1}^{2}\langle\mathbf{e}_i,X\mathbf{e}_i\rangle|\mathbf{e}_j\rangle\langle\mathbf{e}_j|=\frac{1}{2}\operatorname{Tr}[X]\mathbb{1}=\Phi(X).$$

Example 5.25. Let $\Phi : M_2(\mathbb{C}) \to M_2(\mathbb{C})$ be given by $\Phi(X) = X^T$, and note that Φ is trace preserving and positive. We have already seen that Φ is not CP, and now we shall check this by computing its characteristic matrix. As in the previous example, let $\mathbf{e}_1 = (0,1)$ and $\mathbf{e}_2 = (0,1)$ so that $\{\mathbf{e}_1,\mathbf{e}_2\}$ is the standard orthonormal basis of $\mathbb{C}^2$, and define $F_\alpha := |\mathbf{e}_{\alpha_1}\rangle\langle\mathbf{e}_{\alpha_2}|$. Then $(c_\Phi)_{\alpha,\beta} = \langle F_\alpha^{*}\#F_\beta, \Phi\rangle_{\hat{H}_{nm}}$, and we shall also use the $|\mathbf{e}_i\rangle\langle\mathbf{e}_j|$ basis to compute the trace in the Hilbert-Schmidt inner product. Computing as before, we find

$$\langle F_\alpha^{*}\#F_\beta, \Phi\rangle_{\hat{H}_{nm}} = \sum_{i,j=1}^{2}\operatorname{Tr}[E_{j,i}F_\alpha\Phi(E_{i,j})F_\beta^{*}] = \sum_{i,j=1}^{2}\operatorname{Tr}\left[E_{j,i}F_\alpha E_{j,i}F_\beta^{*}\right]$$

$$= \sum_{i,j=1}^{2}\langle\mathbf{e}_i,F_\alpha\mathbf{e}_j\rangle\langle\mathbf{e}_j,F_\beta\mathbf{e}_i\rangle = \sum_{i=1}^{2}\langle\mathbf{e}_i,F_\alpha F_\beta\mathbf{e}_i\rangle = \delta_{\alpha,\beta'},$$

where $(\beta_1,\beta_2)' = (\beta_2,\beta_1)$. If we order the indices as $(1,1),(1,2),(2,1),(2,2)$, then c_Φ is the matrix

$$\begin{bmatrix} 1 & 0 & 0 & 0 \\ 0 & 0 & 1 & 0 \\ 0 & 1 & 0 & 0 \\ 0 & 0 & 0 & 1 \end{bmatrix},$$

which is the same nonpositive semidefinite matrix we encountered in Example 5.11. Since this matrix is not positive definite, Theorem 5.22 says that Φ is not *CP*.

5.4.1. CP maps on von Neumann algebras. So far we have discussed CP maps from $M_n(\mathbb{C})$ to $M_m(\mathbb{C})$, or, what is the same thing, from $\mathcal{B}(\mathcal{H})$ to $\mathcal{B}(\mathcal{K})$ where $\mathcal{H}$ and $\mathcal{K}$ are finite-dimensional Hilbert spaces.

Definition 5.26. Let $\mathcal{H}$ and $\mathcal{K}$ be finite-dimensional Hilbert spaces. Let $\mathcal{M}$ be a von Neumann subalgebra of $\mathcal{B}(\mathcal{H})$, and let $\mathcal{N}$ be a von Neumann subalgebra of $\mathcal{B}(\mathcal{K})$. Then a linear map $\Phi : \mathcal{M} \to \mathcal{N}$ is CP if and only if for all $k \in \mathbb{N}$, the map $\Phi \otimes \mathbb{1} : \mathcal{M} \otimes M_k(\mathbb{C}) \to \mathcal{N} \otimes M_k(\mathbb{C})$ is positive.

Theorem 5.27. *Let $\mathcal{H}$ and $\mathcal{K}$ be finite-dimensional Hilbert spaces. Let $\mathcal{M}$ and $\mathcal{N}$ be von Neumann subalgebras of $\mathcal{B}(\mathcal{H})$ and $\mathcal{B}(\mathcal{K})$, respectively. A linear map $\Phi : \mathcal{M} \to \mathcal{N}$ is CP if and only if $\Phi \circ \mathcal{E}_{\mathcal{M}} : \mathcal{B}(\mathcal{H}) \to \mathcal{B}(\mathcal{K})$ is CP where $\mathcal{E}_{\mathcal{M}}$ is the conditional expectation onto $\mathcal{M}$ in $\mathcal{B}(\mathcal{H})$.*

Proof. Suppose $\Phi : \mathcal{M} \to \mathcal{N}$ is CP. Then for all $k \in \mathbb{N}$, with $\mathbb{1}$ denoting the identity on $\mathcal{H}_k$, $\Phi \otimes \mathbb{1}$ is positive. Since $\mathcal{E}_{\mathcal{M}} : \mathcal{B}(\mathcal{H}) \to \mathcal{B}(\mathcal{H})$ is CP, $\mathcal{E}_{\mathcal{M}} \otimes \mathbb{1}$ is positive. Therefore $\Phi \circ \mathcal{E}_{\mathcal{M}} \otimes \mathbb{1} = (\Phi \otimes \mathbb{1}) \circ (\mathcal{E}_{\mathcal{M}} \otimes \mathbb{1})$ is positive. Hence $\Phi \circ \mathcal{E}_{\mathcal{M}} : \mathcal{B}(\mathcal{H}) \to \mathcal{B}(\mathcal{K})$ is CP.

Now suppose that $\Phi \circ \mathcal{E}_{\mathcal{M}} : \mathcal{B}(\mathcal{H}) \to \mathcal{B}(\mathcal{K})$ is CP. If $\{E_{i,j} : 1 \leq i, j \leq k\}$ is any orthonormal basis for $\mathcal{H}_k$, every $X \in \mathcal{M} \otimes \mathcal{H}_k$ has an expansion

$$X = \sum_{i,j=1}^{k} X_{i,j} \otimes E_{i,j},$$

where each $X_{i,j} \in \mathcal{M}$. Therefore, since $\mathcal{E}_{\mathcal{M}}(X_{i,j}) = X_{i,j}$,

$$(\Phi \circ \mathcal{E}_{\mathcal{M}}) \otimes \mathbb{1}(X) = \sum_{i,j=1}^{k} \Phi \circ \mathcal{E}_{\mathcal{M}}(X_{i,j}) \otimes E_{i,j}$$

$$= \sum_{i,j=1}^{k} \Phi(X_{i,j}) \otimes E_{i,j} = \Phi \otimes \mathbb{1}(X).$$

This shows that for $X \in \mathcal{M} \otimes \mathcal{H}_k$, $(\Phi \circ \mathcal{E}_{\mathcal{M}}) \otimes \mathbb{1}(X) \geq 0$ if and only if $\Phi \otimes \mathbb{1}(X) \geq 0$. Therefore, $\Phi : \mathcal{M} \to \mathcal{N}$ is CP. $\qquad \square$

Theorem 5.27 provides a simple extension of any CP map between von Neumann algebras $\mathcal{M}$ and $\mathcal{N}$ on finite-dimensional Hilbert spaces $\mathcal{H}$ and $\mathcal{K}$ to a CP map from $\mathcal{B}(\mathcal{H})$ to $\mathcal{B}(\mathcal{K})$, or equivalently, from $M_n(\mathbb{C})$ to $M_m(\mathbb{C})$ where n and m are the dimensions of $\mathcal{H}$ and $\mathcal{K}$, respectively. Therefore, every theorem we have proved, and will prove, about CP maps from $M_n(\mathbb{C})$ to $M_m(\mathbb{C})$ applies to CP maps from $\mathcal{M}$ to $\mathcal{N}$. There is no loss of generality in working with full matrix algebras.

The following theorem is a special case of a theorem of Stinespring [**203**] who did not assume the Hilbert spaces involved to be finite dimensional and considered the case in which the range is commutative. Størmer treated the commutative domain case by different means; see his book [**204**]. The two are equivalent in our setting.

Theorem 5.28 (Stinespring's commutative range theorem). *Let $\mathcal{H}$ and $\mathcal{K}$ be finite-dimensional Hilbert spaces. Let $\mathcal{M}$ and $\mathcal{N}$ be von Neumann subalgebras of $\mathcal{B}(\mathcal{H})$ and $\mathcal{B}(\mathcal{K})$, respectively. Suppose that $\mathcal{M}$ is commutative. Then any positive linear map $\Phi : \mathcal{M} \to \mathcal{N}$ is CP. Likewise, suppose that $\mathcal{N}$ is commutative. Then any positive linear map $\Phi : \mathcal{M} \to \mathcal{N}$ is CP.*

Proof. When $\mathcal{M}$ is commutative, it is spanned by a set $\{P_1, \ldots, P_m\}$ of mutually orthogonal projections. Let $\Phi : \mathcal{M} \to \mathcal{N}$ be positive. For $j = 1, \ldots, m$, define $E_j := \Phi(P_j) \geq 0$. Fix any $1 \leq j \leq m$, and choose a unit vector $\mathbf{u}_j \in \mathrm{ran}(P_j)$. Let $\{\mathbf{v}_{j,1}, \ldots, \mathbf{v}_{j,r_j}\}$ be an orthonormal basis of $\mathrm{ran}(E_j)$ consisting of eigenvectors of E_j so that for each $1 \leq k \leq r_j$, $E_j \mathbf{v}_{j,k} = \lambda_{j,k} \mathbf{v}_{j,k}$. Then defining $\mathbf{w}_{j,k} = \sqrt{\lambda_{j,k}} \mathbf{v}_{j,k}$,

$$(5.4.6) \qquad E_j = \sum_{k=1}^{r_j} |\mathbf{w}_{j,k}\rangle\langle\mathbf{w}_{j,k}|.$$

Define $V_{j,k} := |\mathbf{u}_j\rangle\langle\mathbf{w}_{j,k}|$. Using (5.4.6) and the fact that $P_\ell \mathbf{u}_j = \delta_{\ell,j} \mathbf{u}_j$,

$$(5.4.7) \qquad \sum_{k=1}^{r_j} V_{j,k}^* P_\ell V_{j,k} = \delta_{\ell,j} E_j.$$

Define $\Psi_j : \mathcal{M} \to \mathcal{N}$ by $\Psi_j(X) = \sum_{k=1}^{r_j} V_{j,k}^* X V_{j,k}$. Every $X \in \mathcal{M}$ has the form $X = \sum_{j=1}^m z_j P_j$ for $z_1, \ldots, z_m$ in $\mathbb{C}$. It follows from (5.4.7) that $\Psi_j(X) = z_j E_j$, and therefore $\Phi(X) = \sum_{j=1}^m z_j E_j = \sum_{j=1}^m \Psi_j(X)$; that is, $\Phi = \sum_{j=1}^m \Psi_j$. Evidently each Ψ_j is CP, and hence Φ is CP.

This shows that every positive map Φ with a commutative domain $\mathcal{M}$ is CP. Now suppose that $\mathcal{N}$, the range of Φ, is commutative. Then $\Phi^\dagger : \mathcal{N} \to \mathcal{M}$ is positive and has a commutative domain. By what was proved above, $\Phi^\dagger$ is CP, and consequently Φ is CP. $\qquad\qquad\square$

Definition 5.29 (States on a von Neumann algebra). Let $\mathcal{M}$ be a von Neumann algebra on a finite-dimensional Hilbert space. A **state** on a von Neumann algebra $\mathcal{M}$ is a positive linear map ρ from $\mathcal{M}$ to $\mathbb{C}$ such $\rho(\mathbb{1}) = 1$, and ρ is **faithful** in case $\rho(X) = 0$ for $X \in \mathcal{M}^+$ if and only if $X = 0$. We denote the set of states on $\mathcal{M}$ by $\mathcal{S}_\mathcal{M}$.

Regard $\mathbb{C}$ as a one-dimensional Hilbert space, and note that $\mathcal{B}(\mathbb{C})$ is also one dimensional, and may be identified with $\mathbb{C}$. Thus a state ρ on $\mathcal{M}$ is a positive linear transformation from $\mathcal{M}$ to $\mathcal{N} := \mathcal{B}(\mathbb{C})$, which is commutative. By Theorem 5.28 it is CP, though this special case can be seen more simply.

More important for us is the following application of Theorem 5.27:

Theorem 5.30. *Let $\mathcal{M}$ be a von Neumann algebra on a finite-dimensional Hilbert space $\mathcal{H}$. Let ρ be a state on $\mathcal{M}$. Then there is a uniquely determined density matrix, also denoted ρ, belonging to $\mathcal{M}$, so that for all $X \in \mathcal{M}$,*

$$(5.4.8) \qquad \rho(X) = \mathrm{Tr}[\rho X].$$

Conversely, given $\rho \in \mathfrak{S}_\mathcal{H} \cap \mathcal{M}$, the map $X \mapsto \mathrm{Tr}[\rho X]$ from $\mathcal{M}$ to $\mathbb{C}$ is a state on $\mathcal{M}$. In particular, there is a one-to-one correspondence between elements of $\mathcal{S}_\mathcal{M}$ and of $\mathfrak{S}_\mathcal{H} \cap \mathcal{M}$.

Proof. As in Theorem 5.27, let $\mathcal{E}_\mathcal{M}$ denote the conditional expectation onto $\mathcal{M}$ in $\mathcal{B}(\mathcal{H})$. Since the composition of positive maps is positive, $\rho \circ \mathcal{E}_\mathcal{M}$, and since

$\mathcal{E}_{\mathcal{M}}$ is unital, $\rho \circ \mathcal{E}_{\mathcal{M}}(\mathbb{1}) = 1$. Hence $\rho \circ \mathcal{E}_{\mathcal{M}}$ is a positive map from $\mathcal{B}(\mathcal{H})$ to $\mathbb{C}$ such that $\rho \circ \mathcal{E}_{\mathcal{M}}(\mathbb{1}) = 1$, and hence there is a $\hat{\rho} \in \mathfrak{S}_{\mathcal{H}}$ such that for all $Y \in \mathcal{B}(\mathcal{H})$,

$$\rho \circ \mathcal{E}_{\mathcal{M}}(Y) = \mathrm{Tr}[\hat{\rho}Y].$$

Then for all $X \in \mathcal{M}$, since $\mathcal{E}_{\mathcal{M}}(X) = X$ and $\mathcal{E}_{\mathcal{M}}^{\dagger} = \mathcal{E}_{\mathcal{M}}$,

$$\rho(X) = \mathrm{Tr}[\hat{\rho}X] = \mathrm{Tr}[\hat{\rho}\mathcal{E}_{\mathcal{M}}(X)] = \mathrm{Tr}[\mathcal{E}_{\mathcal{M}}(\hat{\rho})X].$$

Since $\mathcal{E}_{\mathcal{M}}$ is trace preserving and CP, $\rho := \mathcal{E}_{\mathcal{M}}(\hat{\rho}) \in \mathfrak{S}_{\mathcal{H}} \cap \mathcal{M}$ is such that (5.4.8) is satisfied for all $X \in \mathcal{M}$. If $\sigma \in \mathfrak{S}_{\mathcal{H}} \cap \mathcal{M}$ and $\rho(X) = \mathrm{Tr}[\sigma X]$ for all $X \in \mathcal{M}$, then $\mathrm{Tr}[(\sigma - \rho)X] = 0$ for all $X \in \mathcal{M}$. Taking $X = \sigma - \rho$ shows that $\sigma = \rho$. Hence there is exactly one $\rho \in \mathfrak{S}_{\mathcal{H}} \cap \mathcal{M}$ such that (5.4.8) is satisfied for all $X \in \mathcal{M}$. Finally, it is evident that for any $\rho \in \mathfrak{S}_{\mathcal{H}} \cap \mathcal{M}, X \mapsto \mathrm{Tr}[\rho X]$ is a state on $\mathcal{M}$. $\qquad\square$

Definition 5.31. Let $\mathcal{M}$ be a von Neumann algebra on a finite-dimensional Hilbert space $\mathcal{H}$. The **support** of a state ρ on $\mathcal{M}$, denoted by $\mathrm{supp}(\rho)$, is defined to be the smallest projection $P \in \mathcal{M}$ such that $\rho(P) = 1$. Thus, a state ρ is faithful if and only if $\mathrm{supp}(\rho) = \mathbb{1}$.

5.5. Stinespring's factorization of CP maps

Let π be a representation of $M_n(\mathbb{C})$ as operators on a finite-dimensional Hilbert space $\mathcal{H}$. Since π is a $*$-homomorphism, it is CP. Let V be a linear transformation from $\mathbb{C}^m$ to $\mathcal{H}$. The map $\Psi(X) : \mathcal{B}(\mathcal{H}) \to M_m(\mathbb{C})$ given by $\Psi(X) = V^*XV$ is CP by Example 5.12. Then $A \mapsto V^*\pi(A)V = \Psi \circ \pi(A)$, being the composition of CP maps, is CP. Stinespring's factorization theorem, specialized to the matrix case, says that every CP map $\Phi : M_n(\mathbb{C}) \to M_m(\mathbb{C})$ may be factorized in this manner; that is, $\Phi(A) = V^*\pi(A)V = \Psi \circ \pi(A)$.

Another name for the same theorem is **Stinespring's dilation theorem**. In many cases of interest, $\mathcal{H}$ has a dimension that is larger than m, and hence $\Phi(A) = V^*\pi(A)V = \Psi \circ \pi(A)$ *dilates* Φ onto a larger space, displaying it as a *projection* of a representation.

Theorem 5.32 (Stinespring's factorization theorem). *For every CP map* $\Phi :$ $M_n(\mathbb{C}) \to M_m(\mathbb{C})$, *there is a Hilbert space* $\mathcal{H}$ *of dimension* $d \leq mn^2$, *a representation* π *of* $M_n(\mathbb{C})$ *on* $\mathcal{H}$, *and a linear map* $V : \mathbb{C}^m \to \mathcal{H}$ *such that*

$$(5.5.1) \qquad\qquad \Phi(A) = V^*\pi(A)V,$$

and Φ *is unital if and only if* V *is an isometry of* $\mathbb{C}^m$ *into* $\mathcal{H}$.

Proof of Theorem 5.32. By Theorem 5.22, for some $p \leq nm$, there exists $\{V_1, \ldots, V_p\} \subset M_{n,m}(\mathbb{C})$ such that

$$\Phi(A) = \sum_{j=1}^{p} V_j^* A V_j.$$

Let $\mathcal{H} := \mathbb{C}^n \otimes \mathbb{C}^p$ with its usual inner product. Define a representation of $M_n(\mathbb{C})$ on $\mathcal{H}$ by

$$\pi(A) := A \otimes \mathbb{1} \in \mathcal{B}(\mathcal{H}).$$

Let $\{\mathbf{u}_1, \ldots, \mathbf{u}_p\}$ be an orthonormal basis for $\mathbb{C}^p$. Define a linear transformation $V : \mathbb{C}^m \to \mathcal{H}$ by

$$(5.5.2) \qquad V\mathbf{x} = \sum_{j=1}^{p} V_j \mathbf{x} \otimes \mathbf{u}_j.$$

Then for $\psi = \sum_{j=1}^{p} \mathbf{y}_j \otimes \mathbf{u}_j \in \mathcal{H}$, $\langle \psi, V\mathbf{x} \rangle_{\mathcal{H}} = \sum_{j=1}^{p} \langle V_j^* \mathbf{y}_j, \mathbf{x} \rangle$, from which it follows that $V^* \psi = \sum_{j=1}^{p} V_j^* \mathbf{y}_j$. Thus for all $\mathbf{x} \in \mathbb{C}^m$,

$$V^* \pi(A) V\mathbf{x} = V^* \left(\sum_{j=1}^{p} AV_j \mathbf{x} \otimes \mathbf{u}_j \right) = \sum_{j=1}^{p} V_j^* AV_j \mathbf{x} = \Phi(A)\mathbf{x}.$$

Finally, note that $\Phi(\mathbb{1}) = V^* V$ so that Φ is unital if and only if V is an isometry from $\mathbb{C}^m$ into $\mathcal{H}$. $\qquad \square$

Definition 5.33. For a CP map Φ, a factorization of Φ of the form (5.5.1) is called a **Stinespring factorization** on $\mathcal{H}$. A Stinespring factorization is **minimal** in case the closed span of

$$(5.5.3) \qquad \{\pi(A) V\mathbf{x} \ : \ A \in M_n(\mathbb{C}), \ \mathbf{x} \in \mathbb{C}^m\}$$

is all of $\mathcal{H}$, the closure being automatic in this finite-dimensional case.

Lemma 5.34. *Stinespring's factorization (5.5.1), where $V : \mathbb{C}^m \to \mathcal{H}$ is given in terms of $\{V_1, \ldots, V_p\} \subset M_{n,m}(\mathbb{C})$ by (5.5.2), is minimal if and only if $\{V_1, \ldots, V_p\}$ is linearly independent.*

Proof. Let $\{\mathbf{u}_1, \ldots, \mathbf{u}_p\}$ be the orthonormal basis for $\mathbb{C}^p$ used to construct V in the proof of Theorem 5.32. The subspace (5.5.3) is a proper subspace of $\mathcal{H}$ if and only if there is a nonzero $\psi = \sum_{j=1}^{p} \mathbf{y}_j \otimes \mathbf{u}_j \in \mathbb{C}^n \otimes \mathbb{C}^p$ such that $\langle \psi, \pi(A) V\mathbf{x} \rangle_{\mathcal{H}} = 0$ for all $A \in M_n(\mathbb{C})$, $\mathbf{x} \in \mathbb{C}^m$. Then since

$$\langle \psi, \pi(A) V\mathbf{x} \rangle_{\mathcal{H}} = \left\langle \sum_{j=1}^{p} V_j^* A^* \mathbf{y}_j, \mathbf{x} \right\rangle,$$

Stinespring's factorization (5.5.1) specified by (5.5.2) is minimal if and only if whenever $(\mathbf{y}_1, \ldots, \mathbf{y}_p) \in \bigoplus^p \mathbb{C}^n$

$$(5.5.4) \qquad \sum_{j=1}^{p} V_j^* A^* \mathbf{y}_j = 0 \quad \text{for all } A \in M_n(\mathbb{C}) \quad \Rightarrow \quad (\mathbf{y}_1, \ldots, \mathbf{y}_p) = 0.$$

Suppose $\{V_1, \ldots, V_p\}$ is linearly independent. Suppose $(\mathbf{y}_1, \ldots, \mathbf{y}_p) \in \bigoplus^p \mathbb{C}^n$ is such that $\sum_{j=1}^p V_j^* A^* \mathbf{y}_j = 0$ for all $A \in M_n(\mathbb{C})$. For any nonzero $\mathbf{y}, \mathbf{z} \in \mathbb{C}^n$, define $A^* \in M_n(\mathbb{C})$ by $A^* = |\mathbf{z}\rangle\langle\mathbf{y}|$. Then

$$\sum_{j=1}^p V_j^* A^* \mathbf{y}_j = \left(\sum_{j=1}^p \langle \mathbf{y}, \mathbf{y}_j \rangle V_j^* \right) \mathbf{z}.$$

Since $\mathbf{z} \in \mathbb{C}^n$ is arbitrary, when the statement on the left side in (5.5.4) is valid, $\sum_{j=1}^p \langle \mathbf{y}, \mathbf{y}_j \rangle V_j^* = 0$. Since $\{V_1^*, \ldots, V_p^*\}$ is linearly independent, $\langle \mathbf{y}, \mathbf{y}_j \rangle = 0$ for all j and all $\mathbf{y}$ in $\mathbb{C}^n$, and and hence $\mathbf{y}_j = 0$ for each $1 \le j \le p$. This proves (5.5.4) when $\{V_1, \ldots, V_p\}$ is linearly independent.

However, if $\{V_1, \ldots, V_p\}$ is not linearly dependent, there is a nonzero $\mathbf{v} \in \mathbb{C}^p$ such that $\sum_{j=1}^p v_j V_j^* = 0$. Let $\mathbf{y} \in \mathbb{C}^n$ be any nonzero vector, and define $\mathbf{y}_j = v_j \mathbf{y}$, $1 \le j \le n$. Let $A = \mathbb{1}$. Then $\sum_{j=1}^p V_j^* A^* \mathbf{y}_j = \left(\sum_{j=1}^p v_j V_j^* \right) \mathbf{y} = 0$, but $(\mathbf{y}_1, \ldots, \mathbf{y}_p) \ne 0$. $\qquad\square$

Example 5.35. Let $\mathcal{H}_1$ and $\mathcal{H}_2$ be two finite-dimensional Hilbert spaces. The partial trace Tr_2 is a CP map from $\mathcal{B}(\mathcal{H}_1 \otimes \mathcal{H}_2)$ to $\mathcal{B}(\mathcal{H}_1)$. A useful Stinespring factorization may be constructed using the Kraus representation from Example 5.16. Let $\{\mathbf{v}_1, \ldots, \mathbf{v}_d\}$ be an orthonormal basis for $\mathcal{H}_2$. For $1 \le j \le d$, define $V_j : \mathcal{H}_1 \to \mathcal{H}_1 \otimes \mathcal{H}_2$ by

$$V_j \mathbf{x} = \mathbf{x} \otimes \mathbf{v}_j.$$

It is easy to check that $\mathrm{Tr}_2[A] = \sum_{j=1}^d V_j^* A V_j$ is a Kraus representation of Tr_2.

Starting from here, the construction used in the proof of Theorem 5.32 yields a Stinespring factorization of Tr_2. Let $\mathcal{H}_3$ be another Hilbert space of dimension d, the dimension of $\mathcal{H}_2$, and let $\{\mathbf{u}_1, \ldots, \mathbf{u}_d\}$ be an orthonormal basis for $\mathcal{H}_3$. Define $\mathcal{K} = \mathcal{H}_1 \otimes \mathcal{H}_2 \otimes \mathcal{H}_3$, and define a representation π of $\mathcal{B}(\mathcal{H}_1 \otimes \mathcal{H}_2)$ on $\mathcal{K}$ by $\pi(X) := X \otimes \mathbb{1}_{\mathcal{H}_3}$. Finally define $V : \mathcal{H}_1 \to \mathcal{K}$ by

$$V\mathbf{x} = \sum_{j=1}^d \mathbf{x} \otimes \mathbf{v}_j \otimes \mathbf{u}_j.$$

Then for all $\mathbf{x}, \mathbf{x}' \in \mathcal{H}_1$ and all $A \in \mathcal{B}(\mathcal{H}_1 \otimes \mathcal{H}_2)$, simple computations yield

$$\langle \mathbf{x}', V^* \pi(A) V \mathbf{x} \rangle = \langle \mathbf{x}', \mathrm{Tr}_2[A] \mathbf{x} \rangle,$$

and this verifies that $A \mapsto V^* \pi(A) V$ is a Stinespring factorization of Tr_2. Moreover, since $\{\mathbf{v}_1, \ldots, \mathbf{v}_d\}$ is orthonormal, $\{V_1, \ldots, V_d\}$ is linearly independent, and then by Lemma 5.34 this Stinespring factorization is minimal.

5.5.1. The structure of quantum operations.

When $\Phi : M_n(\mathbb{C}) \to M_m(\mathbb{C})$ is unital and CP, the Strinespring factorization $\Phi(A) = V^* \pi(A) V$ can be written in a form that is more useful for quantum mechanical applications. By Theorem 5.32, $V : \mathbb{C}^m \to \mathcal{H} = \mathbb{C}^n \otimes \mathbb{C}^p$ is an isometry and $p \le mn$. For present purposes, it is convenient to take $p = mn$, so that Φ has a Kraus representation of the form

$$\Phi(X) = \sum_{i=1}^{m} \sum_{j=1}^{n} V_{i,j}^* X V_{i,j},$$

which is exactly what we would get using all of the eigenvectors of the characteristic matrix C_Φ to construct the Kraus operators, even if the corresponding eigenvalues are zero. Then some of the Kraus operators will be zero, but they cause no harm.

Proceeding as in the proof of Theorem 5.32, define $\mathcal{H} := \mathbb{C}^n \otimes \mathbb{C}^n \otimes \mathbb{C}^m$. Let $\{\mathbf{u}_1, \ldots, \mathbf{u}_m\}$ and $\{\mathbf{v}_1, \ldots, \mathbf{v}_n\}$ be orthonormal bases for $\mathbb{C}^m$ and C^n, respectively. Define a map $V : \mathbb{C}^m \to \mathcal{H}$ by

$$V\mathbf{x} = \sum_{i=1}^{n} \sum_{j=1}^{m} V_{i,j}\mathbf{x} \otimes \mathbf{v}_i \otimes \mathbf{u}_j.$$

Define a representation π of $M_m(\mathbb{C})$ on $\mathcal{H}$ by $\pi(A) = A \otimes \mathbb{1} \otimes \mathbb{1}$. Then exactly as in the proof of Theorem 5.32, one finds that $V^*\pi(A)V = \Phi(A)$. Now suppose that Φ is unital. Then V is an isometry.

Now let $\{\phi_1, \ldots, \phi_{n^2}\}$ be an orthonormal basis of $\mathbb{C}^n \otimes \mathbb{C}^n$, and define an isometry $\widehat{V}$ from $\mathbb{C}^m$ into $\mathcal{H}$, namely $\widehat{V}\mathbf{x} = \phi_1 \otimes \mathbf{x}$. The general vector $\eta \in \mathcal{H}$ can be written as

$$(5.5.5) \qquad \eta = \sum_{j=1}^{n^2} \phi_j \otimes \eta_j$$

for uniquely determined $\eta_j \in \mathbb{C}^m$. Then $\langle \widehat{V}^*\eta, \mathbf{x} \rangle = \langle \eta, \widehat{V}\mathbf{x} \rangle = \langle \eta_1, \mathbf{x} \rangle$, and hence $\widehat{V}^*\eta = \eta_1$.

Since V and $\widehat{V}$ have the same rank, there is a (nonunique) unitary $\mathcal{U} \in \mathcal{B}(\mathcal{H})$ such that $V = \mathcal{U}\widehat{V}$, so that

$$\Phi(A) = \widehat{V}^*\big(\mathcal{U}^*(A \otimes \mathbb{1} \otimes \mathbb{1})\mathcal{U}\big)\widehat{V}.$$

This expresses Φ as the composition of three maps:

(1) the natural embedding $A \mapsto A \otimes \mathbb{1} \otimes \mathbb{1}$,

(2) the unitary conjugation $X \mapsto \mathcal{U}^*X\mathcal{U}$, and finally

(3) $X \mapsto \widehat{V}^*X\widehat{V}$.

Therefore, $\Phi^\dagger : M_m(\mathbb{C}) \to M_n(\mathbb{C})$ is the composition of the Hermitian conjugates of these three factors in the reverse order. In particular, the first factor in $\Phi^\dagger$ is the map $Y \mapsto \widehat{V}Y\widehat{V}^*$ from $M_m(\mathbb{C})$ to $\mathcal{B}(\mathcal{H})$. With $\eta \in \mathcal{H}$ given by (5.5.5),

$$\widehat{V}Y\widehat{V}^*\eta = \phi_1 \otimes Y\eta_1 = \big(|\phi_1\rangle\langle\phi_1| \otimes Y\big)\eta.$$

The adjoint of $A \mapsto A \otimes \mathbb{1} \otimes \mathbb{1}$ is Tr_{23}, the partial trace over the final two factors. Altogether,

$$(5.5.6) \qquad \Phi^\dagger(Y) := \mathrm{Tr}_{23}\left[\mathcal{U}(|\phi_1\rangle\langle\phi_1| \otimes Y)\mathcal{U}^* \right],$$

since the choice of the orthonormal basis $\{\phi_1, \ldots, \phi_{n^2}\}$ was arbitrary, ϕ_1 can be any unit vector in $\mathbb{C}^n \otimes \mathbb{C}^n$, though of course the unitary $\mathcal{U}$ depends on the choice of the orthonormal basis.

As discussed earlier, this factorization of $\Phi^\dagger : M_m(\mathbb{C}) \to M_n(\mathbb{C})$ may be understood in physical terms. Think of $\mathbb{C}^m$ as the Hilbert space for the *input system*, and $\mathbb{C}^n \otimes \mathbb{C}^n$ is the environment of the input system, Likewise, after applying the unitary operation $\mathcal{U}$, we may think of the first factor of $\mathbb{C}^n$ in $\mathcal{H}$ as the Hilbert space for the *output system*, and the remaining $\mathbb{C}^n \otimes \mathbb{C}^m$ in $\mathcal{H}$ as the environment of the output system. If the initial state of the input system is given by ρ and the initial state of its environment is given by $|\phi\rangle\langle\phi|$, the initial state of the combined system is $|\phi\rangle\langle\phi| \otimes \rho$. This state is then made to undergo some unitary evolution, resulting in $\mathcal{U}(|\phi\rangle\langle\phi| \otimes \rho)\mathcal{U}^*$. Finally, this state is restricted to to the output system, resulting in $\Phi^\dagger(\rho)$. Thus, every CP trace preserving map represents a quantum operation in this physical sense.

Theorem 5.36. *Let $\mathcal{H}$ and $\mathcal{K}$ be finite-dimensional Hilbert spaces, and let $\Phi^\dagger :$ $\mathcal{B}(\mathcal{H}) \to \mathcal{B}(\mathcal{K})$ be quantum operation. Then there is another Hilbert space $\widehat{\mathcal{K}}$ with same dimension as $\mathcal{K}$, a pure state $\omega \in \mathcal{B}(\mathcal{H} \otimes \widehat{\mathcal{K}})$, and a $\mathcal{U}$ unitary in $\mathcal{B}(\mathcal{K} \otimes \mathcal{H} \otimes \widehat{\mathcal{K}})$ such that*

$$(5.5.7) \qquad \Phi^\dagger(X) := \mathrm{Tr}_{\mathcal{H} \otimes \widehat{\mathcal{K}}}\left[\mathcal{U}(\omega \otimes X)\mathcal{U}^* \right].$$

Alternatively, there is a finite set $\{U_1, \ldots, U_N\}$ of unitaries in $\mathcal{B}(\mathcal{K} \otimes \mathcal{H} \otimes \widehat{\mathcal{K}})$ such that for all $X \in \mathcal{B}(\mathcal{H})$

$$(5.5.8) \qquad \frac{1}{d}\Phi^\dagger(X) \otimes \mathbb{1}_{\mathcal{H} \otimes \widehat{\mathcal{K}}} = \frac{1}{N}\sum_{j=1}^{N} U_j(\omega \otimes X)U_j,$$

where d is the dimension of $\mathcal{H} \otimes \widehat{\mathcal{K}}$.

Proof. The first statement is a simple rephrasing of (5.5.6) in Hilbert space terms with $\omega := |\phi\rangle\langle\phi|$ where ϕ is a unit vector in $\mathcal{H} \otimes \widehat{\mathcal{K}}$.

Then by Example 4.38 and Theorem 4.40, there is a set $\{G_1, \ldots, G_N\}$ of unitaries in $\mathcal{K} \otimes \mathcal{H} \otimes \widehat{\mathcal{K}}$ such that for all $A \in \mathcal{B}(\mathcal{K} \otimes \mathcal{H} \otimes \widehat{\mathcal{K}})$,

$$(5.5.9) \qquad \frac{1}{d}\mathrm{Tr}_{\mathcal{H} \otimes \widehat{\mathcal{K}}}[A] \otimes \mathbb{1}_{\mathcal{H} \otimes \widehat{\mathcal{K}}} = \frac{1}{N}\sum_{j=1}^{N} G_j A G_j^*.$$

Defining $U_j = \mathcal{U}G_j$ and combining (5.5.7) and (5.5.9) yields (5.5.8). $\qquad\square$

The physical significance of Theorem 5.36 is the following: every quantum operation of the form (5.5.7) corresponds to a sequence of operations one might perform in a laboratory. Moreover, the composition of any finite number of such quantum operations also corresponds to a sequence of operations that one might perform in a laboratory, but if the number of composition factors is large, the expression in terms of repeated attachment of ancilla, unitary transformations, and partial traces, over and over again, would be very complicated. Fortunately, Theorem 5.36 says that any such composition can be expressed in terms of a single attachment of ancilla, followed by a single unitary transformation, followed by a single partial trace.

5.6. Inequalities for 2-positive maps

A theorem giving necessary and sufficient conditions for a block matrix $\left[\begin{smallmatrix} X & K \\ K^* & Y \end{smallmatrix}\right]$ to be positive semidefinite was proved in [**3**, Theorem 1]. The conditions involve the Moore-Penrose generalized inverse A^+ of the matrix A; see Theorem 2.34.

Lemma 5.37. *For $X \in M_m^+(\mathbb{C})$, $Y \in M_n^+(\mathbb{C})$, and $K \in M_{m,n}(\mathbb{C})$ the following are equivalent.*

$$(5.6.1) \qquad \begin{bmatrix} X & K \\ K^* & Y \end{bmatrix} \geq 0.$$

$$(5.6.2) \qquad \ker(X) \subseteq \ker(K^*) \quad \text{and} \quad Y \geq K^* X^+ K.$$

$$(5.6.3) \qquad \ker(Y) \subseteq \ker(K) \quad \text{and} \quad X \geq K Y^+ K^*.$$

Proof. Suppose that (5.6.1) is satisfied. If the first condition in (5.6.2) is not satisfied, there exists $\mathbf{w} \in \mathbb{C}^m$ such that $X\mathbf{w} = 0$, but $\mathbf{v} := K^*\mathbf{w} \neq 0$. Then for all $t > 0$,

$$\left\langle \begin{pmatrix} -t\mathbf{w} \\ \mathbf{v} \end{pmatrix}, \begin{bmatrix} X & K \\ K^* & Y \end{bmatrix} \begin{pmatrix} -t\mathbf{w} \\ \mathbf{v} \end{pmatrix} \right\rangle = -2t\|\mathbf{v}\|^2 + \langle \mathbf{v}, Y\mathbf{v} \rangle.$$

For sufficiently large t, this is negative, which is impossible since (5.6.1) is satisfied. Hence $\ker(X) \subseteq \ker(K^*)$. Next, for any $\mathbf{v} \in \mathbb{C}^n$, define the vector $\begin{pmatrix} -X^+ K\mathbf{v} \\ \mathbf{v} \end{pmatrix} \in \mathbb{C}^{m+n}$. Then since $X^+ X X^+ = X^+$,

$$\left\langle \begin{pmatrix} -X^+ K\mathbf{v} \\ \mathbf{v} \end{pmatrix}, \begin{bmatrix} X & K \\ K^* & Y \end{bmatrix} \begin{pmatrix} -X^+ K\mathbf{v} \\ \mathbf{v} \end{pmatrix} \right\rangle = \langle \mathbf{v}, Y\mathbf{v} \rangle - \langle \mathbf{v}, K^* X^+ K\mathbf{v} \rangle.$$

The left side is nonnegative because of (5.6.1), and then since $\mathbf{v} \in \mathbb{C}^n$ is arbitrary, $Y \geq K^* X^+ K$. This completes the proof that (5.6.1) implies (5.6.2).

To show that (5.6.2) implies (5.6.1), define $Z := X^{1/2}$, and compute

$$(5.6.4) \qquad \begin{bmatrix} Z & 0 \\ K^*Z^+ & 0 \end{bmatrix} \begin{bmatrix} Z & Z^+K \\ 0 & 0 \end{bmatrix} = \begin{bmatrix} X & ZZ^+K \\ K^*Z^+Z & K^*X^+K \end{bmatrix}.$$

Note that $P := ZZ^+ = XX^+$ is the orthogonal projection onto the range of X, so that $P^\perp$ is the orthogonal projection onto $\ker(X)$. Since $\ker(X) \subseteq \ker(K^*)$, $K^*P^\perp = 0$, and hence $K^* = K^*P$, so that (5.6.4) becomes

$$\begin{bmatrix} Z & 0 \\ K^*Z^+ & 0 \end{bmatrix} \begin{bmatrix} Z & Z^+K \\ 0 & 0 \end{bmatrix} = \begin{bmatrix} X & K \\ K^* & K^*X^+K \end{bmatrix},$$

which proves that $\begin{bmatrix} X & K \\ K^* & K^*X^+K \end{bmatrix} \geq 0$. Evidently if $Y \geq K^*X^+K$, (5.6.1) is satisfied.

The upper left and lower right corners of $\begin{bmatrix} X & K \\ K^* & Y \end{bmatrix}$ are on an equal footing and hence the equivalence of (5.6.1) with (5.6.2) implies the equivalence of (5.6.1) with (5.6.3). $\qquad\square$

The realization that the characterization of positivity of 2×2 block matrices given by Lemma 5.37 is a useful source of matrix inequalities goes back to Anderson [6] who used Lemma 5.37 for $X, Y > 0$ to study the harmonic mean of two positive operators; see the discussion just above Definition 5.70 later in this chapter.

Our first application of Lemma 5.37 is to a fundamental theorem due to Lieb and Ruskai [144]. They proved that if $\Phi : M_n(\mathbb{C}) \to M_m(\mathbb{C})$ is CP with $\Phi(\mathbb{1}) > 0$, then for all $K \in M_n(\mathbb{C})$ and all $X \in M_n^{++}(\mathbb{C})$,

$$(5.6.5) \qquad \Phi\left(K^*\frac{1}{X}K\right) \geq \Phi(K)^*\frac{1}{\Phi(X)}\Phi(K).$$

Some years later, Choi [63, Proposition 4.1] extended this result: He relaxed the requirement that Φ be CP to the weaker requirement that Φ be 2-positive, keeping the requirement that $\Phi(\mathbb{1}) > 0$. He also proved that if $\Phi : M_n(\mathbb{C}) \to M_m(\mathbb{C})$ is a positive map such that $\Phi(\mathbb{1}) > 0$, then Φ is 2-positive if and only if (5.6.5) is valid for all $K \in M_n(\mathbb{C})$, and all $X \in M_n^{++}(\mathbb{C})$. Choi's proof was based on Lemma 5.37 in the special case in which $\ker(X) = 0$ and $\ker(Y) = 0$ so that conditions (5.6.2) and (5.6.3) are vacuous. The version of this result without any condition on $\Phi(\mathbb{1})$ is the following [52]:

Theorem 5.38. *A positive map $\Phi : M_n(\mathbb{C}) \to M_n(\mathbb{C})$ is 2-positive if and only if for all $K \in M_n(\mathbb{C})$ and all $X \in M_n^+(\mathbb{C})$ such that $\ker(X) \subseteq \ker(K^*)$,*

$$(5.6.6) \qquad \Phi(K^*X^+K) \geq \Phi(K)^*\Phi(X)^+\Phi(K).$$

The proof will use the following lemma from [52].

Lemma 5.39. *Let* $\Phi : M_n(\mathbb{C}) \to M_m(\mathbb{C})$ *be a positive map satisfying* (5.6.6) *for all* $X \in M_n^+(\mathbb{C})$ *and all* $K \in M_n(\mathbb{C})$ *such that* $\ker(X) \subseteq \ker(K^*)$. *Then for all such* X *and* K, $\ker(\Phi(X)) \subseteq \ker(\Phi(K)^*)$.

Proof. First consider the case in which $X = \mathbb{1}$ so that for all $L \in M_n(\mathbb{C})$, $0 = \ker(X) \subseteq \ker(L^*)$. Then by (5.6.6)

$$\Phi(L^*L) \geq \Phi(L)^*\Phi(\mathbb{1})^+\Phi(L) = |(\Phi(\mathbb{1})^+)^{1/2}\Phi(L)|^2.$$

By Lemma 5.2, $\mathrm{ran}(\Phi(L)) \subseteq \mathrm{ran}(\Phi(\mathbb{1})) = \ker(\Phi(\mathbb{1}))^\perp = \ker(\Phi(\mathbb{1})^+)^\perp$. Therefore,

$$(5.6.7) \qquad \ker(\Phi(L^*L)) \subseteq \ker(\Phi(L)) \qquad \text{for all} \quad L \in M_n(\mathbb{C}).$$

With this established, consider any $X \in M_n^+(\mathbb{C})$, and any $K \in M_n(\mathbb{C})$ such that $\ker(X) \subseteq \ker(K^*)$. Because of this last condition, $X \geq \lambda KK^*$ for some $\lambda > 0$, and then since Φ is positive $\Phi(X) \geq \lambda\Phi(KK^*)$, and then by (5.6.7) with K^* in place of L, $\ker(\Phi(X)) \subseteq \ker(\Phi(KK^*)) \subseteq \ker(\Phi(K)^*)$. $\qquad\square$

Proof of Theorem 5.38. Suppose that Φ is 2-positive, and $\ker(X) \subseteq \ker(K^*)$. By Lemma 5.37 $\begin{bmatrix} X & K \\ K^* & K^*X^+K \end{bmatrix} \geq 0$ and then since Φ is 2 positive, $\begin{bmatrix} \Phi(X) & \Phi(K) \\ \Phi(K)^* & \Phi(K^*X^+K) \end{bmatrix} \geq 0$. Then by Lemma 5.37 once again, (5.6.6) is valid.

For the converse, suppose that $\begin{bmatrix} X & K \\ K^* & Y \end{bmatrix} \geq 0$. By Lemma 5.37, $\ker(X) \subseteq \ker(K^*)$ and $Y \geq K^*X^+K$. By (5.6.6) and the positivity of Φ,

$$\Phi(Y) \geq \Phi(K^*X^+K) \geq \Phi(K)^*\Phi(X)^+\Phi(K),$$

and by Lemma 5.39, $\ker(\Phi(X)) \subseteq \ker(\Phi(K)^*)$. Therefore,

$$\begin{bmatrix} \Phi(X) & \Phi(K) \\ \Phi(K)^* & \Phi(Y) \end{bmatrix} \geq \begin{bmatrix} \Phi(X) & \Phi(K) \\ \Phi(K)^* & \Phi(K)^*\Phi(X)^+\Phi(K) \end{bmatrix} \geq 0,$$

where the last inequality is from By Lemma 5.37 once again. This shows that Φ is 2-positive. $\qquad\square$

Corollary 5.40 (Kadison-Choi inequality)**.** *Let* $\Phi : M_m(\mathbb{C}) \to M_n(\mathbb{C})$ *be 2-positive and quasi unital; that is,* $\Phi(\mathbb{1})$ *is an orthogonal projection. Then for all* $K \in M_m(\mathbb{C})$,

$$(5.6.8) \qquad\qquad \Phi(K^*K) \geq \Phi(K)^*\Phi(K).$$

Proof. Taking $X = \mathbb{1}$, the condition $\ker(X) \subseteq \ker(K^*)$ of Theorem 5.38 is trivially satisfied so that $\Phi(K^*K) \geq \Phi(K)^*\Phi(\mathbb{1})^+\Phi(K)$. Then by Lemma 5.6 and the fact that $P^+ = P$ for any orthogonal projection P, $\Phi(\mathbb{1})^+\Phi(K) = \Phi(K)$. This proves (5.6.8). $\qquad\square$

Kadison [**125**] proved the inequality (5.6.8) for all *self-adjoint K* and all *positive* Φ, and he referred to this inequality as a Schwarz inequality.

Definition 5.41 (Schwarz map). A linear map $\Phi : M_n(\mathbb{C}) \to M_m(\mathbb{C})$ is a **Schwarz map** in case

$$\Phi(A^*A) \geq \Phi(A)^*\Phi(A)$$

is satisfied for all $A \in M_n(\mathbb{C})$.

Example 5.42. By Corollary 5.40, the class of Schwarz maps includes all quasi unital 2-positive maps, but Choi proved in [**63**, Appendix A] that there exist Schwarz maps that are not 2-positive, using the following example: Given $0 \leq \lambda \leq 1$, define $\Phi_\lambda : M_2(\mathbb{C}) \to M_2(\mathbb{C})$ by

$$\Phi_\lambda(A) = \lambda A^T + (1 - \lambda)\frac{1}{2}\operatorname{Tr}[A]\mathbb{1}.$$

Evidently, each Φ_λ is a positive unital map. Φ_0 is the conditional expectation onto $\mathcal{N} = \mathbb{C}\mathbb{1}$, and hence is CP, while Φ_1 is the map considered in Example 5.11 that we have seen is not even 2-positive.

Because Φ_λ is unital, for all $z \in \mathbb{C}$ and all $A \in M_2(\mathbb{C})$,

$$\Phi_\lambda(A^*A) - \Phi_\lambda(A)^*\Phi_\lambda(A) = \Phi_\lambda((A - z\mathbb{1})^*(A - z\mathbb{1})) - \Phi_\lambda(A - z\mathbb{1})^*\Phi_\lambda(A - z\mathbb{1}),$$

and hence Φ_λ is a Schwarz map if and only if $\Phi_\lambda(A^*A) \geq \Phi_\lambda(A)^*\Phi_\lambda(A)$ for all traceless $A \in M_2(\mathbb{C})$.

For traceless A, $\Phi_\lambda(A) = \lambda A^T$, and hence

$$\Phi_\lambda(A)^*\Phi_\lambda(A) = \lambda^2(A^*)^T A^T = \lambda^2\overline{AA^*},$$

while

$$\Phi_\lambda(A^*A) = \lambda(A^*A)^T + (1 - \lambda)\frac{1}{2}\operatorname{Tr}[A^*A]\mathbb{1}.$$

Since $\sigma(\overline{AA^*}) = \sigma(AA^*) = \sigma(A^*A)$, $\operatorname{Tr}[A^*A]\mathbb{1} \geq \overline{AA^*}$, and hence

$$\Phi_\lambda(A^*A) - \Phi_\lambda(A)^*\Phi_\lambda(A) \geq \lambda(A^*A)^T + \left(\frac{1-\lambda}{2} - \lambda^2\right)\overline{AA^*}.$$

Thus Φ_λ is a Schwarz map for $0 \leq \lambda \leq 1/2$.

However, with A, B, C, and D as in Example 5.11,

$$\begin{bmatrix} \Phi_\lambda(A) & \Phi_\lambda(B) \\ \Phi_\lambda(C) & \Phi_\lambda(D) \end{bmatrix} = \begin{bmatrix} (1+\lambda)/2 & 0 & 0 & 0 \\ 0 & (1-\lambda)/2 & \lambda & 0 \\ 0 & \lambda & (1-\lambda)/2 & 0 \\ 0 & 0 & 0 & (1+\lambda)/2 \end{bmatrix},$$

and this matrix is positive semidefinite if and only if $\lambda \leq 1/3$. Hence Φ_λ is not 2-positive for $\lambda > 1/3$. In summary, for $1/3 < \lambda \leq 1/2$, the unital map Φ_λ is a Schwarz map, but is not 2-positive. This construction of Choi's was further developed in [**209**] and [**106**, Example 3.6].

As a consequence of Theorem 5.38, we obtain a theorem that extends an important convexity result first proved by Kiefer [**127**].

Theorem 5.43 (Kiefer's theorem). *Define $D \subset M_{m,n}(\mathbb{C}) \times M_m^+(\mathbb{C})$ by*

$$D = \{(K,X) \ : \ \ker(X) \subset \ker(K^*)\}.$$

Then D is a convex cone, and the function $(K,X) \mapsto K^*X^+K$ is convex from D to $M_n^+(C)$.

Proof. Let $(K_0,X_0),(K_1,X_1) \in D$. For $0 < \lambda < 1$, $\ker((1-\lambda)X_0 + \lambda X_1) = \ker(X_0) \cap \ker(X_1)$, and this is contained in

$$\ker(K_0^*) \cap \ker(K_1^*) \subseteq \ker((1-\lambda)K_0^* + \lambda K_1^*).$$

This shows that D is convex, and evidently if $(K,X) \in D$ and $\lambda > 0$, $(\lambda K, \lambda X) \in D$. Therefore D is a convex cone.

Next, again for $(K_0,X_0),(K_1,X_1) \in D, 0 < \lambda < 1$, define

$$\widehat{K} := \begin{bmatrix} (1-\lambda)K_0 & 0 \\ 0 & \lambda K_1 \end{bmatrix} \quad \text{and} \quad \widehat{X} := \begin{bmatrix} (1-\lambda)X_0 & 0 \\ 0 & \lambda X_1 \end{bmatrix}.$$

Then $\ker(\widehat{X}) \subset \ker(\widehat{K}^*)$. We now apply Theorem 5.38 taking Φ to be the partial trace $\Phi\left(\begin{bmatrix} A & B \\ C & D \end{bmatrix}\right) = A + D$ as in Example 5.16. Since Φ is CP , Theorem 5.38 gives

$$\Phi(\widehat{K}^*\widehat{X}^+\widehat{K}) \geq \Phi(\widehat{K})^*\Phi(\widehat{X})^+\Phi(\widehat{K}).$$

Writing this out,

$$(1-\lambda)K_0^*X_0^+K_0 + \lambda K_1^*X_1^+K_1$$
$$\geq ((1-\lambda)K_0 + \lambda K_1)^*((1-\lambda)X_0 + \lambda X_1)^+((1-\lambda)K_0 + \lambda K_1),$$

which proves the convexity. $\qquad\qquad\square$

Remark 5.44. The function $(K,X) \mapsto K^*X^+K$ is homogeneous of degree one, and then as a consequence (see the simple Lemma 6.7 below), the convexity inequality is equivalent to the *subadditivity inequality*,

$$(5.6.9) \qquad (K_0 + K_1)^*(X_0 + X_1)^+(K_0 + K_1) \leq K_0^*X_0^+K_0 + K_1^*X_1^+K_1$$

for all $(K_0,X_0),(K_1,X_1) \in D$.

The next theorem, concerning cases of equality in the Schwarz inequality (5.6.8), is due to Choi [**63**].

Theorem 5.45. *Let $\mathcal{M}$ and $\mathcal{N}$ be von Neumann subalgebras of $\mathcal{B}(\mathcal{H})$ and $\mathcal{B}(\mathcal{K})$, respectively, and let $\Phi \ : \ \mathcal{M} \to \mathcal{N}$ be CP and unital. Then*

$$\mathcal{C} := \{X \in \mathcal{M} \ : \ \Phi(X^*X) = \Phi(X)^*\Phi(X) \quad and \quad \Phi(XX^*) = \Phi(X)\Phi(X)^*\}$$

is a von Neumann subalgebra of $\mathcal{M}$, and Φ is a $*$-homomorphism when restricted to $\mathcal{C}$. Moreover, for all $X \in \mathcal{C}$ and all $Y \in \mathcal{M}$,

$$(5.6.10) \qquad \Phi(XY) = \Phi(X)\Phi(Y) \quad and \quad \Phi(YX) = \Phi(Y)\Phi(X).$$

Proof. Note that

$$\begin{bmatrix} A^*A & A^*B^* \\ BA & BB^* \end{bmatrix} = \begin{bmatrix} A & B^* \\ 0 & 0 \end{bmatrix}^* \begin{bmatrix} A & B^* \\ 0 & 0 \end{bmatrix}.$$

Therefore, since Φ is CP and unital, so is $\Phi \otimes \mathbb{1}$, and we have the Schwarz inequality

$$\begin{bmatrix} \Phi(A^*A) & \Phi(A^*B^*) \\ \Phi(BA) & \Phi(BB^*) \end{bmatrix} \geq \begin{bmatrix} \Phi(A) & \Phi(B^*) \\ 0 & 0 \end{bmatrix}^* \begin{bmatrix} \Phi(A) & \Phi(B^*) \\ 0 & 0 \end{bmatrix},$$

and this is

$$(5.6.11) \qquad \begin{bmatrix} \Phi(A^*A) - \Phi(A)^*\Phi(A) & \Phi(A^*B^*) - \Phi(A)^*\Phi(B)^* \\ \Phi(BA) - \Phi(B)\Phi(A) & \Phi(BB^*) - \Phi(B)\Phi(B)^* \end{bmatrix} \geq 0.$$

Evidently, $\mathcal{C}$ is closed under the involution and $\mathbb{1} \in \mathcal{C}$. Now suppose $A \in \mathcal{C}$. Then for all $B \in \mathcal{M}$, the upper left entry of the block matrix in (5.6.11) is zero, and hence by Lemma 5.37 the off-diagonal entries must also be zero. In particular for all $B \in \mathcal{M}$, $\Phi(BA) = \Phi(B)\Phi(A)$. Next, suppose $B \in \mathcal{C}$. Then for all $A \in \mathcal{M}$, the lower right entry of the block matrix in (5.6.11) is zero, and hence by Lemma 5.37, the off-diagonal entries must also be zero. In particular for all $A \in \mathcal{M}$, $\Phi(BA) = \Phi(B)\Phi(A)$. This proves (5.6.10).

In particular, when both $X, Y \in \mathcal{C}$, repeatedly using (5.6.10),

$$\Phi(XYY^*X^*) = \Phi(X)\Phi(Y)\Phi(Y^*)\Phi(X^*) = \Phi(XY)\Phi((XY)^*),$$

and likewise $\Phi(Y^*X^*XY) = \Phi((XY)^*)\Phi(XY)$. This shows that $XY \in \mathcal{C}$. Also by (5.6.10),

$$\Phi((X+Y)(X+Y)^*) = \Phi(X+Y)\Phi((X+Y)^*)$$

and

$$\Phi((X+Y)^*(X+Y)) = \Phi((X+Y)^*)\Phi(X+Y).$$

This completes the proof that $\mathcal{C}$ is a subalgebra of $\mathcal{M}$, and the fact that Φ restricted to $\mathcal{C}$ is a $*$-homomorphism is immediate from (5.6.10). $\qquad \square$

Definition 5.46 (Multiplicative domain). The von Neumann algebra $\mathcal{C}$ in Theorem 5.45 is called the **multiplicative domain of** Φ.

Example 5.47. Let $\mathcal{M}$ be a von Neumann algebra and let $\mathcal{N}$ be a von Neumann subalgebra of $\mathcal{M}$. Let $\mathcal{E}_{\mathcal{N}}$ denote the conditional expectation onto $\mathcal{N}$, and let $\mathcal{C}$ be the multiplicative domain of $\mathcal{E}_{\mathcal{N}}$. For all $X \in \mathcal{N}$, X^*, X^*X, and $XX^* \in \mathcal{N}$, and hence

$$\mathcal{E}_{\mathcal{N}}(X^*X) = X^*X = \mathcal{E}_{\mathcal{N}}(X^*)\mathcal{E}_{\mathcal{N}}(X),$$

and likewise $\mathcal{E}_{\mathcal{N}}(XX^*) = \mathcal{E}_{\mathcal{N}}(X)\mathcal{E}_{\mathcal{N}}(X^*)$. Hence $\mathcal{N} \subseteq \mathcal{C}$.

Now let $X \in \mathcal{C}$ of $\mathcal{E}_{\mathcal{N}}$, and define $Y := X - \mathcal{E}_{\mathcal{N}} X$. Then

$$Y^*Y = X^*X - X^*\mathcal{E}_{\mathcal{N}}(X) - \mathcal{E}_{\mathcal{N}}(X)^*X + \mathcal{E}_{\mathcal{N}}(X)^*\mathcal{E}_{\mathcal{N}}(X).$$

Since $X \in \mathcal{C}$ and $\mathcal{E}_{\mathcal{N}} X \in \mathcal{N} \subseteq \mathcal{C}$, applying $\mathcal{E}_{\mathcal{N}}$ to both sides and using (5.6.10) yields

$$\mathcal{E}_{\mathcal{N}}(Y^*Y) = \mathcal{E}_{\mathcal{N}}(X)^*\mathcal{E}_{\mathcal{N}}(X) - 2\mathcal{E}_{\mathcal{N}}(X)^*\mathcal{E}_{\mathcal{N}}(X) + \mathcal{E}_{\mathcal{N}}(X)^*\mathcal{E}_{\mathcal{N}}(X) = 0.$$

Since $\mathcal{E}_{\mathcal{N}}$ is trace preserving, $\mathrm{Tr}[Y^*Y] = 0$, and hence $Y = 0$. Therefore $X = \mathcal{E}_{\mathcal{N}} X \in \mathcal{N}$. This proves that $\mathcal{C} \subseteq \mathcal{N}$, and hence $\mathcal{C} = \mathcal{N}$.

5.7. The mean ergodic theorem

The following theorem is fundamental to the rest of this chapter.

Theorem 5.48 (Mean ergodic theorem). *Let $\mathcal{M}$ be a von Neumann subalgebra of $\mathcal{B}(\mathcal{H})$ for a finite-dimensional Hilbert space $\mathcal{H}$. Let $\Phi : \mathcal{M} \to \mathcal{M}$ be unital and CP. Then the following limit exists:*

$$(5.7.1) \qquad \mathcal{P}_\Phi := \lim_{N \to \infty} \frac{1}{N} \sum_{n=1}^{N} \Phi^n$$

and $\mathcal{P}_\Phi$ is a unital and CP transformation on $\mathcal{M}$. Moreover, $\mathcal{P}_\Phi^\dagger$, the Hermitian adjoint of $\mathcal{P}_\Phi$, is given by

$$(5.7.2) \qquad \mathcal{P}_\Phi^\dagger := \lim_{N \to \infty} \frac{1}{N} \sum_{n=1}^{N} (\Phi^\dagger)^n$$

and $\mathcal{P}_\Phi^\dagger$ is a trace preserving and CP transformation on $\mathcal{M}$.

For a general von Neumann algebra version, see [**137**]. In our finite-dimensional setting, all of the natural topologies on the space of linear transformations on $\mathcal{M}$ are equivalent, and we need not specify the sense of convergence. Before proving Theorem 5.48, we discuss an application to *invariant states*. Recall that a state on $\mathcal{M}$ is a positive linear functional ρ from $\mathcal{M}$ to $\mathbb{C}$ such that $\rho(\mathbb{1}) = 1$. If $\Phi : \mathcal{M} \to \mathcal{M}$ is unital and CP (even unital and positive), then $\rho \circ \Phi$ is also a state on $\mathcal{M}$. Thus the map $\rho \mapsto \rho \circ \Phi$ takes $\mathcal{S}_{\mathcal{M}}$ into itself. The fixed points of this map on $\mathcal{S}_{\mathcal{M}}$ are the *invariant states* under Φ.

Definition 5.49. Let $\mathcal{M}$ be a von Neumann algebra on a finite-dimensional Hilbert space $\mathcal{H}$. Let $\Phi : \mathcal{M} \to \mathcal{M}$ be unital and CP. A state ρ on $\mathcal{M}$ is an **invariant state for** Φ in case $\rho \circ \Phi = \rho$. The set of states on $\mathcal{M}$ that are invariant under Φ is denoted $\mathcal{S}_{\mathcal{M},\Phi}^{\mathrm{INV}}$.

Remark 5.50. By Theorem 5.30, there is a one-to-one correspondence between states ρ on $\mathcal{M}$ and density matrices $\rho \in \mathcal{M}$, and the correspondence is such that $\rho(X) = \text{Tr}[\rho X]$ for all $X \in \mathcal{M}$. The state ρ is invariant under Φ if and only if

$$\rho(X) = \rho \circ \Phi(X) = \text{Tr}[\rho\Phi(X)] = \text{Tr}[\Phi^\dagger \rho X]$$

for all $X \in \mathcal{M}$. Hence the states in $\mathcal{S}^{\text{INV}}_{\mathcal{M},\Phi}$ are in one-to-one correspondence with the density matrices $\rho \in \mathcal{M}$ such that

$$(5.7.3) \qquad\qquad\qquad \Phi^\dagger \rho = \rho.$$

Since $\mathcal{S}_{\mathcal{M}}$ is a compact convex set, and since $\rho \mapsto \rho \circ \Phi$ is continuous from $\mathcal{S}_{\mathcal{M}}$ into itself, the Brouwer fixed point theorem can be, and often is, used to show that $\mathcal{S}^{\text{INV}}_{\mathcal{M},\Phi}$ is nonempty. However, a proof using Theorem 5.48 yields other important information about $\mathcal{S}^{\text{INV}}_{\mathcal{M},\Phi}$.

Theorem 5.51. *Let $\mathcal{M}$ be a von Neumann algebra on a Hilbert space $\mathcal{H}$ of dimension d. Let $\Phi : \mathcal{M} \to \mathcal{M}$ be unital and CP. Then:*

(1) $\mathcal{S}^{\text{INV}}_{\mathcal{M},\Phi} \neq \emptyset$.

(2) *The exists a state $\bar\rho \in \mathcal{S}^{\text{INV}}_{\mathcal{M},\Phi}$ that has maximal support in the sense that* $\text{supp}(\rho) \leq \text{supp}(\bar\rho)$ *for all* $\rho \in \mathcal{S}^{\text{INV}}_{\mathcal{M},\Phi}$.

(3) $\mathcal{S}^{\text{INV}}_{\mathcal{M},\Phi} = \mathcal{S}^{\text{INV}}_{\mathcal{M},\mathcal{P}_\Phi}$ *where $\mathcal{P}_\Phi$ is the unital CP map defined in* (5.7.1).

Proof. To prove (1), it suffices by Remark 5.50 to find $\rho \in \mathfrak{S}_{\mathcal{H}} \cap \mathcal{M}$ such that (5.7.3) is satisfied. Since $\mathcal{P}^\dagger_\Phi$ is trace preserving and CP, for any choice of $\rho_0 \in \mathfrak{S}_{\mathcal{H}} \cap \mathcal{M}$, $\mathcal{P}^\dagger_\Phi \rho_0 \in \mathfrak{S}_{\mathcal{H}} \cap \mathcal{M}$. Also since the identity $\Phi^\dagger \circ \mathcal{P}^\dagger_\Phi = \mathcal{P}^\dagger_\Phi$ follows directly from (5.7.2), $\Phi^\dagger(\mathcal{P}^\dagger_\Phi \rho_0) = \mathcal{P}^\dagger_\Phi \rho_0$. Thus, for any choice of $\rho_0 \in \mathfrak{S}_{\mathcal{H}} \cap \mathcal{M}$, $\mathcal{P}^\dagger_\Phi \rho_0$ satisfies (5.7.3), and hence $\mathcal{S}^{\text{INV}}_{\mathcal{M},\Phi}$ is not empty.

To prove (2), we make a particular choice of ρ_0. Since every $\rho \in \mathfrak{S}_{\mathcal{H}}$ satisfies $\rho \leq \mathbb{1}$, every $\rho \in \mathfrak{S}_{\mathcal{H}}$ satisfies $\rho \leq d\rho_0$. Since $\mathcal{P}^\dagger_\Phi$ is positive, $\mathcal{P}^\dagger_\Phi \rho \leq d\mathcal{P}^\dagger_\Phi \rho_0$. If ρ is the density matrix of an invariant state, then $\rho = \Phi^\dagger \rho$, and hence $\rho = \mathcal{P}^\dagger_\Phi \rho$, so that

$$\rho = \mathcal{P}^\dagger_\Phi \rho \leq d\mathcal{P}^\dagger_\Phi \rho_0,$$

and this implies that $\text{supp}(\rho) \subseteq \text{supp}(\mathcal{P}^\dagger_\Phi \rho_0)$. Defining $\bar\rho := \mathcal{P}^\dagger_\Phi \rho_0$ for this choice of ρ_0, gives us a density matrix $\bar\rho \in \mathcal{M}$ such that $\Phi^\dagger \bar\rho = \bar\rho$, such that $\text{supp}(\bar\rho)$ contains the support of every other density matrix ρ in $\mathcal{M}$ that satisfies (5.7.3).

To prove (3), note that if $\rho = \rho \circ \Phi$, then $\rho = \rho \circ \Phi^m$ for all positive m, and hence by (5.7.1), $\rho = \rho \circ \mathcal{P}_\Phi$. Conversely, suppose that $\rho = \rho \circ \mathcal{P}_\Phi$. Again by (5.7.1), $\Phi_0 \circ \mathcal{P}_{\Phi_0} = \mathcal{P}_{\Phi_0} = \mathcal{P}_{\Phi_0} \circ \Phi_0$. Therefore,

$$\rho \circ \Phi_0 = (\rho \circ \mathcal{P}_{\Phi_0}) \circ \Phi_0 = \rho \circ (\mathcal{P}_{\Phi_0} \circ \Phi_0) = \rho \circ \mathcal{P}_{\Phi_0} = \rho.$$

Thus, ρ is invariant under Φ_0. $\qquad\square$

It is important in many problems in quantum physics to have detailed information about the structure of $\mathcal{S}_{\mathcal{M},\Phi}^{\mathrm{INV}}$, and we return to this topic in the next section after first proving Theorem 5.48, which we deduce from a very general result in linear algebra, the **ergodic theorem for contractions**, proved here for completeness.

Let $\mathcal{V}$ be any finite-dimensional vector space equipped with a norm $\lvert\!\lvert\!\lvert \cdot \rvert\!\rvert\!\rvert$. That is, $\mathbf{v} \mapsto \lvert\!\lvert\!\lvert \mathbf{v} \rvert\!\rvert\!\rvert$ is any function on $\mathcal{V}$ with values in $[0, \infty)$ such that $\lvert\!\lvert\!\lvert \mathbf{v} \rvert\!\rvert\!\rvert = 0$ only if $\mathbf{v} = 0$, $\lvert\!\lvert\!\lvert z\mathbf{v} \rvert\!\rvert\!\rvert = |z| \lvert\!\lvert\!\lvert \mathbf{v} \rvert\!\rvert\!\rvert$ for all $z \in \mathbb{C}$ and $\mathbf{v} \in \mathcal{V}$, and finally, $\lvert\!\lvert\!\lvert \mathbf{v} + \mathbf{w} \rvert\!\rvert\!\rvert \leq \lvert\!\lvert\!\lvert \mathbf{v} \rvert\!\rvert\!\rvert + \lvert\!\lvert\!\lvert \mathbf{w} \rvert\!\rvert\!\rvert$ for all $\mathbf{v}, \mathbf{w} \in \mathcal{V}$.

A linear transformation T on $\mathcal{V}$ is a **contraction** in case $\lvert\!\lvert\!\lvert T\mathbf{v} \rvert\!\rvert\!\rvert \leq \lvert\!\lvert\!\lvert \mathbf{v} \rvert\!\rvert\!\rvert$ for all $\mathbf{v} \in \mathcal{V}$. Recall that a nonzero $\mathbf{v} \in \mathcal{V}$ is a *generalized eigenvector of T* for the eigenvalue λ in case for some $m \in \mathbb{N}$, $(T - \lambda\mathbb{1})^m\mathbf{v} = 0$. Then $\mathbf{v}$ is an eigenvector if the least value of m for which this is true is $m = 1$, but otherwise not, and in any case, if $\mathbf{v}$ is a generalized eigenvector with eigenvalue λ, the least value of m for which $(T - \lambda\mathbb{1})^m\mathbf{v} = 0$ is no more than the dimension of $\mathcal{V}$. (Note that the notion of eigenvalue is not generalized. Even when we are discussing generalized eigenvectors, $\sigma(T)$, the spectrum of T, is still the set of all eigenvalues of T.)

For any linear transformation on a finite-dimensional vector space $\mathcal{V}$, the space $\mathcal{V}$ is the direct sum of the generalized eigenspaces $\mathcal{V}_\lambda$, $\lambda \in \sigma(T)$. That is, $\mathcal{V} = \bigoplus_{\lambda \in \sigma(T)} \mathcal{V}_\lambda$. (This is usually discussed in linear algebra texts in connection with the Jordan normal form). Therefore, there exist well defined projections onto each of the generalized eigenspaces. (This is one of the few contexts in which we are dealing with nonorthogonal projections. In fact, here we are using the theory of linear transformations without reference to any inner product.) For an eigenvalue λ of T, the *geometric multiplicity* of λ is the dimension of the corresponding eigenspace, and the *algebraic multiplicity* of λ is the dimension of the corresponding generalized eigenspace. Thus, the algebraic multiplicity is greater than or equal to the geometric multiplicity, with equality if and only if all generalized eigenvectors are actually eigenvectors.

Lemma 5.52. *Let $\mathcal{V}$ be any finite-dimensional vector space equipped with a norm $\lvert\!\lvert\!\lvert \cdot \rvert\!\rvert\!\rvert$, and let T be a contraction on $\mathcal{V}$. Then all of the eigenvalues λ of T satisfy $|\lambda| \leq 1$. Moreover, if λ is an eigenvalue of T with $|\lambda| = 1$, then the algebraic and geometric multiplicities of λ are equal.*

Proof. If $\mathbf{v}$ is an eigenvector with eigenvalue λ, then $|\lambda| \lvert\!\lvert\!\lvert \mathbf{v} \rvert\!\rvert\!\rvert = \lvert\!\lvert\!\lvert T\mathbf{v} \rvert\!\rvert\!\rvert \leq \lvert\!\lvert\!\lvert \mathbf{v} \rvert\!\rvert\!\rvert$, and hence $|\lambda| \leq 1$.

Next, suppose λ is an eigenvalue of T with $|\lambda| = 1$. If there exist generalized eigenvectors for λ that are not eigenvectors, then there exists a vector $\mathbf{v} \in \mathcal{V}$ such that $(T - \lambda\mathbb{1})^2\mathbf{v} = 0$ but $\mathbf{w} = T\mathbf{v} - \lambda\mathbf{v} \neq 0$. By the binomial formula,

$T^n \mathbf{v} = (\lambda \mathbb{1} + (T - \lambda \mathbb{1}))^n (\mathbf{v}) = \lambda^n \mathbf{v} + n \lambda^{n-1} \mathbf{w}$ so that

$$\||T^n \mathbf{v}\|| \geq n \||\mathbf{w}\|| - \||\mathbf{v}\||.$$

For sufficiently large n, $\||T^n \mathbf{v}\|| > \||\mathbf{v}\||$, which is impossible. Therefore, if λ is an eigenvalue of T, and $|\lambda| = 1$, $\mathcal{V}_\lambda$ is actually the eigenspace of T for eigenvalue λ. $\qquad\square$

Theorem 5.53 (Ergodic theorem for contractions). *Let $\mathcal{V}$ be a finite-dimensional vector space, and let T be a linear transformation on $\mathcal{V}$ that is a contraction with respect to some norm $\|| \cdot \||$ on $\mathcal{V}$. If $1 \in \sigma(T)$, define P_T to be the projection onto $\mathcal{V}_1$ corresponding to the decomposition of $\mathcal{V}$ into the direct sum $\mathcal{V} = \bigoplus_{\lambda \in \sigma(T)} \mathcal{V}_\lambda$ of the generalized eigenspaces of T. If $1 \notin \sigma(T)$, define $P_T = 0$. Then $P_T = \lim_{N \to \infty} \frac{1}{N} \sum_{n=1}^{N} T^n$.*

Proof. By Lemma 5.52, provided $1 \in \sigma(T)$, every $\mathbf{v} \in \mathcal{V}_1$ satisfies $T^n \mathbf{v} = \mathbf{v}$ for all $n \in \mathbb{N}$, and hence $\lim_{N \to \infty} \frac{1}{N} \sum_{n=1}^{N} T^n \mathbf{v} = \mathbf{v}$.

Then, because $\mathcal{V} = \bigoplus_{\lambda \in \sigma(T)} \mathcal{V}_\lambda$, if suffices to prove that for all $\mathbf{v} \in \mathcal{V}_\lambda$, $\lambda \neq 1$,

$$(5.7.4) \qquad\qquad \lim_{N \to \infty} \frac{1}{N} \sum_{j=1}^{N} T^n \mathbf{v} = 0.$$

Suppose $\lambda \in \sigma(T)$, and $|\lambda| = 1$ but $\lambda \neq 1$. By Lemma 5.52, every $\mathbf{v} \in \mathcal{V}_\lambda$ is an eigenvector of T, so that $T\mathbf{v} = \lambda \mathbf{v}$, and hence

$$\lim_{N \to \infty} \frac{1}{N} \sum_{j=1}^{N} T^n \mathbf{v} = \lim_{N \to \infty} \frac{1}{N} \frac{\lambda^{N+1} - \lambda}{\lambda - 1} \mathbf{v} = 0.$$

Again by Lemma 5.52, if $\lambda \in \sigma(T)$ and $|\lambda| \neq 1$, then $|\lambda| < 1$. Let $\mathbf{v} \in \mathcal{V}_\lambda$. Since $(T - \lambda \mathbb{1})^k \mathbf{v} = 0$ for $k > \dim(\mathcal{V})$, applying the binomial formula to $T^n \mathbf{v} = (\lambda \mathbb{1} + (T - \lambda \mathbb{1}))^n \mathbf{v}$ yields a sum of at most $\dim(\mathcal{V}) + 1$ terms, each of which goes to zero as n goes to infinity. This shows that $\lim_{n \to \infty} T^n(\mathbf{v}) = 0$, and hence (5.7.4) is proved in this case too. $\qquad\square$

Proof of Theorem 5.48. We equip $\mathcal{M}$ with the operator norm as usual. When $\Phi : \mathcal{M} \to \mathcal{M}$ is unital and CP, it is a contraction in the operator norm. Indeed, by the Schwarz inequality for unital CP maps, a special case of Corollary 5.40, and the fact that for all $X \in \mathcal{M}, X^*X \leq \|X^*X\| \mathbb{1}$,

$$\Phi(X)^* \Phi(X) \leq \Phi(X^*X) \leq \|X^*X\| \Phi(\mathbb{1}) \leq \|X^*X\| \mathbb{1}.$$

Therefore, $\|\Phi(X)\| \leq \|X\|$. Then by Theorem 5.53, the limit in (5.7.1) exists, and $\mathcal{P}_\Phi$ is the projection onto the eigenspace of Φ with eigenvalue 1. Since the set of unital CP maps is closed and convex, $\mathcal{P}_\Phi$ is unital and CP.

Then since $\Phi \mapsto \Phi^\dagger$ is continuous in the space of linear transformations of $\mathcal{M}$, and since $(\Phi^m)^\dagger = (\Phi^\dagger)^m$ for all positive m, the limit in (5.7.2) exists because the limit in (5.7.1) exists. Since the set of trace preserving CP maps is closed and convex, $\mathcal{P}_\Phi$ is trace preserving and CP. $\qquad\square$

5.8. The set of invariant states for unital CP maps

Let $\mathcal{M}$ be a von Neumann algebra on a finite-dimensional Hilbert space $\mathcal{H}$, and let $\Phi : \mathcal{M} \to \mathcal{M}$ be unital and CP. The set of states on $\mathcal{M}$ that are invariant under Φ, $\mathcal{S}_{\mathcal{M},\Phi}^{\mathrm{INV}}$, is closely connected with the set of X in $\mathcal{M}$ such that $\Phi(X) = X$; that is the eigenspace of Φ with eigenvalue 1.

Theorem 5.54. *Let $\mathcal{M}$ be a von Neumann algebra on a finite-dimensional Hilbert space, and let $\Phi : \mathcal{M} \to \mathcal{M}$ be CP and unital with Kraus representation $\Phi(X) = \sum_{j=1}^m V_j^* X V_j$. Suppose that there exists a faithful state ρ on $\mathcal{M}$ that is invariant under Φ. Define*

$$\mathcal{C}_\Phi = \{X \in \mathcal{M} \; : \; \Phi(X) = X\}.$$

Then $\mathcal{C}_\Phi$ is the commutant of the set of the Kraus operators:

$$(5.8.1) \qquad\qquad \mathcal{C}_\Phi = \{V_1, \ldots, V_m, V_1^*, \ldots, V_m^*\}'.$$

In particular, $\mathcal{C}_\Phi$ is a von Neumann subalgebra of $\mathcal{M}$, and for all $A, B \in \mathcal{C}_\Phi$ and all $X \in \mathcal{M}$,

$$(5.8.2) \qquad\qquad \Phi(AXB) = A\Phi(X)B.$$

Proof. Let $X \in \mathcal{C}_\Phi$. Since Φ is Hermitian, $\Phi(X^*) = X^*$. Then

$$[X, V_j]^*[X, V_j] = V_j^* X^* X V_j + X^* V_j^* V_j X - V_j^* X^* V_j X - X^* V_j^* X V_j.$$

Summing on j, and using unitality of Φ so that $\sum_{j=1}^m V_j^* V_j = \mathbb{1}$,

$$\sum_{j=1}^m [X, V_j]^*[X, V_j] = \Phi(X^* X) + X^* X - \Phi(X)^* X - X^* \Phi(X)$$

$$= \Phi(X^* X) - X^* X.$$

Since ρ is invariant under Φ, $\rho(\Phi(X^* X) - X^* X) = 0$ and therefore

$$\rho\left(\sum_{j=1}^m [X, V_j]^*[X, V_j]\right) = 0.$$

Since ρ is faithful, $\sum_{j=1}^m [X, V_j]^*[X, V_j] = 0$. Since $X \in \mathcal{C}_\Phi$ is arbitrary, this proves that $\mathcal{C}_\Phi \subseteq \{V_1, \ldots, V_m, V_1^*, \ldots, V_m^*\}'$ and the opposite inclusion is trivial since $\sum_{j=1}^m V_j^* V_j = \mathbb{1}$. Now (5.8.2) follows from (5.8.1). $\qquad\square$

While by Theorem 5.51, $S_{\mathcal{M},\Phi}^{\text{INV}}$ is never empty, it may contain no faithful states. See Exercise 19 for an example which shows moreover that when no faithful state exists in $S_{\mathcal{M},\Phi}^{\text{INV}}$, none of the conclusions of Theorem 5.54 are valid. However, the next theorem provides a simple reduction to the case in which faithful invariant states exist.

Theorem 5.55. *Let $\mathcal{M}$ be a von Neumann algebra on a finite-dimensional Hilbert space $\mathcal{H}$. Let $\Phi : \mathcal{M} \to \mathcal{M}$ be unital and CP. Suppose that $S_{\mathcal{M},\Phi}^{\text{INV}}$ does not contain any faithful state. Let $\bar{\rho} \in S_{\mathcal{M},\Phi}^{\text{INV}}$ be a state of maximal support, which exists by Theorem 5.51. Let P be the support of $\bar{\rho}$. Define*

$$\mathcal{N} := \{PXP \ : \ X \in \mathcal{M}\}.$$

Let $\widetilde{\Phi}^{\dagger}$ denote the restriction of $\Phi^{\dagger}$ to $\mathcal{N}$.

(1) *For all $Y \in \mathcal{N}$, $\widetilde{\Phi}^{\dagger}(Y) \in \mathcal{N}$ and $\text{Tr}[\widetilde{\Phi}^{\dagger}(Y)] = \text{Tr}[Y]$. Thus, $\widetilde{\Phi}^{\dagger}$ is a trace preserving CP map from $\mathcal{N}$ to $\mathcal{N}$, and hence $\widetilde{\Phi} = (\widetilde{\Phi}^{\dagger})^{\dagger}$ is a untial CP map from $\mathcal{N}$ to $\mathcal{N}$.*

(2) *Define $\Psi : \mathcal{M} \to \mathcal{N}$ by $\Psi(X) = PXP$. Then the map $\rho \mapsto \rho \circ \Psi$ is a one-to-one map from $S_{\mathcal{N},\widetilde{\Phi}}^{\text{INV}}$ onto $S_{\mathcal{M},\Phi}^{\text{INV}}$.*

(3) *$S_{\mathcal{N},\widetilde{\Phi}}^{\text{INV}}$ contains at least one faithful state.*

According to Theorem 5.55, if we can determine the structure of the set of invariant states under the assumption that it includes at least one faithful state, we can determine it in general.

Proof of Theorem 5.55. Let $\bar{\rho} \in \mathfrak{S}_{\mathcal{H}} \cap \mathcal{M}$ denote the density matrix corresponding to an invariant state $\bar{\rho}$ with maximal support, so that the projection P is the projection onto the range of $\bar{\rho}$. Therefore there exists some $\lambda > 0$ (the smallest nonzero eigenvalue of $\bar{\rho}$) such that $\bar{\rho} \geq \lambda P$. Then for any $Y \in \mathcal{N}^{+}$, $Y = PYP \leq \|Y\|P$, and hence

$$\Phi^{\dagger}(Y) \leq \|Y\|\Phi^{\dagger}(P) \leq \|Y\|\lambda^{-1}\Phi^{\dagger}(\bar{\rho}) = \|Y\|\lambda^{-1}\bar{\rho}.$$

Therefore, $P^{\perp}\Phi^{\dagger}(Y)P^{\perp} = 0$. Then for all $Z \in \mathcal{M}$,

$$\text{Tr}[ZP\Phi^{\dagger}(Y)P^{\perp}] = \langle (ZP(\Phi^{\dagger}(Y))^{1/2})^{*}, (\Phi^{\dagger}(Y))^{1/2}P^{\perp} \rangle$$

$$\leq \|ZP(\Phi^{\dagger}(Y))^{1/2}\|_{2} \left(\text{Tr}[P^{\perp}\Phi^{\dagger}(Y)P^{\perp}] \right)^{1/2} = 0.$$

Since Z is arbitrary, $P\Phi^{\dagger}(Y)P^{\perp} = 0$, and then $P^{\perp}\Phi^{\dagger}(Y)P = 0$. Altogether, for all $Y \in \mathcal{N}^{+}$,

$$\Phi^{\dagger}(Y) = P\Phi^{\dagger}(Y)P.$$

Since every operator in $\mathcal{N}$ is a linear combination of four operators in $\mathbb{N}^{+}$, this identity holds for all $Y \in \mathcal{N}$. Since $\widetilde{\Phi}^{\dagger}$ is the restriction of $\Phi^{\dagger}$ to $\mathcal{N}$, it is trace preserving and CP since $\Phi^{\dagger}$ has these properties. This proves (1).

By the maximality property of $\bar{\rho}$, for all $\rho \in \mathfrak{S}_{\mathcal{H}} \cap \mathcal{M}$ such that $\Phi^{\dagger}\rho = \rho$, $P\rho P = \rho$. Hence the density matrix of every invariant state belongs to $\mathcal{N}$, and since on $\mathcal{N}$, $\widetilde{\Phi}^{\dagger} = \Phi^{\dagger}$, a density matrix $\rho \in \mathcal{N}$ satisfies $\widetilde{\Phi}^{\dagger}(\rho) = \rho$ if and only if $\Phi^{\dagger}(\rho) = \rho$. Next, note that for a density matrix $\rho \in \mathcal{N}$, and all $X \in \mathcal{M}$,

$$\mathrm{Tr}[X\rho] = \mathrm{Tr}[XP\rho P] = \mathrm{Tr}[\Psi(X)\rho],$$

so that if ρ is the state on $\mathcal{N}$ with density matrix ρ, $\rho \circ \Psi$ is the state on $\mathcal{M}$ with density matrix ρ. This proves (2).

Finally, as in the proof of (1), for some $\lambda > 0$, $\bar{\rho} \geq \lambda P$. Let $X \in \mathcal{N}$ and suppose that $\bar{\rho}(X^*X) = 0$. Then

$$0 = \bar{\rho}(X^*X) = \mathrm{Tr}[\bar{\rho}X^*X] \geq \lambda\,\mathrm{Tr}[PX^*X] = \lambda\,\mathrm{Tr}[X^*X].$$

Therefore, $X = 0$, and $\bar{\rho}$ is faithful. This proves (3). $\qquad\square$

Throughout the rest of this section, we use the following notation: $\mathcal{M}$ is a von Neumann algebra on a finite-dimensional Hilbert space $\mathcal{H}$, and S is a set of states on $\mathcal{M}$ (identified with density matrices in $\mathcal{M}$) that includes at least one faithful state. Define $\mathcal{I}_S$ to be the set of all unital CP maps Φ that leave every state in S invariant.

Lemma 5.56. *For each $\Phi \in \mathcal{I}_S$, let $\mathcal{C}_{\Phi}$ denote the corresponding fixed point algebra. Then there exists a $\Phi_0 \in \mathcal{I}_S$ such that*

$$(5.8.3) \qquad\qquad \mathcal{C}_{\Phi_0} \subseteq \mathcal{C}_{\Phi} \quad \text{for all} \quad \Phi \in \mathcal{I}_S.$$

Proof. For $\Phi_1, \Phi_2 \in \mathcal{I}_S$, and $\lambda_1, \lambda_2 \in (0,1)$ with $\lambda_1 + \lambda_2 = 1$, let $\Psi := \lambda_1 \Phi_1 + \lambda_2 \Phi_2$. It is clear that $\mathcal{C}_{\Phi_1} \cap \mathcal{C}_{\Phi_2} \subseteq \mathcal{C}_{\Psi}$. In fact, $\mathcal{C}_{\Psi} = \mathcal{C}_{\Phi_1} \cap \mathcal{C}_{\Phi_2}$. To see this, let $X \in \mathcal{C}_{\Psi}$, and define $Q(X) := \sum_{j=1}^{2} \lambda_j |X - \Phi_j(X)|^2$. Then

$$Q(X) = X^*X - X^*\Psi(X) - \Psi(X)^*X + \sum_{j=1}^{2} \lambda_j \Phi_j(X)^*\Phi_j(X)$$

$$= \sum_{j=1}^{2} \lambda_j (\Phi_j(X)^*\Phi_j(X) - X^*X) \leq \sum_{j=1}^{2} \lambda_j (\Phi_j(X^*X) - X^*X).$$

Applying the faithful invariant state ρ_0 to both sides shows that $Q(X) = 0$. Therefore $\Phi_j(X) = X$ for $j = 1, 2$, and hence $\mathcal{C}_{\Phi_1} \cap \mathcal{C}_{\Phi_2} \supseteq \mathcal{C}_{\Psi}$.

Now consider an arbitrary finite subset $\{\Phi_1, \dots, \Phi_n\}$ of $\mathcal{I}_S$, and along with it, $\bigcap_{j=1}^{n} \mathcal{C}_{\Phi_j}$. The dimension of this subalgebra is bounded below by 1 since the identity is always included in it. Make a choice of the finite set $\{\Phi_1, \dots, \Phi_n\}$ that yields the minimal dimension of $\bigcap_{j=1}^{n} \mathcal{C}_{\Phi_j}$. Then evidently $\bigcap_{j=1}^{n} \mathcal{C}_{\Phi_j} = \bigcap_{\Phi \in \mathcal{I}_S} \mathcal{C}_{\Phi}$. Now a simple induction using the first part shows that with $\Phi_0 := \frac{1}{n} \sum_{j=1}^{n} \Phi_j$, $\mathcal{C}_{\Phi_0} = \bigcap_{\Phi \in \mathcal{I}_S} \mathcal{C}_{\Phi}$. $\qquad\square$

Now let $\mathcal{P}_{\Phi_0}$ be the unital CP map defined in terms Φ_0 in Theorem 5.48 by taking the limit in (5.7.1). By (3) of Theorem 5.51, since every $\rho \in S$ is invariant under Φ_0, every $\rho \in S$ is invariant under $\mathcal{P}_{\Phi_0}$. We have also seen that $\mathcal{P}_{\Phi_0}$ is the projector onto the eigenspace of Φ_0 with eigenvalue 1, and as such, we can expect it to have a particularly simple structure, which we now investigate.

Let $\{P_1, \ldots, P_m\}$ be a set of mutually orthogonal projections that span the center of $\mathcal{C}_{\Phi_0}$. For $j = 1, \ldots, m$, let $\mathcal{H}^{(j)}$ denote the range of P_j, so that

$$(5.8.4) \qquad \mathcal{H} = \bigoplus_{j=1}^{m} \mathcal{H}^{(j)}.$$

Then by Theorem 4.28, each $\mathcal{H}^{(j)}$ has a decomposition

$$(5.8.5) \qquad \mathcal{H}^{(j)} = \mathcal{H}^{(j)}_\ell \otimes \mathcal{H}^{(j)}_r,$$

and $\mathcal{C}_{\Phi_0}$ has the form

$$(5.8.6) \qquad \mathcal{C}_{\Phi_0} = \bigoplus_{j=1}^{m} \mathcal{C}_j,$$

where for each j,

$$(5.8.7) \qquad \mathcal{C}_j = P_j \mathcal{C}_{\Phi_0} P_j = \mathcal{B}(\mathcal{H}^{(j)}_\ell) \otimes \mathbb{1}.$$

Theorem 5.57. *Let $\Phi_0 \in \mathcal{I}_S$ satisfy (5.8.3). Let $\mathcal{P}_{\Phi_0}$ be the projection of $\mathcal{M}$ onto $\mathcal{C}_{\Phi_0}$ defined by (5.7.1). Then with respect to the decomposition of $\mathcal{H}$ and $\mathcal{C}_{\Phi_0}$ given by (5.8.4), (5.8.5), (5.8.6), and (5.8.7), and:*

   (1) *There exists a set $\{\omega_1, \ldots, \omega_m\}$, where for each j, $\omega_j \in \mathfrak{S}_{\mathcal{H}^{(j)}_r}$, such that for all $X \in \mathcal{B}(\mathcal{H})$, the projection $\mathcal{P}_{\Phi_0}$ is given by*

$$(5.8.8) \qquad \mathcal{P}_{\Phi_0}(X) = \sum_{j=1}^{m} \mathrm{Tr}_{\mathcal{H}^{(j)}_r}[P_j X P_j (\mathbb{1} \otimes \omega_j)] \otimes \mathbb{1},$$

   and consequently,

$$(5.8.9) \qquad \mathcal{P}^{\dagger}_{\Phi_0}(\rho) = \sum_{j=1}^{m} \mathrm{Tr}_{\mathcal{H}^{(j)}_r}[P_j \rho P_j] \otimes \omega_j.$$

   In particular, a state ρ is invariant under Φ_0 if and only if for some set $\{\sigma_i, \ldots \sigma_m\}$ where for each j, $\sigma_j \in \mathfrak{S}_{\mathcal{H}^{(j)}_\ell}$, and probabilities $p_1, \ldots, p_m$ on $\{1, \ldots, m\}$,

$$(5.8.10) \qquad \rho = \bigoplus_{j=1}^{m} p_j \sigma_j \otimes \omega_j.$$

(2) *A CP unital map Ψ on $\mathcal{M}$ belongs to $\mathcal{I}_S$ if and only if Ψ has the form*

$$(5.8.11) \qquad \Psi(X) = \bigoplus_{j=1}^{m} (\mathbb{1} \otimes \Psi_j)(P_j X P_j),$$

where for each $j = 1,\ldots,m$, $\Psi_j \in \mathcal{B}(\mathcal{B}(\mathcal{H}_r^{(j)}))$ satisfies $\Psi_j^\dagger(\omega_j) = \omega_j$. Consequently, for all density matrices $\rho \in \mathcal{M}$,

$$(5.8.12) \qquad \Psi^\dagger(\rho) = \bigoplus_{j=1}^{m} (\mathbb{1} \otimes \Psi_j^\dagger)(P_j \rho P_j),$$

so that Ψ leaves invariant every state ρ that is invariant under Φ_0.

Proof. As the first step we prove part of (2), namely that every $\Psi \in \mathcal{I}_S$ has the form (5.8.11) where each $\Psi_j \in \mathcal{B}(\mathcal{B}(\mathcal{H}_r^{(j)}))$.

Let $\Psi \in \mathcal{I}_S$. Since for each $1 \leq i,j,k \leq m$, $P_i, P_k \in \mathcal{C}_{\Phi_0} \subseteq \mathcal{C}_\Psi$, it follows from (5.8.2) that for each $X \in \mathcal{M}$,

$$P_i \Psi(P_j X P_j) P_k = \Psi(P_i P_j X P_j P_k) = \delta_{i,j}\delta_{k,j}\Psi(P_j X P_j),$$

and therefore

$$P_j \Psi(P_j X P_j) P_j = \Psi(P_j X P_j).$$

That is, $\mathcal{B}(\mathcal{H}_j)$ is invariant under Ψ.

Now consider any $A \in \mathcal{B}(\mathcal{H}_\ell^{(j)})$ and any $B \in \mathcal{B}(\mathcal{H}_r^{(j)})$. Then $A \otimes B = (A \otimes \mathbb{1})(\mathbb{1} \otimes B) = (\mathbb{1} \otimes B)(A \otimes \mathbb{1})$. Since $A \otimes \mathbb{1} \in \mathcal{C}_{\Phi_0} \subseteq \mathcal{C}_\Psi$, by (5.8.2) again

$$(5.8.13) \qquad \Psi(A \otimes B) = (A \otimes \mathbb{1})\Psi(\mathbb{1} \otimes B) = \Psi(\mathbb{1} \otimes B)(A \otimes \mathbb{1}).$$

Therefore, $\Psi(\mathbb{1} \otimes B) \in (\mathcal{B}(\mathcal{H}_\ell^{(j)}) \otimes \mathbb{1})' = \mathbb{1} \otimes \mathcal{B}(\mathcal{H}_r^{(j)})$. It follows that for some $\Psi_j \in \mathcal{B}(\mathcal{B}(\mathcal{H}_r^{(j)}))$, $\Psi(\mathbb{1} \otimes B) = \mathbb{1} \otimes \Psi_j(B)$. Going back to (5.8.13),

$$\Psi(A \otimes B) = (A \otimes \mathbb{1})\Psi(\mathbb{1} \otimes B) = A \otimes \Psi_j(B).$$

By linearity, as an operator on $\mathcal{H}_j$, $\Psi = \mathbb{1} \otimes \Psi_j$. This completes the first step.

In the second step, we prove (1). By part (3) of Theorem 5.51, $\mathcal{P}_{\Phi_0} \in \mathcal{I}_S$. By what was proved in the first step, for all $B \in \mathcal{B}(\mathcal{H}_r^{(j)})$ and all $\Psi \in \mathcal{I}_S$,

$$\Psi(\mathbb{1} \otimes B) \in \mathbb{1} \otimes \mathcal{B}(\mathcal{H}_r^{(j)}) = \mathcal{C}'_{\Phi_0}.$$

Since $\mathcal{P}_{\Phi_0}$ is a projection onto $\mathcal{C}_{\Phi_0}$, $\mathcal{P}_{\Phi_0}(\mathbb{1} \otimes B)$ lies in the center of $\mathcal{C}_{\Phi_0}$, and specifically, the center of $\mathcal{C}_j$, which is a factor. Hence for some linear functional f_j on $\mathcal{B}(\mathcal{H}_r^{(j)})$,

$$\mathcal{P}_{\Phi_0}(\mathbb{1} \otimes B) = f_j(B)\mathbb{1}.$$

Since Φ_0 is positive and unital, f_j is a state on $\mathcal{B}(\mathcal{H}_r^{(j)})$, and hence there is a density matrix ω_j on $\mathcal{H}_r^{(j)}$ such that $\mathcal{P}_{\Phi_0}(\mathbb{1} \otimes B) = \mathrm{Tr}[B\omega_j]\mathbb{1}$. Altogether,

$$\mathcal{P}_{\Phi_0}(A \otimes B) = \mathrm{Tr}_{\mathcal{H}_r^{(j)}}(A \otimes B\omega_j) \otimes \mathbb{1}.$$

By linearity, for all $X \in \mathcal{B}(\mathcal{H}_j)$, $\mathcal{P}_{\Phi_0}(X) = \mathrm{Tr}_{\mathcal{H}_r^{(j)}}[X(\mathbb{1} \otimes \omega_j)] \otimes \mathbb{1}$. Putting the pieces together yields (5.8.8), from which (5.8.9) follows by taking the adjoint. By (3) of Theorem 5.51, ρ is invariant under Φ_0 if and only ρ is invariant under $\mathcal{P}_{\Phi_0}$, and $\mathcal{P}_{\Phi_0}(\rho)$ is given by (5.8.10) where $p_j\sigma_j = \mathrm{Tr}_{\mathcal{H}_r^{(j)}}[P_j\rho P_j]$. Conversely, any ρ of the form (5.8.10) is invariant under $\mathcal{P}_{\Phi_0}$ by (5.8.9). This completes the proof of (1).

The third step completes the proof of (2). In the first step we proved that every $\Psi \in \mathcal{J}_S$ has the form (5.8.11). Taking the adjoint, we see that $\Psi^\dagger$ has the form specified in (5.8.12).

Then since $\Psi^\dagger(\rho) = \rho$ for all $\rho \in S$, and since by (1) such density matrices ρ have the form $\rho_0 = \sum_{j=1}^m p_j\sigma_{0,j} \otimes \omega_j$, we must have $\Psi_j^\dagger(\omega_j) = \omega_j$. Therefore, if ρ is of the form (5.8.10), and $\Psi^\dagger$ has the form specified in (2), then $\Psi^\dagger\rho_0 = \rho_0$. $\quad\square$

5.9. The no-cloning theorem

The inputs and outputs of any device that operates on quantum states, from one quantum system to another, are related by a completely positive trace preserving map $\Phi^\dagger$; that is, a quantum operation. A quantum cloning device would take as input a quantum state ρ on a Hilbert space $\mathcal{H}$, and yield as output the quantum state $\Phi^\dagger(\rho) = \rho \otimes \rho$. This is too much to ask (see Exercise 32). We therefore consider a weaker form of cloning, called *broadcasting* in [24], and we shall see that one can broadcast a classical set of states, but no more.

Consider a quantum operation $\Phi^\dagger : \mathcal{B}(\mathcal{H}) \to \mathcal{B}(\mathcal{H} \otimes \mathcal{H})$. If it is the case that

$$\mathrm{Tr}_1[\Phi^\dagger(\rho)] = \mathrm{Tr}_2[\Phi^\dagger(\rho)] = \rho,$$

then we shall say that the device has *broadcasted* the state ρ. This is certainly the case if $\Phi^\dagger(\rho) = \rho \otimes \rho$, in which case we say the the device has *cloned* the state ρ.

Broadcasting is is weaker than cloning; if a device clones the state ρ, then it broadcasts it. This weaker notion arises naturally in the consideration of cloning of density matrices; see Exercsie 32.

Definition 5.58. Let $\mathcal{H}$ be a finite-dimensional Hilbert space. A **cloning device** is a CP unital map $\Phi : \mathcal{B}(\mathcal{H} \otimes \mathcal{H}) \to \mathcal{B}(\mathcal{H})$. For $j = 1, 2$, define $\Phi_j : \mathcal{B}(\mathcal{H}) \to \mathcal{B}(\mathcal{H})$ by

$$\Phi_1(X) = \Phi(X \otimes \mathbb{1}) \quad \text{and} \quad \Phi_2(X) = \Phi(\mathbb{1} \otimes X).$$

The set of **cloneable states** is the set of states ρ such that $\rho = \Phi_1^\dagger \rho = \Phi_2^\dagger \rho$. Note that $\Phi_1^\dagger = \mathrm{Tr}_2 \circ \Phi^\dagger$ and $\Phi_2^\dagger = \mathrm{Tr}_1 \circ \Phi^\dagger$, so this is the same as saying $\rho = \mathrm{Tr}_2 \circ \Phi^\dagger \rho = \mathrm{Tr}_1 \circ \Phi^\dagger \rho$. Note that the terminology is such that a state ρ is cloneable by Φ if and only if $\Phi^\dagger$ broadcasts ρ.

At the end of Section 2.3, we examined a communication scheme in which Hal and Katy use a shared pair of atoms in an entangled state to instantly transmit a message from Hal's lab to Katy's. Depending on the message Hal intends to send, he measures one of two observables agreed upon in advance. In the process, he *steers* the state of Katy's atom into one of four possible states. We have seen that if Katy has a device that can clone all four of these states, then she can read the message. The four states are real density matrices on $\mathbb{C}^2$ with its usual inner product, and they span the space real self-adjoint 2×2 matrices. Hence, any cloning device that could clone these four states, would have to be able to clone all real 2×2 density matrices. However, no such cloning device can exist, even in the wide sense of cloning. The following theorem is due Barmum, Caves, Fuchs, Josza, and Schumacher [24], and it was somewhat extended and simplified by Lindblad [150].

Theorem 5.59 (The general no-cloning theorem). *Let $\mathcal{H}$ be a finite-dimensional Hilbert space. Let $\Phi : \mathcal{B}(\mathcal{H} \otimes \mathcal{H}) \to \mathcal{B}(\mathcal{H})$ be a cloning device. Suppose that at least one faithful state ρ_0 is cloneable. The set of cloneable states is commutative so that the corresponding density matrices may all be simultaneously diagonalized. In this sense, the set of cloneable states is classical.*

Example 5.60. Let $\mathcal{H} = \mathbb{C}^2$ with its usual inner product. The scheme for faster-than-light communication discussed the end of Section 2.3 requires a cloning device $\Phi : \mathcal{B}(\mathcal{H} \otimes \mathcal{H}) \to \mathcal{B}(\mathcal{H})$ that strictly clones the possible states that Katy can have in her lab after Hal has sent the message. Since these states span the real 2×2 density matrices, it would have to be the case that $\Phi_1^\dagger(\rho) = \Phi_2^\dagger(\rho) = \rho$ for all real density matrices on $\mathcal{H}$. Note that $\rho_0 = \frac{1}{2}\mathbb{1}$ is real, self-adjoint, and strictly positive. Hence if the set of cloneable states includes all real density matrices on $\mathcal{H}$, it includes ρ_0. But then by part (1) of Theorem 5.59, the sets of cloneable states, even in the wide sense, is commutative. However, the set of real density matrices on $\mathcal{H}$ is not commutative. Hence no such device can exist. The communication scheme does not require a cloning device that can clone every state, so a no-cloning theorem precluding only this is not enough.

It will be useful to formulate the key lemma in greater generality than required for proving Theorem 5.59.

Lemma 5.61. *Let $\mathcal{H}_1$ and $\mathcal{H}_2$ be two finite-dimensional Hilbert spaces. Let $\Phi : \mathcal{B}(\mathcal{H}_1 \otimes \mathcal{H}_2) \to \mathcal{B}(\mathcal{H}_1)$ be CP and unital. Define $\Phi_1 : \mathcal{B}(\mathcal{H}_1) \to \mathcal{B}(\mathcal{H}_1)$ and $\Phi_2 : \mathcal{B}(\mathcal{H}_2) \to \mathcal{B}(\mathcal{H}_1)$ by $\Phi_1(X) = \Phi(X \otimes \mathbb{1})$ and $\Phi_2(Y) = \Phi(\mathbb{1} \otimes Y)$. Suppose there exists a faithful state ρ_0 that is invariant under Φ_1.*

Then for all $Y \in \mathcal{B}(\mathcal{H}_2)$, all $X \in \mathcal{C}_{\Phi_1}$,

$$X\Phi(\mathbb{1} \otimes Y) = \Phi(X \otimes Y) = \Phi(\mathbb{1} \otimes Y)X,$$

so that, in particular, $\Phi_2(Y) \in \mathcal{C}'_{\Phi_1}$.

Proof. Let $\Phi(A) = \sum_{j=1}^{m} V_j^* A V_j$ be a Kraus representation of Φ. For each j, $V_j \in \mathcal{B}(\mathcal{H}_1, \mathcal{H}_1 \otimes \mathcal{H}_2)$, and

$$\begin{aligned}
((X \otimes \mathbb{1})V_j - V_jX)^*((X \otimes \mathbb{1})V_j - V_jX) &= V_j^*(X^*X \otimes \mathbb{1})V_j + X^*V_j^*V_jX \\
&\quad - V_j^*(X^* \otimes \mathbb{1})V_jX \\
&\quad - X^*V_j^*(X \otimes \mathbb{1})V_j.
\end{aligned}$$

Summing on j yields, because $X \in \mathcal{C}_{\Phi_1}$,

$$\Phi_1(X^*X) + X^*X - \Phi_1(X^*)X - X^*\Phi_1(X) = \Phi_1(X^*X) - X^*X$$

$$= \sum_{j=1}^{m} |(X \otimes \mathbb{1})V_j - V_jX|^2.$$

Applying the faithful invariant state ρ, $\rho(\Phi_1(X^*X) - X^*X) = 0$, and therefore $\sum_{j=1}^{m} \rho(|(X \otimes \mathbb{1})V_j - V_jX|^2)$. It follows that for each j, $(X \otimes \mathbb{1})V_j = V_jX$. Replacing X by X^*, also in $\mathcal{C}_{\Phi_1}$, and taking adjoints yields $V_j^*(X \otimes \mathbb{1}) = XV_j^*$ for each j.

Then for any $X \in \mathcal{C}_{\Phi_1}$, $Y \in \mathcal{B}(\mathcal{H}_2)$,

$$\begin{aligned}
(5.9.1) \qquad \Phi(X \otimes Y) &= \Phi((X \otimes \mathbb{1})(\mathbb{1} \otimes Y)) = X\Phi(\mathbb{1} \otimes Y) = X\Phi_2(Y) \\
&= \Phi((\mathbb{1} \otimes Y)(X \otimes \mathbb{1})) = \Phi(\mathbb{1} \otimes Y)X = \Phi_2(Y)X. \qquad \square
\end{aligned}$$

Proof of Theorem 5.59. Let $\Psi := \frac{1}{2}(\Phi_1 + \Phi_2)$. By Lemma 5.56, since there is a faithful state ρ_0 invariant under Φ_1 and Φ_2, $\mathcal{C}_\Psi = \mathcal{C}_{\Phi_1} \cap \mathcal{C}_{\Phi_2}$. By Lemma 5.61 applied with $\mathcal{H}_1 = \mathcal{H}_2 = \mathcal{H}$, $\Phi_2(\mathcal{B}(\mathcal{H})) \subseteq \mathcal{C}'_{\Phi_1}$, and therefore, $\mathcal{C}_{\Phi_2} \subseteq \mathcal{C}'_{\Phi_1}$. Since $\mathcal{C}_\Psi = \mathcal{C}_{\Phi_1} \cap \mathcal{C}_{\Phi_2}$, $\mathcal{C}_\Psi$ is commutative.

Let S be the set of cloneable states. Then each $\rho \in S$ is invariant under Φ_1 and Φ_2 and hence under Ψ, and hence again under $\mathcal{R}_\Psi$, the projection onto $\mathcal{C}_\Psi$ given by (5.7.1). That is, $\mathcal{R}_\Psi^\dagger \rho = \rho$. We now apply Theorem 5.57.

Let $\mathcal{C}_\Psi = \bigoplus_{j=1}^{m} \mathcal{C}_j$ be the decomposition of $\mathcal{C}_\Psi$ as in (5.8.6). Since $\mathcal{C}_\Psi$ is commutative, each $\mathcal{C}_j$ is commutative, and therefore the Hilbert space $\mathcal{H}_\ell^{(j)}$ in (5.8.7) is one dimensional. Therefore, $\mathcal{H}_j = \mathcal{H}_\ell^{(j)} \otimes \mathcal{H}_r^{(j)}$ simplifies to $\mathcal{H}_j = \mathcal{H}_r^{(j)}$, and the one-dimensional density matrix σ_j in (5.8.10) is simply the number 1, and (5.8.10) simplifies to

$$(5.9.2) \qquad \rho = \bigoplus_{j=1}^{m} p_j \omega_j.$$

Since ω_j acts nontrivially only on $\mathcal{H}_j$, $\{\omega_1,\ldots,\omega_m\}$ is a commutative set of density matrices. Since every $\rho \in S$ has the form specified in (5.9.2), S is commutative. $\qquad\square$

The no-cloning theorem is not only physically important, the ideas going into its proof will be useful later on. We close this section by drawing some useful consequences from Lemma 5.61.

Lemma 5.62. *Let $\mathcal{H}_1$ and $\mathcal{H}_2$ be finite-dimensional Hilbert spaces.*

    (1) *Let $\Phi : \mathcal{B}(\mathcal{H}_1 \otimes \mathcal{H}_2) \to \mathcal{B}(\mathcal{H}_1)$ be CP and unital. Then Φ has the property that for all $X \in \mathcal{B}(\mathcal{H}_1)$,*

$$(5.9.3) \qquad\qquad \Phi(X \otimes \mathbb{1}) = X.$$

if and only if there exists $\omega \in \mathfrak{S}_{\mathcal{H}_2}$ such that for all $A \in \mathcal{B}(\mathcal{H}_1 \otimes \mathcal{H}_2)$,

$$(5.9.4) \qquad\qquad \Phi(A) = \mathrm{Tr}_2[(\mathbb{1} \otimes \omega)A].$$

    (2) *Let $\Phi^\dagger : \mathcal{B}(\mathcal{H}_1) \to \mathcal{B}(\mathcal{H}_1 \otimes \mathcal{H}_2)$ be CP and trace preserving. Then*

$$(5.9.5) \qquad\qquad \mathrm{Tr}_2[\Phi^\dagger(\rho)] = \rho$$

for all $\rho \in \mathfrak{S}_{\mathcal{H}_1}$ if and only if there is an $\omega \in \mathfrak{S}_{\mathcal{H}_2}$ such that for all $\rho \in \mathfrak{S}_{\mathcal{H}_1}$,

$$(5.9.6) \qquad\qquad \Phi^\dagger(\rho) = \rho \otimes \omega.$$

Proof. Suppose Φ has the property (5.9.3). Define $\Phi_1(X) = \Phi(X \otimes \mathbb{1})$. Then Φ_1 is the identity, and every faithful state is invariant under Φ_1, and $\mathcal{C}_{\Phi_1} = \mathcal{B}(\mathcal{H}_1)$. By Lemma 5.61, for all $Y \in \mathcal{B}(\mathcal{H}_2)$,

$$\Phi(\mathbb{1} \otimes Y) \in \mathcal{B}(\mathcal{H}_1)' = \mathbb{C}\mathbb{1}.$$

Hence, there is a linear functional φ on $\mathcal{B}(\mathcal{H}_2)$ such that $\Phi(\mathbb{1} \otimes Y) = \varphi(Y)\mathbb{1}$. Since Φ is positive, φ is positive, and since Φ is unital, $\varphi(\mathbb{1}) = 1$. Hence φ is a state, and there is a uniquely determined density matrix ω such that $\varphi(Y) = \mathrm{Tr}[\omega Y]$ for all $Y \in \mathcal{B}(\mathcal{H}_2)$. Consequently, for all $X \in \mathcal{B}(\mathcal{H}_1)$ and all $Y \in \mathcal{B}(\mathcal{H}_2)$, again using Lemma 5.61,

$$\Phi(X \otimes Y) = \Phi(\mathbb{1} \otimes Y)X = \mathrm{Tr}[\omega Y]X = \mathrm{Tr}_2[(\mathbb{1} \otimes \omega)X \otimes Y].$$

Now (5.9.3) follows by linearity.

Conversely, if Φ has the form specified in (5.9.4), it evidently has the property (5.9.3). This proves (1).

Now observe that by duality, (1) and (2) of Lemma 5.62 are equivalent. Since Tr_2 is the adjoint of the inclusion map $X \mapsto X \otimes \mathbb{1}$, Φ satisfies (5.9.3) for all X if and only if $\Phi^\dagger$ satisfies (5.9.5) for all $\rho \in \mathfrak{S}_{\mathcal{H}_1}$. Also, Φ has the form specified in (5.9.4) if and only if $\Phi^\dagger$ has the form specified in (5.9.6). $\qquad\square$

Corollary 5.63. *Let $\mathcal{K}_1$, $\mathcal{K}_2$, $\mathcal{K}_3$ be finite-dimensional Hilbert spaces. Let $\Phi_2^\dagger : \mathcal{B}(\mathcal{K}_2) \to \mathcal{B}(\mathcal{K}_2)$ be a quantum operation. Define $\Phi_{12}^\dagger : \mathcal{B}(\mathcal{K}_1 \otimes \mathcal{K}_2) \to \mathcal{B}(\mathcal{K}_1 \otimes \mathcal{K}_2)$ by*

$$(5.9.7) \qquad \Phi_{12}^\dagger = \mathbb{1} \otimes \Phi_2^\dagger.$$

Let $\Psi^\dagger : \mathcal{B}(\mathcal{K}_1 \otimes \mathcal{K}_2) \to \mathcal{B}(\mathcal{K}_1 \otimes \mathcal{K}_2 \otimes \mathcal{K}_3)$ be a quantum operation such that

$$(5.9.8) \qquad \Phi_{12}^\dagger = \mathrm{Tr}_3 \circ \Psi^\dagger.$$

Then there is a quantum operation $\Psi_{23}^\dagger : \mathcal{B}(\mathcal{K}_2) \to \mathcal{B}(\mathcal{K}_2 \otimes \mathcal{K}_3)$ such that

$$(5.9.9) \qquad \Psi^\dagger = \mathbb{1} \otimes \Psi_{23}^\dagger.$$

Proof. Fix $Y \in \mathfrak{S}_{\mathcal{K}_2}$, and define a map $\widehat{\Psi}_Y^\dagger : \mathcal{B}(\mathcal{K}_1) \to \mathcal{B}(\mathcal{K}_1 \otimes \mathcal{H}_2 \otimes \mathcal{K}_3)$ by

$$(5.9.10) \qquad \widehat{\Psi}_Y^\dagger(X) = \Psi^\dagger(X \otimes Y)$$

for $X \in \mathcal{B}(\mathcal{K}_1)$. Since $Y \in \mathfrak{S}_{\mathcal{K}_2}$, $\widehat{\Psi}_Y^\dagger$ is a quantum operation. Moreover, by (5.9.8), for all $X \in \mathcal{B}(\mathcal{K}_1)$,

$$\mathrm{Tr}_{23}[\widehat{\Psi}_Y^\dagger(X)] = \mathrm{Tr}_{23}[\Psi^\dagger(X \otimes Y)] = \mathrm{Tr}_2[\Phi_{12}^\dagger(X \otimes Y)] = X,$$

where the final equality is from (5.9.7). By part (2) of Theorem 5.59, there exists $\omega_Y \in \mathfrak{S}_{\mathcal{K}_2 \otimes \mathcal{K}_3}$ such that $\widehat{\Psi}_Y^\dagger(X) = X \otimes \omega_Y$. Then by (5.9.10),

$$\Psi^\dagger(X \otimes Y) = X \otimes \omega_Y.$$

The map $Y \mapsto \omega_Y$ extends to a uniquely determined linear map Ψ_{23} from $\mathcal{B}(\mathcal{K}_2)$ to $\mathcal{B}(\mathcal{K}_2 \otimes \mathcal{K}_3)$ such that (5.9.9) is satisfied. Since $\Psi^\dagger$ is a quantum operation, so is $\Psi_{23}^\dagger$. $\qquad\square$

5.10. Inequalities for positive maps

In this section we study the properties of positive maps that are merely positive, not 2-positive or Schwarz, let alone completely positive. The basis of much of this is Theorem 5.43 which has a number of important consequences for maps that are positive and nothing more. The first is the Schwarz inequality of Kadison [125] and Choi [63], which will be shown to hold for all *normal* $A \in M_n(\mathbb{C})$ assuming no more than positivity of Φ.

Theorem 5.64 (Kadison-Choi inequality for positive maps). *Let* $\Phi : M_n(\mathbb{C}) \to M_m(\mathbb{C})$ *be a positive linear map. Then for all normal operators* $A \in M_n(\mathbb{C})$,

$$(5.10.1) \qquad \Phi(A^*A) \geq \Phi(A)^* \Phi(\mathbb{1})^+ \Phi(A).$$

In particular, if Φ *is quasi unital*

$$\Phi(A^*A) \geq \Phi(A)^* \Phi(A).$$

Proof. By the spectral theorem, A has the spectral representation $A = \sum_{j=1}^{k} z_k P_k$ where $\sigma(A) = \{z_1, \ldots, z_k\} \subset \mathbb{C}$ and $\{P_1, \ldots, P_k\}$ is a set of orthogonal projections with $\sum_{j=1} P_j = \mathbb{1}$.

Note that for any $X \in M_n^+(\mathbb{C})$, $X = XX^+X$. Then since for each j, $\Phi(P_j) \in M_n^+(\mathbb{C})$,

$$(5.10.2) \qquad \Phi(P_j) = \Phi(P_j)\Phi(P_j)^+\Phi(P_j).$$

Since $A^*A = \sum_{j=1}^{k} |z_j|^2 P_j$, the subadditivity form of Kiefer's inequality, (5.6.9), yields

$$\Phi(A^*A) = \sum_{j=1}^{k} |z_j|^2 \Phi(P_j) = \sum_{j=1}^{k} (z_j\Phi(P_j))^* \Phi(P_j)^+ (z_j\Phi(P_j))$$

$$\geq \left(\sum_{j=1}^{k} z_j\Phi(P_j)\right)^* \left(\sum_{j=1}^{k} \Phi(P_j)\right)^+ \left(\sum_{j=1}^{k} z_j\Phi(P_j)\right)$$

$$= \Phi(A^*)\Phi(\mathbb{1})^+\Phi(A).$$

Since Φ is quasi unital, $\Phi(\mathbb{1})^+ = \Phi(\mathbb{1})$, and then $\Phi^+(\mathbb{1})\Phi(A) = \Phi(A)$ follows from Lemma 5.6. $\qquad\square$

Kadison proved [125] the inequality (5.10.1) for A self-adjoint, and Choi [63] extended this to the case where A is normal. The same argument that was used here to prove Theorem 5.64 can be used to prove another theorem of Choi [61]:

Theorem 5.65. *Let* $\Phi : M_n(\mathbb{C}) \to M_m(\mathbb{C})$ *be a quasi unital positive linear map. Then for all* $X \in M_n^+(\mathbb{C})$,

$$\Phi(X^+) \geq (\Phi(X))^+.$$

Proof. Let $X = \sum_{j=1}^{k} \lambda_j P_j$, $\sum_{j=1}^{k} P_j = \mathbb{1}$ be the spectral representation of X. For $\lambda \geq 0$, write $\lambda^+ = \lambda^{-1}$ is $\lambda > 0$, and $\lambda^+ = 0$, otherwise. Again using (5.10.2) and (5.6.9),

$$
\begin{aligned}
\Phi(X^+) &= \sum_{j=1}^{k} \lambda_j^+ \Phi(P_j) \\
&= \sum_{j=1}^{k} \lambda_j^+ \Phi(P_j)\Phi(P_j)^+ \Phi(P_j) \\
&\geq \sum_{j=1}^{k} \Phi\left(\sum_{j=1}^{k} P_j\right) \Phi\left(\sum_{j=1}^{k} \lambda_j P_j\right)^+ \Phi\left(\sum_{j=1}^{k} P_j\right) \\
&= \Phi(\mathbb{1})\Phi(X)^+ \Phi(\mathbb{1}) = \Phi(X)^+. \qquad\qquad \square
\end{aligned}
$$

For $n \in \mathbb{N}$, the **closed unit ball** in $M_n(\mathbb{C})$ is defined to be the set

$$
\mathbf{B}_n := \{X \in M_n(\mathbb{C}) \ : \ \|X\| \leq 1\}.
$$

For a linear map $\Phi : M_n(\mathbb{C}) \to M_m(\mathbb{C})$, define

$$
\|\Phi\| := \sup\{\|\Phi(X)\| \ : \ X \in \mathbf{B}_n\},
$$

which is the operator norm of Φ with respect to the operator norms on $M_n(\mathbb{C})$ and $M_m(\mathbb{C})$.

Lemma 5.66. *For all $A \in \mathbf{B}_n$, there exist two unitaries U_1 and U_2 such that*

$$
(5.10.3) \qquad\qquad A = \tfrac{1}{2}(U_1 + U_2).
$$

Moreover, the set of extreme points of $\mathbf{B}_n$ is exactly the set of $n \times n$ unitary matrices.

Proof. Let $A = U|A|$ be a polar decomposition of A with U unitary. Then $|A| = (A^*A)^{1/2} \in \mathbf{B}_n$. Define $V_{\pm} = |A| \pm i\sqrt{\mathbb{1} - |A|^2}$ and note that V_+ and V_- are both unitary, and of course $|A| = \tfrac{1}{2}(V_+ + V_-)$. Therefore, (5.10.3) is satisfied with $U_1 = UV_+$ and $U_2 = UV_-$.

For the second part, let A be extreme in $\mathbf{B}_n$, and let U_1, U_2 be two unitaries such that (5.10.3) is satisfied. Since A is extreme, $U_1 = U_2 = A$, and hence A is unitary.

It remains to show that every unitary U is extreme. We first show that $\mathbb{1}$ is extreme. By Remark A.10 it suffices to show that if $X, Y \in \mathbf{B}_n$ and $\mathbb{1} = \tfrac{1}{2}(X+Y)$, then $X = Y = \mathbb{1}$. Let $A := \tfrac{1}{2}(X+X^*)$ and $B = \tfrac{1}{2}(Y+Y^*)$. Then A and B are self-adjoint, and $\mathbb{1} = \tfrac{1}{2}(A+B)$. Let λ be an eigenvlaue of A. Then since $B = 2\mathbb{1} - A$, $2 - \lambda$ is an eigenvalue of B, but since $A, B \in \mathbf{B}_n$, $|\lambda| \leq 1$ and $|2 - \lambda| \leq 1$. Therefore, $\lambda = 1$. Hence, $\sigma(A) = \{1\}$, and since A is self-adjoint, $A = \mathbb{1}$, and

$X = \mathbb{1} + iC$, where C is self-adjoint. Therefore, $X^*X = \mathbb{1} + C^2 \in \mathbf{B}_n$, and hence $C = 0$. Thus, $X = \mathbb{1}$, and then $Y = \mathbb{1}$. Thus $\mathbb{1}$ is extreme in $\mathbf{B}_n$.

Now let U be any unitary in $M_n(\mathbb{C})$ and suppose that $U = \frac{1}{2}(X + Y), X, Y \in \mathbf{B}_n$. Then $\mathbb{1} = \frac{1}{2}(U^*X + U^*Y)$, and since $U^*X, U^*Y \in \mathbf{B}_n$, $U^*X = U^*Y = \mathbb{1}$, and this is the same as $X = Y = U$. Hence, every unitary U in $M_n(\mathbb{C})$ is extreme. $\square$

Russo and Dye [**189**] showed that in any C^* algebra, the convex hull of the unitary elements are dense in the unit ball, which is rather trivial in our setting, but then they showed in [**189**, Corollary 1] that for a positive unital map Φ from one unital C^* algebra to another, $\|\Phi\| = \|\Phi(\mathbb{1})\| = 1$. The following is commonly called the **Russo-Dye theorem** in the matrix algebra setting.

Theorem 5.67 (Russo-Dye theorem). *Let* $\Phi : M_n(\mathbb{C}) \to M_m(\mathbb{C})$ *be a quasi unital positive linear map. Then for all* $A \in \mathbf{B}_n$,

$$\Phi(A)^*\Phi(A) \le \Phi(\mathbb{1}) \le \mathbb{1},$$

so that, in particular, $\|\Phi\| = \|\Phi(\mathbb{1})\| = 1$.

Proof. Suppose Φ is positive. Let U_1 and U_2 be two unitaries such that (5.10.3) is satisfied. By the convexity of $X \mapsto X^*X$ and then Theorem 5.64,

$$\begin{aligned}
\Phi(A)^*\Phi(A) = \Phi\left(\frac{1}{2}U_1 + \frac{1}{2}U_2\right)^* \Phi\left(\frac{1}{2}U_1 + \frac{1}{2}U_2\right) \\
\le \frac{1}{2}\Phi(U_1)^*\Phi(U_1) + \frac{1}{2}\Phi(U_2)^*\Phi(U_2) \\
\le \frac{1}{2}\Phi(U_1^*U_1) + \frac{1}{2}\Phi(U_2^*U_2) = \Phi(\mathbb{1}) \le \mathbb{1}. \qquad \square
\end{aligned}$$

Theorem 5.68. *Let* $\Phi : M_n(\mathbb{C}) \to M_m(\mathbb{C})$ *be a positive linear map. Then for all* $A \in \mathbf{B}_n$,

$$\Phi(A)^*\Phi(A) \le \|\Phi(\mathbb{1})\|\Phi(\mathbb{1}),$$

so that, in particular, $\|\Phi\| = \|\Phi(\mathbb{1})\|$.

Proof. If Φ is quasi unital, this follows from Theorem 5.67. Otherwise, define Ψ in terms of Φ by

$$\Psi(X) = (\Phi(\mathbb{1})^+)^{1/2}\Phi(X)(\Phi(\mathbb{1})^+)^{1/2},$$

and note that Ψ is positive and quasi unital. Hence, by Theorem 5.67,

$$(5.10.4) \qquad\qquad \Psi(A)^*\Psi(A) \le \Psi(\mathbb{1}) = P,$$

where P is the orthogonal projection onto $\operatorname{ran}(\Phi(\mathbb{1}))$. However,

$$\Psi(A)^*\Psi(A) = (\Phi(\mathbb{1})^+)^{1/2}\Phi(A)^*\Phi(\mathbb{1})^+\Phi(A)(\Phi(\mathbb{1})^+)^{1/2}.$$

Multiplying through on both sides in (5.10.4) by $\Phi(\mathbb{1})^{1/2}$, and using Lemma 5.2,

$$\Phi(A)^*\Phi(\mathbb{1})^+\Phi(A) \le \Phi(\mathbb{1}).$$

Since $\|\Phi(\mathbb{1})\|\Phi(\mathbb{1})^+ \geq P$, we have from Lemma 5.2 once more that

$$\Phi(A)^*\Phi(A) = \Phi(A)^*P\Phi(A) \leq \|\Phi(\mathbb{1})\|\Phi(A)^*\Phi(\mathbb{1})^+\Phi(A) \leq \|\Phi(\mathbb{1})\|\Phi(\mathbb{1}). \quad \square$$

Theorem 5.69 (Ando). *Let* $\Phi : M_n(\mathbb{C}) \to M_m(\mathbb{C})$ *be a positive map. Then for all* $H \in M_n^{\text{s.a.}}(\mathbb{C})$ *and all* $X \in M_n^+(\mathbb{C})$ *such that* $\ker(X) \subseteq \ker(H)$,

$$\Phi(HX^+H) \geq \Phi(H)\Phi(X)^+\Phi(H).$$

Proof. Define $\Psi : M_n(\mathbb{C}) \to M_m(\mathbb{C})$ by

$$\Psi(Y) = (\Phi(X)^+)^{1/2}\Phi(X^{1/2}YX^{1/2})(\Phi(X)^+)^{1/2},$$

and note that Ψ is positive and quasi unital. Then taking $Y := (X^+)^{1/2}H(X^+)^{1/2}$, which is self-adjoint, Theorem 5.64 yields

$$(5.10.5) \qquad\qquad \Psi(Y^2) \geq \Psi(Y)^2.$$

Since $\ker(X) \subseteq \ker(H)$, $X^{1/2}Y^2X^{1/2} = HX^+H$, and hence

$$\Psi(Y^2) = (\Phi(X)^+)^{1/2}\Phi(HX^+H)(\Phi(X)^+)^{1/2}$$

and

$$\Psi(Y)^2 = (\Phi(X)^+)^{1/2}\Phi(H)\Phi(X)^+\Phi(H)(\Phi(X)^+)^{1/2}.$$

Multiplying through in (5.10.5) by $\Phi(X)^{1/2}$ on both the left and right, and using the fact that $\Phi(X)^{1/2}(\Phi(X)^+)^{1/2} =: P$ is the orthogonal projection onto the range of $\Phi(X)$, we find

$$P\Phi(HX^+H)P \geq P\Phi(H)\Phi(X)^+\Phi(H)P.$$

Since $\ker(X) \subseteq \ker(H) = \ker(H^2)$, there is some $c > 0$ such that $H^2 \leq cX$, and hence $\Phi(H^2) \leq c\Phi(X)$. Also, $HX^+H \leq \|X^+\|H^2$ and hence $\Phi(HX^+H) \leq \|X^+\|\Phi(H^2)$. Altogether, these inequalities yield

$$\operatorname{ran}(\Phi(HX^+H)) \subseteq \operatorname{ran}(\Phi(H^2)) \subseteq \operatorname{ran}(\Phi(X)).$$

This already shows that $P\Phi(HX^+H)P = \Phi(HX^+H)$.

Next, we show that $\operatorname{ran}(\Phi(H^2)) = \operatorname{ran}(\Phi(H))$. Then since $\operatorname{ran}(\Phi(H^2)) \subseteq \operatorname{ran}(\Phi(X))$, $P\Phi(H)\Phi(X)^+\Phi(H)P = \Phi(H)\Phi(X)^+\Phi(H)$.

It remains to show that $\operatorname{ran}(\Phi(H^2)) = \operatorname{ran}(\Phi(H))$, or equivalently $\ker(\Phi(H^2)) = \ker(\Phi(H))$. By (5.10.1),

$$\Phi(H^2) \geq \Phi(H)\Phi(\mathbb{1})^+\Phi(H),$$

from which it is obvious that $\ker(\Phi(H)) \subseteq \ker(\Phi(H^2))$. To see that equality holds, let Q denote the orthogonal projection onto $\operatorname{ran}(\Phi(\mathbb{1}))$. Then for some $c > 0$, $\Phi(\mathbb{1})^+ \geq cQ$, and if $\mathbf{v} \in \ker(\Phi(H^2)) \subseteq \ker(\Phi(H)\Phi(\mathbb{1})^+\Phi(H))$,

$$0 = \langle \Phi(H)\mathbf{v}, \Phi(\mathbb{1})^+\Phi(H)\mathbf{v}\rangle \geq c\langle \Phi(H)\mathbf{v}, Q\Phi(H)\mathbf{v}\rangle = c\|\Phi(H)\mathbf{v}\|^2. \quad \square$$

Ando's theorem has an application to the *harmonic mean* of positive definite matrices that was studied by Anderson and Duffin [7] who used a slightly different formula.

Definition 5.70 (Harmonic mean). For $X, Y \in M_n^{++}(\mathbb{C})$, the **harmonic mean** of X and Y, denoted $X : Y$, is defined by

$$X : Y := 2(X^{-1} + Y^{-1})^{-1}.$$

As Ando observed, since $X^{-1} + Y^{-1} = X^{-1}(X + Y)Y^{-1}$,

$$(X^{-1} + Y^{-1})^{-1} = Y(X + Y)^{-1}X = ((X + Y) - X)(X + Y)^{-1}X$$

$$\tag{5.10.6} = X - X(X + Y)^{-1}X.$$

It is left as an exercise to show, using this, that for all $X, Y \in M_n^{++}(\mathbb{C})$, and all positive maps $\Phi : M_n(\mathbb{C}) \to M_m(\mathbb{C})$,

$$\tag{5.10.7} \Phi(X : Y) \geq \Phi(X) : \Phi(Y).$$

A general theory of **operator means**, including the harmonic mean, the geometric mean (see Exercise 11) and the arithmetic mean as special cases, has been developed by Kubo and Ando [**134**]. A geometric development of this theory has been given by Bhatia and Holbrook [**37**].

Exercises

(1) For $A, X \in M_n(\mathbb{C})$, the Hadamard product of A and X, $A \circ X$ is the entry-wise product of A and X. That is, for $1 \leq i, j \leq n$, $(A \circ X)_{i,j} = A_{i,j}X_{i,j}$. Fix $A \in M_n(\mathbb{C})$ and define a linear map Φ on $M_n(\mathbb{C})$ by $\Phi(X) = A \circ X$. Show that if $A \in M_n^+(\mathbb{C})$, then Φ is CP , and find a Kraus representation for Φ in terms of a spectral decomposition of A.

(2) Let $\rho \in M_n^{++}(\mathbb{C})$ be a density matrix. Define an operator Ψ on $M_n(\mathbb{C})$ by

$$\Psi(X) = \int_0^\infty \frac{\sqrt{\rho}}{\rho + t\mathbb{1}} X \frac{\sqrt{\rho}}{\rho + t\mathbb{1}} \, dt.$$

Show that Ψ is unital, CP, and trace preserving. Show that there is a Kraus representation of Ψ involving only n Kraus operators.

(3) Let Φ be a linear map from $M_n(\mathbb{C})$ to $M_m(\mathbb{C})$. Recall that Φ is said to be Hermitian in case $\Phi(A^*) = \Phi(A)^*$ for all A. Let C_Φ be the characteristic matrix of Φ as defined in Definition 5.21 using any orthonormal basis $\{F_\alpha\}_{\alpha \in \mathcal{I}_{n,m}}$ of $\mathcal{H}_{n,m}$. Show that Φ is Hermitian if and only if C_Φ is self-adjoint.

(4) Let Φ be a Hermitian linear map from $M_n(\mathbb{C})$ to $M_m(\mathbb{C})$. Show that there exists a set $\{V_1, \ldots, V_p\} \subset M_{n,m}(\mathbb{C})$ and a set $\{\lambda_1, \ldots, \lambda_p\} \subset \mathbb{R}$ with $p \leq mn$ such that for all $X \in M_n(\mathbb{C})$, $\Phi(X) = \sum_{j=1}^p \lambda_j V_j^* X V_j$.

(5) Let Φ be a Hermitian linear map from $M_n(\mathbb{C})$ to $M_m(\mathbb{C})$. Show that Φ is the difference of two CP maps. Write the transpose map $X \mapsto X^T$ on $M_2(\mathbb{C})$ as a difference of CP maps.

(6) Let $\rho \in M_n^{++}(\mathbb{C})$ be a density matrix. The *GNS inner product* on $M_n(\mathbb{C})$ with respect to ρ is defined by $\langle X, Y \rangle_\rho = \mathrm{Tr}[X^*Y\rho]$. For a linear map $\Phi : M_n(\mathbb{C}) \to M_n(\mathbb{C})$, let $\Phi^{*,\rho}$ denote its adjoint with respect to $\langle \cdot, \cdot \rangle_\rho$. Compute a formula for $\Phi^{*,\rho}$ in terms of $\Phi^\dagger$ and ρ. Show that it may be that $\Phi^{*,\rho}$ is not CP when Φ is CP. However, if $\Phi(X) = \sum_{j=1}^m V_j^* X V_j$ and for each j, $\rho V_j \rho^{-1} = \lambda_j V_j$ for some number λ_j, then $\Phi^{*,\rho}$ is CP.

(7) Define a map $\Phi : M_{mn}(\mathbb{C}) \to M_m(\mathbb{C})$ by identifying $X \in M_{mn}(\mathbb{C})$ with the $n \times n$ block matrix with entries in $X_{i,j} \in M_m(\mathbb{C})$, and setting $\Phi(X) = \frac{1}{n} \sum_{i,j=1}^n X_{i,j}$. Show that Φ is CP and unital.

(8) Show that $\Phi : \mathcal{M}_m(\mathbb{C}) \to M_n(\mathbb{C})$ is CP if and only if for all $k \in \mathbb{N}$, all $\{A_1, \ldots, A_k\} \subset M_m(\mathbb{C})$ and all $\{\mathbf{v}_1, \ldots, \mathbf{v}_k\} \subset \mathbb{C}^n$, $\sum_{i,j=1}^n \langle \mathbf{v}_i, \Phi(A_i^* A_j) \mathbf{v}_j \rangle \geq 0$.

(9) Define $\Phi_\lambda : M_3(\mathbb{C}) \to M_3(\mathbb{C})$ by $\Phi_\lambda(A) = \lambda A^T + (1 - \lambda)\frac{1}{2} \mathrm{Tr}[A]\mathbb{1}$. For which values of λ is Φ_λ 2-positive? For which values of λ is Φ_λ 3-positive (and hence CP)?

(10) (Bhatia, Davis, and Mathias) Let $\Phi : M_n(\mathbb{C}) \to M_n(\mathbb{C})$ be 3-positive. Show that for all $A, B \in M_n(\mathbb{C})$,

$$\begin{bmatrix} \Phi(A^*A) & \Phi(A^*B) & \Phi(A^*) \\ \Phi(B^*A) & \Phi(B^*B) & \Phi(B^*) \\ \Phi(A) & \Phi(B) & \Phi(\mathbb{1}) \end{bmatrix} \geq 0.$$

Then apply Lemma 5.37 to prove that

$$\begin{bmatrix} \Phi(A^*A) & \Phi(A^*B) \\ \Phi(B^*A) & \Phi(B^*B) \end{bmatrix} \geq \begin{bmatrix} \Phi(A)^*\Phi(A) & \Phi(A)^*\Phi(B) \\ \Phi(B)^*\Phi(A) & \Phi(B)^*\Phi(B) \end{bmatrix}.$$

Show that this last inequality need not be true if Φ is only 2-positive and not 3-positive.

(11) (Ando) Fix $A, B \in M_n^{++}(\mathbb{C})$. Define

$$K_{AB} := \left\{ C \in M_n^+(\mathbb{C}) : \begin{bmatrix} A & C \\ C & B \end{bmatrix} \geq 0 \right\}.$$

(a) Show that K_{AB} is convex. Use Lemma 5.37 to show that $C \in K_{AB}$ if and only if $M_G(A, B) := A^{1/2}(A^{-1/2}BA^{-1/2})^{1/2}A^{1/2} \geq C$. That is, the set K_{AB} has a maximal element, which is $M_G(A, B)$.

(b) Show that $M_G(A, B) = M_G(B, A)$, and that if $[A, B] = 0$, $M_G(A, B) = A^{1/2}B^{1/2}$. $M_G(A, B)$ is called the **geometric mean** of A and B, originally defined and studied in [**183**].

(c) Show that the geometric mean is jointly concave: if $A_0, A_1, B_0, B_1 \in \mathcal{M}_n^{++}(\mathbb{C})$ and for $0 < \lambda < 1$, $A_\lambda := (1 - \lambda)A_0 + \lambda A_1$ and $B_\lambda := (1 - \lambda)B_0 + \lambda B_1$, then

$$M_G(A_\lambda, B_\lambda) \geq (1 - \lambda)M_G(A_0, B_0) + \lambda M_G(A_1, B_1).$$

(12) Fix $A \in M_n^{++}(\mathbb{C})$, $K \in M_n(\mathbb{C})$, and $0 < r \leq 1$. Define

$$C_{A,K} := \left\{ CX \in M_n^+(\mathbb{C}) \; : \; \begin{bmatrix} X & K \\ K^* & A^r \end{bmatrix} \geq 0 \right\}.$$

(a) Show that $C_{A,K}$ is convex. Use Lemma 5.37 to show that $X \in C_{AK}$ if and only if $K^* A^{-r} K \leq X$. That is the set C_{AK} has a minimal element, which is $K^* A^{-r} K$.

(b) Show that $(A, K) \mapsto K^* A^{-r} K$ jointly convex. That is, if $A_0, A_1 \in M_n^{++}(\mathbb{C})$, $K_0, K_1 \in M_n(\mathbb{C})$ and for $0 < \lambda < 1$, $A_\lambda := (1 - \lambda)A_0 + \lambda A_1$ and $K_\lambda := (1 - \lambda)K_0 + \lambda K_1$, then

$$K_\lambda A_\lambda^{-r} K_\lambda \leq (1 - \lambda)K_0^* A_0^{-r} K_0 + \lambda K_1^* A_1^{-r} K_1.$$

(13) This exercise yields a proof of Naimark's theorem via Stinespring's factorization. Let X be a finite set of cardinality m. Let $\mathcal{E}$ be a POVM on X with values in $\mathcal{B}(\mathcal{H})$, and let $\mathcal{H}$ be a finite-dimensional Hilbert space. Let $\mathcal{K}$ be a Hilbert space of dimension m, and let $\{e_x\}_{x \in X}$ be an orthonormal basis of $\mathcal{K}$. Let $\mathcal{M}$ be the von Neumann subalgebra of $\mathcal{B}(\mathcal{K})$ spanned by $\{|e_x\rangle\langle e_x|\}_{x \in X}$. Note that $\mathcal{M}$ is commutative. Define $\Phi : \mathcal{M} \to \mathcal{B}(\mathcal{H})$ by $\Phi\left(\sum_{x \in X}^m z_x |e_x\rangle\langle e_x|\right) = \sum_{j=1}^m z_x E_x$.

(a) Show that Φ is CP and unital.

(b) Find a Stinespring factorization of Φ, so that $\Phi(X) = V^* \pi(X) V$ where V is an isometry and π is a representation. For each $x \in X$, define $\widehat{\mathcal{E}}(x) := \pi(|e_x\rangle\langle e_x|)$. Show that $\widehat{\mathcal{E}}$ is a PVM on X with values in $\mathcal{B}(\mathcal{H} \otimes \mathcal{K})$.

(c) Show that $\mathcal{E}(x) = V^* \widehat{\mathcal{E}}(x) V$ is a Naimark dilation of $\mathcal{E}$.

(14) Use the operator Jensen inequality to show that if a unital map $\Phi : M_n(\mathbb{C}) \to M_n(\mathbb{C})$ is CP, and f is operator convex on $(0, \infty)$, then for all $X \in M_n^{++}(\mathbb{C})$, $f(\Phi(X)) \leq \Phi(f(X))$. Then use Theorem 5.28 to show that the inequality remains valid if Φ is only assumed to be positive.

(15) The map from $\mathcal{B}(\mathcal{H}_1 \otimes \mathcal{H}_2)$ to $\mathcal{B}(\mathcal{H}_1 \otimes \mathcal{H}_2)$ given by $X \mapsto D_2^{-1} \operatorname{Tr}_2[X] \otimes \mathbb{1}_{\mathcal{H}_2}$, where d_2 is the dimension of $\mathcal{H}_2$, is the conditional expectation in $\mathcal{B}(\mathcal{H}_1 \otimes \mathcal{H}_2)$ onto the subalgebra $\mathcal{B}(\mathcal{H}_1) \otimes \mathbb{1}_{\mathcal{H}_2}$, and therefore is CP. Find a minimal Stinespring factorization of this map.

(16) Let $\Phi : M_m(\mathbb{C}) \to M_n(\mathbb{C})$ be a completely positive and unital map, and let $\Phi(x) = V^* \pi(X) V$ be a Stinespring factorization of it. Show that $X \in M_m(\mathbb{C})$ satisfies $\Phi(X^*)\Phi(X) = \Phi(X^*X)$ if and only if $\operatorname{ran}(V)$ is an invariant subspace for $\pi(X)$, and that in this case, for all such X, and all $Y \in M_m(\mathbb{C})$, $\Phi(YX) = \Phi(Y)\Phi(X)$.

(17) Define the map $\Phi : M_{mn}(\mathbb{C}) \to M_m(\mathbb{C})$ obtained by identifying $X \in M_{mn}(\mathbb{C})$ with the $n \times n$ block matrix with entries in $X_{i,j} \in M_m(\mathbb{C})$, and setting $\Phi(X) = \frac{1}{n} \sum_{i,j=1}^n X_{i,j}$ is CP and unital. Find a minimal Stinespring factorization.

(18) Let $\mathcal{H}$ be a finite-dimensional Hilbert space, and let Φ be a unital CP map on $\mathcal{B}(\mathcal{H})$ with the Kraus representation $\Phi(X) = \sum_{j=1}^m V_j^* X V_j$. Let P be an orthogonal projection on $\mathcal{H}$.

(a) Show that if $\Phi(P) = P$, then $\sum_{j=1}^m \|[V_j, P]\|^2 = 0$, and hence that $P \in \{V_1, \ldots, V_m, V_1^*, \ldots, V_m^*\}'$.

(b) The orthogonal projection P is said to **strongly reduce** Φ in case $\Phi(PXP) = P\Phi(X)P$ for all $X \in \mathcal{B}(\mathcal{H})$. Show that P strongly reduces Φ if and only if $P \in \{V_1, \ldots, V_m, V_1^*, \ldots, V_m^*\}'$.

(c) A CP unital map Φ is **strongly reducible** if and only if there exists a nontrivial orthogonal projection P that strongly reduces Φ. Show that Φ is strongly reducible if and only if $\{V_1, \ldots, V_m, V_1^*, \ldots, V_m^*\}' \neq \mathbb{C}\mathbb{1}$.

(d) An orthogonal projection P **reduces** Φ in case $P\Phi(P^\perp X P^\perp)P = 0$ for all $X \in \mathcal{B}(\mathcal{H})$. Show that P reduces Φ if and only if $P\Phi(PXP)P = P\Phi(X)P$ for all $X \in \mathcal{B}(\mathcal{H})$, and that this is the case if and only if $P\Phi^\dagger(PXP)P = \Phi^\dagger(PXP)$ for all $X \in \mathcal{B}(\mathcal{H})$.

(e) Suppose that P is a nontrival projection that reduces Φ, and $\mathcal{K}$ is the range of P. Define $\widetilde{\Phi}$ on $\mathcal{B}(\mathcal{K})$ by $\widetilde{\Phi}(Y) := P\Phi(Y)P$ for all $Y \in \mathcal{B}(\mathcal{K})$. Show that $\widetilde{\Phi}$ is a unital CP map on $\mathcal{B}(\mathcal{K})$.

(19) Let $\mathcal{H}$ be a three-dimensional Hilbert space. Let $\{\mathbf{u}_1, \mathbf{u}_2, \mathbf{u}_3\}$ be an orthonormal basis of $\mathcal{H}$, and for $1 \leq i, j \leq 3$, define $E_{i,j} := |\mathbf{u}_i\rangle\langle\mathbf{u}_j|$. Define

$$V_1 := E_{1,1}, \quad V_2 := E_{2,2}, \quad V_3 := 2^{-1/2}E_{1,3}, \quad V_4 := 2^{-1/2}E_{2,3}.$$

Define a CP map Φ on $\mathcal{B}(\mathcal{H})$ by $\Phi(X) = \sum_{j=1}^{4} V_j^* X V_j$.

(a) Show that Φ is unital, so that $\Phi^\dagger$ is a quantum operation.

(b) Show that the eigenspace of $\Phi^\dagger$ for the eigenvalue 1 is two dimensional and spanned by $E_{1,1}$ and $E_{2,2}$, so that ρ is an invariant state for Φ if and only if $\rho = pE_{1,1} + (1-p)E_{2,2}$ for some $0 \leq p \leq 1$. In particular, Φ has no faithful invariant state.

(c) Let $\mathcal{C}_\Phi$ be the set of fixed points of Φ, as in Theorem 5.54. Show that $\mathcal{C}_\Phi = \text{span}(\{E_{1,1} + \tfrac{1}{2}E_{3,3}, E_{2,2} + \tfrac{1}{2}E_{3,3}\})$. Show that $\mathcal{C}_\Phi$ is not an algebra, and hence that the assumption of the existence of a faithful invariant state cannot be dropped from Theorem 5.54.

(d) Show that $\{V_1, V_2, V_3, V_4, V_1^*, V_2^*, V_3^*, V_4^*\}' = \mathbb{C}\mathbb{1}$ so that also the conclusion (5.8.1) of Theorem 5.54 need not be valid when a faithful invariant state does not exist. Also show that (5.8.2) of Theorem 5.54 is not generally valid for this map Φ.

(20) Use (5.10.6) to prove (5.10.7).

(21) (Choi) For $n \geq 2$, define $\Phi : M_n(\mathbb{C}) \to M_n(\mathbb{C})$ by

$$\Phi(X) := \frac{1}{n^2 - n - 1}((n-1)\operatorname{Tr}[X]\mathbb{1} - X).$$

Show that Φ is positive, unital, and trace preserving, but is not completely positive.

(22) Let $\mathcal{H}$ be a finite-dimensional Hilbert space. Let $\Phi : \mathcal{B}(\mathcal{H}) \to \mathcal{B}(\mathcal{H})$. Suppose that K belongs to the multiplicative domain of Φ. Show that for all $X \in \mathcal{B}(\mathcal{H})$,

$$\Psi^\dagger(KX) = K\Psi^\dagger(X) \quad \text{and} \quad \Psi^\dagger(XK) = \Psi^\dagger(X)K.$$

(23) With the harmonic mean $X : Y$ specified as in Definition 5.70, show using (5.10.6) that for all $X, Y \in M_n^{++}(\mathbb{C})$, and all positive maps $\Phi : M_n(\mathbb{C}) \to M_m(\mathbb{C})$, $\Phi(X : Y) \geq \Phi(X) : \Phi(Y)$.

(24) Let $\Phi : \mathcal{B}(\mathcal{H}) \to \mathcal{B}(\mathcal{H})$ be unital and CP. Suppose that $K \in \mathcal{B}(\mathcal{H})$ belongs to the multiplicative domain of Φ. Show that for all $X \in \mathcal{B}(\mathcal{H})$, $\Phi^{\dagger}(XK) = \Phi^{\dagger}(X)K$ and $\Phi^{\dagger}(KX) = K\Phi^{\dagger}(X)$.

(25) For $A, B \in M_n^{++}(\mathbb{C})$, and $0 \leq t \leq 1$, define

$$A\#_t B := A^{1/2}(A^{-1/2}BA^{-1/2})^t A^{1/2},$$

and note that $A\#_{1/2}B = M_G(A, B)$, the geometric mean of A and B from Exercise 11. Note that id A and B commute, $A\#_t B = A^{1-t}B^t$, so that $A\#_t B$ may be viewed as a *weighted* geometric mean of A and B. The following exercises develop this point of view.

Use the integral representation for X^{-t}, $0 < t < 1$, to show that

$$A\#_t B = \frac{\sin(\pi t)}{\pi} \int_0^{\infty} \lambda^t \frac{1}{A^{-1} + \lambda B^{-1}} d\lambda = \frac{\sin(\pi t)}{2\pi} \int_0^{\infty} X : (\lambda Y)\lambda^t d\lambda,$$

thus expressing $A\#_t B$ for $0 < t < 1$ in terms of the harmonic mean.

(26) Continuing with Exercise 24, show that for all $0 < t < 1$ and all positive maps $\Phi : M_n(\mathbb{C}) \to M_m(\mathbb{C})$,

$$\Phi(A\#_t B) \geq \Phi(A)\#_t \Phi(B).$$

Then show that if $C \in M_n(\mathbb{C})$ is invertible,

$$C^*(A\#_t B)C = (C^*AC)\#_t(C^*BC).$$

(27) (Bhatia and Holbrook) Let $X(t)$ be a smooth path in $M_n^{++}(C)$.
   (a) For invertible $A \in M_n(C)$, define $Z(t) := A^*X(t)A$. Show that

$$\|Z(t)^{-1/2}Z'(t)Z(t)^{-1/2}\|_2^2 = \|X(t)^{-1/2}X'(t)X(t)^{-1/2}\|_2^2.$$

Then make $M_n^{++}(\mathbb{C})$ into a Riemannain manifold by equipping it with the distance $\delta(A, B)$ given by

$$\delta(A, B) = \inf_{X(\cdot) \in \mathcal{P}_{X,Y}} \left\{ \int_0^1 \|X(t)^{-1/2}X'(t)X(t)^{-1/2}\|_2 dt \right\},$$

where $\mathcal{P}_{X,Y}$ is the set of smooth paths $X(t) \in M_n^{++}(\mathbb{C})$, $t \in (0, 1)$ such that $X(0) = A$ and $X(1) = B$. Show that this metric is **conjugation invariant**. That is, for all invertible C, $\delta(C^*AC, C^*BC) = \delta(A, B)$.
   (b) Given any smooth path $t \mapsto X(t)$, define $H(t) := \log(X(t))$ so that $X(t) = e^{H(t)}$, and show that $H'(t) = \Psi_t(X(t)^{-1/2}X'(t)X(t)^{-1/2})$, where $\Psi_t : M_n(\mathbb{C}) \to M_n(\mathbb{C})$ is unital and CP. Show that, as a consequence,

$$\|H'(t)\|_2 \leq \|X(t)^{-1/2}X'(t)X(t)^{-1/2}\|_2.$$

   (c) Let $X(t)$ be a smooth path in $M_n(\mathbb{C})$ with $X(0) = A$ and $X(1) = B$. Show that Then, with $H(t) = \log X(t)$

$$\|\log B - \log A\|_2 \leq \delta(A, B),$$

and that when A and B commute, there is equality, and the unique minimal length path from A to B in unit time is $A^{1-t}B^t$.

(d) By the conjugation invariance $\delta(A, B) = \delta(\mathbb{1}, A^{-1/2}BA^{-1/2})$, and then by part (c), the unique minimal length path from $\mathbb{1}$ to $A^{-1/2}BA^{-1/2}$ in unit time is $X(t) = (A^{-1/2}BA^{-1/2})^t$. Show that the unique minimal length path from A to B in unit time is

$$X(t) = A\#_t B,$$

where $A\#_t B$ is the weighted geometric mean of Exercise 24. In fact, the formula $A\#_t B = A^{1/2}(A^{-1/2}BA^{-1/2})^t A^{1/2}$ makes sense for all $t \in \mathbb{R}$. Show that this entire path is a geodesic, along which the speed has the constant value $\|\log(A^{-1/2}BA^{-1/2})\|_2$, so that, in particular, $\delta(A, B) = \|\log(A^{-1/2}BA^{-1/2})\|_2$.

(28) For $A, B \in M_n^{++}(\mathbb{C})$, let $A\#_t B$ be defined for all $t \in \mathbb{R}$ as in Exercise 27. Then show that for all $s, t \in \mathbb{R}$ and $0 < \lambda < 1$,

$$A\#_{(1-\lambda)s+\lambda t}B = (A\#_s B)\#_\lambda(A\#_t B).$$

(29) (Kubo and Ando) For $0 < t < 1$, define the weighted geometric mean $A\#_t B$ on $M_n^{++}(\mathbb{C})$ as in Exercise 24. show that $(X, Y) \mapsto X\#_t Y$ is jointly concave, and monotone increasing in X and in Y.

(30) For $A, B \in M_n^{++}(\mathbb{C})$, let $A\#_t B$ be defined for all $t \in \mathbb{R}$ as in Exercise 26. Show that

$$A\#_{-1}B = AB^{-1}A \quad \text{and} \quad A\#_2 B = BA^{-1}B,$$

which are both convex in A and B. Show that in fact for all $t \in [-1, 0] \cup [1, 2]$, the map $(A, B) \mapsto A\#_t B$ is jointly convex.

(31) (Fujii and Kamei [**89**]) For $A, B \in M_n^{++}(\mathbb{C})$, show that

$$A^{1/2}\log(A^{1/2}B^{-1}A^{1/2})A^{1/2} = \int_0^\infty \left(\frac{A}{\lambda+1} - \frac{1}{A^{-1}+(\lambda B)^{-1}}\right)d\lambda,$$

and then show that $(A, B) \mapsto A^{1/2}\log(A^{1/2}B^{-1}A^{1/2})A^{1/2}$ is jointly convex on $M_n^{++}(\mathbb{C}) \times M_n^{++}(\mathbb{C})$.

(32) Let ρ be a density matrix with rank two and spectral decomposition $\rho = \sum_{j=1}^2 \lambda_j |\mathbf{u}_j\rangle\langle\mathbf{u}_j|$. Let $\Phi^\dagger$ be a quantum operation such that $\Phi^\dagger(|\mathbf{u}_j\rangle\langle\mathbf{u}_j|) = |\mathbf{u}_j\rangle\langle\mathbf{u}_j| \otimes |\mathbf{u}_j\rangle\langle\mathbf{u}_j|$ for $j = 1, 2$. Show that it is not the case that $\Phi^\dagger(\rho) = \rho \otimes \rho$, but that $\mathrm{Tr}_1[\Phi^\dagger(\rho)] = \mathrm{Tr}_2[\Phi^\dagger(\rho)] = \rho$.

(33) Let $\alpha \in (0, 1)$, and let

$$E_1 := \begin{bmatrix} 1-\alpha & 0 \\ 0 & \alpha \end{bmatrix} \quad \text{and} \quad E_2 := \begin{bmatrix} \alpha & 0 \\ 0 & 1-\alpha \end{bmatrix}.$$

Let $\mathcal{X} = \{1, 2\}$, and define a function $\mathcal{E}$ from $\mathcal{X}$ to the 2×2 matrices, regarded as operators on $\mathcal{H} = \mathbb{C}^2$, by $\mathcal{E}(x) = E_x$.

(a) Show that $\mathcal{E}$ is a POVM.

(b) Let $\mathcal{K}$ be another copy of $\mathbb{C}^2$. Find an isometry V from $\mathcal{H}$ to $\mathcal{H} \otimes \mathcal{K}$ and PVM $\widehat{\mathcal{E}}$ on $\mathcal{X}$ with values in $\mathcal{B}(\mathcal{H} \otimes \mathcal{K})$ such that for $x = 1, 2$, $\mathcal{E}(x) = V^* \widehat{\mathcal{E}} V$.

(c) Find a unitary $\mathcal{U}$ on $\mathcal{H} \otimes \mathcal{K}$ and a unit vector $\phi \in \mathcal{K}$ so that the isometry V from part (b) has the form $V\mathbf{x} = \mathcal{U}(\mathbf{x} \otimes \phi)$.

Chapter 6

Some basic trace function inequalities

6.1. The above the tangent plane inequality and subgradients

A number of theorems proved in this chapter concern functions F on $M_n^{\text{s.a.}}(\mathbb{C})$ or $M_n(\mathbb{C})$ defined by $F(X) = \text{Tr}[f(X)]$ or $F(X) = \text{Tr}[f(|X|)]$, where f is a convex function on $\mathbb{R}$. We give $M_n^{\text{s.a.}}(\mathbb{C})$ or $M_n(\mathbb{C})$ the structure of real Hilbert spaces by the following device: If $\mathcal{H}$ is a complex Hilbert space with inner product $\langle \cdot, \cdot \rangle$, let $\mathcal{K}$ denote $\mathcal{H}$ considered as a vector space over $\mathbb{R}$ (so that for all $\mathbf{v} \in \mathcal{H}$, $\mathbf{v}$ and $i\mathbf{v}$ are linearly independent in $\mathcal{K}$), and equip $\mathcal{K}$ with the inner product

$$\langle \mathbf{v}, \mathbf{w} \rangle_{\mathcal{K}} = \Re\langle \mathbf{v}, \mathbf{w} \rangle.$$

In particular, $M_n(\mathbb{C})$ equipped with the inner product $\Re\langle X, Y \rangle_{\text{HS}}$ is a real Hilbert space although the vectors in it are complex matrices. The important point is that the inner product of two vectors is always a real number.

A function f on a real Hilbert space $\mathcal{K}$ with values in $(-\infty, \infty]$ is **convex** in case for all $\lambda \in (0, 1)$ and all $\mathbf{v}, \mathbf{w} \in \mathcal{K}$,

$$f((1 - \lambda)\mathbf{v} + \lambda\mathbf{w}) \leq (1 - \lambda)f(\mathbf{v}) + \lambda f(\mathbf{w}).$$

A convex function f on $\mathcal{K}$ is **proper** in case it is not identically equal to ∞, and it is **lower semicontinuous** in case

$$(6.1.1) \qquad \{\mathbf{v} \in \mathcal{K} \,:\, f(\mathbf{v}) > t\} \quad \text{is open for all } t \in \mathbb{R}.$$

Evidently, in this case, $\{\mathbf{v} \in \mathcal{K} \;:\; f(\mathbf{v}) \le t\}$ is closed for all $t \in \mathbb{R}$. Therefore, if $\{\mathbf{v}_n\}$ is a sequence in $\mathcal{K}$ such that $\lim_{n\to\infty} \mathbf{v}_n = \mathbf{v}$, then

$$(6.1.2) \qquad\qquad f(\mathbf{v}) \le \liminf_{n\to\infty} f(\mathbf{v}_n).$$

It is left as an exercise to show that if (6.1.2) is true for all convergent sequences in $\mathcal{K}$, then (6.1.1) is true for all t, and thus (6.1.2) provides an alternate characterization of lower semicontinuity.

Lemma 6.1. *Let $\mathcal{F}$ be an arbitrary family of functions on $\mathcal{K}$ with values in $(-\infty, \infty]$. Define*

$$\bigvee_{\mathcal{F}} F(\mathbf{v}) = \sup\{F(\mathbf{v}) \;:\; F \in \mathcal{F}\}.$$

If each $F \in \mathcal{F}$ is convex, then $\bigvee_{\mathcal{F}} F$ is convex. If each $F \in \mathcal{F}$ is lower semicontinuous, then $\bigvee_{\mathcal{F}} F$ is lower semicontinuous.

Proof. Let $\mathbf{v}, \mathbf{w} \in \mathcal{K}$ and $0 < \lambda < 1$. If $\bigvee_{\mathcal{F}} F((1-\lambda)\mathbf{v} + \lambda\mathbf{w}) < \infty$, let $\epsilon > 0$ and $F \in \mathcal{F}$ be such that $F((1-\lambda)\mathbf{v} + \lambda\mathbf{w}) \ge \bigvee_{\mathcal{F}} F((1-\lambda)\mathbf{v} + \lambda\mathbf{w}) - \epsilon$. Since F is convex,

$$\bigvee_{\mathcal{F}} F((1-\lambda)\mathbf{v} + \lambda\mathbf{w}) \le (1-\lambda)F(\mathbf{v}) + \lambda F(\mathbf{w}) + \epsilon$$

$$\le (1-\lambda)\bigvee_{\mathcal{F}} F(\mathbf{v}) + \lambda \bigvee_{\mathcal{F}} F(\mathbf{w}) + \epsilon.$$

Since $\epsilon > 0$ is arbitrary, the inequality defining convexity is proved when $\bigvee_{\mathcal{F}} F((1-\lambda)\mathbf{v} + \lambda\mathbf{w}) < \infty$. The same sort of reasoning shows that when $\bigvee_{\mathcal{F}} F((1-\lambda)\mathbf{v} + \lambda\mathbf{w}) = \infty$, at least one of $\bigvee_{\mathcal{F}} F(\mathbf{v})$ or $\bigvee_{\mathcal{F}} F(\mathbf{w})$ is infinite, and this proves the convexity.

Likewise, if each $F \in \mathcal{F}$ is lower semicontinuous, so is $\bigvee_{\mathcal{F}} F$ since

$$\left\{\mathbf{v} \in \mathcal{K} \;:\; \bigvee_{\mathcal{F}} F(\mathbf{v}) > t\right\} = \bigcup_{F \in \mathcal{F}}\{\mathbf{v} \in \mathcal{K} \;:\; F(\mathbf{v}) > t\},$$

and an arbitrary union of open sets is open. $\square$

Definition 6.2 (Legendre transform). Let $\mathcal{K}$ be a real Hilbert space with inner product $\langle \cdot, \cdot \rangle_{\mathcal{K}}$. Let F be a proper function on $\mathcal{K}$ with values in $(-\infty, \infty]$. Then the **Legendre transform** of F on $\mathcal{K}$ is the function F^* defined by

$$F^*(\mathbf{y}) = \sup_{\mathbf{x} \in \mathcal{K}}\{\langle \mathbf{y}, \mathbf{x}\rangle_{\mathcal{K}} - F(\mathbf{x})\}.$$

By the very definition of F^* one has that for all $\mathbf{x}, \mathbf{y} \in \mathcal{K}$,

$$(6.1.3) \qquad\qquad \langle \mathbf{y}, \mathbf{x}\rangle \le F^*(\mathbf{y}) + F(\mathbf{x}).$$

This is known as **Young's inequality**; see Appendix A for more information. To describe the cases of equality, we need the following definition.

Definition 6.3 (Subgradient of a function on a real Hilbert space). Let $\mathcal{K}$ be a real Hilbert space. Let F be a proper function on $\mathcal{K}$ with values in $(-\infty, \infty]$. At each $\mathbf{x}$ such that $F(\mathbf{x}) < \infty$, the **subgradient** of F at $\mathbf{x}$, $\partial F(\mathbf{x})$, is defined to be the set of all $\mathbf{y} \in \mathcal{K}$ such that for all $\mathbf{w} \in \mathcal{K}$,

$$(6.1.4) \qquad F(\mathbf{w}) \geq F(\mathbf{x}) + \langle \mathbf{y}, \mathbf{w} - \mathbf{x} \rangle_{\mathcal{K}}.$$

Remark 6.4. It is left as an exercise to show that if F is convex and differentiable at $\mathbf{x}$, then $\partial F(\mathbf{x}) = \{\nabla F(\mathbf{x})\}$. In Appendix A it is shown that if F is a proper convex lower semicontinuous function, and F is finite on an open set containing $\mathbf{x}$, then $\partial F(\mathbf{x})$ is a nonempty compact convex set. The graph of the function $\mathbf{y} \mapsto F(\mathbf{x}) + \langle \mathbf{w}, \mathbf{y} - \mathbf{x} \rangle_{\mathcal{K}}$ is a hyperplane in $\mathcal{K}$, and when the inequality (6.1.4) is satisfied, the graph of F lies above this hyperplane, but touches it at $\mathbf{y} = \mathbf{x}$. In this case, the hyperplane is called a **supporting hyperplane**. Supporting hyperplanes exist at all points $\mathbf{x}$ such that $\partial F(\mathbf{x})$ is nonempty. Inequality (6.1.4) is called the **above the tangent plane inequality**.

Theorem 6.5 (Cases of equality in Young's inequality). *Let $\mathcal{K}$ be a real Hilbert space. Let F be a proper function on $\mathcal{K}$ with values in $(-\infty, \infty]$. If $F(\mathbf{x}) < \infty$, there is equality in (6.1.3) if and only if $\mathbf{y} \in \partial F(\mathbf{x})$.*

Proof. By Young's inequality, $\langle \mathbf{y}, \mathbf{w} \rangle_{\mathcal{K}} \leq F^*(\mathbf{y}) + F(\mathbf{w})$ for all $\mathbf{w} \in \mathcal{K}$. By assumption, $\langle \mathbf{y}, \mathbf{x} \rangle_{\mathcal{K}} = F^*(\mathbf{y}) + F(\mathbf{x})$, and hence

$$(F^*(\mathbf{y}) + F(\mathbf{w})) - (F^*(\mathbf{y}) + F(\mathbf{x})) \geq \langle \mathbf{y}, \mathbf{w} \rangle_{\mathcal{K}} - \langle \mathbf{y}, \mathbf{x} \rangle_{\mathcal{K}},$$

which is the same as (6.1.4). Thus $\mathbf{y} \in \partial F(\mathbf{x})$.

Conversely, suppose $\mathbf{y} \in \partial F(\mathbf{x})$. Then by (6.1.4), for all $\mathbf{w} \in \mathcal{K}$

$$\langle \mathbf{y}, \mathbf{x} \rangle \geq F(\mathbf{x}) + \langle \mathbf{y}, \mathbf{w} \rangle_{\mathcal{K}} - F(\mathbf{w}).$$

Taking the supremum over $\mathbf{w}$ yields $\langle \mathbf{y}, \mathbf{x} \rangle \geq F(\mathbf{x}) + F^*(\mathbf{y})$, and hence by (6.1.3), $\langle \mathbf{y}, \mathbf{x} \rangle \leq F(\mathbf{x}) + F^*(\mathbf{y})$. Therefore, $\langle \mathbf{y}, \mathbf{x} \rangle = F(\mathbf{x}) + F^*(\mathbf{y})$. $\qquad\square$

6.1.1. Homogeneity and convexity.

Definition 6.6 (Homogeneity). A function $F : \mathcal{K} \to (-\infty, \infty]$ is **homogeneous of degree p**, $p \in \mathbb{R}$, in case for all $t \in \mathbb{R}$ and $\mathbf{v} \in \mathcal{K}$,

$$F(t\mathbf{v}) = |t|^p F(\mathbf{v}).$$

We shall frequently encounter such functions, especially those that are homogeneous of degree one.

Lemma 6.7. *Let F be a function on a real Hilbert space $\mathcal{K}$ with values in $[0, \infty)$ that is homogeneous of degree one. Then F is convex if and only if F is subadditive; that is*

$$F(\mathbf{v} + \mathbf{w}) \leq F(\mathbf{v}) + F(\mathbf{w}).$$

Proof. Since F is homogeneous of degree one, for all $\lambda \in (0,1)$ and all $\mathbf{v}, \mathbf{w} \in \mathcal{K}$,

$$(6.1.5) \qquad F((1-\lambda)\mathbf{v}) + F(\lambda\mathbf{w}) = (1-\lambda)F(\mathbf{v}) + \lambda F(\mathbf{w}).$$

If F is convex, the right side of (6.1.5) is at least $F((1-\lambda)\mathbf{v}+\lambda\mathbf{w})$, so that F is subadditive. If F is subadditive, the left side of (6.1.5) is at least $F((1-\lambda)\mathbf{v}+\lambda\mathbf{w})$, so that F is convex. $\qquad\square$

Remark 6.8 (Convexity and level sets). Let F be any real valued function on $\mathcal{K}$. For each $t \in R$, the t-**level set of** F is the set $\{\mathbf{x} : F(\mathbf{x}) \leq t\}$. If F is convex, then each of its level sets is convex. To see this, fix $t \in \mathbb{R}$ and suppose $\mathbf{v}, \mathbf{w}$ are such that $F(\mathbf{v}) \leq t$ and $F(\mathbf{w}) \leq t$. Then for any $0 \leq \lambda \leq 1$,

$$F((1-\lambda)\mathbf{v} + \lambda\mathbf{w}) \leq (1-\lambda)F(\mathbf{v}) + \lambda F(\mathbf{w}) \leq (1-\lambda)t + \lambda t = t.$$

This shows that for all $t \in \mathbb{R}$, $\{\mathbf{x} : F(\mathbf{x}) \leq t\}$ is convex whenever F is convex.

Convexity of the level sets of F is not in general enough to guarantee that F is convex. However, convexity of the level sets together with homogeneity of degree one does guarantee convexity.

Lemma 6.9. *Let F be a function on a real Hilbert space $\mathcal{K}$ with values in $[0, \infty)$ that is homogeneous of degree one. If every level set of F is convex, then F is convex. Let G be a convex function on a real Hilbert space $\mathcal{K}$ with values in $[0, \infty)$ that is homogeneous of degree $p > 0$. That is*

$$G(\lambda\mathbf{v}) = \lambda^p G(\mathbf{v})$$

for all $\lambda > 0$ and all $\mathbf{v} \in \mathcal{K}$. Then $F(\mathbf{v}) := (G(\mathbf{v}))^{1/p}$ is convex.

Proof. For the first part, by Lemma 6.7 it suffices to show that $F(\mathbf{v} + \mathbf{w}) \leq F(\mathbf{v}) + F(\mathbf{w})$ for all $\mathbf{v}, \mathbf{w} \in \mathcal{K}$. If $F(\mathbf{v})$ or $F(\mathbf{w})$ is infinite, this is trivial, so we may suppose both are finite. Let $\epsilon > 0$ and define

$$\mathbf{v}' := \frac{1}{F(\mathbf{v}) + \epsilon}\mathbf{v} \qquad \text{and} \qquad \mathbf{w}' := \frac{1}{F(\mathbf{w}) + \epsilon}\mathbf{w}.$$

Since both $\mathbf{v}'$ and $\mathbf{w}'$ belong to the level set corresponding to $t = 1$,

$$\frac{F(\mathbf{v}) + \epsilon}{F(\mathbf{v}) + F(\mathbf{w}) + 2\epsilon}\mathbf{v}' + \frac{F(\mathbf{w}) + \epsilon}{F(\mathbf{v}) + F(\mathbf{w}) + 2\epsilon}\mathbf{w}' \in \{\mathbf{x} : F(\mathbf{x}) \leq 1\}.$$

That is,

$$F\left(\frac{1}{F(\mathbf{v}) + F(\mathbf{w}) + 2\epsilon}\mathbf{v} + \frac{1}{F(\mathbf{v}) + F(\mathbf{w}) + 2\epsilon}\mathbf{w}\right) \leq 1,$$

and by the homogeneity of F, $F(\mathbf{v} + \mathbf{w}) \leq F(\mathbf{v}) + F(\mathbf{w}) + 2\epsilon$. Since $\epsilon > 0$ is arbitrary, this proves that F is subadditive, and therefore convex.

For the final part, if G is convex, then the level sets of G are convex by Remark 6.8. But every level set of G is a level set of $F = G^{1/p}$ and vice-versa.

Hence every level set of F is convex, and since G is homogeneous of degree p, F is homogeneous of degree one. Therefore by Lemma 6.7, F is convex. $\square$

Legendre transforms of convex functions that are homogeneous of degree one have a very special form:

Theorem 6.10. *Suppose that F is homogeneous of degree one on a real Hilbert space $\mathcal{K}$ that is also proper, lower semicontinuous, and convex. Then there is a closed convex set $\Omega \subset \mathcal{K}$ such that the Legendre transform F^* of F is given by*

$$(6.1.6) \qquad F^*(\mathbf{v}) = \begin{cases} 0 & \mathbf{v} \in \Omega \\ \infty & \mathbf{v} \notin \Omega. \end{cases}$$

Proof. By definition, $F^*(\mathbf{v}) = \sup\{\langle \mathbf{v}, \mathbf{w}\rangle_{\mathcal{K}} - F(\mathbf{w})\}$. Replacing $\mathbf{w}$ by $\lambda\mathbf{w}$, $\lambda > 0$, and using the homogeneity,

$$(6.1.7) \qquad \langle \mathbf{v}, \lambda\mathbf{w}\rangle_{\mathcal{K}} - F(\lambda\mathbf{w}) = \lambda\left(\langle \mathbf{v}, \mathbf{w}\rangle_{\mathcal{K}} - F(\mathbf{w})\right).$$

If $\langle \mathbf{v}, \mathbf{w}\rangle_{\mathcal{K}} - F(\mathbf{w}) > 0$, the supremum over $\lambda > 0$ of the quantity in (6.1.7) is ∞, and hence $F^*(\mathbf{v}) = \infty$. On the other hand, if $\langle \mathbf{v}, \mathbf{w}\rangle_{\mathcal{K}} - F(\mathbf{w}) \le 0$, the supremum over $\lambda > 0$ of the quantity in (6.1.7) is 0, and hence the only other possible value for $F^*(\mathbf{v})$ is 0. Thus F^* has the form specified in (6.1.6). Since F^* is a Legendre transform, it is convex and lower semicontinuous, and therefore Ω is convex and closed. $\square$

Another lemma relating convexity and homogeneity, due to Lieb [**141**], will be used several times in what follows.

Lemma 6.11. *Let $\mathcal{C}$ be a convex cone in a finite-dimensional Hilbert space, and let $F : \mathcal{C} \to \mathbb{R}$ be homogeneous of degree one. Define*

$$G(x, y) := \lim_{t\downarrow 0} \frac{F(x + ty) - F(x)}{t}.$$

If F is convex, then for all $x, y \in \mathcal{C}$, $G(x, y) \le F(y)$. Conversely, if F is continuously differentiable and $G(x, y) \le F(y)$ for all $x, y \in \mathcal{C}$, then F is convex.

Proof. Assume first that F is convex, and compute

$$\begin{aligned} F(x + ty) &= (1 + t)F\left(\frac{1}{1+t}x + \frac{t}{1+t}y\right) \\ &\le (1 + t)\left(\frac{1}{1+t}F(x) + \frac{t}{1+t}F(y)\right) \\ &= F(x) + tF(y). \end{aligned}$$

Next assume F is continuously differentiable and $G(x, y) \leq F(y)$ for all $x, y \in \mathcal{C}$. For $0 < \lambda < 1$, $x, y \in \mathcal{C}$, let $z := \lambda x + (1 - \lambda)y$. Then

$$F(z) - \lambda F(x) = F(\lambda x + (1 - \lambda)y) - F(\lambda x)$$

$$= \int_0^{1-\lambda} G(\lambda x + sy, y)\mathrm{d}s \leq (1 - \lambda)F(y). \qquad \square$$

6.2. The Fenchel-Moreau theorem

Affine functions are evidently convex and lower semicontinuous. Since F^* is the supremum over a family of affine functions, F^* is automatically convex and lower semicontinuous by Lemma 6.1. It is not hard to see that if F is proper, so is F^*. Thus, the Legendre transform takes the class of proper, lower semicontinuous convex functions into itself. The Fenchel-Moreau theorem says that in fact, not only does it map this class of functions onto itself, it is involutive:

Theorem 6.12 (Fenchel-Moreau theorem). *Let F be a proper lower semicontinuous convex function on a real Hilbert space $\mathcal{K}$. Then F^*, the Legendre transform of F, is also a proper lower semicontinuous convex function on $\mathcal{K}$. Moreover, if F^{**} denotes the Legendre transform of F^*, then $F^{**} = F$.*

A full proof of Theorem 6.12 is provided Appendix A, which can also be read as a self-contained introduction to all of the tools of convex geometry employed in this book.

For a first application, let $\mathcal{K}$ be the real Hilbert space consisting of $M_n^{\mathrm{s.a.}}(\mathbb{C})$ with the inner product $\langle X, Y \rangle_{\mathcal{K}} = \mathrm{Tr}[XY]$, noting that when $X, Y \in M_n^{\mathrm{s.a.}}(\mathbb{C})$, $\mathrm{Tr}[XY]$ is always real. Given a convex function f on the real line, define the function F on $\mathcal{K}$ by

(6.2.1) $F(X) = \mathrm{Tr}[f(X)],$

where $f(X)$ is defined using the spectral theorem. According to Theorem 3.3, F is convex on $\mathcal{K}$. Then next lemma specifies the Legendre transform of F in terms of the Legendre transform of f.

Theorem 6.13. *Let f be a proper, lower semicontinuous convex function on $\mathbb{R}$, and let f^* be its Legendre transform. Let $\mathcal{K}$ be the real Hilbert space consisting of $M_n^{\mathrm{s.a.}}(\mathbb{C})$ with the inner product $\langle X, Y \rangle_{\mathcal{K}} = \mathrm{Tr}[XY]$. Let F be the function of $\mathcal{K}$ specified in (6.2.1). Then F is a proper lower semicontinuous convex function on $\mathcal{K}$, and its Legendre transform F^* is given by $F^*(Y) = \mathrm{Tr}[f^*(Y)]$.*

If f and f^* are both strictly convex and differentiable on $\mathbb{R}$, then

$$\langle Y, X \rangle_{\mathcal{K}} = F^*(Y) + F(X)$$

if and only if

$$Y = f'(X) \quad \text{and} \quad X = (f^*)'(Y).$$

Proof. Let $X, Y \in M_n^{\text{s.a.}}(\mathbb{C})$, and let $\{\mathbf{u}_1, \ldots, \mathbf{u}_n\}$ be an orthonormal basis of $\mathbb{C}^n$ consisting of eigenvectors of X with $X\mathbf{u}_j = \lambda_j \mathbf{u}_j$. Then

$$\langle X, Y \rangle - \text{Tr}[f(X)] = \sum_{j=1}^{n} \left(\lambda_j \langle \mathbf{u}_j, Y\mathbf{u}_j \rangle - f(\lambda_j) \right)$$

$$(6.2.2) \qquad \leq \sum_{j=1}^{n} f^*(\langle \mathbf{u}_j, Y\mathbf{u}_j \rangle)$$

$$(6.2.3) \qquad \leq \sum_{j=1}^{n} \langle \mathbf{u}_j, f^*(Y)\mathbf{u}_j \rangle = \text{Tr}[f^*(Y)],$$

where the inequality in (6.2.2) is from the definition of f^*; that is, $f^*(y) = \sup_{x \in \mathbb{R}} \{xy - f(x)\}$, and the inequality in (6.2.3) is from Lemma 3.2.

This upper bound is saturated. Suppose first that $\text{Tr}[f^*(Y)] < \infty$. Let $\{\mathbf{v}_1, \ldots, \mathbf{v}_n\}$ be an orthonormal basis of $\mathbb{C}^n$ consisting of eigenvectors of Y with $Y\mathbf{v}_j = \mu_j \mathbf{v}_j$, and choose $\epsilon > 0$. Then each $f^*(\mu_j)$ is finite and there exists $x_j \in \mathbb{R}$ such that $\mu_j x_j - f(x_j) \geq f^*(\mu_j) - \epsilon$. Define $X := \sum_{j=1}^{n} x_j |\mathbf{v}_j\rangle\langle\mathbf{v}_j|$. Then

$$\text{Tr}[XY] - \text{Tr}[f(X)] \geq \sum_{j=1}^{n} (f^*(\mu_j) - \epsilon) = \text{Tr}[f^*(Y)] - n\epsilon.$$

Since $\epsilon > 0$ is arbitrary, this proves that

$$(6.2.4) \qquad \text{Tr}[f^*(Y)] = \sup\{\text{Tr}[XY] - \text{Tr}[f(X)]\},$$

when $\text{Tr}[f^*(Y)] < \infty$. When $\text{Tr}[f^*(Y)] = \infty$, then $f^*(\mu) = \infty$ for some eigenvalue μ of Y, and then $x\mu - f(x)$ can be made arbitrarily large with an appropriate choice of x. Let $\mathbf{v}$ be a unit vector with $Y\mathbf{v} = \mu\mathbf{v}$, and take $X = x|\mathbf{v}\rangle\langle\mathbf{v}|$ to see that $\langle X, Y \rangle - \text{Tr}[f(X)]$ takes on arbitrarily large values, completing the proof of (6.2.4) in general.

So far we have not used the lower semicontinuity of f, only that it is convex and proper. But when f is also lower semicontinuous, then there is a perfect symmetry between f and f^*: then by the Fenchel-Moreau theorem, f is the Legendre transform of f^*, and thus the roles of f and f^* may be interchanged. It follows that $\text{Tr}[f(X)]$ is the Legendre transform of $\text{Tr}[f^*(Y)]$, and hence is proper, lower semicontinuous and convex.

For the final part, since f^* is strictly convex, there is equality in (6.2.3) if and only if each $\mathbf{u}_j$ is an eigenvector of Y, and then in this case

$$Y = \sum_{j=1}^{n} \langle \mathbf{u}_j, Y\mathbf{u}_j \rangle |\mathbf{u}_j\rangle\langle\mathbf{u}_j|.$$

Furthermore, by Young's inequality, for each j,

$$(6.2.5) \qquad \lambda_j \langle \mathbf{u}_j, Y\mathbf{u}_j \rangle \leq f^*(\langle \mathbf{u}_j, Y\mathbf{u}_j \rangle) + f(\lambda_j).$$

Hence when there is equality in (6.2.3), there must be equality in (6.2.5) for each j, and this means that $\langle \mathbf{u}_j, Y\mathbf{u}_j \rangle \in \partial f(\lambda_j)$. By the differentiability of f, this means $\langle \mathbf{u}_j, Y\mathbf{u}_j \rangle = f'(\lambda_j)$, and hence $Y = f'(X)$. For the rest, we again invoke the symmetry. $\qquad\square$

We shall often encounter trace functions on $M_n(\mathbb{C})$ of the form

$$X \mapsto \mathrm{Tr}[f(|X|)],$$

where f is convex and monotone increasing. If it were the case that for all $X, Y \in M_n(\mathbb{C})$, $|X + Y| \leq |X| + |Y|$, it would be the case that

$$\mathrm{Tr}\left[f\left(\left|\frac{X+Y}{2}\right|\right)\right] \leq \mathrm{Tr}\left[f\left(\frac{|X|+|Y|}{2}\right)\right] \leq \frac{1}{2}\mathrm{Tr}[f(|X|) + f(|Y|)].$$

However, it is *not true* that $|X + Y| \leq |X| + |Y|$ for all $X, Y \in M_n(\mathbb{C})$. (It is left as an exercise for the reader to find simple counterexamples in $M_2(\mathbb{C})$.) Therefore, we need a different extension strategy.

The simplest way to proceed is via the following doubling argument: For $X \in M_n(\mathbb{C})$ define $\widehat{X} \in M_{2n}^{\mathrm{s.a.}}(\mathbb{C})$ by

$$\widehat{X} = \begin{bmatrix} 0 & X \\ X^* & 0 \end{bmatrix}.$$

Let $X = \sum_{j=1}^n \sigma_j |\mathbf{v}_j\rangle\langle\mathbf{u}_j|$ be an SVD of X as in Theorem 2.25. For each j define

$$\mathbf{w}_j := \frac{1}{\sqrt{2}}\begin{pmatrix} \mathbf{v}_j \\ \mathbf{u}_j \end{pmatrix} \quad \text{and} \quad \mathbf{w}_{n+j} := \frac{1}{\sqrt{2}}\begin{pmatrix} -\mathbf{v}_j \\ \mathbf{u}_j \end{pmatrix}.$$

Then for each $1 \leq j \leq n$, $\widehat{X}\mathbf{w}_j = \sigma_j\mathbf{w}_j$ and $\widehat{X}\mathbf{w}_{n+j} = -\sigma_j\mathbf{w}_{n+j}$, and

$$(6.2.6) \qquad \widehat{X} = \sum_{j=1}^n \sigma_j |\mathbf{w}_j\rangle\langle\mathbf{w}_j| - \sum_{j=1}^n \sigma_j |\mathbf{w}_{n+j}\rangle\langle\mathbf{w}_{n+j}|$$

gives a spectral decomposition of $\widehat{X}$.

We are ready to extend the scope of Theorem 6.13 from $M_n^{\mathrm{s.a.}}(\mathbb{C})$ to all of $M_n(\mathbb{C})$. To take Legendre transforms of functions on $M_n(\mathbb{C})$, we regard $M_n(\mathbb{C})$ as a real vector space and equip it with the inner product $\Re\langle X, Y \rangle = \Re(\mathrm{Tr}[X^*Y])$.

Theorem 6.14. *Let f be any even proper lower semicontinuous convex function on $\mathbb{R}$, and let f^* be its Legendre transform. $X \mapsto \mathrm{Tr}[f(|X|)]$ is a proper lower semicontinuous convex function on $M_n(\mathbb{C})$, and for all $Y \in M_n(\mathbb{C})$,*

$$(6.2.7) \quad \mathrm{Tr}[f^*(|Y|)] = \sup_{X \in M_n(\mathbb{C})} \{\Re\langle Y, X\rangle - \mathrm{Tr}[f(|X|)]\} \quad \text{for all } Y \in M_n(\mathbb{C}),$$

which identifies the Legendre transform of $\mathrm{Tr}[f(|X|)]$ as $\mathrm{Tr}[f^*(|Y|)]$.

Furthermore, suppose that f and f^* are both strictly convex and differentiable, and let $X = U|X|$ and $Y = V|Y|$ be the polar decomposition of X and Y. Then

$$(6.2.8) \qquad \Re\langle Y, X\rangle = \mathrm{Tr}[f^*(|Y|)] + \mathrm{Tr}[f(|X|)]$$

if and only if

$$Y = Uf'(|X|) \quad \text{and} \quad X = V(f^*)'(|Y|).$$

Proof. Fix $X, Y \in M_n(\mathbb{C})$ and let $\widehat{X}, \widehat{Y} \in M_{2n}^{\mathrm{s.a.}}$ be defined as above. By (6.2.6),

$$\mathrm{Tr}[f(\widehat{X})] = 2\,\mathrm{Tr}[f(|X|)] \quad \text{and} \quad \mathrm{Tr}[f(\widehat{Y})] = 2\,\mathrm{Tr}[f(|Y|)],$$

and by the definitions of $\widehat{X}$ and $\widehat{Y}$,

$$\langle \widehat{X}, \widehat{Y}\rangle = \mathrm{Tr}[X^*Y] + \mathrm{Tr}[XY^*] = 2\Re\langle X, Y\rangle.$$

Thus (6.2.7) follows from Theorem 6.13 which also says that when there is equality in (6.2.8), then $\widehat{Y} = f'(\widehat{X})$, and since f' is odd, (6.2.6) yields

$$f'(\widehat{X}) = \sum_{j=1}^{n} f'(\sigma_j)|\mathbf{w}_j\rangle\langle\mathbf{w}_j| - \sum_{j=1}^{n} f'(\sigma_j)|\mathbf{w}_{n+j}\rangle\langle\mathbf{w}_{n+j}|,$$

and the upper right block of $f'(\widehat{X})$ is $\sum_{j=1}^{n} f'(\sigma_j)|\mathbf{v}_j\rangle\langle\mathbf{u}_j| = Uf'(|X|)$. The relation $X = V(f^*)'(|Y|)$ follows the same way. $\qquad \square$

6.3. The Schatten p-norms

Our first application of Theorem 6.14 involves the functions $f(x) = \frac{1}{p}|x|^p$ for $1 \leq p < \infty$. A simple computation shows that $f^*(y) = \frac{1}{q}|y|^q$, where $1/p + 1/q = 1$. From the convexity of f and Theorem 6.13, $X \mapsto \mathrm{Tr}[|X|^p]$ is convex on $M_n^{\mathrm{s.a.}}(\mathbb{C})$.

Definition 6.15 (Schatten p-norms). For $1 \leq p < \infty$, define the **Schatten p norm** on $M_n(\mathbb{C})$ by

$$\|X\|_p := (\mathrm{Tr}[|X|^p])^{1/p}.$$

For $p = \infty$, we define $\|X\|_\infty$ to be the operator norm of X, i.e., the largest singular value of X.

The Schatten p-norm of $A \in M_n(\mathbb{C})$ depends only on the singular values of A: if $A = \sum_{j=1}^{r} \sigma_j |\mathbf{u}_j\rangle\langle\mathbf{v}_j|$ is an SVD of A, then for all $1 \leq p < \infty$, $|A|^p = \sum_{j=1}^{r} \sigma_j^p |\mathbf{v}_j\rangle\langle\mathbf{v}_j|$, and then since $\{\mathbf{v}_1, \ldots, \mathbf{v}_r\}$ is orthonormal,

$$\|A\|_p := \left(\sum_{j=1}^{r} \sigma_j^p \right)^{1/p}.$$

Of course, for $p = \infty$, $\|A\|_\infty = \max\{\sigma_1, \ldots, \sigma_r\}$.

Theorem 6.16. *For all $A \in M_n(\mathbb{C})$, all $1 \leq p \leq \infty$, and all unitaries $U, V \in M_n(\mathbb{C})$,*

$$\|UAV\|_p = \|A\|_p$$

and

$$\|A^*\|_p = \|A\|_p.$$

Moreover, for all $A \in M_n(\mathbb{C})$,

(6.3.1) $$|\mathrm{Tr}[A]| \leq \|A\|_1.$$

In particular, for all $A, B \in M_n(\mathbb{C})$,

(6.3.2) $$|\langle A, B \rangle| \leq \|A^*B\|_1.$$

Proof. UAV and A^* have the same singular values as A, and this proves the first two parts. For the third, let $A = U|A|$ be the polar decomposition of A. Then by the Cauchy-Schwarz inequality

$$|\mathrm{Tr}[A]| = |\mathrm{Tr}[U|A|^{1/2}|A|^{1/2}]| = \langle |A|^{1/2}U^*, |A|^{1/2} \rangle$$

$$\leq (\mathrm{Tr}[U|A|U^*])^{1/2}(\mathrm{Tr}[|A|])^{1/2} = \|A\|_1.$$

Finally, $\langle A, B \rangle = \mathrm{Tr}[A^*B]$, and (6.3.2) follows from (6.3.1). $\qquad\square$

Theorem 6.17 (Minkowski and Hölder inequalities for traces). *For $X, Y \in M_n(\mathbb{C})$, and $1 \leq p \leq \infty$,*

(6.3.3) $$\|X + Y\|_p \leq \|X\|_p + \|Y\|_p,$$

and, with $1/q + 1/p = 1$,

(6.3.4) $$|\langle X, Y \rangle| \leq \|X\|_p\|Y\|_q$$

and

(6.3.5) $$\|XY\|_1 \leq \|X\|_p\|Y\|_q.$$

Moreover, let $1 < p < \infty$, and let $X = U|X|$ be the polar decomposition of X. Then $|\langle X, Y \rangle| = \|X\|_p\|Y\|_q$ if and only if for some $z \in \mathbb{C}$,

(6.3.6) $$Y = zU|X|^{p-1}.$$

Proof of Theorem 6.17. Since for $1 \leq p \leq \infty$, the function $f(x) := |x|^p$ is even and convex on $\mathbb{R}$, the function $G(X) := \text{Tr}[|X|^p] = (F(X))^p$ is convex on $M_n(\mathbb{C})$. It also is homogeneous of degree p. Therefore by Lemma 6.9 the function $F(X) := \|X\|_p = G(X)^{1/p}$ is convex and homogeneous of degree one. Then by Lemma 6.7, F is subadditive. Subadditivity of F is exactly (6.3.3). We have already seen that (6.3.3) is satisfied for $p = \infty$ since $\|X\|_\infty$ is the operator norm of X.

Next, by Young's inequality (6.1.3) and (6.2.7),

$$(6.3.7) \qquad \mathfrak{R}\langle X, Y\rangle \leq \frac{1}{q}\text{Tr}[|Y|^q] + \frac{1}{p}\text{Tr}[|X|^p].$$

Choose $\theta \in \mathbb{R}$ so that $\mathfrak{R}(e^{i\theta}\langle X, Y\rangle) = |\langle X, Y\rangle|$. Replacing Y with $e^{i\theta}Y$ does not change the right side of (6.3.7), and hence we have

$$(6.3.8) \qquad |\langle X, Y\rangle| \leq \frac{1}{q}\text{Tr}[|Y|^q] + \frac{1}{p}\text{Tr}[|X|^p].$$

Now for $t > 0$, replace X by tX and Y by $t^{-1}Y$ in (6.3.8) to obtain

$$(6.3.9) \qquad |\langle X, Y\rangle| \leq \frac{1}{q}\text{Tr}[|Y|^q]t^{-q} + \frac{1}{p}\text{Tr}[|X|^p]t^p$$

for all $t > 0$. Now choose t to minimize the right side of (6.3.9). This yields (6.3.4).

Let $U|XY|$ be the polar decomposition of XY, so that $|XY| = U^*XY$,

$$\|XY\|_1 = \text{Tr}[U^*XY] = \langle X^*U, Y\rangle \leq \|X^*U\|_p\|Y\|_q = \|X\|_p\|Y\|_q,$$

where we used (6.3.4) and Theorem 6.16.

Suppose there is equality in (6.3.4) and neither $X = 0$ nor $Y = 0$. Define $A = \|X\|_p^{-1}X$ and $B = e^{i\theta}\|Y\|_q^{-1}Y$, where $\theta \in [0, 2\pi)$. Then $|\langle A, B\rangle| = 1$. Choose θ so that $\langle A, B\rangle = 1$. Then since $1 = 1/p + 1/q$,

$$\mathfrak{R}\langle A, B\rangle = \frac{1}{p}\|A\|_p^p + \frac{1}{q}\|B\|_q^q = 1.$$

Since $f_t(x) = |x|^t/t$ is both strictly convex and differentiable on $\mathbb{R}$ for all $1 < t < \infty$, Theorem 6.14 says that with $A = U|A|$, so that $X = U|X|$ with the same U,

$$B = Uf_p'(|A|) = \|X\|_p^{1-p}U|X|^{p-1}.$$

which is the case if and only if (6.3.6) is satisfied for some $z \in \mathbb{C}$. $\qquad\square$

The inequality (6.3.3) is known as **Minkowski's inequality for the Schatten norms**, and the inequalities (6.3.4) and (6.3.5) are known as **Hölder's inequality for traces**.

The fact that the Minkowski inequality is satisfied for the Schatten p-norms justifies our use of "norm" in naming them, since they are obviously homogeneous of degree one and $\|X\|_p = 0$ if and only if $X = 0$.

As a consequence of Minkowski's inequality, for $1 \le p < \infty$, the function d_p on $M_n(\mathbb{C}) \times M_n(\mathbb{C})$ defined by $d_p(X, Y) := \|X - Y\|_p$ satisfies the triangle inequality, $d_p(X, Z) \le d_p(X, Y) + d_p(Y, Z)$. It also evidently satisfies $d_p(X, Y) = d_p(Y, X)$ and $d_p(X, Y) = 0$ if and only if $X = Y$. Thus, d_p is a metric on $M_n(\mathbb{C})$ for all $1 \le p \le \infty$: we already have seen this for $p = \infty$, and also for $p = 2$ since $\|\cdot\|_2$ is simply the Hilbert-Schmidt norm. The norm $\|\cdot\|_1$, the largest in the family, is especially important; see Example 6.19. It too has a name, the **trace norm**.

Lemma 6.18. *Let $1 \le p_1 < p_2 \le \infty$. Then for all $X \in M_n(\mathbb{C})$,*

$$\|X\|_{p_2} \le \|X\|_{p_1},$$

and there is equality if and only if X is rank one. Moreover, as $p \uparrow \infty$, $\|X\|_p \downarrow \|X\|_\infty$.

Proof. Assume $X \ne 0$. By the homogeneity of the norms, we may suppose that $\|X\|_{p_1} = 1$. Let $X = \sum_{j=1}^{r} \sigma_j |\mathbf{v}_j\rangle\langle\mathbf{u}_j|$ be a singular value decomposition of X. Suppose first that $p_2 < \infty$. Then since for each j, $0 < \sigma_j \le 1$,

$$1 = \|X\|_{p_1}^{p_1} = \sum_{j=1}^{r} \sigma_j^{p_1} \ge \sum_{j=1}^{r} \sigma_j^{p_2} = \|X\|_{p_2}^{p_2}.$$

Hence $\|X\|_{p_2} \le 1 = \|X\|_{p_1}$. There is equality if and only for each j, $\sigma_j^{p_2} = \sigma_j^{p_1}$, and this is the case if and only if $r = 1$ and $\sigma_1 = 1$.

Now let $p_2 = \infty$. Suppose that the singular values are arranged in decreasing order so that $\|X\|_\infty = \sigma_1$. Then evidently,

$$\sigma_1 \le \left(\sum_{j=1}^{r} \sigma_j^{p_1} \right)^{1/p_1} \le (r\sigma_1^{p_1})^{1/p_1},$$

and hence $\|X\|_\infty \le \|X\|_{p_1} \le r^{1/p_1}\|X\|_\infty$. Again, there is equality on the left if and only if $r = 1$, and the final statement is now clear. $\qquad\square$

Example 6.19 (The trace norm and quantum states). Let $A \in \mathcal{B}(\mathcal{H})$ be self-adjoint and regarded as an observable. Let ρ and σ be two density matrices representing two states of the quantum system. Then the expected values of observations of A in the two states are $\mathrm{Tr}[A\rho]$ and $\mathrm{Tr}[A\sigma]$, respectively. The difference is bounded by

$$|\mathrm{Tr}[A\rho] - \mathrm{Tr}[A\sigma]| = |\mathrm{Tr}[A(\rho - \sigma)]| \le \|A\|_\infty \|\rho - \sigma\|_1.$$

Hence if two quantum states differ by at most ϵ in the trace norm, the difference in the expected values assigned to any observable A is bounded by $\|A\|\epsilon$, where $\|A\|$ is the operator norm of A.

6.3.1. Legendre transforms of the Schatten p-norms. We next compute the Legendre transforms of the norms $\| \cdot \|_p$, which are proper lower semicontinuous functions on $\mathcal{K}$. Since these norms are homogenous of degree one, their Legendre transforms have the form specified in Theorem 6.10.

Theorem 6.20 (Duality for Schatten p-norms). *For $1 \leq p \leq \infty$, let $1/q = 1 - 1/p$. Define*
$$\mathbf{B}_q := \{Y \in M_n(\mathbb{C}) \ : \ \|Y\|_q \leq 1\}.$$
Then for all $X \in M_n(C)$,

(6.3.10) $\qquad \|X\|_p = \sup\{\Re\langle Y, X\rangle \ : \ Y \in \mathbf{B}_q\} = \max\{\Re\langle Y, X\rangle \ : \ Y \in \mathbf{B}_q\}.$

If $X \in M_n^+(\mathbb{C})$, there exists $Y \in M_n^+(\mathbb{C}) \cap \mathbf{B}_q$ such that $\|X\|_p = \langle Y, X\rangle$.

Remark 6.21. Define
$$G_q(Y) := \begin{cases} 0 & Y \in \mathbf{B}_q \\ \infty & Y \notin \mathbf{B}_q, \end{cases}$$
and note that $\mathbf{B}_q$ is convex (by Minkowski's inequality) and closed, so that G_q is proper, lower semicontinuous and convex. Then (6.3.10) can be written as
$$\|X\|_p = \sup\{\Re\langle Y, X\rangle - G_q(Y)\},$$
and this shows that $\| \cdot \|_p$ is the Legendre transform of G_q, and then by the Fenchel-Moreau theorem, G_q is the Legendre transform of $\| \cdot \|_p$. For any $\theta \in \mathbb{R}$, $G_q(e^{i\theta}A) = G_q(A)$, and by an appropriate choice of θ we have $\Re\langle Y, X\rangle = |\langle Y, X\rangle|$. Likewise, $G_q(Y^*) = G_q(Y)$ and $\Re\langle Y, X\rangle = \mathrm{Tr}[Y^*X]$. Hence an equivalent formulation of (6.3.10) is
$$\|X\|_p = \max\{|\langle Y, X\rangle| \ : \ Y \in B_q\} = \max\{\Re(\mathrm{Tr}[YX]) \ : \ Y \in B_q\}.$$

Proof of Theorem 6.20. By Hölder's inequality,

(6.3.11) $\qquad \sup\{\Re\langle Y, X\rangle \ : \ Y \in B_q\} \leq \|X\|_p.$

The second part of Theorem 6.17 says that equality holds in (6.3.11) and hence the supremum is a maximum. Moreover, if $p < \infty$, let $X = U|X|$ be the polar decomposition of X. By Theorem 6.17, the supremum is attained at, and only at,

(6.3.12) $\qquad Y = \|X\|_p^{1-p} U|X|^{p-1}.$

Note that if $X \geq 0$, and Y is given by (6.3.12), then $Y \geq 0$. The case $p = \infty$ is left as an exercise. $\qquad\square$

Remark 6.22. There are several other ways to prove Hölder's inequality for traces, and we shall give two more quite different proofs: one using complex interpolation later in this chapter, and the other using majorization methods in Chapter 11. The proof using Legendre transforms and duality illustrates the use of tools that will be applied extensively in what follows, already in the next chapter in which we discuss entropy.

6.3.2. Schatten norm bounds for operators $\Phi : \mathcal{B}(\mathcal{H}) \to \mathcal{B}(\mathcal{K})$.

Definition 6.23. Let $\mathcal{H}, \mathcal{K}$ be finite-dimensional Hilbert spaces, and let $\Phi :$ $\mathcal{B}(\mathcal{H}) \to \mathcal{B}(\mathcal{K})$. Let $1 \leq p, q \leq \infty$. Then the quantity $\|\Phi\|_{p \to q}$ is defined by

$$(6.3.13) \qquad \|\Phi\|_{p \to q} := \sup\{\|\Phi(X)\|_q \; : \; X \in \mathcal{B}(\mathcal{H}) \text{ such that } \|X\|_p \leq 1\}.$$

Example 6.24. Let $\Phi : \mathcal{B}(\mathcal{H}) \to \mathcal{B}(\mathcal{K})$ be positive and unital. By the Russo-Dye theorem, Theorem 5.67, $\|\Phi(A)\| \leq \|A\|$ with equality if $A = \mathbb{1}$, and hence $\|\Phi\|_{\infty \to \infty} = 1$.

Lemma 6.25. *Let $\mathcal{H}, \mathcal{K}$ be finite-dimensional Hilbert spaces, and let $\Phi : \mathcal{B}(\mathcal{H}) \to \mathcal{B}(\mathcal{K})$. Let $1 \leq p, q \leq \infty$, and define p', q' so that $\frac{1}{p} + \frac{1}{p'} = 1$ and $\frac{1}{q} + \frac{1}{q'} = 1$. Then*

$$\|\Phi^{\dagger}\|_{q' \to p'} = \|\Phi\|_{p \to q}.$$

Proof. Let $X \in \mathcal{B}(\mathcal{H})$ satisfy $\|X\|_p = 1$ and $\|\Phi\|_{p \to q} = \|\Phi(X)\|_q$. Let $Y \in \mathcal{B}(\mathcal{K})$ satisfy $\|Y\|_{q'} = 1$ and $\langle Y, \Phi(X) \rangle = \|\Phi(X)\|_q$. By Hölder's inequality and Definition 6.23

$$\|\Phi\|_{p \to q} = \langle Y, \Phi(X) \rangle = \langle \Phi^{\dagger}(Y), X \rangle \leq \|\Phi^{\dagger}(Y)\|_{p'} \|X\|_p$$
$$= \|\Phi^{\dagger}(Y)\|_{p'} \leq \|\Phi^{\dagger}\|_{q' \to p'} \|Y\|_{q'} = \|\Phi^{\dagger}\|_{q' \to p'}.$$

Therefore $\|\Phi\|_{p \to q} \leq \|\Phi^{\dagger}\|_{q' \to p'}$, and by symmetry, $\|\Phi^{\dagger}\|_{q' \to p'} \leq \|\Phi\|_{p \to q}$. $\square$

Theorem 6.26. *Let $\Phi^{\dagger} : \mathcal{B}(\mathcal{K}) \to \mathcal{B}(\mathcal{H})$ be a positive trace preserving map. Then*

$$(6.3.14) \qquad\qquad \|\Phi^{\dagger}\|_{1 \to 1} = 1.$$

As a special case, if $\Phi^{\dagger}$ is any quantum operation, then $\Phi^{\dagger}$ is contractive in the trace norm.

Proof. Since Φ is positive and unital, (6.3.14) is an immediate consequence of Example 6.24 and Lemma 6.25. $\square$

The following is from [**46**, Theorem 2], where it is proved for conditional expectations, though the proof, which is given here, extends directly to the case of general CP maps.

Theorem 6.27. *Let $\mathcal{H}, \mathcal{K}$ be finite-dimensional Hilbert spaces, and let $\Phi : \mathcal{B}(\mathcal{H}) \to \mathcal{B}(\mathcal{K})$ be CP. Then for all $X \in \mathcal{M}$ and all $1 \leq q \leq \infty$,*

$$\|\Phi(X)\|_q \leq \|\Phi(|X|)\|_q^{1/2} \|\Phi(|X^*|)\|_q^{1/2}.$$

Proof. Let $\Phi(X) := \sum_{j=1}^{m} V_j^* X V_j$ be a Kraus representation of Φ. Let $X = U|X|$ be the polar decomposition of X. Then $\|\Phi(X)\|_q = \mathrm{Tr}[C\Phi(U|X|)]$ for some $C \in \mathcal{B}(\mathcal{K})$ such that $\|C\|_{q'} = 1$, where $1/q' = 1 - 1/q$. Let $C = W|C|$ be the polar decomposition of C. Note that $U|X|U^* = |X^*|$ and $W|C|W^* = |C^*|$.

$$\|\Phi(X)\|_q = \mathrm{Tr}\left[\sum_{j=1}^{m} CV_j^* U|X|^{1/2}|X|^{1/2}V_j \right]$$

$$= \sum_{j=1}^{m} \mathrm{Tr}\left[|C|^{1/2}V_j^* U|X|^{1/2}|X|^{1/2}V_j W|C|^{1/2} \right]$$

$$\leq \left(\sum_{j=1}^{m} \mathrm{Tr}[|C|V_j^*|X^*|V_j] \right)^{1/2} \left(\sum_{j=1}^{m} \mathrm{Tr}[|C^*|V_j^*|X|V_j] \right)^{1/2}$$

$$= (\mathrm{Tr}[|C|\Phi(|X^*|)])^{1/2} (\mathrm{Tr}[(|C^*|)\Phi(|X|)])^{1/2}$$

$$\leq \|\Phi(|X^*|)\|_q^{1/2}\|\Phi(|X|)\|_q^{1/2}. \qquad \square$$

The following theorem of Watrous [225] can be seen as a corollary of Theorem 6.27. It says that when Φ is CP, the supremum in (6.3.13) is equal to the supremum over the smaller set of of *positive X* such that $\|X\|_p \leq 1$. This fact is often useful in bounding $\|\Phi\|_{p\to q}$.

Theorem 6.28. *Let $\mathcal{H}, \mathcal{K}$ be finite-dimensional Hilbert spaces, and let $\Phi : \mathcal{B}(\mathcal{H}) \to \mathcal{B}(\mathcal{K})$ be CP. Then for all $1 \leq p, q \leq \infty$,*

$$\|\Phi\|_{p\to q} := \sup\{\|\Phi(X)\|_q \; : \; X \in \mathcal{B}^+(\mathcal{H}) \text{ such that } \|X\|_p \leq 1\}.$$

Proof. Since $\||X^*|\|_p = \||X|\|_p = \|X\|_p$, this is immediate from Theorem 6.27.
$$\square$$

Another proof of Theorem 6.28, valid more generally for 2-positive maps, has been given by Audenaert [18].

By now, it is clear that the theory of Schatten norms for operators closely parallels that of the L^p norms for functions. The point of view that the Schatten spaces $\mathcal{S}_p$ are *noncommutative L^p spaces*, was put forward and extensively developed by Dixmier [75] and Segal [196]. A number of important results about operators on L^p spaces follow from a classical theorem due to Marcel Reisz (the younger brother of Frigyes Riesz for whom the Riesz representation theorem is named) and his student G. Olof Thorin. A simple modification of Thorin's proof, due to Elias Stein [200] and presented below, yields a much more powerful theorem for operators on L^p spaces. As Dixmier showed, Thorin's proof extends readily to the noncommutative setting, and Stein's extension carries over as well.

Definition 6.29. Let $\mathcal{H}, \mathcal{K}$ be finite-dimensional Hilbert spaces. Let $\mathcal{O}$ be an open set in $\mathbb{C}$. A function $z \mapsto \Phi_z \in \mathcal{L}(\mathcal{B}(\mathcal{H}), \mathcal{B}(\mathcal{K}))$ is **analytic** in $\mathcal{O}$ in case for all $A \in \mathcal{B}(\mathcal{H})$ and all $B \in \mathcal{B}(\mathcal{K})$, $g(z) := \mathrm{Tr}[B\Phi_z(A)]$ is an analytic function of $z \in \mathcal{O}$.

Theorem 6.30 (Riesz-Thorin-Stein theorem for Schatten norms). *Let* $1 \leq p_0, p_1, q_0, q_1 \leq \infty$. *For* $0 < \theta < 1$, *define*

$$\frac{1}{p_\theta} = \frac{1-\theta}{p_0} + \frac{\theta}{p_1} \quad and \quad \frac{1}{q_\theta} = \frac{1-\theta}{q_0} + \frac{\theta}{q_1}.$$

Let $S := \{x + iy \ : \ 0 < x < 1\}$, *and suppose that* Φ_z *is analytic on* S *and continuous on* $\overline{S}$ *with values in* $\mathcal{L}(\mathcal{B}(\mathcal{H}), \mathcal{B}(\mathcal{K}))$ *such that for some finite constant* C,

$$(6.3.15) \qquad\qquad \sup_{z \in \overline{S}}\{\|\Phi_z\|_{\infty \to \infty}\} \leq C.$$

Then

$$(6.3.16) \qquad \|\Phi\|_{p_\theta \to q_\theta} \leq \left(\sup_{y \in \mathbb{R}}\{\|\Phi_{iy}\|_{p_0 \to q_0}\}\right)^{1-\theta} \left(\sup_{y \in \mathbb{R}}\{\|\Phi_{1+iy}\|_{p_1 \to q_1}\}\right)^{\theta}.$$

Remark 6.31. Often in applications, $\sup_{y \in \mathbb{R}}\{\|\Phi_{iy}\|_{p_0 \to q_0}\} = \|\Phi_0\|_{p_0 \to q_0}$ and $\sup_{y \in \mathbb{R}}\{\|\Phi_{1+iy}\|_{p_1 \to q_1}\} = \|\Phi_1\|_{p_1 \to q_1}$ so that (6.3.16) simplifies to

$$(6.3.17) \qquad\qquad \|\Phi\|_{p_\theta \to q_\theta} \leq \|\Phi_0\|_{p_0 \to q_0}^{1-\theta}\|\Phi_1\|_{p_1 \to q_1}^{\theta}.$$

In 1927, M. Riesz proved the L^p space analogue of Theorem 6.30 under the restriction that $\frac{1}{p_0} + \frac{1}{p_1} \leq 1$ and $\frac{1}{q_0} + \frac{1}{q_1} \leq 1$, for Φ independent of z. His proof [**186**] used real variable techniques. A little over a decade later, his student Thorin [**208**] proved the L^p space analogue for all $0 \leq \frac{1}{p_0}, \frac{1}{p_1}, \frac{1}{q_0}, \frac{1}{q_1} \leq 1$, again for Φ independent of z. His very simple argument used Hadamard's three lines theorem about analytic functions. Stein's extension of Thorin's proof [**200**] to the case in which Φ_z is analytic in z was part of his thesis, published in 1956. In 1953, Dixmier showed [**75**] how the Riesz-Thorin theorem for L^p spaces adapts to the Schatten norm setting, and Stein's extension carries over in the same way.

Hadamard's three lines theorem is explained in Appendix B, and the reader unfamiliar with it is advised to consult this appendix.

Before presenting the proof, we present a first application.

Theorem 6.32. *Let* $\mathcal{H}$ *be a finite-dimensional Hilbert space, and let* $\Psi : \mathcal{B}(\mathcal{H})$ *be positive, unital, and trace preserving. Then* $\|\Psi\|_{p \to p} = 1$ *for all* $1 \leq p \leq \infty$.

Proof. Since Ψ is positive and trace preserving, $\|\Psi\|_{1 \to 1} = 1$ by Theorem 6.26. Since Ψ is positive and unital, $\|\Psi\|_{\infty \to \infty} = 1$ by Example 6.24. Then by Theorem 6.30, with $\Psi_z := \Psi$ for all z, $\|\Psi\|_{p \to p} \leq (\|\Psi\|_{1 \to 1})^{1/p}(\|\Psi\|_{\infty \to \infty})^0 = 1$. $\qquad \square$

Proof of Theorem 6.30. Let $X \in \mathcal{B}(\mathcal{H})$ be such that $\|X\|_{p_\theta} = 1$. Let $Y \in \mathcal{B}(\mathcal{K})$ be such that $\|Y\|_{q'_\theta} = 1$ and furthermore,

$$\mathrm{Tr}[\Phi_\theta(X)Y] = \|\Phi_\theta(X)\|_{q_\theta},$$

noting that such a Y exists by Theorem 6.20. Write the polar decompositions of X and Y^* in the form $X = U_X D_X^{\frac{1}{p_\theta}}$ and $Y^* = U_Y^* D_Y^{\frac{1}{q'_\theta}}$ so that $D_X \in \mathfrak{S}_{\mathcal{H}}$ and $D_Y \in \mathfrak{S}_{\mathcal{K}}$. Note that $Y = D_Y^{\frac{1}{q'_\theta}} U_Y$. It will be convenient to extend U_X and U_Y to be unitary if they happen to be only partial isometries. Define

$$X(z) = U_X D_X^{\frac{1-z}{p_0} + \frac{z}{p_1}} \quad \text{and} \quad Y(z) = D_Y^{\frac{1-z}{q'_0} + \frac{z}{q'_1}} U_Y.$$

For $z = x + iy$ with $0 \le x \le 1$ so that $\frac{1-x}{p_0} + \frac{x}{p_1} \ge 0$,

$$\|X(z)\|_\infty = \|D_X^{\frac{1-z}{p_0} + \frac{z}{p_1}}\|_\infty = \|D_X\|_\infty^{\frac{1-x}{p_0} + \frac{x}{p_1}} \le 1,$$

since $\|\rho\|_\infty \le 1$ for any density matrix ρ. Likewise, $\|Y(z)\|_\infty \le 1$ for all $z \in \overline{S}$.

Define the function $f(z)$ by $f(z) := \mathrm{Tr}[Y(z)\Phi_z(X(z))]$. Then f is analytic on the whole complex plane, and since $X(\theta) = X$ and $Y(\theta) = Y$,

$$(6.3.18) \qquad f(\theta) = \mathrm{Tr}[\Phi_\theta(X)Y] = \|\Phi_\theta(X)\|_{q_\theta}.$$

To apply Hadamard's three lines theorem, Theorem B.1, we need to know that $|f(z)|$ is bounded on the strip $\overline{S}$. Let $d_{\mathcal{K}}$ denote the dimension of $\mathcal{K}$. By the Cauchy-Schwarz inequality, Lemma 1.42, and finally $\|X(z)\|_\infty, \|Y(z)\|_\infty \le 1$,

$$|f(z)| \le \|Y(z)\|_2 \|\Phi_z(X(z))\|_2 \le d_{\mathcal{K}} \|Y(z)\|_\infty \|\Phi_z(X(z))\|_\infty$$
$$\le d_{\mathcal{K}} \|\Phi_z\|_{\infty \to \infty}.$$

Therefore, by (6.3.15), $|f(z)| \le d_{\mathcal{K}} C$ for all $z \in \overline{S}$.

Next, we need a more precise bound on $|f(z)|$ on the boundary of S. For all $y \in \mathbb{R}$,

$$X(iy) = U_X D_X^{1/p_0} D_X^{-iy/p_0 + iy/p_1}.$$

By the unitary invariance of the Schatten p-norms,

$$\|X(iy)\|_{p_0} = \|U_X D_X^{1/p_0} D_X^{-iy/p_0 + iy/p_1}\|_{p_0} = \|D_X^{1/p_0}\|_{p_0} = 1,$$

and therefore, $\|\Phi_{iy}(X(iy))\|_{q_0} \le \|\Phi_{iy}\|_{p_0 \to q_0}$. The same sort of calculation also yields $\|Y(iy)\|_{q'_0} = 1$. Then by Hölder's inequality,

$$|f(iy)| \le \|Y(iy)\|_{q'_0} \|\Phi_{iy}(X(iy))\|_{q_0} = \|\Phi_{iy}(X(iy))\|_{q_0}$$

and hence

$$(6.3.19) \qquad |f(iy)| \le \|\Phi_{iy}\|_{p_0 \to q_0}.$$

The same sort of reasoning yields

$$(6.3.20) \qquad |f(1+iy)| \le \|\Phi_{1+iy}\|_{p_1 \to q_1}$$

for all $y \in \mathbb{R}$. Therefore, using (6.3.18), (6.3.19), and (6.3.20) in Hadamard's three lines theorem yields

$$\|\Phi_\theta(X)\|_{q_\theta} \le \left(\sup_{y \in \mathbb{R}} \{\|\Phi_{iy}\|_{p_0 \to q_0}\} \right)^{1-\theta} \left(\sup_{y \in \mathbb{R}} \{\|\Phi_{1+iy}\|_{p_1 \to q_1}\} \right)^{\theta}.$$

Since this is valid for all X with $\|X\|_{p_\theta} = 1$, (6.3.16) is proved. $\qquad \square$

The next section contains an important application of Theorem 6.30, but we close this section by giving another proof of Hölder's inequality that illustrates the application of Theorem 6.30. The inequality to be proved, namely that for all $1 \le p, q, r$ with $\frac{1}{p} + \frac{1}{q} = \frac{1}{r}$, and all $X, Y \in \mathcal{B}(\mathcal{H})$,

$$(6.3.21) \qquad \|XY\|_r \le \|X\|_p \|Y\|_q,$$

can be viewed as a bound on $\|\Phi\|_{p \to r}$ where Φ denotes right multiplication by Y; that is $\Phi = R_Y$. For general $Z \in \mathcal{B}(\mathcal{H})$, $\|R_Z\|_{s \to t}$ is easy to compute when $t = \infty$ or $t = 1$. For $t = \infty$, note that for all $X \in \mathcal{B}(\mathcal{H})$, $\|R_Z X\|_\infty = \|XZ\|_\infty \le \|X\|_\infty \|Z\|_\infty$, and $R_Z \mathbb{1} = Z$. Therefore

$$(6.3.22) \qquad \|R_Z\|_{\infty \to \infty} = \|Z\|_\infty.$$

For $t = 1$, note that if $U|XZ|$ is the polar decomposition of XZ, $\|R_Z X\|_1 = \|XZ\|_1 = \mathrm{Tr}[U^*XZ] \le \|X\|_s \|Z\|_{s/(s-1)}$ using the elementary Hölder's inequality for traces, and there is equality for an appropriate choice of X. Therefore, for all $1 \le s \le \infty$,

$$(6.3.23) \qquad \|R_Z\|_{s \to 1} = \|Z\|_{s/(s-1)}.$$

We shall use Theorem 6.30 to *interpolate* between (6.3.22) and (6.3.23) for an appropriate choice of $Z(z)$. In fact, Theorem 6.30 is often called the **Riesz-Thorin-Stein interpolation theorem**.

Proof of (6.3.21) **using Theorem 6.30.** To order to use (6.3.22) and (6.3.23) to bound $\|XY\|_r$, we must choose θ so that $\frac{1}{r} = \frac{1}{q_\theta} = \frac{1-\theta}{\infty} + \frac{\theta}{1}$. Therefore, we take $\theta := \frac{1}{r}$.

Now construct an analytic operator valued function Φ_z such that $\Phi_\theta = R_Y$: Let $Y^* = U^*|Y^*|$ be the polar decomposition of Y^* so that $Y = |Y^*|U$. Define $Z(z) := |Y^*|^{rz}U$, and then define Φ_z to be right multiplication by $Z(z)$. Then $\Phi_{1/r}(X) = R_Y X = XY$. Moreover,

$$\|XY\|_r = \|\Phi_\theta(X)\|_{q_\theta} \le \|X\|_{p_\theta} \|\Phi_\theta\|_{p_\theta \to q_\theta}.$$

For $z = x + iy$, by the unitary invariance of the operator norm,

$$\|\Phi_z(X)\| \leq \|X\|\|Z(z)\| = \|X\|\|Y\|^{rx} \leq (1 + \|Y\|)^r\|X\|$$

for all $z \in \overline{S}$. Hence with this choice of Φ_z, condition (6.3.15) of Theorem 6.30 is satisfied.

For any $0 \leq x \leq 1$ and $1 \leq t, y \leq \infty$, again by the unitary invariance of the Schatten t norm,

$$\|\Phi_{x+iy}(A)\|_t = \||A|Y^*|^{rx}|Y^*|^{iry}U\|_t = \||A|Y^*|^{rx}\|_t = \|\Phi_x(A)\|_t.$$

Hence for all $0 \leq x \leq 1$, and all $1 \leq s, t \leq \infty$,

$$(6.3.24) \qquad \sup_{y \in \mathbb{R}}\{\|\Phi_{x+iy}\|_{s \to t}\} = \|\Phi_x\|_{s \to t}.$$

Therefore, the conclusion of Theorem 6.30 simplifies to (6.3.17).

We have already fixed $\theta = \frac{1}{r}$ and $q_\theta = r$. In (6.3.22), $p_0 = \infty$, and in (6.3.23), $p_1 = s$, to be specified. Now choose s so that $p_\theta = p$: $\frac{1}{p_\theta} = \frac{1-\theta}{\infty} + \frac{\theta}{s} = \frac{\theta}{s}$. Therefore, we apply (6.3.23) with $s = \frac{r}{p}$. Since $\frac{p}{r} + \frac{q}{r} = 1$, $\frac{s}{s-1} = \frac{q}{r}$ and since $Z(1) = |Y^*|^r U$, (6.3.23) becomes

$$\|\Phi_1\|_{p_1 \to q_1} = \|R_{Z(1)}\|_{s \to 1} = \|Y^r\|_{q/r} = \|Y\|_q^r.$$

Since Φ_0 is right multiplication by a unitary,

$$\|\Phi_0\|_{p_0 \to q_0} = 1.$$

By Theorem 6.30 and (6.3.24),

$$\|XY\|_r \leq \|X\|_p \|\Phi_0\|_{p_0 \to q_0}^{1-\theta} \|\Phi_1\|_{p_1 \to q_1}^{\theta} = \|X\|_p (\|Y\|_q^r)^{1/r} = \|X\|_p \|Y\|_q. \qquad \square$$

6.4. Noncommutative L^p norms associated to a density matrix

Definition 6.33. Let $\mathcal{H}$ be a finite-dimensional Hilbert space, and let $\sigma \in \mathfrak{S}_{\mathcal{H}}^\circ$, the interior of $\mathfrak{S}_{\mathcal{H}}$, so that $\sigma > 0$. The corresponding **Kubo-Martin-Schwinger (KMS) inner product** on $\mathcal{B}(\mathcal{H})$, $\langle \cdot, \cdot \rangle_\sigma$ is defined by

$$(6.4.1) \qquad \langle X, Y \rangle_\sigma := \mathrm{Tr}[X^*\sigma^{1/2}Y\sigma^{1/2}].$$

Define the operator $\Psi_\sigma : \mathcal{B}(\mathcal{H}) \to \mathcal{B}(\mathcal{H})$ by

$$\Psi_\sigma(Y) := \sigma^{1/2}Y\sigma^{1/2},$$

and note that

$$(6.4.2) \qquad \langle X, Y \rangle_\sigma = \langle X, \Psi_\sigma(Y) \rangle_{\mathrm{HS}}.$$

The operator Ψ_σ is a positive operator on the Hilbert space $\widehat{\mathcal{H}}$ consisting of $\mathcal{B}(\mathcal{H})$ equipped with the Hilbert-Schmidt inner product: Let $\{\mathbf{u}_1, \ldots, \mathbf{u}_d\}$ be an orthonormal basis of $\mathcal{H}$ consisting of eigenvectors of σ, $\sigma\mathbf{u}_j = \lambda_j\mathbf{u}_j$, $j = 1, \ldots, d$. For $1 \leq i, j \leq d$, define $E_{i,j} := |\mathbf{u}_i\rangle\langle\mathbf{u}_j|$. Then $\{E_{i,j}\}_{1 \leq i,j \leq d}$ is an orthonormal basis of $\widehat{\mathcal{H}}$, and $\Psi_\sigma(E_{i,j}) = \lambda_i\lambda_j E_{i,j}$. Since all of the eigenvalues are positive, this proves that $\Psi_\sigma \in \mathcal{B}^{++}(\widehat{\mathcal{H}})$. It then follows from (6.4.2) that $\langle X, X\rangle_\sigma \geq 0$ with equality if and only if $X = 0$, and thus $\langle \cdot, \cdot \rangle_\sigma$ is indeed an inner product on $\mathcal{B}(\mathcal{H})$.

Let $\mathcal{X}$ be a finite set, and let σ be a probability density on $\mathcal{X}$. Define an inner product $\langle \cdot, \cdot \rangle_\sigma$ on the space of complex valued functions on $\mathcal{X}$ by

$$\langle f, g\rangle_\sigma := \sum_{x \in \mathcal{X}} \overline{f(x)}g(x)\sigma(x).$$

In the noncommutative setting, there are many possible analogues of this inner product that are distinct because of noncommutativity. The distribution of powers of σ in (6.4.1) is particularly symmetric and simple; it is also physically natural. We shall see some justification for this later, but we note now that Kubo, Martin, and Schwinger were all physicists.

Associated to the KMS inner product is the corresponding Hilbert space norm $\| \cdot \|_{2,\sigma}$ given by

$$\|X\|_{2,\sigma} := \langle X, X\rangle_\sigma^{1/2} = \langle X, \Psi_\sigma(X)\rangle_{\mathrm{HS}}^{1/2}.$$

There is also a corresponding family of noncommutative L^p norms.

Definition 6.34 (Noncommutative L^p norms). Let $\mathcal{H}$ be a finite-dimensional Hilbert space and let $\sigma \in \mathfrak{S}_{\mathcal{H}}^\circ$. For $1 \leq p \leq \infty$, define a function $X \mapsto \|X\|_{p,\sigma}$ on $X \in \mathcal{B}(\mathcal{H})$ by

$$(6.4.3) \qquad \|X\|_{p,\sigma} := \|\Psi_\sigma^{1/p}(X)\|_p = \|\sigma^{\frac{1}{2p}} X \sigma^{\frac{1}{2p}}\|_p,$$

where $\| \cdot \|_p$ denotes the Schatten p-norm.

Lemma 6.35. *Let $\mathcal{H}$ be a finite-dimensional Hilbert space and let $\sigma \in \mathfrak{S}_{\mathcal{H}}^\circ$. For $1 \leq p \leq \infty$, $\| \cdot \|_{p,\sigma}$ is a norm on $\mathcal{B}(\mathcal{H})$.*

Proof. For $p = \infty$, there is nothing to prove: for all $\sigma \in \mathfrak{S}_{\mathcal{H}}^\circ$, $\| \cdot \|_{\infty,\sigma} = \| \cdot \|_\infty$ which is a norm on $\mathcal{B}(\mathcal{H})$. Now consider $1 \leq p < \infty$. Evidently, $\|\lambda X\|_{p,\sigma} = |\lambda|\|X\|_{p,\sigma}$, and since $\sigma > 0$, $\|\lambda X\|_{p,\sigma} = 0$ if and only if $X = 0$. Finally, for all $X, Y \in \mathcal{B}(\mathcal{H})$,

$$\|X + Y\|_{p,\sigma} = \|\Psi_\sigma^{1/p}(X) + \Psi_\sigma^{1/p}(X)\|_p$$

$$\leq \|\Psi_\sigma^{1/p}(X)\|_p + \|\Psi_\sigma^{1/p}(X)\|_p = \|X\|_{p,\sigma} + \|Y\|_{p,\sigma},$$

so that the triangle inequality is satisfied. $\qquad\qquad\qquad\qquad\qquad\qquad\qquad\square$

By construction, $\Psi_\sigma^{1/p}$ is an isometry between $\mathcal{B}(\mathcal{H})$ equipped with the norm $\|\cdot\|_{p,\sigma}$ and $\mathcal{S}_p$; that is, $\mathcal{B}(\mathcal{H})$ is equipped with the Schatten p-norm. For this reason, the norms $\|\cdot\|_{p,\sigma}$ readily inherit a number of properties from the Schatten norms.

Theorem 6.36. *Let $\mathcal{H}$ be a finite-dimensional Hilbert space and let $\sigma \in \mathfrak{S}_{\mathcal{H}}^\circ$. For $1 \le p, q \le \infty$, $\frac{1}{p} + \frac{1}{q} = 1$, and $X, Y \in \mathcal{B}(\mathcal{H})$*

$$(6.4.4) \qquad |\langle X, Y\rangle_\sigma| \le \|X\|_{p,\sigma}\|Y\|_{q,\sigma}$$

and

$$(6.4.5) \qquad \|X\|_{p,\sigma} = \max\{\mathfrak{R}\langle X, Y\rangle_\sigma \ : \ \|Y\|_{q,\sigma} = 1\},$$

and if $X \in \mathcal{B}^+(\mathcal{H})$, there is a unique $Y \in \mathcal{B}^+(\mathcal{H})$ such that $\|Y\|_{q,\sigma} = 1$ and $\|X\|_{p,\sigma} = \langle X, Y\rangle_\sigma$.

Proof. From the definition (6.4.3), the positivity of Ψ_σ, and $\frac{1}{p} + \frac{1}{q} = 1$,

$$\langle X, Y\rangle_\sigma = \langle X, \Psi_\sigma(Y)\rangle_{\mathrm{HS}} = \langle \Psi_\sigma^{1/p}(X), \Psi_\sigma^{1/q}(Y)\rangle_{\mathrm{HS}}.$$

Then by Hölder's inequality,

$$|\langle X, Y\rangle_\sigma| \le \|\Psi_\sigma^{1/p}(X)\|_p\|\Psi_\sigma^{1/q}(Y)\|_q = \|X\|_{p,\sigma}\|Y\|_{p,\sigma},$$

which proves (6.4.4). Now suppose that $p < \infty$. By the statement about cases of equality in Theorem 6.17, $|\langle \Psi_\sigma^{1/p}(X), \Psi_\sigma^{1/q}(Y)\rangle_{\mathrm{HS}}| = \|\Psi_\sigma^{1/p}(X)\|_p\|\Psi_\sigma^{1/q}(Y)\|_q$ if and only if $Y = zU|\Psi_\sigma^{1/p}(X)|^{p-1}$, where $\Psi_\sigma^{1/p}(X) = U|\Psi_\sigma^{1/p}(X)|$ is the polar decompositions of $\Psi_\sigma^{1/p}(X)$, and $z \in \mathbb{C}$. If we further require that $\|Y\|_{q,\sigma} = 1$ and that $\mathfrak{R}\langle X, Y\rangle_\sigma = |\langle X, Y\rangle_\sigma|$, this uniquely determines z, and there is equality in (6.4.5) only in case $Y = Y_X := \|X\|_{p,\sigma}^{1-p}U|\Psi_\sigma^{1/p}(X)|^{p-1}$. Note that if $X \ge 0$, then $Y \ge 0$. The case $p = \infty$ is left as an exercise. $\qquad\square$

Definition 6.37. Let $\mathcal{H}, \mathcal{K}$ be finite-dimensional Hilbert spaces. Let $\rho \in \mathfrak{S}_{\mathcal{H}}^\circ$ and $\sigma \in \mathfrak{S}_{\mathcal{K}}^\circ$. For $1 \le p, q \le \infty$ define a norm $\Phi \mapsto \|\Phi\|_{(\rho,p)\to(\sigma,q)}$ on the space $\mathcal{L}(\mathcal{B}(\mathcal{H}), \mathcal{B}(\mathcal{K}))$ of linear transformations from $\mathcal{B}(\mathcal{H})$ to $\mathcal{B}(\mathcal{K})$ by

$$\|\Phi\|_{(\rho,p)\to(\sigma,q)} = \sup\{\|\Phi(X)\|_{\sigma,q} \ : \ \|X\|_{\rho,p} = 1\}.$$

Lemma 6.38. *Let $\mathcal{H}, \mathcal{K}$ be finite-dimensional Hilbert spaces, and let $\Phi : \mathcal{B}(\mathcal{H}) \to \mathcal{B}(\mathcal{K})$ be a linear transformation. Let $\rho \in \mathfrak{S}_{\mathcal{H}}^\circ$ and $\sigma \in \mathfrak{S}_{\mathcal{K}}^\circ$. Then for all $X \in \mathcal{B}(\mathcal{H})$ and all $Y \in \mathcal{B}(\mathcal{K})$,*

$$\langle Y, \Phi(X)\rangle_\sigma = \langle \Psi_\rho^{-1} \circ \Phi^\dagger \circ \Psi_\sigma(Y), X\rangle_\rho.$$

Proof. By definition,

$$\langle Y, \Phi(X)\rangle_\sigma = \mathrm{Tr}[Y^*\Psi_\sigma(\Phi(X))] = \mathrm{Tr}[(\Phi^\dagger(\Psi_\sigma(Y)))^*X]$$

$$= \mathrm{Tr}[(\Psi_\rho^{-1}(\Phi^\dagger(\Psi_\sigma(Y))))^*\Psi_\rho(X)] = \langle \Psi_\rho^{-1} \circ \Phi^\dagger \circ \Psi_\sigma(Y), X\rangle_\rho. \quad\square$$

Lemma 6.39. *Let $\mathcal{H}, \mathcal{K}$ be finite-dimensional Hilbert spaces, and let $\Phi : \mathcal{B}(\mathcal{H}) \to \mathcal{B}(\mathcal{K})$ be a linear transformation. Let $1 \leq p, q \leq \infty$, and define p', q' so that $\frac{1}{p} + \frac{1}{p'} = 1$ and $\frac{1}{q} + \frac{1}{q'} = 1$. Then for $\rho \in \mathfrak{S}^{\circ}_{\mathcal{H}}$ and $\sigma \in \mathfrak{S}^{\circ}_{\mathcal{K}}$,*

$$\|\Psi_\rho^{-1} \circ \Phi^\dagger \circ \Psi_\sigma\|_{(q',\sigma) \to (p',\rho)} = \|\Phi\|_{(p,\rho) \to (q,\sigma)}.$$

Proof. Let $Y \in \mathcal{B}(\mathcal{K})$ be such that $\|Y\|_{q',\sigma} = 1$. By Theorem 6.36, there exists $X \in \mathcal{B}(\mathcal{H})$ such that $\|X\|_{q,\rho} = 1$ and

$$\|\Psi_\rho^{-1} \circ \Phi^\dagger \circ \Psi_\sigma(Y)\|_{\rho,p'} = \langle \Psi_\rho^{-1} \circ \Phi^\dagger \circ \Psi_\sigma(Y), X \rangle_\rho.$$

Therefore, by Lemma 6.38,

$$\|\Psi_\rho^{-1} \circ \Phi^\dagger \circ \Psi_\sigma(Y)\|_{\rho,p'} = \langle Y, \Phi(X) \rangle_\sigma \leq \|Y\|_{q',\sigma} \|\Phi(X)\|_{q,\sigma} \leq \|\Phi\|_{(p,\rho) \to (q,\sigma)}.$$

This shows that $\|\Psi_\rho^{-1} \circ \Phi^\dagger \circ \Psi_\sigma\|_{(q',\sigma) \to (p',\rho)} \leq \|\Phi\|_{(p,\rho) \to (q,\sigma)}$, and the inequality is saturated for appropriate X and Y. $\qquad\square$

The following theorem is due to Beigi [26].

Theorem 6.40 (Riesz-Thorin theorem for maps from $L^p(\sigma)$ to $L^q(\sigma)$). *Let $1 \leq p_0, p_1, q_0, q_1 \leq \infty$. For $0 < \theta < 1$, define*

$$\frac{1}{p_\theta} = \frac{1-\theta}{p_0} + \frac{\theta}{p_1} \quad \text{and} \quad \frac{1}{q_\theta} = \frac{1-\theta}{q_0} + \frac{\theta}{q_1}.$$

Then for all $\Phi \in \mathcal{L}(\mathcal{B}(\mathcal{H}), \mathcal{B}(\mathcal{K}))$,

$$(6.4.6) \qquad \|\Phi\|_{(\rho,p_\theta) \to (\sigma,q_\theta)} \leq \|\Phi\|^{1-\theta}_{(\rho,p_0) \to (\sigma,q_0)} \|\Phi\|^{\theta}_{(\rho,p_1) \to (\sigma,q_1)}.$$

Proof. Define a map $\Phi_z : \mathcal{B}(\mathcal{H}) \to \mathcal{B}(\mathcal{K})$ by

$$\Phi_z(X) := \Psi_\sigma^{\frac{1-z}{q_0} + \frac{z}{q_1}} \circ \Phi \circ \Psi_\rho^{-\frac{1-z}{p_0} - \frac{z}{p_1}}.$$

Then Φ_z is analytic in the entire complex plane and for $z = x + iy, 0 \leq x \leq 1$, and $X \in \mathcal{B}(\mathcal{H})$, since $\|\sigma\| \leq 1$,

$$\|\Phi_z(X)\| = \|\Psi_\sigma^{\frac{1-x}{q_0} + \frac{x}{q_1}}(\Phi(\Psi_\rho^{-\frac{1-z}{p_0} - \frac{z}{p_1}}(X)))\|$$

$$\leq \|\Phi\|_{\infty \to \infty} \|\Psi_\rho^{-\frac{1-x}{p_0} - \frac{x}{p_1}}(X)\|$$

$$\leq \|\Phi\|_{\infty \to \infty} \|X\| \|\rho^{-1}\|.$$

Therefore, $\|\Phi_z\|_{\infty \to \infty} \leq \|\Phi\|_{\infty \to \infty} \|\rho^{-1}\|$ for all $z \in \overline{S}$ so that condition (6.3.15) of Theorem 6.30 is satisfied.

Next, for all $y \in \mathbb{R}$,

$$\|\Phi_{iy}(X)\|_{q_0} = \|\Psi_\sigma^{\frac{1}{q_0}} \Phi(\Psi_\rho^{-\frac{1-iy}{p_0} - \frac{iy}{p_1}}(X))\|_{q_0} = \|\Phi(\Psi_\rho^{-\frac{1-iy}{p_0} - \frac{iy}{p_1}}(X))\|_{q_0,\sigma}$$

$$\leq \|\Phi\|_{(p_0,\rho) \to (q_0,\sigma)} \|\Psi_\rho^{-\frac{1}{p_0}}(X)\|_{p_0,\rho} = \|\Phi\|_{(p_0,\rho) \to (q_0,\sigma)} \|X\|_{p_0}.$$

Therefore,

$$\sup_{y\in\mathbb{R}}\{\|\Phi_{iy}\|_{p_0\to q_0}\} \leq \|\Phi\|_{(p_0,\rho)\to(q_0,\sigma)}.$$

The same sort of reasoning also yields

$$\sup_{y\in\mathbb{R}}\{\|\Phi_{1+iy}\|_{p_1\to q_1}\} \leq \|\Phi\|_{(p_1,\rho)\to(q_1,\sigma)}.$$

Then (6.4.6) follows from Theorem 6.30 because $\Phi_\theta(X) = \Psi_\sigma^{1/q_\theta}(\Phi(\Psi_\rho^{-1/p_\theta}(X)))$.

$\square$

We now turn to some important applications of Theorem 6.40 that center on the following operator:

Definition 6.41. Let $\mathcal{H}, \mathcal{K}$ be finite-dimensional Hilbert spaces, and let $\Phi :$ $\mathcal{B}(\mathcal{H}) \to \mathcal{B}(\mathcal{K})$ be a positive linear map. Let $\sigma \in \mathfrak{S}_{\mathcal{K}}^\circ$, and suppose that $\Phi^\dagger(\sigma) > 0$. The **Accardi-Cecchini coarse graining operator** is the map

$$\mathcal{A}_{\Phi,\sigma*} := \Psi_{\Phi^\dagger(\sigma)}^{-1} \circ \Phi^\dagger \circ \Psi_\sigma,$$

which, by Lemma 6.38, is the adjoint of Φ considered as a map from $L^2(\Phi^\dagger(\sigma))$ to $L^2(\sigma)$.

Remark 6.42. Note that the operator $\mathcal{A}_{\Phi,\sigma}$ is positive, and moreover it is unital since $\mathcal{A}_{\Phi,\sigma}(\mathbb{1}) = \Psi_{\Phi^\dagger(\sigma)}(\Phi^\dagger(\sigma)) = \mathbb{1}$. By the Russo-Dye theorem,

$$(6.4.7) \qquad \|\mathcal{A}_{\Phi,\sigma}\|_{\infty\to\infty} = 1.$$

The map $\mathcal{A}_{\Phi,\sigma}$ was introduced by Accardi [1] and further studied by Accardi and Cecchini [2] in the context of conditional expectations in noncommutative probability theory. Some aspects of their investigation are developed in the exercises.

The following theorem is due to Beigi [26] in the case in which Φ is CP and unital, but as observed by Müller-Hermes and Reeb [160], the complete positivity plays no essential role in Beigi's proof.

Theorem 6.43. *Let $\mathcal{H}, \mathcal{K}$ be finite-dimensional Hilbert spaces, and let $\Phi :$ $\mathcal{B}(\mathcal{H}) \to \mathcal{B}(\mathcal{K})$ be positive and unital. Let $\sigma \in \mathfrak{S}_{\mathcal{H}}^\circ$, and suppose that $\Phi^\dagger(\sigma) > 0$. Then for all $1 \leq p \leq \infty$,*

$$(6.4.8) \qquad \|\mathcal{A}_{\Phi,\sigma}\|_{(p,\sigma)\to(p,\Phi^\dagger(\sigma))} = 1$$

and

$$(6.4.9) \qquad \|\Phi\|_{(p,\Phi^\dagger(\sigma))\to(p,\sigma)} = 1.$$

Proof. Since for any invertible density matrix ρ, $\|\cdot\|_{\infty,\rho} = \|\cdot\|_\infty$, (6.4.7) is equivalent to

$$(6.4.10) \qquad \|\mathcal{A}_{\Phi,\sigma}\|_{(\infty,\sigma)\to(\infty,\Phi^\dagger(\sigma))} = 1.$$

Next, for $X \in \mathcal{B}(\mathcal{H})$,

$$\|\mathcal{A}_{\Phi,\sigma}(X)\|_{1,\Phi^\dagger(\sigma)} = \left\|\Phi^\dagger(\Psi_\sigma(X))\right\|_1 \leq \|\Phi^\dagger\|_{1\to 1}\|X\|_{1,\sigma}.$$

Therefore, $\left\|\mathcal{A}_{\Phi,\sigma}\right\|_{(1,\sigma)\to(1,\Phi^\dagger(\sigma))} \leq \|\Phi^\dagger\|_{1\to 1} = 1$, where the last equality is due to Theorem 6.26. Since $\mathcal{A}_{\Phi,\sigma}$ is unital, there is even equality, so that

$$(6.4.11) \qquad\qquad \|\mathcal{A}_{\Phi,\sigma}\|_{(1,\sigma)\to(1,\Phi^\dagger(\sigma))} = 1.$$

Using Theorem 6.40 to interpolate between (6.4.10) and (6.4.11) yields (6.4.8), and then (6.4.9) follows from Lemma 6.39. $\qquad\qquad\qquad\qquad\square$

Theorem 6.43 has an important consequence that does not require the hypothesis that $\Phi^\dagger(\sigma) > 0$. The following approximation will be used:

Let $\Phi^\dagger : \mathcal{B}(\mathcal{K}) \to \mathcal{B}(\mathcal{H})$ be positive and trace preserving. Take $0 < \epsilon < 1$ and define

$$(6.4.12) \qquad \Phi^\dagger_\epsilon(A) := (1-\epsilon)\Phi^\dagger(A) + \frac{\epsilon}{\dim(\mathcal{H})}\,\mathrm{Tr}[A]\mathbb{1}_{\mathcal{H}}.$$

Then $\Phi^\dagger_\epsilon$ is positive and trace preserving and $\Phi^\dagger_\epsilon(\mathbb{1}) > 0$. We recover Φ from Φ_ϵ in the limit $\epsilon \downarrow 0$.

Theorem 6.44. *Let $\mathcal{H}, \mathcal{K}$ be finite-dimensional Hilbert spaces, and let $\Phi : \mathcal{B}(\mathcal{H}) \to \mathcal{B}(\mathcal{K})$ be positive and unital. Then for all $A \in \mathcal{B}^+(\mathcal{H})$ and all $B \in \mathcal{B}^+(\mathcal{K})$, and all $1 \leq p \leq \infty$*

$$(6.4.13) \qquad \mathrm{Tr}\left[\left((\Phi^\dagger(A))^{\frac{1-p}{2p}}\,\Phi^\dagger(B)(\Phi^\dagger(A))^{\frac{1-p}{2p}}\right)^p\right] \leq \mathrm{Tr}\left[\left(A^{\frac{1-p}{2p}}BA^{\frac{1-p}{2p}}\right)^p\right]$$

and

$$(6.4.14) \qquad \mathrm{Tr}\left[\left(A^{\frac{1}{2p}}\Phi(B)A^{\frac{1}{2p}}\right)^p\right] \leq \mathrm{Tr}\left[\left((\Phi^\dagger(A))^{\frac{1}{2p}}B(\Phi^\dagger(A))^{\frac{1}{2p}}\right)^p\right].$$

Proof. By homogeneity, we may assume that $A = \sigma \in \mathfrak{S}_{\mathcal{K}}$. Suppose that $\sigma > 0$ and that $\Phi^\dagger(\mathbb{1}) > 0$, so that $\Phi^\dagger(\sigma) > 0$. By Theorem 6.43, we have both (6.4.8) and (6.4.9). Starting from (6.4.8), for all $B \in \mathcal{B}^+(\mathcal{K})$,

$$\|\Psi^{-1}_{\Phi^\dagger(\sigma)}\Phi^\dagger(B)\|_{p,\Phi^\dagger(\sigma)} = \|\mathcal{A}_{\Phi,\sigma}(\Psi^{-1}_\sigma(B))\|_{p,\Phi^\dagger(\sigma)} \leq \|\Psi^{-1}_\sigma(B)\|_{p,\sigma}.$$

Taking the pth power of both sides and replacing σ by A yields (6.4.13). To remove the hypothesis that $\Phi^\dagger(\sigma) > 0$, use the approximation (6.4.12) and take limits to obtain (6.4.14). The same reasoning starting from (6.4.9) yields (6.4.14). $\qquad\qquad\qquad\qquad\square$

Theorem 6.44 has important applications. Let $\rho, \sigma \in \mathfrak{S}_{\mathcal{H}}$ with $\mathrm{ran}(\rho) \subseteq \mathrm{ran}(\sigma)$. For $1 < p < \infty$, define

$$(6.4.15) \qquad Q_p(\rho\|\sigma) := \left\|\sigma^{-1/2}\rho\sigma^{-1/2}\right\|^p_{p,\sigma} = \mathrm{Tr}\left[\left(\sigma^{\frac{1-p}{2p}}\rho\sigma^{\frac{1-p}{2p}}\right)^p\right].$$

The quantity $Q_p(\rho\|\sigma)$ is closely related to a quantity called the *sandwiched Rényi relative entropy of ρ with respect to σ* that will be studied in Chapter 10. Theorem 6.43 says that $Q_p(\rho\|\sigma)$ is **monotone under quantum operations** and more generally, is **monotone under positive trace preserving maps**. Corollary 6.45 was drawn in [26] for the completely positive case; see [160] for the general case.

Corollary 6.45. *Let $\mathcal{H}, \mathcal{K}$ be finite-dimensional Hilbert spaces, and let $\Phi : \mathcal{B}(\mathcal{H}) \to \mathcal{B}(\mathcal{K})$ be positive and unital. Then for all $\rho, \sigma \in \mathfrak{S}_{\mathcal{K}}$ such that $\sigma > 0$,*

$$(6.4.16) \qquad Q_p(\Phi^\dagger \rho \| \Phi^\dagger \sigma) \leq Q_p(\rho\|\sigma).$$

and

$$(6.4.17) \qquad \mathrm{Tr}[\Phi^\dagger(\rho)(\log(\Phi^\dagger(\rho)) - \log(\Phi^\dagger(\sigma)))] \leq \mathrm{Tr}[\rho(\log(\rho) - \log(\sigma))].$$

Proof. The monotonicity relation (6.4.16) is an immediate consequence of (6.4.13). Note that for $p = 1$, $Q_p(\rho\|\sigma) = \mathrm{Tr}[\rho] = 1$ and hence for all $p > 1$,

$$\frac{Q_p(\Phi^\dagger \rho \| \Phi^\dagger \sigma) - 1}{p - 1} \leq \frac{Q_p(\rho\|\sigma) - 1}{p - 1},$$

and taking the limit $p \downarrow 1$ yields (6.4.17). (See Exercise 15.) $\qquad\square$

The quantity $D(\rho\|\sigma) := \mathrm{Tr}[\rho(\log(\rho) - \log(\sigma))]$ is known as the *relative entropy of ρ with respect to σ*, and will be studied extensively in the next chapter. The fact that it is monotone decreasing under positive trace preserving maps, and therefore under quantum operations, is fundamentally important in quantum information theory.

The inequality (6.4.14) of Theorem 6.44 has important consequences.

Corollary 6.46. *Let $\mathcal{H}$ be a finite-dimensional Hilbert space, and let $\Phi : \mathcal{B}(\mathcal{H}) \to \mathcal{B}(\mathcal{H})$ be CP and unital. Let $\rho \in \mathfrak{S}_{\mathcal{H}}$ be invariant for Φ, meaning that $\Phi^\dagger(\rho) = \rho$. Then for all $1 \leq p \leq \infty$,*

$$(6.4.18) \qquad \|\Phi\|_{(p,\rho)\to(p,\rho)} \leq 1.$$

Proof. When $\Phi^\dagger(\rho) = \rho$, (6.4.18) is simply (6.4.9) with ρ in place of σ. $\qquad\square$

6.5. The Powers-Størmer inequality

By Lemma 6.18, for all $A, B \in M_n(\mathbb{C})$, $\|A - B\|_1 \geq \|A - B\|_2$. There is another lower bound on $\|A - B\|_1$ involving the Hilbert-Schmidt norm when $A, B \in M_n^+(\mathbb{C})$, due to Powers and Størmer [181].

Theorem 6.47 (The Powers-Størmer inequality). *For all $A, B \in M_n^+(\mathbb{C})$,*

$$(6.5.1) \qquad \|A^{1/2} - B^{1/2}\|_2^2 \leq \|A - B\|_1.$$

Example 6.48 (The Powers-Størmer inequality and purification). Let ρ and σ be density matrices on a finite-dimensional Hilbert space $\mathcal{H}$. Then by (6.5.1)

$$\|\rho^{1/2} - \sigma^{1/2}\|_2^2 \leq \|\rho - \sigma\|_1.$$

Let $\{\mathbf{u}_1, \ldots, \mathbf{u}_n\}$ be any orthonormal basis of $\mathcal{H}$. As in (2.4.4), define ψ and ϕ in $\mathcal{H} \otimes \mathcal{H}$ by

$$(6.5.2) \qquad \psi = \sum_{i,j=1}^{n} \langle \mathbf{u}_i, \rho^{1/2} \mathbf{u}_j \rangle \mathbf{u}_i \otimes \mathbf{u}_j \quad \text{and} \quad \phi = \sum_{i,j=1}^{n} \langle \mathbf{u}_i, \sigma^{1/2} \mathbf{u}_j \rangle \mathbf{u}_i \otimes \mathbf{u}_j.$$

Then ψ and ϕ are purifications of ρ and σ, respectively, and

$$(6.5.3) \qquad \langle \psi, \phi \rangle = \sum_{i,j=1}^{n} \langle \rho^{1/2} \mathbf{u}_j, \mathbf{u}_i \rangle \langle \mathbf{u}_i, \sigma^{1/2} \mathbf{u}_j \rangle = \mathrm{Tr}[\rho^{1/2} \sigma^{1/2}] \geq 0,$$

so that $\|\psi - \phi\|_{\mathcal{H} \otimes \mathcal{H}}^2 = \|\rho^{1/2} - \sigma^{1/2}\|_2^2$. Therefore, since $\langle \psi, \phi \rangle \geq 0$,

$$(6.5.4) \qquad \|\rho - \sigma\|_1 \geq \|\psi - \phi\|_{\mathcal{H} \otimes \mathcal{H}}^2 = 2(1 - |\langle \psi, \phi \rangle|).$$

Thus, when the trace norm of $\rho - \sigma$ is small, there exist purifications of ρ and σ that are close in the $\mathcal{H} \otimes \mathcal{H}$ norm.

We give proofs of two different and useful extensions of the Powers-Størmer inequality. For the first, rewrite (6.5.1) as

$$(6.5.5) \qquad \mathrm{Tr}[|A - B| - A - B + 2A^{1/2}B^{1/2}] \geq 0.$$

Motivated by a problem in quantum information theory, the following generalization of (6.5.5) has been proved in [**20, 21**].

Theorem 6.49. *For all $A, B \in M_n^+(\mathbb{C})$ and all $0 \leq s \leq 1$,*

$$\mathrm{Tr}[|A - B| - A - B + 2A^s B^{1-s}] \geq 0.$$

A simple proof of this has been found by Ozawa, and it was used by Ogata [**171**] in a general von Neumann algebra setting. Ozawa's unpublished argument is indicated in her paper.

Lemma 6.50 (Ozawa's lemma). *Let $A, B, C \in M_n^+(\mathbb{C})$ with $C \geq A$ and $C \geq B$. Then for all $0 \leq s \leq 1$,*

$$(6.5.6) \qquad \mathrm{Tr}[C^s B^{1-s}] - \mathrm{Tr}[A^s B^{1-s}] \leq \mathrm{Tr}[C] - \mathrm{Tr}[A^s C^{1-s}]$$

and

$$(6.5.7) \qquad \mathrm{Tr}[A^s B^{1-s}] \leq \mathrm{Tr}[C^s B^{1-s}] \quad \text{and} \quad \mathrm{Tr}[A^s B^{1-s}] \leq \mathrm{Tr}[A^s C^{1-s}].$$

Proof. By the operator monotonicity of $X \mapsto X^p$ on $M_n^+(\mathbb{C})$ for $0 \le p \le 1$, $C^s - A^s \ge 0$ and $C^{1-s} - B^{1-s} \ge 0$. Therefore,

$$0 \le \mathrm{Tr}[(C^s - A^s)(C^{1-s} - B^{1-s}]$$
$$= \big(\mathrm{Tr}[C] - \mathrm{Tr}[A^s C^{1-s}]\big) - \big(\mathrm{Tr}[C^s B^{1-s}] - \mathrm{Tr}[A^s B^{1-s}]\big),$$

which proves (6.5.6). Next,

$$\mathrm{Tr}[A^s B^{1-s}] = \mathrm{Tr}[B^{(1-s)/2} A^s B^{(1-s)/2}]$$
$$\le \mathrm{Tr}[B^{(1-s)/2} C^s B^{(1-s)/2}] = \mathrm{Tr}[C^s B^{1-s}],$$

once more using the operator monotonicity of $X \mapsto X^p$, $0 \le p \le 1$. This proves the first inequality in (6.5.7) and the proof of the second is the same. $\qquad\square$

Proof of Theorem 6.49. Define $C := B + (B - A)_-$, and note that $C \ge B$ for trivial reasons and also $C - A = (B - A) + (B - A)_- = (B - A)_+$ so that $C \ge A$. Therefore, by (6.5.7), then (6.5.6), and then (6.5.7) again,

$$\mathrm{Tr}[B] - \mathrm{Tr}[A^s B^{1-s}] = \mathrm{Tr}[B^s B^{1-s}] - \mathrm{Tr}[A^s B^{1-s}]$$
$$\le \mathrm{Tr}[C^s B^{1-s}] - \mathrm{Tr}[A^s B^{1-s}]$$
$$\le \mathrm{Tr}[C] - \mathrm{Tr}[A^s C^{1-s}]$$
$$\le \mathrm{Tr}[C] - \mathrm{Tr}[A] = \mathrm{Tr}[(B - A)_+].$$

Altogether,

$$2\,\mathrm{Tr}[A^s B^{1-s}] \ge \mathrm{Tr}[2B - 2(B - A)_+]$$
$$= \mathrm{Tr}[(A + B) - ((A - B) + 2(A - B)_-)]$$
$$= \mathrm{Tr}[A + B - |A - B|]. \qquad\square$$

The second generalization of the Powers-Størmer inequality is due to Caspers, Parcet, Perrin, and Ricard [**55**]:

Theorem 6.51. *For all $A, B \in M_n^+(\mathbb{C})$, all $p \ge 1$, and all $0 < \alpha < 1$,*

$$\|A^\alpha - B^\alpha\|_p \le \|A - B\|_{p\alpha}^\alpha.$$

Proof. Consider the case $C = B - A \le 0$. Using the integral representation

$$(B + tC)^\alpha = c_\alpha \int_0^\infty \lambda^{\alpha-1}(\mathbb{1} - \lambda(\lambda + (B + tC))^{-1})\mathrm{d}\lambda,$$

$$A^\alpha - B^\alpha = c_\alpha \int_0^1 \left(\int_0^\infty \lambda^\alpha(\lambda + (B + tC))^{-1} C(\lambda + (B + tC))^{-1}\mathrm{d}\lambda \right) \mathrm{d}t.$$

For $X \in M_n^+(\mathbb{C})$, define $\Psi_X : M_n(\mathbb{C}) \to M_n(\mathbb{C})$ by

$$\Psi_X(K) := \frac{c_\alpha}{\alpha} \int_0^\infty \lambda^\alpha(\lambda + X)^{-1}(X^{(1-\alpha)/2} K X^{(1-\alpha)/2})(\lambda + X)^{-1}\mathrm{d}\lambda.$$

Since $c_\alpha \int_0^\infty \frac{t^{\alpha-1}}{t+1} dt = 1$, a simple calculation shows that $\Psi(\mathbb{1}) = \mathbb{1}$. Since evidently $\Psi^\dagger = \Psi$, Ψ is also trace preserving. Since Ψ is evidently positive, it follows from Theorem 6.32 that $\|\Psi\|_{p \to p} = 1$ for all $p \geq 1$. Then by the operator monotonicity of $x^{\alpha-1}$,

$$\|A^\alpha - B^\alpha\|_p \leq \alpha \int_0^1 \|(B + tC)^{(\alpha-1)/2} C (B + tC)^{(\alpha-1)/2}\|_p dt$$

$$= \alpha \int_0^1 \|C^{1/2}(B + tC)^{(\alpha-1)} C^{1/2}\|_p dt$$

$$\leq \alpha \int_0^1 t^{\alpha-1}\|C^\alpha\|_p dt = \|C^\alpha\|_p = \|A - B\|_{\alpha p}^\alpha.$$

It remains to remove the assumption that $A > B$. Therefore, let $A, B \in M_n^+(\mathbb{C})$. Let $C = A - B$, and let $C = C_+ - C_-$ be the decomposition of C into its positive and negative parts. Define

$$X := (A + C_-)^\alpha - A^\alpha \quad \text{and} \quad Y := (B + C_+)^\alpha - B^\alpha.$$

By the operator monotonicity of x^α, $A, Y \in M_n^+(\mathbb{C})$, and since $A + C_- = B + C_+$, $A^\alpha - B^\alpha = X - Y$. By Exercise 10, $\|X - Y\|_p^p \leq \|X\|_p^p + \|Y\|_p^p$. Therefore, applying what we have proved above in the second inequality below,

$$\|A^\alpha - B^\alpha\|_p^p = \|X - Y\|_p^p \leq \|X\|_p^p + \|Y\|_p^p \leq \|C_-\|_{\alpha p}^\alpha + \|C_+\|_{\alpha p}^\alpha = \|A - B\|_{\alpha p}^\alpha. \quad \square$$

There is a complement to the Powers-Størmer inequality. The following is a refinement of a lemma in [53].

Theorem 6.52. *For $X, Y \in M_{m,n}(\mathbb{C})$,*

$$(6.5.8) \qquad \|X^*X - Y^*Y\|_1 \leq \|X - Y\|_2 \|X + Y\|_2.$$

Proof. By Theorem 6.20 and Remark 6.21, for any $A \in M_m(\mathbb{C})$, there exists $U \in M_m(\mathbb{C})$ with $\|U\| = 1$ such that $\|A\|_1 = \text{Tr}[UA]$. When A is self-adjoint, one can say more. Let $A = \sum_{\lambda \in \sigma(A)} \lambda P_\lambda$ be the spectral decomposition of A. Define $U := \sum_{\lambda \in \sigma(A),\, \lambda > 0} P_\lambda - \sum_{\lambda \in \sigma(A),\, \lambda < 0} P_\lambda$. Then U is a self-adjoint partial isometry and $\text{Tr}[UA] = \|A\|_1$. In fact, $A = U|A|$ is the polar decomposition of A.

Therefore, there is a self-adjoint partial isometry U such that

$$\|X^*X - Y^*Y\|_1 = \text{Tr}[U(X^*X - Y^*Y)].$$

Using the fact that $\Re \operatorname{Tr}[A^*B] = \frac{1}{2}(\operatorname{Tr}[A^*B] + \operatorname{Tr}[AB^*])$ and cyclicity of the trace,

$$
\begin{aligned}
\operatorname{Tr}[U(X^*X - Y^*Y)] &= \operatorname{Tr}[U(X^* - Y^*)X + UY^*(X - Y)] \\
&= \operatorname{Tr}[(X^* - Y^*)XU + UY^*(X - Y)] \\
&= \tfrac{1}{2}\operatorname{Tr}[(X^* - Y^*)XU + UY^*(X - Y)] \\
&\quad + \tfrac{1}{2}\operatorname{Tr}[(X - Y)UX^* + YU(X^* - Y^*)] \\
&= \tfrac{1}{2}\operatorname{Tr}[(X^* - Y^*)(X + Y)U] \\
&\quad + \tfrac{1}{2}\operatorname{Tr}[(X - Y)U(X^* + Y^*)] \\
&\leq \|X - Y\|_2 \|X + Y\|_2. \qquad \square
\end{aligned}
$$

Theorem 6.53. *Let $\mathcal{H}$ be a finite-dimensional Hilbert space. For all $\rho, \sigma \in \mathfrak{S}_{\mathcal{H}}$, there exist purifications ψ and ϕ of ρ and σ, respectively, in $\mathcal{H} \otimes \mathcal{H}$ such that*

$$
\| \, |\psi\rangle\langle\psi| - |\phi\rangle\langle\phi| \, \|_1 \leq \sqrt{2\|\rho - \sigma\|_1}.
$$

Proof. Let $\mathbf{u}$ be a unit vector in $\mathcal{H} \otimes \mathcal{H}$. For any two purifications ψ and ϕ of ρ and σ, respectively, define $X := |\mathbf{u}\rangle\langle\psi|$ and $Y := |\mathbf{u}\rangle\langle\phi|$. Then $X^*X = |\psi\rangle\langle\psi|$ and $Y^*Y = |\phi\rangle\langle\phi|$, while $\|X \pm Y\|_2 = \| \, |\mathbf{u}\rangle\langle\psi \pm \phi| \, \|_2 = \|\psi \pm \phi\|$. Therefore, by Theorem 6.52,

$$
\| \, |\psi\rangle\langle\psi| - |\phi\rangle\langle\phi| \, \|_1 \leq 2^{1/2}\|\psi - \phi\|.
$$

Choose ψ and ϕ as in (6.5.2). From (6.5.4), $\|\psi - \phi\| \leq \sqrt{\|\rho - \sigma\|_1}$. $\qquad \square$

Example 6.54. Let $\rho, \sigma \in \mathfrak{S}_{\mathcal{H}}$. Let ψ and ϕ be their respective purifications as defined in (6.5.2) so that $\langle\psi, \phi\rangle = |\langle\psi, \phi\rangle|$. Let $X = \rho^{1/2}$ and $Y = \sigma^{1/2}$. Then $\|X^*X - Y^*Y\|_1 = \|\rho - \sigma\|_1$ and $\|X \pm Y\|_2^2 = 2(1 \pm \operatorname{Tr}[\rho^{1/2}\sigma^{1/2}])$. By Theorem 6.52 and (6.5.3)

$$
\|\rho - \sigma\|_1 \leq 2\sqrt{1 - (\operatorname{Tr}[\rho^{1/2}\sigma^{1/2}])^2} = 2\sqrt{1 - |\langle\psi, \phi\rangle|^2}.
$$

Combining this with (6.5.4) yields

$$
(6.5.9) \qquad 2(1 - |\langle\psi, \phi\rangle|) \leq \|\rho - \sigma\|_1 \leq 2\sqrt{1 - |\langle\psi, \phi\rangle|^2}.
$$

Using (6.5.3), we may rewrite (6.5.9) as

$$
(6.5.10) \qquad (1 - \operatorname{Tr}[\rho^{1/2}\sigma^{1/2}]) \leq \frac{1}{2}\|\rho - \sigma\|_1 \leq \sqrt{1 - \operatorname{Tr}[\rho^{1/2}\sigma^{1/2}]^2}.
$$

In deriving (6.5.9), we have made a simple but special choice of the purifications of ρ and σ. As we shall soon see, one can do slightly better.

6.6. Fidelity and Uhlmann's theorem

Let ψ and ϕ be two unit vectors in a Hilbert space $\mathcal{H}$. Consider the corresponding rank one projections $|\psi\rangle\langle\psi|$ and $|\phi\rangle\langle\phi|$. These may be regarded both as quantum pure states and as quantum observables. Let us regard $|\psi\rangle\langle\psi|$ as a state and $A := |\phi\rangle\langle\phi|$ as an observable. If a measurement of A is made with the system initially in the pure state $|\psi\rangle\langle\psi|$, then with probability

$$(6.6.1) \qquad p = \mathrm{Tr}[|\psi\rangle\langle\psi||\phi\rangle\langle\phi|] = |\langle\psi,\phi\rangle|^2$$

after the measurement, the system is in the state $|\phi\rangle\langle\phi|$. The quantity p in (6.6.1) is called the **transition probability** between the two quantum states, and is symmetric in the two states; it does not matter which one is regarded as the observable. It is of interest to extend the notion of transition probability from pure states to general states; such a generalization was made by Bures [42] to define a metric on states on a von Neumann algebra. Uhlmann further developed these ideas and proved the following theorem [214].

Theorem 6.55. *Let $\rho, \sigma \in \mathfrak{S}_{\mathcal{H}}$, and let $\mathcal{H}$ be a finite-dimensional Hilbert space. Let ψ and ϕ be any two such purifications of ρ and σ, respectively in $\mathcal{H} \otimes \mathcal{H}$. Then the transition probability between ψ and ϕ is bounded above by*

$$(6.6.2) \qquad |\langle\psi,\phi\rangle|^2 \le \|\rho^{1/2}\sigma^{1/2}\|_1^2 = (\mathrm{Tr}[(\sigma^{1/2}\rho\sigma^{1/2})^{1/2}])^2.$$

Moreover, there exists a choice of the purifications ψ and ϕ for which equality holds in (6.6.2).

Definition 6.56 (Uhlmann fidelity). For $\rho, \sigma \in \mathfrak{S}_{\mathcal{H}}$, the **Uhlmann fidelity** of ρ and σ, denoted $F(\rho,\sigma)$, is the least upper bound of the possible transition probabilities between any two purifications of ρ and σ in $\mathcal{H} \otimes \mathcal{H}$. That is,

$$(6.6.3) \qquad F(\rho,\sigma) := \sup\{|\langle\psi,\phi\rangle|^2 \ : \ \psi, \phi \text{ purify } \rho, \sigma\}.$$

By Theorem 6.55, $F(\rho,\sigma)$ the supremum in (6.6.3) is attained, so that it is a maximum, and $F(\rho,\sigma) = \|\rho^{1/2}\sigma^{1/2}\|_1^2$. The Uhlmann fidelity of ρ and σ generalizes the notion of transition probability to mixed states. Since for all A, B, $\|AB\|_1 = \|BA\|_1$,

$$F(\rho,\sigma) = \|\rho^{1/2}\sigma^{1/2}\|_1^2 = \|\sigma^{1/2}\rho^{1/2}\|_1^2 = F(\sigma,\rho),$$

and hence $F(\rho,\sigma)$ is symmetric in ρ and σ.

Before proving Uhlmann's theorem, we discuss the relation between $F(\rho,\sigma)$ and $\|\rho - \sigma\|_1$.

Theorem 6.57. *For all $\rho, \sigma \in \mathfrak{S}_{\mathcal{H}}$. Then*

$$(6.6.4) \qquad 1 - \sqrt{F(\rho,\sigma)} \le \frac{1}{2}\|\rho - \sigma\|_1 \le \sqrt{1 - F(\rho,\sigma)}.$$

Proof. Let ψ and ϕ be purifications in $\mathcal{H} \otimes \mathcal{H}$ of ρ and σ, respectively, such that $\langle \psi, \phi \rangle = \sqrt{F(\rho, \sigma)}$. Such purifications exist by Uhlmann's theorem. Let $X, Y \in \mathcal{B}(\mathcal{H})$ be defined in terms of ψ and ϕ as in Theorem 2.22 so that $|\langle \psi, \phi \rangle| = \mathrm{Tr}[XY^*]$, $\rho = XX^*$ and $\sigma = YY^*$. Then applying (6.5.8) with X^*, Y^* in place of X, Y yields

$$\|\rho - \sigma\|_1 \le 2(1 - |\langle \psi, \phi \rangle|)^{1/2}(1 + |\langle \psi, \phi \rangle|)^{1/2}$$
$$= 2\sqrt{1 - F(\rho, \sigma)}.$$

This proves the inequality on the right in (6.6.4). The inequality on the left is (6.5.4), a direct consequence of the Powers-Størmer inequality. $\qquad\square$

Inequalities (6.6.4) are the **Fuchs–van de Graaf inequalities** [88]. Since $\mathrm{Tr}[\rho^{1/2}\sigma^{1/2}] \le \|\rho^{1/2}\sigma^{1/2}\|_1$, the inequality on the right in (6.6.4) is stronger than the inequality on the right in (6.5.10), but the opposite is true of the inequalities on the left.

It turns out the Uhlmann's theorem is one half of a theorem showing how the fidelity $F(\rho, \sigma)$ arises as the answer to both a maximization problem and a minimization problem, and it is efficient to prove the two theorems together. We shall do this after introducing the minimization problem. (See Exercises 22 and 23 for a direct approach.)

The **Hellinger distance** $d_H(p, q)$ between two probability measures p and q on $\{1, \dots, n\}$ is defined to be

$$(6.6.5) \qquad d_H(p, q) := \left(\sum_{j=1}^{n} \left(\sqrt{p(j)} - \sqrt{q(j)} \right)^2 \right)^{1/2}.$$

The right side of (6.6.5) is the Euclidean distance between the vectors

$$\left(\sqrt{p(1)}, \dots, \sqrt{p(n)} \right) \quad \text{and} \quad \left(\sqrt{q(1)}, \dots, \sqrt{q(n)} \right),$$

and hence the Hellinger distance is indeed a metric.

By the definition (6.6.5), $d_H^2(p, q) := 2\left(1 - \sum_{j=1}^{n} \sqrt{p(j)q(j)}\right)$, Given two classical probability p, q measures on $\{1, \dots, n\}$, their **Bhattacharyya coefficient** $B(p, q)$ is defined by

$$B(p, q) := \sum_{j=1}^{n} \sqrt{p(j)q(j)}.$$

Recall the discussion of measurement procedures and POVM from Section 4.7. We would like to choose a POVM, that is, a measurement procedure, to make $d_H(\mathcal{E}(\rho), \mathcal{E}(\sigma))$ as large as possible, which hopefully makes it as easy as possible to distinguish between ρ and σ. Since $d_H^2(p, q) = 2(1 - B(p, q))$, maximizing the Hellinger distance is the same as minimizing the Bhattacharyya coefficient. This motivates the following question. Given $\rho, \sigma \in \mathfrak{S}_{\mathcal{H}}$, is there a POVM in $\mathcal{H}$ that minimizes $B(\mathcal{E}(\rho), \mathcal{E}(\sigma))$ over all POVM's $\mathcal{E}$? The answer is yes, and the minimum value turns out to be $\sqrt{F(\rho, \sigma)}$. Moreover, the optimal POVM turns out to be a PVM corresponding to the measurement of a particular self-adjoint operator H constructed out of ρ and σ. This information tells us which measurement $\mathcal{E}$ we should make to maximize the Hellinger distance between $\mathcal{E}(\rho)$ and $\mathcal{E}(\sigma)$. The following theorem is due to Fuchs and Caves [87].

Theorem 6.58. *Let $\mathcal{H}$ be a finite-dimensional Hilbert space, and let $\rho, \sigma \in \mathfrak{S}_{\mathcal{H}}$ with $\sigma > 0$. Let H be the self-adjoint operator on $\mathcal{H}$ defined by*

$$(6.6.6) \qquad H := \sigma^{-1/2}(\sigma^{1/2}\rho\sigma^{1/2})^{1/2}\sigma^{-1/2},$$

and let $\mathcal{E}_0$ denote the PVM given by the spectral decomposition of H. Then for any POVM $\mathcal{E}$ in $\mathcal{H}$,

$$B(\mathcal{E}_0(\rho), \mathcal{E}_0(\sigma)) \leq B(\mathcal{E}(\rho), \mathcal{E}(\sigma))$$

and

$$(6.6.7) \qquad B(\mathcal{E}_0(\rho), \mathcal{E}_0(\sigma)) = \|\rho^{1/2}\sigma^{1/2}\|_1.$$

By (6.6.6) and (6.6.7), since $F(\rho, \sigma) = \|\rho^{1/2}\sigma^{1/2}\|_1^2$,

$$\sqrt{F(\rho, \sigma)} = \inf\{B(\mathcal{E}(\rho), \mathcal{E}(\sigma)) \; : \; \mathcal{E} \text{ is a POVM}\}.$$

Combining this with (6.6.3),

$$\sup\{|\langle\psi, \phi\rangle| \; : \; \psi, \phi \text{ purify } \rho, \sigma\}$$
$$= \|\rho^{1/2}\sigma^{1/2}\|_1 = \inf\{B(\mathcal{E}(\rho), \mathcal{E}(\sigma)) \; : \; \mathcal{E} \text{ is a POVM}\}.$$

To prove this, it suffices to show that

$$(6.6.8) \quad \sup\{|\langle\psi, \phi\rangle| \; : \; \psi, \phi \text{ purify } \rho, \sigma\}$$
$$\leq \inf\{B(\mathcal{E}(\rho), \mathcal{E}(\sigma)) \; : \; \mathcal{E} \text{ is a POVM}\},$$

and then to show that there is a choice of the purifications ψ and ϕ and of the POVM $\mathcal{E}$ such that

$$(6.6.9) \qquad |\langle\psi, \phi\rangle| = B(\mathcal{E}(\rho), \mathcal{E}(\sigma)) = \|\rho^{1/2}\sigma^{1/2}\|_1.$$

We now proceed to show (6.6.8) and (6.6.9). In doing so, we find explicit choices for ψ and ϕ and $\mathcal{E}$ such that (6.6.9) is satisfied, and this yields a simultaneous proof of Theorem 6.55 and Theorem 6.58.

Proof of Theorems 6.55 and 6.58. We first prove (6.6.8). Let $\mathcal{E} = \{E_1, \dots, E_n\}$ be any POVM for $\mathcal{H}$, and let ψ and ϕ be any purifications of ρ and σ, respectively, in $c H \otimes \mathcal{K}$.

By Theorem 2.33, there are partial isometries $U, V \in \mathcal{B}(\mathcal{H})$ whose final spaces are $\operatorname{ran}(\rho)$ and $\operatorname{ran}(\sigma)$, respectively, so that $\psi = \mathcal{U}(\rho^{1/2}U)$ and $\phi = \mathcal{U}(\sigma^{1/2}V)$, where $\mathcal{U}$ is the unitary map defined in Theorem 2.22 (determined by the choice of an orthonormal basis for $\mathcal{H}$). Moreover, by Theorem 2.22,

$$(6.6.10) \qquad \operatorname{Tr}_{\mathcal{K}}[|\phi\rangle\langle\psi|] = \sigma^{1/2}VU^*\rho^{1/2}.$$

Define $\widehat{\mathcal{E}} := \{E_1 \otimes \mathbb{1}, \dots, E_n \otimes \mathbb{1}\}$ which is a POVM in $\mathcal{H} \otimes \mathcal{H}$. Replace V by $e^{i\theta}V$ as needed to ensure that $\langle\psi, \phi\rangle \geq 0$, noting that this does not affect the state $|\phi\rangle\langle\phi|$. Then by (6.6.10),

$$\langle\psi, \phi\rangle = \sum_{j=1}^{n}\langle\psi, (E_j \otimes \mathbb{1})\phi\rangle = \sum_{j=1}^{n}\operatorname{Tr}[(E_j \otimes \mathbb{1})|\phi\rangle\langle\psi|]$$

$$= \sum_{j=1}^{n}\operatorname{Tr}[E_j \operatorname{Tr}_{\mathcal{K}}[|\phi\rangle\langle\psi|]] = \sum_{j=1}^{n}\operatorname{Tr}[E_j\sigma^{1/2}VU^*\rho^{1/2}]$$

$$(6.6.11) \qquad = \sum_{j=1}^{n}\langle E_j^{1/2}\rho^{1/2}U, E_j^{1/2}\sigma^{1/2}V\rangle.$$

By the Cauchy-Schwarz inequality in $\mathcal{B}(\mathcal{K}, \mathcal{H})$, for each j,

$$(6.6.12) \qquad |\langle E_j^{1/2}\rho^{1/2}U, E_j^{1/2}\sigma^{1/2}V\rangle| \leq \left(\operatorname{Tr}[\rho E_j]\right)^{1/2}\left(\operatorname{Tr}[\sigma E_j]\right)^{1/2}.$$

Together, (6.6.11), the elementary inequality $|\sum_{j=1}^{n} z_j| \leq \sum_{j=1}^{n}|z_j|$ and (6.6.12) prove that

$$(6.6.13) \qquad |\langle\psi, \phi\rangle| \leq \sum_{j=1}^{n}\left(\operatorname{Tr}[\rho E_j]\right)^{1/2}\left(\operatorname{Tr}[\sigma E_j]\right)^{1/2} = B(\mathcal{E}(\rho), \mathcal{E}(\sigma)).$$

Since ψ and ϕ are arbitrary purifications of ρ and σ, and since $\mathcal{E}$ is any POVM, this proves (6.6.8).

We now proceed to construct purifications ψ and ϕ and a POVM $\mathcal{E}$ for which (6.6.9) is satisfied.

To do this, examine the cases of equality in the two inequalities that were used to arrive at (6.6.13). Since we have chosen the phase on ϕ so that $\langle\psi, \phi\rangle \geq 0$, there is equality in (6.6.13) if and only if for each j,

$$(6.6.14) \qquad \langle E_j^{1/2}\rho^{1/2}U, E_j^{1/2}\sigma^{1/2}V\rangle \geq 0,$$

and for some number c_j,

$$(6.6.15) \qquad E_j^{1/2}\rho^{1/2}U = c_j E_j^{1/2}\sigma^{1/2}V,$$

which is the condition for equality in the Cauchy-Schwarz inequality. When (6.6.15) is satisfied, $\langle E_j^{1/2}\rho^{1/2}U, E_j^{1/2}\sigma^{1/2}V\rangle = \overline{c_j}\,\mathrm{Tr}[E_j\sigma]$, so that (6.6.14) is satisfied if and only if $c_j > 0$. Then, since σ is invertible, (6.6.15) is satisfied if and only if

$$(6.6.16) \qquad E_j^{1/2}\rho^{1/2}UV^*\sigma^{1/2} = E_j^{1/2}c_j\sigma.$$

Multiply on the left by $E_j^{1/2}$ and sum on j. This yields $\rho^{1/2}UV^*\sigma^{1/2} = H\sigma$, where H is the the positive operator $H := \sum_{j=1}^n c_j E_j$. Since σ is invertible, $\rho^{1/2}UV^*\sigma^{1/2} = H\sigma$ is equivalent to

$$(6.6.17) \qquad \sigma^{1/2}\rho^{1/2}UV^* = \sigma^{1/2}H\sigma^{1/2}.$$

The right side is positive semidefinite. For general choices of ψ and ϕ, and hence of U and V, the left side will not be even self-adjoint, and (6.6.17) will not be satisfied. Define $W := UV^*$, which is a partial isometry since V is unitary. The final space of U is $\mathrm{supp}(\rho)$ and again since V is unitary, the final space of W is $\mathrm{supp}(\rho)$.

Taking adjoints in (6.6.17) yields $W^*\rho^{1/2}\sigma^{1/2} = \sigma^{1/2}H\sigma^{1/2}$, and therefore, since WW^* is the orthogonal projection onto $\mathrm{supp}(\rho)$,

$$\rho^{1/2}\sigma^{1/2} = W\left(\sigma^{1/2}H\sigma^{1/2}\right).$$

By the uniqueness of the polar decomposition, when (6.6.17) is satisfied, $\sigma^{1/2}H\sigma^{1/2} = |\rho^{1/2}\sigma^{1/2}|$, and hence H is given by (6.6.6).

Now it is clear how to choose the purification $|\phi\rangle\langle\phi|$ so that (6.6.17) is satisfied. Make *any* choice $|\psi\rangle\langle\psi|$ that purifies ρ. Let W be the partial isometry in the polar decomposition of $\rho^{1/2}\sigma^{1/2}$. Now choose V, and hence ϕ, so that $W = UV^*$, which is possible since U and W have the same final space and V is unitary. This fixes our choice of purifications.

Now we specify the POVM $\mathcal{E}$. Let $H = \sum_{\lambda\in\sigma(H)} \lambda P_\lambda$ be the spectral decomposition of H. Let the spectrum of H be $\{\lambda_1,\ldots,\lambda_m\}$ and define a PVM $\mathcal{E}_0$ by $E_j := P_{\lambda_j}$, $j = 1,\ldots,m$. In fact, this is a PVM, and since each E_j is a projection, $E_j^{1/2} = P_{\lambda_j}$. Writing (6.6.17) in the equivalent form $\rho^{1/2}UV^*\sigma^{1/2} = H\sigma$, we have

$$E_j^{1/2}\rho^{1/2}UV^*\sigma^{1/2} = E_j^{1/2}H\sigma = P_{\lambda_j}H\sigma = \lambda_j E_j^{1/2}\sigma.$$

Therefore, (6.6.16) is satisfied for each j with $E_j = P_{\lambda_j}$ and $c_j = \lambda_j \geq 0$. Hence both conditions for equality in (6.6.13) are satisfied for our choices.

Returning to (6.6.11) for this choice of purifications and $\mathcal{E}$ yields

$$\langle\psi,\phi\rangle = \sum_{j=1}^m \mathrm{Tr}[E_j\sigma^{1/2}VU^*\rho^{1/2}] = \mathrm{Tr}[\rho^{1/2}\sigma^{1/2}VU^*] = \mathrm{Tr}[|\rho^{1/2}\sigma^{1/2}|],$$

where the final equality comes from $\rho^{1/2}\sigma^{1/2} = UV^*|\rho^{1/2}\sigma^{1/2}|$, as shown above. This shows that the common value of the infimum and supremum in (6.6.8) is $\|\rho^{1/2}\sigma^{1/2}\|_1 = \sqrt{F(\rho,\sigma)}$. $\qquad\square$

6.7. Klein's inequality and the Peierls-Bogoliubov inequality

This section introduces two important inequalities. We begin with **Klein's inequality** [130] which is a direct consequence of the *above the tangent plane* inequality (see Remark 6.4) and the spectral theorem.

Theorem 6.59 (Klein's inequality). *Let $A, B \in M_n^{\mathrm{s.a.}}(\mathbb{C})$. Let f be convex on an open interval containing $\sigma(A) \cup \sigma(B)$ and finite on an open set containing $\sigma(A)$. Let g be any function on $\mathbb{R}$ such that for each $\lambda \in \sigma(A)$, $g(\lambda) \in \partial f(\lambda)$. Then*

$$(6.7.1) \qquad \mathrm{Tr}[f(B)] \geq \mathrm{Tr}[f(A)] + \mathrm{Tr}[g(A)(B - A)].$$

If f is strictly convex on an interval containing $\sigma(A) \cup \sigma(B)$, then there is equality in (6.7.1) if and only if $A = B$.

Proof. Let $A = \sum_{\lambda \in \sigma(A)} \lambda P_\lambda$ and $B = \sum_{\mu \in \sigma(B)} \mu Q_\mu$ be the spectral decompositions of A and B. Since $\sum_{\lambda \in \sigma(A)} P_\lambda = \sum_{\mu \in \sigma(B)} Q_\mu = \mathbb{1}$, for any function g on $\mathbb{R}$,

$$
\begin{aligned}
(6.7.2) \quad \mathrm{Tr}&[f(B) - f(A) - g(A)(B - A)] \\
&= \sum_{\lambda \in \sigma(A)} \sum_{\mu \in \sigma(B)} [f(\mu) - f(\lambda) - g(\lambda)(\mu - \lambda)] \, \mathrm{Tr}[P_\lambda Q_\mu].
\end{aligned}
$$

Since $g(\lambda) \in \partial f(\lambda)$, $[f(\mu) - f(\lambda) - g(\lambda)(\mu - \lambda)] \geq 0$ for all μ and λ by the definition of the subgradient. Also, since $P_\lambda, Q_\mu \geq 0$, $\mathrm{Tr}[P_\lambda Q_\mu] \geq 0$. Hence the right side of (6.7.2) is nonnegative. This proves (6.7.1).

Now suppose that there is equality in (6.7.1) and that f is strictly convex on an interval containing $\sigma(A) \cup \sigma(B)$. Then for $\mu \neq \lambda$, and all $s \in \partial f(\lambda)$, $[f(\mu) - f(\lambda) - s(\mu - \lambda)] > 0$, and hence for each $\lambda \in \sigma(A)$ and $\mu \in \sigma(B)$ such that $\lambda \neq \mu$, $\mathrm{Tr}[P_\lambda Q_\mu] = 0$. Since $\sum_{\mu \in \sigma(B)} \mathrm{Tr}[P_\lambda Q_\mu] = \mathrm{Tr}[P_\lambda] > 0$, $\lambda \in \sigma(B)$ and $P_\lambda \leq Q_\lambda$. The same reasoning shows that for each $\mu \in \sigma(B)$, $\mu \in \sigma(A)$ and $Q_\mu \leq P_\mu$. Thus, $A = B$. $\qquad\square$

Klein's inequality leads to the **Peierls-Bogoliubov inequality** [40]:

Theorem 6.60 (Peierls-Bogoliubov inequality). *For all $H, K \in M_n^{\mathrm{s.a.}}(\mathbb{C})$,*

$$(6.7.3) \qquad \frac{\mathrm{Tr}[e^H K]}{\mathrm{Tr}[e^H]} \leq \log\left(\frac{\mathrm{Tr}[e^{H+K}]}{\mathrm{Tr}[e^H]}\right),$$

and there is equality if and only if K is a multiple of the identity.

Remark 6.61. The Golden-Thompson inequality, Theorem 3.27, says that for all $H, K \in M_n^{\text{s.a.}}(\mathbb{C})$, $\text{Tr}[e^{H+K}] \leq \text{Tr}[e^H e^K]$. Writing $X = e^K$ and defining the density matrix $\rho := \frac{1}{\text{Tr}[e^H]} e^H$, we obtain a weakened form of (6.7.3):

$$(6.7.4) \qquad\qquad \text{Tr}[\log X \rho] \leq \log \text{Tr}[X \rho],$$

valid for all $H \in M_n^{\text{s.a.}}(\mathbb{C})$ and all $X > 0$ in $M_n(\mathbb{C})$.

This weakened version of the Peierls-Bogoliubov inequality is thus a version of Jensen's inequality for the logarithm function and density matrices, and it is easy to prove using Lemma 3.2; this is left as an exercise. The stronger inequality (6.7.3) will have a number of applications in what follows where (6.7.4) would not be of use. The Peierls-Bogoliubov inequality may be regarded as a strong tracial form of Jensen's inequality for the logarithm function.

Proof of Theorem 6.60. Taking $f(t) = e^t$, (6.7.1) becomes $\text{Tr}[e^B] \geq \text{Tr}[e^A] + \text{Tr}[e^A(B-A)]$. Since the exponential function is strictly convex, there is equality if and only if $A = B$. For $c \in \mathbb{R}$ and $H, K \in M_n^{\text{s.a.}}(\mathbb{C})$, choose $A = c + H$ and $B = H + K$ to obtain

$$\text{Tr}[e^{H+K}] \geq e^c \, \text{Tr}[e^H] + e^c \, \text{Tr}[e^H(K - c)].$$

In this case, $A = B$ is the same as $K = c\mathbb{1}$, so there is equality if and only if K is a multiple of the identity. Finally, choosing $c = \text{Tr}[e^H K]/\text{Tr}[e^H]$, we obtain $\text{Tr}[e^{H+K}] \geq e^{\text{Tr}[e^H K]/\text{Tr}[e^H]} \, \text{Tr}[e^H]$ which yields (6.7.3). $\qquad\square$

6.8. Hölder's inequality for Schatten norms in general form

We have seen that Hölder's inequality for traces can be written in the form $\|AB\|_1 \leq \|A\|_p \|B\|_q$ for all $A, B \in M_n(\mathbb{C})$ and all $1 \leq p, q \leq \infty$ such that $1/p + 1/q = 1$. The following is an important generalization.

Theorem 6.62. *For all $A, B \in M_n(\mathbb{C})$, all $1 \leq p, q, r$ such that $1/p + 1/q = 1/r$,*

$$(6.8.1) \qquad\qquad \|AB\|_r \leq \|A\|_p \|B\|_q.$$

Under the further condition that $p, q < \infty$, there is equality in (6.8.1), if and only if for some $c \geq 0$, $|A|^p = c|B^*|^q$.

Remark 6.63. Later, in Theorem 11.28, it will be shown that the condition $1 \leq p, q, r$ can be relaxed to $0 \leq p, q, r$. In the more general setting, (6.8.1) is no longer a statement about norms, but nonetheless, it has important applications discovered by Zhang [237] to be discussed later. For present purposes, Theorem 6.62 suffices, and the proof illustrates the methods introduced so far.

Lemma 6.64. *For all $A, B \in M_n(\mathbb{C})$, all $1 \leq p, q, r$ such that that $1/p + 1/q = 1/r$, and such that $p, q \geq 2$, the inequality (6.8.1) is satisfied.*

Proof. Let $AB = \sum_{j=1}^{k} \sigma_j |\mathbf{u}_j\rangle\langle\mathbf{v}_j|$ be a singular value decomposition as in Theorem 2.25 with $k := \operatorname{rank}(AB)$.

For $1 \le j \le k$, $\sigma_j = \langle \mathbf{u}_j, AB\mathbf{v}_j\rangle$, and by the Cauchy-Schwarz inequality,

$$\|AB\|_r^r = \sum_{j=1}^{k} \langle A^*\mathbf{u}_j, B\mathbf{v}_j\rangle^r \le \sum_{j=1}^{k} \langle \mathbf{u}_j, AA^*\mathbf{u}_j\rangle^{r/2}\langle\mathbf{v}_j, B^*B\mathbf{v}_j\rangle^{r/2}.$$

Define $s := p/r$ and $t = q/r$, and note that $1/s + 1/t = 1$. Then by Hölder's inequality for sequences (or diagonal matrices),

$$\|AB\|_r^r \le \left(\sum_{j=1}^{k}\langle\mathbf{u}_j, |A^*|^2\mathbf{u}_j\rangle^{rs/2}\right)^{1/s}\left(\sum_{j=1}^{k}\langle\mathbf{v}_j, |B|^2\mathbf{v}_j\rangle^{rt/2}\right)^{1/t}.$$

Since $p = rs \ge 2$ and $q = rt \ge 2$, for all j, Lemma 3.2 yields

$$\langle\mathbf{u}_j, |A^*|^2\mathbf{u}_j\rangle^{rs/2} \le \langle\mathbf{u}_j, |A^*|^p\mathbf{u}_j\rangle \quad \text{and} \quad \langle\mathbf{v}_j, |B|^2\mathbf{v}_j\rangle^{rt/2} \le \langle\mathbf{v}_j, |B|^q\mathbf{v}_j\rangle,$$

and thus $\|AB\|_r^r \le \|A^*\|_p^r\|B\|_q^r = \|A\|_p^r\|B\|_q^r$. This proves (6.8.1). $\qquad\square$

Lemma 6.65. *For all $A, B, C \in M_n(\mathbb{C})$, all $p, q, t \ge 1$ such that*

$$\text{(6.8.2)} \qquad \frac{1}{p} + \frac{1}{q} + \frac{1}{t} = 1,$$

$$\text{(6.8.3)} \qquad |\operatorname{Tr}[ABC]| \le \|A\|_p\|B\|_q\|C\|_t.$$

Proof. The condition (6.8.2) is symmetric in p, q, t. Suppose these indices are ordered so that $p \ge q \ge t$. If $t > 2$, then also $p, q > 2$. If $t \le 2$, then $1 - 1/t \le 1/2$, and hence $1/p + 1/q \le 1/2$. In this case, $p, q \ge 2$. Hence in any case, the two largest indices are at least 2.

By cyclicity of the trace, $\operatorname{Tr}[ABC] = \operatorname{Tr}[BCA] = \operatorname{Tr}[CAB]$, and hence whichever pair of matrices is to be computed in the norms with index at least 2 may be put in the first two places. Therefore, without loss of generality, relabeling as needed, we may assume that $p, q \ge 2$ in (6.8.3).

Then with r defined by $1/r = 1 - 1/t$, Hölder's inequality (6.3.4) yields

$$|\operatorname{Tr}[ABC]| = |\langle C^*, AB\rangle| \le \|AB\|_r\|C\|_t,$$

and then by what has been proved for $p, q \ge 2$, $\|AB\|_r \le \|A\|_p\|B\|_q$. This proves (6.8.3) whenever (6.8.2) is satisfied. $\qquad\square$

Proof of Theorem 6.62. We show that the inequality (6.8.1) in general follows from Lemmas 6.64 and 6.65.

Let t be defined by $1/t := 1 - 1/r$. Then there exists $C \in M_n(\mathbb{C})$ such that $\|C\|_t = 1$ and $\|AB\|_r = \operatorname{Tr}[ABC]$. By Lemma 6.65, $\operatorname{Tr}[ABC] \le \|A\|_p\|B\|_q\|C\|_t = \|A\|_p\|B\|_q$, and this proves (6.8.1).

It is left as an exercise to prove the statement about the cases of equality. $\qquad\square$

Theorem 6.66 (Hölder's inequality for Schatten norms in general form). *For all $1 \leq r \leq \infty$, or all $m \geq 2$, all $A_1, \ldots, A_m \in M_n(\mathbb{C})$, and all $0 \leq p_1, \ldots, p_m \leq \infty$ such that*

$$(6.8.4) \qquad \sum_{j=1}^{m} \frac{1}{p_j} = \frac{1}{r},$$

$$\|A_1 \cdots A_m\|_r \leq \prod_{j=1}^{m} \|A_j\|_{p_j}.$$

Proof. By Theorem 6.62, $\|A_1 A_2\|_r \leq \|A_1\|_{p_1} \|A_2\|_{p_2}$ whenever $1/p_1 + 1/p_2 = 1/r$, which proves the $m = 2$ case. The general case follows by induction: By (6.8.4) for $m = 2$, $\|A_1 \cdots A_m\|_r \leq \|A_1 \cdots A_{m-1}\|_s \|A_m\|_{p_m}$ where $1/s + 1/p_m = 1/r'$. By (6.8.4), this is the same as $\sum_{j=1}^{m-1} \frac{1}{p_j} = \frac{1}{s}$. By the inductive hypothesis, $\|A_1 \cdots A_{m-1}\|_s \leq \prod_{j=1}^{m-1} \|A_j\|_{p_j}$. $\qquad\square$

Corollary 6.67. *Let $A, B \in M_n^+(\mathbb{C})$, and let $m = 2^\ell$, $\ell \in \mathbb{N}$. Then*

$$(6.8.5) \qquad \mathrm{Tr}[A^m B^m] \geq \mathrm{Tr}[(AB)^m].$$

In fact, we have the stronger inequality

$$(6.8.6) \qquad \mathrm{Tr}[A^m B^m] - \mathrm{Tr}[(AB)^m] \geq \tfrac{1}{2} \|[A^{m/2}, B^{m/2}]\|_2^2,$$

where for $X, Y \in M_n(\mathbb{C})$, $[X, Y] = XY - YX$ is the commutant of X and Y. In particular, there is equality in (6.8.5) if and only if A and B commute.

Remark 6.68. Define $X := A^m$ and $Y := B^m$. Then (6.8.5) can be written as $\mathrm{Tr}[(X^{1/m} Y^{1/m})^m] \leq \mathrm{Tr}[XY]$. This is a very special case of the *Araki-Lieb-Thirring inequality*, to be proved in Chapter 11, which says that for $0 < s < t$,

$$(6.8.7) \qquad \mathrm{Tr}[(X^{1/t} Y^{1/t})^t] \leq \mathrm{Tr}[(X^{1/s} Y^{1/s})^s].$$

The inequality (6.8.7) for $s = 1$ is the *Lieb-Thirring inequality*, proved in Chapter 10. With the methods so far at our disposal, we can only give a short proof of a very special case. However, this special case has important consequences.

Proof. By Theorem 6.66 with $r = 1$ and each $p_j = m$,

$$\mathrm{Tr}[(AB)^m] \leq \|(AB)^m\|_1 \leq \|AB\|_m^m.$$

Since $|AB| = (BA^2 B)^{1/2}$, $\|AB\|_m^m = \mathrm{Tr}[(BA^2 B)^{m/2}]$. This proves that

$$(6.8.8) \qquad \mathrm{Tr}[(AB)^m] \leq \mathrm{Tr}[(BA^2 B)^{m/2}] = \mathrm{Tr}[(A^2 B^2)^{m/2}],$$

where the final equality used cyclicity of the trace and the fact that $m/2 \in \mathbb{N}$.

For $0 \le j \le \ell$, define $f(j) = \mathrm{Tr}\big[(A^{2^j} B^{2^j})^{2^{\ell-j}}\big]$. Then $f(0) = \mathrm{Tr}[(AB)^m]$ and $f(\ell) = \mathrm{Tr}[A^m B^m]$. By the telescoping sum formula,

$$(6.8.9) \qquad \mathrm{Tr}[A^m B^m] - \mathrm{Tr}[(AB)^m] = f(\ell) - f(0) = \sum_{j=1}^{\ell} (f(j) - f(j-1)).$$

Defining $C := A^{2^{j-1}}$ and $D := B^{2^{j-1}}$ and $n := 2^{\ell-j+1}$,

$$f(j) - f(j-1) = \mathrm{Tr}[(C^2 D^2)^{n/2}] - \mathrm{Tr}[(CD)^n].$$

By (6.8.8), $f(j) - f(j-1) \ge 0$ for all $0 \le j \le \ell$. Then (6.8.5) follows from this and (6.8.9). Better yet, from (6.8.9),

$$\mathrm{Tr}[A^m B^m] - \mathrm{Tr}[(AB)^m] \ge f(\ell) - f(\ell-1) = \mathrm{Tr}[A^m B^m] - \mathrm{Tr}[(A^{m/2} B^{m/2})^2].$$

Not only is this particular term nonnegative, it can be written as a perfect square. Let $X := A^{m/2}$ and $Y := B^{m/2}$. Then

$$[X, Y]^*[X, Y] = XY^2 X + YX^2 Y - XYXY - YXYX,$$

and therefore $\mathrm{Tr}[A^m B^m] - \mathrm{Tr}[(A^{m/2} B^{m/2})^2] = \frac{1}{2}\|[A^{m/2}, B^{m/2}]\|_2^2$. This proves (6.8.6). $\qquad\qquad\square$

We now give a second proof of the Golden-Thompson inequality, and more—this version specifies the cases of equality.

Theorem 6.69. *For all self-adjoint $H, K \in M_n(\mathbb{C})$, $\mathrm{Tr}[e^{H+K}] \le \mathrm{Tr}[e^H e^K]$, and there is equality if and only if H and K commute.*

Proof. Let H and K be self-adjoint in $M_n(\mathbb{C})$. Define $A := e^{H/m}$ and $B := e^{K/m}$. By the cyclicity of the trace and (6.8.6), when $m = 2^\ell$, ℓ a positive integer,

$$\mathrm{Tr}[e^H e^K] - \mathrm{Tr}[(e^{H/m} e^{K/m})^m] \ge \frac{1}{2}\|[e^{H/2}, e^{K/2}]\|_2^2.$$

By the Lie product formula, Lemma 3.16,

$$\mathrm{Tr}[e^H e^K] - \mathrm{Tr}[e^{H+K}] \ge \frac{1}{2}\|[e^{H/2}, e^{K/2}]\|_2^2.$$

Therefore, there is equality in the Golden-Thompson inequality if and only if $[e^{H/2}, e^{K/2}] = 0$, and this is the case if and only if H and K commute. $\qquad\square$

The statement about cases of equality in the Golden-Thompson inequality was first explicily proved in [**86**]. The simple argument here is almost implicit in the original proof of Thompson, but the more intricate proof in [**86**] yields much more—the cases of equality in the Araki-Lieb-Thirring inequality.

6.8.1. Reverse Hölder inequalities. As a further application of the general form of Hölder's inequality and Corollary 6.67, we prove some reverse Hölder inequalities for traces; see [59] and [106, Appendix]. For $X \in M_n(\mathbb{C})$, let r denote the rank of X, and let $X = \sum_{j=1}^{r} \sigma_j |\mathbf{v}_j\rangle\langle\mathbf{u}_j|$ be a singular value decomposition of X. Then for $1 \leq j \leq r, \sigma_j > 0$. For all $-\infty < p < 0$, and $0 < p < 1$, define $\|X\|_p = \left(\sum_{j=1}^{r} \sigma_j^p\right)^{1/p}$ and note that for $1 \leq p < \infty$, this same formula gives the Schatten p-norm of X. Now we simply let p take on any nonzero real value.

For $-\infty < p < 0$ and $0 < p < 1$, $X \mapsto \|X\|_p$ is not a norm because it does not satisfy the triangle inequality. However, for these ranges of p, there are reverse forms of Hölder inequalities.

In the following, for $Y \in M_n^+(\mathbb{C})$, the *generalized inverse* of Y is the matrix Y^+ defined by $Y^+ := \sum_{\lambda \in \sigma(Y), \, \lambda > 0} \lambda^{-1} P_\lambda$, where $Y = \sum_{\lambda \in \sigma(Y)} \lambda P_\lambda$ is the spectral decomposition of Y. Note that the singular values of Y^+ are simply the inverses of the singular values of Y. Moreover,

$$YY^+ = Y^+ Y = P,$$

where P is the orthogonal projection on to the range of Y; that is, $P = \sum_{\lambda \in \sigma(Y), \, \lambda > 0} P_\lambda$ in terms of the spectral decomposition of Y.

Lemma 6.70. *Let* $X, Y \in M_n^+(\mathbb{C})$, *and suppose that* $\mathrm{ran}(X) \subseteq \mathrm{ran}(Y)$. *Let* $0 < r < 1$, *and define* $r' := r/(r-1) < 0$. *Then for all* $0 < \alpha < r, \alpha = 2^{-\ell}, \ell \in \mathbb{N}$,

$$(6.8.10) \qquad \mathrm{Tr}\left[(Y^\alpha X^{2\alpha} Y^\alpha)^{1/2\alpha}\right] \geq \|X\|_r \|Y\|_{r'}.$$

Proof. Let P be the projection onto the range of Y, so that $P = Y^\alpha (Y^+)^\alpha$. Since $\mathrm{ran}(X) \subseteq \mathrm{ran}(Y)$, $X^\alpha = X^\alpha P = (X^\alpha Y^\alpha)(Y^+)^\alpha$. We shall apply the general form of Hölder's inequality to estimate $\|(X^\alpha Y^\alpha)(Y^+)^\alpha\|_s$ for appropriate $s > 1$. Define

$$\frac{1}{s} := \frac{\alpha}{r}, \quad \frac{1}{p} := \alpha \quad \text{and} \quad \frac{1}{q} := -\frac{\alpha}{r'}.$$

Then since $1/r = 1 - 1/r'$, $1/s = 1/p + 1/q$ with $p, q, s > 0$. Since $\alpha < r, s > 1$, and then of course $p, q > 1$. By the general form of Hölder's inequality,

$$\|X^\alpha\|_s \leq \|X^\alpha Y^\alpha\|_p \|(Y^+)^\alpha\|_q,$$

and this can be rewritten as

$$\|X\|_r^\alpha \leq \|X^\alpha Y^\alpha\|_{1/\alpha} \|Y\|_{r'}^{-\alpha}.$$

Since $|X^\alpha Y^\alpha| = (Y^\alpha X^{2\alpha} Y^\alpha)^{1/2}$, this proves (6.8.10). $\qquad\square$

Theorem 6.71. *Let* $X, Y \in M_n^+(\mathbb{C})$, *and suppose that* $\mathrm{ran}(X) \subseteq \mathrm{ran}(Y)$. *Let* $0 < r < 1$, *and define* $r' := r/(r-1) < 0$. *Then*

$$(6.8.11) \qquad \mathrm{Tr}[XY] \geq \|X\|_r \|Y\|_{r'},$$

and there is equality in case

$$(6.8.12) \qquad Y = c\|X\|_r^{1-r}(X^+)^{1-r}$$

for any constant $c \geq 0$.

Proof. Define $m = 1/2\alpha$, with $\alpha = 2^{-\ell}$, $\ell \in N$ chosen large enough that $\alpha < r$. Then m is a nonnegative integer power of 2, and

$$\mathrm{Tr}[(X^\alpha Y^{2\alpha} X^\alpha)^{1/2\alpha}] = \mathrm{Tr}[(X^{1/2m} Y^{1/m} X^{1/2m})^m].$$

By Corollary 6.67, specifically (6.8.5) with $B := X^{1/m}$ and $A := Y^{1/m}$,

$$\mathrm{Tr}[(X^{1/2m} Y^{1/m} X^{1/2m})^m] \leq \mathrm{Tr}[XY].$$

This together with Lemma 6.70 proves (6.8.11). Finally it is easy to check that if Y is defined in terms of X by (6.8.12) for $c = 1$, then $\|Y\|_{r'} = 1$ and $\mathrm{Tr}[XY] = \|X\|_r$, and this proves the final statement. $\qquad\square$

Corollary 6.72. *Let $X \in M_n^+(\mathbb{C})$, and let $0 < r < 1$ or $-\infty < r < 0$. Then with $r' = r/(r-1)$,*

$$(6.8.13) \qquad \|X\|_r = \min\{\mathrm{Tr}[XY] \ : \ Y \in M_n^+(\mathbb{C}), \|Y\|_{r'} = 1\}.$$

Proof. For $Y \geq 0$, $\|Y\|_{r'} = 1$, (6.8.11) yields $\mathrm{Tr}[XY] \geq \|X\|_r$, and then Y is defined in terms of X by (6.8.12) for $c = 1$, $Y \geq 0$, $\|Y\|_{r'} = 1$ and $\mathrm{Tr}[XY] = \|X\|_r$. This proves the corollary for $0 < r < 1$. Switching the roles of r and r' yields the result for $-\infty < r < 0$ in the same way. The details are left as an exercise. $\qquad\square$

Corollary 6.73. *For all $A, B \in M_n(\mathbb{C})$ such that $\mathrm{ran}(A^*) \subseteq \mathrm{ran}(B)$, and all $0 < r < 1$,*

$$(6.8.14) \qquad \|AB\|_1 \geq \|A\|_r \|B\|_{r'},$$

where $r' = r/(1-r) < 0$.

Proof. Let $A = U|A|$ and $B^* = V|B^*|$ be the polar decompositions of A and B^* so that $AB = U|A||B^*|V^*$. Then by the unitary invariance of the Schatten norms,

$$\|AB\|_1 = \|U|A||B^*|V\|_1 = \||A||B^*|\|_1 \geq \mathrm{Tr}[|A||B^*|].$$

Since $\mathrm{ran}(|B^*|) = \mathrm{ran}(B)$ and $\mathrm{ran}(|A|) = \mathrm{ran}(A^*)$, we may apply Theorem 6.71 to estimate $\mathrm{Tr}[|A||B^*|]$, and this yields (6.8.14). $\qquad\square$

Exercises

(1) Let $f(x) := \log(\cosh(x))$, $x \in \mathbb{R}$. Show that f is convex, and compute the Legendre transform of f.

(2) Let $g(y) := \frac{1}{2}(1+y)\log(1+y) + \frac{1}{2}(1-y)\log(1-y)$, $y \in [-1,1]$, with $g(y) = \infty$ for $|y| \geq 1$. Show that g is convex, and compute the Legendre transform of g.

(3) Let $\mathcal{H}$ be a finite-dimensional Hilbert space. With f and g defined as in Exercises (1) and (2) above, show that for all $X, Y \in \mathcal{B}^{\text{s.a.}}(\mathcal{H})$ with $\|Y\| \leq 1$,

$$\mathrm{Tr}[XY] \leq \mathrm{Tr}[f(X)] + \mathrm{Tr}[g(Y)].$$

When is there equality in this inequality?

(4) Let $F : \mathcal{B}^+(\mathcal{H}) \to \mathbb{R}$ be homogeneous of degree one. Let G be the restriction of F to $\mathfrak{S}_{\mathcal{H}}$. Show that F is convex (concave) if and only if G is convex (concave).

(5) Let $XY \in M_n(\mathbb{C})$, and $1 < p < \infty$, $q = p/(p-1)$. Suppose $\langle X, Y \rangle = \|X\|_p \|Y\|_q$. Show that $|X|$ and $|Y|$ commute.

(6) Give another proof of Theorem 6.26 using Theorem 6.32 and Theorem 5.36.

(7) Let ψ and ϕ be two unit vectors in $\mathcal{H}$. Consider the two pure states $\rho := |\psi\rangle\langle\psi|$ and $\sigma := |\phi\rangle\langle\phi|$. Show that

$$2 \min_{\theta \in [0,2\pi)} \|\psi - e^{i\theta}\phi\| \geq \|\rho - \sigma\|_1 \geq \min_{\theta \in [0,2\pi)} \|\psi - e^{i\theta}\phi\|.$$

(8) Prove $\mathrm{Tr}[\log(B)] \leq \mathrm{Tr}[\log(A)] + \mathrm{Tr}[A^{-1}B] - n$ for $A, B \in M_n^{++}(\mathbb{C})$.

(9) Let A be a positive operator on the tensor product of two finite-dimensional Hilbert spaces $\mathcal{H}_1 \otimes \mathcal{H}_2$. Show that for all $p \geq 1$,

$$\left(\mathrm{Tr}[\mathrm{Tr}_1[A]]^p\right)^{1/p} \leq \mathrm{Tr}\left[\left(\mathrm{Tr}_2[A^p]\right)^{1/p}\right]$$

and that the inequality reverses for $0 < p \leq 1$.

(10) Let $X, Y \in M_n^+(\mathbb{C})$, and $p \geq 1$. Show that $\|X - Y\|_p^p \leq \|X\|_p^p + \|Y\|_p^p$.

(11) (Beighi) Let $\sigma \in \mathfrak{S}_{\mathcal{H}}^\circ$, and let $\mathcal{H}$ be a finite-dimensional Hilbert space. For $1 \leq p_0, p_1 \leq \infty$, $0 < \theta < 1$, define $p_\theta := \dfrac{1-\theta}{p_0} + \dfrac{\theta}{p_1}$. Use the three lines theorem to prove that for all $X \in \mathcal{B}(\mathcal{H})$, $\|X\|_{\rho,p_\theta} \leq \|X\|_{\rho,p_0}^{1-\theta} \|X\|_{\rho,p_0}^\theta$.

(12) Let $\mathcal{H}, \mathcal{K}$ be finite-dimensional Hilbert spaces, and let $\Phi : \mathcal{B}(\mathcal{H}) \to \mathcal{B}(\mathcal{K})$ be a positive linear map. Let $\sigma \in \mathfrak{S}_{\mathcal{K}}^\circ$, and suppose that $\Phi^\dagger(\sigma) > 0$. Let $A_{\Phi,\sigma}$ be defined as in Definition 6.41. Show that for all $Y \in \mathcal{B}(\mathcal{K})$, $\mathrm{Tr}[\Phi^\dagger(\sigma)A_{\Phi,\sigma}(Y)] = \mathrm{Tr}[\sigma Y]$.

(13) Show that the function $\| \cdot \|_{p \to q}$, given in Definiton 6.23, is a norm on $\mathcal{L}(\mathcal{B}(\mathcal{H}), \mathcal{B}(\mathcal{K}))$.

(14) Let $\mathcal{H}$ be a finite-dimensional Hilbert space, and let $\sigma \in \mathfrak{S}_{\mathcal{H}}^\circ$. For $0 \leq \alpha \leq 1$, and $X, Y \in \mathcal{B}(\mathcal{H})$, define $\langle X, Y \rangle_{\sigma,\alpha} = \mathrm{Tr}[X^* \sigma^\alpha Y \sigma^{1-\alpha}]$.
    (a) Show that $\langle X, Y \rangle_{\sigma,\alpha} \geq 0$ with equality if and only if $X = 0$, so that $\langle \cdot, \cdot \rangle_{\sigma,\alpha}$ is an inner product on $\mathcal{B}(\mathcal{H})$.

(b) Let $\mathcal{K}$ be another finite-dimensional Hilbert space, and let $\Phi \in \mathcal{L}(\mathcal{B}(\mathcal{H}), \mathcal{B}(\mathcal{K}))$. Let $\rho \in \mathfrak{S}_{\mathcal{H}}^{\circ}$ and $\sigma \in \mathfrak{S}\mathcal{K}^{\circ}$. Let $\widehat{\Phi}$ denote the adjoint of Φ with respect to the inner products $\langle \cdot, \cdot \rangle_{\rho,\alpha}$ and $\langle \cdot, \cdot \rangle_{\sigma,\alpha}$. That is, for all $X \in \mathcal{B}(\mathcal{H})$ and all $Y \in \mathcal{B}(\mathcal{K})$, $\langle Y, \Phi(X) \rangle_{\sigma,\alpha} = \langle \widehat{\Phi}(Y), X \rangle_{\rho,\alpha}$. Compute an explicit formula for $\widehat{\Phi}$ in terms of $\Phi^{\dagger}$, ρ and σ. Show that for $\alpha \neq \frac{1}{2}$, it is not generally true that $\widehat{\Phi}$ is positive when Φ is positive.

(15) Let $A(p) \in \mathcal{B}^{++}(\mathcal{H})$ depend smoothly on the parameter p near $p = 1$. Define $f(p) := \operatorname{Tr}[e^{A(p)}]$. Use Lemma 3.13 to show that $f'(1) = \operatorname{Tr}[e^{A(1)} A'(1)]$. Then for $B(p) \in \mathcal{B}^{++}(\mathcal{H})$ depending smoothly on the parameter p near $p = 1$, define $g(p) = \operatorname{Tr}[B(p)^p] = \operatorname{Tr}[e^{p \log B(p)}]$. Show that

$$g'(1) = \operatorname{Tr}[B(1)(\log B(1) + (\log B)'(1))],$$

where $(\log B)'(1) = \int_0^{\infty} \frac{1}{t+B(1)} B'(1) \frac{1}{t+B(1)} \, dt$.

Now apply this with $B(p) := \sigma^{\frac{1-p}{2p}} \rho \sigma^{\frac{1-p}{2p}}$ for two strictly positive density matrices ρ and σ, noting that $B(1) = \rho$, and $B'(1) = -\frac{1}{2}(\log \sigma \rho + \rho \log \sigma)$ to show that with $Q_p(\rho \| \sigma)$ defined as in (6.4.15),

$$\frac{d}{dp} Q_p(\rho \| \sigma) \Big|_{p=1} = \operatorname{Tr}[\rho(\log \rho - \log \sigma)],$$

thus providing the details of the computation in Corollary 6.45.

(16) Let ρ_{12} be a density matrix on the tensor product of two finite-dimensional Hilbert spaces. Apply the Pieirls-Bogoliubov inequality with $H = \log \rho_{12}$ and $K = \frac{1}{2}(\log \rho_1 + \log \rho_2 - \log \rho_{12})$ to show that

$$\tfrac{1}{2}(S_{12} - S_1 - S_2) \leq \log\left(\operatorname{Tr}\left[\rho_{12}^{1/2}(\rho_1 \otimes \rho_2)^{1/2}\right]\right),$$

and from here show that

$$S_1 + S_2 - S_{12} \geq -2 \ln\left(1 - \tfrac{1}{2} \operatorname{Tr}\left[\sqrt{\rho_{12}} - \sqrt{\rho_1 \otimes \rho_2}\right]^2\right).$$

(17) Apply the Pieirls-Bogoliubov inequality to prove

$$D(\rho \| \sigma) \geq -2 \ln \operatorname{Tr}\left[\rho^{1/2} \sigma^{1/2}\right].$$

(18) Prove the statement about cases of equality in Theorem 6.62.

(19) Let $\mathcal{H}$ be a finite-dimensional Hilbert space. Define the real Hilbert space $\mathcal{K} := \mathcal{H} \oplus \mathcal{B}^{\text{s.a.}}(\mathcal{H})$ with the inner product

$$\langle (\mathbf{w}, Y), (\mathbf{v}, X) \rangle_{\mathcal{K}} = 2\Re\langle \mathbf{w}, \mathbf{v} \rangle + \operatorname{Tr}[YX].$$

Define the function F on $\mathcal{K}$ by

$$F(\mathbf{v}, A) := \begin{cases} \langle \mathbf{v}, A^+ \mathbf{v} \rangle & A \in \mathcal{B}^+(\mathcal{H}), \quad \mathbf{v} \in \operatorname{ran}(A), \\ \infty & \text{otherwise.} \end{cases}$$

(a) For $A \in \mathcal{B}^+(\mathcal{H})$, let P denote the orthogonal projection onto $\operatorname{ran}(A)$. Prove the identity

$$F(\mathbf{v}, A) + \operatorname{Tr}[A(B + |\mathbf{w}\rangle\langle\mathbf{w}|)] = \langle (\mathbf{v}, A), (\mathbf{w}, B) \rangle_{\mathcal{K}}$$
$$+ \|A^{1/2}(A^+\mathbf{v} - \mathbf{w})\|^2.$$

(b) Define a function G on $\mathcal{K}$ by $G(\mathbf{w}, B) := \begin{cases} 0 & B \leq -|\mathbf{w}\rangle\langle\mathbf{w}| \\ \infty & \text{otherwise} \end{cases}$. Show that $F(\mathbf{v}, A)$ $:= \sup\{\langle(\mathbf{v}, A), (\mathbf{w}, B)\rangle_{\mathcal{K}} - G(\mathbf{w}, B)\}$, and then that F and G are proper, lower semicontinuous convex functions on $\mathcal{K}$, and are Legendre transforms of one another. Apply Theorem 6.10 to show that $\{(B, \mathbf{w}) : B \leq -|\mathbf{w}\rangle\langle\mathbf{w}|\}$ is a closed convex set in $\mathcal{K}$, and also give a direct proof of this.

(c) Show that if $F(\mathbf{v}, A) < \infty$, the subgradient of F at $(\mathbf{v}, A)$, $\partial F(\mathbf{v}, A)$, is given by

$$\partial F(\mathbf{v}, A) = \{(A^+\mathbf{v}, -|A^+\mathbf{v}\rangle\langle A^+\mathbf{v}|)\},$$

and that for all $(\mathbf{w}, B)$ with $G(\mathbf{w}, B) < \infty$,

$$F(\mathbf{v}, A) \geq \langle(\mathbf{v}, A), (\mathbf{w}, B)\rangle_{\mathcal{K}} + \|A^{1/2}(A^+\mathbf{v} - \mathbf{w})\|^2.$$

(20) Use the fact that $(X, A) \mapsto X^* A^{-1} X$ is convex on $\mathcal{B}(\mathcal{H}) \times \mathcal{B}^{++}(\mathcal{H})$ if and only for all $\mathbf{u} \in \mathcal{H}$, $(X, A) \mapsto \langle X\mathbf{u}, A^{-1} X\mathbf{u}\rangle$ is convex on $\mathcal{B}(\mathcal{H}) \times \mathcal{B}^{++}(\mathcal{H})$, together with the results of Exercise 19 to give another proof of Kiefer's theorem. Much more can be done along these lines:

(a) First show that for $x, y > 0$,

$$g(x, y) := x(\log x - \log y) + y - x = \int_0^1 \frac{(y - x)^2}{(1 - t)x + ty} t\, dt.$$

Let $\mathcal{H}$ denote $M_n(\mathbb{C})$ equipped with the Hilbert-Schmidt inner product. For $X, Y \in M_n^{++}$, let R_X and L_Y be the operators of right and left multiplication on $\mathcal{H}$, as in Definition 4.52. Define a function $\widetilde{D}(X\|Y)$ on $M_n^{++}(\mathbb{C}) \times M_n^{++}(\mathbb{C})$ by $\widetilde{D}(X\|Y) := \text{Tr}[X(\log X - \log Y) + Y - X]$. By Theorem 4.54, R_X and L_Y are commuting strictly positive operators, we may substitute R_X and L_Y for x and y in the function $g(x, y)$ using the spectral calculus. Using this, show that $\widetilde{D}(X\|Y) = \langle\mathbb{1}, g(R_X, L_Y)\mathbb{1}\rangle_{\mathcal{H}}$. Now use the integral representation of $g(x, y)$ proved above to write $\widetilde{D}(X\|Y)$ in terms of the function $F(A\mathbf{v})$ studied in Exercise 19: Show that

$$\widetilde{D}(X\|Y) = \int_0^1 F(A_{X,Y,t}\mathbf{v}_{X,Y})t\, dt, \tag{$*$}$$

where $A_{X,Y,t} := (1 - t)R_X + tL_Y$ $\mathbf{v}_{X,Y} := (L_Y - R_X)\mathbb{1} = Y - X$.

(b) Use the fact that $(X, Y) \mapsto A_{X,Y,t}$ and $(X, Y) \mapsto \mathbf{v}_{X,Y}$ are linear to conclude that $\widetilde{D}(X\|Y)$ is jointly convex in X and Y, that is for $X_0, X_1, Y_0, Y_1 \in M_n^{++}(\mathbb{C})$ and $0 < \lambda < 1$, with $X_\lambda := (1 - \lambda)X_0 + \lambda X_1$, and likewise for Y_λ,

$$\widetilde{D}(X_\lambda\|Y_\lambda) \leq (1 - \lambda)\widetilde{D}(X_0\|Y_0) + \lambda\widetilde{D}(X_1\|Y_1). \tag{$**$}$$

The inequality $(**)$ plays an important role in the next chapter.

(21) This exercise presents a variation on a result of Pusz and Woronowicz [**184**], later simplified by Donald [**77**], and further simplified here. Let $\mathcal{H}$ be $M_n(\mathbb{C})$ equipped with the Hilbert-Schmidt inner product. Let $\mathcal{K}$ be the real Hilbert space defined in terms of $\mathcal{H}$ in Exercise 19. Let Λ denote the set of all continuous functions $t \mapsto (\mathbf{w}(t), B(t)) \in \mathcal{K}$ such that for each t, $B(t) \leq -|\mathbf{w}(t)\rangle\langle\mathbf{w}(t)|$.

(a) Prove that Λ is a convex set.

(b) Show that for all $B \in \mathcal{B}(\mathcal{H})$, there exist uniquely determined B_L and B_R in $M_n(\mathbb{C})$ such that for all $H \in M_n^{\text{s.a.}}(\mathbb{C})$,

$$\text{Tr}[L_H B] = \text{Tr}[H B_L] \quad \text{and} \quad \text{Tr}[R_H B] = \text{Tr}[H B_R].$$

(c) For $(\mathbf{w}(\cdot), B(\cdot)) \in \Lambda$, define $H(\mathbf{w}(\cdot), B(\cdot)), K(\mathbf{w}(\cdot), B(\cdot)) \in M_n(\mathbb{C})$ by

$$H(\mathbf{w}(\cdot), B(\cdot)) := \int_0^1 (-\mathbf{w}(t) - \mathbf{w}^*(t) + (1-t)B_R(t))t\,dt$$

and

$$K(\mathbf{w}(\cdot), B(\cdot)) := \int_0^1 (\mathbf{w}(t) + \mathbf{w}^*(t) + tB_L(t))t\,dt.$$

(d) Define $\Omega \subset M_n^{\text{s.a.}}(\mathbb{C}) \oplus M_n^{\text{s.a.}}(\mathbb{C})$ to be the closure of the set of all (H, K) such that for some $(\mathbf{w}(\cdot), B(\cdot)) \in \Lambda$, $H = H(\mathbf{w}(\cdot), B(\cdot))$ and $K = K(\mathbf{w}(\cdot), B(\cdot))$. Then define the function R on $M_n^{\text{s.a.}}(\mathbb{C}) \oplus M_n^{\text{s.a.}}(\mathbb{C})$ by

$$R(H, K) := \begin{cases} 0 & (H, K) \in \Omega \\ \infty & (H, K) \notin \Omega. \end{cases}$$

Prove that

$$\widetilde{D}(X \| Y) = \sup_{H, K \in M_n^{\text{s.a.}}(\mathbb{C})} \{\mathrm{Tr}[HX] + \mathrm{Tr}[KY] - R(H, K)\},$$

and hence that $(X, Y) \mapsto \widetilde{D}(X, Y)$ is a proper lower semicontinuous convex function. Finally, show that the supremum is actually a maximum, realized uniquely at $H(\mathbf{w}(\cdot), B(\cdot)), K(\mathbf{w}(\cdot), B(\cdot)) \in M_n(\mathbb{C})$, where $\mathbf{w}(t) := A^{-1}(t)\mathbf{v}$ and $B(t) = -|\mathbf{w}(t)\rangle\langle\mathbf{w}(t)|$, with $A(t) := (1-t)R_X + tL_Y$ and $\mathbf{v} = Y - X$.

(e) Show that the function $R(H, K)$ defined in part (d) is the Legendre transform of $\widetilde{D}(X \| Y)$.

(22) Let $\mathcal{H}$ be a finite-dimensional Hilbert space and let ρ and σ be density matrices on $\mathcal{H}$ with ranks r and s, respectively. Let $\mathcal{K}$ be another Hilbert space whose dimension is at least $\max\{r, s\}$. Let ψ and ϕ be purifications of ρ and σ in $\mathcal{H} \otimes \mathcal{K}$, with Schrödinger decompositions

$$\rho = \sum_{j=1}^r |\mathbf{u}_j\rangle\langle\mathbf{v}_j| \quad \text{and} \quad \sigma = \sum_{k=1}^s |\tilde{\mathbf{u}}_k\rangle\langle\tilde{\mathbf{v}}_k|.$$

(a) Show that

$$\langle\psi, \phi\rangle = \mathrm{Tr}[\rho^{1/2}\sigma^{1/2}K] \quad \text{where} \quad K := \sum_{j,k}\langle\mathbf{v}_j, \tilde{\mathbf{v}}_k\rangle|\tilde{\mathbf{u}}_k\rangle\langle\mathbf{u}_j|.$$

(b) Show that $|\langle\psi, \phi\rangle| \le \|\rho^{1/2}\sigma^{1/2}\|_1$, which is the upper bound from Uhlmann's fidelity theorem, except that here we do not require that $\mathcal{K} = \mathcal{H}$.

(23) This exercise builds on the previous exercise, using the notation introduced there.

(a) Define partial isometries $U : \mathcal{H} \to \mathbb{C}^r$ and $\tilde{U} : \mathcal{H} \to \mathbb{C}^s$ by

$$(U\mathbf{z})_j := \langle\mathbf{u}_j, \mathbf{z}\rangle \quad \text{and} \quad (\tilde{U}\mathbf{w})_k = \langle\tilde{\mathbf{u}}_k, \mathbf{w}\rangle.$$

Define $L \in M_{s,r}(\mathbb{C})$ by $L_{k,j} := \langle\mathbf{v}_j, \tilde{\mathbf{v}}_k\rangle$. Show that $K = \tilde{U}^*LU$ and that $\|K\| = \|L\| \le 1$.

(b) It is of interest to determine when K is a partial isometry. Show that K is a partial isometry if and only if L is a partial isometry.

(c) Define orthogonal projections P and Q in $\mathcal{B}(\mathcal{H})$ by

$$P = \sum_{j=1}^{r} |\mathbf{v}_j\rangle\langle\mathbf{v}_j| \quad \text{and} \quad Q = \sum_{k=1}^{s} |\tilde{\mathbf{v}}_k\rangle\langle\tilde{\mathbf{v}}_k| \, .$$

Show that $(L^*L)_{k,\ell} = \langle\tilde{\mathbf{v}}_k, QPQ\tilde{\mathbf{v}}_\ell\rangle$ and $((L^*L)^2)_{k,\ell} = \langle\tilde{\mathbf{v}}_k, QPQPQ\tilde{\mathbf{v}}_\ell\rangle$. Using Excercise 2 of Chapter 4, show that L, and hence K, is a partial isometry if and only if $QPQ = PQP = P \wedge Q$.

(d) Show that if K is an isometry, $K^*K = \sum_{k,\ell}\langle\mathbf{v}_k, Q \wedge P\mathbf{v}_\ell\rangle|\mathbf{u}_\ell\rangle\langle\mathbf{u}_k|$, and find an analogous formula for KK^*. Now note that if we replace $\{\mathbf{v}_1,\ldots,\mathbf{v}_r\}$ with another orthonormal set $\{\mathbf{w}_1,\ldots,\mathbf{w}_r\}$ that has the same span in $\mathcal{K}$, this gives us another purification of ρ but does not change P or Q. Let Π be any orthogonal projection on $\mathcal{H}$ whose range lies in the span of $\{\mathbf{u}_1,\ldots,\mathbf{u}_r\}$, and rank$(\Pi)$ = rank$(Q \wedge P)$. Show that there is an orthonormal basis $\{\mathbf{w}_1,\ldots,\mathbf{w}_r\}$ of span$(\{\mathbf{v}_1,\ldots,\mathbf{v}_r\})$ such that $\Pi = \sum_{k,\ell}\langle\mathbf{w}_k, Q \wedge P\mathbf{w}_\ell\rangle|\mathbf{u}_\ell\rangle\langle\mathbf{u}_k|$.

(e) Show without any assumptions on the rank of either ρ or σ that there exist purifications ψ and ϕ of ρ and σ in $\mathcal{H} \otimes \mathcal{H}$ such that $|\langle\psi,\phi\rangle| = \|\rho^{1/2}\sigma^{1/2}\|_1$.

Chapter 7

Fundamental entropy inequalities

7.1. Classical entropy

Classical entropy functionals, and inequalities pertaining to them, are fundamentally important in statistical mechanics and information theory. Our main concern is with their quantum counterparts, but it is important to understand the classical inequalities. One reason is that some of these inequalities transfer fairly directly to the quantum setting. Another is that other inequalities, true in the classical case, are simply false in the quantum case, and it is important to understand why.

We begin by reviewing some classical probability theory on finite sets. Correlations are fundamental to statistical mechanics and information theory, and so before doing anything else, we discuss the joint distribution of two random variables.

Let $\mathcal{X}$ and $\mathcal{Y}$ be two finite sets. Define the functions $X : \mathcal{X} \times \mathcal{Y} \to \mathcal{X}$ and $Y : \mathcal{X} \times \mathcal{Y} \to \mathcal{Y}$ by

$$X(x,y) = x \quad \text{and} \quad Y(x,y) = y.$$

Let $\rho_{XY}(x,y)$ be a probability density on $\mathcal{X} \times \mathcal{Y}$; that is, for all (x,y), $\rho_{XY}(x,y) \geq 0$, and $\sum_{(x,y)\in\mathcal{X}\times\mathcal{Y}} \rho_{XY}(x,y) = 1$. Then X and Y become random variables because we can now talk about the probability that they take on certain values. The probability density $\rho_{XY}(x,y)$ specifies their **joint distribution**: For $E \subseteq \mathcal{X} \times \mathcal{Y}$, the probability of E, $\mathbb{P}(E)$ is given by $\mathbb{P}(E) = \sum_{(x,y)\in E} \rho_{XY}(x,y)$. For $A \subseteq \mathcal{X}$, $\{X \in A\}$ denotes $\{(x,y) : x \in A\}$, and hence $\mathbb{P}(\{X \in A\}) = \sum_{(x,y)\in A\times\mathcal{Y}} \rho_{XY}(x,y)$. The **marginal density** of X is the probability density ρ_X

on $\mathcal{X}$ given by $\rho_X(x) := \sum_{y\in\mathcal{Y}}\rho_{XY}(x,y)$. Then $\mathbb{P}(\{X \in A\}) = \sum_{x\in A}\rho_X(x)$. Likewise, define $\rho_Y(y) = \sum_{x\in\mathcal{X}}\rho_{XY}(x,y)$, the marginal density of Y, and then for $B \subseteq \mathcal{Y}$, $\mathbb{P}(\{Y \in B\}) = \sum_{y\in B}\rho_Y(y)$.

Definition 7.1. Let $\mathcal{Z}$ be a finite set, and let ρ_Z be a probability density on $\mathcal{Z}$. Then the **Shannon entropy** of ρ_Z, $H(\rho_Z)$, is the quantity defined by

$$(7.1.1) \qquad H(\rho_Z) = - \sum_{z\in\mathcal{Z}} \rho_Z(z)\log\rho_Z(z).$$

The Shannon entropy takes its name from the fundamental role it plays in Shannon's information theory [**197**]. For brevity, we shall generally simply refer to it as entropy. It will often be convenient to write $H(Z)$ to denote the entropy specified in (7.1.1), putting the emphasis on the random variable Z instead of on the density, but $H(Z)$ depends on Z only through the *set* of probabilities $\{\rho_Z(z) \; : \; z \in \mathcal{Z}\}$.

It is standard notation in probability and information theory to use H for the classical entropy.

The set $\mathcal{P}_{\mathcal{Z}}$ of probability densities on $\mathcal{Z}$ is convex: if $\rho,\sigma \in \mathcal{P}_{\mathcal{Z}}$ and $0 \leq \lambda \leq 1$, $(1-\lambda)\rho + \lambda\sigma \in \mathcal{P}_{\mathcal{Z}}$.

Lemma 7.2. *Let $\mathcal{Z}$ be a finite set of cardinality d, and let $\mathcal{P}_{\mathcal{Z}}$ denote the set of probability densities on $\mathcal{Z}$ with respect to counting measure. Then the entropy H is a strictly concave function on $\mathcal{P}_{\mathcal{Z}}$, and for all $\rho \in \mathcal{P}_{\mathcal{Z}}$,*

$$(7.1.2) \qquad 0 \leq H(\rho) \leq \log(d).$$

There is equality on the left in (7.1.2) if and only if for some $z' \in \mathcal{Z}$, $\rho(z) = \delta_{z,z'}$ for all $z \in \mathcal{Z}$. There is equality on the right in (7.1.2) if and only if ρ is the uniform density on $\mathcal{Z}$; that is, $\rho(z) = 1/d$ for all $z \in \mathcal{Z}$.

Proof. Since the function $f(t) = -t\log t$ is strictly concave, for any two probability densities ρ and σ and any $0 < \lambda < 1$, $f((1-\lambda)\rho(z) + \lambda\sigma(z)) \geq (1-\lambda)f(\rho(z))+\lambda f(\sigma(z))$, with equality if and only if $\rho(z) = \sigma(z)$. Summing on z, $H((1-\lambda)\rho + \lambda\sigma) \geq (1-\lambda)H(\rho) + \lambda H(\sigma)$, with equality if and only if $\rho = \sigma$. This proves the concavity.

To see that H is nonnegative, observe that since $0 \leq \rho(z) \leq 1$ for all z, $-\rho(z)\log\rho(z) \geq 0$ for all z, with equality if and only if $\rho(z) = 0$ or $\rho(z) = 1$. Therefore, $H(\rho) \geq 0$ with equality if and only if $\rho(z') = 1$ for some $z' \in \mathcal{Z}$, and hence $\rho(z) = 0$ for all $z \neq z'$.

For the upper bound we use the strict concavity. Since counting measure is permutation invariant, permutations act on the set of probability densities on $\mathcal{Z}$ in a simple way: let the elements of $\mathcal{Z}$ be $z_1,\ldots,z_d$, and for $\pi \in \mathcal{S}_d$, the symmetric group on d letters, define $\pi : \mathcal{Z} \to \mathcal{Z}$ by $\pi(z_j) = z_{\pi(j)}$. Then for any probability density ρ on $\mathcal{Z}$, $\rho \circ \pi$ is also a probability density, and evidently

$H(\rho \circ \pi) = H(\rho)$. Let σ_u denote the uniform probability density, $\sigma_u(z) = 1/d$ for all z. Then for every probability density ρ

$$\sigma_u = \frac{1}{d!} \sum_{\pi \in \mathcal{S}_d} \rho \circ \pi.$$

Therefore, by the strict concavity of the entropy and Jensen's inequality,

$$\log(d) = H(\sigma_u) = H\left(\frac{1}{d!} \sum_{\pi \in \mathcal{S}_d} \rho \circ \pi\right) \geq \frac{1}{d!} \sum_{\pi \in \mathcal{S}_d} H(\rho \circ \pi) = H(\rho),$$

and the inequality is strict unless ρ is uniform, in which case $\rho \circ \pi$ is independent of π. $\qquad\square$

Lemma 7.2 is a first indication that the entropy is a good measure of *randomness*: the minimum value of the entropy is attained only at the nonrandom densities that assign all of the probability to a single outcome x', and the maximum value of the entropy is attained only at the uniform density, for which any outcome is as likely as any other.

So far, we have defined H as a functional on probability densities on a finite set $\mathcal{X}$. Extend it to a functional on all real valued functions of $\mathcal{X}$ as follows:

$$H(f) := \begin{cases} -\sum_{x \in \mathcal{X}} f(x) \log f(x) & f \geq 0 \text{ and } \sum_{x \in \mathcal{X}} f(x) = 1, \\ \infty & \text{otherwise.} \end{cases}$$

Let $\mathcal{K}$ denote the real Hilbert space consisting of real valued functions h on $\mathcal{X}$ with inner product $\langle h, g \rangle = \sum_{x \in \mathcal{X}} h(x)g(x)$. Then $-H$ is a convex function of $\mathcal{K}$. It is both simple and useful to compute its Legendre transform. Let $\mathcal{P}_x$ denote the set of all probability densities on $\mathcal{X}$ which is the subset of $\mathcal{K}$ on which $-H$ is finite.

For $\rho \in \mathcal{P}_x \subset \mathcal{K}$ such that $\rho(x) > 0$ for all x, and $h \in \mathcal{K}$, by Jensen's inequality for the logarithm,

$$\langle \rho, h \rangle - (-H(\rho)) = \sum_{x \in \mathcal{X}} \rho(x)(h(x) - \log \rho(x))$$

$$= \sum_{x \in \mathcal{X}} \rho(x) \log e^{(h(x) - \log \rho(x))}$$

$$\leq \log\left(\sum_{x \in \mathcal{X}} \rho(x) e^{(h(x) - \log \rho(x))}\right) = \log\left(\sum_{x \in \mathcal{X}} e^{h(x)}\right),$$

and since the logarithm is strictly concave, there is equality if and only if for some constant c, $h(x) - \log \rho(x) = c$ for all x.

This proves that the Legendre transform of $-H$ is given by

$$(-H)^*(h) = \log\left(\sum_{x \in \mathcal{X}} e^{h(x)}\right).$$

The function $(-H)^*$ is differentiable on $\mathcal{K}$ and its derivative at h is given by

$$D_{(-H^*)}(h)(x) = \left(\sum_{x \in \mathcal{X}} e^{h(x)} \right)^{-1} e^{h(x)},$$

so that $D_{(-H^*)}(h)$ is the unique element of the subgradient of $(-H)^*$ at h. By Young's inequality and the conditions for equality in it, this proves:

Theorem 7.3 (Classical Gibbs variational principle). *Let $\mathcal{X}$ be any finite set. Let $\rho \in \mathcal{P}_{\mathcal{X}} \subset \mathcal{K}$, and let $h \in \mathcal{K}$. Then*

$$(7.1.3) \qquad \langle \rho, h \rangle \leq \sum_{x \in \mathcal{X}} \rho(x) \log \rho(x) + \log \left(\sum_{x \in \mathcal{X}} e^{h(x)} \right),$$

and there is equality if and only if $\rho(x) = \left(\sum_{x \in \mathcal{X}} e^{h(x)} \right)^{-1} e^{h(x)}$ for all x. In particular, if there is equality, $\rho(x) > 0$ for all x.

Corollary 7.4. *Let $\mathcal{X}$ be any finite set. Let $\rho, \sigma \in \mathcal{P}_{\mathcal{X}} \subset \mathcal{K}$, with $\sigma(x) > 0$ for all $x \in \mathcal{X}$, and let $h \in \mathcal{K}$. Then*

$$\langle \rho, h \rangle \leq \sum_{x \in \mathcal{X}} \rho(x)(\log \rho(x) - \log \sigma(x)) + \log \left(\sum_{x \in \mathcal{X}} e^{h(x)} \sigma(x) \right),$$

and there is equality if and only if for some constant c, $h = \log \rho - \log \sigma + c$.

Proof. Simply replace h by $h + \log \sigma$ in (7.1.3). $\qquad\qquad\qquad\square$

7.1.1. Conditional entropy. Returning to our two random variables X and Y, we have a number of entropies to consider. The first three are $H(\rho_{XY})$, $H(\rho_X)$, and $H(\rho_Y)$, for which we also write:

$$H(XY) := H(\rho_{XY}), \quad H(X) := H(\rho_X), \quad \text{and} \quad H(Y) := H(\rho_Y).$$

In addition to these three entropies, there are also entropies of conditional probabilities:

Definition 7.5. Let X and Y be two random variables with values in finite sets $\mathcal{X}$ and $\mathcal{Y}$, respectively. Let $\rho_{XY}(x, y)$ be the probability distribution of $\mathcal{X} \times \mathcal{Y}$ specifying their joint distribution. Then the **conditional probability of X given $Y = y$** is the probability density on $\mathcal{X}$ given by

$$\rho(x|y) := \begin{cases} \frac{\rho_{XY}(x,y)}{\rho_Y(y)} & \rho_Y(y) > 0, \\ 0 & \rho_Y(y) = 0. \end{cases}$$

Note that if $\rho_Y(y) = 0$, then $\rho_{XY}(x, y) = 0$ for all x, and if $\rho_Y(y) > 0$, $\rho(x|y)$ is a probability density on $\mathcal{X}$. The **(classical discrete) conditional entropy** $H(X|Y)$ of X with respect to Y is defined by

$$(7.1.4) \quad H(X|Y) = - \sum_{(x,y) \in \mathcal{X} \times \mathcal{Y}} \rho_{XY}(x, y) \log(\rho(x|y)) = \sum_{y \in \mathcal{Y}} H(\rho(\cdot|y)) \rho_Y(y).$$

Lemma 7.6. *Let X and Y be random variables with values in finite sets $\mathcal{X}$ and $\mathcal{Y}$, respectively. Then*

$$(7.1.5) \qquad H(X|Y) = H(XY) - H(Y)$$

and

$$(7.1.6) \qquad 0 \leq H(X|Y) \leq H(X).$$

Moreover, there is equality on the left if and only if $X = f(Y)$ for some function $f : \mathcal{Y} \to \mathcal{X}$, so that X is perfectly correlated with Y, and there is equality on the right if and only if X and Y are independent; that is $\rho_{XY}(x,y) = \rho_X(x)\rho_Y(y)$. Consequently,

$$(7.1.7) \qquad H(XY) \leq H(X) + H(Y)$$

with equality if and only if X and Y are independent.

Proof. The formula (7.1.5) is an immediate consequence of the definition. Since $H(X|Y) = \sum_{y \in \mathcal{Y}} H(\rho(\cdot|y))\rho_Y(y)$ the lower bound (7.1.6) is a consequence of the lower bound in (7.1.2). Moreover, since each $H(\rho(\cdot|y)) \geq 0$, $H(X|Y) = 0$ if and only if $H(\rho(\cdot|y)) = 0$ for each y. By the conditions for equality in Lemma 7.2, this means that $X = f(Y)$.

To prove the upper bound, note that by the strict convexity of $x \log x$, application of Jensen's inequality yields

$$0 \leq \sum_{(x,y) \in \mathcal{X} \times \mathcal{Y}} \left(\frac{\rho_{XY}(x,y)}{\rho_X(x)\rho_Y(y)} \right) \log\left(\frac{\rho_{XY}(x,y)}{\rho_X(x)\rho_Y(y)} \right) \rho_X(x)\rho_Y(y)$$

$$= \sum_{(x,y) \in \mathcal{X} \times \mathcal{Y}} \rho_{XY}(x,y) \log\left(\frac{\rho_{XY}(x,y)}{\rho_X(x)\rho_Y(y)} \right) = H(X) - H(X|Y)$$

with equality if and only if $\rho_{XY}(x,y) = \rho_X(x)\rho_Y(y)$. Combining (7.1.5) and (7.1.6) yields (7.1.7). $\qquad\square$

Remark 7.7. The inequality (7.1.7) is known as the **subadditivity of the entropy**.

The conditional information $H(X|Y)$ provides a measure of the uncertainty in X when Y is known. In the extreme case $X = f(Y)$ for some function f, once Y is known, X is also known, and there is no uncertainty in X. However, if for all x, y, $\rho_{XY}(x,y) = \rho_X(x)\rho_Y(y)$ so that X and Y are independent, $H(X|Y)$ takes on its maximum value $H(X)$: knowledge of Y does nothing to reduce the uncertainty about the value of the random variable X.

Let $\mathcal{Y}'$ be another finite set, and let $f : \mathcal{Y} \to \mathcal{Y}'$ be a function, and define the random variable $\widehat{Y} := f(Y)$. If f is not one-to-one, outcomes that were distinct for Y are now lumped together in $\widehat{Y}$ and hence the value of $\widehat{Y}$ tells us less about the value of X, or anything else, than does the value of Y. One can therefore expect that $H(X|Y) \leq H(X|\widehat{Y})$. This is the case.

Lemma 7.8. *Let X and Y be random variables with values in finite sets $\mathcal{X}$ and $\mathcal{Y}$, respectively. Let $\mathcal{Y}'$ be another finite set, and let $f : \mathcal{Y} \to \mathcal{Y}'$ be a function, and define the random variable $\widehat{Y} := f(Y)$. Then*

$$H(X|Y) \leq H(X|\widehat{Y}).$$

Proof. Let ρ_{XY} denote the joint density of X and Y, and let $\sigma_{X\widehat{Y}}$ denote the joint density of X and $\widehat{Y}$. Then for $x \in \mathcal{X}$ and $y' \in \mathcal{Y}'$,

$$\sigma_{X\widehat{Y}}(x,y') = \sum_{\{y \,:\, f(y)=y'\}} \rho_{XY}(x,y) \quad \text{and} \quad \sigma_{\widehat{Y}}(y') = \sum_{\{y \,:\, f(y)=y'\}} \rho_Y(y).$$

Consequently, the conditional probability $\sigma(x|y')$ is given by

$$\sigma(x|y') = \frac{\sigma_{X\widehat{Y}}(x,y')}{\sigma_{\widehat{Y}}(y')} = \sum_{\{y \,:\, f(y)=y'\}} \frac{\rho_{XY}(x,y)}{\rho_Y(y)} \frac{\rho_Y(y)}{\sigma_{\widehat{Y}}(y')}.$$

Define $\lambda(y|y') := \frac{\rho_Y(y)}{\sigma_{\widehat{Y}}(y')}$, and note that $\sum_{\{y \,:\, f(y)=y'\}} \lambda(y|y') = 1$. Therefore, $\sigma(x|y') = \sum_{\{y \,:\, f(y)=y'\}} \lambda(y|y')\rho(x|y)$, and this identity expresses $\sigma(x|y')$ as a convex combination of the conditional probabilities $\rho(x|y)$ for which $f(y) = y'$. Then by (7.1.4),

$$H(X|\widehat{Y}) = \sum_{y' \in \mathcal{Y}'} H(\sigma(\cdot|y'))\sigma_{\widehat{Y}}(y')$$

$$\geq \sum_{y' \in \mathcal{Y}'} \left(\sum_{\{y \,:\, f(y)=y'\}} \lambda(y|y')H(\rho(\cdot|y)) \right) \sigma_{\widehat{Y}}(y')$$

$$= \sum_{y \in \mathcal{Y}} H(\rho(\cdot|y))\rho_Y(y) = H(X|Y). \qquad \square$$

Lemma 7.6 says that if $X = f(Y)$, then $H(X|Y) = 0$. It turns out that if $\mathbb{P}(\{X \neq f(Y)\})$ is small for some $f : \mathcal{Y} \to \mathcal{X}$, then $H(X|Y)$ is also small, and the relation between the two notions of *smallness* depends only on the cardinality of $\mathcal{X}$.

Theorem 7.9 (Fano's theorem). *Let X and Y be random variables with values in finite sets $\mathcal{X}$ and $\mathcal{Y}$, respectively. Let d denote the cardinality of $\mathcal{X}$. Let $0 < \epsilon < 1$. Suppose that for some $f : \mathcal{Y} \to \mathcal{X}$, $\mathbb{P}(\{X \neq f(Y)\}) \leq \epsilon$. Then*

$$(7.1.8) \qquad H(X|Y) \leq \epsilon \log(d-1) + \eta(\epsilon),$$

where $\eta(\epsilon) := -\epsilon \log \epsilon - (1-\epsilon)\log(1-\epsilon) > 0$.

Proof. Let $\widehat{Y}$ denote the random variable $f(Y)$, which has values in $\mathcal{X}$. Let $\sigma_{X\widehat{Y}}(x, x')$ denote the joint distribution of X and $\widehat{Y}$. Let $E \subseteq \mathcal{X} \times \mathcal{X}$ denote the set

$$E := \{(x, x') \; : \; x \neq x'\},$$

which is the *off-diagonal* part of the cartesian product $\mathcal{X} \times \mathcal{X}$. For $x' \in \mathcal{X}$, define

$$(7.1.9) \qquad q(x') := \sum_{x \neq x'} \frac{\sigma_{X\widehat{Y}}(x, x')}{\sigma_{\widehat{Y}}(x')} \quad \text{and} \quad p(x') = \sum_{x \neq x'} \sigma_{X\widehat{Y}}(x, x').$$

Then

$$\begin{aligned}
H(X|\widehat{Y}) &= - \sum_{(x,x') \in \mathcal{X} \times \mathcal{X}} \left(\frac{\sigma_{X\widehat{Y}}(x, x')}{\sigma_{\widehat{Y}}(x')} \right) \log\left(\frac{\sigma_{X\widehat{Y}}(x, x')}{\sigma_{\widehat{Y}}(x')} \right) \sigma_{\widehat{Y}}(x') \\
&= - \sum_{(x,x') \in E} \left(\frac{\sigma_{X\widehat{Y}}(x, x')q(x')}{\sigma_{\widehat{Y}}(x')q(x')} \right) \log\left(\frac{\sigma_{X\widehat{Y}}(x, x')q(x')}{\sigma_{\widehat{Y}}(x')q(x')} \right) \sigma_{\widehat{Y}}(x') \\
&\quad + - \sum_{(x,x') \in E^c} \left(\frac{\sigma_{X\widehat{Y}}(x, x')}{\sigma_{\widehat{Y}}(x')} \right) \log\left(\frac{\sigma_{X\widehat{Y}}(x, x')}{\sigma_{\widehat{Y}}(x')} \right) \sigma_{\widehat{Y}}(x') \\
&= I + II.
\end{aligned}$$

Since $q(x')\sigma_{\widehat{Y}}(x') = p(x')$,

$$I = - \sum_{(x,x') \in E} \left(\frac{\sigma_{X\widehat{Y}}(x, x')}{p(x')} \right) \left(\log\left(\frac{\sigma_{X\widehat{Y}}(x, x')}{p(x')} \right) + \log q(x') \right) p(x').$$

Since for each x', $x \mapsto \frac{\sigma_{X\widehat{Y}}(x,x')}{p(x')}$ is a probability density on $X \backslash \{x'\}$, the estimate (7.1.2) yields

$$- \sum_{x \neq x'} \left(\frac{\sigma_{X\widehat{Y}}(x, x')}{p(x')} \right) \log\left(\frac{\sigma_{X\widehat{Y}}(x, x')}{p(x')} \right) \leq \log(d - 1),$$

and hence $I \leq \mathbb{P}(E) \log(d - 1) - \sum_{x'} \log q(x') p(x')$. Finally, by Jensen's inequality,

$$- \sum_{x'} \log q(x') p(x') = - \sum_{x'} \log q(x') q(x') \sigma_{\widehat{Y}}(x') \leq -\mathbb{P}(E) \log \mathbb{P}(E).$$

Next, to estimate II, note that by (7.1.9)

$$\begin{aligned}
II &= - \sum_{(x,x') \in E^c} \left(\frac{\sigma_{X\widehat{Y}}(x, x')}{\sigma_{\widehat{Y}}(x')} \right) \log\left(\frac{\sigma_{X\widehat{Y}}(x, x')}{\sigma_{\widehat{Y}}(x')} \right) \sigma_{\widehat{Y}}(x') \\
&= - \sum_{x' \in \mathcal{X}} (1 - q(x')) \log(1 - q(x')) \sigma_{\widehat{Y}}(x') \\
&\leq -(1 - \mathbb{P}(E)) \log(1 - \mathbb{P}(E)),
\end{aligned}$$

where the inequality is Jensen's inequality.

Combining the estimates on I and II yields

$$H(X|\widehat{Y}) \leq \mathbb{P}(E)\log(d-1) - \mathbb{P}(E)\log\mathbb{P}(E) - (1-\mathbb{P}(E))\log(1-\mathbb{P}(E)).$$

By Lemma 7.8, $H(X|Y) \leq H(X|\widehat{Y})$, and this yields (7.1.8). $\qquad\square$

7.1.2. Quantitative continuity of entropy. We now turn to the continuity properties of entropy. Since $x\log x$ is continuous on $[0,1]$, it is evident from the definition that $H(\rho)$ depends continuously on $\rho \in \mathcal{P}_{\mathcal{X}}$ for any finite set $\mathcal{X}$. However, in many applications we need to know more: Given $\epsilon > 0$, how small must $\sum_{x\in\mathcal{X}} |\rho(x) - \sigma(x)|$ be to ensure that $|H(\rho) - H(\sigma)| \leq \epsilon$?

The answer to this question depends on the cardinality of $\mathcal{X}$:

Theorem 7.10. *Let $\mathcal{X}$ be a set of finite cardinality $d \geq 2$. For any probability densities $\rho, \sigma \in \mathcal{P}_{\mathcal{X}}$,*

$$(7.1.10) \qquad |H(\rho) - H(\sigma)| \leq \frac{1}{2}\|\rho - \sigma\|_1 \log(d-1) + \eta\left(\frac{1}{2}\|\rho - \sigma\|_1\right),$$

where for $t \in [0,1]$, $\eta(t) := -t\log t - (1-t)\log(1-t)$.

Remark 7.11. The bound (7.1.10) is sharp: Let $\mathcal{X} = \{1, \dots, d\}$, and define

$$\sigma(j) := \delta_{1,j} \quad \text{and} \quad \rho(j) := (1-\epsilon)\delta_{1,j} + \frac{\epsilon}{d-1}(1-\delta_{1,j}).$$

Then $\|\rho - \sigma\|_1 = 2\epsilon$, $H(\sigma) = 0$ and $H(\rho) = \epsilon\log(d-1) + \eta(\epsilon)$. However, the function $f(t) := t\log(d-1) + \eta(t)$ is not monotone on $[0,1]$. It is concave, with a maximum value of $\log d$ occurring at $t = (d-1)/d$. Thus if we define

$$g(t) := \begin{cases} t\log(d-1) + \eta(t) & 0 \leq t \leq (d-1)/d \\ \log d & (d-1)/d \leq t \leq 1, \end{cases}$$

then $f \leq g$, and g is monotone nondecreasing. Therefore, we obtain the following modulus of continuity estimate for the entropy:

$$\frac{1}{2}\|\rho - \sigma\|_1 \leq \delta \quad \Rightarrow \quad |H(\rho) - H(\sigma)| \leq g(\delta).$$

Definition 7.12. Let $\mathcal{X}$ be a finite set, and let $\rho, \sigma \in \mathcal{P}_{\mathcal{X}}$. A **coupling** of ρ and σ is a probability density γ on $\mathcal{X} \times \mathcal{X}$ whose left and right marginals are ρ and σ, respectively. That is,

$$\rho(x) = \sum_{x'\in\mathcal{X}} \gamma(x, x') \quad \text{and} \quad \sigma(x') = \sum_{x\in\mathcal{X}} \gamma(x, x').$$

Equivalently, a coupling of ρ and σ may be thought of as a pair of random variables X and X' with values in $\mathcal{X}$ such that $\mathbb{P}(\{X = x\}) = \rho(x)$ and $\mathbb{P}(\{X' = x'\}) = \sigma(x')$. Then the joint density γ of X and X' is a coupling of ρ and σ, and of course, associated to any coupling such joint density, we have the pair of random variables X and X'. The set of all couplings of ρ and σ is denoted $\Gamma_{\rho,\sigma}$.

For example, the product density $\gamma(x, x') = \rho(x)\sigma(x')$ is a coupling of ρ and σ, but the couplings that turn out to be most useful are those in which the corresponding random variables X and X' are more closely correlated. For example, if $\rho = \sigma$, then $\gamma(x, x') = \delta_{x,x'}\rho(x)$ is a coupling of ρ and σ, and then X and X' are perfectly correlated. One might expect that if ρ and σ are close in the sense that $\sum_{x\in\mathcal{X}} |\rho(x) - \sigma(x)|$ is small, then there exists a coupling with X and X' strongly correlated. This is indeed the case.

Lemma 7.13. *Let $\mathcal{X}$ be a finite set, and let $\rho, \sigma \in \mathcal{P}_\mathcal{X}$. Let X, X' be any coupling of ρ and σ. Then*

$$(7.1.11) \qquad \mathbb{P}(\{X = X'\}) \leq \sum_{x\in\mathcal{X}} \min\{\rho(x), \sigma(x)\}.$$

Moreover, there exists a coupling for which there is equality in (7.1.11).

Proof. Define $A \subseteq \mathcal{X}$ by $A := \{x \ : \ \rho(x) \geq \sigma(x)\}$, and define $\gamma(x) := \min\{\rho(x), \sigma(x)\}$. Then

$$\begin{aligned}
\mathbb{P}(\{X = X'\}) &= \mathbb{P}(\{X = X'\} \cap \{X \in A\}) + \mathbb{P}(\{X = X'\} \cap \{X \in A^c\}) \\
&\leq \mathbb{P}(\{X' \in A\}) + \mathbb{P}(\{X \in A^c\}) \\
&= \sum_{x\in A} \sigma(x) + \sum_{x\in A^c} \rho(x) = \sum_{x\in\mathcal{X}} \gamma(x).
\end{aligned}$$

For the second part, we may suppose that $\rho \neq \sigma$ since otherwise $\gamma(x, x') = \delta_{x,x'}\rho(x)$ is a coupling for which there is equality in (7.1.11). Define $p := \sum_{x\in\mathcal{X}} \gamma(x)$. Since $\rho \neq \sigma$, $p < 1$, and so we may define

$$\gamma(x, x') := \delta_{x,x'}\gamma(x) + \frac{1}{1-p}(\rho(x) - \gamma(x))(\sigma(x') - \gamma(x')).$$

Then simple computations show that $\gamma(x, x')$ is a coupling of ρ and σ for which there is equality in (7.1.11). $\qquad\square$

Definition 7.14. A coupling of $\rho, \sigma \in \mathcal{P}_\mathcal{X}$ such that there is equality in (7.1.11) is a **maximal coupling of ρ and σ**.

Lemma 7.15. *Let $\mathcal{X}$ be a finite set, and let $\rho, \sigma \in \mathcal{P}_\mathcal{X}$. Let X, X' be a maximal coupling of ρ and σ. Then*

$$(7.1.12) \qquad \mathbb{P}(\{X \neq X'\}) = \frac{1}{2}\sum_{x\in\mathcal{X}} |\rho(x) - \sigma(x)| = \frac{1}{2}\|\rho - \sigma\|_1.$$

Proof. Again define $\gamma(x) := \min\{\rho(x), \sigma(x)\}$. Then

$$\gamma(x) = \frac{\rho(x) + \sigma(x) - |\rho(x) - \sigma(x)|}{2}.$$

Summing over x yields $\sum_{x\in\mathcal{X}} \gamma(x) = 1 - \frac{1}{2}\sum_{x\in\mathcal{X}} |\rho(x) - \sigma(x)|$. Then (7.1.12) follows since X, X' is a maximal coupling of ρ and σ. $\qquad\square$

Proof of Theorem 7.10. Let X and Y be a maximal coupling of ρ and σ so that if ρ_{XY} is their joint distribution, $\rho_X = \rho$ and $\rho_Y = \sigma$. Without loss of generality, we may assume that $H(X) \geq H(Y)$. Since $H(XY) - H(X) = H(X|Y) \geq 0$, $H(X) \leq H(XY)$. Therefore,

$$H(X) - H(Y) \leq H(XY) - H(Y) = H(X|Y).$$

Since the coupling is maximal, the claim now follows from Fano's theorem and Lemma 7.15. $\qquad\qquad\qquad\qquad\qquad\qquad\qquad\qquad\qquad\qquad\qquad\qquad\qquad\quad$ $\square$

7.1.3. Relative entropy.

Definition 7.16. Given two probability densities ρ and σ on a finite set $\mathcal{X}$, the **relative entropy of ρ with respect to** σ, $D(\rho\|\sigma)$, is

$$(7.1.13) \quad D(\rho\|\sigma) = \begin{cases} \sum_{x\in\mathcal{X}} \rho(x)(\log\rho(x) - \log\sigma(x)) & \text{if } \sigma(x) = 0 \Rightarrow \rho(x) = 0 \\ \infty & \text{otherwise.} \end{cases}$$

Relative entropy is a measure of the distinguishability of two probability densities. In the information theoretic literature, it is sometimes called the **Kullback-Liebler divergence**, and the paper [136] of Kullback and Liebler can be consulted for more information on how this quantity measures the degree to which ρ *diverges* from σ, but the first fundamental fact is that $D(\rho\|\sigma) \geq 0$ unless $\rho = \sigma$.

To see this, assume that $\sigma(x) > 0$ whenever $\rho(x) > 0$. Then (7.1.13) can be written as

$$D(\rho\|\sigma) = \sum_{x\in\mathcal{X}} \sigma(x)\left(\frac{\rho(x)}{\sigma(x)}\right)\log\left(\frac{\rho(x)}{\sigma(x)}\right).$$

By the convexity of $t \mapsto t\log t$ and Jensen's inequality, $D(\rho\|\sigma) \geq 0$ with equality if and only if $\rho = \sigma$.

Lemma 7.17. *The function* $(\rho,\sigma) \mapsto D(\rho\|\sigma)$ *is jointly convex on* $\mathcal{P}(\mathcal{X}) \times \mathcal{P}(\mathcal{X})$.

Proof. Corollary 7.4 to the Gibbs variational principle can be written in terms of the relative entropy as

$$(7.1.14) \qquad D(\rho\|\sigma) = \sup_{h:\mathcal{X}\to\mathbb{R}} \left\{ \sum_{x\in\mathcal{X}} \rho(x)h(x) - \log\left(\sum_{x\in\mathcal{X}} e^{h(x)}\sigma(x) \right) \right\}$$

since the inequality (7.1.14) can be rewritten to say that for all h,

$$\sum_{x\in\mathcal{X}} \rho(x)h(x) - \log\left(\sum_{x\in\mathcal{X}} e^{h(x)}\sigma(x) \right) \leq D(\rho\|\sigma).$$

Moroever, there is equality if and only if $h = \log\rho - \log\sigma + c$, c any constant.

Observe that $\sum_{x \in \mathcal{X}} \rho(x) h(x)$ is linear in ρ, and $-\log\left(\sum_{x \in \mathcal{X}} e^{h(x)} \sigma(x)\right)$ is convex in σ. The sum of these two functions is jointly convex in (ρ, σ), and hence (7.1.14) expresses $D(\rho \| \sigma)$ as the supremum of a family of jointly convex functions. $\qquad \square$

We have already observed that $D(\rho \| \sigma) \geq 0$ with equality if and only if $\rho = \sigma$. However, more is true:

Theorem 7.18 (Pinsker's inequality). *Let ρ and σ be two probability densities. Then*

$$(7.1.15) \qquad D(\rho \| \sigma) \geq \frac{1}{2}\left(\sum_{x \in \mathcal{X}} |\rho(x) - \sigma(x)|\right)^2.$$

This inequality is known as Pinsker's inequality [180], though he did not have the sharp constant $\frac{1}{2}$ which is due to Csiszár and Kullback [65,135]. The proof of Theorem 7.18 proceeds in two steps. The first step is to prove it in the simplest case in which $\mathcal{X} = \{0, 1\}$. Every $\rho \in \mathcal{P}_x$ is determined by specifying $p := \rho(0)$ since then $\rho(1) = 1 - p$: for this simple choice of $\mathcal{X}$, $\rho \in \mathcal{P}_x$ may be identified with $[0, 1]$.

Lemma 7.19. *Let $\mathcal{X} = \{0, 1\}$. Then (7.1.15) is valid for all $\rho, \sigma \in \mathcal{P}_x$.*

Proof. Let $p := \rho(0)$ and $q := \sigma(0)$. We may assume $0 < q < 1$ since the endpoint cases are trivial. Then expressing both sides of (7.1.15) in terms of p and q, it becomes

$$(7.1.16) \qquad q\left(\frac{p}{q}\right)\log\left(\frac{p}{q}\right) + (1-q)\left(\frac{1-p}{1-q}\right)\log\left(\frac{1-p}{1-q}\right) \geq 2(p-q)^2.$$

Define $t := p/q - 1$, and define the function

$$\phi_q(t) := q(1+t)\log(1+t) + (1-q)(1 - tq/(1-q))\log(1 - tq/(1-q))$$

for $-1 < t < (1-q)/q$. Evidently, $\phi_q(0) = \phi_q'(0) = 0$, and

$$\phi_q''(t) = q^2\left(\frac{1}{q+qt} + \frac{1}{1-q-qt}\right).$$

Note that ϕ_q'' is strictly convex on $(-1, (1-q)/q)$, and it tends to infinity at both boundary points. Therefore, ϕ_q'' is minimized at its unique critical point in $(-1, (1-q)/q)$. Simple computations show that $\phi_q''(t) \geq 4q^2$, and hence $\phi_q(t) \geq 2q^2t^2$ for all $(-1, (1-q)/q)$. Recalling that $t = p/q - 1$, (7.1.16) is proved. $\qquad \square$

Proof of Theorem 7.18. We may assume $\sigma(x) > 0$ for all x. Define $f(x) := \rho(x)/\sigma(x)$ so that $D(\rho\|\sigma) = \sum_{x\in\mathcal{X}} f(x)\log f(x)\sigma(x)$. Define $A \subset \mathcal{X}$ by $A := \{x \in \mathcal{X} : f(x) > 1\}$, and define

$$P := \sum_{x\in A} f(x)\sigma(x) = \sum_{x\in A} \rho(x) \quad\text{and}\quad Q := \sum_{x\in A} \sigma(x).$$

Then

$$\sum_{x\in\mathcal{X}} |\rho(x) - \sigma(x)| = \sum_{x\in A}(\rho(x) - \sigma(x)) + \sum_{x\in A^c}(\sigma(x) - \rho(x))$$
$$= 2(P - Q),$$

and hence

$$(7.1.17)\qquad\qquad \frac{1}{2}\left(\sum_{x\in\mathcal{X}} |\rho(x) - \sigma(x)|\right)^2 = 2(P - Q)^2.$$

By Jensen's inequality once more,

$$D(\rho\|\sigma) = Q\frac{1}{Q}\sum_{x\in A} f(x)\log f(x)\sigma(x)$$
$$+ (1 - Q)\frac{1}{1-Q}\sum_{x\in A^c} f(x)\log f(x)\sigma(x)$$
$$\geq Q\left(\frac{1}{Q}\sum_{x\in A} f(x)\sigma(x)\right)\log\left(\frac{1}{Q}\sum_{x\in A} f(x)\sigma(x)\right)$$
$$+ (1-Q)\left(\frac{1}{1-Q}\sum_{x\in A^c} f(x)\sigma(x)\right)\log\left(\frac{1}{1-Q}\sum_{x\in A^c} f(x)\sigma(x)\right)$$
$$= Q\left(\frac{P}{Q}\right)\log\left(\frac{P}{Q}\right) + (1-Q)\left(\frac{1-P}{1-Q}\right)\log\left(\frac{1-P}{1-Q}\right),$$

and by Lemma 7.19, this last quantity is at least $2(P-Q)^2$. Combining this with (7.1.17), the proof is complete. $\qquad\square$

7.1.4. Mutual information. Given three finite sets $\mathcal{X}$, $\mathcal{Y}$, and $\mathcal{Z}$ and a probability density ρ on $\mathcal{X} \times \mathcal{Y} \times \mathcal{Z}$, form the conditional joint and marginal distributions of X and Y, given $Z = z$:

$$\rho(x, y|z) := \frac{\rho(x, y, z)}{\rho(z)}, \qquad \rho(x|z) := \frac{\rho(x, z)}{\rho(z)}, \quad\text{and}\quad \rho(y|z) := \frac{\rho(y, z)}{\rho(z)}.$$

Then for each $z \in \mathcal{Z}$, $\rho(x, y|z)$ and $\rho(x|z)\rho(y|z)$ are two probability densities in $\mathcal{X} \times \mathcal{Y}$. We can thus form their relative entropy, and by Theorem 7.18

$$D(\rho(x, y|z)\|\rho(x|z)\rho(y|z)) \geq 0$$

with equality if and only if $\rho(x, y|z) = \rho(x|z)\rho(y|z)$.

Definition 7.20 (Conditional mutual information). The **conditional mutual information of** X **and** Y **given** Z is the quantity $I(X, Y|Z)$ defined by

$$I(X, Y|Z) = \sum_{z \in \mathcal{Z}} \rho(z) D(\rho(x, y|z) \| \rho(x|z)\rho(y|z)).$$

The random variables X and Y are **conditionally independent, given** Z in case $\rho(x, y|z)$ is a product; that is $\rho(x, y|z) = \rho(x|z)\rho(y|z)$.

It is an immediate consequence of the nonnegativity of the relative entropy that $I(X, Y|Z) \geq 0$, with equality if and only if X and Y are conditionally independent given Z. In this case, once Z is known, the value of X gives no further information about the value of Y, and the value of Y gives no further information about the value of X. For this reason, the **Markov property** of a discrete time stochastic process may be expressed in terms of conditional information.

Definition 7.21 (Markov chain). Let $\mathcal{S}$ be a finite set. Let $N \in \mathbb{N}$, and $\mathbb{P}$ be a probability measure on the cartesian product $\mathcal{S}^N$. For each $1 \leq j \leq N$, define $S_j : \mathcal{S}^N \to \mathcal{S}$ by $S_j(s_1, \ldots, s_N) = s_j$. Then S_j is a random variable whose distribution is determined by $\mathbb{P}$. Regard this sequence of random variables as a discrete time stochastic process in $\mathcal{S}$, which is then the **state space** of the process. For each $1 < j \leq N$, define the **past at time** j to be the random variable $P_j(s_1, \ldots, s_N) = (s_1, \ldots, s_{j-1})$ which takes values in $\mathcal{S}^{j-1}$ and specifies the history of the process prior to time j. Likewise, for $1 \leq j < N$, define the **future at time** j to be the random variable $F_j(s_1, \ldots, s_N) = (s_{j+1}, \ldots, s_N)$ which takes values in $\mathcal{S}^{N-j}$ and specifies the future of the process after time j. Then the stochastic process $S_1, \ldots, S_N$ is a **Markov chain** in case for each $1 < j < N$, P_j and F_j are conditionally independent given S_j. By what we have noted above, this is the case if and only if $I(P_j, F_j|S_j) = 0$ for each $1 < j < N$.

When $S_1, \ldots, S_N$ is a Markov chain, the probability measure $\mathbb{P}$ has a particularly simple structure,

$$(7.1.18) \quad \mathbb{P}(s_1, \ldots, s_N)$$
$$= \mathbb{P}(P_N = (s_1, \ldots, s_{N-1}))\mathbb{P}(s_1, \ldots, s_N) \mid P_N = (s_1, \ldots, s_{N-1}))$$
$$= \mathbb{P}(P_N = (s_1, \ldots, s_{N-1}))\mathbb{P}(S_N = s_N \mid S_{N-1} = s_{N-1}),$$

where the second equality is valid on account of the Markov property.

For each $j = 1, \ldots, N - 1$ define a function $P_{j,j+1}(s, s')$ on $\mathcal{S} \times \mathcal{S}$ by

$$P_{j,j+1}(s, s') := \mathbb{P}(S_{j+1} = s' \mid S_j = s).$$

Evidently, for each s such that $\mathbb{P}(S_j = s) \neq 0$, $\sum_{s' \in \mathcal{S}} P_{j,j+1}(s, s') = 1$, and $P_{j,j+1}(s, s')$ gives the probability of a transition from s to s' at time j, given that the state at time j is s. Returning to (7.1.18), we may write this as

$$\mathbb{P}(s_1, \ldots, s_N) = \mathbb{P}(P_N = (s_1, \ldots, s_{N-1}))P_{N-1,N}(s_{N-1}, s_N).$$

When $S_1, \ldots, S_N$ is a Markov chain, so is $S_1, \ldots, S_{N-1}$, and we may express $\mathbb{P}(P_{N-1} = (s_1, \ldots, s_N))$ in terms of $\mathbb{P}(P_{N-1} = (s_1, \ldots, s_{N-2}))$ and the transition probability $P_{N-2,N-1}(s_{N-2}, s_{N-1})$ using the same reasoning. Continuing in this way, a simple induction yields

$$\mathbb{P}(s_1, \ldots, s_N) = \mathbb{P}(S_1 = s_1) \prod_{j=1}^{N-1} P_{j,j+1}(s_j, s_{j+1}).$$

That is, the probability measure $\mathbb{P}$ on the path space S^N of a Markov chain is completely determined by the initial probability distribution $\mathbb{P}(S_1 = s)$ and the transition probability functions $P_{j,j+1}(s, s')$, $j = 1, \ldots, N - 1$.

It is a remarkable fact that although the notion of conditional probability is problematic in quantum mechanics, as explained in the next section, there is a natural notion of a quantum Markov chain, and again the Markov property is characterized by zero conditional information. The key to extending the notion of conditional information to the quantum setting is to write it in terms of entropies of marginals without making reference to conditional probabilities. This is easy to do. A simple calculation shows

$$(7.1.19) \qquad I(X, Y|Z) = H(XZ) + H(YZ) - H(XYZ) - H(Z),$$

and this proves

$$(7.1.20) \qquad H(XYZ) + H(Z) \leq H(XZ) + H(YZ)$$

with equality if and only if $I(X, Y|Z) = 0$.

Specializing to the case in which Z consists of a single point, $H(Z) = 0$, $H(XZ) = H(X)$, $H(YZ) = H(Y)$, and $H(XYZ) = H(XY)$, and we obtain the classical subadditivity of the entropy (7.1.7). The stronger inequality (7.1.20) is known in the mathematical physics literature as the **classical strong subadditivity of the entropy**.

7.1.5. Other relative entropies and variational formulas.

The relative entropy provides an important measure of the *divergence* between two classical probability densities. There are other closely related functionals that provide other measures of the divergence that, depending on the application, may be more incisive or relevant.

Definition 7.22. A **relative entropy generating function** is a function $g : (0, \infty) \to \mathbb{R}$ that is twice continuously differentiable and strictly convex such that $g(1) = 0$.

The **perspective function** [66] of g is the function $\hat{g} : (0, \infty) \times (0, \infty) \to \mathbb{R}$ defined by

$$(7.1.21) \qquad \hat{g}(s, t) := tg(s/t).$$

The **quasi relative entropy** D_g associated to g is the functional of $\mathcal{P}_X \times \mathcal{P}_X$, X a finite set, given by

$$(7.1.22) \qquad D_g(\rho\|\sigma) = \sum_{x \in X} \hat{g}(\rho(x), \sigma(x)).$$

Example 7.23. Taking $g(t) = t \log t$, $\hat{g}(s,t) = s(\log s - \log t)$ and $D_g(\rho\|\sigma) = \sum_{x \in X} \rho(x)(\log \rho(x) - \log \sigma(x))$. Thus, the standard relative entropy is a quasi entropy.

For different examples, take $g(t) = 1 - t^p$, $0 < p < 1$. Then $\hat{g}(s,t) = t - t^{1-p}s^p$ and $D_g(\rho\|\sigma) = 1 - \sum_{x \in X} \sigma^{1-p}(x)\rho^p(x)$.

Lemma 7.24. *Let g be a relative entropy generating function. Then the corresponding perspective function $\hat{g}$ is homogeneous of degree one and convex.*

Proof. It is evident that $\hat{g}$ is homogeneous of degree one. Therefore, to prove the convexity of $\hat{g}$ it suffices to prove that $\hat{g}$ is subadditive. Let $s_1, s_2, t_1, t_2 > 0$. Then

$$g\left(\frac{s_1 + s_2}{t_1 + t_2}\right) = g\left(\frac{t_1}{t_1 + t_2}\frac{s_1}{t_1} + \frac{t_2}{t_1 + t_2}\frac{s_2}{t_2}\right)$$
$$\leq \frac{t_1}{t_1 + t_2}g\left(\frac{s_1}{t_1}\right) + \frac{t_2}{t_1 + t_2}g\left(\frac{s_2}{t_2}\right),$$

and since g is strictly convex, there is equality if and only if $s_1/t_1 = s_2/t_2$. Then by the definition of the perspective function,

$$\hat{g}(s_1 + s_2, t_1 + t_2) \leq \hat{g}(s_1, t_1) + \hat{g}(s_2, t_2)$$

with equality if and only if $s_1/t_1 = s_2/t_2$. $\qquad\square$

By Theorem A.32, g^*, the Legendre transform of g, is twice continuously differentiable and strictly convex on (c,d) where $c := \lim_{t \to 0} g'(t)$ and $d := \lim_{t \to \infty} g'(t)$. Also by Theorem A.32, $(g^*)'(g'(t)) = t$ for all $t > 0$. Therefore, g^* is strictly monotone increasing on (c,d).

Define $a := \lim_{u \downarrow c} g^*(u)$ and $b := \lim_{u \uparrow d} g^*(u)$. Define w to be the inverse function of g^*, so that for all $v \in (a,b)$, $g^*(w(v)) = v$, and for all $u \in (-\infty, c)$, $w(g^*(u)) = u$.

Lemma 7.25. *Let g be a relative entropy generating function, and let w be the inverse function to its Legendre transform g^*. Then w is concave.*

Proof. By the convexity of g^*, and the fact that it is inverse to w,

$$g^*((1 - \lambda)w(v_1) + \lambda w(v_2)) \leq (1 - \lambda)v_1 + \lambda v_2.$$

Since w is the inverse of g^*, w is also monotone increasing, and applying w to both sides yields $(1 - \lambda)w(v_1) + \lambda w(v_2) \leq w((1 - \lambda)v_1 + \lambda v_2)$. $\qquad\square$

Theorem 7.26. *Let $\mathcal{X}$ be a finite set, and let $\mathcal{P}_{\mathcal{X}}$ be the set of probability densities on $\mathcal{X}$. Let g be a relative entropy generating function. Then the corresponding quasi relative entropy satisfies*

$$D_g(\rho\|\sigma) \geq 0$$

with equality if and only if $\rho = \sigma$. Moreover, $D_g(\rho,\sigma)$ is a jointly convex function of ρ and σ, as displayed in the following two variational formulas for $D_g(\rho,\sigma)$. First,

$$(7.1.23) \qquad D_g(\rho\|\sigma) = \sup_{h:\mathcal{X}\to\mathbb{R}} \sum_{x\in\mathcal{X}} \big(\rho(x)h(x) - \sigma(x)g^*(h(x))\big),$$

with the supremum attained only at $h(x) = g'(\rho(x)/\sigma(x))$ for all x.

Second, with w denoting the inverse function to g^*, and I being the open interval on which it is defined,

$$(7.1.24) \qquad D_g(\rho\|\sigma) = \sup_{h:\mathcal{X}\to I} \sum_{x\in\mathcal{X}} (\rho(x)w(h(x)) - \sigma(x)h(x)),$$

and the supremum is attained only at $h(x) = g^*(g'(\rho(x)/\sigma(x)))$.

Proof. Let $\hat{g}$ be the perspective function of g. Then by (7.1.21), (7.1.22), and Jensen's inequality,

$$D_g(\rho\|\sigma) = \sum_{x\in\mathcal{X}} \sigma(x)g(\rho(x)/\sigma(x)) \geq g\left(\sum_{x\in\mathcal{X}} \rho(x)\right) = g(1) = 0,$$

and since g is strictly convex, there is equality if and only if $\rho = \sigma$.

Next, by the Fenchel-Moreau theorem, $g(t) = \sup_{u\in\mathbb{R}}\{ut - g^*(u)\}$, and therefore, for each x,

$$\hat{g}(\rho(x),\sigma(x)) = \sup_{u\in\mathbb{R}}\{\rho(x)u - \sigma(x)g^*(u)\},$$

with the supremum being attained only for $(g^*)'(u) = \rho(x)/\sigma(x)$, from which (7.1.23) follows directly. Then since g' is the inverse function to $(g^*)'$, the unique optimizing function h is given by $h(x) = g'(\rho(x)/\sigma(x))$.

Then replacing $h(x)$ by $w(h(x))$, where w is the inverse function to g^*, one deduces the equivalent formulation (7.1.24). $\qquad\square$

Example 7.27. Take $g(t) = t\log t$ for $t \geq 0$. Then $g^*(u) = e^{u-1}$, and the functions inverse to g^* is the function w on $(0,\infty)$ given by $w(v) = \log(v) + 1$. Hence for the regular relative entropy, we have the two variational formulas,

$$D(\rho\|\sigma) = \sup_{h:\mathcal{X}\to\mathbb{R}} \sum_{x\in\mathcal{X}} \big(\rho(x)h(x) - \sigma(x)e^{h(x)-1}\big)$$

and, with w denoting the inverse function to g^*,

$$(7.1.25) \qquad D(\rho\|\sigma) = \sup_{h:\mathcal{X}\to(0,\infty)} \sum_{x\in\mathcal{X}} (\rho(x)(\log(h(x)) + 1) - \sigma(x)h(x)).$$

These are both different from the variational formula (7.1.14) coming from the Gibbs variational principle. The variational formula (7.1.25) is particularly useful in quantum applications, as shown in [**33**], because of the operator concavity of the logarithm. We shall return to this after discussing quantum relative entropy.

Definition 7.28. For $0 < \alpha < 1$, the α-**Rényi relative entropy**, or α-**Rényi divergence** is the functional D_α on $\mathcal{P}_\mathcal{X} \times \mathcal{P}_\mathcal{X}$ given by

$$D_\alpha(\rho\|\sigma) := \frac{1}{\alpha - 1} \log\left(\sum_{x\in\mathcal{X}} \rho^\alpha(x)\sigma^{1-\alpha}(x)\right).$$

For $\alpha > 1$, $D_\alpha(\rho\|\sigma)$ is defined by the same formula except we set $D_\alpha(\rho\|\sigma) = \infty$ if for some $x \in \mathcal{X}$, $\sigma(x) = 0$ and $\rho(x) > 0$.

The α-Rényi relative entropy can be written in terms of the quasi relative entropy for the function $g_\alpha(t) = t^\alpha$ for $t > 0$. Indeed,

$$D_\alpha(\rho\|\sigma) = \frac{1}{\alpha - 1} \log\big(1 - D_{g_\alpha}(\rho\|\sigma)\big).$$

Then Theorem 7.26 gives variational formulas for the α-Rényi relative entropy.

The standard relative entropy arises as a limit of the α-Rényi relative entropy:

$$D(\rho\|\sigma) = \lim_{\alpha\to 1} D_\alpha(\rho\|\sigma).$$

Other aspects of α-Rényi relative entropy are explored in the exercises.

We have focused here on classical entropy on discrete probability spaces because that is what is relevant here. See [**142**] for a parallel treatment of classical discrete, classical continuous, and quantum entropy.

7.2. Quantum entropy

The von Neumann entropy is a direct generalization of the classical entropy function:

Definition 7.29. The **von Neumann entropy** of $\rho \in \mathfrak{S}$, $S(\rho)$, is defined by

$$S(\rho) = -\operatorname{Tr}[\rho \log \rho].$$

$S(\rho)$ is often called the **quantum entropy**. Note that $S(\rho)$ depends on ρ only through the eigenvalues of ρ, repeated according to multiplicity, and in particular, if U is any unitary, $S(U^*\rho U) = S(\rho)$.

If ρ is a density matrix on a Hilbert space of finite dimension d with the spectral decomposition $\rho = \sum_{j=1}^{d} \lambda_j |\mathbf{u}_j\rangle\langle\mathbf{u}_j|$, then $S(\rho) = -\sum_{j=1}^{n} \lambda_j \log \lambda_j$. In particular, if $X = \{1,\ldots,d\}$ and ρ_X is the classical probability density $\rho_X(j) :=\lambda_j$, then

$$(7.2.1) \qquad\qquad S(\rho) = H(\rho_X).$$

Lemma 7.30. *For any finite-dimensional Hilbert space $\mathcal{H}$, the von Neumann entropy S is a strictly concave, unitarily invariant function on $\mathfrak{S}_{\mathcal{H}}$, and for all $\rho \in \mathfrak{S}_{\mathcal{H}}$,*

$$(7.2.2) \qquad\qquad 0 \le S(\rho) \le \log(\dim(\mathcal{H})).$$

Moreover, $S(\rho) = 0$ if and only if ρ is a pure state; that is, if and only if for some unit vector ψ, $\rho = |\psi\rangle\langle\psi|$, and $S(\rho) = \log(\dim(\mathcal{H}))$ if and only if $\rho = (\dim(\mathcal{H}))^{-1}\mathbb{1}$.

Proof. Since

$$(7.2.3) \qquad\qquad f(x) := \begin{cases} x\log x & x \ge 0 \\ +\infty & x < 0 \end{cases}$$

is a strictly convex function on $\mathbb{R}$, $\mathrm{Tr}[f(X)]$ is a strictly convex function on $\mathcal{B}^{\mathrm{s.a.}}(\mathcal{H})$ by Theorem 3.3. Hence its restriction to the set of density matrices $\mathfrak{S}$ is strictly convex.

Let $\rho = \sum_{j=1}^{d}\lambda_j|\mathbf{u}_j\rangle\langle\mathbf{u}_j|$ be a spectral expansion of ρ, and let ρ_X be the classical probability density $\rho_X(j) = \lambda_j$ on $X = \{1,\ldots,d\}$. Then (7.2.2) follows from the identity (7.2.1) and Lemma 7.2. Moreover, by Lemma 7.2, $H(\rho_X) = d$ if and only if $\rho_X(j) = \frac{1}{d}$ for all j, in which case $\rho = \frac{1}{d}\mathbb{1}$. Likewise, $H(\rho_X) = 0$ if and only if for some $k \in \{1,\ldots,d\}$, $\rho_X(j) = \delta_{j,k}$, and in this case $\rho = |\mathbf{u}_k\rangle\langle\mathbf{u}_k|$. $\qquad\square$

Lemma 7.30 can be used to prove an important monotonicity property of the quantum entropy.

Theorem 7.31. *Let $\mathcal{H}$ be a Hilbert space of finite dimension d. Let $\{P_1,\ldots,P_k\}$ be a PVM; i.e, a set of mutually orthogonal projections with $\sum_{j=1}^{k}P_j = \mathbb{1}$. Let $\mathcal{P}$ be corresponding pinching operation, $\mathcal{P}(A) = \sum_{j=1}^{d}P_jAP_j$. Then for all density matrices ρ on $\mathcal{H}$,*

$$S(\rho) \le S(\mathcal{P}(\rho)).$$

Proof. Since the pinching operation is a conditional expectation, by Theorem 4.40, there exists a set $\{U_1,\ldots,U_m\}$ of unitaries on $\mathcal{H}$ such that $\mathcal{P}(X) = \frac{1}{m}\sum_{k=1}^{m} UXU^*$. Then since $S(\cdot)$ is concave and unitarily invariant,

$$S(\mathcal{P}(\rho)) = S\left(\frac{1}{m}\sum_{k=1}^{m} U\rho U^*\right) \ge \frac{1}{m}\sum_{k=1}^{m} S(U\rho U^*) = S(\rho). \qquad\square$$

It is now a simple matter to extend Theorem 7.10 on the quantitative continuity of the classical entropy to the quantum setting [17, 234]:

Theorem 7.32. *Let $\mathcal{H}$ be a set of finite dimension $d \geq 2$. For any probability densities $\rho, \sigma \in \mathfrak{S}_{\mathcal{H}}$ such that $\|\rho - \sigma\|_1 \leq 2(d-1)/d$,*

$$|S(\rho) - S(\sigma)| \leq \left(\frac{1}{2}\|\rho - \sigma\|_1\right)\log(d-1) + \eta\left(\frac{1}{2}\|\rho - \sigma\|_1\right),$$

where for $t \in [0,1]$, $\eta(t) := -t\log t - (1-t)\log(1-t) \geq 0$.

Proof. We may suppose that $S(\sigma) \geq S(\rho)$. Let $\rho = \sum_{j=1}^{d} \lambda_j |\mathbf{u}_j\rangle\langle\mathbf{u}_j|$ be a spectral decomposition of ρ. Define $P_j := |\mathbf{u}_j\rangle\langle\mathbf{u}_j|$, $j = 1, \ldots, d$, and the pinching operation $\mathcal{P}$ by $\mathcal{P}(X) := \sum_{j=1}^{d} P_j X P_j$, and note that $\mathcal{P}(\rho) = \rho$. Let ρ_{cl} and σ_{cl} denote the classical probability densities on $\{1, \ldots, d\}$ associated to $\mathcal{P}(\rho)$ and $\mathcal{P}(\sigma)$, respectively. Then $S(\rho) = H(\rho_{\mathrm{cl}})$, and by Theorem 7.31, $S(\sigma) \leq S(\mathcal{P}(\sigma)) = H(\sigma_{\mathrm{cl}})$,

$$S(\sigma) - S(\rho) \leq S(\mathcal{P}(\sigma)) - S(\mathcal{P}(\rho)) = H(\sigma_{\mathrm{cl}}) - H(\rho_{\mathrm{cl}}).$$

Then by Theorem 7.10,

$$H(\sigma_{\mathrm{cl}}) - H(\rho_{\mathrm{cl}}) \leq \frac{1}{2}\|\rho_{\mathrm{cl}} - \sigma_{\mathrm{cl}}\|_1 \log(d-1) + \eta\left(\frac{1}{2}\|\rho_{\mathrm{cl}} - \sigma_{\mathrm{cl}}\|_1\right).$$

By Theorem 6.32, $\|\rho_{\mathrm{cl}} - \sigma_{\mathrm{cl}}\|_1 = \|\mathcal{P}(\rho) - \mathcal{P}(\sigma)\|_1 \leq \|\rho - \sigma\|_1$. $\qquad\square$

Remark 7.33. The bound provided by Theorem 7.32 is sharp because it is sharp in the classical case; see Remark 7.11.

We close this subsection with a complement to the concavity of the entropy. Let $\{\rho^{(1)}, \ldots, \rho^{(m)}\}$ be a set of m density matrices, and let $\{p_1, \ldots, p_m\}$ be a set of m nonnegative numbers with $\sum_{j=1}^{m} p_j = 1$. Then the concavity of the entropy says that

$$(7.2.4) \qquad \sum_{j=1}^{m} p_j S(\rho^{(j)}) \leq S\left(\sum_{j=1}^{m} p_j \rho^{(j)}\right).$$

Evidently there is equality in (7.2.4) if for some j, $p_j = 1$, which is the case if and only if $H(p) := -\sum_{j=1}^{m} p_j \log p_j = 0$, independent of what the density matrices are. It turns out that there is almost equality in (7.2.4) whenever $H(p)$ is small. This was discovered and used in the classical case first, but the proof extends directly to the quantum case; see [72, 128].

Lemma 7.34. *Let $\{\rho^{(1)}, \ldots, \rho^{(m)}\}$ be a set of m density matrices, and let $\{p_1, \ldots, p_m\}$ be a set of m nonnegative numbers with $\sum_{j=1}^{m} p_j = 1$. Then*

$$(7.2.5) \qquad S\left(\sum_{j=1}^{m} p_j \rho^{(j)}\right) \leq \sum_{j=1}^{m} p_j S(\rho^{(j)}) - \sum_{j=1}^{m} p_j \log p_j.$$

Proof. Since the logarithm is operator monotone, for each k,

$$-\log\left(\sum_{j=1}^{m} p_j \rho^{(j)}\right) \leq -\log\left(p_k \rho^{(k)}\right).$$

Taking the trace against $\sum_{j=1}^{m} p_j \rho^{(j)}$ yields (7.2.5). $\qquad\square$

7.2.1. von Neumann entropy on bipartite systems. Let $\mathcal{H}_1$ and $\mathcal{H}_2$ be two Hilbert spaces. We may think of these as the state spaces for two parts of a bipartite system whose state space is $\mathcal{H}_1 \otimes \mathcal{H}_2$. Let ρ_{12} denote a density matrix on $\mathcal{H}_1 \otimes \mathcal{H}_2$. Then we write ρ_1 and ρ_2 to denote the two partial traces:

$$\rho_1 := \mathrm{Tr}_2[\rho_{12}] \quad \text{and} \quad \rho_2 := \mathrm{Tr}_1[\rho_{12}].$$

There are then three entropies associated to ρ_{12}, namely $S(\rho_{12})$, $S(\rho_1)$, and $S(\rho_2)$. The following brief notation will be useful:

$$S_{12} = S(\rho_{12}), \qquad S_1 = S(\rho_1), \quad \text{and} \quad S_2 = S(\rho_2).$$

Several important inequalities relate these three entropies. The first is the **sub-additivity of the entropy**.

Theorem 7.35. *Let ρ_{12} be a density matrix on the tensor product of finite-dimensional Hilbert spaces $\mathcal{H}_1 \otimes \mathcal{H}_2$. Then*

$$S(\rho_{12}) \leq S(\rho_1) + S(\rho_2),$$

and there is equality if and only $\rho_{12} = \rho_1 \otimes \rho_2$.

We will use the following lemma:

Lemma 7.36. *Let ρ and σ be two density matrices on a finite-dimensional Hilbert space $\mathcal{H}$. Suppose that $\ker(\sigma) \subseteq \ker\rho$. Then*

$$(7.2.6) \qquad\qquad \mathrm{Tr}[\rho(\log\rho - \log\sigma)] \geq 0,$$

and there is equality if and only if $\rho = \sigma$.

Proof. Apply Klein's inequality with $f(x) = x\log x$ to obtain

$$\mathrm{Tr}[\rho\log\rho] \geq \mathrm{Tr}[\sigma\log\sigma] + \mathrm{Tr}[(1+\log\sigma)(\rho-\sigma)] = \mathrm{Tr}[\rho] - \mathrm{Tr}[\sigma] + \mathrm{Tr}[\rho\log\sigma].$$

Since $\mathrm{Tr}[\rho] = \mathrm{Tr}[\sigma] = 1$, this reduces to (7.2.6). Since $f(x) = x\log x$ is strictly convex, the statement about cases of equality follows from the statement about cases of inequality in Klein's inequality. $\qquad\square$

Proof of Theorem 7.35. Apply Lemma 7.36 with $\mathcal{H} = \mathcal{H}_1 \otimes \mathcal{H}_2$, $\rho = \rho_{12}$, and $\sigma = \rho_1 \otimes \rho_2$, and by Lemma 7.36, there is equality if and only if $\rho_{12} = \rho_1 \otimes \rho_2$. $\qquad\square$

The second of these inequalities is the **Araki-Lieb triangle inequality**. By Lemma 7.30, ρ_{12} is a pure state if and only if $S_{12} = 0$, and then by Corollary 2.17, $S(\rho_1) = S(\rho_2)$. Note that the dimensions of $\mathcal{H}_1$ and $\mathcal{H}_2$ are unrelated and can be very different, but if ρ_1 and ρ_2 are the partial traces of a pure state on $\mathcal{H}_1 \otimes \mathcal{H}_2$, they have the same entropy, and hence $S(\rho_2) \leq \log(\dim(\mathcal{H}_1))$, no matter how large the dimension of $\mathcal{H}_2$ may be.

A theorem of Araki and Lieb [**15**] says that if ρ_{12} is almost pure, in the sense that $S(\rho_{12})$ is small, then $S(\rho_1)$ and $S(\rho_2)$ are almost equal.

Theorem 7.37 (Araki-Lieb triangle inequality). *Let ρ_{12} be a density matrix on the tensor product of Hilbert spaces $\mathcal{H}_1 \otimes \mathcal{H}_2$. Then*

$$S(\rho_{12}) \geq |S(\rho_1) - S(\rho_2)|.$$

Proof. By Theorem 2.21, there exists a Hilbert space $\mathcal{H}_3$, which we can take to have the same dimension as $\mathcal{H}_1 \otimes \mathcal{H}_2$, and a pure state

$$\rho_{123} := |\psi\rangle\langle\psi|$$

on $\mathcal{H}_1 \otimes \mathcal{H}_2 \otimes \mathcal{H}_3$ such that $\mathrm{Tr}_3[\rho_{123}] = \rho_{12}$. Let $\rho_{23} = \mathrm{Tr}_1[\rho_{123}]$, etc. Then since ρ_{123} is pure, Corollary 2.17 says that $S(\rho_1) = S(\rho_{23})$ and $S(\rho_{12}) = S(\rho_3)$. Hence

$$\begin{aligned} S(\rho_{12}) - S(\rho_1) &= S(\rho_3) - S(\rho_{23}) \\ &\geq -S(\rho_2), \end{aligned}$$

where the inequality comes from Theorem 7.35. Therefore, $S(\rho_{12}) \geq S(\rho_1) - S(\rho_2)$, and then by symmetry, also $S(\rho_{12}) \geq S(\rho_2) - S(\rho_1)$. $\qquad\square$

7.2.2. Quantum conditional entropy. There simply is not a meaningful way to define an analogue of the conditional probability densities $\rho(y|x) = \rho_{XY}(x, y)/\rho_X(x)$ and $\rho(x|y) = \rho_{XY}(x, y)/\rho_Y(y)$ for a quantum bipartite state ρ_{12}. What is missing is an analogue of the set $\mathcal{Y}$ of all possible outcomes of an observation of the second system, because in the quantum case we must choose an observable A to observe, and different choices of A will have different spectrum $\sigma(A)$, and hence different possible outcomes.

However, given two classical random variables X and Y, the classical conditional probability $H(Y|X)$ is given by $H(Y|X) = H(XY) - H(X)$, a formula that does not *directly* involve conditional probabilities. The quantum analogues of $H(XY)$ and $H(X)$ are S_{12} and S_1, respectively, and hence we define:

Definition 7.38. Consider a bipartite quantum system on $\mathcal{H}_1 \otimes \mathcal{H}_2$ where the Hilbert spaces $\mathcal{H}_1$ and $\mathcal{H}_2$ are the state spaces of the two component systems.

Let $\rho_{12} \in \mathfrak{S}_{\mathcal{H}_1 \otimes \mathcal{H}_2}$. Then the **quantum conditional entropy** of ρ_{12} conditioned on system 1 is the quantity $S_{2|1}(\rho_{12})$ given by

$$S_{2|1}(\rho_{12}) := S_{12} - S_1.$$

The quantum conditional entropy of ρ_{12} conditioned on system 2, $S_{1|2}(\rho_{12})$, is given by $S_{1|2}(\rho_{12}) = S_{12} - S_2$.

As a consequence of Theorem 7.35, $S_{12} - S_1 \leq S_2$, and hence the quantum conditional entropy satisfies

$$(7.2.7) \qquad\qquad S_{2|1}(\rho_{12}) \leq S_2,$$

which is the analogue of the classical upper bound $H(Y|X) \leq H(Y)$.

One might hope that the quantum analogue of the lower bound $H(Y|X) \geq 0$ would also be valid. That is, one might hope that the inequality $S_{12} - S_1 > 0$ would be valid. *However, it is not.* In fact, let $\mathcal{H}$ be a Hilbert space of finite dimension d, and let $\{\mathbf{u}_1, \ldots, \mathbf{u}_d\}$ be an orthonormal basis for $\mathcal{H}$. Define a state ψ on $\mathcal{H} \otimes \mathcal{H}$ by

$$\psi := \frac{1}{\sqrt{d}} \sum_{j=1}^{d} \mathbf{u}_j \otimes \mathbf{u}_j,$$

and then define $\rho_{12} := |\psi\rangle\langle\psi|$. Since ρ_{12} is a rank one projection, i.e., a pure state, $S(\rho_{12}) = 0$. However, simple computations show that $\rho_1 = \rho_2 = \frac{1}{d}\mathbb{1}$, and hence $S_1 = S_2 = \log d$. Hence $S_{2|1}(\rho_{12}) = S_{1|2}(\rho_{12}) = -\log d$. However, we do have a priori lower bounds showing that in this example the conditional entropy is as negative as possible. An interesting discussion of the meaning of negative quantum conditional entropy can be found in [**111**].

Lemma 7.39. *Let $\mathcal{H}_1$ and $\mathcal{H}_2$ be Hilbert spaces, and let ρ_{12} be a density matrix on $\mathcal{H}_1 \otimes \mathcal{H}_2$. Then*

$$(7.2.8) \qquad\qquad S_{2|1}(\rho_{12}) \geq -\min\{S_1, S_2\}.$$

Proof. Since $S_{2|1}(\rho_{12}) = S_{12} - S_1$, it follows from the nonnegativity of the entropy that $S_{2|1}(\rho_{12}) \geq -S_1$. Next, by the Araki-Lieb triangle inequality, $S_1 - S_2 \leq S_{12}$, and hence $-S_2 \leq S_{12} - S_1 = S_{2|1}(\rho_{12})$. $\qquad\square$

By Theorem 6.32 $\|\rho_1 - \sigma_1\|_1 \leq \|\rho_{12} - \sigma_{12}\|_1$. Consequently, two applications of Theorem 7.32 give a bound on the modulus of continuity of the conditional entropy $S_{2|1}(\rho_{12})$ as a function of ρ_{12}. However, this bound depends on the product of the dimensions of $\mathcal{H}_1$ and $\mathcal{H}_2$. Alicki and Fannes proved in [**5**] a much stronger bound, giving an estimate that depends only on the dimension of $\mathcal{H}_2$; this will be proved in the next chapter.

7.2.3. Quantum relative entropy.

Closely related to the von Neumann entropy is the Umegaki relative entropy [**217**]:

Definition 7.40. For two density matrices $\rho, \sigma \in \mathfrak{S}$, the **Umegaki relative entropy of ρ with respect to σ** is the quantity

$$(7.2.9) \qquad D(\rho\|\sigma) = \begin{cases} \mathrm{Tr}[\rho(\log\rho - \log\sigma)] & \ker(\sigma) \subseteq \ker(\rho) \\ \infty & \ker(\sigma) \nsubseteq \ker(\rho). \end{cases}$$

Definition 7.40 is a direct transcription of its classical version, Definition 7.16. Moreover, if ρ and σ commute, we can compute the trace in (7.2.9) using an orthonormal basis $\{\mathbf{u}_1, \dots, \mathbf{u}_d\}$ consisting of simultaneous eigenvectors of ρ and σ. Define ρ_{cl} and σ_{cl} to be the classical probability densities on $\{1, \dots, d\}$ given by $\rho_{\mathrm{cl}}(j) = \langle \mathbf{u}_j, \rho\mathbf{u}_j \rangle$ and $\sigma_{\mathrm{cl}}(j) = \langle \mathbf{u}_j, \sigma\mathbf{u}_j \rangle$, respectively. Then

$$D(\rho\|\sigma) = D(\rho_{\mathrm{cl}}\|\sigma_{\mathrm{cl}}).$$

Therefore, every theorem about Umegaki relative entropy includes as a special case a theorem about the classical relative entropy. The special case is that in which ρ and σ commute, and frequently this is called the *classical case*, and the general case is called the *quantum case*. It is a highly nontrivial fact that analogues of many theorems for the classical relative entropy are valid, as we shall see, when extended to the Umegaki relative entropy.

For the classical relative entropy, $\log\rho - \log\sigma$ is of course $\log(\rho/\sigma)$. For noncommutative densities matrices ρ and σ, there are other possibles interpretations of "ρ divided by σ". For example, one might consider $\rho^{1/2}\sigma^{-1}\rho^{1/2}$ leading to an alternate formula $\mathrm{Tr}[\rho\log(\rho^{1/2}\sigma^{-1}\rho^{1/2})]$, which also has the property that it reduces to the classical relative entropy when ρ and σ commute. This functional is known as the **Belavkin-Stasewski relative entropy** [**27**]. It turns out to be a "good" extension of the classical relative entropy. However, it is the definition of Umegaki that will be the most important in much of what follows, and we simply refer to it as the relative entropy of ρ with respect to σ.

Lemma 7.41. *For any finite-dimensional Hilbert space $\mathcal{H}$, the relative entropy is a unitarily invariant function on $\mathfrak{S}_{\mathcal{H}} \times \mathfrak{S}_{\mathcal{H}}$; that is for all unitary U on $\mathcal{H}$, and all $\rho, \sigma \in \mathfrak{S}_{\mathcal{H}}$,*

$$(7.2.10) \qquad D(U\rho U^*\|U\sigma U^*) = D(\rho\|\sigma),$$

and $D(\rho\|\sigma) \geq 0$ with equality if and only if $\rho = \sigma$. Furthermore, for each fixed $\sigma \in \mathfrak{S}_{\mathcal{H}}$, $D(\rho\|\sigma)$ is strictly convex as a function of ρ, and for each fixed $\rho \in \mathfrak{S}_{\mathcal{H}}$, $D(\rho\|\sigma)$ is strictly convex as a function of σ.

Proof of Lemma 7.41. The unitary invariance (7.2.10) is a consequence of the definition (7.2.9) and cyclicity of the trace. For the positivity, apply Klein's inequality with $f(x) = x\log x$ to obtain

$$\mathrm{Tr}[\rho\log\rho] \geq \mathrm{Tr}[\sigma\log\sigma] + \mathrm{Tr}[(1+\log\sigma)(\rho-\sigma)] = \mathrm{Tr}[\rho] - \mathrm{Tr}[\sigma] + \mathrm{Tr}[\rho\log\sigma].$$

Rearranging terms yields $\mathrm{Tr}[\rho(\log \rho - \log \sigma)] + \mathrm{Tr}[\sigma] - \mathrm{Tr}[\rho] \geq 0$; that is, $D(\rho\|\sigma) \geq 0$. Since $f(x) = x \log x$ is strictly convex on $[0, \infty)$, there is equality if and only if $\rho = \sigma$.

Since $f(x) = x \log x$ is strictly convex on $[0, \infty)$, $\rho \mapsto \mathrm{Tr}[\rho \log \rho]$ is strictly convex by Theorem 3.3, and the subset of $\mathfrak{S}_{\mathcal{H}}$ consisting of ρ such that $\ker(\rho) \subseteq \ker(\sigma)$ is convex. This proves the strict convexity in ρ for fixed σ. The function $\sigma \mapsto -\mathrm{Tr}[\rho \log(\sigma)]$ is strictly convex because of the strict operator convexity of the logarithm, and this proves the strict convexity in σ for fixed ρ. $\qquad\square$

It is also worth observing that the quantum conditional entropy may be expressed in terms of the quantum relative entropy as follows: Let ρ_{12} a bipartite state on $\mathcal{H}_1 \otimes \mathcal{H}_2$. Let $\tau_1 := \frac{1}{d_1}\mathbb{1}$ where d_1 is the dimension of $\mathcal{H}_1$. Then

$$D(\rho_{12}\|\tau_1 \otimes \rho_2) = \mathrm{Tr}[\rho_{12} \log \rho_{12}] - \mathrm{Tr}[\rho_2 \log \rho_2] + \log d_1$$
$$(7.2.11) \qquad\qquad = -(S_{12} - S_2) + \log d_1 = S_{1|2}(\rho_{12}) + \log d_1.$$

7.2.4. The chain rule for relative entropy. Let $\mathcal{M}$ be a commutative von Neumann subalgebra of $\mathcal{B}(\mathcal{H})$, and let $\mathcal{H}$ be a finite-dimensional Hilbert space. Let $\{P_1, \ldots, P_m\}$ be a set of mutually orthogonal projections whose span is $\mathcal{M}$. Let $\mathcal{H}_j$ denote $\mathrm{ran}(P_j)$. Every density matrix $\rho \in \mathcal{M}'$ has the block form

$$(7.2.12) \qquad\qquad \rho = \bigoplus_{j=1}^{m} p_j \rho_j,$$

where $p_j = \mathrm{Tr}[P_j \rho]$ and $P_j \rho P_j = p_j \rho_j$. For two density matrices $\rho, \sigma \in \mathcal{M}'$ there is a useful formula for $D(\rho\|\sigma)$ in terms of the constituents of the decompositions of ρ and σ as in (7.2.12).

Theorem 7.42 (Chain rule for relative entropy). *Let $\mathcal{M}$ be a commutative von Neumann subalgebra of $\mathcal{B}(\mathcal{H})$, and let $\mathcal{H}$ be a finite-dimensional Hilbert space. Let ρ, σ be density matrices in $\mathcal{M}'$. Let $\rho = \bigoplus_{j=1}^{m} p_j \rho_j$ and $\sigma = \bigoplus_{j=1}^{m} q_j \sigma_j$ be their decompositions as in (7.2.12). Then*

$$(7.2.13) \qquad D(\rho\|\sigma) = \sum_{j=1}^{m} p_j(\log p_j - \log q_j) + \sum_{j=1}^{m} p_j D(\rho_j\|\sigma_j).$$

In particular if $p_j = q_j$ for each j, $D(\rho\|\sigma) = \sum_{j=1}^{m} p_j D(\rho_j\|\sigma_j)$.

Proof. A simple calculation yields

$$(7.2.14) \qquad \log \rho - \log \sigma = \sum_{j=1}^{m} \left((\log p_j - \log q_j)P_j + (\log \rho_j - \log \sigma_j)P_j\right).$$

Since (7.2.12) is equivalent to $\rho = \sum_{j=1}^{m} p_j \rho_j P_j$, (7.2.13) follows from (7.2.14).

$\qquad\square$

Theorem 7.42 describes the additivity of relative entropy over direct sums. The relative entropy is also additive over tensor products:

Theorem 7.43. *Let $\mathcal{H}_1, \mathcal{H}_2$ be two finite-dimensional Hilbert spaces. For $\rho_1, \sigma_1 \in \mathfrak{S}_{\mathcal{H}_1}$ and $\rho_2, \sigma_2 \in \mathfrak{S}_{\mathcal{H}_2}$*

$$D(\rho_1 \otimes \rho_2 \| \sigma_1 \otimes \sigma_2) = D(\rho_1 \| \sigma_1) + D(\rho_2 \| \sigma_2).$$

Proof. Note that

$$\log(\rho_1 \otimes \rho_2) - \log(\sigma_1 \otimes \sigma_2)$$
$$= (\log \rho_1 - \log \sigma_1) \otimes \mathbb{1} + \mathbb{1} \otimes (\log \rho_2 - \log \sigma_2).$$

$\square$

Evidently Theorem 7.43 extends directly to tensor products of more than two factors.

7.3. Gibbs variational principle

A simple computation shows that with f defined by (7.2.3), its Legendre transform f^* is given by $f^*(y) = e^{y-1}$, $y \in \mathbb{R}$. By either Theorem 3.3 or Theorem 6.13, we have that $\mathrm{Tr}[e^Y]$ is convex on $M_n^{\mathrm{s.a.}}(\mathbb{C})$. However, more is true: The function $Y \mapsto \log \mathrm{Tr}[e^Y]$ is convex on $M_n^{\mathrm{s.a.}}(\mathbb{C})$. This is a stronger statement since while the exponential of a convex function is always convex, the logarithm of a positive convex function need not be convex.

The convexity of $Y \mapsto \log \mathrm{Tr}[e^Y]$ may be seen by computing the Legendre transform of a different convex function on $M_n^{\mathrm{s.a.}}(\mathbb{C})$ whose restriction to $\mathfrak{S}$ agrees with $-\mathrm{Tr}[X \log X]$. Define a function F on $M_n^{\mathrm{s.a.}}(\mathbb{C})$ by

$$(7.3.1) \qquad F(X) := \begin{cases} -S(X) & X \in \mathfrak{S} \\ +\infty & X \notin \mathfrak{S}. \end{cases}$$

Then for $X \in M_n^{\mathrm{s.a.}}(\mathbb{C})$, define $S(X) = -F(X)$. Evidently, the convexity of $-S$ on $\mathfrak{S}$ implies the convexity of F on $M_n^{\mathrm{s.a.}}(\mathbb{C})$, and it is clear that F is proper, lower semicontinuous and convex.

Equip the real vector space $M_n^{\mathrm{s.a.}}(\mathbb{C})$ with the inner product $\langle X, Y \rangle$ that it inherits from $M_n(\mathbb{C})$. The next lemma shows that the Legendre transform of F on this Hilbert space is the function $G(Y)$ given by

$$(7.3.2) \qquad G(Y) := \log(\mathrm{Tr}[e^Y]).$$

For all $Y, H \in M_n^{\mathrm{s.a.}}(\mathbb{C})$, by Lemma 3.13 and cyclicity of the trace, G is differentiable and

$$\frac{d}{dt} \log(\mathrm{Tr}[e^{Y+tH}]) \Big|_{t=0} = \frac{\mathrm{Tr}[e^Y H]}{\mathrm{Tr}[e^Y]}.$$

Therefore, the subgradient of G at Y contains the single element $\frac{e^Y}{\text{Tr}[e^Y]}$, which belongs to $\mathfrak{S}$ for all Y. Being differentiable, G is certainly lower semicontinuous, and $G(Y)$ is finite at all Y so it is certainly proper. It is less trivial to show by direct computation that G is convex, but this will follow once we show that $G = F^*$.

Lemma 7.44. *The Legendre transform F^* of the function F defined by (7.3.1), is the function G on $M_n^{\text{s.a.}}(\mathbb{C})$ defined by (7.3.2) so that for all $X, Y \in M_n^{\text{s.a.}}(\mathbb{C})$,*

$$(7.3.3) \qquad\qquad \text{Tr}[XY] \leq F(X) + G(Y).$$

Moreover, there is equality in (7.3.3) if and only if $X = \frac{e^Y}{\text{Tr}[e^Y]}$. In particular the function $Y \mapsto \log \text{Tr}[e^Y]$ is a proper lower semicontinuous convex function on $M_n^{\text{s.a.}}(\mathbb{C})$.

Proof. Let $X \in \mathfrak{S}$ with $X > 0$, and define $H := \log(X)$ so that $\text{Tr}[e^H] = 1$. By the Peierls-Bogoliubov inequality, for all $K \in M_n^{\text{s.a.}}(\mathbb{C})$

$$(7.3.4) \qquad\qquad \text{Tr}[XK] = \text{Tr}[e^H K] \leq \log(\text{Tr}[e^{H+K}]),$$

with equality if and only if K is a multiple of the identity. Replacing K in (7.3.4) with $K - H$ yields

$$(7.3.5) \qquad\qquad \text{Tr}[XK] \leq \log(\text{Tr}[e^K]) + \text{Tr}[X \log X].$$

Then there is equality in (7.3.5) if and only if $K - \log X = c\mathbb{1}$ for some constant c, in which case $X = e^{-c}e^K$. By continuity of $\text{Tr}[X \log X]$ and $\text{Tr}[XK]$ in X, the inequality holds for all $X \in \mathfrak{S}$, whether or not $X > 0$. This proves that

$$\log(\text{Tr}[e^Y]) = \sup_{X \in \mathfrak{S}}\{\langle X, Y \rangle + S(X)\}$$
$$= \sup_{X \in M_n^{\text{s.a.}}(\mathbb{C})} \{\langle X, Y \rangle - F(X)\},$$

and hence $F^* = G$. We have already seen that the subgradient of G at Y contains the single element $\frac{e^Y}{\text{Tr}[e^Y]}$, and by the conditions for equality in Young's inequality, there is equality in (7.3.3) if and only if $X := e^K / \text{Tr}[e^K]$. $\qquad\square$

Remark 7.45. The relative interior of the set $\mathfrak{S}$, $\mathfrak{S}^\circ$, consists of the density matrices ρ such that $\rho > 0$, and the relative boundary of $\mathfrak{S}$ consists of all others. (See Appendix A for the terminology.) Since for all $Y \in M_n^{\text{s.a.}}(\mathbb{C})$, $(\text{Tr}[e^Y])^{-1}e^Y \in \mathfrak{S}^\circ$, there cannot be equality in (7.3.3) unless $Y \in \mathfrak{S}^\circ$. By the conditions for equality in Young's inequality, if X has zero as an eigenvalue, the subgradient of F at X is empty, and otherwise that subgradient consists of the single element $Y = \log X$.

Lemma 7.44 leads almost directly to the **Gibbs variational principle of quantum statistical mechanics**.

Theorem 7.46 (Gibbs variational principle). *For all $\rho \in \mathfrak{S}$,*

$$(7.3.6) \qquad -S(\rho) = \sup_{K \in M_n^{\mathrm{s.a.}}(\mathbb{C})} \{\mathrm{Tr}[\rho K] - \log(\mathrm{Tr}[e^K])\},$$

and for all $K \in M_n^{\mathrm{s.a.}}(\mathbb{C})$,

$$(7.3.7) \qquad \log(\mathrm{Tr}[e^K]) = \sup_{\rho \in \mathfrak{S}} \{\mathrm{Tr}[\rho K] + S(\rho)]\}.$$

Moreover, for all $\rho, \sigma \in \mathfrak{S}$, $\sigma > 0$.

$$(7.3.8) \qquad D(\rho\|\sigma) = \sup_{K \in M_n^{\mathrm{s.a.}}(\mathbb{C})} \{\mathrm{Tr}[\rho K] - \log(\mathrm{Tr}[e^{K + \log \sigma}])\},$$

and for all $K \in M_n^{\mathrm{s.a.}}(\mathbb{C})$,

$$(7.3.9) \qquad \log(\mathrm{Tr}[e^{K + \log \sigma}]) = \sup_{\rho \in \mathfrak{S}} \{\mathrm{Tr}[\rho K] - D(\rho\|\sigma)\}.$$

Proof. Evidently, (7.3.7) is simply a restatement of the main conclusion of Lemma 7.44. We have already observed that the function F defined in (7.3.1) is proper, lower-semicontinuous, and convex. Hence by the Fenchel-Moreau theorem $F = F^{**}$, and then since $F^* = G$, with G defined in (7.3.2), $F = G^*$, and this is a restatement of (7.3.6).

Now let $\sigma \in \mathfrak{S}$, $\sigma > 0$, and replace K with $K + \log \sigma$ in (7.3.6) and (7.3.7) to obtain (7.3.8) and (7.3.9). $\qquad \square$

Theorem 7.46 is important in quantum statistical mechanics. We briefly explain the physical context of an important variational problem involving the von Neumann entropy, referring to [206] for more information. Consider a finite-dimensional quantum system with Hamiltonian $H \in M_n^{\mathrm{s.a.}}(\mathbb{C})$ on $\mathbb{C}^n$ equipped with the usual inner product. The Hamiltonian H is the quantum observable corresponding to the energy, and so if the state of the system is $\rho \in \mathfrak{S}$, the energy E is given by $E := \mathrm{Tr}[H\rho]$. More precisely, suppose that $e_1, \ldots, e_r$ are the distinct eigenvalues of H arranged in increasing order, and that $P_1, \ldots, P_r$ are the corresponding spectral projections, so that

$$H = \sum_{j=1}^{r} e_j P_j.$$

Then the rules of quantum mechanics stipulate that any measurement of the energy of the system will yield one of the eigenvalues of H, and the probability that e_j is the outcome is given by $\mathrm{Tr}[\rho P_j]$. Thus, $E = \mathrm{Tr}[H\rho]$ gives the limiting value for the average of a long sequence of independent repeated measurements of the energy in the state ρ. By the spectral theorem, $e_1 \leq \mathrm{Tr}[H\rho] \leq e_r$ for all $\rho \in \mathfrak{S}$.

It is of interest to compute the maximum value of the entropy on the set of states $\rho \in \mathfrak{S}$ with energy E; that is, to determine

$$\sup\{S(\rho) \ : \ \rho \in \mathfrak{S}, \mathrm{Tr}(H\rho) = E\},$$

where to avoid trivialities, we assume that $e_1 \leq E \leq e_r$.

Theorem 7.47. *Let $H \in M_n^{\mathrm{s.a.}}(\mathbb{C})$, and let e_1 and e_r be the least and largest eigenvalues of H. For $\beta \in \mathbb{R}$, define*

$$\rho_{\beta,H} := \frac{1}{\mathrm{Tr}\left[e^{-\beta H}\right]} e^{-\beta H}.$$

Then for all E with $e_1 < E < e_r$, there is a unique β_E such that $\mathrm{Tr}[H\rho_{\beta_E,H}] = E$, and

$$\sup\{S(\rho) \ : \ \rho \in \mathfrak{S}, \mathrm{Tr}(H\rho) = E\}$$

is attained uniquely at $\rho = \rho_{\beta_E,H}$.

Proof. Note that

$$(7.3.10) \qquad \lim_{\beta \to -\infty} \rho_{\beta,H} = (\mathrm{Tr}[P_r])^{-1}P_r \quad \text{and} \quad \lim_{\beta \to \infty} \rho_{\beta,H} = (\mathrm{Tr}[P_1])^{-1}P_1.$$

Define $f_H(\beta) := \log \mathrm{Tr}[e^{-\beta H}]$. Differentiating yields $f_H'(\beta) = -\mathrm{Tr}[H\rho_{\beta,H}]$. It then follows from (7.3.10) that

$$(7.3.11) \qquad \lim_{\beta \to -\infty} f_H'(\beta) = -e_r \quad \text{and} \quad \lim_{\beta \to \infty} f_H'(\beta) = -e_1.$$

By Lemma 7.44, $K \mapsto \log(\mathrm{Tr}[e^K])$ is a convex function of K, and hence f_H is a convex function function of β, and hence f_H' is an increasing function. In fact, it is strictly increasing, as one can check by differentiating twice in β:

$$f_H''(\beta) = \mathrm{Tr}[(H - \mathrm{Tr}[H\rho_{\beta,H}]\mathbb{1})^2 \rho_{\beta,H}] > 0.$$

Hence $f_H'(\beta) = -\mathrm{Tr}[H\rho_{\beta,H}]$ is a strictly increasing function of β, and by (7.3.11) its range is the interval $(-e_r, -e_1)$. This proves that for $e_1 < E < e_r$, there is a unique solution β_E of the equation $f_H'(\beta) = -\mathrm{Tr}[H\rho_{\beta,H}] = E$.

By Lemma 7.44, for all $\rho \in \mathfrak{S}$, $\log \mathrm{Tr}[e^{-\beta H}] - S(\rho) \geq -\beta \mathrm{Tr}[H\rho]$, which can be rewritten as $S(\rho) \leq \log \mathrm{Tr}[e^{-\beta H}] + \beta \mathrm{Tr}[H\rho]$ with equality if and only if $\rho = \rho_{\beta,H}$. Thus for $\beta = \beta_E$ and all $\rho \in \mathfrak{S}$ with $\mathrm{Tr}[H\rho] = E$,

$$S(\rho) \leq \log \mathrm{Tr}[e^{-\beta_E H}] + \beta E$$

with equality if and only if $\rho = \rho_{\beta_E,H}$. $\qquad \square$

Density matrices of the form $\rho_{\beta,H}$ are known as **Gibbs states**, and this extremal property which characterizes them —that they, and only they, maximize entropy given the expected value of the energy—is directly connected with their physical significance.

Even more directly from Theorem 7.46, for $\beta > 0$,

$$\log \operatorname{Tr}[e^{-\beta H}] = \sup\{\operatorname{Tr}[(-\beta H)\rho] + S(\rho) \; : \; \rho \in \mathfrak{S}\}$$

$$= -\beta \inf\left\{\left(\operatorname{Tr}[H\rho] - \frac{1}{\beta}S(\rho)\right) \; : \; \rho \in \mathfrak{S}\right\},$$

and the infimum is uniquely attained at the Gibbs state $\rho_{\beta,H}$. The functional $F_{\beta,H}$ on $\mathfrak{S}$ defined by $F_{\beta,H}(\rho) := \operatorname{Tr}[H\rho] - \frac{1}{\beta}S(\rho)$ is the **Helmholtz free energy at inverse temperature** β (see [**206**]). Thus the Gibbs state $\rho_{\beta,H}$ may also be characterized as the minimizer of the Helmholtz free energy functional $F_{\beta,H}$, and it represents the equilibrium state of the system when it is in contact with a heat bath at inverse temperature β.

7.4. Measured relative entropies

Let $\mathcal{H}$ be a finite-dimensional Hilbert space. Let $\rho, \sigma \in \mathfrak{S}_{\mathcal{H}}$. To avoid trivialities, we assume $\sigma > 0$. Suppose we have been sent a quantum state that is either ρ or σ, and we would like to perform a measurement to decide which it is as reliably as possible. One way to proceed is as follows.

Let $\mathcal{X}$ be a finite set, and let $\mathcal{E}$ be a POVM on $\mathcal{X}$ with values in $\mathcal{B}^+(\mathcal{H})$. Then as in Definition 4.43, there is a map, also denoted $\mathcal{E}$, from $\mathfrak{S}_{\mathcal{H}}$ to $\mathcal{P}_{\mathcal{X}}$, the set of probability densities on $\mathcal{X}$.

We may then apply any classical relative entropy function to the pair $\mathcal{E}(\rho)$ and $\mathcal{E}(\sigma)$. The larger the relative entropy, the easier it will be to distinguish between ρ and σ.

Let g be a relative entropy generating function on $(0, \infty)$ in the sense of Definition 7.22, and let D_g denote the corresponding quasi relative entropy function. We now seek to choose the POVM $\mathcal{E}$ to make $D_g(\mathcal{E}(\rho)\|\mathcal{E}(\sigma))$ as large as possible.

Definition 7.48. Given a relative entropy generating function g, define the corresponding **measured g-relative entropy** $D_g^M(\rho\|\sigma)$ on $\mathfrak{S}_{\mathcal{H}} \times \mathfrak{S}_{\mathcal{H}}$ by

$$D_g^M(\rho\|\sigma) = \sup\{D_g(\mathcal{E}(\rho)\|\mathcal{E}(\sigma)) \; : \; \mathcal{E} \text{ is a POVM with values in } \mathcal{B}^+(\mathcal{H})\}.$$

For $g(t) = t \log t$, so that D_g is the standard relative entropy, we simply write $D^M(\rho\|\sigma)$, and refer to this as the **measured relative entropy**.

The following theorem is based on results from [33].

Theorem 7.49. *Let g be a relative entropy generating function in the sense of Definition 7.22. Let* $\mathrm{dom}(g^*)$ *and* $\mathrm{ran}(g^*)$ *denote the domain and range of its Legendre transform* g^*. *Let* $\rho, \sigma \in \mathfrak{S}_{\mathcal{H}}$ *with* $\sigma > 0$.

(1) *Suppose that the Legendre transform* g^* *is operator convex. Then*

$$(7.4.1) \qquad D_g^M(\rho\|\sigma) = \sup_{K \in \mathcal{B}^{\mathrm{s.a.}}(\mathcal{H}),\ \sigma(K) \subset \mathrm{dom}(g^*)} \{\mathrm{Tr}[\rho K - \sigma g^*(K)]\}.$$

The supremum in (7.4.1) *is attained and when* g^* *is strictly operator convex, there is a unique PVM* $\mathcal{E}_0$ *such that*

$$D_g^M(\rho\|\sigma) = D_g(\mathcal{E}_0(\rho)\|\mathcal{E}_0(\sigma)).$$

(2) *Suppose that* w, *the inverse function to* g^*, *is operator concave. Then*

$$(7.4.2) \qquad D_g^M(\rho\|\sigma) = \sup_{K \in \mathcal{B}^{\mathrm{s.a.}}(\mathcal{H}),\ \sigma(K) \subset \mathrm{ran}(g^*)} \{\mathrm{Tr}[\rho w(K) - \sigma K]\}.$$

The supremum in (7.4.2) *is attained and when* w *is strictly operator concave, there is a unique PVM* $\mathcal{E}_0$ *such that*

$$D_g^M(\rho\|\sigma) = D_g(\mathcal{E}_0(\rho)\|\mathcal{E}_0(\sigma)).$$

Proof. Let $\mathcal{E}$ be any POVM on a finite set $\mathcal{X}$ with values in $\mathcal{B}^+(\mathcal{H})$. Define $E_x := \mathcal{E}(x)$, so that $\mathcal{E}(\rho)(x) = \mathrm{Tr}[\rho E_x]$. Suppose that g^* is operator convex. In this case, we use the variational formula (7.1.23), and have

$$D_g(\mathcal{E}(\rho)\|\mathcal{E}(\sigma)) = \sup_{h:\mathcal{X}\to\mathbb{R}} \sum_{x\in\mathcal{X}} \left(\mathrm{Tr}[E_x\rho]h(x) - \mathrm{Tr}[E_x\sigma]g^*(h(x)) \right)$$

$$(7.4.3) \qquad = \sup_{h:\mathcal{X}\to\mathbb{R}} \mathrm{Tr}\left[\sum_{x\in\mathcal{X}} \left(E_x\rho h(x) - \sigma E_x^{1/2} g^*(h(x)\mathbb{1}) E_x^{1/2} \right) \right].$$

By the operator Jensen inequality,

$$(7.4.4) \qquad \sum_{x\in\mathcal{X}} E_x^{1/2} g^*(h(x)\mathbb{1}) E_x^{1/2} \geq g^*\left(\sum_{x\in\mathcal{X}} h(x)E_x \right).$$

Define the self-adjoint operator $H := \sum_{x\in\mathcal{X}} h(x)E_x$. Then (7.4.3) and (7.4.4) yield $D_g(\mathcal{E}(\rho)\|\mathcal{E}(\sigma)) \leq \mathrm{Tr}[\rho H - \sigma g^*(H)]$. This shows that

$$D_g^M(\rho\|\sigma) \leq \sup_{K \in \mathcal{B}^{\mathrm{s.a.}}(\mathcal{H})} \{\mathrm{Tr}[\rho K - \sigma g^*(K)]\}.$$

For $K \in \mathcal{B}^{\text{s.a.}}$, let $K = \sum_{\lambda \in \sigma(K)} \lambda P_\lambda$ be the spectral decomposition of K. Define a PVM $\mathcal{E}_0$ on $\sigma(K)$ by $\mathcal{E}_0(\lambda) = P_\lambda$. Then

$$\text{Tr}[\rho K - \sigma g^*(K)] = \sum_{\lambda \in \sigma(K)} (\text{Tr}[\rho P_\lambda]\lambda - g^*(\lambda)\text{Tr}[\sigma P_\lambda])$$

$$(7.4.5) \qquad \leq \sum_{\lambda \in \sigma(K)} \text{Tr}[\sigma P_\lambda] g\left(\frac{\text{Tr}[\rho P_\lambda]}{\text{Tr}[\sigma P_\lambda]}\right)$$

$$= D_g(\mathcal{E}_0(\rho)\|\mathcal{E}_0(\sigma)) \leq D_g^M(\rho\|\sigma).$$

This proves (7.4.1). By the condition for equality in Young's inequality, there is equality in (7.4.5) if and only if for each λ, $\lambda = g'\left(\frac{\text{Tr}[\rho P_\lambda]}{\text{Tr}[\sigma P_\lambda]}\right)$.

Therefore, when $\rho, \sigma > 0$, we need only consider $K \in \mathcal{B}^{\text{s.a.}}(\mathcal{H})$ such that $g'(\|\rho^{-1}\|^{-1}) \leq K \leq g'(\|\sigma^{-1}\|)$. This set is compact, and hence the supremum is achieved. Since g^* is strictly operator convex, there is a unique maximizer K, and the PVM $\mathcal{E}_0$ corresponding to the spectral decomposition of K satisfies $D_g^M(\rho\|\sigma) = D_g(\mathcal{E}_0(\rho)\|\mathcal{E}_0(\sigma))$.

If w is operator concave, we proceed in a similar manner, this time using the variational formula (7.1.24) with I denoting the open interval that is the interior of the set on which w is finite. Then the operator Jensen inequality yields

$$D_g(\mathcal{E}(\rho)\|\mathcal{E}(\sigma)) = \sup_{h:\mathcal{X}\to I} \text{Tr}\left[\sum_{x\in\mathcal{X}} \left(\rho E_x^{1/2} w(h(x)\mathbb{1})E_x^{1/2} - \sigma E_x h(x)\right)\right]$$

$$\leq \text{Tr}[\rho w(H) - \sigma H],$$

where H is again defined below (7.4.4). Again, letting $\mathcal{E}_0$ denote the PVM corresponding to the spectral decomposition of H,

$$\text{Tr}[\rho w(H) - \sigma H] = \sum_{\lambda \in \sigma(H)} (w(\lambda)\text{Tr}[\rho P_\lambda] - \lambda\text{Tr}[\sigma P_\lambda])$$

$$(7.4.6) \qquad \leq \sum_{\lambda \in \sigma(K)} \text{Tr}[\sigma P_\lambda]g\left(\frac{\text{Tr}[\rho P_\lambda]}{\text{Tr}[\sigma P_\lambda]}\right)$$

$$= D_g(\mathcal{E}_0(\rho)\|\mathcal{E}_0(\sigma)) \leq D_g^M(\rho\|\sigma).$$

There is equality in (7.4.6) if and only if for each λ, $w'(\lambda) = \frac{\text{Tr}[\sigma P_\lambda]}{\text{Tr}[\rho P_\lambda]}$. We need only consider H in the compact set $\|\sigma^{-1}\|^{-1} \leq w'(H) \leq \|\rho^{-1}\|$ and hence the supremum is attained, and when w is strictly operator concave, the maximizer is unique. $\qquad \square$

Example 7.50. Let $g(t) = t\log t$ so that D_g is the standard relative entropy. Then $g^*(u) = e^{u-1}$, with $\text{dom}(g^*) = (-\infty, \infty)$ and $\text{ran}(g^*) = (0, \infty)$. While g^* is not operator convex, its inverse function $w(v) = \log v + 1$ is strictly operator concave on $(0, \infty)$. Therefore (7.4.2) yields the following variational formula for the the standard measured relative entropy,

$$(7.4.7) \qquad D^M(\rho\|\sigma) = \sup_{K \in \mathcal{B}^{++}(\mathcal{H})} \{\mathrm{Tr}[\rho \log K - \sigma(K - 1)]\}.$$

There is an alternate variational formula for the measured relative entropy that will prove useful. By the concavity of the logarithm $\mathrm{Tr}[\sigma K] - 1 \geq \log(\mathrm{Tr}[\sigma K])$, and hence from (7.4.7), for all $K > 0$,

$$\mathrm{Tr}[\rho \log K - \sigma(K - 1)] \leq \mathrm{Tr}[\rho \log K] - \log(\mathrm{Tr}[\sigma K]).$$

However, the right side is invariant under the replacement $K \to tK$, $t > 0$. Choosing t so that $\mathrm{Tr}[\sigma(tK)] = 1$,

$$\mathrm{Tr}[\rho \log K] - \log(\mathrm{Tr}[\sigma K]) = \mathrm{Tr}[\rho \log(tK)] = \mathrm{Tr}[\rho \log(tK) - \sigma(tK - 1)].$$

Therefore,

$$\sup_{K \in \mathcal{B}^{++}(\mathcal{H})} \{\mathrm{Tr}[\rho \log K - \sigma(K - 1)] = \sup_{K \in \mathcal{B}^{++}(\mathcal{H})} \{\mathrm{Tr}[\rho \log K - \log(\mathrm{Tr}[\sigma K])],$$

and hence we also have

$$(7.4.8) \qquad D^M(\rho\|\sigma) = \sup_{K \in \mathcal{B}^{++}(\mathcal{H})} \{\mathrm{Tr}[\rho \log K - \log(\mathrm{Tr}[\sigma K])]\}.$$

Moreover, again by Theorem 7.49, for $\rho, \sigma > 0$, there is a uniquely determined $K \in \mathcal{B}^{++}(\mathcal{H})$ at which the supremum in (7.4.7) is attained. This operator K is the unique solution (by the strict concavity of the logarithm) of

$$0 = \frac{d}{dt} \mathrm{Tr}[\rho w(K + tH) - \sigma(K + tH)]\Big|_{t=0}$$

$$= \mathrm{Tr}\left[\left(\int_0^\infty \frac{1}{\lambda + K}\rho\frac{1}{\lambda + K}d\lambda - \sigma\right)H\right],$$

and therefore

$$(7.4.9) \qquad \rho = \int_0^1 K^{1-s}\sigma K^s ds.$$

If ρ and σ commute, this equation is solved by $K = \rho\sigma^{-1}$. Otherwise, it seems difficult to solve this equation for K, given $\rho, \sigma > 0$.

Theorem 7.51. *Let* $\rho, \sigma > 0$ *in* $\mathfrak{S}_{\mathcal{H}}$, *and let* $\mathcal{H}$ *be a finite-dimensional Hilbert space. Then*

$$(7.4.10) \qquad \frac{1}{2}\|\rho - \sigma\|_1^2 \leq D^M(\rho\|\sigma) \leq D(\rho\|\sigma),$$

and there is equality on the right if and only if ρ *and* σ *commute.*

Proof. We first prove the inequality on the right. As shown in Example 7.50, there is a unique $K > 0$ such that

$$(7.4.11) \qquad D^M(\rho\|\sigma) = \mathrm{Tr}\left[\rho \log(K) + 1 - \sigma K\right].$$

Making the change of variable $K = e^H$, (7.4.11) becomes

$$D^M(\rho\|\sigma) = \mathrm{Tr}\left[\rho H + 1 - \sigma e^H\right] \leq \mathrm{Tr}[\rho H] - \left(\mathrm{Tr}\left[e^{H+\log\sigma}\right] - 1\right)$$
$$\leq \mathrm{Tr}[\rho H] - \log\left(\mathrm{Tr}\left[e^{H+\log\sigma}\right]\right) \leq D(\rho\|\sigma),$$

where we have used the Golden-Thompson inequality, the elementary inequality $t - 1 \geq \log t$, and the Gibbs variational principle (7.3.8).

There is equality in the Golden-Thompson inequality if and only if H and σ commute, and this is the case if and only if K and σ commute. In this case (7.4.9) reduces to $\rho = K\sigma$, and hence ρ commutes with σ. Thus, when ρ and σ do not commute, $D^M(\rho\|\sigma) < D(\rho\|\sigma)$.

We now prove the inequality on the left which is known as the **quantum Pinsker inequality**. Define $H := \rho - \sigma$ which is self-adjoint. Let $H = \sum_{\lambda\in\sigma(H)} \lambda P_\lambda$ be the spectral decomposition of H. Define $\mathcal{X} := \sigma(H)$, and define a POVM on $\mathcal{X}$ by $\mathcal{E}(\lambda) := P_\lambda$. Define classical probability densities on $\mathcal{X}$ that we also denote by ρ and σ as follows: $\rho(\lambda) := \mathrm{Tr}[\rho P_\lambda]$ and $\sigma(\lambda) := \mathrm{Tr}[\sigma P_\lambda]$. Then $|\rho(\lambda) - \sigma(\lambda)| = |\mathrm{Tr}(\rho - \sigma)P_\lambda| = |\lambda|\,\mathrm{Tr}[P_\lambda]$. Therefore,

$$(7.4.12) \qquad \sum_{\lambda\in\mathcal{X}} |\rho(\lambda) - \sigma(\lambda)| = \|H\|_1 = \|\rho - \sigma\|_1.$$

By Theorem 7.51, and the definition of the measured relative entropy,

$$(7.4.13) \qquad D(\rho\|\sigma) \geq D^M(\rho\|\sigma) \geq \sum_{\lambda\in\mathcal{X}} \rho(\lambda)(\log\rho(\lambda) - \log\sigma(\lambda)).$$

By Theorem 7.18,

$$(7.4.14) \qquad \sum_{\lambda\in\mathcal{X}} \rho(\lambda)(\log\rho(\lambda) - \log\sigma(\lambda)) \geq \frac{1}{2}\left(\sum_{\lambda\in\mathcal{X}} |\rho(\lambda) - \sigma(\lambda)|\right)^2.$$

Combining (7.4.12), (7.4.13), and (7.4.14) yields the quantum Pinsker inequality. $\qquad\square$

There is a simple but important complement to (7.4.10).

Theorem 7.52. *Let $\mathcal{H}$ be a Hilbert space of finite dimension d. Let $\rho, \sigma \in \mathfrak{S}_\mathcal{H}$, $\sigma > 0$. Let $|\sigma(\sigma)|$ denote the cardinality of the spectrum of σ. Then*

$$(7.4.15) \qquad D^M(\rho\|\sigma) \geq D(\rho\|\sigma) - \log|\sigma(\sigma)|.$$

Consequently, for all $n \in \mathbb{N}$,

$$(7.4.16) \qquad \frac{1}{n}D^M(\rho^{\otimes n}\|\sigma^{\otimes n}) \geq D(\rho\|\sigma) - \frac{1}{n}\log\left(\frac{(n+d-1)^{d-1}}{(d-1)!}\right),$$

and in particular

$$(7.4.17) \qquad \lim_{n\to\infty} \frac{1}{n} D^M(\rho^{\otimes n}\|\sigma^{\otimes n}) = D(\rho\|\sigma).$$

Hiai and Petz [107] proved (7.4.17), which is known as the **asymptotic achievability** of the relative entropy.

Proof. Let $\sigma = \sum_{\lambda\in\sigma(\sigma)} \lambda P_\lambda$ be the spectral decomposition of σ. Let $\mathcal{N}$ denote the von Neumann algebra of all polynomials in σ, and let $\mathcal{E}_{\mathcal{N}}$ be the conditional expectation onto $\mathcal{N}$. Since $t\log t$ is convex,

$$\mathrm{Tr}[\rho\log\rho] \geq \mathrm{Tr}[(\mathcal{E}_{\mathcal{N}}\rho)\log(\mathcal{E}_{\mathcal{N}}(\rho))] = \mathrm{Tr}[\rho\log(\mathcal{E}_{\mathcal{N}}(\rho))].$$

Since $\mathcal{E}_{\mathcal{N}}(\rho) = \sum_{\lambda\in\sigma(\sigma)} P_\lambda\rho P_\lambda$, the pinching inequality yields

$$\mathcal{E}_{\mathcal{N}}\rho \geq \frac{1}{|\sigma(\sigma)|}\rho.$$

Together with operator monotonicity of the logarithm function, this yields

$$\mathrm{Tr}[\rho\log(\mathcal{E}_{\mathcal{N}}(\rho))] \geq \mathrm{Tr}[\rho\log\rho] - \log|\sigma(\sigma)|,$$

and hence $\mathrm{Tr}[\rho(\log(\mathcal{E}_{\mathcal{N}}(\rho)) - \log\sigma)] \geq D(\rho\|\sigma) - \log|\sigma(\sigma)|$. However,

$$\mathrm{Tr}[\rho(\log(\mathcal{E}_{\mathcal{N}}(\rho)) - \log\sigma)] = \sum_{\lambda\in\sigma(\sigma)} (\mathrm{Tr}[\rho P_\lambda](\log(\mathrm{Tr}[\rho P_\lambda] - \mathrm{Tr}[\sigma P_\lambda]))$$

$$\leq D^M(\rho\|\sigma).$$

This proves (7.4.15). Then since $D(\rho^{\otimes n}\|\sigma^{\otimes n}) = nD(\rho\|\sigma)$, (7.4.16) follows from (7.4.15) and the bound on $|\sigma(\sigma)|$ provided by Lemma 2.43. $\square$

Example 7.53. For $0 < \alpha < 1$, let

$$g_\alpha(t) = \frac{1}{\alpha} - \frac{1}{\alpha}t^\alpha.$$

Then $g^*(u) = \infty$ for all $u > 0$, and for $u < 0$,

$$(7.4.18) \qquad g^*(u) = \begin{cases} \frac{1-\alpha}{\alpha}(-u)^{\alpha/(\alpha-1)} - \frac{1}{\alpha} & u < 0 \\ \infty & u \geq 0. \end{cases}$$

This is operator convex if and only if $-1 \leq \alpha/(\alpha - 1)$, which is the case if and only if $0 < \alpha \leq 1/2$.

However, the inverse function $w(v)$, $v \in (0,\infty)$ is given by

$$w(v) = -\left(\frac{\alpha}{1-\alpha}v + \frac{1}{1-\alpha}\right)^{(\alpha-1)/\alpha}.$$

This is operator concave if and only if $-1 \leq (\alpha - 1)/\alpha$, which is the case if and only if $1/2 \leq \alpha < 1$. Thus, for all $\alpha \in (0,1)$, we have either the requisite operator convexity or concavity for application of Theorem 7.49.

Example 7.54. We continue with Example 7.53, but focus on the case $\alpha = 1/2$, for which more can be said. That is, we take $g_{1/2} := 2 - 2t^{1/2}$. Then by (7.4.18), $g_{1/2}^*(u) = (-u)^{-1} - 2$. This is strictly operator convex on $(-\infty, 0)$. For $t \in (0, \infty)$, $g_{1/2}'(t) = -t^{-1/2}$ takes values in $(-\infty, 0)$.

In seeking to maximize $\mathrm{Tr}[\rho K - \sigma g^*(K)]$ we need only consider K such that

$$-\|\rho^{-1}\|^{1/2} \leq K \leq -\|\sigma^{-1}\|^{-1/2}.$$

Therefore, there is a unique $K < 0$ such that

$$(7.4.19) \qquad D_{g_{1/2}}^M(\rho\|\sigma) = 2 + \mathrm{Tr}[\rho K + \sigma K^{-1}],$$

and the Euler-Lagrange equation satisfied by K is

$$(7.4.20) \qquad \rho = K^{-1}\sigma K^{-1},$$

and this is equivalent to $\sigma^{1/2}\rho\sigma^{1/2} = (\sigma^{1/2}K^{-1}\sigma^{1/2})^2$ so that

$$(7.4.21) \qquad K^{-1} = -\sigma^{-1/2}(\sigma^{1/2}\rho\sigma^{1/2})^{1/2}\sigma^{-1/2}.$$

Writing (7.4.20) in the from $K\rho K = \sigma$, the same reasoning leads to

$$(7.4.22) \qquad K = -\rho^{-1/2}(\rho^{1/2}\sigma\rho^{1/2})^{1/2}\rho^{-1/2}.$$

Inserting (7.4.21) and (7.4.22) into (7.4.19) yields

$$D_{g_{1/2}}^M(\rho\|\sigma) = 2 - 2\|\rho^{1/2}\sigma^{1/2}\|_1.$$

Since for any POVM on a finite set $\mathcal{X}$ with values in $\mathcal{B}^+(\mathcal{H})$,

$$D_{g_{1/2}}(\rho\|\sigma) = 2 - \sum_{x \in \mathcal{X}} \sqrt{\mathrm{Tr}[\rho\mathcal{E}(x)]\,\mathrm{Tr}[\sigma\mathcal{E}(x)]},$$

this analysis yields another proof of the theorem of Fuchs and Caves relating the fidelity to measurements of ρ and σ, Theorem 6.58.

7.5. **Strong subadditivity of the von Neumann entropy**

A **tripartite density matrix** is a density matrix ρ_{123} on $\mathcal{H}_1 \otimes \mathcal{H}_2 \otimes \mathcal{H}_3$, the tensor product of three Hilbert spaces. One may think of each Hilbert space as corresponding to observables localized in three regions. In some statistical mechanical applications, two of the regions are spatially separated and are sufficiently far apart that they do not directly interact with one another, but do interact with the third, which is in the middle.

In this context we use the same sort of notation that was used in Section 7.2.1: For a density matrix ρ_{123} on $\mathcal{H}_1 \otimes \mathcal{H}_2 \otimes \mathcal{H}_3$, $\rho_{12} := \mathrm{Tr}_3[\rho_{123}]$, $\rho_1 := \mathrm{Tr}_{23}[\rho_{123}]$, etc. Commas will not be needed since in the theorems that follow no double digit subscripts will arise. Also we will denote $\mathcal{H}_1 \otimes \mathcal{H}_2 \otimes \mathcal{H}_3$ by $\mathcal{H}_{123}$ and $\mathcal{H}_1 \otimes \mathcal{H}_3$ by $\mathcal{H}_{13}$, for example.

As before, although ρ_2 acts on $\mathcal{H}_2$ by definition, we may identify it with the operator $\rho_2 \otimes \mathbb{1}_{13} \in \mathcal{B}(\mathcal{H}_{123})$, acting trivially on the first and third factors. By the spectral theorem

$$\log(\rho_2 \otimes \mathbb{1}_{13}) = \log(\rho_2) \otimes \mathbb{1}_{13},$$

and hence it is convenient to simply write $\log \rho_2$ in place of $\log \rho_2 \otimes \mathbb{1}_{13}$, making the same sort of identification. We treat the other marginal densities of ρ_{123} is the same way. We also use the following notation for the von Neumann entropies of these density matrices: $S_{123} := S(\rho_{123})$, $S_{12} = S(\rho_{12})$, $S_1 = S(\rho_1)$, etc.

The **strong subadditivity** of the von Neumman entropy (SSA) is an inequality relating marginal entropies of a tripartite density matrix ρ_{123}. In 1968 Lanford and Robinson [**138**] conjectured that

(7.5.1) $S_{12} + S_{23} \geq S_{123} + S_2,$

in analogy with the classical case. If $\mathcal{H}_2$ is one dimensional, then (7.5.1) reduces to $S_1 + S_3 \geq S_{13}$ (i.e., to subadditivity), but when $\dim(\mathcal{H}_2) > 1$, then there is nontrivial *overlap* in the spaces on which ρ_{12} and ρ_{23} act, and because of this, one cannot bound $S_{12} + S_{23}$ by S_{123}. In fact, suppose $\rho_{123} = \rho_1 \otimes \rho_{23}$. Then $\rho_{12} = \rho_1 \otimes \rho_2$, and a simple calculation gives

$$S_{12} + S_{23} = (S_1 + S_2) + S_{23} = S_{123} + S_2.$$

In 1968, Lanford and Robinson [**138**] conjectured that for general ρ_{123} this identity is replaced by the inequality (7.5.1) and they referred to this conjectured inequality as the strong subadditivty (SSA) of the quantum entropy.

The term SSA is used in statistical mechanics for what is known in classical information theory as positivity of the conditional mutual information, as discussed in Section 7.1. The classical version of SSA was used by Robinson and Ruelle [**187**, Proposition 1] to show the existence of the thermodynamic limit in classical statistical mechanics. In 1968, Lanford and Robinson attempted to extend this result to the quantum setting, but did not fully succeed because their approach required quantum SSA which they could not prove. They wrote in [**138**, p. 1125] referring to classical SSA, that: "One could believe, and even support one's belief by heuristic physical arguments that the same condition holds for the quantum entropy". For an overview of the physical context, see the review article of Wehrl [**227**].

The classical analogue is elementary to prove, as was done in Section 7.1, but the **usual** classical proof uses conditional probabilities. (See [**47**] for a proof that does not.) Recall that if $\rho(x, y)$ is a joint probability distribution for two random variables, and $\rho(y)$ is the marginal distribution for y, the corresponding conditional probability density $\rho(x|y)$ is $\rho(x, y)/\rho(y)$. Unfortunately, in the quantum setting there are many ways one might try to divide ρ_{12} by ρ_2, but none of them yields a satisfactory notion of conditional quantum probability, and none of them provides a basis for adapting the easy proof of classical SSA to the quantum case.

Five years after it was made, the conjecture of Lanford and Robinson was proved by Lieb and Ruskai [143] using results obtained in the same year by Lieb [141]. We now give an elementary proof of SSA. The starting point is a reformulation of SSA provided by Lieb and Araki [15]:

Lemma 7.55. *The inequality*

$$(7.5.2) \qquad S(\rho_{12}) + S(\rho_{23}) - S(\rho_1) - S(\rho_3) \geq 0$$

is valid for all tripartite density matrices ρ_{123} if and only if SSA is valid for all tripartite density matrices.

Proof. Let $\mathcal{H}_4$ be a fourth Hilbert space, and let ψ be a unit vector in $\mathcal{H}_1 \otimes \mathcal{H}_2 \otimes \mathcal{H}_3 \otimes \mathcal{H}_4$ such that

$$\mathrm{Tr}_4[|\psi\rangle\langle\psi|] = \rho_{123}.$$

One can consider $\mathcal{H}_1 \otimes \mathcal{H}_2 \otimes \mathcal{H}_3 \otimes \mathcal{H}_4$ as a bipartite space $\mathcal{H} \otimes \mathcal{K}$ in various ways, and since the two marginal entropies of a pure bipartite state are equal,

$$(7.5.3) \qquad S(\rho_{123}) = S(\rho_4) \quad \text{and} \quad S(\rho_{23}) = S(\rho_{14}).$$

Thus, (7.5.1) becomes

$$(7.5.4) \qquad S(\rho_{12}) + S(\rho_{14}) \geq S(\rho_4) + S(\rho_2),$$

which, up to relabeling, is the same as (7.5.2). Conversely, because of (7.5.3), once one has proved (7.5.4), one has proved (7.5.1). $\qquad\square$

Lin, Kim, and Hsieh [146] have proved a remarkable operator inequality that has (7.5.2) as a direct corollary.

Theorem 7.56 (Lin, Kim, and Hsieh). *Let ρ_{12} be an invertible density matrix on $\mathcal{H}_1 \otimes \mathcal{H}_2$, and let σ_{23} be any density matrix on $\mathcal{H}_2 \otimes \mathcal{H}_3$. Then*

$$(7.5.5) \qquad \rho_1^{-1} \otimes \sigma_{23} \leq \rho_{12}^{-1} \otimes \sigma_3.$$

Remark 7.57. The inequality (7.5.5) extends to an inequality for positive operators X on $\mathcal{H}_1 \otimes \mathcal{H}_2$ and Y on $\mathcal{H}_2 \otimes \mathcal{H}_3$ since we may normalize these to form $\rho_{12} = (\mathrm{Tr}[X])^{-1}X$ and $\sigma_{23} := (\mathrm{Tr}[Y])^{-1}Y$, and the normalization factors may be cancelled from both sides. The normalization is only relevant to the application.

Next, while there is no restriction on the finite dimensions of the Hilbert spaces d_1, d_2, and d_3, we may freely assume that $d_2 \geq \max\{d_1, d_3\}$. To see this, let $\widetilde{\mathcal{H}}_2$ be a Hilbert space that contains $\mathcal{H}_2$ as a proper subspace. Let P be the projection onto $\mathcal{H}_2$ in $\widetilde{\mathcal{H}}_2$. If ρ_{12} is invertible on $\mathcal{H}_1 \otimes \mathcal{H}_2$, $\rho_{12}(\epsilon) := \rho_{12} + \epsilon \mathbb{1}_1 \otimes P$ is invertible on $\mathcal{H}_1 \otimes \widetilde{\mathcal{H}}_2$ for all $\epsilon > 0$. Of course, any density matrix σ_{23} on

$\mathcal{H}_2 \otimes \mathcal{H}_3$ may be regarded as a density matrix on $\widetilde{\mathcal{H}}_2 \otimes \mathcal{H}_3$. Then the restriction of $\rho_{12}^{-1}(\epsilon) \otimes \sigma_3 \geq \rho_1^{-1}(\epsilon) \otimes \sigma_{23}$ to $\mathcal{H}_1 \otimes \mathcal{H}_2 \otimes \mathcal{H}_3$ reduces to (7.5.5) as ϵ decreases to zero.

As a direct corollary of Theorem 7.56, one obtains (7.5.2), and hence strong subadditivity:

Theorem 7.58 (Lieb and Ruskai). *The inequality* (7.5.1) *is valid for all tripartite density matrices.*

Proof. Since the logarithm is operator monotone, (7.5.5) implies

$$(7.5.6) \qquad \log \rho_{12} + \log \sigma_{23} - \log \rho_1 - \log \sigma_3 \leq 0.$$

Now let ρ_{123} be a density matrix on $\mathcal{H}_1 \otimes \mathcal{H}_2 \otimes \mathcal{H}_3$, and apply (7.5.6) with $\rho_{12} = \mathrm{Tr}_3[\rho_{123}]$ and $\sigma_{23} = \mathrm{Tr}_1[\rho_{123}] = \rho_{23}$. Then taking the trace against $-\rho_{123}$ yields (7.5.2), and hence (7.5.1) by Lemma 7.55. The condition that ρ_{12} be invertible is removed using continuity of the conditional entropy. $\qquad\square$

Theorem 7.56 has an elementary proof [**50**].

Lemma 7.59. *Let Ψ and Φ be unit vectors in $\mathcal{H}_1 \otimes \mathcal{H}_2$ and $\mathcal{H}_2 \otimes \mathcal{H}_3$, respectively, and let $\rho_{12} = |\Psi\rangle\langle\Psi|$ and $\sigma_{23} = |\Phi\rangle\langle\Phi|$. Suppose that σ_3 is invertible. Let P denote the projection into the orthogonal complement of Φ in $\mathcal{H}_2 \otimes \mathcal{H}_3$. For $\epsilon > 0$, define*

$$X_{23}(\epsilon) = |\Phi\rangle\langle\Phi| + \epsilon P.$$

Then for all ϵ sufficiently small,

$$(7.5.7) \qquad \rho_{12} \otimes \sigma_3^{-1} \leq \left(1 + \sqrt{\epsilon}\right) \rho_1 \otimes X_{23}(\epsilon)^{-1}.$$

Remark 7.60. Since σ_2 and σ_3 have the same rank, σ_3 can only be invertible if $d_2 \geq d_3$.

Proof of Theorem 7.56. By Remark 7.57, we may assume $d_2 \geq d_3$. Since both sides of (7.5.7) are linear in ρ_{12}, and since any density matrix is a convex combination of pure states, (7.5.7) is valid for all density matrices ρ_{12} on $\mathcal{H}_1 \otimes \mathcal{H}_2$. Then by the monotonicity of the inverse function, if ρ_{12} is invertible,

$$\rho_{12}^{-1} \otimes \sigma_3 \geq \left(1 + \sqrt{\epsilon}\right)^{-1} \rho_1^{-1} \otimes X_{23}(\epsilon).$$

Now taking ϵ to zero, we recover (7.5.5) for pure states σ_{23} such that σ_3 is invertible. Such pure states are dense by Schrödinger's theorem, and again, since any density matrix on $\mathcal{H}_2 \otimes \mathcal{H}_3$ is a convex combination of pure states, it follows that (7.5.5) is valid for all density matrices σ_{23}. $\qquad\square$

Proof of Lemma 7.59. Let $\{\phi_j\}_{1 \leq j \leq d_2 d_3}$ be an orthonormal basis for $\mathcal{H}_2 \otimes \mathcal{H}_3$ in which $\phi_1 = \Phi$. The general unit vector $\varphi \in \mathcal{H}_1 \otimes \mathcal{H}_2 \otimes \mathcal{H}_3$ has the form $\varphi = \sum_{\ell=1}^{d_2 d_3} \mathbf{w}_\ell \otimes \phi_\ell$, where $\sum_{\ell=1}^{d_2 d_3} \|\mathbf{w}_\ell\|^2 = 1$. Then with $B := \rho_1 \otimes X_{23}(\epsilon)^{-1}$,

$$(7.5.8) \qquad \langle \varphi, B\varphi \rangle = \langle \mathbf{w}_1, \rho_1 \mathbf{w}_1 \rangle + \frac{1}{\epsilon} \sum_{\ell=2}^{d_2 d_3} \langle \mathbf{w}_\ell, \rho_1 \mathbf{w}_\ell \rangle.$$

We must compare this with $\langle \varphi, A\varphi \rangle$, where $A := \rho_{12} \otimes \sigma_3^{-1}$. Recall that for all $a, b, \delta > 0$, $ab \leq \frac{1}{2}(\delta a^2 + \delta^{-1} b^2)$. Hence for any $\delta > 0$,

$$\langle \varphi, A\varphi \rangle = \sum_{\ell,\ell'=1}^{d_2 d_3} \langle \mathbf{w}_\ell \otimes \phi_\ell, A\mathbf{w}_{\ell'} \otimes \phi_{\ell'} \rangle$$

$$= \langle \mathbf{w}_1 \otimes \Phi, A\mathbf{w}_1 \otimes \Phi \rangle + 2 \sum_{\ell=2}^{d_2 d_3} \langle \mathbf{w}_1 \otimes \Phi, A\mathbf{w}_\ell \otimes \phi_\ell \rangle$$

$$+ \sum_{\ell,\ell'=2}^{d_2 d_3} \langle \mathbf{w}_\ell \otimes \phi_\ell, A\mathbf{w}_{\ell'} \otimes \phi_{\ell'} \rangle$$

$$\leq (1 + \delta d_2 d_3) \langle \mathbf{w}_1 \otimes \Phi, A\mathbf{w}_1 \otimes \Phi \rangle$$

$$+ \left(\frac{1}{2\delta} + d_2 d_3 \right) \sum_{\ell=2}^{d_2 d_3} \langle \mathbf{w}_\ell \otimes \phi_\ell, A\mathbf{w}_\ell \otimes \phi_\ell \rangle.$$

Next, with μ denoting the least eigenvalue of σ_3, $A \leq \mu^{-1} \rho_{12} \otimes \mathbb{1}$. Therefore,

$$\langle \mathbf{w}_\ell \otimes \phi_\ell, A\mathbf{w}_\ell \otimes \phi_\ell \rangle \leq \mu^{-1} \langle \mathbf{w}_\ell \otimes \phi_\ell, \rho_{12} \otimes \mathbb{1} \mathbf{w}_\ell \otimes \phi_\ell \rangle$$

$$\leq \mu^{-1} \sum_{j=1}^{d_2 d_3} \langle \mathbf{w}_\ell \otimes \phi_j, \rho_{12} \otimes \mathbb{1} \mathbf{w}_\ell \otimes \phi_j \rangle$$

$$\leq \mu^{-1} \langle \mathbf{w}_\ell, \mathrm{Tr}_{23}[\rho_{12} \otimes \mathbb{1}] \mathbf{w}_\ell \rangle = d_3 \mu^{-1} \langle \mathbf{w}_\ell, \rho_1 \mathbf{w}_\ell \rangle.$$

Then our previous estimate on $\langle \varphi, A\varphi \rangle$ becomes
$$(7.5.9)$$
$$\langle \varphi, A\varphi \rangle \leq (1 + \delta d_2 d_3) \langle \mathbf{w}_1 \otimes \Phi, A\mathbf{w}_1 \otimes \Phi \rangle + d_3 \mu^{-1} \left(\frac{1}{2\delta} + d_2 d_3 \right) \sum_{\ell=2}^{d_2 d_3} \langle \mathbf{w}_\ell, \rho_1 \mathbf{w}_\ell \rangle.$$

We claim that

$$(7.5.10) \qquad \langle \mathbf{w}_1 \otimes \Phi, A\mathbf{w}_1 \otimes \Phi \rangle \leq \langle \mathbf{w}_1, \rho_1 \mathbf{w}_1 \rangle.$$

Once this is established, choose $\delta = \sqrt{\epsilon}/d_2 d_3$ and then (7.5.9) becomes

$$\langle \varphi, A\varphi \rangle \leq (1 + \sqrt{\epsilon})\langle \mathbf{w}_1, \rho_1 \mathbf{w}_1 \rangle + \frac{d_2 d_3^2}{\mu}\left(\frac{1}{\sqrt{\epsilon}} + 1\right)\sum_{\ell=2}^{d_2 d_3}\langle \mathbf{w}_\ell, \rho_1 \mathbf{w}_\ell \rangle.$$

Then from (7.5.8), $\langle \varphi, A\varphi \rangle \leq (1 + \sqrt{\epsilon})\langle \varphi, B\varphi \rangle$ for all sufficiently small ϵ.

To prove (7.5.10), let $\Psi = \sum_{j=1}^{d_1} \lambda_j^{\frac{1}{2}} \mathbf{u}_j \otimes \mathbf{v}_j$ and $\Phi = \sum_{k=1}^{d_3} \mu_k^{\frac{1}{2}} \mathbf{x}_k \otimes \mathbf{y}_k$ be decompositions of Ψ and Φ provided by Schrödinger's theorem. Then $\rho_1 = \sum_{j=1}^{d_1} \lambda_j |\mathbf{u}_j\rangle\langle\mathbf{u}_j|$, $\sigma_3 = \sum_{k=1}^{d_3} \mu_k |\mathbf{y}_k\rangle\langle\mathbf{y}_k|$ and

$$A = \sum_{j,j',k} \lambda_j^{\frac{1}{2}} \lambda_{j'}^{\frac{1}{2}} \mu_k^{-1} |\mathbf{u}_j \otimes \mathbf{v}_j \otimes \mathbf{y}_k\rangle\langle\mathbf{u}_{j'} \otimes \mathbf{v}_{j'} \otimes \mathbf{y}_k|.$$

Writing out $\langle \mathbf{w}_1 \otimes \Phi, A\mathbf{w}_1 \otimes \Phi \rangle$ yields the sum

$$\sum_{i,i'j,j',k} \mu_k^{-1}(\mu_i \mu_{i'} \lambda_j \lambda_{j'})^{\frac{1}{2}}\langle \mathbf{u}_{j'} \otimes \mathbf{v}_{j'} \otimes \mathbf{y}_k, \mathbf{w}_1 \otimes \mathbf{x}_i \otimes \mathbf{y}_i\rangle\langle\mathbf{w}_1 \otimes \mathbf{x}_{i'} \otimes \mathbf{y}_{i'}, \mathbf{u}_j \otimes \mathbf{v}_j \otimes \mathbf{y}_k\rangle,$$

in which only the terms with $i = i' = k$ can be nonzero, and hence the sum reduces to $\sum_{j,j',k}(\lambda_j \lambda_{j'})^{\frac{1}{2}}\langle \mathbf{w}_1, \mathbf{u}_j\rangle\langle\mathbf{u}_{j'}, \mathbf{w}_1\rangle\langle\mathbf{v}_{j'}, \mathbf{x}_k\rangle\langle\mathbf{x}_k, \mathbf{v}_j\rangle$. Define the matrix $M \in M_{d_1}(\mathbb{C})$ with entries $M_{j',j} := \sum_{k=1}^{d_3} \mu_k \langle \mathbf{v}_{j'}, \mathbf{x}_k\rangle\langle\mathbf{x}_k, \mathbf{v}_j\rangle$ and the vector $\mathbf{a} \in \mathbb{C}^{d_1}$ with entries $a_j := \lambda_j^{\frac{1}{2}}\langle \mathbf{w}_1, \mathbf{u}_j\rangle$. Note that $0 < M \leq \mathbb{1}$ (with equality on the right if $d_3 = d_2$, so that $\{\mathbf{x}_1, \ldots, \mathbf{x}_{d_3}\}$ is an orthonormal basis for $\mathcal{H}_2$), and therefore, $\langle \mathbf{w}_1 \otimes \Phi, A\mathbf{w}_1 \otimes \Phi \rangle = \langle \mathbf{a}, M\mathbf{a} \rangle \leq \|\mathbf{a}\|^2 = \langle \mathbf{w}_1, \rho_1 \mathbf{w}_1 \rangle$. $\qquad\square$

Lieb and Ruskai gave several equivalent reformulations of SSA that have turned out to be very important in their own right [**143**].

Theorem 7.61 (Lieb and Ruskai, 1973, four equivalent formulations of SSA). *The following statements are all true and equivalent in that once any one of them is proven, simple arguments yield the other three:*

(1) *For all density matrices ρ_{123} on $\mathcal{H}_1 \otimes \mathcal{H}_2 \otimes \mathcal{H}_3$, (7.5.1) is satisfied.*

(2) *The conditional entropy is concave. That is, the map $\rho_{12} \mapsto S_{1|2}(\rho_{12}) = S_{12} - S_2$ is concave on the set of density matrices ρ_{12} on $\mathcal{H}_1 \otimes \mathcal{H}_2$.*

(3) *The relative entropy functional $(\rho, \sigma) \mapsto D(\rho\|\sigma)$ is montone under partial traces; i.e., for all density matrices ρ_{12}, σ_{12} on $\mathcal{H}_1 \otimes \mathcal{H}_2$, $D(\rho_1\|\sigma_1) \leq D(\rho_{12}\|\sigma_{12})$.*

(4) *The relative entropy functional $(\rho, \sigma) \mapsto D(\rho\|\sigma)$ is jointly convex.*

Proof. We first prove that $(3) \Rightarrow (1)$ making use of the relation (7.2.11) between conditional and relative entropy. Let $\tau_3 := d_3^{-1}\mathbb{1}_3$, where d_3 is the dimension of $\mathcal{H}_3$. Since $D(X\|Y) = \text{Tr}[X(\log X - \log Y)]$,

$$D(\rho_{123}\|\rho_{12} \otimes \tau_3) = -S_{123} + S_{12} + \log d_3$$

and

$$D(\rho_{23}\|\rho_2 \otimes \tau_3) = -S_{23} + S_2 + \log d_3.$$

Therefore,

$$(7.5.11) \qquad S_{12} + S_{23} - S_{123} - S_2 = D(\rho_{123}\|\rho_{12} \otimes \tau_3) - D(\rho_{23}\|\rho_2 \otimes \tau_3).$$

Then since $\mathrm{Tr}_1[\rho_{123}] = \rho_{23}$ and $\mathrm{Tr}_1[\rho_{12} \otimes \tau_3] = \rho_2 \otimes \tau_3$, $D(\rho_{123}\|\rho_{12} \otimes \tau_3) - D(\rho_{23}\|\rho_2 \otimes \tau_3) \geq 0$ as a consequence of the monotonicity of the relative entropy under partial traces, and then by (7.5.11), this implies (7.5.1).

We next prove that $(1) \Rightarrow (2)$. For two density matrices ρ_{12} and σ_{12} on $\mathcal{H}_1 \otimes \mathcal{H}_2$, take $\mathcal{H}_3$ to be $\mathbb{C}^2$ with the usual inner product and define the tripartite state $\frac{1}{2}\begin{bmatrix} \rho_{12} & 0 \\ 0 & \sigma_{12} \end{bmatrix}$. Then

$$S_{123} = \log 2 - \frac{1}{2}\mathrm{Tr}[\rho_{12}\log\rho_{12}] - \frac{1}{2}\mathrm{Tr}[\sigma_{12}\log\sigma_{12}],$$

$$S_{23} = \log 2 - \frac{1}{2}\mathrm{Tr}[\rho_2\log\rho_2] - \frac{1}{2}\mathrm{Tr}[\sigma_2\log\sigma_2],$$

$$S_{12} = -\mathrm{Tr}\left[\frac{\rho_{12} + \sigma_{12}}{2}\log\left(\frac{\rho_{12} + \sigma_{12}}{2}\right)\right],$$

and

$$S_2 = -\mathrm{Tr}\left[\frac{\rho_2 + \sigma_2}{2}\log\left(\frac{\rho_2 + \sigma_2}{2}\right)\right].$$

Now (7.5.1) yields

$$S\left(\frac{\rho_{12} + \sigma_{12}}{2}\right) - S\left(\frac{\rho_2 + \sigma_2}{2}\right) \geq \frac{1}{2}\left(S(\rho_{12}) - S(\rho_2)\right) + \frac{1}{2}\left(S(\sigma_{12}) - S(\sigma_2)\right).$$

We next prove that $(2) \Rightarrow (3)$. For this, it is useful to consider the conditional entropy as a function on all of $\mathcal{B}^{++}(\mathcal{H}_1 \otimes \mathcal{H}_2)$ and not only on density matrices. For $X_{12} \in \mathcal{B}^{++}(\mathcal{H}_1 \otimes \mathcal{H}_2)$, let $X_1 = \mathrm{Tr}_2[X_{12}]$ and define $F(X_{12}) = -\mathrm{Tr}[X_{12}\log X_{12}] + \mathrm{Tr}[X_1\log X_1]$. Then F restricted to $\mathfrak{S}_{\mathcal{H}_1 \otimes \mathcal{H}_2}$ is the conditional entropy. Clearly, F is homogeneous of degree one. For $X_{12}, Y_{12} \in \mathcal{B}^{++}(\mathcal{H}_1 \otimes \mathcal{H}_2)$, define $\rho_{12} := (\mathrm{Tr}[X_{12}])^{-1}X_{12}$ and $\sigma_{12} := (\mathrm{Tr}[Y_{12}])^{-1}Y_{12}$. Then for $0 < \lambda < 1$ and $\mu := \frac{\lambda\,\mathrm{Tr}[Y_{12}]}{(1-\lambda)\,\mathrm{Tr}[X_{12}]+\lambda\,\mathrm{Tr}[Y_{12}]}$, simple computations give

$$F((1-\lambda)X_{12} + \lambda Y_{12}) = ((1-\lambda)\,\mathrm{Tr}[X_{12}] + \lambda\,\mathrm{Tr}[Y_{12}])F((1-\mu)\rho_{12} + \mu\sigma_{12})$$

$$\geq (1-\lambda)F(X_{12}) + \lambda F(Y_{12})$$

where the inequality comes from the concavity of F on $\mathfrak{S}_{\mathcal{H}_1 \otimes \mathcal{H}_2}$. Thus, the extended function F is concave on $\mathcal{B}^{++}(\mathcal{H}_1 \otimes \mathcal{H}_2)$. We are now in a position to apply Lemma 6.11 on the derivatives of convex functions that are homogeneous of degree one.

Note that if G is the function on $\mathcal{B}^{++}(\mathcal{H})$, $\mathcal{H}$ a finite-dimensional Hilbert space, defined by $G(X) := -\operatorname{Tr}[X \log X]$, then for all $X \in \mathcal{B}^{++}(\mathcal{H})$ and all self-adjoint H

$$\lim_{t \downarrow 0} \frac{1}{t}(G(X + tH) - G(X)) = -\operatorname{Tr}[H] - \operatorname{Tr}[H \log X],$$

and from here we readily obtain the derivative of F.

Fix two density matrices ρ_{12} and σ_{12} on $\mathcal{H}_1 \otimes \mathcal{H}_2$. By Lemma 6.11,

$$(-1 - \operatorname{Tr}[\rho_2 \log \sigma_2]) - (-1 - \operatorname{Tr}[\rho_{12} \log \sigma_{12}])$$
$$\leq -\operatorname{Tr}[\rho_2 \log \rho_2] + \operatorname{Tr}[\rho_{12} \log \rho_{12}]$$

Rearranging terms, $D(\rho_{12}\|\sigma_{12}) \geq D(\rho_2\|\sigma_2)$, which is the monotonicity of the relative entropy under partial traces.

At this point, we have proved the equivalence of (1), (2), and (3). To see that (3) implies (4), Let ρ_1, ρ_2 and σ_1, σ_2 be two pairs density matrices on $\mathcal{H}$, and let $0 < \lambda < 1$. Define the density matrices ρ and σ on $\mathcal{H} \otimes \mathbb{C}^2$ by

$$\rho := \begin{bmatrix} (1-\lambda)\rho_1 & 0 \\ 0 & \lambda\rho_2 \end{bmatrix} \quad \text{and} \quad \sigma := \begin{bmatrix} (1-\lambda)\sigma_1 & 0 \\ 0 & \lambda\sigma_2 \end{bmatrix},$$

where we have written ρ and σ in block form. Finally define $\rho_\lambda := (1-\lambda)\rho_1 + \lambda\rho_2$ and $\sigma_\lambda := (1-\lambda)\sigma_1 + \lambda\sigma_2$. Note that $\rho_\lambda = \operatorname{Tr}_2[\rho]$ and $\sigma_\lambda = \operatorname{Tr}_2[\sigma]$. Therefore, by (3),

$$D(\rho_\lambda\|\sigma_\lambda) \leq D(\rho\|\sigma) = (1+\lambda)D(\rho_1\|\sigma_1) + \lambda D(\rho_2\|\sigma_2),$$

and hence the relative entropy is jointly convex. Finally, we show that (4) implies (2). In fact, this is an immediate consequence of the formula (7.2.11) relating relative and conditional entropy. Finally, since SSA is valid, all four of the equivalent statements are valid. $\square$

7.6. The data processing inequality

The **data processing inequality (DPI)**, proved by Lindblad [149], is one of the cornerstones of quantum information theory. He had earlier treated a special case in [147], and obtained the complete result after the proof of SSA.

Theorem 7.62 (Data processing inequality). *Let $\mathcal{H}$ and $\mathcal{K}$ be finite-dimensional Hilbert spaces, and let $\Phi^\dagger$ be a quantum operation from $\mathcal{B}(\mathcal{H})$ to $\mathcal{B}(\mathcal{K})$. Let $\rho, \sigma \in \mathfrak{S}_{\mathcal{H}}$. Then*

$$\tag{7.6.1} D(\Phi^\dagger(\rho)\|\Phi^\dagger(\sigma)) \leq D(\rho\|\sigma).$$

Since $D(\rho\|\sigma)$ measures the difficulty of experimentally distinguishing between ρ and σ, the DPI says that applying any quantum operation to ρ and σ only makes them more difficult to distinguish. We have already seen one proof of Theorem 7.62. In fact, in Corollary 6.45 we have seen that (7.6.1) is valid for all positive trace preserving maps—complete positivity is not needed. However, Lindblad's original proof, which does require complete positivity, is instructive, and is given here.

Proof of Theorem 7.62. Observe that as a direct consequence of the definition of the relative entropy, for any density matrix ω on any other Hilbert space, and any density matrices ρ, σ on $\mathcal{H}$,

$$(7.6.2) \qquad D(\omega \otimes \rho \| \omega \otimes \sigma) = D(\rho \| \sigma).$$

That is, adjoining ancilla does not change the relative entropy.

By Theorem 5.36 and (5.5.8) in particular, there is a Hilbert space $\hat{\mathcal{K}}$ of the same dimension as $\mathcal{K}$, and a density matrix ω on $\mathcal{K} \otimes \hat{\mathcal{K}}$, such that

$$(7.6.3) \qquad \Phi^\dagger(\rho) \otimes \frac{1}{d}\mathbb{1}_{\hat{\mathcal{K}} \otimes \mathcal{H}} = \frac{1}{N}\sum_{j=1}^{N} U_j(\omega \otimes \rho)U_j^*,$$

where $d = \dim(\hat{\mathcal{K}} \otimes \mathcal{H})$. The same is true with σ in place of ρ. By (7.6.2), $D(\Phi^\dagger(\rho)\|\Phi^\dagger(\sigma)) = D\left(\Phi^\dagger(\rho) \otimes \frac{1}{d}\mathbb{1}_{\mathcal{H} \otimes \hat{\mathcal{K}}}\|\Phi^\dagger(\sigma) \otimes \frac{1}{d}\mathbb{1}_{\mathcal{H} \otimes \hat{\mathcal{K}}}\right)$. Combining this with (7.6.3),

$$D(\Phi^\dagger(\rho)\|\Phi^\dagger(\sigma)) = D\left(\frac{1}{N}\sum_{j=1}^{N} U_j(\omega \otimes \rho)U_j^* \middle\| \frac{1}{N}\sum_{j=1}^{N} U_j(\omega \otimes \sigma)U_j^*\right)$$

$$\leq \frac{1}{N}\sum_{j=1}^{N} D(U_j(\omega \otimes \rho)U_j^* \| U_j(\omega \otimes \sigma)U_j^*),$$

where the inequality is the joint convexity of the relative entropy.

By Lemma 7.41, the relative entropy is unitarily invariant, and hence for each j, $D(U_j(\omega \otimes \rho)U_j^* \| U_j(\omega \otimes \sigma)U_j^*) = D(\omega \otimes \rho \| \omega \otimes \sigma) = D(\rho \| \sigma)$. $\qquad \square$

Exercises

(1) Let ρ and σ be probability densities on a finite set $\mathcal{X}$. Define $d^2(\rho, \sigma)$ by

$$d^2(\rho, \sigma) = H(\tfrac{1}{2}(\rho + \sigma)) - \tfrac{1}{2}(H(\rho) + H(\sigma)).$$

Then by the strict concavity of the entropy, $d(\rho, \sigma) > 0$ for $\rho \neq \sigma$. Show that $d(\rho, \sigma) \geq \frac{1}{2}\|\rho - \sigma\|_1$. (In fact, d satisfies the triangle inequality and thus defines a metric on $\mathcal{P}(\mathcal{X})$.)

(2) Prove the chain rule for the classical relative entropy: Let ρ_{XY} and σ_{XY} be joint probability distributions on $\mathcal{X} \times \mathcal{Y}$. Then

$$D(\rho_{XY}\|\sigma_{XY}) = D(\rho_X\|\sigma_Y) + \sum_{x \in \mathcal{X}} \rho_X(x)D(\rho_{Y|X=x}\|\sigma_{Y|X=x}).$$

(3) Let $\mathcal{X}$ be a finite set, and let $\rho, \sigma \in \mathcal{P}(\mathcal{X})$. For $A \subset \mathcal{X}$, define $p := \sum_{x \in A} \rho(x)$ and $q := \sum_{x \in A} \sigma(x)$. If $D(\rho\|\sigma) < \infty$ and $q = 0$, then $p = 0$. The point of this exercise is to show, moreover, that if q is small, then so is p. Use the first variational formula from Example 7.27 to prove that for all $t > 0$,

$$D(\rho\|\sigma) \geq tp - e^{t-1}q.$$

Now make an appropriate choice of t to show that for all $\epsilon > 0$,

$$q \leq \frac{\epsilon}{2} e^{-2D(\rho\|\sigma)/\epsilon} \quad \Rightarrow \quad p \leq \epsilon.$$

(4) For $\rho, \sigma \in \mathfrak{S}$ such that $D(\rho\|\sigma) < \infty$, for all $\epsilon > 0$, and all orthogonal projections P, show that

$$\mathrm{Tr}[P\sigma] \leq \frac{\epsilon}{2} e^{-2D(\rho\|\sigma)/\epsilon} \quad \Rightarrow \quad \mathrm{Tr}[P\rho] \leq \epsilon.$$

(5) Let $\mathcal{H}_1, \mathcal{H}_2$ be Hilbert spaces, and let $\rho_{12} \in \mathfrak{S}_{\mathcal{H}_1 \otimes \mathcal{H}_2}$. Show that $S_1 + S_2 - S_{12} \leq 2\min\{S_1, S_2\}$.

(6) Let $\mathcal{H}_1, \ldots, \mathcal{H}_N$ be finite-dimensional Hilbert spaces. Let ρ be a density matrix on $\mathcal{H}_1 \otimes \cdots \otimes \mathcal{H}_N$. Let ρ_j denote its marginal density on the jth factor $\mathcal{H}_j$. Show that

$$\sum_{j=1}^{N} S(\rho_j) \geq S(\rho)$$

with equality if and only if $\rho = \rho_1 \otimes \cdots \otimes \rho_N$.

(7) One relation between relative and conditional entropy is given in (7.2.11). Here is another. Let $\mathcal{H}_1, \mathcal{H}_2$ be Hilbert spaces of finite dimensions d_1, d_2, respectively. Let $\tau_1 := d_1^{-1} \mathbb{1}$. Show that

$$-S_{1|2}(\rho_{12}) = \inf\{D(\rho_{12}\|\tau_1 \otimes \sigma_2) \; : \; \sigma_2 \in \mathfrak{S}_{\mathcal{H}_2}\} + \log d_1.$$

Use this together with Lemma A.25 to give another proof that $(4) \Rightarrow (2)$ in Theorem 7.61.

(8) This exercise develops a geometric proof of Pinsker's theorem. The following inequality for the Schatten p-norms was proved in [23]: For all $X, Y \in M_n(\mathbb{C})$, all $n \in \mathbb{N}$, and all $1 < p \leq 2$,

$$\frac{\|X + Y\|_p^2 + \|X - Y\|_p^2}{2} \geq \|X\|_p^2 + (p-1)\|Y\|_p^2. \tag{$*$}$$

(This is a consequence of [23, (1.7)] since for $a, b \geq 0$ and $1 \leq p \leq 2$, $p \mapsto \left(\frac{a^p + b^p}{2}\right)^{1/p}$ is increasing in p on $[1, \infty)$.)

(a) Show that for $p = 2$, this inequality is an identity, namely the parallelogram law for Hilbert space norms. In the rest of the exercise we assume $1 < p \leq 2$.

(b) Let A, B be unit vectors in S_p; that is $\|A\|_p = \|B\|_p = 1$. Define $X := \frac{1}{2}(A+B)$ and $Y := \frac{1}{2}(A - B)$. Show that $(*)$ becomes

$$\left\|\frac{A+B}{2}\right\|_p \leq 1 - \frac{p-1}{8}\|A - B\|_p^2. \tag{$**$}$$

Since the unit ball in S_p is convex for $1 < p \leq 2$, the midpoint $\frac{1}{2}(A + B)$ of the segment connecting any two unit vectors A and B lies inside the unit ball; that is $\frac{1}{2}(A - B)\|_p \le 1$. Equation $(**)$ quantifies this, it says that in fact, $1 - \|\frac{1}{2}(A - B)\|_p$ is at least as large as a constant, namely $\frac{p-1}{8}$, times the square of the distance between A and B. Normed spaces with this property are said to be **2-uniformly convex**, and they share a number of properties possessed by Hilbert spaces.

(c) Replace Y by tY, $t > 0$, in $(*)$ to deduce that

$$\frac{1}{2t^2}(\|X + tY\|_p^2 + \|X - tY\|_p^2 - 2\|X\|_p^2) \ge (p - 1)\|Y\|_p^2$$

to deduce that for $X \in M_n^{++}(\mathbb{C})$ and $Y \in M_n^{\mathrm{s.a.}}(\mathbb{C})$,

$$\left.\frac{\mathrm{d}^2}{\mathrm{d}t^2}\|X + tY\|_p^2\right|_{t=0} \ge 2(p - 1)\|Y\|_p^2 \qquad (***)$$

and

$$\left.\frac{\mathrm{d}}{\mathrm{d}t}\|X + tY\|_p^2\right|_{t=0} = 2\|X\|_p^{2-p}\,\mathrm{Tr}[X^{p-1}Y],$$

and then prove that

$$\|X + Y\|_p^2 \ge \|X\|_p^2 + 2\|X\|_p^{2-p}\,\mathrm{Tr}[X^{p-1}Y] + (p - 1)\|X - Y\|_p^2.$$

(In fact, $(*)$ is proved in **[23]** by directly proving $(***)$.)

Let ρ, σ be $n \times n$ density matrices with $\sigma \in M_n^{++}$. Apply the results of part (c) with $X = \sigma$ and $Y = \rho - \sigma$ to conclude that

$$\frac{\|\rho\|_p^2 - 1}{p - 1} \ge \frac{\|\sigma\|_p^2 - 1}{p - 1} + 2\|\sigma\|_p^{2-p}\,\mathrm{Tr}\left[\frac{\sigma^{p-1} - \mathbb{1}}{p - 1}(\rho - \sigma)\right] + \|\rho - \sigma\|_p^2,$$

and then take the limit $p \downarrow 1$ to conclude

$$\mathrm{Tr}[\rho \log \rho] \ge \mathrm{Tr}[\sigma \log \sigma] + \mathrm{Tr}[\log \sigma(\rho - \sigma)] + \tfrac{1}{2}\|\rho - \sigma\|_1^2,$$

which is the same as Pinsker's inequality, $D(\rho\|\sigma) \ge \tfrac{1}{2}\|\rho - \sigma\|_1^2$.

(9) Use the result of Exercise 30 of Chapter 5 to prove that Belavkin-Stasewski relative entropy, $D_{\mathrm{BS}}(\rho\|\sigma) := \mathrm{Tr}[\rho \log(\rho^{1/2}\sigma^{-1}\rho^{1/2})]$, is jointly convex in ρ and σ, and then use this to show that is satisfies the data processing inequality; that is, for all quantum operations $\Phi^\dagger$, $D_{\mathrm{BS}}(\Phi^\dagger(\rho)\|\Phi^\dagger(\sigma)) \le D_{\mathrm{BS}}(\rho\|\sigma)$.

Chapter 8

Consequences and refinements of SSA

8.1. The Alicki-Fannes inequality

Christandl and Winter conjectured a bound on the modulus of continuity of the quantum conditional entropy that depends only on the dimension of one of the Hilbert spaces involved [**64**, Conjecture 14]. A bound depending on the dimensions of both Hilbert spaces is trivial to obtain from Theorem 7.32, but this is not enough for the application proposed in [**64**] which is discussed in the next chapter.

Alicki and Fannes [**5**] independently arrived at the conjecture and proved it. Their proof makes use of the concavity of the conditional entropy, one of the equivalent forms of SSA; see Theorem 7.61. Winter [**234**] refined their proof to obtain a nearly optimal bound and showed that the general structure of the argument could be adapted to prove nearly optimal continuity bounds for a number of quantities involving entropy.

Theorem 8.1. *Let ρ and σ be two density matrices of the tensor product $\mathcal{H}_1 \otimes \mathcal{H}_2$ of two Hilbert spaces with $\mathcal{H}_1$ having finite dimension d_1. Then with $\eta(t) := -t \log t$,*

$$\tfrac{1}{2}\|\rho - \sigma\|_1 \leq \delta \quad \Rightarrow \quad |S_{2|1}(\sigma) - S_{2|1}(\rho)| \leq 2 \log d_1 \delta + (1 + \delta)\eta\left(\frac{\delta}{1 + \delta}\right).$$

Proof. Let $t := \tfrac{1}{2}\|\rho - \sigma\|_1$. The first step of the proof is to introduce three new density matrices ω, τ, and τ' such that

$$(8.1.1) \qquad \omega = \frac{1}{1+t}\sigma + \frac{t}{1+t}\tau \quad \text{and} \quad \omega = \frac{1}{1+t}\rho + \frac{t}{1+t}\tau'.$$

Note that if t is small ρ, σ, and ω are all close to one another. We shall then apply convexity inequalities to get lower and upper bounds on $S_{2|1}(\omega)$ in terms of the conditional entropies of ρ, σ, τ, and τ' and this will prove Theorem 8.1.

Define $\tau := \frac{1}{t}(\rho - \sigma)_+ \in \mathfrak{S}$ so that

$$\rho = \sigma + (\rho - \sigma) \leq \sigma + t\tau$$

$$= (1 + t)\left(\frac{1}{1+t}\sigma + \frac{t}{1+t}\tau\right) =: (1 + t)\omega.$$

Then since $(1 + t)\omega - \rho \geq 0$ and $\mathrm{Tr}[(1+t)\omega - \rho] = t$, $\tau' := \frac{1}{t}((1+t)\omega - \rho) \in \mathfrak{S}$. This yields ω, τ, and τ' satisfying (8.1.1).

Everything is symmetric in ρ and σ up to this point. Now suppose that $S_{2|1}(\sigma) \geq S_{2|1}(\rho)$. Then we seek a lower bound on $S_{2|1}(\omega)$ involving $S_{2|1}(\sigma)$, and an upper bound on $S_{2|1}(\omega)$ involving $S_{2|1}(\rho)$.

To get the lower bound, combine the first expression in (8.1.1) with the concavity of the conditional entropy to obtain

$$(8.1.2) \qquad S_{2|1}(\omega) \geq \frac{1}{1+t}S_{2|1}(\sigma) + \frac{t}{1+t}S_{2|1}(\tau).$$

Now use the second expression in (8.1.1) to produce an upper bound on $S_{2|1}(\omega)$ as follows:

$$S_{2|1}(\omega) = S(\omega) - S(\omega_1)$$

$$= S\left(\frac{1}{1+t}\rho + \frac{t}{1+t}\tau'\right)$$

$$(8.1.3) \qquad + \frac{1}{1+t}\,\mathrm{Tr}[\rho_1 \log(\omega_1)] + \frac{t}{1+t}\,\mathrm{Tr}[\tau_1' \log(\omega_1)].$$

By Lemma 7.34,

$$S\left(\frac{1}{1+t}\rho + \frac{t}{1+t}\tau'\right) \leq \frac{1}{1+t}S(\rho) + \frac{t}{1+t}S(\tau') + \eta\left(\frac{t}{1+t}\right).$$

By the Gibbs variational principle, Theorem 7.46,

$$\mathrm{Tr}[\rho_1 \log \omega_1] \leq \log\left(\mathrm{Tr}[e^{\log \omega_1}]\right) - S(\rho_1) = -S(\rho_1).$$

Likewise, $\mathrm{Tr}[\tau_1' \log \omega_1] \leq -S(\tau')$. Combining the last three estimates with (8.1.3) yields

$$(8.1.4) \qquad S_{2|1}(\omega) \leq \frac{1}{1+t}S_{2|1}(\rho) + \frac{t}{1+t}S_{2|1}(\tau') + \eta\left(\frac{t}{1+t}\right).$$

Combining (8.1.2) and (8.1.4),

$$(8.1.5) \qquad S_{2|1}(\sigma) - S_{2|1}(\rho) \leq t\left[S_{2|1}(\tau') - S_{2|1}(\tau)\right] + (1 + t)\eta\left(\frac{t}{1+t}\right).$$

We do not have much information on τ and τ' except that they are density matrices on $\mathcal{H}_1$, but this is all we need: by (7.2.7) and (7.2.8), and then (7.2.2) for any $\tau \in \mathfrak{S}_{\mathcal{H}_1}$, $|S_{2|1}(\tau)| \leq S_1(\tau) \leq \log d_1$. Since we have assumed that $S_{2|1}(\sigma) \geq S_{2|1}(\rho)$, (8.1.5) becomes

$$(8.1.6) \qquad |S_{2|1}(\sigma) - S_{2|1}(\rho)| \leq 2t \log d_1 + (1+t)\eta\left(\frac{t}{1+t}\right).$$

Finally, $f(t) := (1+t)\eta\left(\frac{t}{1+t}\right)$ is monotone increasing in t; hence the right side of (8.1.6) is monotone increasing in t. Therefore, Theorem 8.1 follows from (8.1.6). $\qquad\square$

8.2. SSA by a method of Petz

We have seen that SSA is equivalent to the monotonicity of the relative entropy under partial traces, or equivalently, under conditional expectations. Petz found a very simple proof of this monotonicity [177] that furthermore opens the way to an analysis of the cases of equality in SSA.

Throughout this section $\mathcal{M}$ is a von Neumann algebra on a finite-dimensional Hilbert space $\mathcal{H}$, and $\mathcal{N}$ is a von Neumman subalgebra of $\mathcal{M}$.

Definition 8.2 (Relative modular operator). Let ρ and σ be two density matrices belonging to $\mathcal{M}$ with $\rho > 0$. The **relative modular operator** $\Delta_{\rho,\sigma}$ is the operator on $\mathcal{M}$ defined by

$$\Delta_{\sigma,\rho}(X) = \sigma X \rho^{-1}$$

for all $X \in \mathcal{M}$.

Note that if $\{\mathbf{u}_1, \ldots, \mathbf{u}_d\}$ is an orthonormal basis of $\mathcal{H}$ consisting of eigenvectors of ρ, $\rho\mathbf{u}_j = \lambda_j\mathbf{u}_j$, and if $\{\mathbf{v}_1, \ldots, \mathbf{v}_d\}$ is an orthonormal basis of $\mathcal{H}$ consisting of eigenvectors of σ, $\sigma\mathbf{v}_i = \mu_i\mathbf{v}_i$, then $\{|\mathbf{v}_i\rangle\langle\mathbf{u}_j|\}_{1 \leq i,j \leq d}$ is an orthonormal basis of $\mathcal{B}(\mathcal{H})$ equipped with the Hilbert-Schmidt inner product $\langle \cdot, \cdot \rangle_{\mathrm{HS}}$ and

$$\Delta_{\sigma,\rho}(|\mathbf{v}_i\rangle\langle\mathbf{u}_j|) = \mu_i\lambda_j^{-1}|\mathbf{v}_i\rangle\langle\mathbf{u}_j|.$$

This shows that $\Delta_{\sigma,\rho}$ is diagonal in this basis with nonnegative entries, and hence $\Delta_{\sigma,\rho}$ is a positive operator on $\mathcal{B}(\mathcal{H})$, and therefore also on $\mathcal{M}$ since it evidently leaves $\mathcal{M}$ invariant. Moreover, if we also have $\sigma > 0$, then all of the eigenvalues of $\Delta_{\sigma,\rho}$ are strictly positive, so that $\Delta_{\sigma,\rho} > 0$.

The relative modular operator is the matricial version of an operator introduced in a more general von Neumann algebra context by Araki [13]. It will be studied in this context in the final chapter. Petz had the insight that the ideas of Araki, though developed for a von Neumann algebra setting in which there are no traces, could be very useful for proving finite-dimensional trace inequalities. The next definition is due to Petz [176, 177].

Definition 8.3 (Petz quasi relative entropy). Let $f : (0, +\infty) \to \mathbb{R}$ be an operator convex function, and let ρ, σ be strictly positive density matrices in $\mathcal{M}$. The f-**relative quasi entropy** is defined by

$$D_f(\rho\|\sigma) = \langle \rho^{1/2}, f(\Delta_{\sigma,\rho})\rho^{1/2}\rangle_{\mathrm{HS}}.$$

Since $-\log(\Delta_{\sigma,\rho})(X) = X\log\rho - \log\sigma X$, the choice $f(s) = -\log s$ yields the Umegaki relative entropy $D(\rho\|\sigma)$. Likewise, for $f(s) = s\log s$,

$$\Delta_{\sigma,\rho}\log(\Delta_{\sigma,\rho})(X) = \sigma\log\sigma X\rho^{-1} - \sigma X\log\rho\rho^{-1}$$

and hence

$$\Delta_{\sigma,\rho}\log(\Delta_{\sigma,\rho})(\rho^{1/2}) = \sigma\log\sigma\rho^{-1/2} - \sigma\log\rho\rho^{-1/2}.$$

Therefore, the choice $f(x) = s\log s$ yields $D(\sigma\|\rho)$.

Let $\mathcal{E}_{\mathcal{N}}$ denote conditional expectation onto $\mathcal{N}$ as in Definition 4.36. The following compact notation will be useful: define $\rho_{\mathcal{N}} := \mathcal{E}_{\mathcal{N}}\rho$ and $\sigma_{\mathcal{N}} := \mathcal{E}_{\mathcal{N}}\sigma$.

Theorem 8.4. *Let ρ and σ be strictly positive density matrices in $\mathcal{M}$. Let f be operator convex, and let $D_f(\rho\|\sigma)$ denote the corresponding quasi relative entropy. Then,*

(8.2.1)
$$D_f(\rho_{\mathcal{N}}\|\sigma_{\mathcal{N}}) \le D_f(\rho\|\sigma).$$

Proof. The proof makes use of several operators on the Hilbert spaces $(\mathcal{M}, \langle\cdot,\cdot\rangle_{\mathrm{HS}})$ and $(\mathcal{N}, \langle\cdot,\cdot\rangle_{\mathrm{HS}})$, the latter being a subspace of the former. Define $U \in \mathcal{B}((\mathcal{M}, \langle\cdot,\cdot\rangle_{\mathrm{HS}})$ by

$$U(X) = \mathcal{E}_{\mathcal{N}}(X)\rho_{\mathcal{N}}^{-1/2}\rho^{1/2}.$$

Since by definition, $\mathcal{E}_{\mathcal{N}}^* = \mathcal{E}_{\mathcal{N}}$, the adjoint operator U^* is given by $U^*(Y) = \mathcal{E}_{\mathcal{N}}(Y\rho^{1/2})\rho_{\mathcal{N}}^{-1/2}$, and hence for $X \in \mathcal{M}$,

$$U^*U(X) = \mathcal{E}_{\mathcal{N}}(\mathcal{E}_{\mathcal{N}}(X)\rho_{\mathcal{N}}^{-1/2}\rho)\rho_{\mathcal{N}}^{-1/2} = \mathcal{E}_{\mathcal{N}}(X),$$

where part (i) of Theorem 4.39 has been used. Since $\mathcal{E}_{\mathcal{N}}$ is an orthogonal projection, U is a partial isometry on $(\mathcal{M}, \langle\cdot,\cdot\rangle_{\mathrm{HS}})$ whose initial space is $\mathcal{N}$. Therefore by the operator Jensen inequality, Theorem 3.25, and in particular, (3.4.7) of Remark 3.26,

(8.2.2)
$$f(U^*\Delta_{\sigma,\rho}U) \le U^*f(\Delta_{\sigma,\rho})U$$

as operators on $(\mathcal{N}, \langle\cdot,\cdot\rangle_{\mathrm{HS}})$. Next we claim that

(8.2.3)
$$U^*\Delta_{\sigma,\rho}U = \Delta_{\sigma_{\mathcal{N}},\rho_{\mathcal{N}}},$$

again as operators on $(\mathcal{N}, \langle\cdot,\cdot\rangle_{\mathrm{HS}})$.

To see this, note that for $X \in \mathcal{N}$, $\Delta_{\sigma,\rho}^{1/2}(U(X)) = \sigma^{1/2}X\rho_{\mathcal{N}}^{-1/2}$, and hence

$$\langle \Delta_{\sigma,\rho}^{1/2}(U(X)), \Delta_{\sigma,\rho}^{1/2}(U(X))\rangle = \mathrm{Tr}((\rho_{\mathcal{N}})^{-1/2}X^*\sigma X(\rho_{\mathcal{N}})^{-1/2})$$
$$= \mathrm{Tr}((\rho_{\mathcal{N}})^{-1/2}X^*\sigma_{\mathcal{N}}X(\rho_{\mathcal{N}})^{-1/2})$$
$$= \langle \Delta_{\sigma_{\mathcal{N}},\rho_{\mathcal{N}}}^{1/2}(X), \Delta_{\sigma_{\mathcal{N}},\rho_{\mathcal{N}}}^{1/2}(X)\rangle.$$

Combining (8.2.2), (8.2.3), and the identity $U(\rho_{\mathcal{N}})^{1/2} = \rho^{1/2}$,

$$D_f(\rho_{\mathcal{N}}\|\sigma_{\mathcal{N}}) = \langle (\rho_{\mathcal{N}})^{1/2}, f(\Delta_{\sigma_{\mathcal{N}},\rho_{\mathcal{N}}})(\rho_{\mathcal{N}})^{1/2}\rangle$$
$$\le \langle U(\rho_{\mathcal{N}})^{1/2}, f(\Delta_{\sigma,\rho})U(\rho_{\mathcal{N}})^{1/2}\rangle$$
$$= \langle \rho^{1/2}, f(\Delta_{\sigma,\rho})\rho^{1/2}\rangle = D_f(\rho\|\sigma). \qquad \square$$

We now examine the cases of equality in (8.2.1) following the strategy of [**53**]. The only inequality used in the proof was the application (8.2.2) of the operator Jensen inequality. We seek an explicit expression for the divergence from equality, and naturally this will depend on the choice of f. We are mainly interested in $f(s) = -\log s$ or $f(s) = s\log s$. However, defining $f_t(s)$ for $t > 0$ by $f_t(s) = (s+t)^{-1}$, we have the integral representation

$$-\log(s) = \int_0^\infty \left(\frac{1}{t+s} - \frac{1}{1+t}\right) dt.$$

Therefore,

$$(8.2.4) \qquad S(\rho\|\sigma) = \int_0^\infty \left(S_{f_t}(\rho\|\sigma) - \frac{1}{1+t}\right) dt,$$

and it suffices to examine the cases of equality in (8.2.1) for $f = f_t$. For this choice of f, it is possible to sharpen the application (8.2.2) of the operator Jensen inequality.

Lemma 8.5. *Let U be a partial isometry on Hilbert space $\mathcal{K}$ whose initial space is the subspace $\mathcal{K}_0$. Let $B \in \mathcal{B}^{++}(\mathcal{K}_0)$, $A\mathcal{B}^{++}(\mathcal{K})$, and suppose that $U^*AU = B$. Then for all $\mathbf{v} \in \mathcal{K}_0$,*

$$\langle \mathbf{v}, U^*A^{-1}U\mathbf{v}\rangle = \langle \mathbf{v}, B^{-1}\mathbf{v}\rangle + \langle \mathbf{w}, A\mathbf{w}\rangle,$$

where

$$\mathbf{w} := UB^{-1}\mathbf{v} - A^{-1}U\mathbf{v}.$$

Proof. We compute, using $U^*U = \mathbb{1}_{\mathcal{K}_0}$,

$$\langle \mathbf{w}, A\mathbf{w}\rangle = \langle UB^{-1}\mathbf{v} - A^{-1}U\mathbf{v}, AUB^{-1}\mathbf{v} - U\mathbf{v}\rangle$$
$$= \langle \mathbf{v}, B^{-1}U^*AUB^{-1}\mathbf{v}\rangle - 2\langle \mathbf{v}, B^{-1}\mathbf{v}\rangle + \langle \mathbf{v}, U^*A^{-1}U\mathbf{v}\rangle$$
$$= -\langle \mathbf{v}, B^{-1}\mathbf{v}\rangle + \langle \mathbf{v}, U^*A^{-1}U\mathbf{v}\rangle. \qquad \square$$

Definition 8.6 (Petz recovery map). Given a density matrix $\rho \in \mathcal{M}, \rho > 0$, the **Petz recovery map** $\mathcal{R}_\rho$ is the map from $\mathcal{N}$ to $\mathcal{M}$ given by

$$(8.2.5) \qquad \mathcal{R}_\rho(X) = \rho^{1/2}(\rho_\mathcal{N}^{-1/2} X \rho_\mathcal{N}^{-1/2})\rho^{1/2}$$

for all $X \in \mathcal{N}$.

Note that $\mathcal{R}(\rho_\mathcal{N}) = \rho$, so that $\mathcal{R}_\rho$ recovers ρ from $\rho_\mathcal{N}$. Evidently, $\mathcal{R}_\rho$ is completely positive and

$$\mathrm{Tr}[\mathcal{R}_\rho(X)] = \mathrm{Tr}[\rho^{1/2}(\rho_\mathcal{N}^{-1/2} X \rho_\mathcal{N}^{-1/2})\rho^{1/2}] = \mathrm{Tr}[\rho(\rho_\mathcal{N}^{-1/2} X \rho_\mathcal{N}^{-1/2})] = \mathrm{Tr}[X].$$

Therefore, $\mathcal{R}_\rho$ is a quantum operation. We are ready to prove Petz's theorem on the cases of equality in the inequality $D(\rho_\mathcal{N} \| \sigma_\mathcal{N}) \leq D(\rho \| \sigma)$.

Theorem 8.7. *Let ρ and σ be strictly positive density matrices in $\mathcal{M}$. Then $D(\rho_\mathcal{N} \| \sigma_\mathcal{N}) = D(\rho \| \sigma)$ if and only if*

$$\mathcal{R}_\rho(\sigma_\mathcal{N}) = \sigma \quad and \quad \mathcal{R}_\sigma(\rho_\mathcal{N}) = \rho.$$

Proof. Apply Lemma 8.5 with $A := (t + \Delta_{\sigma,\rho})$, $B = (t + \Delta_{\sigma_\mathcal{N},\rho_\mathcal{N}})$, and $\mathbf{v} := (\rho_\mathcal{N})^{1/2}$, and with U defined as above. The lemma's condition, $U^*AU = B$, follows from (8.2.3) and the fact that $U^*U = \mathbb{1}_\mathcal{N}$. Therefore, since $U(\rho_\mathcal{N})^{1/2} = \rho^{1/2}$, Lemma 8.5 yields

$$\begin{aligned}
D_{f_t}(\rho \| \sigma) - D_{f_t}(\rho_\mathcal{N} \| \sigma_\mathcal{N}) &= \langle \rho^{1/2}, (t + \Delta_{\sigma,\rho})^{-1} \rho^{1/2} \rangle \\
&\quad - \langle \rho_\mathcal{N}^{1/2}, (t + \Delta_{\sigma_\mathcal{N},\rho_\mathcal{N}})^{-1} \rho_\mathcal{N}^{1/2} \rangle \\
&= \langle \mathbf{w}_t, (t + \Delta_{\sigma,\rho})\mathbf{w}_t \rangle \geq t \|\mathbf{w}_t\|^2,
\end{aligned}$$

where, recalling that $U(\rho_\mathcal{N})^{1/2} = \rho^{1/2}$,

$$\mathbf{w}_t := U(t + \Delta_{\sigma_\mathcal{N},\rho_\mathcal{N}})^{-1}(\rho_\mathcal{N})^{1/2} - (t + \Delta_{\sigma,\rho})^{-1}\rho^{1/2}.$$

By (8.2.4), $D(\rho \| \sigma) = D(\rho_\mathcal{N} \| \sigma_\mathcal{N})$ if and only if $\mathbf{w}_t = 0$ for all t.

Using the integral representation of the square root function,

$$X^{1/2} = \frac{1}{\pi} \int_0^\infty t^{1/2} \left(\frac{1}{t} - \frac{1}{t + X} \right) dt,$$

and $U(\mathcal{N}\rho)^{1/2} = \rho^{1/2}$ once more, we conclude that

$$U(\Delta_{\sigma_\mathcal{N},\rho_\mathcal{N}})^{1/2}(\rho_\mathcal{N})^{1/2} - (\Delta_{\sigma,\rho})^{1/2}\rho^{1/2} = \frac{1}{\pi} \int_0^\infty t^{1/2}\mathbf{w}_t dt.$$

On the other hand,

$$\begin{aligned}
U(\Delta_{\sigma_\mathcal{N},\rho_\mathcal{N}})^{1/2}(\rho_\mathcal{N})^{1/2} - (\Delta_{\sigma,\rho})^{1/2}\rho^{1/2} &= U(\sigma_\mathcal{N})^{1/2} - \sigma^{1/2} \\
&= (\sigma_\mathcal{N})^{1/2}(\rho_\mathcal{N})^{-1/2}\rho^{1/2} - \sigma^{1/2}.
\end{aligned}$$

Therefore, if $D(\rho \| \sigma) = D(\rho_\mathcal{N} \| \sigma_\mathcal{N})$, then

$$(8.2.6) \qquad \sigma^{1/2} = (\sigma_\mathcal{N})^{1/2}(\rho_\mathcal{N})^{-1/2}\rho^{1/2}.$$

Then with $X := (\sigma_{\mathcal{N}})^{1/2}(\rho_{\mathcal{N}})^{-1/2}\rho^{1/2}$ and $Y = \sigma^{1/2}$, $X = Y$ implies $X^*X = Y^*Y$ which is

$$\sigma = \rho^{1/2}(\rho_{\mathcal{N}})^{-1/2}\sigma_{\mathcal{N}}(\rho_{\mathcal{N}})^{-1/2}\rho^{1/2} = \mathcal{R}_\rho(\sigma_{\mathcal{N}}).$$

Next, (8.2.6) can be rewritten as $\rho^{1/2} = \sigma^{1/2}(\sigma_{\mathcal{N}})^{-1/2}(\rho_{\mathcal{N}})^{1/2}$, and hence

$$\rho = \sigma^{1/2}(\sigma_{\mathcal{N}})^{-1/2}\rho_{\mathcal{N}}(\sigma_{\mathcal{N}})^{-1/2}\sigma^{1/2} = \mathcal{R}_\sigma(\rho_{\mathcal{N}}). \qquad \square$$

So far in this section we have discussed the monotonicity of relative entropy under conditional expectations. It is a simple matter to extend our conclusions to general quantum operations.

Let $\mathcal{H}$ and $\mathcal{K}$ be finite-dimensional Hilbert spaces, and let $\Phi^\dagger : \mathcal{B}(\mathcal{H}) \to \mathcal{B}(\mathcal{K})$ be a quantum operation. By Theorem 5.36, there is another Hilbert space $\widehat{\mathcal{K}}$ with same dimension as $\mathcal{K}$, a density matrix $\omega \in \mathcal{B}(\mathcal{H} \otimes \widehat{\mathcal{K}})$, and a $\mathcal{U}$ unitary in $\mathcal{B}(\mathcal{K} \otimes \mathcal{H} \otimes \widehat{\mathcal{K}})$,

$$(8.2.7) \qquad \Phi^\dagger(X) := \mathrm{Tr}_{\mathcal{H} \otimes \widehat{\mathcal{K}}}\left[\mathcal{U}(X \otimes \omega)\mathcal{U}^* \right].$$

Define $\mathcal{M} := \mathcal{B}(\mathcal{K} \otimes \mathcal{H} \otimes \widehat{\mathcal{K}})$ and $\mathcal{N} := \mathcal{B}(\mathcal{K}) \otimes \mathbb{1}_{\mathcal{H} \otimes \widehat{\mathcal{K}}}$. Also define $\tau := d^{-1}\mathbb{1}$, where $d := \dim(\mathcal{H})\dim(\mathcal{K})$ so that τ is the uniform density matrix on $\mathcal{H} \otimes \widehat{\mathcal{K}}$. By (4.6.2) of Example 4.38, we may rewrite (8.2.7) as

$$\Phi^\dagger(X) \otimes \tau := \mathcal{E}_{\mathcal{N}}\left[\mathcal{U}(X \otimes \omega)\mathcal{U}^* \right].$$

Then for any two $\rho, \sigma \in \mathfrak{S}_{\mathcal{H}}$,

$$
\begin{aligned}
D(\Phi^\dagger(\rho)\|\Phi^\dagger(\sigma)) &= D(\Phi^\dagger(\rho) \otimes \tau\|\Phi^\dagger(\sigma) \otimes \tau) \\
&= D\big(\mathcal{E}_{\mathcal{N}}\left[\mathcal{U}(\rho \otimes \omega)\mathcal{U}^* \right] \| \mathcal{E}_{\mathcal{N}}\left[\mathcal{U}(\sigma \otimes \omega)\mathcal{U}^* \right]\big) \\
&\leq D\big(\mathcal{U}(\rho \otimes \omega)\mathcal{U}^* \| \mathcal{U}(\sigma \otimes \omega)\mathcal{U}^*\big) \\
&= D(\rho \otimes \omega \| \sigma \otimes \omega) = D(\rho \| \sigma),
\end{aligned}
$$

(8.2.8)

where the only inequality comes from Theorem 8.4. Otherwise, we have used only the unitary invariance of the relative entropy and trivial identities.

By Theorem 8.7 there is equality in (8.2.8) if and only if the Petz recovery map that recovers $\mathcal{U}(\rho \otimes \omega)\mathcal{U}^*$ from $\mathcal{E}_{\mathcal{N}}\left[\mathcal{U}(\rho \otimes \omega)\mathcal{U}^* \right] = \Phi^\dagger(\rho) \otimes \tau$ also recovers $\mathcal{U}(\sigma \otimes \omega)\mathcal{U}^*$ from $\Phi^\dagger(\sigma) \otimes \tau$. Assume for now that $\Phi^\dagger(\rho) > 0$. Then for all $X \in \mathcal{B}(\mathcal{K})$, $X \otimes \tau \in \mathcal{N}$, letting $\mathcal{R}$ denote this recovery map,

$$\mathcal{R}(X \otimes \tau)$$

$$= \mathcal{U}(\rho \otimes \omega)^{1/2} \mathcal{U}^* \left[(\Phi^\dagger(\rho) \otimes \tau)^{-1/2}X \otimes \tau(\Phi^\dagger(\rho) \otimes \tau)^{-1/2} \right] \mathcal{U}(\rho \otimes \omega)^{1/2} \mathcal{U}^*.$$

Therefore, defining $Z := (\Phi^\dagger(\rho))^{-1/2}\Phi^\dagger(\sigma)(\Phi^\dagger(\rho))^{-1/2}$,

$$\sigma \otimes \omega = (\rho \otimes \omega)^{1/2} \mathcal{U}^* \left[Z \otimes \mathbb{1} \right] \mathcal{U}(\rho \otimes \omega)^{1/2},$$

and hence

$$\sigma = \mathrm{Tr}_{\mathcal{H} \otimes \widetilde{\mathcal{K}}}[\sigma \otimes \omega]$$
$$= \rho^{1/2} \left(\mathrm{Tr}_{\mathcal{H} \otimes \widetilde{\mathcal{K}}}[\mathcal{U}^*[Z \otimes \mathbb{1}]\,\mathcal{U}(\mathbb{1} \otimes \omega)]\right) \rho^{1/2} = \rho^{1/2} \left(\Phi(Z)\right) \rho^{1/2}.$$

Summarizing, there is equality in (8.2.8) if and only if

$$(8.2.9) \qquad\qquad \mathcal{R}_{\Phi,\rho}(\Phi^\dagger(\sigma)) = \sigma,$$

where

$$(8.2.10) \qquad \mathcal{R}_{\Phi,\rho}(X) := \rho^{1/2} \left(\Phi\left((\Phi^\dagger(\rho))^{-1/2} X (\Phi^\dagger(\rho))^{-1/2}\right)\right) \rho^{1/2}.$$

Definition 8.8. Let $\mathcal{H}$ and $\mathcal{K}$ be finite-dimensional Hilbert spaces, and let $\Phi : \mathcal{B}(\mathcal{K}) \to \mathcal{B}(\mathcal{H})$ be completely positive and unital. Let $\rho \in \mathfrak{S}_{\mathcal{H}}$. Then the **Petz recovery map** $\mathcal{R}_{\Phi,\rho} : \mathcal{B}(\mathcal{K}) \to \mathcal{B}(\mathcal{H})$ is the quantum operation defined by (8.2.10)

Remark 8.9. Definition 8.8 is an extension of Definition 8.6 to which it reduces in case $\Phi^\dagger$ is a conditional expectation. Moreover, we have already encountered $\mathcal{R}_{\Phi,\rho}$ in a sense: it is simply the Hilbert-Schmidt adjoint of the Accardi-Cecchini coarse graining operator $\mathcal{A}_{\Phi,\rho}$ defined in Definition 6.41. Hence, we have seen this operator arise naturally in connection with the DPI in two different ways.

Theorem 8.10 (Petz). *Let $\mathcal{H}$ and $\mathcal{K}$ be finite-dimensional Hilbert spaces, and let $\Phi : \mathcal{B}(\mathcal{K}) \to \mathcal{B}(\mathcal{H})$ be completely positive and unital. Then for any two invertible $\rho, \sigma \in \mathfrak{S}_{\mathcal{H}}$,*

$$(8.2.11) \qquad\qquad D(\Phi^\dagger(\rho)\|\Phi^\dagger(\sigma)) = D(\rho\|\sigma)$$

if and only if either

$$(8.2.12) \qquad \mathcal{R}_{\Phi,\rho}(\Phi^\dagger(\sigma)) = \sigma \quad or \quad \mathcal{R}_{\Phi,\sigma}(\Phi^\dagger(\rho)) = \rho,$$

and in this case, both are valid.

Proof. Suppose $\mathcal{R}_{\Phi,\rho}(\Phi^\dagger(\sigma)) = \sigma$. By definition, $\mathcal{R}_{\Phi,\rho}, \mathcal{R}_{\Phi,\rho}(\Phi^\dagger(\rho)) = \rho$. Since $\mathcal{R}_{\Phi,\rho}$ is a quantum operation, the DPI yields

$$D(\rho\|\sigma) = D(\mathcal{R}_{\Phi,\rho}(\Phi^\dagger(\rho))\|\mathcal{R}_{\Phi,\rho}(\Phi^\dagger(\sigma))) \leq D(\Phi^\dagger(\rho)\|\Phi^\dagger(\sigma)).$$

However, by the DPI once more, $D(\Phi^\dagger(\rho)\|\Phi^\dagger(\sigma)) \leq D(\rho\|\sigma)$. Hence equality holds in (8.2.11). The same reasoning shows that if $\mathcal{R}_{\Phi,\sigma}(\Phi^\dagger(\rho)) = \rho$, then equality holds in (8.2.11). Thus, either condition in (8.2.12) is sufficient for (8.2.11).

Now suppose that (8.2.11) is satisfied. Then we have seen in the computations leading up to (8.2.9) that the first equation in (8.2.12) is satisfied, and the same sort of computation shows that the second is also satisfied. $\qquad\square$

8.3. SSA and the quantum Markov property

In Definition 7.21, the Markov property for a classical discrete-time stochastic process is expressed in terms of conditional mutual information, and following the definition, it was shown how this leads to the description of a Markovian measure on the path space in terms of one-step transition operators. We now define quantum conditional mutual information in analogy with the classical formula (7.1.19), and use this to define quantum Markov chains. We then show that, as in the classical case, quantum Markov chains can be described in terms of one step transition operators.

Definition 8.11. For any tripartite density matrix ρ_{123} on $\mathcal{H}_1 \otimes \mathcal{H}_2 \otimes \mathcal{H}_3$, the **conditional mutual information of** 1 **and** 3 **given** 2, $I_{13|2}$, is defined by

$$I_{13|2}(\rho_{123}) := S_{12} + S_{23} - S_{123} - S_2 \geq 0.$$

Theorem 8.12. *For any tripartite density matrix ρ_{123} on $\mathcal{H}_1 \otimes \mathcal{H}_2 \otimes \mathcal{H}_3$,*

$$(8.3.1) \qquad\qquad I_{13|2}(\rho_{123}) \geq 0,$$

and there is equality in (8.3.1) if and only if there is a quantum operation T : $\mathcal{B}(\mathcal{H}_2) \to \mathcal{B}(\mathcal{H}_2 \otimes \mathcal{H}_3)$ such that

$$(8.3.2) \qquad\qquad (\mathbb{1} \otimes T)\rho_{12} = \rho_{123}.$$

Proof. By Definition 8.11, the inequality (8.3.1) is an equivalent expression of SSA. Next, express $I_{13|2}(\rho_{123})$ as the difference of two relative entropies, as done in (7.5.11) in the proof of Theorem 7.61: Let τ_1 denote the normalized identity on $\mathcal{H}_1$, so that

$$\begin{aligned} I_{12|3}(\rho_{123}) &= S_{12} + S_{23} - S_{123} - S_2 \\ &= D(\rho_{123}\|\tau_1 \otimes \rho_{23}) - D(\rho_{12}\|\tau_1 \otimes \rho_2). \end{aligned}$$

When $I_{13|2}(\rho_{123}) = 0$, $D(\rho_{123}\|\tau_1 \otimes \rho_{23}) = D(\rho_{12}\|\tau_1 \otimes \rho_2)$, and then by Theorem 8.7, the Petz recovery map $\mathcal{R}$ that recovers $\tau_1 \otimes \rho_{23}$ from $\tau_1 \otimes \rho_2$ must also recover ρ_{123} from ρ_{12}. By (8.2.5), $\mathcal{R}$ is given by

$$\mathcal{R}(\gamma) = (\tau_1 \otimes \rho_{23})^{1/2}(\tau_1 \otimes \rho_2)^{-1/2}\gamma(\tau_1 \otimes \rho_2)^{-1/2}(\tau_1 \otimes \rho_{23})^{1/2}$$

for all $\gamma \in \mathfrak{S}_{\mathcal{H}_1 \otimes \mathcal{H}_2}$. Evidently, $\mathcal{R}$ has the form $\mathbb{1} \otimes T$, where T : $\mathcal{B}(\mathcal{H}_2) \to \mathcal{B}(\mathcal{H}_2 \otimes \mathcal{H}_3)$ is given by

$$T(\omega) = (\rho_{23})^{1/2}(\rho_2)^{-1/2}\omega(\rho_2)^{-1/2}(\rho_{23})^{1/2},$$

and then with this T, (8.3.2) is satisfied.

Now suppose that (8.3.2) is satisfied for some quantum operation T. Applying Tr_1 to both sides yields $T\rho_2 = \rho_{23}$. Therefore, with $R := \mathbb{1} \otimes T$,

$$D(\rho_{123}\|\tau_1 \otimes \rho_{23}) = D(R\rho_{12}\|R\tau_1 \otimes \rho_2) \leq D(\rho_{12}\|\tau_1 \otimes \rho_2)$$

by the DPI since R is a quantum operation. However, by SSA,

$$D(\rho_{123}\|\tau_1 \otimes \rho_{23}) \geq D(\rho_{12}\|\tau_1 \otimes \rho_2),$$

and hence there is equality in (8.3.1). $\square$

Definition 8.13. Let ρ be a density matrix on the tensor product of N copies of a finite-dimensional Hilbert space $\mathcal{K}$, that is, on $\mathcal{K}_N := \mathcal{K}^{\otimes N}$. For each $j = 1,\ldots,N$, define $\mathcal{K}_{\leq j}$ to be the tensor product of the first j factors of $\mathcal{K}$ in $\mathcal{K}_N$, and define $\mathcal{K}_{\geq j}$ to be the tensor product of the final $N - j + 1$ factors of $\mathcal{K}$ in $\mathcal{K}_N$. Finally, define $\mathcal{K}_{\{j\}}$ to be the jth factor of $\mathcal{K}$ in $\mathcal{K}_N$. Then for $2 \leq j \leq N - 1$, we have

$$\mathcal{K}_N = \mathcal{K}_{\leq j-1} \otimes \mathcal{K}_{\{j\}} \otimes \mathcal{K}_{\geq j+1},$$

displaying $\mathcal{H}_N$ as a tripartite space in $N - 2$ different ways.

The density matrix ρ has **Markov property** in case for each $2 \leq j \leq N-1$, $I_{13|2}(\rho) = 0$, where

$$\mathcal{H}_1 = \mathcal{K}_{\leq j-1}, \quad \mathcal{H}_2 = \mathcal{K}_{\{j\}}, \quad \text{and} \quad \mathcal{H}_3 = \mathcal{K}_{\geq j+1}.$$

Theorem 8.14. *Let $\mathcal{K}$ be a finite-dimensional Hilbert space, and let ρ be a density matrix on $\mathcal{K}_N := \mathcal{K}^{\otimes N}$. For each $1 \leq j \leq N$, define $\rho_{\leq j}$ to be the marginal density on the first j factors of $\mathcal{K}$ in $\mathcal{K}_N$. Then ρ has the Markov property if and only if for each $j = 1,\ldots,N-1$, there is a quantum operation $T_j : \mathcal{B}(\mathcal{K}) \to \mathcal{B}(\mathcal{K} \otimes \mathcal{K})$ such that*

$$\rho_{\leq j+1} = \mathbb{1} \otimes T_j\rho_{\leq j},$$

where $\mathbb{1}$ denote the identity on $\mathcal{B}(\mathcal{K}_{\leq j-1})$.

Proof. This is an immediate consequence of Theorem 8.12 and Definition 8.13.
                                                                          $\square$

Theorem 8.14 gives a first indication, but only that, of the importance of the conditional mutual information, which will figure prominently in the next chapter. We conclude this section by proving some simple but useful properties of conditional mutual information.

Lemma 8.15 (Additivity). *Suppose that $\mathcal{H}_1 = \mathcal{H}_0 \otimes \mathbb{C}^2$ and ρ_{123} has the block representation*

$$\rho_{123} = \begin{bmatrix} \lambda\sigma_{023}^{(1)} & 0 \\ 0 & (1-\lambda)\sigma_{023}^{(2)} \end{bmatrix},$$

where for $j = 1, 2$, $\sigma_{023}^{(j)} \in \mathfrak{S}(\mathcal{H}_0 \otimes \mathcal{H}_2 \otimes \mathcal{H}_3)$. Then

$$I_{13|2}(\rho_{123}) = \lambda I_{03|2}(\sigma_{023}^{(1)}) + (1-\lambda)I_{03|2}(\sigma_{023}^{(2)}).$$

Proof. By familiar computations,

$$S_{123}(\rho_{123}) = \lambda S_{023}(\sigma_{023}^{(1)}) + (1 - \lambda)S(\sigma_{023}^{(2)}) - (\lambda \log \lambda + (1 - \lambda)\log(1 - \lambda)),$$

and likewise for the other three terms. The part depending only on λ shows up four times, twice with a positive sign, and twice with a negative sign, and hence it cancels in the sum. $\qquad\square$

Obviously, the above lemma extends to direct sums of any finite number of terms. The next lemma proves a monotonicity property of the conditional mutual information.

Lemma 8.16 (Monotonicity). *Let* Φ *be a quantum operation on* $\mathcal{B}(\mathcal{H}_1)$. *Then for all* $\rho_{123} \in \mathfrak{S}_{\mathcal{H}_1 \otimes \mathcal{H}_2 \otimes \mathcal{H}_3}$,

$$I_{13|2}(\Phi \otimes \mathbb{1}_{23}(\rho_{123})) \leq I_{13|2}(\rho_{123}).$$

Proof. Recall that with τ_3 denoting the trace normalized identity on $\mathcal{H}_3$,

$$I_{13|2}(\Phi \otimes \mathbb{1}_{23}(\rho_{123})) = D(\Phi \otimes \mathbb{1}_{23}(\rho_{123})\|\Phi \otimes \mathbb{1}_{23}(\rho_{12} \otimes \tau_3)) - D(\rho_{23}\|\rho_2 \otimes \tau_3),$$

where we have used the identity $\mathrm{Tr}_1[\Phi \otimes \mathbb{1}_{23}(\rho_{123})] = \mathrm{Tr}_1[\rho_{123}] = \rho_{23}$. By the DPI, $D(\Phi \otimes \mathbb{1}_{23}(\rho_{123})\|\Phi \otimes \mathbb{1}_{23}(\rho_{12} \otimes \tau_3)) \leq D(\rho_{123}\|\rho_{12} \otimes \tau_3)$. $\qquad\square$

8.4. Cases of equality in SSA

Let ρ_{123} be a density matrix on $\mathcal{H}_1 \otimes \mathcal{H}_2 \otimes \mathcal{H}_3$. Suppose that $\rho_{123} = \rho_1 \otimes \rho_{23}$. Then $S_{123} = S_1 + S_{23}$ and $S_{12} = S_1 + S_2$. Therefore,

$$(8.4.1) \qquad I_{13|2}(\rho_{123}) = S_{12} + S_{23} - S_{123} - S_2 = 0.$$

Likewise, (8.4.1) is satisfied if $\rho_{123} = \rho_{12} \otimes \rho_3$. More generally, one can put "part of ρ_2" together with ρ_1, and "part of ρ_2" together with ρ_3. That is, suppose $\mathcal{H}_2$ is a tensor product

$$\mathcal{H}_2 = \mathcal{H}_{\ell_2} \otimes \mathcal{H}_{r_2} \quad \text{and} \quad \rho_{123} = \rho_{1\ell_2} \otimes \rho_{r_2 3},$$

where ℓ and r stand for "left" and "right". Then $S_{123} = S_{1\ell_2} + S_{r_2 3}$, $S_{23} = S_{\ell_2} + S_{r_2 3}$ and $S_2 = S_{\ell_2} + S_{r_2}$. Once again, (8.4.1) is satisfied.

Examples of this type may be patched together taking direct sums to yield further examples. Let σ_{123} be a density matrix on $\mathcal{H}_1 \otimes \mathcal{K}_2 \otimes \mathcal{H}_3$ for some Hilbert space $\mathcal{K}_2$ and the same $\mathcal{H}_1$ and $\mathcal{H}_3$. Then for any $0 < p < 1$, $p\rho_{123} \oplus (1 - p)\sigma_{123}$ is a tripartite density matrix on $\mathcal{H}_1 \otimes (\mathcal{H}_2 \oplus \mathcal{K}_2) \otimes \mathcal{H}_3$, and clearly

$$I_{13|2}(p\rho_{123} \oplus (1 - p)\sigma_{123}) = pI_{13|2}(\rho_{123}) + (1 - p)I_{13|2}(\sigma_{123}).$$

It has been shown in [99] that this type of construction yields all cases of equality in SSA. (Necessary and sufficient conditions for equality in SSA are given in [190], but it does not seem easy to use these conditions to verify that this description of cases of equality is complete.)

Theorem 8.17. *Let ρ_{123} be a tripartite density matrix on the tensor product $\mathcal{H}_1 \otimes \mathcal{H}_2 \otimes \mathcal{H}_3$ of finite-dimensional Hilbert spaces. Assume $\rho_2 > 0$. Then $I_{13|2}(\rho_{123}) = 0$ if and only if $\mathcal{H}_2$ has the form*

$$(8.4.2) \qquad \mathcal{H}_2 = \bigoplus_{j=1}^{m} \mathcal{H}_{\ell_2}^{(j)} \otimes \mathcal{H}_{r_2}^{(j)},$$

and ρ_{123} has the form

$$(8.4.3) \qquad \rho_{123} = \bigoplus_{j=1}^{m} p_j \rho_{12_\ell}^{(j)} \otimes \rho_{r_2 3}^{(j)}$$

for probabilities $p_1, \ldots, p_k$ and density matrices $\rho_{12_\ell}^{(j)}, \rho_{r_2 3}^{(j)}$, $j = 1, \ldots, m$ where for each j, $\rho_{12_\ell}^{(j)} \in \mathcal{S}_{\mathcal{H}_1 \otimes \mathcal{H}_{\ell_2}^{(j)}}$ and $\rho_{r_2,3}^{(j)} \in \mathcal{S}_{\mathcal{H}_{r_2}^{(j)} \otimes \mathcal{H}_3}$.

Proof. We have already seen that if ρ_{123} has the form (8.4.3), then $I_{13|2}(\rho_{123}) = 0$. Now suppose that $I_{13|2}(\rho_{123}) = 0$. By Theorem 8.12, there is a quantum operation $T : \mathcal{B}(\mathcal{H}_2) \to \mathcal{B}(\mathcal{H}_2 \otimes \mathcal{H}_3)$ such that

$$(8.4.4) \qquad (\mathbb{1} \otimes T)\rho_{12} = \rho_{123}.$$

Applying Tr_3 to both sides,

$$(8.4.5) \qquad (\mathbb{1} \otimes \Phi^\dagger)\rho_{12} = \rho_{12}, \quad \text{where} \quad \Phi^\dagger := \mathrm{Tr}_3 \circ T.$$

Applying Tr_1 to both sides of (8.4.5) yields $\Phi^\dagger \rho_2 = (\mathrm{Tr}_3 \circ T)\rho_2 = \rho_2$, so that ρ_2 is an invariant state for Φ, and since $\rho_2 > 0$, Φ has a faithful invariant state.

Let $\mathcal{I}_{\rho_2}$ be the set of all completely positive unital maps Ψ on $\mathcal{B}(\mathcal{H}^2)$ that leave ρ_2 invariant. By Lemma 5.56, there exists $\Phi_0 \in \mathcal{I}_{\rho_2}$ such that the fixed point algebra $\mathcal{C}_{\Phi_0}$ satisfies $\mathcal{C}_{\Phi_0} \subseteq \mathcal{C}_\Psi$ for all $\Psi \in \mathcal{I}_{\rho_2}$.

Then by Theorem 5.57, $\mathcal{H}_2$ has a decomposition of the form (8.4.2) and there exists a set of density matrices $\{\omega_1, \ldots, \omega_m\}$, with $\omega_j \in \mathcal{S}_{\mathcal{H}_{r_2}^{(j)}}$ for each j, such that for all ρ that are invariant under every $\Psi \in \mathcal{I}_{\rho_2}$,

$$\rho = \sum_{j=1}^{m} p_j \sigma_j \otimes \omega_j$$

for some set $\{p_1, \ldots, p_m\}$ of probabilities and some set of density matrices $\{\sigma_1, \ldots, \sigma_m\}$, with $\sigma_j \in \mathcal{S}_{\mathcal{H}_{\ell_2}^{(j)}}$ for each j; in particular, ρ_2 has this form.

Furthermore, also by Theorem 5.57, for every $\Psi \in \mathcal{I}_{\rho_2}$, if Ψ_j denotes the restriction of $\Phi^\dagger$ to $\mathcal{B}(\mathcal{H}_2^{(j)})$, $\Phi_j^\dagger$ has the form $\Phi_j^\dagger = \left(\mathbb{1}_{\mathcal{H}_{\ell_2}^{(j)}} \otimes \Psi_j^\dagger \right)$, where $\Psi_j^\dagger$ is a quantum operation on $\mathcal{B}(\mathcal{H}_{2_r}^{(j)})$ such that $\Psi_j^\dagger(\omega_r^{(j)}) = \omega_r^{(j)}$. In particular, this applies to the map $\Phi^\dagger$ defined in (8.4.5), and then it follows that

$$\rho_{12} = (\mathbb{1} \otimes \Phi^\dagger)\rho_{12} = \sum_{j=1}^{m} \mathrm{Tr}_{\mathcal{H}_r^{(j)}}[(\mathbb{1} \otimes P_j)\rho_{12}(\mathbb{1} \otimes P_j)] \otimes \omega_r^{(j)}.$$

For $j = 1, \ldots, m$, define p_j and $\rho_{12_\ell}^{(j)} \in \mathfrak{S}_{\mathcal{H}_1 \otimes \mathcal{H}_{2_\ell}^{(j)}}$,

$$p_j = \mathrm{Tr}[[(\mathbb{1} \otimes P_j)\rho_{12}(\mathbb{1} \otimes P_j)]] \quad \text{and} \quad p_j\rho_{12_\ell}^{(j)} := \mathrm{Tr}_{\mathcal{H}_r^{(j)}}[(\mathbb{1} \otimes P_j)\rho_{12}(\mathbb{1} \otimes P_j)].$$

Then $\rho_{12} = \sum_{j=1}^{m} p_j\rho_{12_\ell}^{(j)} \otimes \omega_r^{(j)}$.

Let $T : \mathcal{B}(\mathcal{H}_2) \to \mathcal{B}(\mathcal{H}_2 \otimes \mathcal{H}_3)$ be the quantum operation in (8.4.4), and let T_j denote the restriction of T to $\mathcal{B}(\mathcal{H}_2^{(j)})$. Then since $\mathrm{Tr}_3 \circ T_j = \Phi_j^\dagger$, it follows from Corollary 5.63 that T_j has the form $T_j = \left(\mathbb{1}_{\mathcal{H}_{\ell_2}^{(j)}} \otimes \widehat{\Psi}_j^\dagger\right)$, where $\widehat{\Psi}_j^\dagger : \mathcal{B}(\mathcal{G}_{2_r}^{(j)}) \to \mathcal{B}(\mathcal{G}_{2_r}^{(j)} \otimes \mathcal{H}_3)$ is a quantum operation. Therefore,

$$(\mathbb{1} \otimes T_j)(\rho_{12_\ell}^{(j)} \otimes \omega_r^{(j)}) = \rho_{12_\ell}^{(j)} \otimes \widehat{\Psi}_j^\dagger(\omega_r^{(j)}).$$

Defining $\rho_{r_2,3}^{(j)} := \widehat{\Psi}_j^\dagger(\omega_r^{(j)})$ yields the decomposition (8.4.3). $\qquad\square$

8.5. SSA and the Peierls-Bogoliubov and Golden-Thompson inequalities

The conjecture of Lanford and Robinson that the classical SSA inequality was also valid in the quantum case attracted attention, and many people had worked to prove it during the half decade it was open. A paper of Araki and Lieb [**15**] is the first publication on the subject. They proved:

Theorem 8.18 (Araki and Lieb 1970). *Let ρ_{123} be a tripartite density matrix. Then*

$$(8.5.1) \qquad\qquad S_{12} + S_{23} \geq S_{123} - \log \mathrm{Tr}[\rho_2^2].$$

If ρ_2 commutes with ρ_{12} or ρ_{23}, then (7.5.1) is valid.

Inequality (8.5.1) is *almost* SSA. By Jensen's inequality, and the concavity of the logarithm, if $\{\lambda_1, \ldots, \lambda_n\}$ are the eigenvalues of ρ_2, repeated according to multiplicity,

$$\log \mathrm{Tr}[\rho_2^2] = \log\left(\sum_{j=1}^{n} \lambda_j\lambda_j\right) \geq \sum_{j=1}^{n} \lambda_j \log \lambda_j.$$

Therefore, $-\log \mathrm{Tr}[\rho_2^2] \leq S(\rho_2)$, and hence (8.5.1) is weaker than (7.5.1).

Nonetheless, Theorem 8.18, together with the Araki-Lieb triangle inequality, also proved in [**15**], were sufficient to solve the quantum thermodynamic limit problem that had motivated the conjecture of Lanford and Robinson in the first place. Despite this, interest in proving the original conjecture remained strong, and the progress of Araki and Lieb influenced future work, and it has been the starting point for some recent refinements of SSA. Hence it is worthwhile to recall the simple proof of Theorem 8.18. The proof uses the Golden-Thompson inequality, Theorem 3.27, which states that for all self-adjoint $H, K \in M_n(\mathbb{C})$,

$$(8.5.2) \qquad \mathrm{Tr}[e^{H+K}] \le \mathrm{Tr}[e^H e^K],$$

and the Peierls-Bogoliubov inequality, Theorem 6.60, which states that for self-adjoint H and K with $\mathrm{Tr}[e^H] = 1$,

$$(8.5.3) \qquad \mathrm{Tr}[Ke^H] \le \log \mathrm{Tr}[e^{H+K}].$$

Proof of Theorem 8.18. Assume that ρ_{123} is positive definite, and define

$$\Delta := S_{123} - S_{12} - S_{23} = \mathrm{Tr}[\rho_{123}(-\log\rho_{123} + \log\rho_{12} + \log\rho_{23})].$$

Now apply (8.5.3) taking $H = \log\rho_{123}$ and $K = -\log\rho_{123} + \log\rho_{12} + \log\rho_{23}$,

$$e^{\Delta} \le \mathrm{Tr}[\exp(\log\rho_{123} - \log\rho_{123} + \log\rho_{12} + \log\rho_{23})]$$
$$= \mathrm{Tr}[\exp(\log\rho_{12} + \log\rho_{23}))] \le \mathrm{Tr}[\rho_{12}\rho_{23}] = \mathrm{Tr}[\rho_2^2],$$

where the second inequality is (8.5.2). This proves (8.5.1).

Now suppose ρ_2 commutes with ρ_{23}, and redefine Δ:

$$\Delta := S_{123} + S_2 - S_{12} - S_{23} = \mathrm{Tr}[\rho_{123}(-\log\rho_{123} - \log\rho_2 + \log\rho_{12} + \log\rho_{23})].$$

Using the Peierls-Bogoliubov inequality as before,

$$(8.5.4) \qquad e^{\Delta} \le \mathrm{Tr}[\exp(-\log\rho_2 + \log\rho_{12} + \log\rho_{23})].$$

Then again by the Golden-Thompson inequality, and then the fact that ρ_2 commutes with ρ_{23},

$$e^{\Delta} \le \mathrm{Tr}[\exp(-\log\rho_2 + \rho_{12} + \log\rho_{23}))]$$
$$\le \mathrm{Tr}[\rho_{12}\exp(-\log\rho_2 + \log\rho_{23})] = \mathrm{Tr}[\rho_{12}\rho_2^{-1}\rho_{23}] = 1.$$

Therefore, $\Delta \le 0$, and hence (7.5.1) is valid in this case. $\qquad\qquad\square$

This result focused attention on the search for a stronger form of the Golden-Thompson inequality that would provide an upper bound on $\mathrm{Tr}[e^{H+K+L}]$ for arbitrary self-adjoint matrices H, K, and L without assuming that any of them commute. Uhlmann in particular took up this line of investigation, and he proved in [**213**, Satz 8.2] a mild generalization of part of Theorem 8.18, namely that (7.5.1) is valid when any two of the matrices ρ_{12}, ρ_{23}, and ρ_2 commute, but this fell short of proving SSA.

Lieb's triple matrix inequality [141, Theorem 5], to be proved below in Theorem 10.15, provided the missing generalization of the Golden-Thompson inequality. Theorem 10.15 says that

$$\mathrm{Tr}[e^{H+K+L}] \leq \mathrm{Tr}[e^H T_{e^{-K}}(e^L)],$$

where

$$T_A(B) = \int_0^\infty \frac{1}{s+A} B \frac{1}{s+A} \, ds.$$

Armed with this, return to (8.5.4), and observe that, using Lieb's triple matrix inequality in the second line, and taking the trace over $\mathcal{H}_3$ in the next step, and then using cyclicity of the trace,

$$e^\Delta \leq \mathrm{Tr}[\exp(-\log \rho_2 + \log \rho_{12} + \log \rho_{23})]$$

$$\leq \int_0^\infty \mathrm{Tr}\left[\frac{1}{s+\rho_2}\rho_{12}\frac{1}{s+\rho_2}\rho_{23}\right] ds$$

$$= \int_0^\infty \mathrm{Tr}\left[\frac{1}{s+\rho_2}\rho_{12}\frac{1}{s+\rho_2}\rho_2\right] ds$$

$$= \int_0^\infty \mathrm{Tr}\left[\rho_{12}\rho_2\frac{1}{(s+\rho_2)^2}\right] ds = \mathrm{Tr}[\rho_{12}] = 1.$$

This proves SSA following the line of the investigation started in [15].

Later, in Section 11.8, we shall present the proof of a more recent extension of the Golden-Thompson inequality to four matrices, due to Sutter, Berta, and Tomamichel [205].

Theorem 8.19. *For all self-adjoint H_1, H_2, H_3, and H_4 in $\mathcal{B}^{\mathrm{s.a.}}(\mathcal{H})$,*

$$\mathrm{Tr}[e^{H_1+H_2+H_3+H_4}] \leq \int_\mathbb{R} \mathrm{Tr}\left[e^{H_1} e^{\frac{1+it}{2}H_2} e^{\frac{1+it}{2}H_3} e^{H_4} e^{\frac{1-it}{2}H_3} e^{\frac{1-it}{2}H_2}\right] d\mu_0(t).$$

Sutter, Berta, and Tomamichel [205] applied Theorem 8.19 to prove an important refinement of SSA or, what is the same thing, the monotonicity of the relative entropy under partial traces.

Theorem 8.20. *Let $\mathcal{H}_1, \mathcal{H}_2$ be two finite-dimensional Hilbert spaces, and let $\rho_{12}, \sigma_{12} \in \mathcal{S}_{\mathcal{H}_1 \otimes \mathcal{H}_2}$. Assume that both ρ_{12} and σ_{12} are strictly positive. Then*

(8.5.5) $$D(\rho_{12}\|\sigma_{12}) - D(\rho_1\|\sigma_1) \geq D^M(\rho_{12}\|\widetilde{R}_\sigma(\rho_1)),$$

where

(8.5.6) $$\widetilde{R}_\sigma(X) := \int_\mathbb{R} \sigma_{12}^{\frac{1+it}{2}} \sigma_1^{-\frac{1+it}{2}} X \sigma_1^{-\frac{1-it}{2}} \sigma_{12}^{\frac{1-it}{2}} \, d\mu_0(t),$$

and D^M denotes the measured relative entropy.

Proof. Recall Gibbs variational principle (7.3.6): For all $\rho \in \mathfrak{S}$,

$$\mathrm{Tr}[\rho \log \rho] = \sup_{K \in M_n^{\mathrm{s.a.}}(\mathbb{C})} \{\mathrm{Tr}[\rho K] - \log(\mathrm{Tr}[e^K])\}.$$

Let $X > 0$, and replace K with $K + \log X$ to obtain

$$\mathrm{Tr}[\rho(\log \rho - \log X)] = \sup_{K \in M_n^{\mathrm{s.a.}}(\mathbb{C})} \{\mathrm{Tr}[\rho K] - \log(\mathrm{Tr}[e^{K + \log X}])\}.$$

This is simply the Gibbs variational principle for the relative entropy (7.3.8) except that now we observe that the formula is valid for any $X > 0$; there is no need for X to be a density matrix. Then

$$\begin{aligned}
\Delta &:= D(\rho_{12}\|\sigma_{12}) - D(\rho_1\|\sigma_1) \\
&= \mathrm{Tr}[\rho_{12}(\log \rho_{12} - \log \sigma_{12} + \log \sigma_1 - \log \rho_1)] \\
&= \mathrm{Tr}[\rho_{12}(\rho_{12} - \log X)],
\end{aligned}$$

where $X := e^{\log \sigma_{12} - \log \sigma_1 + \log \rho_1}$. Therefore, for any $K \in M_n^{\mathrm{s.a.}}(\mathbb{C})$,

$$\Delta \geq \mathrm{Tr}[\rho_{12}K] - \log(\mathrm{Tr}[e^{K + \log \sigma_{12} - \log \sigma_1 + \log \rho_1}]).$$

Now apply Theorem 8.19 with $H_1 := K$, $H_4 := \rho_1$, $H_2 := \sigma_{12}$ and $H_4 := \sigma_1$. The result is

$$\Delta \geq \mathrm{Tr}[\rho_{12}K] - \log(\mathrm{Tr}[e^K \widetilde{R}_\sigma(\rho_1)]).$$

Then by (7.4.8), optimizing over K yields the measured relative entropy $D^M(\rho_{12}\|\widetilde{R}_\sigma(\rho_1))$. $\qquad\square$

By Theorem 7.51, $D^M(\rho_{12}\|\widetilde{R}_\sigma(\rho_1)) \geq \frac{1}{2}\|\rho_{12} - \widetilde{R}_\sigma(\rho_1)\|_1^2$, and hence (8.5.5) yields

$$(8.5.7) \qquad D(\rho_{12}\|\sigma_{12}) - D(\rho_1\|\sigma_1) \geq \frac{1}{2}\|\rho_{12} - \widetilde{R}_\sigma(\rho_1)\|_1^2.$$

This is the bound we shall need in the next chapter. The breakthrough result in this direction is due to Fawzi and Renner [82] who gave a different lower bound in terms of fidelity, which also yields a bound of the type (8.5.7). The topic has been actively investigated in recent years; see [34] for a discussion of this work.

Exercises

(1) Let $\mathcal{M}$ be a von Neumann algebra on a finite-dimensional Hilbert space $\mathcal{H}$. Let $f(s) = s^p$, $-1 < p < 0$. For invertible density matrices $\rho, \sigma \in \mathcal{M}$, compute an explicit formula for $D_f(\rho \| \sigma)$, the Petz quasi entropy determined by f.

(2) Let $\mathcal{M}$ be a von Neumann algebra on a finite-dimensional Hilbert space $\mathcal{H}$. Let $f(s) = -s^p$, $0 < p < 1$. For invertible density matrices $\rho, \sigma \in \mathcal{M}$, compute an explicit formula for $D_f(\rho \| \sigma)$, the Petz quasi entropy determined by f.

(3) Let $\mathcal{M}$ be a von Neumann algebra on a finite-dimensional Hilbert space $\mathcal{H}$. Let $f(s) = s^p$, $1 < p \leq 2$. For invertible density matrices $\rho, \sigma \in \mathcal{M}$, compute an explicit formula for $D_f(\rho \| \sigma)$, the Petz quasi entropy determined by f.

(4) Let $\mathcal{M}$ be a von Neumann algebra on a finite-dimensional Hilbert space $\mathcal{H}$, and let $\mathcal{N}$ be a von Neumann subalgebra of $\mathcal{M}$. For each of the quasi entropies considered in the first three exercises, determine the condition for cases of equality in $D_f(\rho_{\mathcal{N}} \| \sigma_{\mathcal{N}}) \leq D_f(\rho \| \sigma)$.

(5) Let ρ_{12} be a density matrix on the tensor product $\mathcal{H}_1 \otimes \mathcal{H}_2$ of two finite-dimensional Hilbert spaces. Suppose that $S_1 \geq S_2$. By the Araki-Lieb triangle inequality $S_{12} \geq S_1 - S_2$. Show that there is equality in this inequality if and only if $\operatorname{rank}(\rho_1) = \operatorname{rank}(\rho_2)\operatorname{rank}(\rho_{12})$ and ρ_{12} has a spectral decomposition of the form

$$\rho_{12} = \sum_{j=1}^{\operatorname{rank}(\rho_{12})} \kappa_j |\phi_j\rangle\langle\phi_j|,$$

where for each i, j, $\operatorname{Tr}_1 |\phi_i\rangle\langle\phi_j| = \delta_{i,j}\rho_2$.

Chapter 9

Quantification of entanglement

9.1. Quantum teleportation and entanglement as a resource

Let $\mathcal{H}_A$ and $\mathcal{H}_B$ be two finite-dimensional Hilbert spaces, and let ψ be a unit vector in $\mathcal{H}_A \otimes \mathcal{H}_B$. As we have seen, ψ has a Schmidt decomposition

$$(9.1.1) \qquad \psi = \sum_{j=1}^{r} \lambda_j \mathbf{u}_j \otimes \mathbf{v}_j,$$

where each $\lambda_j > 0$ with $\sum_{j=1}^{r} \lambda_j = 1$ and where $\{\mathbf{u}_1, \ldots, \mathbf{u}_r\}$ and $\{\mathbf{v}_1, \ldots, \mathbf{v}_r\}$ are orthonormal in $\mathcal{H}_A$ and $\mathcal{H}_B$, respectively. The number r is the Schmidt rank of ψ, and ψ, or properly speaking $|\psi\rangle\langle\psi|$, which is entangled if and only if $r > 1$. It is intuitive that in some sense the larger r is, and the more uniformly distributed the numbers $\lambda_1, \ldots, \lambda_r$ are, the *more entangled* the state ψ. Evidently r can be no larger than $\min\{d_A, d_B\}$, where d_A and d_B are the dimensions of $\mathcal{H}_A$ and $\mathcal{H}_B$, respectively. We now make a definition that will be shown to be useful below.

Definition 9.1. Let $\mathcal{H}_A$ and $\mathcal{H}_B$ be two Hilbert spaces with finite dimensions d_A and d_B, respectively. A unit vector $\psi \in \mathcal{H}_A \otimes \mathcal{H}_B$ with Schmidt decomposition (9.1.1) is **maximally entangled** in case $r = \min\{d_A, d_B\}$, and $\lambda_j = 1/r$ for each $j = 1, \ldots, r$.

If $\mathcal{H}_A$ and $\mathcal{H}_B$ have the same dimension d, a unit vector $\psi \in \mathcal{H}_A \otimes \mathcal{H}_B$ is maximally entangled if and only if

$$\mathrm{Tr}_B[|\psi\rangle\langle\psi|] = \frac{1}{d}\mathbb{1}_A \quad \text{and} \quad \mathrm{Tr}_A[|\psi\rangle\langle\psi|] = \frac{1}{d}\mathbb{1}_B.$$

In the 1990s, people began to develop quantum communication protocols that made essential use of entanglement, and this sparked a strong interest in the quantification of entanglement for pure and mixed states in multipartite systems with two or more components. One of these quantum communication protocols is **quantum teleportation** [30]. This provides a means to exactly communicate a quantum state from one laboratory to another using a *shared maximally entangled state* and a classical communication line to relay two bits of information on the results of an experiment in one laboratory to the other.

To explain this in the simplest terms, all single particle Hilbert spaces will be copies of $\mathbb{C}^2$ with the usual inner product. In $\mathcal{H}_A$, we use the notation

$$|0\rangle_A := \begin{pmatrix} 0 \\ 1 \end{pmatrix} \quad \text{and} \quad |1\rangle_A := \begin{pmatrix} 1 \\ 0 \end{pmatrix}.$$

We use similar notation for $\mathcal{H}_B$ and $\mathcal{H}_C$, a third two-dimensional Hilbert space that will enter the discussion. States $|\psi\rangle\langle\psi|$, where $\psi \in \mathbb{C}^2$, are called **qbits**. They represent one basic unit of quantum information.

Suppose Alice has access to observables on $\mathcal{H}_A$ and $\mathcal{H}_C$, while Bob has access to observables on $\mathcal{H}_B$. We suppose that an entangled state $\psi_{AB} \in \mathcal{H}_A \otimes \mathcal{H}_B$ has been prepared, and is of the form

$$\psi_{AB} = \frac{1}{\sqrt{2}}(|0\rangle_A \otimes |0\rangle_B + |1\rangle_A \otimes |1\rangle_B),$$

which is a maximally entangled state.

Suppose that in addition, Alice has another state ϕ_C on $\mathcal{H}_C$, and this is the state that she wishes to communicate to Bob:

$$\phi_C = \phi_0|0\rangle_C + \phi_1|1\rangle_C.$$

The combined state in the two laboratories is

$$\phi_C \otimes \psi_{AB} = \frac{\phi_0}{\sqrt{2}}(|0\rangle_C \otimes |0\rangle_A \otimes |0\rangle_B + |0\rangle_C \otimes |1\rangle_A \otimes |1\rangle_B)$$

$$(9.1.2) \qquad = \frac{\phi_1}{\sqrt{2}}(|1\rangle_C \otimes |0\rangle_A \otimes |0\rangle_B + |1\rangle_C \otimes |1\rangle_A \otimes |1\rangle_B).$$

To communicate the state ϕ_C to Bob, Alice performs a von Neumann measurement on $\mathcal{H}_C \otimes \mathcal{H}_A$, generating two bits of classical information, which she sends to Bob. Bob then applies a unitary transformation acting on $\mathcal{H}_B$, depending on the message received, to the state he now has. After this, the state will be $\phi_0|0\rangle_B + \phi_1|1\rangle_B$. The state ϕ_C has been communicated to Bob.

Here is how this is done. The measurement Alice will make can be described using the **Bell basis** which which is an orthonormal basis of $\mathcal{H}_C \otimes \mathcal{H}_A$ consisting of maximally entangled unit vectors $\{\Phi_{CA}^+, \Psi_{CA}^+, \Phi_{CA}^-, \Psi_{CA}^-\}$, where

$$\Phi_{CA}^+ := 2^{-1/2}(|0\rangle_C \otimes |0\rangle_A + |1\rangle_C \otimes |1\rangle_A),$$

$$\Psi_{CA}^+ := 2^{-1/2}(|0\rangle_C \otimes |1\rangle_A + |1\rangle_C \otimes |0\rangle_A),$$

$$\Phi_{CA}^- := 2^{-1/2}(|0\rangle_C \otimes |0\rangle_A - |1\rangle_C \otimes |1\rangle_A),$$

$$\Psi_{CA}^- := 2^{-1/2}(|0\rangle_C \otimes |1\rangle_A - |1\rangle_C \otimes |0\rangle_A).$$

Since $|0\rangle_C \otimes |0\rangle_A = \frac{1}{\sqrt{2}}(\Phi_{CA}^+ + \Phi_{CA}^-)$, $|1\rangle_C \otimes |1\rangle_A = \frac{1}{\sqrt{2}}(\Phi_{CA}^+ - \Phi_{CA}^-)$, etc., it is easy to write $\phi_C \otimes \psi_{AB}$ in terms of the Bell basis by making these substitutions in (9.1.2). One finds:

$$\phi_C \otimes \psi_{AB} = \tfrac{1}{2}\Phi_{CA}^+ \otimes (\phi_0|0\rangle_B + \phi_1|1\rangle_B) + \tfrac{1}{2}\Psi_{CA}^+ \otimes (\phi_0|1\rangle_B + \phi_1|0\rangle_B)$$
$$+ \tfrac{1}{2}\Phi_{CA}^- \otimes (\phi_0|0\rangle_B - \phi_1|1\rangle_B) + \tfrac{1}{2}\Psi_{CA}^- \otimes (\phi_0|1\rangle_B - \phi_1|0\rangle_B).$$

Alice will use the PVM on $\{0, 1, 2, 3\}$ with values in $\mathcal{B}^+(\mathcal{H}_C \otimes \mathcal{H}_A)$ given by

$$E_0 := |\Phi_{CA}^+\rangle\langle\Phi_{CA}^+|, \quad E_1 := |\Psi_{CA}^+\rangle\langle\Psi_{CA}^+|,$$
$$E_2 := |\Phi_{CA}^-\rangle\langle\Phi_{CA}^-|, \quad E_3 := |\Psi_{CA}^-\rangle\langle\Psi_{CA}^-|.$$

After making her measurement, she communicates the result, which is one of the numbers $0, 1, 2, 3$ to Bob. This requires two classical bits.

If Alice measured 0, the state of the system has collapsed to $\Phi_{CA}^+ \otimes (\phi_0|0\rangle_B + \phi_1|1\rangle_B)$. In this case, after receiving Alice's message, Bob does nothing. He already has a copy of ϕ_C.

If Alice measured 1, the state of the system has collapsed to $\Psi_{CA}^+ \otimes (\phi_0|1\rangle_B + \phi_1|0\rangle_B)$. After receiving Alice's message, Bob applies the unitary $U_1 := |1\rangle_B\langle 0|_B + |0\rangle_B\langle 1|_B$, and since $U_1(\phi_0|1\rangle_B + \phi_1|0\rangle_B) = (\phi_0|0\rangle_B + \phi_1|1\rangle_B)$, Bob now has a copy a copy of ϕ_C.

The remaining cases are similar. If Bob learns that Alice measured 2, he applies $U_2 := |1\rangle_B\langle 1|_B - |1\rangle_B\langle 1|_B$, and if Bob learns that Alice measured 3, he applies $U_3 = |0\rangle_B\langle 1|_B - |1\rangle_B\langle 0|_B$. In all cases, at the end of the procedure, the state he has is $\phi_0|0\rangle_B + \phi_1|1\rangle_B$.

This example explains the notion of entanglement as a resource. Maximally entangled pairs of qbits are the "coin of the realm" in quantum information theory. The teleportation protocol depends on Alice and Bob having a shared maximally entangled pair of qbits distributed across their laboratories. The pair of particles that Alice and Bob shared and used in the teleportation protocol were entangled because they had interacted in some way in the past, although later in the two laboratories there is no interaction between them.

Entanglement is not something Alice and Bob can produce using only local operations and classical communication in a pair of particles that have never interacted. They can of course produce classical correlations in this manner, using the classical communication to coordinate their operations, as they did in the teleportation protocol. But although quantum entanglement entails correlation, it is something else; classical correlations are useless for implementing the teleportation protocol. Therefore, one absolute requirement on any quantification of entanglement is *the quantity of entanglement must not increase under the action of local operations and classical communication.* We now formalize this.

9.1.1. Local operations and classical communication (LOCC). Given a bipartite system $\mathcal{H}_A \otimes \mathcal{H}_B$, let $\mathcal{X}$ be a finite set, and let $\mathcal{E}$ be a POVM on $\mathcal{X}$ with values in $\mathcal{B}^+(\mathcal{H}_A)$. Then defining $\widetilde{\mathcal{E}}$ by $\widetilde{\mathcal{E}}(x) := \mathcal{E}(x) \otimes \mathbb{1}_B$, $\widetilde{\mathcal{E}}$ is a POVM on $\mathcal{X}$ with values in $\mathcal{B}^+(\mathcal{H}_A \otimes \mathcal{H}_B)$. Such a POVM is said to be a **local POVM acting on the system** A. Local POVMs acting on the system B are defined in an analogous manner.

Now suppose we have two laboratories. Alice in the first laboratory has access only to observables in $\mathcal{B}^{\text{s.a.}}(\mathcal{H}_A)$, while Bob in the second has access only to observables in $\mathcal{B}^{\text{s.a.}}(\mathcal{H}_B)$. The two laboratories are connected by a classical telephone line. Alice makes a measurement using some POVM on some set $\mathcal{X}$ with values in $\mathcal{B}^+(\mathcal{H}_A)$ agreed upon in advance by the two experimentalists, using their telephone line. Again using the telephone line, Alice tells Bob what value $x \in \mathcal{X}$ was observed. Based on that information, Bob chooses a POVM, depending on x, and makes a local measurement in his laboratory.

It will be helpful to rephrase this in terms of quantum operations. Let Φ_A be a quantum operation on $\mathcal{B}(\mathcal{H}_A)$. Associated to any Kraus representation of Φ_A,

$$(9.1.3) \qquad \Phi_A(X) = \sum_{j=1}^m A_j X A_j^*, \quad \sum_{j=1}^m A_j^* A_j = \mathbb{1}_A,$$

is a POVM $\mathcal{E}_A$ on $\mathcal{X} = \{1, \ldots, m\}$ with values in $\mathcal{B}^+(\mathcal{H}_A)$ given by $\mathcal{E}(x) = A_x^* A_x$. If the initial state of the combined system on $\mathcal{H}_A \otimes \mathcal{H}_B$ is ρ, after Alice has done her measurement and found x, the state of the system is

$$p_x^{-1}(A_x \otimes \mathbb{1}_B)\rho(A_x^* \otimes \mathbb{1}_B) \quad \text{where} \quad p_x := \mathrm{Tr}[(A_x \otimes \mathbb{1}_B)\rho(A_x^* \otimes \mathbb{1}_B)],$$

and $p_x > 0$ or else Alice would not have observed this value. Bob has a set quantum operations $\{\Phi_B^{(x)} : x \in \mathcal{X}\}$, and based on Alice's communication, he applies $\mathbb{1}_A \otimes \Phi_B^{(x)}$ to the state $p_x^{-1}(A_x \otimes \mathbb{1}_B)\rho(A_x^* \otimes \mathbb{1}_B)$. This is exactly what was done in the teleportation protocol, and in this case each $\Phi_B^{(x)}$ was simply a local unitary conjugation.

What we have described so far is **one-way LOCC**, where LOCC stands for local operations and classical control.

Definition 9.2. Let $\mathcal{H}_A, \mathcal{H}_B$ be finite-dimensional Hilbert spaces. Let **LOCC1** denote the class of quantum operations Ψ on $\mathcal{B}(\mathcal{H}_A \otimes \mathcal{H}_B)$ of the following form: There is a quantum operation Φ_A on $\mathcal{B}(\mathcal{H}_A)$ with a preferred Kraus representation of the form (9.1.3). With $\mathcal{X} = \{1, \dots, m\}$, there is also a set $\{\Phi_B^{(x)} : x \in \mathcal{X}\}$ of quantum operations on $\mathcal{B}(\mathcal{H}_B)$. Then for all $X \in \mathcal{B}(\mathcal{H}_A \otimes \mathcal{H}_B)$,

$$\Psi(X) := \sum_{x \in \mathcal{X}} \Psi_x(X) := \sum_{x \in \mathcal{X}} \mathbb{1}_A \otimes \Phi_B^{(x)}((A_x \otimes \mathbb{1}_B)X(A_x^* \otimes \mathbb{1}_B)).$$

Note that each of the maps Ψ_x is CP, and each satisfies $\mathrm{Tr}[\Psi_x(X)] \leq \mathrm{Tr}[X]$ for all $X \in \mathcal{B}^+(\mathcal{H}_A \otimes \mathcal{H}_B)$. The sum $\Psi = \sum_{x \in \mathcal{X}}$ is CP and trace preserving.

A POVM $\mathcal{E}$ on $\mathcal{H}_A \otimes \mathcal{H}_B$ is a **LOCC1 POVM** in case it is a POVM on the cartesian product $\mathcal{X} \times \mathcal{Y}$ of two finite sets with values in $\mathcal{B}^+(\mathcal{H}_A \otimes \mathcal{H}_B)$ such that

$$\mathcal{E}(x, y) = E_x \otimes F_{(y|x)},$$

where for each x, y, $E_x \in \mathcal{B}^+(\mathcal{H}_A)$, $F_{(y|x)} \in \mathcal{B}^+(\mathcal{H}_B)$, and furthermore, $\sum_{x \in \mathcal{X}} E_x = \mathbb{1}_A$ and for each $x \in \mathcal{X}$, $\sum_{y \in \mathcal{Y}} F_{(y|x)} = \mathbb{1}_B$.

Evidently, given Kraus representations of all of the local quantum operations Φ_A, $\Phi_B^{(x)}$, $x \in \mathcal{X}$, figuring in an LOCC1 quantum operation, one could construct in the obvious manner an associated LOCC1 POVM. Conversely, given preferred factorizations $E_x = A_x^* A_x$ and $F_{(y|x)} = (B_y^{(x)})^* B_y^{(x)}$, one can construct an associated quantum operation.

In the definition, we have specified that the communication in LOCC1 is one way from A to B. Of course we can interchange the roles of A and B and obtain the class of LOCC1 operations and POVMs with one way communication from B to A. In some settings it is natural to take the union of these as LOCC1, but in others it is natural to maintain the asymmetry, which we do. By concatenating arbitrarily many LOCC1 operations in with alternating directions of classical communication, we obtain the class of **LOCC** operations.

9.2. Entanglement of mixed states

So far we have considered only the entanglement of pure states. But even if Alice and Bob start with a pure state, their operations on it may change that. Before we quantify entanglement, we must extend the notion to mixed states.

A density matrix ρ_{AB} on the tensor product of two Hilbert spaces $\mathcal{H}_A \otimes \mathcal{H}_B$ is **separable** if and only if it has a decomposition as a convex combination of tensor products,

$$(9.2.1) \qquad \rho_{AB} = \sum_{k=1}^{n} \lambda_k \rho_A^{(k)} \otimes \rho_B^{(k)},$$

where the λ_k are positive and sum to 1, $\rho_A^{(k)} \in \mathfrak{S}_{\mathcal{H}_A}$, and $\rho_B^{(k)} \in \mathfrak{S}_{\mathcal{H}_B}$. A bipartite state that is not separable is **entangled**. We write $\mathfrak{S}_{\mathcal{H}_A \otimes \mathcal{H}_B}^{\mathbf{sep}}$ to denote the set of separable states on $\mathcal{H}_A \otimes \mathcal{H}_B$.

The set $\mathfrak{S}_{\mathcal{H}_A \otimes \mathcal{H}_B}^{\mathbf{sep}}$ is evidently convex. Going forward, we suppose $\mathcal{H}_A$ and $\mathcal{H}_B$ have finite dimensions d_A and d_B, respectively. Then by the spectral theorem, every product state $\rho_A \otimes \rho_B$ can be decomposed as a convex combination of states of the form

$$|\psi\rangle\langle\psi| \otimes |\phi\rangle\langle\phi| = |\psi \otimes \phi\rangle\langle\psi \otimes \phi|.$$

Such states are extreme in $\mathfrak{S}_{\mathcal{H}_A \otimes \mathcal{H}_B}$, and hence in $\mathfrak{S}_{\mathcal{H}_A \otimes \mathcal{H}_B}^{\mathbf{sep}}$. Since every state in $\mathfrak{S}_{\mathcal{H}_A \otimes \mathcal{H}_B}^{\mathbf{sep}}$ is a convex combination of such states, the set of extreme states in $\mathfrak{S}_{\mathcal{H}_A \otimes \mathcal{H}_B}^{\mathbf{sep}}$, $E(\mathfrak{S}_{\mathcal{H}_A \otimes \mathcal{H}_B}^{\mathbf{sep}})$, is given by

$$\{|\psi \otimes \phi\rangle\langle\psi \otimes \phi| \; : \; \|\psi\| = \|\phi\| = 1\}.$$

Since $\mathfrak{S}_{\mathcal{H}_A \otimes \mathcal{H}_B}$ lies in a $(d_A^2 d_B^2 - 1)$-dimensional real affine space, and since $E(\mathfrak{S}_{\mathcal{H}_A \otimes \mathcal{H}_B}^{\mathbf{sep}})$ is closed, it follows from Carathéodory's theorem, Theorem A.15, and Corllary A.16, that every state in $\mathfrak{S}_{\mathcal{H}_A \otimes \mathcal{H}_B}^{\mathbf{sep}}$ can be written as a convex combination of at most $(d_A d_B)^2$ extreme elements of $\mathfrak{S}_{\mathcal{H}_A \otimes \mathcal{H}_B}^{\mathbf{sep}}$. It then follows that $\mathfrak{S}_{\mathcal{H}_A \otimes \mathcal{H}_B}^{\mathbf{sep}}$ is a closed subset of $\mathfrak{S}_{\mathcal{H}_A \otimes \mathcal{H}_B}$. Finally then, $\mathfrak{S}_{\mathcal{H}_A \otimes \mathcal{H}_B}^{\mathbf{sep}}$ is a compact, convex set, and hence the entangled states are the complement in $\mathfrak{S}_{\mathcal{H}_A \otimes \mathcal{H}_B}$ of a compact, convex set. We record that fact that ρ_{AB} is separable if and only if it has an expression of the form

$$(9.2.2) \qquad \rho_{AB} = \sum_{k=1}^{m} \lambda_k |\psi^{(k)} \otimes \phi^{(k)}\rangle\langle\psi^{(k)} \otimes \phi^{(k)}|$$

as a convex combination of separable pure states.

Lemma 9.3. *The set $\mathfrak{S}_{\mathcal{H}_A \otimes \mathcal{H}_B}^{\mathbf{sep}}$ of separable states is invariant under LOCC operations.*

Proof. It suffices to consider LOCC1 operations Ψ with either direction of communication. Let Ψ have the form specified in Definition 9.2. Let ρ_{AB} have the form specified in (9.2.1). Then $\Psi(\rho_{AB}) := \sum_{x \in \mathcal{X}} \sum_{k=1}^{n} \lambda_k A_x \rho_A^{(k)} A_x^* \otimes \Phi_B^{(x)}(\rho_B^{(k)})$. $\qquad\square$

The next definition is taken from Vedral and Plenio [220]. Other authors have discussed various modifications of the definition, and indeed this definition is a modification of one proposed earlier [219] by two of the same authors and two others.

Definition 9.4 (Entanglement monotone). An **entanglement monotone** is a function E on $\mathfrak{S}_{\mathcal{H}_A \otimes \mathcal{H}_B}$ with values in $[0, \infty)$ such that

(1) $E(\rho) = 0$ if and only if $\rho \in \mathfrak{S}^{\mathbf{sep}}_{\mathcal{H}_A \otimes \mathcal{H}_B}$.

(2) If Ψ is any LOCC1 quantum operation as in Definition 9.2 (but with the communication from either A to B or from B to A), then for all $\rho \in \mathfrak{S}_{\mathcal{H}_A \otimes \mathcal{H}_B}$ with $p_x := \mathrm{Tr}[\Psi_x(\rho)]$,

$$(9.2.3) \qquad \sum_{x \in \mathcal{X}} p_x E\left(p_x^{-1} \Psi_x(\rho)\right) \le E(\rho).$$

Condition (1) says that E is a **faithful** measure of entanglement.

The more complex conditon (2) requires more explanation. As noted in [220], if E is convex, as it will be in the main examples to be discussed below, since $\Psi(\rho) = \sum_{x \in \mathcal{X}} p_x \Psi_x(\rho)$,

$$(9.2.4) \qquad E(\Psi(\rho)) \le E(\rho)$$

for all LOCC1 operations, and hence for all LOCC operations. In [219], condition (9.2.4) was used in place of (9.2.3). However, the motivation for imposing both conditions is the same: We seek measures of entanglement that do not increase under LOCC, for if they did, part of what they would be measuring would be classical correlations, which can be increased by LOCC. The simpler condition (9.2.4) captures this idea. In the presence of convexity, which we will have in our main examples, (9.2.3) is a stronger expression of this. It says that *on average*, entanglement cannot be increased by an LOCC measurement procedure. However, there are useful ways to *concentrate entanglement*, originally discussed in [31] and further discussed in [112, 220], and many other works. In the words of [31],

> though Alice and Bob cannot by local actions increase their expected entanglement, they can gamble with it, spending their initial amount on a chance of obtaining a greater amount.

The inequality (9.2.3) expresses exactly the fact that Alice and Bob cannot increase their expected entanglement, but if their protocol is well chosen, certain output states $p_x^{-1} \Psi_x(\rho)$ may be more entangled than ρ.

Lemma 9.5. *An entanglement monotone E has the property that it is invariant under local unitary transformations. That is, for all $\rho \in \mathfrak{S}^{\mathbf{sep}}_{\mathcal{H}_A \otimes \mathcal{H}_B}$,*

$$(9.2.5) \qquad E(\rho) = E(U_A \otimes U_B \, \rho \, U_A^* \otimes U_B^*).$$

Proof. Let U_B be unitary on $\mathcal{H}_B$, and let Ψ have the form specified in (9.1.3) where $\mathcal{X} = \{x\}$ has a single point, $A_x = \mathbb{1}_A$, and $\Phi_B^{(x)}(Y) = U_B Y U_B^*$ for all $Y \in \mathcal{B}(\mathcal{H}_B)$. Then by condition (2) (in which the averaging is trivial in this case),

$$E(\mathbb{1}_A \otimes U_B \, \rho \, \mathbb{1}_A^* \otimes U_B^*) \leq E(\rho)$$

for all $\rho \in \mathfrak{S}_{\mathcal{H}_A \otimes \mathcal{H}_B}^{\text{sep}}$. Replacing U_B by U_B^* and ρ by $\mathbb{1}_A \otimes U_B \rho \mathbb{1}_A^* \otimes U_B^*$ yields

$$E(\rho) \leq E(\mathbb{1}_A \otimes U_B \, \rho \, \mathbb{1}_A^* \otimes U_B^*).$$

Altogether, $E(\rho) \leq E(\mathbb{1}_A \otimes U_B \, \rho \, \mathbb{1}_A^* \otimes U_B^*) \leq E(\rho)$. The same reasoning applies to unitaries acting locally on $\mathcal{H}_A$, and concatenating such local unitary operations on $\mathcal{H}_B$, and then on $\mathcal{H}_A$, yields (9.2.5). $\qquad\square$

It is convenient in some proofs to "unpack" condition (2) into the composition of two simpler operations in which only one laboratory at a time is active. This is possible [221] when the function E is convex. The following definition involves ensembles of mixed states which differ from the ensembles of pure state introduced much earlier in Definition 2.23 only in that now the states in the ensemble may be mixed.

Definition 9.6. Let $\mathcal{H}_A$ and $\mathcal{H}_B$ be finite-dimensional Hilbert spaces. A **unilocal operation** on $\mathfrak{S}_{\mathcal{H}_A \otimes \mathcal{H}_B}$ is a map

$$\rho_{AB} \to \{p_k, \rho_k\}_{1 \leq k \leq n}$$

from density matrices on $\mathfrak{S}_{\mathcal{H}_A \otimes \mathcal{H}_B}$ to ensembles on $\mathcal{H}_A \otimes \mathcal{H}_B$, where either there are operators $\{A_1, \dots, A_n\}$ in $\mathcal{B}(\mathcal{H}_A)$ such that $\sum_{k=1}^n A_k^* A_k = \mathbb{1}_A$ and

$$p_k := \operatorname{Tr}[(A_k \otimes \mathbb{1}_B)\rho_{AB}(A_k^* \otimes \mathbb{1}_B)] \quad \text{and} \quad \rho_k := p_k^{-1}(A_k \otimes \mathbb{1}_B)\rho_{AB}(A_k^* \otimes \mathbb{1}_B),$$

or $\{p_k, \rho_k\}_{1 \leq k \leq n}$ is given by the corresponding expression with the operations acting on $\mathcal{H}_B$ instead of $\mathcal{H}_A$.

Lemma 9.7. *Suppose that a function E on $\mathfrak{S}_{\mathcal{H}_A \otimes \mathcal{H}_B}$ is convex and has the property that for any unilocal operation*

$$\rho_{AB} \to \{p_k, \rho_k\}_{1 \leq k \leq n},$$

$\sum_{k=1}^n p_k E(\rho_k) \leq E(\rho_{AB})$. *Then E satisfies condition (2) in the definition of an entanglement monotone.*

Proof. Let Ψ have the form (9.1.3),

$$\Psi(X) := \sum_{x \in \mathcal{X}} \Psi_x(X) := \sum_{x \in \mathcal{X}} \mathbb{1}_A \otimes \Phi_B^{(x)}((A_x \otimes \mathbb{1}_B)X(A_x^* \otimes \mathbb{1}_B)).$$

Define

$$p_x := \operatorname{Tr}[(A_x \otimes \mathbb{1}_B)\rho_{AB}(A_x^* \otimes \mathbb{1}_B)] \quad \text{and} \quad \hat{\rho}_x := p_x^{-1}(A_x \otimes \mathbb{1}_B)\rho_{AB}(A_x^* \otimes \mathbb{1}_B).$$

Since E is monotone under unilocal operations, $\sum_{y \in \mathcal{Y}} p_x E(\hat{\rho}_x) \leq E(\rho_{AB})$.

For a sufficiently large finite set $\mathcal{Y}$, we may use elements of $\mathcal{Y}$ as indices for a Kraus representation of each $\Phi_B^{(x)}$:

$$\Phi_B^{(x)}(X) = \sum_{y \in \mathcal{Y}} (\mathbb{1}_A \otimes B_{(y|x)}) X (\mathbb{1}_A \otimes B_{(y|x)}^*).$$

Define

$$p_{(y|x)} := p_x \operatorname{Tr}[(\mathbb{1}_A \otimes B_{(y|x)}) \hat{\rho}_x (\mathbb{1}_A \otimes B_{(y|x)}^*)],$$
$$\rho_{(y|x)} := p_{(y|x)}^{-1}((\mathbb{1}_A \otimes B_{(y|x)}) \hat{\rho}_x (\mathbb{1}_A \otimes B_{(y|x)}^*)).$$

Then $p_x^{-1} \Psi_x(\rho_{AB}) = \sum_{y \in \mathcal{Y}} p_x^{-1} p_{(y|x)} \rho_{(y|x)}$ and $\sum_{y \in \mathcal{Y}} p_x^{-1} p_{(y|x)} = 1$, so that by the convexity of E, $E\big(p_x^{-1} \Psi_x(\rho_{AB})\big) \leq \sum_{y \in \mathcal{Y}} p_x^{-1} p_{(y|x)} E(\rho_{(y|x)})$. Therefore,

$$\sum_{x \in \mathcal{X}} p_x E\big(p_x^{-1} \Psi_x(\rho_{AB})\big) \leq \sum_{x \in \mathcal{X}} p_x \sum_{y \in \mathcal{Y}} p_x^{-1} p_{(y|x)} E(\rho_{(y|x)})$$
$$\leq \sum_{x \in \mathcal{X}} p_x E(\hat{\rho}_x)$$
$$= \sum_{x \in \mathcal{X}} p_x E(p_x^{-1}(A_x \otimes \mathbb{1}_B) \rho_{AB} (A_x^* \otimes \mathbb{1}_B))$$
$$\leq E(\rho_{AB}). \qquad \square$$

The next section introduces one important entanglement monotone.

9.3. Relative entropy of entanglement

Definition 9.8. Let $\mathcal{H}_A, \mathcal{H}_B$ be finite-dimensional Hilbert spaces. For a bipartite state ρ_{AB} on $\mathcal{H}_A \otimes \mathcal{H}_B$, its **relative entropy of entanglement** is the quantity

$$E_R(\rho_{AB}) = \min\{D(\rho_{AB} \| \sigma_{AB}) \ : \ \sigma_{AB} \in \mathfrak{S}_{\mathcal{H}_A \otimes \mathcal{H}_B}^{\mathbf{sep}}\},$$

where the minimum exists and is unique since $\mathfrak{S}_{\mathcal{H}_A \otimes \mathcal{H}_B}^{\mathbf{sep}}$ is compact and the relative entropy is continuous and strictly convex in each argument.

Theorem 9.9. *The relative entropy of entanglement is an entanglement monotone, and moreover is a convex functional on $\mathfrak{S}_{\mathcal{H}_A \otimes \mathcal{H}_B}$.*

Proof. By Pinsker's inequality $E_R(\rho_{AB}) \geq \frac{1}{2} \|\rho_{AB} - \sigma_{AB}\|_1^2$ for all $\sigma_{AB} \in \mathfrak{S}_{\mathcal{H}_A \otimes \mathcal{H}_B}^{\mathbf{sep}}$, and thus $E_R(\rho_{AB})$ is a faithful quantifier of the extent to which ρ_{AB} is nonseparable. Thus, condition (1) in Definition 9.4 is satisfied.

We now show that (2) is satisfied. Given $\rho_{AB} \in \mathfrak{S}_{\mathcal{H}_A \otimes \mathcal{H}_B}$, let $\sigma_{AB} \in \mathfrak{S}^{\mathbf{sep}}_{\mathcal{H}_A \otimes \mathcal{H}_B}$ be such that $E_R(\rho_{AB}) = D(\rho_{AB}\|\sigma_{AB})$. Then

$$\sum_{x \in \mathcal{X}} p_x E_R(\Psi_x(\rho_{AB})) \leq \sum_{x \in \mathcal{X}} p_x D(\Psi_x(\rho_{AB})\|\Psi_x(\sigma_{AB}))$$

$$\leq \sum_{x \in \mathcal{X}} p_x D(\rho_{AB}\|\sigma_{AB}) = D(\rho_{AB}\|\sigma_{AB}) = E_R(\rho_{AB}),$$

where the first inequality comes from Lemma 9.3, and the second inequality is the DPI.

The fact that E_R is convex is an immediate consequence of the joint convexity of the relative entropy and Lemma A.25. $\qquad\square$

When an entanglement monotone E is convex, E cannot be increased simply by taking mixtures; that is, convex combinations, which is a very natural requirement to impose on a measure of entanglement, and some authors include convexity as part of the definition.

The next theorem is due to Vedral and Plenio [**220**].

Theorem 9.10. *Let $\mathcal{H}_A, \mathcal{H}_B$ be finite-dimensional Hilbert spaces. Let $\rho_{AB} := |\psi_{AB}\rangle\langle\psi_{AB}|$ be a pure state on $\mathcal{H}_A \otimes \mathcal{H}_B$. Let $\psi_{AB} = \sum_{j=1}^{r} \lambda_j^{1/2} \mathbf{u}_j \otimes \mathbf{v}_j$ be a Schmidt decomposition of ψ_{AB}. Define*

$$\hat{\rho}_{AB} := \sum_{j=1}^{r} \lambda_j |\mathbf{u}_j\rangle\langle\mathbf{u}_j| \otimes |\mathbf{v}_j\rangle\langle\mathbf{v}_j| \in \mathfrak{S}^{\mathbf{sep}}_{\mathcal{H}_A \otimes \mathcal{H}_B}.$$

Then

$$(9.3.1) \qquad E_R(\rho_{AB}) = D(\rho_{AB}\|\hat{\rho}_{AB}) = S(\rho_A) = S(\rho_B),$$

where ρ_A and ρ_B are the two marginal densities of ρ_{AB}.

Proof. Let $\sigma_{AB} \in \mathfrak{S}^{\mathbf{sep}}_{\mathcal{H}_A \otimes \mathcal{H}_B}$, and $t \in (0, 1)$. Define

$$f(t) := D(\rho_{AB}\|(1 - t)\hat{\rho}_{AB} + t\sigma_{AB}).$$

Then f is convex on $[0, 1]$, and therefore if $f'_+(0) \geq 0$ where

$$f'_+(0) = \lim_{t \to 0} \frac{1}{t}(f(t) - f(0)),$$

then $f(1) \geq f(0)$, and this is the same as

$$D(\rho_{AB}\|\sigma_{AB}) \geq D(\rho_{AB}\|\hat{\rho}_{AB}).$$

Thus, to prove the first equation in (9.3.1) it suffices to prove that $f'_+(0) \geq 0$.

Since ρ_{AB} is pure, $f(t) := -\mathrm{Tr}[\rho_{AB}\log((1-t)\hat{\rho}_{AB} + t\sigma_{AB})]$. Therefore,

$$f'(0) = \int_0^\infty \mathrm{Tr}\left[\rho_{AB}\frac{1}{\lambda+\hat{\rho}_{AB}}(\hat{\rho}_{AB}-\sigma_{AB})\frac{1}{\lambda+\hat{\rho}_{AB}}\right]d\lambda$$

$$= 1 - \int_0^\infty \mathrm{Tr}\left[\sigma_{AB}\frac{1}{\lambda+\hat{\rho}_{AB}}\rho_{AB}\frac{1}{\lambda+\hat{\rho}_{AB}}\right]d\lambda.$$

Using the Schmidt decomposition of ψ_{AB},

$$\int_0^\infty \frac{1}{\lambda+\hat{\rho}_{AB}}\rho_{AB}\frac{1}{\lambda+\hat{\rho}_{AB}}d\lambda = \sum_{j,k=1}^r \frac{\sqrt{\lambda_j\lambda_k}}{\Lambda(\lambda_j,\lambda_k)}|\mathbf{u}_j\otimes\mathbf{v}_j\rangle\langle\mathbf{u}_k\otimes\mathbf{v}_k|,$$

where for $a,b > 0$, $\Lambda(a,b)$ is the logarithmic mean of a,b (see Definition 3.14). Since $\Lambda(a,b) \geq \sqrt{ab}$ by (3.2.12),

$$f'(0) \geq 1 - \sum_{j,k=1}^r \langle\mathbf{u}_j\otimes\mathbf{v}_j,\sigma_{AB}\,\mathbf{u}_k\otimes\mathbf{v}_k\rangle.$$

Then for $\sigma_{AB} = \sum_{\ell=1}^s p_\ell\sigma_A^{(\ell)}\otimes\sigma_B^{(\ell)}$, using the Cauchy-Schwarz inequality,

$$\sum_{j,k=1}^r \langle\mathbf{u}_j\otimes\mathbf{v}_j,\sigma_{AB}\,\mathbf{u}_k\otimes\mathbf{v}_k\rangle = \sum_{\ell=1}^s p_\ell\sum_{j,k=1}^r \langle\mathbf{u}_j,\sigma_A^{(\ell)}\mathbf{u}_k\rangle\langle\mathbf{v}_j,\sigma_B^{(\ell)}\mathbf{v}_k\rangle$$

$$\leq \sum_{\ell=1}^s p_\ell\|\sigma_A^{(\ell)}\|_2\|\sigma_B^{(\ell)}\|_2 \leq 1.$$

This shows that $f'(0) \geq 0$, proving the first part of (9.3.1). Now a simple computation yields $D(\rho_{AB}\|\hat{\rho}_{AB}) = -\sum_{j=1}^r \lambda_j\log\lambda_j$, which proves the second part of (9.3.1). $\square$

For a pure state ψ_{AB}, the **entanglement entropy** of ρ_{AB}, $E(\rho_{AB})$, was defined in **[31]** to be the von Neumann entropy of the marginals, ρ_A, ρ_B of $\rho_{AB} := |\psi_{AB}\rangle\langle\psi_{AB}|$, that is

$$(9.3.2) \qquad E(\psi_{AB}) = S(\rho_A) = S(\rho_B).$$

By Theorem 9.10, the entanglement entropy is the restriction of the relative entropy of entanglement to the pure states on $\mathcal{H}_A\otimes\mathcal{H}_B$. What we have called *maximally entangled* pure states at the beginning of this chapter are pure states that maximize the entanglement entropy.

9.3.1. Nonadditivity of E_R. By Theorem 7.43, the relative entropy is additive over tensor products. This leads directly to a **subadditivity over tensor products** property of the relative entropy of entanglement—but nothing more.

Let $\mathcal{K}_A, \mathcal{K}_B$ be finite-dimensional Hilbert spaces, and let ρ_{AB} and $\hat{\rho}_{AB}$ be entangled states on $\mathcal{K}_A \otimes \mathcal{K}_B$. Let σ_{AB} and $\hat{\sigma}_{AB}$ be the separable states so that $E_R(\rho_{AB}) = D(\rho_{AB}\|\sigma_{AB})$ and $E_R(\hat{\rho}_{AB}) = D(\hat{\rho}_{AB}\|\hat{\sigma}_{AB})$. Then $\sigma_{AB} \otimes \hat{\sigma}_{AB}$ is separable in $(\mathcal{H}_A \otimes \mathcal{K}_A) \otimes (\mathcal{H}_B \otimes \mathcal{K}_B)$, and by Theorem 7.43,

$$E_R(\rho_{AB}) + E_R(\hat{\rho}_{AB}) = D(\rho_{AB}\|\sigma_{AB}) + D(\hat{\rho}_{AB}\|\hat{\sigma}_{AB})$$
$$(9.3.3) \qquad\qquad = D(\rho_{AB} \otimes \hat{\rho}_{AB}\|\sigma_{AB} \otimes \hat{\sigma}_{AB}) \geq E_R(\rho_{AB} \otimes \hat{\rho}_{AB}).$$

For a long time it was conjectured, with support from numerical evidence, that in fact $E_R(\rho_{AB} \otimes \hat{\rho}_{AB}) = E_R(\rho_{AB}) + E_R(\hat{\rho}_{AB})$, so that like the relative entropy, the relative entropy of entanglement would be additive over tensor products. (See, e.g., [**220**, Section V].) However, Vollbrecht and Werner provided a counterexample [**222**]. Take $\mathcal{H}_A = \mathcal{H}_B = \mathbb{C}^d$, $d > 2$. Take ρ_{AB} to be the normalized projection onto the antisymmetric subspace $\mathbb{C}^d \wedge \mathbb{C}^d$. Vollbrecht and Werner used the symmetry properties of this state to compute that

$$E_R(\rho_{AB}) = \log 2 \quad \text{and} \quad E_R(\rho_{AB} \otimes \rho_{AB}) = 2\log 2 - \log\left(2\frac{d-1}{d}\right).$$

Additivity fails in the rather spectacular sense that for all $\epsilon > 0$, there are entangled states ρ_{AB} such that $E_R(\rho_{AB} \otimes \rho_{AB}) \leq (1 + \epsilon)E_R(\rho_{AB})$, at least in a sufficiently high dimension d.

As a direct consequence of the subadditivity inequality (9.3.3),

$$(9.3.4) \qquad\qquad E_R^\infty(\rho_{AB}) := \lim_{n\to\infty} \frac{1}{n} E_R(\rho_{AB}^{\otimes n})$$

exists. This is a natural *regularization* of the function E_R. However, it is not a priori clear that it is a *faithful* measure of entanglement. Could it be that $E_R^\infty(\rho_{AB}) = 0$ for an entangled state ρ_{AB}?

The answer, due to Piani [**179**], is "no". Piani's key idea is to bring into the discussion the measured relative entropy, but defined not in terms of a supremum over all POVMs, but only over a restricted class of POVMs.

Definition 9.11. Let $\mathcal{H}_A, \mathcal{H}_B$ be finite-dimensional Hilbert spaces, and let $\rho_{AB}, \sigma_{AB} \in \mathfrak{S}_{\mathcal{H}_A \otimes \mathcal{H}_B}$. Their **measured relative entropy with respect to LOCC1** is the quantity defined by

$$D_{\text{LOCC1}}^M(\rho_{AB}\|\sigma_{AB}) := \sup\{D(\mathcal{E}(\rho_{AB})\|\mathcal{E}(\sigma_{AB})) \; : \; \mathcal{E} \in \text{LOCC1}\},$$

and the **LOCC1 relative entropy of entanglement** is defined by

$$E_{R,\text{LOCC1}}(\rho_{AB}) := \min\{D_{\text{LOCC1}}^M(\rho_{AB}\|\sigma_{AB}) \; : \; \sigma_{AB} \in \mathfrak{S}_{\mathcal{H}_A \otimes \mathcal{H}_B}^{\text{sep}}\}.$$

Lemma 9.12. *Let $\mathcal{H}_A, \mathcal{H}_B$ be finite-dimensional Hilbert spaces. For all $\rho_{AB} \in \mathfrak{S}_{\mathcal{H}_A \otimes \mathcal{H}_B}$,*

$$E_{R,\text{LOCC1}}(\rho_{AB}) = 0 \quad \Longleftrightarrow \quad \rho_{AB} \in \mathfrak{S}_{\mathcal{H}_A \otimes \mathcal{H}_B}^{\text{sep}}.$$

Proof. We first prove that for all $\rho_{AB}, \sigma_{AB} \in \mathfrak{S}_{\mathcal{H}_A \otimes \mathcal{H}_B}$.

$$(9.3.5) \qquad D^M_{\text{LOCC1}}(\rho_{AB} \| \sigma_{AB}) = 0 \quad \Longleftrightarrow \quad \rho_{AB} = \sigma_{AB}.$$

Let $\mathcal{E}$ and $\mathcal{F}$ be minimally informationally complete POVMs on $\mathcal{X}$ and $\mathcal{Y}$ with values in $\mathcal{B}^+(\mathcal{H}_A)$ and $\mathcal{B}^+(\mathcal{H}_B)$, respectively, as defined in Section 4.7.2. Then their tensor product is a minimally informationally complete POVM on $\mathcal{X} \times \mathcal{Y}$ with values in $\mathcal{B}^+(\mathcal{H}_A \times \mathcal{H}_B)$, and it belongs to LOCC1. In fact, it does not even involve classical communication. This proves (9.3.5).

Since $D^M_{\text{LOCC1}}(\rho_{AB} \| \sigma_{AB})$ is, by Definition 9.11, the supremum of a family of jointly convex continuous functions of ρ_{AB} and σ_{AB}, it is jointly convex and lower-semicontinuous. Then since $\mathfrak{S}^{\text{sep}}_{\mathcal{H}_A \otimes \mathcal{H}_B}$ is compact, the function $\sigma_{AB} \mapsto D^M_{\text{LOCC1}}(\rho_{AB} \| \sigma_{AB})$ has a minimizer $\hat{\rho}_{AB}$ in $\mathfrak{S}^{\text{sep}}_{\mathcal{H}_A \otimes \mathcal{H}_B}$, so that $E_{R,\text{LOCC1}}(\rho_{AB}) = D^M_{\text{LOCC1}}(\rho_{AB} \| \hat{\rho}_{AB})$, which, by the first part, is zero if and only if $\rho_{AB} = \hat{\rho}_{AB}$. $\quad\square$

The following theorem is due to Piani [**179**].

Theorem 9.13. *Let $\mathcal{H}_A, \mathcal{H}_B, \mathcal{K}_A, \mathcal{K}_B$ be finite-dimensional Hilbert spaces. Let $\rho \in \mathfrak{S}_{(\mathcal{H}_A \otimes \mathcal{K}_A) \otimes (\mathcal{H}_B \otimes \mathcal{K}_B)}$. Let $\rho_{\mathcal{H}} := \text{Tr}_{\mathcal{K}_A \otimes \mathcal{K}_B}[\rho]$ and $\rho_{\mathcal{K}} := \text{Tr}_{\mathcal{H}_A \otimes \mathcal{H}_B}[\rho]$. Then*

$$(9.3.6) \qquad E_R(\rho) \geq E_{R,\text{LOCC1}}(\rho_{\mathcal{H}}) + E_R(\rho_{\mathcal{K}}).$$

Consequently, for all $\rho_{12} \in \mathfrak{S}_{\mathcal{H}_1 \otimes \mathcal{H}_2}$, where $\mathcal{H}_1, \mathcal{H}_2$ are finite-dimensional Hilbert spaces, $E_R^\infty(\rho_{12}) \geq E_{R,\text{LOCC1}}(\rho_{12})$, so that $E_R^\infty(\rho_{12}) = 0$ if and only if ρ_{12} is separable.

We preface the proof with a preliminary construction and a lemma. The key to the proof is the chain rule for the relative entropy, Theorem 7.42, which expresses relative entropy of two density matrices with the same block-diagonal structure as a sum of two relative entropies, one classical and one quantum. The classical term will yield the $E_{R,\text{LOCC1}}$ component on the right in (9.3.6), while the quantum term will yield the E_R component on the right in (9.3.6). The construction in the next paragraph enables us to write an LOCC1 operation in block form, opening the way to the application of Theorem 7.42.

Consider the bipartite space $(\mathcal{H}_A \otimes \mathcal{K}_A) \otimes (\mathcal{H}_B \otimes \mathcal{K}_B)$. A density matrix on this space is separable if and only if it has the form

$$(9.3.7) \qquad \sigma = \sum_{j=1}^n s_j \omega^{(j)} \otimes \tau^{(j)},$$

where for each j, $\omega^{(j)} \in \mathfrak{S}_{\mathcal{H}_A \otimes \mathcal{K}_A}$, $\tau^{(j)} \in \mathfrak{S}_{\mathcal{H}_B \otimes \mathcal{K}_B}$, and $s_j > 0$.

Let $\mathcal{E}(x, y) = E_x \otimes F_{(y|x)}$ be a LOCC1 POVM on $\mathcal{X} \times \mathcal{Y}$ with values in $\mathcal{B}^+(\mathcal{H}_A \otimes \mathcal{H}_B)$. Let $\mathcal{L}$ be a Hilbert space whose dimension is the cardinality of $\mathcal{X} \times \mathcal{Y}$ so that there exists an orthonormal basis $\{|x, y\rangle \ : \ (x, y) \in \mathcal{X} \times \mathcal{Y}\}$ indexed by points in $\mathcal{X} \times \mathcal{Y}$.

Define $\Phi : \mathcal{B}((\mathcal{H}_A \times \mathcal{K}_A) \otimes (\mathcal{H}_B \otimes \mathcal{K}_B)) \to \mathcal{B}(\mathcal{K}_A \otimes \mathcal{K}_B \otimes \mathcal{L})$ by

$$(9.3.8) \quad \Phi(A) = \sum_{(x,y)\in\mathcal{X}\times\mathcal{Y}} \mathrm{Tr}_{\mathcal{H}_A\otimes\mathcal{H}_B}[(E_x \otimes F_{(y|x)} \otimes \mathbb{1}_{\mathcal{K}_A\otimes\mathcal{K}_B})A] \otimes |x,y\rangle\langle x,y|.$$

Note that Φ is a quantum operation.

Lemma 9.14. *Let $\sigma \in \mathfrak{S}^{\mathrm{sep}}_{(\mathcal{H}_A\otimes\mathcal{K}_A)\otimes(\mathcal{H}_B\otimes\mathcal{K}_B)}$. Then*

$$(9.3.9) \qquad \Phi(\sigma) := \sum_{(x,y)\in\mathcal{X}\times\mathcal{Y}} q(x,y)\sigma_{x,y} \otimes |x,y\rangle\langle x,y|,$$

where each $\sigma_{x,y} \in \mathfrak{S}^{\mathrm{sep}}_{\mathcal{K}_A\otimes\mathcal{K}_B}$.

Proof. Write σ in the form (9.3.7), and define

$$q^{(j)}(x) := \mathrm{Tr}[E_x \otimes \mathbb{1}_{\mathcal{K}_A}\omega^{(j)}] \quad \text{and} \quad q^{(j)}(y|x) := \mathrm{Tr}[F_{(y|x)} \otimes \mathbb{1}_{\otimes\mathcal{K}_B})\tau^{(j)}].$$

Note that

$$(9.3.10) \qquad q(x,y) := \sum_{j=1}^{n} s_j q^{(j)}(x)q^{(j)}(y|x) = \mathrm{Tr}[\mathcal{E}(x,y)\,\mathrm{Tr}_{\mathcal{K}_A\otimes\mathcal{K}_B}[\sigma]].$$

Then define

$$\omega_x^{(j)} := \frac{1}{q^{(j)}(x)} := \mathrm{Tr}_{\mathcal{H}_A}[E_x \otimes \mathbb{1}_{\mathcal{K}_A}\omega^{(j)}]$$

and

$$\tau_{y|x}^{(j)} := \frac{1}{q^{(j)}(y|x)} \mathrm{Tr}_{\mathcal{H}_B}[F_{(y|x)} \otimes \mathbb{1}_{\otimes\mathcal{K}_B})\tau^{(j)}].$$

Finally, define $\sigma_{x,y} := \sum_{j=1}^{n} s_j\omega_x^{(j)} \otimes \tau_{y|x}^{(j)}$, and note that (9.3.9) is satisfied. $\square$

Proof of Theroem 9.13. Let σ be a separable density matrix on $(\mathcal{H}_A \otimes \mathcal{K}_A) \otimes (\mathcal{H}_B \otimes \mathcal{K}_B)$ such that $E_R(\rho) = D(\rho\|\sigma)$. Let $\mathcal{E}(x,y) = E_x \otimes F_{(y|x)}$ be a LOCC1 POVM on $\mathcal{X} \times \mathcal{Y}$, and let the quantum operation Φ be defined in terms of it by (9.3.8). Then by the DPI, $E_R(\rho) \geq D(\Phi(\rho)\|\Phi(\sigma))$.

Defining

$$(9.3.11) \qquad p(x,y) := [(E_x \otimes F_{(y|x)} \otimes \mathbb{1}_{\mathcal{K}_A\otimes\mathcal{K}_B})\rho] = \mathrm{Tr}[\mathcal{E}(x,y)\rho_{\mathcal{H}}]$$

and

$$\rho_{x,y} := \frac{1}{p(x,y)} \mathrm{Tr}_{\mathcal{H}_A\otimes\mathcal{H}_B}[(E_x \otimes F_{(y|x)} \otimes \mathbb{1}_{\mathcal{K}_A\otimes\mathcal{K}_B})\rho],$$

$$(9.3.12) \qquad \Phi(\rho) = \sum_{(x,y)\in\mathcal{X}\times\mathcal{Y}} p(x,y)\rho_{x,y} \otimes |x,y\rangle\langle x,y|.$$

Therefore $\mathrm{Tr}_{\mathcal{L}}[\Phi(\rho)] = \rho_{\mathcal{K}}$; that is,

$$(9.3.13) \qquad \rho_{\mathcal{K}} = \sum_{(x,y)\in\mathcal{X}\times\mathcal{Y}} p(x,y)\rho_{x,y}.$$

Then by Theorem 7.42, the chain rule for relative entropy, and also (9.3.9) and (9.3.12),

$$D(\Phi(\rho)\|\Phi(\sigma)) = \sum_{(x,y)\in\mathcal{X}\times\mathcal{Y}} p(x,y)(\log p(x,y) - \log q(x,y))$$
$$+ \sum_{(x,y)\in\mathcal{X}\times\mathcal{Y}} p(x,y)D\left(\rho_{x,y}\|\sigma_{x,y}\right).$$

By (9.3.10) and (9.3.11)

$$D(\mathcal{E}(\rho_{\mathcal{H}})\|\mathcal{E}(\sigma_{\mathcal{H}})) = \sum_{(x,y)\in\mathcal{X}\times\mathcal{Y}} p(x,y)(\log p(x,y) - \log q(x,y)).$$

For any $\epsilon > 0$, choose $\mathcal{E}$ to be nearly optimal for $D^M_{\mathrm{LOCC1}}(\rho_{\mathcal{H}}\|\sigma_{\mathcal{H}})$ so that

$$\sum_{(x,y)\in\mathcal{X}\times\mathcal{Y}} p(x,y)(\log p(x,y) - \log q(x,y)) \geq E_{R,\mathrm{LOCC1}}(\rho_{\mathcal{H}}) - \epsilon.$$

Next, by the joint convexity of the relative entropy and (9.3.13),

$$\sum_{(x,y)\in\mathcal{X}\times\mathcal{Y}} p(x,y)D\left(\rho_{x,y}\|\sigma_{x,y}\right) \geq D(\rho_{\mathcal{K}}\|\tilde{\sigma}_{\mathcal{K}}),$$

where $\tilde{\sigma}_{\mathcal{K}} := \sum_{(x,y)\in\mathcal{X}\times\mathcal{Y}} p(x,y)\sigma_{x,y}$ is separable by Lemma 9.14, so that $D(\rho_{\mathcal{K}}\|\tilde{\sigma}_{\mathcal{K}}) \geq E_R(\rho_{\mathcal{K}})$. Since $\epsilon > 0$ is arbitrary, this proves (9.3.6).

For the final statement, let $\rho_{12} \in \mathfrak{S}_{\mathcal{H}_1\otimes\mathcal{H}_2}$. Apply the first part of the theorem with $\mathcal{H}_A = \mathcal{H}_1$, $\mathcal{H}_B := \mathcal{H}_1^{\otimes(n-1)}$, $\mathcal{K}_A = \mathcal{H}_2$, $\mathcal{K}_B := \mathcal{H}_2^{\otimes(n-1)}$, and $\rho := \rho_{12}^{\otimes n}$ to conclude $E_R(\rho_{12}^{\otimes n}) \geq E_{R,\mathrm{LOCC1}}(\rho_{12}) + E_R(\rho_{12}^{\otimes(n-1)})$. Then use induction (and finally the DPI) to conclude that $E_R(\rho_{12}^{\otimes n}) \geq nE_{R,\mathrm{LOCC1}}(\rho_{12})$. The final statement now follows from Lemma 9.12. $\qquad\square$

9.4. Entanglement of formation

We have seen that for pure states $\sigma_{AB} := |\psi\rangle\langle\psi|$ on $\mathcal{H}_A \otimes \mathcal{H}_B$, the relative entropy of entanglement $E_R(\sigma_{AB})$ reduces the von Neumann entropy of the marginals σ_A and σ_B; that is $E_R(\sigma_{AB}) = S(\sigma_A) = S(\sigma_B)$.

Recall from Definition 2.23 that for $\rho_{AB} \in \mathfrak{S}_{\mathcal{H}_A\otimes\mathcal{H}_B}$, $\mathrm{Ens}(\rho_{AB})$ denotes the set of all ensemble decompositions $\{p_j, |\psi_j\rangle\langle\psi_j|\}_{1\leq j\leq s}$ of ρ_{AB}; that is, the set of all expressions of ρ_{AB} as a convex combination

$$(9.4.1) \qquad \rho_{AB} = \sum_{j=1}^{s} p_j|\psi_j\rangle\langle\psi_j|$$

of a finite set of pure states.

Let F be a convex function on $\mathfrak{S}_{\mathcal{H}_A \otimes \mathcal{H}_B}$ such that $F(\sigma_{AB}) = E_R(\sigma_{AB})$ for all pure states σ_{AB}. For $\rho_{AB} \in \mathfrak{S}_{\mathcal{H}_A \otimes \mathcal{H}_B}$, let $\{p_j, |\psi_j\rangle\langle\psi_j|\}_{1 \le j \le s} \in \mathrm{Ens}(\rho_{AB})$.

Since F is concave, (9.4.1) yields

$$F(\rho_{AB}) \le \sum_{j=1}^{s} p_j F(|\psi_j\rangle\langle\psi_j|) = \sum_{j=1}^{s} p_j S\left(\mathrm{Tr}_B\left[|\psi_j\rangle\langle\psi_j|\right]\right).$$

Definition 9.15. The **entanglement of formation** is the function E_F on $\mathfrak{S}_{\mathcal{H}_A \otimes \mathcal{H}_B}$ defined as

$$(9.4.2) \qquad E_F(\rho_{AB}) := \inf\left\{ \sum_{j=1}^{s} p_j E(\psi_j) \; : \; \{p_j, |\psi_j\rangle\langle\psi_j|\} \in \mathrm{Ens}(\rho_{AB})\right\},$$

where $E(\psi_j)$ is the entanglement entropy of ψ_j, as defined in (9.3.2).

It is an easy exercise to see that E_F is convex on $\mathfrak{S}_{\mathcal{H}_A \otimes \mathcal{H}_B}$, and by what has been observed above, it is the *largest* convex function on $\mathfrak{S}_{\mathcal{H}_A \otimes \mathcal{H}_B}$ that agrees with E_R on the pure states. The construction is standard, and in the standard terminology, E_F is the **convex roof function** of E_R on the pure states. Therefore,

$$(9.4.3) \qquad\qquad\qquad E_F \ge E_R.$$

We shall see below that E_R can be much smaller than E_F. The entanglement of formation was introduced, physically motived, and studied in [**32, 100**]. It is shown in [**100**] that $E_F(\rho_{AB}^{\otimes n})$ represents the cost, in terms the number of maximally entangled qbits required, to produce n copies of ρ_{AB} using LOCC operations, asymptotically in n. In this sense, it quantifies entanglement in an operational way.

Although $\mathrm{Ens}(\rho)$ is not closed, let alone compact, the infimum in Definition 9.15 is actually a minimum. The following proof is due to Uhlmann [**216**].

Lemma 9.16. *Let $\mathcal{H}_A, \mathcal{H}_B$ be Hilbert spaces of dimensions d_A, d_B, respectively, and let $\rho_{AB} \in \mathfrak{S}_{\mathcal{H}_A \otimes \mathcal{H}_B}$. There exists an ensemble that achieves the infimum in (9.4.2), and moreover this ensemble $\{p_j, |\psi_j\rangle\langle\psi_j|\}_{1 \le j \le s}$ is such that $s \le d_A^2 d_B^2 + 1$.*

Proof. Let $S \subset \mathcal{B}^{\mathrm{s.a.}}(\mathcal{H}_A \otimes \mathcal{H}_B)$ consist of operators of the form $\omega + E(\omega)\mathbb{1}$ where ω is a pure state $|\psi\rangle\langle\psi|$ on $\mathcal{H}_A \otimes \mathcal{H}_B$, and $E(\omega)$ denotes its entanglement entropy, which is the same as $E_F(\omega)$. Let K denote the closed convex hull of S. The set of extreme points of K is precisely the set S. By Caratheodory's theorem, Theorem A.15, and Corollary A.16, every operator in K is a convex combination of at most $d_A^2 d_B^2 + 1$ operators in S.

We now claim that for all $\rho_{AB} \in \mathfrak{S}_{\mathcal{H}_A \otimes \mathcal{H}_B}$,

$$(9.4.4) \qquad\qquad E_F(\rho_{AB}) = \inf\{\lambda \ge 0 \; : \; \rho_{AB} + \lambda\mathbb{1} \in K\}.$$

First note that if $\rho_{AB} + \lambda \mathbb{1} \in K$, there exists an ensemble $\{p_j, \omega_j\}_{1 \le j \le s}$ of pure states $j = 1, \ldots, s$, $s \le d_A^2 d_B^2 + 1$, such that $\rho_{AB} = \sum_{j=1}^{s} p_j \omega_j$ and $\lambda = \sum_{j=1}^{s} p_j E(\omega_j)$. Therefore, $\lambda \ge E_F(\rho_{AB})$.

However, for any $\lambda > E_F(\rho_{AB})$, there exists an ensemble $\{p_j, \omega_j\}_{1 \le j \le n}$ for some n, not necessarily bounded by $d_A^2 d_B^2 + 1$, such that $E_F(\rho_{AB}) \le \sum_{j=1}^{n} p_j E(\omega_j) \le \lambda$ and $\sum_{j=1}^{n} p_j \omega_j = \rho_{AB}$. Therefore, (9.4.4) is proved.

Since K is closed, the infimum in (9.4.4) is a minimum. Therefore, there exists an ensemble $\{p_j, \omega_j\}_{1 \le j \le s}$ such that $\sum_{j=1}^{s} p_j \omega_j = \rho_{AB}$ and $\sum_{j=1}^{s} p_j E(\omega_j) = E_F(\rho_{AB})$. The bound on s was established in the first paragraph. $\qquad\square$

E_F is also an entanglement monotone, as shown in [**32**]. First, since E_R is faithful, (9.4.3) implies that $E_F(\rho_{AB}) > 0$ for any entangled state. On the other hand, if ρ_{AB} is separable, then it has an expression of the form (9.2.2), and from this it follows that $E_F(\rho_{AB}) = 0$. Hence, $E_F(\rho_{AB}) = 0$ if and only if ρ_{AB} is separable.

The proof that E_F satisfies requirement (2) in the definition of an entanglement monotone is more involved that it was for E_R. We follow [**221**], using Lemma 9.7.

Theorem 9.17. *The entanglement of formation is an entanglement monotone.*

Proof. It remains to prove requirement (2). First, consider any pure state $\rho_{AB} := |\phi\rangle\langle\phi|$, and consider any LOCC1 operation Ψ of the form given in (9.1.3); that is

$$\Psi(X) := \sum_{x \in \mathcal{X}} \Psi_x(X) := \sum_{x \in \mathcal{X}} (\mathbb{1}_A \otimes B_x) X (\mathbb{1}_A \otimes B_x^*).$$

Define $p_x := \mathrm{Tr}[\Psi_x(\rho_{AB})]$ and $\rho_x := p_x^{-1} \Psi_x(|\phi\rangle\langle\phi|)$ for $p_x > 0$. Also define, as usual, $\rho_A := \mathrm{Tr}_B[\rho_{AB}]$ and $\rho_B := \mathrm{Tr}_A[\rho_{AB}]$. By the partial cyclicity of the partial trace,

$$\sum_{x \in \mathcal{X}} p_x \mathrm{Tr}_B[\rho_x] = \mathrm{Tr}_B\left[\sum_{x \in \mathcal{X}} (\mathbb{1}_A \otimes B_x)\rho_{AB}(\mathbb{1}_A \otimes B_x^*)\right]$$

$$(9.4.5) \qquad = \mathrm{Tr}_B\left[\sum_{x \in \mathcal{X}} (\mathbb{1}_A \otimes B_x^*)(\mathbb{1}_A \otimes B_x)\rho_{AB}\right] = \rho_A.$$

For each $x \in \mathcal{X}$, let $\rho_x = \sum_{y \in \mathcal{Y}} q_{(y|x)} |\psi^{(y|x)}\rangle\langle\psi^{(y|x)}|$ express ρ_x as a convex combination of pure states where $\mathcal{Y}$ is some sufficiently large finite set. Then by the definition of E_F, the concavity of the von Neumann entropy, and (9.4.5),

$$\sum_{x\in\mathcal{X}} p_x E_F(\rho_x) \leq \sum_{x\in\mathcal{X}}\sum_{y\in\mathcal{Y}} p_x q_{(y|x)} S(\mathrm{Tr}_B[|\psi^{(y|x)}\rangle\langle\psi^{(y|x)}|])$$

$$\leq S\left(\mathrm{Tr}_B\left[\sum_{x\in\mathcal{X}}\sum_{y\in\mathcal{Y}} p_x q_{(y|x)}|\psi^{(y|x)}\rangle\langle\psi^{(y|x)}|\right]\right)$$

$$= S\left(\mathrm{Tr}_B\left[\sum_{x\in\mathcal{X}} p_x\rho_x\right]\right) = S(\rho_A) = E_F(\rho_{AB}).$$

This proves the theorem for the pure state. For general $\rho_{AB} \in \mathfrak{S}_{\mathcal{H}_A\otimes\mathcal{H}_B}$, let $\rho_{AB} = \sum_{j=1}^{m} \lambda_j|\psi_j\rangle\langle\psi_j|$ be an optimal expression of ρ_{AB} as a convex combination of pure states so that $E_F(\rho_{AB}) = \sum_{j=1}^{m} \lambda_j E_F(|\psi_j\rangle\langle\psi_j|)$, noting that this exists by Lemma 9.16.

Define $p_{x,j} := \mathrm{Tr}[\Psi_x(|\psi_j\rangle\langle\psi_j|)]$ and $\rho_{x,j} := p_{x,j}^{-1}\Psi_x(|\psi_j\rangle\langle\psi_j|)$. Then

$$\sum_{j=1}^{m} \lambda_j p_{x,j}\rho_{x,j} = \Psi_x(\rho_{AB}) \quad\text{and}\quad p_x := \mathrm{Tr}[\Psi_x(|\psi_j\rangle\langle\psi_j|)] = \sum_{j=1}^{m} \lambda_j p_{x,j}.$$

Note that $\sum_{j=1}^{m} \frac{\lambda_j p_{x,j}}{p_x} = 1$. Therefore, with $\rho_x := p_x^{-1}\Psi_x(\rho_{AB})$,

$$\sum_{x\in\mathcal{X}} p_x E_F(\rho_x) = \sum_{x\in\mathcal{X}} p_x E_F\left(\sum_{j=1}^{m} \frac{\lambda_j p_{x,j}}{p_x}\rho_{x,j}\right)$$

$$\leq \sum_{x\in\mathcal{X}}\sum_{j=1}^{m} \frac{\lambda_j p_{x,j}}{p_x} p_x E_F(\rho_{x,j})$$

$$= \sum_{j=1}^{m} \lambda_j\left(\sum_{x\in\mathcal{X}} p_{x,j} E_F(\rho_{x,j})\right)$$

$$\leq \sum_{j=1}^{m} \lambda_j E_F(|\psi_j\rangle\langle\psi_j|) = E_F(\rho_{AB}),$$

where the first inequality is the convexity of E_F, and the second is the inequality proved in the first part for pure states. $\square$

9.4.1. Entanglement cost.

As noted above, $E_F(\rho_{AB}^{\otimes n})$ represents the cost in terms of the number of maximally entangled qbits required to produce n copies of ρ_{AB} using LOCC operations, asymptotically in n. For a long time it was conjectured that entanglement of formation is additive over tensor products, in which case we would have $E_F(\rho_{AB}^{\otimes n}) = nE_F(\rho_{AB})$. Schor proved [192] that this conjecture is equivalent to three other conjectures. One of these other conjectures is that *maximum entropy output of quantum operations* is additive [192].

We shall not describe this here, but simply note that Hastings [98] gave a non-constructive proof that this conjecture is false. Hence entanglement of formation is known not to be additive in general, but as of now, not a single explicit counterexample is known.

However, E_F is at least subadditive. To see this, let $\mathcal{K}_A, \mathcal{K}_B$ be another pair of finite-dimensional Hilbert spaces, and let $\sigma_{AB} \in \mathfrak{S}_{\mathcal{K}_A \otimes \mathcal{K}_B}$. Writing

$$\rho_{AB} = \sum_{j=1}^{m} p_j |\phi_j\rangle\langle\phi_j| \quad \text{and} \quad \sigma_{AB} = \sum_{k=1}^{n} q_k |\psi_k\rangle\langle\psi_k|.$$

Then

$$\rho_{AB} \otimes \sigma_{AB} = \sum_{j=1}^{m}\sum_{k=1}^{n} p_j q_k |\phi_j \otimes \psi_k\rangle\langle\phi_j \otimes \psi_k|$$

and

$$S(\mathrm{Tr}_B[|\phi_j \otimes \psi_k\rangle\langle\phi_j \otimes \psi_k|]) = S(\mathrm{Tr}_B[|\phi_j\rangle\langle\phi_j|]) + S(\mathrm{Tr}_B[|\psi_k\rangle\langle\psi_k|]).$$

Therefore, $E_F(\rho_{AB} \otimes \sigma_{AB}) \leq E_F(\rho_{AB}) + E_F(\sigma_{AB})$. It follows that for all $\ell, m \in \mathbb{N}$, $E_F(\rho_{AB}^{\otimes(\ell+m)}) \leq E_F(\rho_{AB}^{\otimes\ell}) + E_F(\rho_{AB}^{\otimes m})$, and therefore, by a standard property of subadditive sequences, $\lim_{n\to\infty} \frac{1}{n} E_F(\rho_{AB}^{\otimes n})$ exists. (See Exercise 3 of this chapter.)

Definition 9.18. Let $\mathcal{H}_A, \mathcal{H}_B$ be finite-dimensional Hilbert spaces. Let $\rho_{AB} \in \mathfrak{S}_{\mathcal{H}_A \otimes \mathcal{H}_B}$. Then the **entanglement cost** $E_C(\rho_{AB})$ is defined by

$$E_C(\rho_{AB}) = \lim_{n\to\infty} \frac{1}{n} E_F(\rho_{AB}^{\otimes n}),$$

that is, E_C is the regularization of E_F.

9.4.2. Continuity of entanglement of formation. The continuity of entanglement of formation was proved by Nielsen [168]. Winter [234] then gave a sharper bound on the modulus of continuity; his approach is followed here.

Theorem 9.19. *Let $\mathcal{H}_A$ and $\mathcal{H}_B$ be Hilbert spaces of finite dimensions d_A and d_B. Let $d := \min\{d_A, d_B\}$. Then*

$$\frac{1}{\sqrt{2}}\|\rho_{AB} - \sigma_{AB}\|_1 \leq \delta$$

$$\Rightarrow |E_F(\rho_{AB}) - E_F(\sigma_{AB})| \leq 2\log d\delta + (1 + \delta)\eta\left(\frac{\delta}{1+\delta}\right),$$

Proof. By Lemma 9.16 for all $\rho_{AB} \in \mathfrak{S}_{\mathcal{H}_A \otimes \mathcal{H}_B}$, there exists an ensemble $\{p_k, |\psi_k\rangle\langle\psi_k|\}_{1 \leq k \leq m}$ such that $E_F(\rho_{AB}) = \sum_{k=1}^{m} p_k E(|\psi_k\rangle\langle\psi_k|)$. We may assume without loss of generality that $m \geq d_{AB}$, the dimension of $\mathcal{H}_{AB}$ by setting $p_k = 0$ for some values of k if necessary.

Let $\mathcal{H}_C$ be an m-dimensional Hilbert space, and let $\{\mathbf{u}_1,\ldots,\mathbf{u}_m\}$ be an orthonormal basis for $\mathcal{H}_C$. Define $\Psi := \sum_{k=1}^{m} p_k^{1/2}\psi_k \otimes \mathbf{u}_k$, so that Ψ is a purification of ρ_{AB}. Now let $\widetilde{\Psi}$ be any other purification of ρ_{AB} on $\mathcal{H}_{AB} \otimes \mathcal{H}_{C'}$ for some Hilbert space $\mathcal{H}_{C'}$ whose dimension is $\operatorname{rank}(\rho_{AB})$. By Theorem 2.32, there exists a partial isometry in $U \in \mathcal{B}(\mathcal{H}_{C'},\mathcal{H}_C)$ such that

$$\Psi = (\mathbb{1}_{AB} \otimes U)\widetilde{\Psi},$$

and we may assume that the initial space of U is $\mathcal{H}_{C'}$. Define $E_k := U^*|\mathbf{u}_k\rangle\langle\mathbf{u}_k|U$ and note that $\sum_{k=1}^{m} E_k = U^*U = \mathbb{1}_{C'}$. Therefore, with $\mathcal{X} := \{1,\ldots,m\}$, and $\mathcal{E} : \mathcal{X} \to \mathcal{B}^+(\mathcal{H}_{C'})$ defined by $\mathcal{E}(x) = E_x$, $\mathcal{E}$ is a POVM. Define $\Phi : \mathcal{B}(\mathcal{H}_{C'}) \to \mathcal{B}(\mathcal{H}_C)$ by $\Phi(X) := \sum_{k=1}^{m} \operatorname{Tr}[E_k X] \otimes |\mathbf{u}_k\rangle\langle\mathbf{u}_k|$, and note that Φ is CP and trace preserving. By construction,

$$\operatorname{Tr}_{C'}[(\mathbb{1}_{AB} \otimes E_k)|\widetilde{\Psi}\rangle\langle\widetilde{\Psi}|] = \operatorname{Tr}_C[(\mathbb{1}_{AB} \otimes |\mathbf{u}_k\rangle\langle\mathbf{u}_k|)|\Psi\rangle\langle\Psi|] = p_k\omega_k.$$

Therefore, $\rho_{ABC'} := (\mathbb{1}_{AB} \otimes \Phi)(|\widetilde{\Psi}\rangle\langle\widetilde{\Psi}|) = \sum_{k=1}^{m} p_k\omega_k \otimes |\mathbf{u}_k\rangle\langle\mathbf{u}_k|$ is an extension of ρ_{AB} and $E_F(\rho_{AB})$ can now be expressed as a conditional entropy of the partial trace $\rho_{AC'} = \operatorname{Tr}_B[\rho_{ABC'}]$:

$$S_{A|C'}(\rho_{AC'}) = S(\rho_{AC'}) - S(\rho_{C'}) = \sum_{k=1}^{m} p_k S(\operatorname{Tr}_B[\omega_k]) = E_F(\rho_{AB}).$$

Now let $\sigma_{AB} \in \mathfrak{S}_{\mathcal{H}_A \otimes \mathcal{H}_B}$. By Theorem 6.53, there exist purifications $\widetilde{\Psi}$ of ρ_{AB} and $\widetilde{\Phi}$ of σ_{AB} such that

(9.4.6)

$$\||\widetilde{\Psi}\rangle\langle\widetilde{\Psi}| - |\widetilde{\Phi}\rangle\langle\widetilde{\Phi}|\|_1 \le \sqrt{2}\|\rho_{AB} - \sigma_{AB}\|_1.$$

Define

$$\sigma_{ABC'} := (\mathbb{1}_{AB} \otimes \Phi)(|\widetilde{\Phi}\rangle\langle\widetilde{\Phi}|) = \sum_{k=1}^{m} q_k\sigma_k \otimes |\mathbf{u}_k\rangle\langle\mathbf{u}_k|,$$

where each σ_k is a density matrix on $\mathcal{H}_{AB}$. Then $\sigma_{AB} = \sum_{k=1}^{m} q_k\sigma_k$, and

$$S_{A|C'}(\sigma_{AC'}) = \sum_{k=1}^{m} q_k] S(\operatorname{Tr}_B[\sigma_k]) \le E_F(\sigma_{AB}),$$

where the final inequality is due to the concavity of the entropy; we can only lower $S(\operatorname{Tr}_B[\sigma_k])$ by expanding σ_k as a convex combination of pure states. Therefore,

$$E_F(\rho_{AB}) - E_F(\sigma_{AB}) \le S_{A|C'}(\rho_{AC'}) - S_{A|C'}(\sigma_{AC'}).$$

Using the Alicki-Fannes bound, we may estimate the right side in terms of

$$\|\rho_{AC'} - \sigma_{AC'}\|_1 = \|\operatorname{Tr}_B[(\mathbb{1}_{AB} \otimes \Phi)(|\widetilde{\Psi}\rangle\langle\widetilde{\Psi}| - |\widetilde{\Phi}\rangle\langle\widetilde{\Phi}|)]\|_1$$

$$\le \||\widetilde{\Psi}\rangle\langle\widetilde{\Psi}| - |\widetilde{\Phi}\rangle\langle\widetilde{\Phi}|\|_1 \le \sqrt{2}\|\rho_{AB} - \sigma_{AB}\|_1,$$

where the first inequality is the contractivity of the trace norm under quantum operations, Theorem 6.26, and the second is (9.4.6). We may assume $E_F(\rho_{AB}) \geq E_F(\sigma_{AB})$, for if not we may interchange their roles. Therefore, by Theorem 8.1, when $\frac{1}{\sqrt{2}}\|\rho_{AB} - \sigma_{AB}\|_1 \leq \delta$, then

$$|E_F(\rho_{AB}) - E_F(\sigma_{AB})| \leq 2 \log d_A \delta + (1 + \delta)\eta\left(\frac{\delta}{1+\delta}\right),$$

and since $d_A \leq d_B$, this proves the theorem. $\square$

See [234] for a proof along these lines of the continuity of entanglement cost, among other related results.

9.5. Symmetric extendibility

Let ρ_{AB} be separable so that it may be written in the form

$$\rho_{AB} = \sum_{j=1}^{n} p_j |\psi_j\rangle\langle\psi_j| \otimes |\phi_j\rangle\langle\phi_j|,$$

where each ψ_j is unit vector in $\mathcal{H}_A$ and each ϕ_j is a unit vector in $\mathcal{H}_B$. Then for $N > 2$, define

$$\rho_{AB^N} = \sum_{j=1}^{n} p_j |\psi_j\rangle\langle\psi_j| \otimes \left(|\phi_j\rangle\langle\phi_j|\right)^{\otimes N},$$

which is a density matrix on $\mathcal{H}_A \otimes \mathcal{H}_B^{\vee N}$, where $\mathcal{H}_B^{\vee N}$ denotes the N-fold symmetric tensor power of $\mathcal{H}_B$, which is the Bosonic subspace of $\mathcal{H}_B^{\otimes N}$, as discussed in Section 2.8. Let $\mathrm{Tr}_{B^{N-1}}$ denote the partial trace over the last $N - 1$ factors of $\mathcal{H}_B$. Then for each j,

$$\mathrm{Tr}_{B^{N-1}}\left[\left(|\phi_j\rangle\langle\phi_j|\right)^{\otimes N}\right] = |\phi_j\rangle\langle\phi_j|,$$

and hence $\mathrm{Tr}_{B^{N-1}}[\rho_{AB^N}] = \rho_{AB}$ so that ρ_{AB^N} is an extension of ρ_{AB}.

As in Section 2.8, for any permutation $\pi \in \mathcal{S}_N$, define V_π to be the corresponding unitary operator given by

$$(9.5.1) \qquad V_\pi(\mathbf{v}_1 \otimes \cdots \otimes \mathbf{v}_N) = \mathbf{v}_{\pi(1)} \otimes \cdots \otimes \mathbf{v}_{\pi(N)}.$$

It then follows that for each $\pi \in \mathcal{S}_N$, $[\rho_{AB^N}, \mathbb{1}_A \otimes V_\pi] = 0$. As in Section 2.8, let P_{BOS}^N be the orthogonal projection onto the Bosonic subspace $\mathcal{H}_B^{\vee N}$ of $\mathcal{H}_B^{\otimes N}$. Note that

$$P_{\mathrm{BOS}}^N = \frac{1}{N!} \sum_{\pi \in \mathcal{S}_N} V_\pi.$$

Definition 9.20. Let $\mathcal{H}_A$ and $\mathcal{H}_B$ be finite-dimensional Hilbert spaces. Let $\rho_{AB} \in \mathfrak{S}_{\mathcal{H}_A \otimes \mathcal{H}_B}$. Let $N \geq 2$. A density matrix $\rho_{AB^N} \in \mathfrak{S}_{\mathcal{H}_A \otimes \mathcal{H}_B^{\otimes N}}$ is a **symmetric extension of order** N of ρ_{AB} in case

$$\mathrm{Tr}_{B^{N-1}}[\rho_{AB^N}] = \rho_{AB},$$

and for all $\pi \in \mathcal{S}_N$,

$$[\rho_{AB^N}, \mathbb{1}_A \otimes V_\pi] = 0.$$

It is a **Bose symmetric extension** of order N if furthermore

$$\rho_{AB^N} = \rho_{AB^N}(\mathbb{1}_A \otimes P_{\mathrm{BOS}}^N).$$

There is a passage, using purification, from symmetric extensions to Bose symmetric extensions. To see this, let ρ_{AB^N} be an N-fold symmetric extension of ρ_{AB}. Use Theorem 2.22 to purify ρ_{AB}: let $\{\mathbf{u}_1, \ldots, \mathbf{u}_{d_A}\}$ be an orthonormal basis of $\mathcal{H}_A$, and define $\Psi_{AA'} = \sum_{j=1}^{d_A} \mathbf{u}_j \otimes \mathbf{u}_j$. Likewise, let $\Phi_{BB'} \in \mathcal{H}_{BB'}$ be constructed in the same manner. Then

$$(\sqrt{\rho_{AB}} \otimes \mathbb{1}_{A'B'})\Psi_{AA'} \otimes \Phi_{BB'}$$

is a purification of ρ_{AB}, and

$$(\sqrt{\rho_{AB^N}} \otimes \mathbb{1}_{A'B'^N})\Psi_{AA'} \otimes (\Phi_{BB'})^{\otimes N}$$

is a pure state in $\mathcal{H}_{AA'} \otimes \mathcal{H}_{BB'}^{\vee N}$ that, because $\sqrt{\rho_{AB^N}} \otimes \mathbb{1}_{A'B'^N}$ is symmetric under permutations in $\mathcal{H}_{BB'}^{\otimes N}$, is a Bose symmetric purification of $\rho_{AA',BB'}$.

It turns out that $\rho_{AB} \in \mathfrak{S}_{\mathcal{H}_A \otimes \mathcal{H}_B}$ is separable if and only if it has a Bose symmetric extension of every order N. This was proved and used in [**76**] as the basis of an algorithm to check for entanglement. A simple and precise quantitative version of this was provided in [**163**]:

Theorem 9.21. *Let $\mathcal{H}_A$ and $\mathcal{H}_B$ be finite-dimensional Hilbert spaces, and let d denote the dimension of $\mathcal{H}_B$. If $\rho_{AB} \in \mathfrak{S}_{\mathcal{H}_A \otimes \mathcal{H}_B}$ has a Bose symmetric extension of ρ_{AB^N} of order N, then*

$$(9.5.2) \qquad \tilde{\rho}_{AB} := \frac{N}{N+d}\rho_{AB} + \frac{d}{N+d}\rho_A \otimes (\tfrac{1}{d}\mathbb{1}_B)$$

is separable.

As a consequence, suppose that ρ_{AB} has a Bose symmetric extension of ρ_{AB^N} of every order N. By Theorem 9.21,

$$\|\rho_{AB} - \tilde{\rho}_{AB}\|_1 \leq \frac{d}{N+d}\|\rho_{AB} - \rho_A \otimes (\tfrac{1}{d}\mathbb{1}_B)\|_1 \leq \frac{2d}{N+d}.$$

Therefore, there is a sequence of separable states converging to ρ_{AB}, and since the set of separable states is closed, ρ_{AB} is separable. We have already seen that separable states have Bose symmetric extensions of every order N, and hence **this property characterizes separable states.**

One can also draw conclusions about the relative entropy of entanglement from the existence of a Bose symmetric extension of order N. With ρ_{AB} and $\tilde{\rho}_{AB}$ as in Theorem 9.21, by the operator concavity of the logarithm,

$$\begin{aligned}
D(\rho_{AB}\|\tilde{\rho}_{AB}) &= \text{Tr}[\rho_{AB}(\log(\rho_{AB}) - \log(\tfrac{N}{N+d}\rho_{AB} + \tfrac{d}{N+d}\rho_A \otimes (\tfrac{1}{d}\mathbb{1}_B)))] \\
&\leq \text{Tr}[\rho_{AB}(\log(\rho_{AB}) - \tfrac{N}{N+d}\log(\rho_{AB}) - \tfrac{d}{N+d}\log(\rho_A \otimes (\tfrac{1}{d}\mathbb{1}_B)))] \\
&= \tfrac{d}{N+d}\text{Tr}[\rho_{AB}(\log(\rho_{AB} - \log(\rho_A \otimes (\tfrac{1}{d}\mathbb{1}_B))))] \\
&= \tfrac{d}{N+d}(S_A - S_{AB} + \log d).
\end{aligned}$$

In particular, letting d_A denote the dimension of $\mathcal{H}_A$,

$$E_R(\rho_{AB}) \leq D(\rho_{AB}\|\tilde{\rho}_{AB}) \leq \frac{d(\log d_A + \log d)}{N + d}.$$

The following corollary will be useful:

Corollary 9.22. *If $\rho_{AB} \in \mathfrak{S}_{\mathcal{H}_A \otimes \mathcal{H}_B}$ has a symmetric extension of ρ_{AB^N} of order N, not necessarily Bose symmetric, then there exists a separable $\sigma_{AB} \in \mathfrak{S}_{\mathcal{H}_A \otimes \mathcal{H}_B}$ such that*

$$(9.5.3) \qquad \|\rho_{AB} - \sigma_{AB}\|_1 \leq \frac{2d_B^2}{N + d_B^2},$$

where d_B is the dimension of $\mathcal{H}_B$.

Proof. Let ρ_{AB^N} be a symmetric extension of ρ_{AB} of order N, and let $\rho_{AA'BB'^N}$ be its purification as described after Definition 9.20. Then by Theorem 9.21, there is a separable state $\tilde{\rho}_{AA'BB'^N}$ such that

$$\|\rho_{AA'BB'} - \tilde{\rho}_{AA'BB'}\|_1 \leq \frac{d_B^2}{N + d_B^2}$$

since the dimension of $\mathcal{H}_{BB'}$ is d_B^2. Now tracing out of $\mathcal{H}_{A'}$ and $\mathcal{H}_{B'}$ and using the fact that the trace norm decreases under partial traces yields (9.5.3). $\qquad \square$

Proof of Theorem 9.21. Let dU denote normalized Haar measure on $\mathcal{U}(\mathcal{H})$, the unitary group on $\mathcal{H}_B$. By Lemma 2.51, the span of the vectors $\psi \otimes \cdots \otimes \psi$ is all of $\bigvee^N \mathcal{H}_B$, the completely symmetric Bose subspace of $\bigotimes^N \mathcal{H}_B$. By Theorem 2.53, $U \mapsto U \otimes \cdots \otimes U$ is an irreducible representation of $\mathcal{U}(\mathcal{H}_R)$ on $\mathcal{H}_B^{\vee N}$, and for an arbitrary unit vector $\phi_0 \in \mathcal{H}_B$,

$$(9.5.4) \qquad \int_{\mathcal{U}(\mathcal{H}_B)} dU |(U\phi_0)^{\otimes N}\rangle\langle(U\phi_0)^{\otimes N}| = \frac{(d_B - 1)!\, N!}{(N + d_B - 1)!} P_{\text{BOS}},$$

where P_{BOS} is the orthogonal projection in $\bigotimes^N \mathcal{H}_B$ onto $\bigvee^N \mathcal{H}_B$. That is, the left side of (9.5.4) is a partial *continuous* POVM. To express it more plainly as a partial POVM, for $U \in \mathcal{U}(\mathcal{H}_B)$, define

$$E_U := |(U\phi_0)^{\otimes N}\rangle\langle(U\phi_0)^{\otimes N}|.$$

Then $\mathbb{1}_A \otimes E_U$ is also a partial POVM on $\mathcal{U}(\mathcal{H}_B)$ with values in $\mathcal{H}_A \otimes \mathcal{H}_{B^N}$.

Now define a probability p by $p := \int_{\mathcal{U}(d)} dU \, \mathrm{Tr}_{B^N}[E_U \rho_{B^N}]$, noting that this is the probability that the partial POVM returns a result on a measurement of ρ_{B^N}. Since ρ_{AB^N} is a Bose symmetric extension of ρ_{AB}, $\rho_{B^N} P_{\mathrm{BOS}}^N = \rho_{B^N}$. Therefore, by (9.5.4), $p := \frac{(d-1)!N!}{(N+d-1)!}$.

Now define the density matrix $\tilde{\rho}_{AB}$ on $\mathcal{H}_A \otimes \mathcal{H}_B$ by

$$
\begin{aligned}
\tilde{\rho}_{AB} &:= \frac{1}{p} \int_{\mathcal{U}(\mathcal{H}_B)} dU \, \mathrm{Tr}_{B^N}[(\mathbb{1}_A \otimes E_U \rho_{AB^N}] \otimes |U\phi_0\rangle\langle U\phi_0| \\
&= \frac{1}{p} \int_{\mathcal{U}(\mathcal{H}_B)} dU \, \mathrm{Tr}_{B^N}[(\mathbb{1}_A \otimes |(U\phi_0)^{\otimes(N+1)}\rangle\langle(U\phi_0)^{\otimes(N+1)}|\rho_{AB^N} \otimes \mathbb{1}_B].
\end{aligned}
$$

From the first equation above, it is clear that $\tilde{\rho}_{AB}$ is separable. To exploit the second equation, recall from (2.8.3) that another way to write P_{BOS} is

$$(9.5.5) \qquad\qquad P_{\mathrm{BOS}} = \frac{1}{N!} \sum_{\pi \in \mathcal{S}_N} V_\pi.$$

Using (9.5.4) and (9.5.5) for $N + 1$ in place of N, we have

$$\tilde{\rho}_{AB} = \frac{(d_B - 1)!}{(N + d_B)!} \sum_{\pi \in \mathcal{S}_N} \mathrm{Tr}_{B^N}[(\rho_{AB^N} \otimes \mathbb{1}_B)(\mathbb{1}_A \otimes V_\pi)].$$

We now claim that

$$(9.5.6) \quad \mathrm{Tr}_{B^N}[(\rho_{AB^N} \otimes \mathbb{1}_B)(\mathbb{1}_A \otimes V_\pi)] = \begin{cases} \rho_A \otimes \mathbb{1}_B & \pi(N+1) = N+1 \\ \rho_{AB} & \pi(N+1) \neq N+1. \end{cases}$$

Once this is established, (9.5.2) follows directly, and we have seen that $\tilde{\rho}_{AB}$ is separable.

It remains to prove (9.5.6). If $\pi(N+1) = N+1$, then $V_\pi = V_{\pi'} \otimes \mathbb{1}_B$ where $\pi' \in \mathcal{S}_N$. Note that

$$\rho_{AB^N} \otimes \mathbb{1}_B(\mathbb{1}_A \otimes V_\pi) = \rho_{AB^N} \otimes \mathbb{1}_B(\mathbb{1}_A \otimes V_{\pi''} \otimes \mathbb{1}_B) = \rho_{AB^N} \otimes \mathbb{1}_B$$

since for all $\pi'' \in \mathcal{S}_N$, $\rho_{AB^N}(\mathbb{1}_A \otimes V_{\pi''}) = \rho_{AB^N}$ due to the Bose symmetry of ρ_{AB^N} in the B variables.

Next, suppose $\pi(N+1) \neq N+1$, and let $X := (\rho_{AB^N} \otimes \mathbb{1}_B)(\mathbb{1}_A \otimes V_\pi)$. We compute the partial trace $\mathrm{Tr}_{B^N}[X]$ using an orthonormal basis $\{\mathbf{u}_1, \ldots, \mathbf{u}_{d_B}\}$ of $\mathcal{H}_B$. For any $\mathbf{x} \in \mathcal{H}_A$ and $1 \leq \ell \leq d_B$,

$$\langle \mathbf{x} \otimes \mathbf{u}_\ell, \mathrm{Tr}_{B^N}[X]\mathbf{x} \otimes \mathbf{u}_\ell \rangle$$

$$= \sum_{j_1,\ldots,j_n=1}^{d_B} \langle \mathbf{x} \otimes \mathbf{u}_{j_1} \otimes \cdots \otimes \mathbf{u}_{j_n} \otimes \mathbf{u}_\ell, X\mathbf{x} \otimes \mathbf{u}_{j_1} \otimes \cdots \otimes \mathbf{u}_{j_n} \otimes \mathbf{u}_\ell \rangle.$$

Let $1 \leq k \leq N$ be such that $\pi(k) = N+1$. Then the inner product in the sum just above is zero unless $j_k = \ell$, and in this case, due to the Bose symmetry in the B variables, the above sum is the same as

$$\sum_{j_2,\ldots,j_n=1}^{d_B} \langle \mathbf{x} \otimes \mathbf{u}_\ell \otimes \mathbf{u}_{j_2} \otimes \cdots \otimes \mathbf{u}_{j_n}, \rho_{AB}\mathbf{x} \otimes \mathbf{u}_\ell \otimes \mathbf{u}_{j_2} \otimes \cdots \otimes \mathbf{u}_{j_n} \rangle$$

$$= \langle \mathbf{x} \otimes \mathbf{u}_\ell \rho_{AB}\mathbf{x} \otimes \mathbf{u}_\ell \rangle.$$

Since the orthonormal basis is arbitrary, this means that for all $\mathbf{x} \in \mathcal{H}_A$ and all $\mathbf{y} \in \mathcal{H}_B$,

$$\langle \mathbf{x} \otimes \mathbf{y}, \mathrm{Tr}_{B^N}[X]\mathbf{x} \otimes \mathbf{y} \rangle = \langle \mathbf{x} \otimes \mathbf{y}, \rho_{AB}\mathbf{x} \otimes \mathbf{y} \rangle,$$

and hence $\mathrm{Tr}_{B^N}[X] = \rho_{AB}$. $\qquad\square$

9.6. Squashed entanglement

Recall that any tripartite density matrix ρ_{ABE} on $\mathcal{H}_A \otimes \mathcal{H}_B \otimes \mathcal{H}_E$, the conditional mutual information of A and B given E, $I_{AB|E}$, is defined by

$$(9.6.1) \qquad I_{AB|E}(\rho_{ABE}) := S_{AE} + S_{BE} - S_{ABE} - S_E \geq 0.$$

Note that SSA is equivalent to the fact that $I_{AB|E} \geq 0$.

Definition 9.23 (Squashed entanglement). The functional E_{sq} on bipartite density matrices ρ_{AB} defined by
$$(9.6.2)$$
$$\mathrm{E}_{\mathrm{sq}}(\rho_{AB}) = \frac{1}{2}\inf\{I_{AB|E}(\rho_{ABE}) : \rho_{ABE} \text{ is a tripartite extension of } \rho_{AB}\}.$$

is the **squashed entanglement** of ρ_{AB}.

This functional was first introduced by Tucci [212] and was rediscovered by Christandl and Winter [64], who studied it and proved that it has many important properties that qualify it as an especially useful measure of entanglement.

By (9.6.1), $\mathrm{E}_{\mathrm{sq}}(\rho_{AB}) \geq 0$. If $\rho_{AB} = |\psi\rangle\langle\psi|$ is a pure state, then every extension ρ_{ABE} is a product; $\rho_{ABE} = |\psi\rangle\langle\psi| \otimes \rho_E$, and hence

$$\mathrm{E}_{\mathrm{sq}}(|\psi\rangle\langle\psi|) = S(\mathrm{Tr}_B[|\psi\rangle\langle\psi|]) = S(\mathrm{Tr}_A[|\psi\rangle\langle\psi|]) = E(|\psi\rangle\langle\psi|).$$

There is also a general upper bound for $E_{sq}(\rho_{AB})$:

Lemma 9.24. *Let $\mathcal{H}_A$ and $\mathcal{H}_B$ be Hilbert spaces of finite dimensions d_A and d_B, respectively. Then for all $\rho_{AB} \in \mathfrak{S}_{\mathcal{H}_A \otimes \mathcal{H}_B}$,*

$$(9.6.3) \qquad E_{sq}(\rho_{AB}) \leq \min\{\log(d_A), \log(d_B)\}.$$

The inequality is saturated when ρ_{AB} is a maximally entangled pure state.

Proof. Consider any extension ρ_{ABE} of ρ_{AB} and write $I_{AB|E}$ as a difference of conditional entropies. There are two ways to do this. One is

$$I_{AB|E}(\rho_{ABE}) = S_{B|E}(\rho_{BE}) - S_{AE|B}(\rho_{ABE}).$$

By the subadditivity of the entropy, $S_{B|E}(\rho_{BE}) = S_{BE} - S_E \leq S_B$ with equality if and only if $\rho_{BE} = \rho_B \otimes \rho_E$. By Lemma 7.39, $S_{AE|B}(\rho_{ABE}) \geq -\min\{S_{AE}, S_B\}$. Therefore,

$$E_{sq}(\rho_{AB}) \leq S_B \leq -\log(d_B).$$

By symmetry in A and B, $E_{sq}(\rho_{AB}) \leq S_A \leq -\log(d_A)$, and this proves (9.6.3). By what was noted above, if ρ_{AB} is a pure state, $E_{sq}(\rho_{AB}) = S_A = S_B$, and hence there is equality in (9.6.3) when ρ_{AB} is a maximally entangled pure state. $\square$

The squashed entanglement is an entanglement monotone. The proof of faithfulness, which was open for some time, requires some preparation, and we start with monotonicity per se, which was proved in [**64**].

9.6.1. LOCC monotonicity of squashed entanglement. The proof of monotonicity follows that of Christandl and Winter [**64**], and uses the method of Vidal [**221**], and hence requires the convexity of E_{sq}.

Lemma 9.25. *The squashed entanglement E_{sq} is convex on $\mathfrak{S}_{\mathcal{H}_A \otimes \mathcal{H}_B}$.*

Proof. Let $\rho_{AB}, \sigma_{AB} \in \mathfrak{S}_{\mathcal{H}_A \otimes \mathcal{H}_B}$, and let $0 < \lambda < 1$. Let ρ_{ABE}, σ_{ABE} be two extensions of ρ_{AB}, σ_{AB}, respectively, where without loss of generality we may assume the third Hilbert space $\mathcal{H}_E$ to be the same. Now let $\mathcal{H}_{E'} := \mathcal{H}_E \otimes \mathbb{C}^2$, and define a tripartite state $\tau_{ABE'}$ $\mathcal{H}_A \otimes \mathcal{H}_B \otimes \mathcal{H}_E \otimes \mathbb{C}^2$ written in block form as

$$\tau_{ABE'} = \begin{bmatrix} \lambda \rho_{ABE} & 0 \\ 0 & (1-\lambda)\sigma_{ABE} \end{bmatrix}.$$

Then by the additivity property of the conditional mutual information, Lemma 8.15,

$$I_{AB|E'}(\tau_{ABE'}) = \lambda I_{AB|E}(\rho_{ABE}) + (1-\lambda)I_{AB|E}(\sigma_{ABE}).$$

By the monotonicity of conditional mutual information under partial traces, Lemma 8.16,

$$I_{AB|E}(\lambda \rho_{ABE} + (1-\lambda)\sigma_{ABE}) \leq I_{AB|E'}(\tau_{ABE'}).$$

It now follows form the definition of E_{sq} that $E_{sq}(\lambda \rho_{ABE} + (1-\lambda)\sigma_{ABE}) \leq \lambda\, E_{sq}(\rho_{AB}) + (1-\lambda)\, E_{sq}(\sigma_{AB})$. $\square$

In proving monotonicity, we shall also need the **chain rule for conditional mutual information**:

Lemma 9.26. *Let $\mathcal{H}_A$, $\mathcal{H}_{A'}$, $\mathcal{H}_B$, $\mathcal{H}_{B'}$, and $\mathcal{H}_E$ be finite-dimensional Hilbert spaces. Then for any density matrix $\rho_{AA'BE}$,*

$$I_{AA',B|E}(\rho_{AA'BE}) = I_{A',B|AE}(\rho_{AA'BE}) + I_{A,B|E}(\rho_{ABE}).$$

Likewise, for any density matrix $\rho_{ABB'E}$,

$$I_{A,BB'|E}(\rho_{AA'BE}) = I_{A,B'|BE}(\rho_{AA'BE}) + I_{A,B|E}(\rho_{ABE}).$$

Proof. This follows directly from the defintion; simply expand the right hand sides, and after cancellations what remains is the quantity on the left. $\square$

Lemma 9.27. *The squashed entanglement E_{sq} is monotone under LOCC operations so that condition (2) in the definition of an entanglement monotone is satisfied.*

Proof. Since E_{sq} is convex, it suffices by Lemma 9.7 to show that it is monotone under unilocal operations by either Alice or Bob. By symmetry, we need only treat the case in which the unilocal operation is applied by Alice.

Given operators $\{A_1, \ldots, A_m\} \subset \mathcal{B}(\mathcal{H}_A)$ such that $\sum_{k=1}^{m} A_k^* A_k = \mathbb{1}_A$, define a quantum operation Φ on $\mathcal{B}(\mathcal{H}_A)$ by $\Phi(X) = \sum_{k=1}^{m} A_k X A_k^*$, and for each k define $p_k \in [0,1]$ and $\rho_{ABE}^{(k)} \in \mathfrak{S}_{\mathcal{H}_A \otimes \mathcal{H}_B \otimes \mathcal{H}_E}$ by

$$p_k \rho_{ABE}^{(k)} := (A_k \otimes \mathbb{1}_{BE}) \rho_{ABE} (A_k^* \otimes \mathbb{1}_{BE}).$$

Finally, let $\mathcal{H}_{A'}$ be a Hilbert space of dimension m, and let $\{\mathbf{u}_1, \ldots, \mathbf{u}_m\}$ be an orthonormal basis for $\mathcal{H}_{A'}$. Define $\tilde{\rho}_{AA'BE} \in \mathfrak{S}_{\mathcal{H}_A \otimes \mathcal{H}_{A'} \otimes \mathcal{H}_B \otimes \mathcal{H}_E}$ by

$$\tilde{\rho}_{AA'BE} := \sum_{k=1}^{m} p_k \rho_{ABE}^{(k)} \otimes |\mathbf{u}_k\rangle\langle\mathbf{u}_k|.$$

By Lemma 8.15, $\sum_{k=1}^{m} p_k = \sum_{k=1}^{m} I_{AB|E}(\rho_{ABE}^{(k)}) = I_{AA'B|E}(\tilde{\rho}_{AA'BE})$, and then by Lemma 9.26 and SSA, $I_{AA'B|E}(\tilde{\rho}_{AA'BE}) > I_{AB|E}(\tilde{\rho}_{ABE})$. However, $\tilde{\rho}_{ABE} = \mathrm{Tr}_{A'}[\tilde{\rho}_{AA'BE}] = \Phi \otimes \mathbb{1}_{BE}(\rho_{ABE})$. Therefore

$$\sum_{k=1}^{m} p_k 2\, E_{sq}(\rho_{AB}^{(k)}) \leq \sum_{k=1}^{m} p_k I_{AB|E}(\rho_{ABE}^{(k)}) \leq I_{AB|E}(\rho_{ABE}). \qquad \square$$

9.6.2. Monogamy of squashed entanglement. One important way in which quantum entanglement differs from classical correlation is while any number of classical random variables $X_1, \dots, X_N$ can be perfectly correlated (so that knowledge of the value of one of them yields knowledge of the values of all of them) if two quantum particles are in a maximally entangled state, neither can be entangled with any third particle.

To see this, consider two qbits in a maximally entangled state—say one of the Bell states

$$\Phi_{AB} = 2^{-1/2}(|0\rangle_A \otimes |0\rangle_B + |1\rangle_A \otimes |1\rangle_B)$$

introduced at the beginning of this chapter. Any extension of $|\Phi_{AB}\rangle\langle\Phi_{AB}|$ to a tripartite state ρ_{ABC} is necessarily of the from

$$\rho_{ABC} = |\Phi_{AB}\rangle\langle\Phi_{AB}| \otimes \rho_C,$$

and hence ρ_{AC} and ρ_{BC} are separable. This is the paradigm for what is called **monogamy of entanglement**. It was first investigated by Coffman, Kundu, and Wooters in terms of the entanglement of formation for two qbit states, using a closed-form formula for $E_F(\rho_{AB})$ for two qbit states ρ_{AB} due to Wooters [236], building on earlier work of Hill and Wooters [109], and their results were extended to multipartite systems of arbitrarily many qbits in [172]. For our purposes, it is much simpler and more relevant to what follows, to discuss this in terms of squashed entanglement, following Koashi and Winder [131].

Theorem 9.28. *Let $\mathcal{H}_A$, $\mathcal{H}_B$, and $\mathcal{H}_C$ be finite-dimensional Hilbert spaces. For any tripartite state ρ_{ABC} on $\mathcal{H}_{ABC}$, let $\rho_{A(BC)}$ denote ρ_{ABC} considered as a bipartite state on $\mathcal{H}_A \otimes \mathcal{H}_{BC}$. Then*

$$(9.6.4) \qquad E_{sq}(\rho_{AB}) + E_{sq}(\rho_{AC}) \leq E_{sq}(\rho_{A(BC)}).$$

Proof. Consider any extension ρ_{ABCE} of ρ_{ABC}. By the chain rule for conditional mutual information, Lemma 9.26,

$$I_{A(BC)|E}(\rho_{ABCE}) = I_{AB|E}(\rho_{ABE}) + I_{AC|BE}(\rho_{ABCE}),$$

from which (9.6.4) follows immediately. $\qquad\qquad\qquad\qquad\qquad\qquad\qquad\square$

Consider any state ρ_{AB} that is maximally entangled in the sense that ρ_{AB} saturates the upper bound in Lemma 9.24; that is $E_{sq}(\rho_{AB}) = \min\{\log d_A, \log d_B\}$. We fix the notation so that $d_A \leq d_B$, and then of course for any extension ρ_{ABC} of ρ_{AB}, $d_{BC} \geq d_A$. Therefore, by Lemma 9.24,

$$E_{sq}(\rho_{A(BC)}) \leq \log(d_A) = E_{sq}(\rho_{AB}),$$

and then by (9.6.4), $E_{sq}(\rho_{AC}) = 0$. As will be shown below, the squashed entanglement is faithful, and hence ρ_{AC} is separable. In other words, if ρ_{AB} is a state of maximal squashed entanglement $\log(d_A)$, then for any extension ρ_{ABC} of ρ_{AB}, ρ_{AC} is separable. This is a very general form of the monogamy relation introduced above in the context of qbits.

Theorem 9.28 has an important consequence for states ρ_{AB} that are N-symmetrically extensible. Suppose that ρ_{AB^N} is an N-symmetric extension of ρ_{AB}. Then by Theorem 9.28,

$$E_{sq}(\rho_{AB}) + E_{sq}(\rho_{A(B^{N-1})}) \leq E_{sq}(\rho_{A(B^N)}).$$

Now a simple induction yields

$$k\, E_{sq}(\rho_{AB}) \leq E_{sq}(\rho_{A(B^{N-k})}) \leq \log(d_A)$$

for all N and k such that $d_A \leq d_B^{N-k}$. Hence $E_{sq}(\rho_{AB}) \leq \frac{d_A}{k}$ for all such k and N.

As shown by Li and Winter [140], not only do states with N symmetric extensions for large N have small squashed entanglement, the converse is also true: if a bipartite state has small squashed entanglement, then it is close in trace norm to state with an N symmetric extension for large N. This yields a proof of the faithfulness of squashed entanglement, and clarifies that squashed entanglement measures the distance to a highly symmetrically extensible state rather than the distance to a separable state.

9.6.3. Faithfulness of squashed entanglement. The easy part of the faithfulness of squashed entanglement is that if ρ_{AB} is separable, then $E_{sq}(\rho_{AB}) = 0$.

Lemma 9.29. *For all bipartite states* ρ_{AB}, $E_{sq}(\rho_{AB}) \leq E_F(\rho_{AB})$.

Proof. Let $\rho_{AB} = \sum_{k=1}^{m} p_k |\psi_k\rangle\langle\psi_k|$ be a representation of ρ_{AB} as an ensemble average of pure states. By the convexity of the E_{sq},

$$(9.6.5) \qquad E_{sq}(\rho_{AB}) \leq \sum_{k=1}^{m} p_k\, E_{sq}(|\psi_k\rangle\langle\psi_k|).$$

Any extension of a pure state $\sigma_{AB} = |\phi\rangle\langle\phi|$ has the form $\sigma_{AB} \otimes \rho_E$ for some density matrix ρ_E on $\mathcal{H}_E$. Then since $I_{AB|E}(\sigma_{AB} \otimes \rho_E) = S_A + S_B = 2S(\text{Tr}_B[|\phi\rangle\langle\phi|])$,

$$E_{sq}(|\phi\rangle\langle\phi|) = S(\text{Tr}_B[|\phi\rangle\langle\phi|]).$$

Going back to (9.6.5), $E_{sq}(\rho_{AB}) \leq \sum_{k=1}^{m} p_k S(\text{Tr}_B[|\psi_k\rangle\langle\psi_k|])$. $\qquad\square$

Since E_F is faithful, if ρ_{AB} is separable, $E_F(\rho_{AB}) = 0$, and then by Lemma 9.29, $E_{sq}(\rho_{AB}) = 0$. This can also be proved directly by choosing a suitable extension of ρ_{AB}. However, it is more difficult to show that if $E_{sq}(\rho_{AB}) = 0$, then ρ_{AB} is separable.

The cases of equality in SSA have been determined in Theorem 8.17, originally proved in [**99**]. By the results obtained there, if $I_{AB|E}(\rho_{ABE}) = 0$, then $\mathcal{H}_E$ has the form $\mathcal{H}_E = \bigoplus_{j=1}^{m} \mathcal{H}_{E\ell}^j \otimes \mathcal{H}_{Er}^j$ and ρ_{ABE} has the form

$$(9.6.6) \qquad \rho_{ABE} = \bigoplus_{j=1}^{m} p_j \rho_{A,E\ell}^j \otimes \rho_{Er,B}^j.$$

Evidently, for any ρ_{ABE} of the form (9.6.6), $\rho_{AB} := \mathrm{Tr}_3(\rho_{ABE})$ is separable. Thus, if one knew that the infimum in Definition 9.23 was attained, it would follow that if $\mathrm{E}_{\mathrm{sq}}(\rho_{AB}) = 0$, then ρ_{AB} is separable. However, it is not known whether the infimum in (9.6.2) is attained, and hence knowledge of the cases of equality in SSA is not enough to prove that if $\mathrm{E}_{\mathrm{sq}}(\rho_{AB}) = 0$, then ρ_{AB} is separable.

Nonetheless, it has been proved in [**41**] that E_{sq} is a faithful measure of entanglement. Together with Lemma 9.27, this proves that squashed entanglement is an entanglement monotone. Other proofs of faithfulness have been given in [**34**] and [**140**]. Here we follow [**140**].

Theorem 9.30. *Let $\epsilon > 0$ and $\rho_{AB} \in \mathfrak{S}_{\mathcal{H}_A \otimes \mathcal{H}_B}$ be such that $\mathrm{E}_{\mathrm{sq}}(\rho_{AB}) \leq \epsilon$. Then for each $k \in \mathbb{N}$, there exists a k-symmetrically extendable state σ_{AB} such that*

$$\|\rho_{AB} - \sigma_{AB}\|_1 \leq 2k\sqrt{\epsilon}.$$

Proof. Choose any extension ρ_{ABE} of ρ_{AB}. From Theorem 8.20,

$$\begin{aligned}
2\epsilon &\geq I_{AB|E}(\rho_{ABE}) \\
&= D(\rho_{ABE}\|\tau_B \otimes \rho_{AE}) - D(\rho_{BE}\|\tau_B \otimes \rho_E) \\
&\geq D^M(\rho_{ABE}\|(\mathbb{1}_A \otimes R)\rho_{AE}) \\
&\geq \tfrac{1}{2}\|\rho_{ABE} - (\mathbb{1}_A \otimes R)\rho_{AE}\|_1^2,
\end{aligned}$$

where R is a CP trace preserving map specified in (8.5.6). Therefore, defining $\omega_{AEB}^{(1)} := (\mathbb{1}_A \otimes R)\rho_{AE}$,

$$(9.6.7) \qquad \|\rho_{ABE} - \omega_{AEB}^{(1)}\|_1 \leq 2\sqrt{\epsilon}.$$

For each $k \in N$, inductively define a density matrix $\omega_{AEB^k}^{(k)}$ on $\mathcal{H}_A \otimes \mathcal{H}_B^{\otimes k}$ by and for $k > 1$,

$$\omega_{AEB^k}^{(k)} := (\mathbb{1}_{AB^{k-1}} \otimes R)\omega_{AEB^{k-1}}^{(k-1)}.$$

By (9.6.7) and that fact that the trace norm decreases under partial traces,

$$(9.6.8) \qquad \|\rho_{AE} - \omega_{AE}^{(1)}\| \leq 2\sqrt{\epsilon} \quad \text{and} \quad \|\rho_{AB} - \omega_{AB}^{(1)}\| \leq 2\sqrt{\epsilon}.$$

Then since $\omega_{AEB^2}^{(2)} = (\mathbb{1}_{AB} \otimes R)\omega_{AEB}^{(1)}$, tracing out the first factor on B,

$$\begin{aligned}
\omega_{AEB}^{(2)} &= (\mathbb{1}_A \otimes R)\omega_{AE}^{(1)} \\
&= (\mathbb{1}_A \otimes R)\rho_{AE} + (\mathbb{1}_A \otimes R)(\omega_{AE}^{(1)} - \rho_{AE}) \\
&= \rho_{ABE} + (\omega_{AEB}^{(1)} - \rho_{ABE}) + (\mathbb{1}_A \otimes R)(\omega_{AE}^{(1)} - \rho_{AE}).
\end{aligned}$$

Then by (9.6.7) and (9.6.8) and the fact that the trace norm decreases under quantum operations, and hence

$$\|\rho_{ABE} - \omega_{AEB}^{(2)}\|_1 \leq 4\sqrt{\epsilon}.$$

Using once more that the trace norm decreases under partial traces,

$$\|\rho_{AE} - \omega_{AE}^{(2)}\|_1 \leq 4\sqrt{\epsilon} \quad \text{and} \quad \|\rho_{AB} - \omega_{AB}^{(2)}\|_1 \leq 4\sqrt{\epsilon}.$$

Now a simple induction leads to, for all $k \in \mathbb{N}$,

$$\|\rho_{ABE} - \omega_{AEB}^{(k)}\|_1 \leq 2k\sqrt{\epsilon},$$

together with

(9.6.9) $$\|\rho_{AE} - \omega_{AE}^{(k)}\|_1 \leq 2k\sqrt{\epsilon} \quad \text{and} \quad \|\rho_{AB} - \omega_{AB}^{(k)}\|_1 \leq 2k\sqrt{\epsilon}.$$

Now let $\sigma_{AB^k}^{(k)}$ be the symmetrization of $\omega_{AB^k}^{(k)}$ on $\mathcal{H}_B^{\otimes k}$:

$$\sigma_{AB^k}^{(k)} = \frac{1}{k!} \sum_{\pi \in X_k} V_\pi^* \omega_{AB^k}^{(k)} V_\pi,$$

with the unitary permutation matrices defined as in (9.5.1). Finally, define $\sigma_{AB} := \mathrm{Tr}_{B^{k-1}}[\sigma_{AB^k}^{(k)}]$. Then $\sigma_{AB^k}^{(k)}$ is a k-symmetric extension of σ_{AB}, and by (9.6.9), $\|\rho_{AB} - \sigma_{AB}\|_1 \leq 2k\sqrt{\epsilon}$. $\qquad\square$

Corollary 9.31. *Suppose that* $\mathrm{E}_{\mathrm{sq}}(\rho_{AB}) \leq \epsilon < 1$. *Then there is a separable state* $\tilde{\sigma}_{AB}$ *such that*

(9.6.10) $$\|\rho_{AB} - \tilde{\sigma}_{AB}\|_1 \leq (2d_B^2 + 2)\epsilon^{1/4}.$$

In particular, if $\mathrm{E}_{\mathrm{sq}}(\rho_{AB}) = 0$, *then* ρ_{AB} *is separable.*

Proof. Suppose that $\mathrm{E}_{\mathrm{sq}}(\rho_{AB}) = 0$. By Theorem 9.30, for all $\epsilon > 0$ and all $k \in \mathbb{N}$, there exists σ_{AB} that is symmetrically k-extensible, and such that $\|\rho_{AB} - \sigma_{AB}\| \leq 2k\sqrt{\epsilon}$. Next by Corollary 9.22, there is a separable state $\tilde{\sigma}_{AB}$ such $\|\sigma_{AB} - \tilde{\sigma}_{AB}\|_1 \leq \frac{2d_B^2}{k+d_B^2}$. By the triangle inequality

(9.6.11) $$\|\rho_{AB} - \tilde{\sigma}_{AB}\|_1 \leq 2k\sqrt{\epsilon} + \frac{2d_B^2}{k + d_B^2}.$$

Choose k to be the greatest integer less than $\epsilon^{-1/4}$, which for $\epsilon < 1$ is at least 1. With this choice of k, (9.6.11) becomes (9.6.10). $\qquad\square$

9.6.4. Additivity of squashed entanglement.

Lemma 9.32. *Let $\mathcal{H}_A$, $\mathcal{H}_{A'}$, $\mathcal{H}_B$, $\mathcal{H}_{B'}$, and $\mathcal{H}_E$ be finite-dimensional Hilbert spaces. Then for any density matrix $\rho_{AA'BE}$,*

$$I_{AA',BB'|E}(\rho_{AA'BB'E}) \geq I_{AB|E}(\rho_{ABE}) + I_{A'B'|EA}(\rho_{AA'B'E}).$$

Proof. Using the chain rule, Lemma 9.26,

$$\begin{aligned}
I_{AA',BB'|E}(\rho_{AA'BB'E}) &= I_{A',BB'|AE}(\rho_{AA'BB'E}) + I_{A,BB'|E}(\rho_{ABB'E}) \\
&= I_{A',B'|ABE}(\rho_{AA'BE}) + I_{A',B|BE}(\rho_{AA'E}) \\
&\quad + I_{A,B'|BE}(\rho_{AA'BE}) + I_{A,B|E}(\rho_{ABE}) \\
&\geq I_{A',B'|ABE}(\rho_{AA'BE}) + I_{A,B|E}(\rho_{ABE}),
\end{aligned}$$

where the two discarded terms are nonnegative by SSA. $\qquad\square$

Theorem 9.33. *The squashed entanglement is subadditive in general, and additive over tensor products.*

Proof. Fix $\epsilon > 0$, and consider an extension $\rho_{AA'BB'E}$ of $\rho_{AA'BB'}$ such that $I_{AA',BB'|E}(\rho_{AA'BB'E}) \leq 2\,\mathrm{E}_{\mathrm{sq}}(\rho_{AA'BB'}) + \epsilon$. By the final inequality in the lemma above,

$$\begin{aligned}
2\,\mathrm{E}_{\mathrm{sq}}(\rho_{AA'}) + 2\,\mathrm{E}_{\mathrm{sq}}(\rho_{BB'}) &\leq I_{A',B'|ABE}(\rho_{AA'BE}) + I_{A,B|E}(\rho_{ABE}) \\
&\leq 2\,\mathrm{E}_{\mathrm{sq}}(\rho_{AA'BB'}) + \epsilon.
\end{aligned}$$

Now let ρ_{ABE} and $\rho_{A'B'E'}$ be extensions of ρ_{AB} and $\rho_{A'B'}$, respectively. Then $\rho_{ABE} \otimes \rho_{A'B'E'}$ is an extension of $\rho_{AB} \otimes \rho_{A'B'}$. We now estimate $I_{AA',BB'|EE'}(\rho_{ABE} \otimes \rho_{A'B'E'})$, as above. The two terms that we discarded are $I_{A,B'|BEE'}(\rho_{ABE} \otimes \rho_{B'E'})$ and $I_{A',B|AB'EE'}(\rho_{ABE} \otimes \rho_{A'B'E'})$, but now these are zero. $\qquad\square$

9.6.5. Continuity of squashed entanglement.
There is a useful alternate formula for $\mathrm{E}_{\mathrm{sq}}(\rho_{AB})$, also due to Christandl and Winter [64] who introduced it in an investigation of the continuity of the squashed entanglement.

Lemma 9.34. *Let $\mathcal{H}_A$, $\mathcal{H}_B$, and $\mathcal{H}_E$ be finite-dimensional Hilbert spaces. Let $\rho_{AB} \in \mathfrak{S}_{\mathcal{H}_{AB}}$. Let $\mathcal{H}_F$ be a Hilbert space of the same dimension as $\mathcal{H}_{AB}$, and let ψ be any purification of ρ_{AB} in $\mathcal{H}_{ABF}$. That is, $\mathrm{Tr}_F[|\psi\rangle\langle\psi|] = \rho_{AB}$.*

Let $\rho_{ABE} \in \mathfrak{S}_{\mathcal{H}_{ABE}}$ be such that $\mathrm{Tr}[\rho_{ABE}] = \rho_{AB}$. Then there is a quantum operation $\Phi : \mathcal{B}(\mathcal{H}_F) \to \mathcal{B}(\mathcal{H}_E)$ such that

$$\rho_{ABE} := (\mathbb{1} \otimes \Phi)(|\psi\rangle\langle\psi|).$$

Conversely, if $\Phi : \mathcal{B}(\mathcal{H}_F) \to \mathcal{B}(\mathcal{H}_E)$ is a quantum operation, then

$$(9.6.12) \qquad \mathrm{Tr}_E[(\mathbb{1} \otimes \Phi)(|\psi\rangle\langle\psi|)] = \rho_{AB}.$$

Consequently,

$$(9.6.13) \qquad \mathrm{E}_{\mathrm{sq}}(\rho_{AB}) = \frac{1}{2} \inf_{\Phi}\{I_{AB|E}((\mathbb{1} \otimes \Phi)(|\psi\rangle\langle\psi|))\},$$

where the infimum is taken over all quantum operations $\Phi : \mathcal{B}(\mathcal{H}_F) \to \mathcal{B}(\mathcal{H}_E)$.

Proof. Take $\mathcal{H}_G$ to be a Hilbert space of the same dimension as $\mathcal{H}_{ABC}$. Let $\phi \in \mathcal{H}_{ABEG}$ be a purification of ρ_{ABE}. Then since $\mathrm{Tr}_{EG}[|\phi\rangle\langle\phi|] = \rho_{AB}$, ϕ is also a purification of ρ_{AB}. Hence there exists a partial isometry $U \in \mathcal{B}(\mathcal{H}_F, \mathcal{H}_{EG})$ such that

$$(\mathbb{1}_{AB} \otimes U)\psi = \phi,$$

and since the dimension of $\mathcal{H}_{EG}$ is at least the dimension of $\mathcal{H}_F$, we may assume that the initial space of U is $\mathcal{H}_F$. Define $\Phi : \mathcal{B}(\mathcal{H}_F) \to \mathcal{B}(\mathcal{H}_E)$ by

$$\Phi(X) = \mathrm{Tr}_G[UXU^*].$$

Evidently Φ is completely positive, and $\mathrm{Tr}[\Phi(X)] = \mathrm{Tr}[UXU^*] = \mathrm{Tr}[U^*UX]$. Since the initial space of U is $\mathcal{H}_F$, $U^*U = \mathbb{1}_F$, and hence $\mathrm{Tr}[\Phi(X)] = \mathrm{Tr}[X]$ for all $X \in \mathcal{B}(\mathcal{H}_F)$. Thus, Φ is a quantum operation.

Next,

$$(\mathbb{1} \otimes \Phi)(|\psi\rangle\langle\psi|) = \mathrm{Tr}_G[|\mathbb{1}_{AB} \otimes U\psi\rangle\langle\mathbb{1}_{AB} \otimes U\psi|]$$
$$= \mathrm{Tr}_G[|\phi\rangle\langle\phi|] = \rho_{ABE}.$$

For the converse, suppose that $\Phi : \mathcal{B}(\mathcal{H}_F) \to \mathcal{B}(\mathcal{H}_E)$ is any quantum operation. Then for any $A \in \mathcal{B}(\mathcal{H}_{AB})$, since $\phi^\dagger$ is unital,

$$\mathrm{Tr}[(A^* \otimes \mathbb{1}_E)(\mathbb{1}_{AB} \otimes \Phi)(|\psi\rangle\langle\psi|)] = \mathrm{Tr}[(A^* \otimes \Phi^\dagger(\mathbb{1}_E))(|\psi\rangle\langle\psi|)]$$
$$= \mathrm{Tr}[(A^* \otimes \mathbb{1}_F)(|\psi\rangle\langle\psi|)] = \mathrm{Tr}[A^*\rho_{AB}].$$

This proves (9.6.12). Finally, (9.6.13) is an immediate consequence of the first two parts. $\qquad\square$

Lemma 9.35. *Fix* $\epsilon > 0$. *Let* $\rho_{AB}, \sigma_{AB} \in \mathfrak{S}_{\mathcal{H}_{AB}}$ *satisfy* $\|\rho_{AB} - \sigma_{AB}\|_1 \leq \epsilon$. *Let* $\mathcal{H}_E$ *be another finite-dimensional Hilbert space. For any* $\rho_{ABE} \in \mathfrak{S}_{\mathcal{H}_{ABE}}$ *such that* $\mathrm{Tr}_E[\rho_{ABE}] = \rho_{AB}$, *there exists* $\sigma_{ABE} \in \mathfrak{S}_{\mathcal{H}_{ABE}}$ *such that* $\mathrm{Tr}_E[\sigma_{ABE}] = \sigma_{AB}$ *and*

$$\|\rho_{ABE} - \sigma_{ABE}\|_1 \leq \sqrt{2\epsilon}.$$

Proof. Let $\mathcal{H}_F$ be a Hilbert space of the same dimension as $\mathcal{H}_{AB}$. By Theorem 6.53, there exist in $\mathcal{H}_{ABF}$ purifications ψ and ϕ of ρ_{AB} and σ_{AB}, respectively, such that

$$\| |\psi\rangle\langle\psi| - |\phi\rangle\langle\phi| \|_1 \leq \sqrt{2\epsilon}.$$

Since quantum operations are contractions in the trace norm, for any quantum operation $\Phi : \mathcal{B}(\mathcal{H}_F) \to \mathcal{B}(\mathcal{H}_E)$,

$$\| (\mathbb{1}_{AB} \otimes \Phi)(|\psi\rangle\langle\psi|) - (\mathbb{1}_{AB} \otimes \Phi)(|\phi\rangle\langle\phi|) \|_1 \leq \sqrt{2\epsilon}.$$

Now apply Lemma 9.34. $\qquad\qquad\qquad\qquad\qquad\qquad\qquad\qquad\qquad\qquad\qquad\square$

Theorem 9.36 (Continuity of squashed entanglement). *Let $\mathcal{H}_A$ and $\mathcal{H}_B$ be Hilbert spaces of finite dimensions d_A and d_B, respectively. Let $\rho_{AB}, \sigma_{AB} \in \mathfrak{S}_{\mathcal{H}_{AB}}$. Then for all $\epsilon > 0$,*

$$(9.6.14) \quad \|\rho_{AB} - \sigma_{AB}\|_1 \leq \epsilon$$

$$\Rightarrow |E_{\mathrm{sq}}(\rho_{AB}) - E_{\mathrm{sq}}(\sigma_{AB})| \leq 3(\eta(2\sqrt{2\epsilon}) + 3\sqrt{2\epsilon}\log(d_A d_B)).$$

Proof. The conditional information can be written in terms of conditional entropies, so that the Alicki-Fannes theorem can be applied. Since $S_{AB|E}(\rho_{ABE}) = S_{ABE} - S_E$, $S_{A|E}(\rho_{AE}) = S_{AE} - S_E$, and $S_{B|E} = S_{BE} - S_E$,

$$(9.6.15) \qquad I_{AB|E}(\rho_{ABE}) = S_{A|E}(\rho_{AE}) + S_{B|E}(\rho_{BE}) - S_{AB|E}(\rho_{ABE}).$$

Now suppose that $\|\rho_{AB} - \sigma_{AB}\|_1 \leq \epsilon$. By Lemma 9.35, for any $\mathcal{H}_E$ and any extension ρ_{ABE} of ρ_{AB}, there is an extension σ_{ABE} of σ_{AB} such that $\| \rho_{ABE} - \sigma_{ABE} \|_1 \leq \sqrt{2\epsilon}$. Since partial traces are contractions in the trace norm, $\| \rho_{AE} - \sigma_{AE} \|_1 \leq \sqrt{2\epsilon}$ and $\| \rho_{BE} - \sigma_{BE} \|_1 \leq \sqrt{2\epsilon}$ as well.

By the Alicki-Fannes theorem,

$$|S_{AB|E}(\rho_{ABE}) - S_{AB|E}(\sigma_{ABE})| \leq \eta(2\sqrt{2\epsilon}) + 3\sqrt{2\epsilon}\log(d_A d_B),$$

$$|S_{A|E}(\rho_{AE}) - S_{A|E}(\sigma_{AE})| \leq \eta(2\sqrt{2\epsilon}) + 3\sqrt{2\epsilon}\log(d_B),$$

$$|S_{B|E}(\rho_{BE}) - S_{B|E}(\sigma_{BE})| \leq \eta(2\sqrt{2\epsilon}) + 3\sqrt{2\epsilon}\log(d_B).$$

Now using (9.6.15),

$$|I_{AB|E}(\rho_{ABE}) - I_{AB|E}(\sigma_{ABE})| \leq 3(\eta(2\sqrt{2\epsilon}) + 3\sqrt{2\epsilon}\log(d_A d_B)).$$

Now let $\epsilon' > 0$ and chose ρ_{ABE} such that $I_{AB|E}(\rho_{ABE}) \leq \epsilon' + E_{\mathrm{sq}}(\rho_{AB})$. Then there exists an extension σ_{ABE} of σ_{AB} such that

$$I_{AB|E}(\sigma_{ABE}) \leq \epsilon' + E_{\mathrm{sq}}(\rho_{AB}) + 3(\eta(2\sqrt{2\epsilon}) + 3\sqrt{2\epsilon}\log(d_A d_B)).$$

Since $\epsilon' > 0$ is arbitrary,

$$E_{\mathrm{sq}}(\sigma_{AB}) \leq E_{\mathrm{sq}}(\rho_{AB}) + 3(\eta(2\sqrt{2\epsilon}) + 3\sqrt{2\epsilon}\log(d_A d_B)).$$

By symmetry, the same estimate holds with ρ_{AB} and σ_{AB} interchanged, and this proves (9.6.14). $\qquad\qquad\qquad\qquad\qquad\qquad\qquad\qquad\qquad\qquad\square$

9.7. Extended SSA

The following extension of SSA is proved in [**49**].

Theorem 9.37 (Extended SSA). *For all tripartite states* ρ_{123},

$$(9.7.1) \qquad S_{13} + S_{23} - S_{123} - S_3 \geq 2\max\{S_1 - S_{12}, S_2 - S_{12}, 0\}.$$

This has the immediate corollary that

$$(9.7.2) \qquad E_{sq}(\rho_{12}) \geq \max\{S_1 - S_{12}, S_2 - S_{12}, 0\}.$$

Therefore if either of the conditional entropies $S_{12} - S_1$ or $S_{12} - S_2$ is strictly negative, then $E_{sq}(\rho_{12}) > 0$, and ρ_{12} is entangled.

It is also shown in [**49**] that the factor of 2 on the right side cannot be replaced by any larger value. The argument uses a tripartite extension of a bipartite state that saturates the Araki-Lieb triangle inequality. This is natural since the proof of Theorem 9.37, which we give below, makes use of the purification arguments used to prove the Araki-Lieb triangle inequality. The construction yields a tripartite state that, when inserted into the definition (9.6.2), yields the upper bound

$$E_{sq}(\rho_{12}) \leq S_1 - S_{12} \qquad \text{with} \qquad S_1 - S_{12} > 0.$$

Then from the lower bound, (9.7.2), we have $E_{sq}(\rho_{12}) = S_1 - S_{12} > 0$. This proves that (9.7.2), and hence the factor of 2 in Theorem 9.37 is sharp. We refer to [**49**] for this construction. A weaker form of the inequality (9.7.2) was given by Christandl and Winter [**64**]; see [**49**] for a discussion.

Proof of Theorem 9.37. Associated to ρ_{123} are three bipartite density matrices, ρ_{12}, ρ_{23}, and ρ_{13}. Applying Lemma 7.55 together with SSA to the two pairs $\{\rho_{12}, \rho_{23}\}$ and $\{\rho_{13}, \rho_{23}\}$ yields

$$S_{12} + S_{23} \geq S_1 + S_3 \quad \text{and} \quad S_{13} + S_{23} \geq S_1 + S_2.$$

Adding these two inequalities yields

$$S_{12} + S_{13} + 2S_{23} \geq 2S_1 + S_2 + S_3.$$

Again consider any purification ρ_{1234} of ρ_{123}. Then $S_{12} = S_{34}$, yielding $S_{23} = S_{14}$, and $S_2 = S_{134}$,

$$S_{13} + S_{34} - S_{134} - S_3 \geq 2(S_1 - S_{14}),$$

which is (9.7.1) with different indices. $\qquad\square$

Exercises

(1) For a finite-dimensional Hilbert space $\mathcal{H}$, let $|\psi\rangle\langle\psi|$ and $|\phi\rangle\langle\phi|$ be two maximally entangled pure states on $\mathcal{H} \otimes \mathcal{H}$. Show that there exists a unitary U on $\mathcal{H}$ such that $|\psi\rangle\langle\psi| = (U \otimes \mathbb{1})|\phi\rangle\langle\phi|$.

(2) For a finite-dimensional Hilbert space $\mathcal{H}$, let $|\psi\rangle\langle\psi|$ and $|\phi\rangle\langle\phi|$ be two pure states on $\mathcal{H} \otimes \mathcal{H}$. Show that ψ and ϕ have the same Schmidt numbers if and only if there exists a local unitary operation $U \otimes V$ on $\mathcal{H} \otimes \mathcal{H}$ with U and V unitary on $\mathcal{H}$, such that $U \otimes V\psi = \phi$.

(3) Let $\{a_k\}$ be a sequence of nonnegative numbers with the property that for all m and n, $a_{m+n} \leq \frac{m}{m+n}a_m + \frac{n}{m+n}a_n$. (Such sequences are said to be **subadditive**.) Define another sequence $\{b_k\}$ by $b_k = a_{2^{-k}}$. Show that $\{b_k\}$ is monotone non-increasing so that $b := \lim_{k\to\infty} b_k$ exists. Then show $\limsup_{k\to\infty} a_k \leq b$ and $\liminf_{k\to\infty} a_k \geq b$. Finally, apply this to show that the limit in (9.3.4) exists.

(4) Show that E_F is convex on $\mathfrak{S}_{\mathcal{H}_A \otimes \mathcal{H}_B}$, as claimed below Definition 9.15.

(5) Show that for any bipartite state ρ_{12} such that $S_{12} = S_1 - S_2$, $E_{\mathrm{sq}}(\rho_{12}) = S_1 - S_{12}$.

(6) Show that there exist bipartite states ρ_{12} for which $S_1 - S_{12} = S_2$, and for which S_1 and S_{12}, and can have arbitrary nonnegative values.

(7) Let $\mathcal{H}$ be a Hilbert space of dimension at least N. Let $\{\mathbf{u}_1, \ldots, \mathbf{u}_N\}$ be orthonormal in $\mathcal{H}$. Define a density matrix $\gamma \in \mathcal{H}^{\otimes N}$ by

$$\gamma := |\mathbf{u}_1 \wedge \cdots \wedge \mathbf{u}_N\rangle\langle\mathbf{u}_1 \wedge \cdots \wedge \mathbf{u}_N|.$$

(Such a density matrix is known as an N-particle Slater determinant.) Define γ_{12} to be the 2-*particle reduced density matrix of* γ; that is,

$$\gamma_{12} := \mathrm{Tr}_{3,\ldots,N}[\gamma].$$

Using γ as an extension of γ_{12}, show that

$$E_{\mathrm{sq}}(\gamma_{12}) \leq \begin{cases} \ln\frac{N+2}{N} & \text{if } N \text{ is even} \\[2ex] \frac{1}{2}\ln\frac{N+3}{N-1} & \text{if } N \text{ is odd.} \end{cases}$$

Chapter 10

Convexity, concavity, and monotonicity

10.1. A tracial characterization of Schwarz maps

A number of important inequalities for completely positive unital maps are valid for the wider class of Schwarz maps, and their proofs in this general setting turn out to be particularly simple. Just as Choi's characterization of 2-positive maps is a powerful source of inequalities for 2-positive maps, a characterization of Schwarz maps [**52**] that is proved in this section leads directly to many useful inequalities.

By definition, $\Phi : M_n(\mathbb{C}) \to M_m(\mathbb{C})$ is a Schwarz map in case $\Phi(K)^*\Phi(K) \leq \Phi(K^*K)$ for all $K \in M_n(\mathbb{C})$. This can be expressed as a comparison of quadratic forms. Since $\Phi(K)\Phi(K)^* \leq \Phi(KK^*)$ if and only if

$$\mathrm{Tr}[X\Phi(K)^*\Phi(K)] \leq \mathrm{Tr}[X\Phi(K^*K)] = \mathrm{Tr}[\Phi^\dagger(X)K^*K]$$

for all $X \in M_m^+(\mathbb{C})$, Φ is a Schwarx map if and only if for all $K \in M_n(\mathbb{C})$ and all $X \in M_m^+(\mathbb{C})$,

$$(10.1.1) \qquad \langle \Phi(K), \Phi(K)X\rangle_{\mathcal{H}_m} \leq \langle K, K\Phi^\dagger(X)\rangle_{\mathcal{H}_n},$$

where $\mathcal{H}_k$ denotes the Hilbert space of $k\times k$ matrices equipped with the Hilbert-Schmidt inner product.

It will be useful to rewrite this in terms of the operators R_X and $R_{\Phi^\dagger(X)}$ on $\mathcal{H}_m$ and $\mathcal{H}_n$ of left multiplication by X and $\Phi^\dagger(X)$, respectively. By Theorem 4.54, $R_X \in \mathcal{B}^+(\mathcal{H}_m)$ and $R_{\Phi^\dagger(X)} \in \mathcal{B}^+(\mathcal{H}_m)$. Then (10.1.1) becomes

$$(10.1.2) \qquad \langle \Phi(K), R_X\Phi(K)\rangle \leq \langle K, R_{\Phi^\dagger(X)}K\rangle.$$

Of course there is a version of (10.1.3) involving left multiplication:

$$\langle K, R_{\Phi^\dagger(X)}K\rangle = \mathrm{Tr}[K^*K\Phi^\dagger(X)] = \mathrm{Tr}[K\Phi^\dagger(X)K^*] = \langle K^*, L_{\Phi^\dagger(X)}K^*\rangle.$$

Making a similar computation for the first term in (10.1.2), and then replacing K^* by K, leads to the conclusion that Φ is a Schwarz map if and only if for all $K \in M_n(\mathbb{C})$ and all $X \in M_m^+(\mathbb{C})$,

$$(10.1.3) \qquad\qquad \langle \Phi(K), L_X\Phi(K)\rangle \le \langle K, L_{\Phi^\dagger(X)}K\rangle.$$

Note that (10.1.2) and (10.1.3) are comparisons of positive semidefinite quadratic forms on $\mathcal{H}_n$. There is a duality theorem for such comparisons, and this will yield a useful characterization of Schwarz maps.

A form of this duality was first applied in this context by Petz [**178**]. He considered $T \in \mathcal{B}(\mathcal{H}, \mathcal{K})$, $\mathcal{H}, \mathcal{K}$ finite-dimensional Hilbert spaces together with $A \in \mathcal{B}^{++}(\mathcal{H})$ and $B \in \mathcal{B}^{++}(\mathcal{K})$. He observed that

$$\begin{aligned}
T^*BT \le A \;&\Longleftrightarrow\; (A^{-1/2}T^*B^{1/2})(B^{1/2}TA^{-1/2}) \le \mathbb{1}\\
&\Longleftrightarrow\; (B^{1/2}TA^{-1/2})(A^{-1/2}T^*B^{1/2}) \le \mathbb{1}\\
&\Longleftrightarrow\; TA^{-1}T^* \le B^{-1}.
\end{aligned}$$

The following theorem adapted from [**52**] uses duality to eliminate the need for A and B to be invertible.

Theorem 10.1. *Let $\mathcal{H}$ and $\mathcal{K}$ be two finite-dimensional Hilbert spaces, and let T be a linear transformation from $\mathcal{H}$ into $\mathcal{K}$. Let $A \in \mathcal{B}^+(\mathcal{H})$ and $B \in \mathcal{B}^+(\mathcal{K})$. Suppose that*

$$(10.1.4) \qquad\qquad \mathrm{ran}(T^*) \subseteq \mathrm{ran}(A) \quad and \quad \mathrm{ran}(T) \subseteq \mathrm{ran}(B).$$

Then

$$(10.1.5) \qquad\qquad \langle \mathbf{v}, T^*BT\mathbf{v}\rangle \le \langle \mathbf{v}, A\mathbf{v}\rangle \qquad for\ all \quad \mathbf{v} \in \mathcal{H}$$

if and only if

$$(10.1.6) \qquad\quad \langle \mathbf{w}, TA^+T^*\mathbf{w}\rangle \le \langle \mathbf{w}, B^+\mathbf{w}\rangle \qquad for\ all \quad \mathbf{w} \in \mathcal{K}.$$

Proof. In preparation for taking Legendre transforms, rewrite (10.1.5) as follows: for all $\mathbf{v} \in \mathcal{H}$ and $\mathbf{w} \in \mathcal{K}$,

$$(10.1.7) \qquad \mathfrak{R}\langle T^*(\mathbf{w}), \mathbf{v}\rangle - \frac{1}{2}\langle \mathbf{v}, A\mathbf{v}\rangle \le \mathfrak{R}\langle T^*(\mathbf{w}), \mathbf{v}\rangle - \frac{1}{2}\langle \mathbf{v}, T^*BT(\mathbf{v})\rangle.$$

Let P denote the orthogonal projection onto $\mathrm{ran}(A)$. Since $AA^+A = A$ and $A^+A = P$, when $\mathrm{ran}(T^*) \subseteq \mathrm{ran}(A)$, so that $PT^*(\mathbf{w}) = T^*(\mathbf{w})$,

$$(10.1.8) \quad \frac{1}{2}\langle (T^*(\mathbf{w}) - A\mathbf{v}), A^+(T^*(\mathbf{w}) - A\mathbf{v})\rangle$$

$$= \frac{1}{2}\langle T^*(\mathbf{w}), A^+T^*(\mathbf{w})\rangle + \frac{1}{2}\langle \mathbf{v}, A\mathbf{v}\rangle - \mathfrak{R}\langle T^*(\mathbf{w}), \mathbf{v}\rangle.$$

Again since $\mathrm{ran}(T^*) \subseteq \mathrm{ran}(A)$, there is some $\mathbf{v}$ such that $A\mathbf{v} = T^*(\mathbf{w})$ and then the left side of (10.1.8) is zero. Therefore, when $\mathrm{ran}(T^*) \subseteq \mathrm{ran}(A)$,

$$(10.1.9) \qquad \sup_{\mathbf{v}\in\mathcal{H}} \left\{ \Re\langle T^*\mathbf{w}, \mathbf{v}\rangle - \frac{1}{2}\langle \mathbf{v}, A\mathbf{v}\rangle \right\} = \frac{1}{2}\langle T^*\mathbf{w}, A^+ T^*\mathbf{w}\rangle .$$

Next suppose that $\mathrm{ran}(T) \subseteq \mathrm{ran}(B)$. Let Q be the orthogonal projection on $\mathrm{ran}(B)$. Taking adjoints, the right hand side of (10.1.7) becomes $\Re\langle \mathbf{w}, T\mathbf{v}\rangle - \frac{1}{2}\langle T\mathbf{v}, BT\mathbf{v}\rangle$, and since $B^+ B B^+ = B^+$ and $BB^+ = Q$,

$$(10.1.10) \quad \frac{1}{2}\langle (T\mathbf{v} - B^+\mathbf{w}), B(T(\mathbf{v}) - B^+\mathbf{w})\rangle$$

$$= \frac{1}{2}\langle T\mathbf{v}, B^+ T\mathbf{v}\rangle + \frac{1}{2}\langle \mathbf{w}, B^+\mathbf{w}\rangle - \Re\langle \mathbf{w}, T\mathbf{v}\rangle .$$

Since the left side of (10.1.10) is always nonnegative,

$$(10.1.11) \qquad \sup_{\mathbf{v}\in\mathcal{H}} \left\{ \Re\langle T^*\mathbf{w}, \mathbf{v}\rangle - \frac{1}{2}\langle \mathbf{v}, T^* BT\mathbf{v}\rangle \right\} \leq \frac{1}{2}\langle \mathbf{w}, B^+\mathbf{w}\rangle .$$

When (10.1.5) is valid, the supremum in (10.1.9) is no greater than the supremum in (10.1.11), and this proves (10.1.6). The proof in the other direction follows the same lines. $\qquad\square$

In our applications, $\mathcal{H}$ and $\mathcal{K}$ will be $\mathcal{H}_p$ and $\mathcal{H}_q$ for some p and q, using the notation from the beginning of Section 5.4, and T will be either Ψ or $\Psi^\dagger$, where $\Psi : \mathcal{H}_p \to \mathcal{H}_q$ is a positive map. The following simple lemma from [52] will justify application of Theorem 10.1.

Lemma 10.2. *Let* $\Psi : M_n(\mathbb{C}) \to M_m(\mathbb{C})$ *be a positive linear map, and let* $X \in M_n^{++}(\mathbb{C})$. *Let* P *denote the orthogonal projection onto* $\mathrm{ran}(\Psi(\mathbb{1}))$. *Then*

$$(10.1.12) \qquad \mathrm{ran}(R_{\Psi(X)}) = \mathrm{ran}(R_P) \quad and \quad \mathrm{ran}(L_{\Psi(X)}) = \mathrm{ran}(L_P).$$

Moreover,

$$(10.1.13) \qquad \mathrm{ran}(\Psi) \subseteq \mathrm{ran}(R_P) \quad and \quad \mathrm{ran}(\Psi) \subseteq \mathrm{ran}(L_P).$$

Proof. Since $X \in M_n^{++}(\mathbb{C})$, $\lambda\mathbb{1} \leq X \leq \mu\mathbb{1}$, where λ and μ are the least and greatest eigenvalues of X. Since Ψ is positive, $\lambda\Psi(\mathbb{1}) \leq \Psi(X) \leq \mu\Psi(\mathbb{1})$. Therefore $\mathrm{ran}(\Psi(X)) = \mathrm{ran}(\Psi(\mathbb{1}))$, and hence

$$(10.1.14) \qquad\qquad \Psi(X)^+ \Psi(X) = P.$$

Furthermore, by Lemma 5.2, for all $K \in M_n(\mathbb{C})$,

$$(10.1.15) \qquad\qquad P\Psi(K) = \Psi(K) = \Psi(K)P.$$

To see that $\mathrm{ran}(R_{\Psi(X)}) \subseteq \mathrm{ran}(R_P)$, let $K \in M_n(\mathbb{C})$. Then

$$R_{\Psi(X)}(K) = K\Psi(X) = K\Psi(X)P = R_P(K\Psi(X)),$$

where the second equality is from (10.1.15).

To see that $\operatorname{ran}(R_P) \subseteq \operatorname{ran}(R_{\Psi(X)})$, let $K \in M_n(\mathbb{C})$. Then

$$R_P(K) = KP = K\Psi(X)^+\Psi(X) = R_{\Psi(X)}(K\Psi(X)^+),$$

where the second equality is from (10.1.14). This proves the first equation in (10.1.12).

Next, for all $K \in M_n(\mathbb{C})$, again using (10.1.15),

$$\Psi(K) = \Psi(K)P = R_P(K),$$

which shows that $\operatorname{ran}(\Psi) \subseteq \operatorname{ran}(R_{\Psi(X)})$. This proves the first equation in (10.1.13). The statements about left multiplication are proved the same way. $\qquad\square$

The following is proved in [**52**].

Theorem 10.3. *A positive linear map* $\Phi : M_n(\mathbb{C}) \to M_m(\mathbb{C})$ *is a Schwarz map if and only if for all* $K \in M_n(\mathbb{C})$ *and* $X \in M_n^+(\mathbb{C})$ *such that* $\ker(X) \subseteq \ker(K)$,

$$(10.1.16)\qquad \operatorname{Tr}[K^*X^+K] \geq \operatorname{Tr}[\Phi(K)^*\Phi(X)^+\Phi(K)].$$

Proof. Fix $X \in M_m^{++}(\mathbb{C})$. By Theorem 4.54, L_X is an invertible operator on $M_m(\mathbb{C})$. To apply Theorem 10.1 to (10.1.3), let $\mathcal{H} := \mathcal{H}_n$, $\mathcal{K} := \mathcal{H}_m$, $A := L_{\Phi^\dagger(X)}$, $B := L_X$, and $T := \Phi$. Since L_X is invertible, $\operatorname{ran}(B) = M_m(\mathbb{C})$, and the second condition in (10.1.4) is trivially satisfied. The first condition is $\operatorname{ran}(\Phi^\dagger) \subseteq \operatorname{ran}(L_{\Phi^\dagger(X)})$, and this is valid by Lemma 10.2.

Now suppose that (10.1.16) is valid whenever $\ker(X) \subseteq \ker(K)$. Then it is valid whenever $X \in M_m^{++}(\mathbb{C})$, and for our specifications of A, B, and T, this becomes (10.1.5), which by Theorem 10.1 implies (10.1.3). Therefore, Φ is a Schwarz map.

Now suppose that Φ is a Schwarz map. Then (10.1.3) is valid, and this is the same as (10.1.5) for our specifications of A, B, and T. Then by Theorem 10.1, (10.1.6) is valid, and for our specifications of A, B, and T, this becomes

$$(10.1.17)\qquad \langle \Phi^\dagger(K), \Phi^\dagger(X)^+\Phi^\dagger(K)\rangle \leq \langle K, X^{-1}K\rangle \qquad \text{for all}\quad K \in \mathcal{H}_m,$$

since X is invertible. Now for general $X \in M_n^+(\mathbb{C})$ and $\epsilon > 0$, define $X_\epsilon := X + \epsilon\mathbb{1}$. With X_ϵ in place of X, (10.1.17) is valid for all $\epsilon > 0$. Since

$$\lim_{\epsilon \to 0}\langle K, X_\epsilon^{-1}K\rangle = \begin{cases} \langle K, X^+K\rangle & \ker(X) \subseteq \ker(K) \\ +\infty & \text{otherwise,} \end{cases}$$

(10.1.16) is valid whenever $\ker(X) \subseteq \ker(K)$. $\qquad\square$

Corollary 10.4. *A positive linear map* $\Phi : M_n(\mathbb{C}) \to M_m(\mathbb{C})$ *is a Schwarz map if and only if for all* $K \in M_n(\mathbb{C})$ *and* $X \in M_n^+(\mathbb{C})$ *such that* $\ker(X) \subseteq \ker(K)$,

$$(10.1.18) \qquad \langle K, L_X^+ K \rangle_{\mathcal{H}_n} \geq \langle \Phi(K), L_{\Phi(X)}^+ \Phi(K) \rangle_{\mathcal{H}_m}$$

and

$$(10.1.19) \qquad \langle K, R_X^+ K \rangle_{\mathcal{H}_n} \geq \langle \Phi(K), R_{\Phi(X)}^+ \Phi(K) \rangle_{\mathcal{H}_m}.$$

Proof. As explained at the beginning of this section, (10.1.18) and (10.1.19) are simply different ways of writing (10.1.16). $\qquad\square$

10.2. Inequalities for Schwarz maps

10.2.1. Uhlmann's theorem for Schwarz maps.
A theorem proved by Uhlmann [**215**] provides a trace inequality, valid for all Schwarz maps, that has the DPI as an immediate corollary among its other important consequences.

Theorem 10.5 (Uhlmann's theorem). *Let* $\Phi : M_n(\mathbb{C}) \to M_m(\mathbb{C})$ *be a Schwarz map. Let* $0 < t < 1$ *and* $m, n \in \mathbb{N}$.

(1) *For all* $K \in M_n(\mathbb{C})$, $X, Y \in M_m^+(\mathbb{C})$,

$$(10.2.1) \qquad \mathrm{Tr}[\Phi(K)^* Y^{1-t} \Phi(K) X^t] \leq \mathrm{Tr}[K^* \Phi^\dagger(Y)^{1-t} K \Phi^\dagger(X)^t].$$

(2) *For all* $X, Y \in M_m^{++}(\mathbb{C})$, *all* $K \in M_m(\mathbb{C})$,

$$(10.2.2) \qquad \mathrm{Tr}[\Phi^\dagger(K^*)\Phi^\dagger(Y)^{t-1}\Phi^\dagger(K)\Phi^\dagger(X)^{-t}] \leq \mathrm{Tr}[K^* Y^{t-1} K X^{-t}].$$

Proof. Let $\Phi : M_n(\mathbb{C}) \to M_m(\mathbb{C})$ be a Schwarz map. Then by Corollary 10.4, for all $\lambda > 0$, all $X, Y \in M_m^{++}(\mathbb{C})$ and all $K \in M_n(\mathbb{C})$,

$$\langle \Phi(K), (\lambda R_{\Phi(X)}^+ + L_{\Phi(Y)}^+)\Phi(K) \rangle_{\mathcal{H}_m} \leq \langle K, (\lambda R_X^+ + L_Y^+)K \rangle_{\mathcal{H}_n}.$$

By Lemma 10.2 and Theorem 10.1, for all $K \in M_m(\mathbb{C})$,

$$\langle \Phi, (\lambda R_{X^{-1}} + L_{Y^{-1}})^{-1} \Phi(K) \rangle \leq \langle K, (\lambda R_{\Phi^\dagger(X)}^+ + L_{\Phi^\dagger(Y)})^+ K \rangle,$$

and this can be written as

$$(10.2.3) \qquad \left\langle \Phi(K), \left(\frac{R_X L_Y^{-1}}{\lambda + R_X L_Y^{-1}} L_Y \right) \Phi(K) \right\rangle$$

$$\leq \left\langle K, \left(\frac{R_{\Phi^\dagger(X)} L_{\Phi^\dagger(Y)}^+}{\lambda + R_{\Phi^\dagger(X)} L_{\Phi^\dagger(Y)}^+} L_{\Phi^\dagger(Y)} \right) K \right\rangle.$$

Recall the integral representation $x^t = \frac{\sin(\pi t)}{\pi} \int_0^\infty \lambda^{t-1} \left(\frac{x}{\lambda + x} \right) d\lambda$. Multiplying both sides of (10.2.3) by $\frac{\sin(\pi t)}{\lambda}^{t-1}$ and integrating yields

$$\langle \Phi(K), (R_X L_Y^{-1})^t L_Y \Phi(K) \rangle \leq \langle K, (R_{\Phi^\dagger(X)} L_{\Phi^\dagger(Y)}^+)^t L_{\Phi^\dagger(Y)} K \rangle.$$

Since for all $K \in M_m(\mathbb{C})$, $(R_X L_Y^{-1})^t L_Y(K) = Y^{1-t} K X^t$, and similarly for $(R_{\Phi^\dagger(X)} L_{\Phi^\dagger(Y)}^+)^t L_{\Phi^\dagger(Y)}$, this last inequality is the same as (10.2.1). This proves (1).

Next, (2) is obtained by applying Theorem 10.1 to (1) which is a comparison of quadratic forms. To apply Theorem 10.1 to (1), take $\mathcal{H}$ and $\mathcal{K}$ to be $M_m(\mathbb{C})$ and $M_n(\mathbb{C})$, respectively, equipped with the Hilbert–Schmidt inner product. Take $T := \Phi$, and let $A \in \mathcal{B}(\mathcal{H}_n)$ and $B \in \mathcal{B}(\mathcal{H}_m)$ be given by

$$A := L_{\Phi^\dagger(Y)}^{1-t} R_{\Phi^\dagger(X)}^t \quad \text{and} \quad B := L_Y^{1-t} R_X^t.$$

As before, by Lemma 10.2 the condition (10.1.4) is satisfied for these choices, and application of Theorem 10.1 then yields (10.2.2). $\qquad\square$

Part (1) of Theorem 10.5 was proved by Uhlmann [215] in this form. Part (2) is due to Hiai and Petz [108] under slightly more restrictive conditions, and with further restrictions by Petz [178]. For this version, see [52].

One application of Uhlmann's inequality (10.2.1), made by Uhlmann himself, is to prove an extended DPI.

Theorem 10.6 (DPI for Schwarz maps). *Let $\Phi : M_n(\mathbb{C}) \to M_m(\mathbb{C})$ be a unital Schwarz map. Then for all density matrices ρ, σ in $M_m(\mathbb{C})$ with $\sigma > 0$, $D(\Phi^\dagger(\rho)\|\Phi^\dagger(\sigma)) \leq D(\rho\|\sigma)$.*

Proof. Take $K = \mathbb{1}$. Since Φ is unital, (10.2.1) reduces to

$$\mathrm{Tr}[\rho^{1-t}\sigma^t] \leq \mathrm{Tr}[\Phi^\dagger(\rho)^{1-t}\Phi^\dagger(\sigma)^t].$$

Since $\Phi^\dagger$ is trace preserving, $1 = \mathrm{Tr}[\rho] = \mathrm{Tr}[\Phi^\dagger(\rho)]$, and hence for all $t > 0$,

$$\frac{\mathrm{Tr}[\rho^{1-t}\sigma^t] - \mathrm{Tr}[\rho]}{t} \leq \frac{\mathrm{Tr}[\Phi^\dagger(\rho)^{1-t}\Phi^\dagger(\sigma)^t] - \mathrm{Tr}[\Phi^\dagger(\rho)]}{t}.$$

Taking the limit $t \downarrow 0$, we obtain

$$\mathrm{Tr}[\rho(\log\sigma - \log\rho)] \leq \mathrm{Tr}[\Phi^\dagger(\rho)(\log\Phi^\dagger(\sigma) - \log\Phi^\dagger(\rho))],$$

and this is the same as $D(\Phi^\dagger(\rho)\|\Phi^\dagger(\sigma)) \leq D(\rho\|\sigma)$. $\qquad\square$

The following lemma produces another pair of useful inequalities from Theorem 10.5.

Lemma 10.7. *For all $X, Y \in M_n^+(\mathbb{C})$, define $B_{Y,X} \in \mathcal{B}(\mathcal{H}_m)$ by*

$$(10.2.4) \qquad B_{Y,X}(K) := \int_0^1 Y^{1-t} K X^t \mathrm{d}t.$$

Then for $X, Y \in M_n^{++}(\mathbb{C})$, $B \in \mathcal{B}^{++}(\mathcal{H}_n)$ and

$$(10.2.5) \qquad B_{Y,X}^{-1}(K) = \int_0^\infty \frac{1}{\lambda+Y} K \frac{1}{\lambda+X} \mathrm{d}\lambda.$$

Proof. Let $\{\mathbf{u}_1, \ldots, \mathbf{u}_n\}$ and $\{\mathbf{v}_1, \ldots, \mathbf{v}_n\}$ be orthonormal bases of $\mathbb{C}^n$ consisting of eigenvectors of X and Y, respectively, with $X\mathbf{u}_j = \lambda_j \mathbf{u}_j$ and $Y\mathbf{v}_i = \mu_i \mathbf{v}_i$ for all i, j. Define $E_{i,j} := |\mathbf{v}_i\rangle\langle\mathbf{u}_j|$, $1 \leq i, j \leq n$, yielding an orthonormal basis of $\mathcal{H}_n$.

$$B_{Y,X}(E_{i,j}) = \left(\int_0^1 \mu_i^{1-t}\lambda_j^t \, dt\right) E_{i,j} = \Lambda(\lambda_j, \mu_i) E_{i,j},$$

where for $a, b > 0$, $\Lambda(a, b)$ is the logarithmic mean of a and b, which is given by

$$\Lambda(a, b) = \begin{cases} (a - b)/(\log a - \log b) & a \neq b \\ a & a = b. \end{cases}$$

This shows that $B_{Y,X} \in \mathcal{B}^{++}(\mathcal{H}_n)$ when $X, Y \in M_n^{++}(\mathbb{C})$, and then a simple calculation shows that for $\lambda_j, \mu_i > 0$,

$$\int_0^\infty \frac{1}{\lambda + Y} E_{i,j} \frac{1}{\lambda + X} \, d\lambda = \frac{1}{\Lambda(\lambda_j, \mu_i)} E_{i,j},$$

which proves (10.2.5). $\qquad\qquad\square$

Theorem 10.8. *Let* $\Phi : M_n(\mathbb{C}) \to M_m(\mathbb{C})$ *be a Schwarz map. Let* $0 < t < 1$ *and* $m, n \in \mathbb{N}$. *Then with* $B_{Y,X}$ *as defined in* (10.2.4),

(1) *For all* $K \in M_n(\mathbb{C})$, $X, Y \in M_m^+(\mathbb{C})$,

$$(10.2.6) \qquad \mathrm{Tr}[\Phi(K)^* B_{Y,X}(\Phi(K))] \leq \mathrm{Tr}[K^* B_{\Phi^\dagger(Y), \Phi^\dagger(X)}(K)].$$

(2) *For all* $X, Y \in M_m^{++}(\mathbb{C})$, *all* $K \in M_m(\mathbb{C})$,

$$(10.2.7) \qquad \mathrm{Tr}[\Phi^\dagger(K)^* B_{\Phi^\dagger(Y), \Phi^\dagger(X)}(\Phi^\dagger(K))] \leq \mathrm{Tr}[K^* B_{Y,X} K].$$

Proof. Integrating in t over $[0, 1]$ on both sides of (10.2.1) in Uhlmann's theorem yields (10.2.6). Then applying Theorem 10.1 and Lemma 10.7 yields (10.2.7). $\qquad\square$

10.2.2. Close relatives of Uhlmann's theorem. Many additional dual pairs

of inequalities may be deduced from Theorem 10.1 together with Theorem 10.3 by proceeding exactly as in the proof of Theorem 10.5. The full family is discussed in [**108**]; we briefly discuss a key idea from this paper in the present framework.

Simply replace the function $f(x) = x^t$ in the proof of Theorem 10.5 by some other postive function g on $(0, \infty)$ having an integral representation of the form

$$(10.2.8) \qquad g(x) := a + bx + \int_0^\infty \frac{x}{\lambda + x} d\mu(\lambda)$$

for some finite Borel measure μ on $(0, \infty)$ and $b \geq 0$. This differs from the integral representation for $f(x) = x^t$ only in the form of μ and the affine term $a + bx$. Then

$$(10.2.9) \qquad g(R_X L_Y^{-1}) L_Y = a L_Y + b R_X + \int_0^\infty \frac{R_X}{\lambda + R_X L_Y^{-1}} d\mu(\lambda),$$

with a similar expression for $g(R_{\Phi^\dagger(X)} L_{\Phi^\dagger(Y)}^+) L_{\Phi^\dagger(Y)}$.

Then exactly as in the proof of Theorem 10.5, up to the affine term that was not present there, for all $X, Y \in M_m^{++}(\mathbb{C})$, all $K \in M_n(\mathbb{C})$ and all Schwarz maps $\Phi : M_n(\mathbb{C}) \to M_m(\mathbb{C})$,

$$(10.2.10) \qquad \langle \Phi(K)^*, g(R_X L_Y^{-1}) L_Y \Phi(K) \rangle \leq \langle K, g(R_{\Phi^\dagger(X)} L_{\Phi^\dagger(Y)}^+) L_{\Phi^\dagger(Y)} K \rangle.$$

The simple argument that this is true for $g(x) = a + bx$, $a, b \geq 0$, is left as an exercise.

Since $X, Y \in M_m^{++}(\mathbb{C})$, $(R_X L_Y^{-1}) L_Y$ is invertible by Theorem 4.54 since by (10.2.9), $g(x) > 0$ for $x > 0$. Therefore, $\mathrm{ran}((R_X L_Y^{-1}) L_Y) = M_m(\mathbb{C})$. As in the proof of Theorem 10.5, one readily checks that

$$\mathrm{ran}(g(R_{\Phi^\dagger(X)} L_{\Phi^\dagger(Y)}^+) L_{\Phi^\dagger(Y)}) = \mathrm{ran}(R_{\Phi^\dagger(X)}) \cap \mathrm{ran}(L_{\Phi^\dagger(Y)}),$$

and that this contains $\mathrm{ran}(\Phi^\dagger)$. Application of Theorem 10.1 yields the inequality

$$(10.2.11) \ \langle \Phi^\dagger(K), \left(g(R_{\Phi^\dagger(X)} L_{\Phi^\dagger(Y)}^+) L_{\Phi^\dagger(Y)}\right)^+ \Phi^\dagger(K) \rangle \leq \langle K, (g(R_X L_Y^{-1}) L_Y)^{-1} K \rangle,$$

valid for all $X, Y \in M_m^{++}(\mathbb{C})$, $K \in M_n(\mathbb{C})$, and all Schwarz maps $\Phi : M_n(\mathbb{C}) \to M_m(\mathbb{C})$.

Thus one has the dual pair of inequalities (10.2.10) and (10.2.11) whenever the positive function g has an integral representation of the form (10.2.8). Hiai and Petz [**108**] showed, using Löwner's theorem, that this condition on g is not only sufficient, but also necessary.

10.2.3. Monotone metrics. Let $\rho(t)$ be a smooth path in the space $\mathfrak{S}^\circ$ of invertible density matrices on a finite-dimensional Hilbert space $\mathcal{H}$. Let $\mathcal{V}$ denote the real vector space of self-adjoint traceless matrices K. For each t, the derivative $\dot{\rho}(t) = K(t)$ belongs to $\mathcal{V}$, and so we may think of $\mathfrak{S}^\circ$ as a differentiable manifold, and identify the tangent space $\mathcal{V}_\rho$ at each $\rho \in \mathfrak{S}^\circ$ with $\mathcal{V}$. Note that $\mathcal{V}_\rho$ does not depend on ρ. A **Riemannian metric** on $\mathfrak{S}^\circ$ is then a smooth function $\rho \mapsto \gamma_\rho$ from $\mathfrak{S}^\circ$ to the space of positive definite quadratic forms on $\mathcal{V}$. Then the distance between $\rho_1, \rho_2 \in \mathfrak{S}^\circ$, $d_\gamma(\rho_1, \rho_2)$, is

$$d_\gamma(\rho_1, \rho_2) = \inf_{\rho(\cdot) \in \mathcal{P}_{\rho_1, \rho_2}} \left\{ \int_0^1 \sqrt{\gamma_{\rho(t)}(\dot{\rho}(t), \dot{\rho}(t))} dt \right\},$$

where $\mathcal{P}_{\rho_1, \rho_2}$ denotes the set of smooth paths $\rho(\cdot)$ in $\mathfrak{S}^\circ$ with $\rho(0) = \rho_0$ and $\rho(1) = \rho_1$.

A Riemannian metric on $\mathfrak{S}^\circ$ is a **monotone metric** in case for all quantum operations $\Phi^\dagger$ on $\mathcal{B}(\mathcal{H})$,

$$(10.2.12) \qquad \gamma_{\Phi^\dagger(\rho)}(\Phi^\dagger(K), \Phi^\dagger(K)) \leq \gamma_\rho(K, K).$$

Then of course, for all $\rho_1, \rho_2 \in \mathfrak{S}^\circ$, $d(\Phi^\dagger(\rho_1), \Phi^\dagger(\rho_1)) \leq d(\rho_1, \rho_2)$. For such a metric, any quantum operation performed on the states can only bring them closer together.

Such metrics have applications, not only in the quantum setting (e.g., [**51**]) but also the classical setting, in which the analogue of $\mathfrak{S}^\circ$ is the set $\mathfrak{S}^\circ_c$ of strictly positive probability densities p on some finite set $\mathcal{X}$ with cardinality $|\mathcal{X}|$. At each $p \in \mathfrak{S}^\circ_c$, identify the tangent space with the space $\mathcal{V}_c$ of real valued functions k on $\mathcal{X}$ such that $\sum_{x \in \mathcal{X}} k(x) = 0$. The analogues of quantum operations are **stochastic maps** P; i.e., elements of $M_{|\mathcal{X}|}(\mathbb{R})$ with nonnegative entries $P_{x,y}$ and the property that for each $y \in \mathcal{X}$, $\sum_{x \in \mathcal{X}}^n P_{x,y} = 1$. We seek the set of metrics γ on $\mathfrak{S}_c$ such that

$$(10.2.13) \qquad \gamma_{Pp}(Pk, Pk) \leq \gamma_p(k, k)$$

for all stochastic P at each $p \in \mathfrak{S}^\circ_c$ and for each tangent vector k. In 1982, Čencov [**57**], building on earlier work of Fisher [**84**], proved that the unique such metric, up to a constant multiple, is the **Fisher information**,

$$(10.2.14) \qquad \gamma_p(k, k) = \sum_{x \in \mathcal{X}} \frac{k^2(x)}{p(x)}.$$

A decade later, Čencov and Morozova [**156**] took up the quantum problem. It is natural to expect that in the quantum setting, monotone metrics would be given by some quantum version of (10.2.14), although in the quantum setting, there are many possible ways to *divide by* $\rho \in \mathfrak{S}^\circ$. Čencov and Morozova came up with several conjectures, all eventually shown to be correct, that certain explicit metrics were in fact monotone metrics, although they did not resolve any of these conjectures. An account of their work can be found in [**178**], where a large class of examples is given. We have several examples at hand. The first is $\gamma_\rho(K, K) = \mathrm{Tr}[K\rho^{-1}K]$, which has the property (10.2.12) by Theorem 10.3.

More generally, for each $t \in [0,1]$,

$$\tilde{\gamma}_\rho^{(t)}(K,K) = \mathrm{Tr}[K\rho^{t-1}K\rho^{-t}]$$

has the property (10.2.12) by part (2) of Theorem 10.5. See [139] for the connection between monotone metrics and quantum entropy. This topic is further developed in the exercises.

10.3. The Wigner-Yanase theorem and the conjectures it inspired

A number of important developments in the theory of trace function inequalities have their roots in work of Wigner and Yanase [230,231], which in turn has its origins in a paper of Araki and Yanase [16]. Araki and Yanase showed that the precision with which a quantum observable can be measured is limited by the extent to which it fails to commute with conserved observables such as the energy.

Wigner and Yanase investigated the implications of this for measuring the amount of information in a quantum state ρ. They developed a list of axioms for a measure of information, and in the case of a quantum mechanical density matrix ρ and a self-adjoint operator H, representing some conserved quantity such as the energy, they proposed [230, Equation 2] which they called the **skew information**,

$$I(\rho,H) := -\frac{1}{2}\mathrm{Tr}[[\rho^{1/2},H]^2] = \mathrm{Tr}[H^2\rho] - \mathrm{Tr}[H\rho^{1/2}H\rho^{1/2}].$$

Wigner and Yanase proved that for fixed H, $\rho \mapsto I(\rho,H)$ is convex, or equivalently that $\rho \mapsto \mathrm{Tr}[H\rho^{1/2}H\rho^{1/2}]$ is concave. Since this function is homogenous of degree one, its concavity does not depend on the normalization of ρ (see Exercise 4, Chapter 6), and we may consider arbitrary $X, Y \in M_n^{++}$ and $K \in M_n(\mathbb{C})$, and form the block matrices

$$(10.3.1) \qquad \rho := \begin{bmatrix} Y & 0 \\ 0 & X \end{bmatrix} \quad \text{and} \quad H := \begin{bmatrix} 0 & K \\ K^* & 0 \end{bmatrix}$$

and then $\mathrm{Tr}[H\rho^{1/2}H\rho^{1/2}] = 2\,\mathrm{Tr}[K^*Y^{1/2}KX^{1/2}]$. Thus, the following is an equivalent formulation of the original Wigner-Yanase theorem:

Theorem 10.9 (Wigner-Yanase theorem). *For all $K \in M_n(\mathbb{C})$, the function*

$$(X,Y) \mapsto \mathrm{Tr}[K^*Y^{1/2}KX^{1/2}]$$

is concave on $M_n^+(\mathbb{C}) \times M_n^+(\mathbb{C})$.

In the final paragraph of [230], Wigner and Yanase wrote that they were not sure that their proposed measure of information is the only one satisfying the axioms they specified, and they reported that they had also considered $-\frac{1}{2}\mathrm{Tr}[[\log\rho,H][\rho,H]]$. They also remark that Freeman Dyson pointed out that these cases fall into a one parameter family of candidates,

$$-\frac{1}{2}\operatorname{Tr}[[\rho^t, H][\rho^{1-t}, H]],$$

where $0 < t \le 1/2$, since $\operatorname{Tr}[[\log\rho, H][\rho, H]] = \lim_{t\downarrow 0}\frac{1}{t}\operatorname{Tr}[[\rho^t, H][\rho^{1-t}, H]]$. Then because

$$-\frac{1}{2}\operatorname{Tr}[[\rho^t, H][\rho^{1-t}, H]] = \operatorname{Tr}[H^2\rho] - \operatorname{Tr}[H\rho^{1-t}H\rho^t],$$

the implicit conjecture suggested by Dyson is that for fixed H, $\rho\mapsto\operatorname{Tr}[H\rho^{1-t}H\rho^t]$ is concave. The *doubling argument* used near (10.3.1) shows that it is the same to conjecture that for all $K \in M_n(\mathbb{C})$, the function $(X, Y) \mapsto \operatorname{Tr}[K^*Y^{1-t}KX^t]$ is concave on $M_n^+(\mathbb{C}) \times M_n^+(\mathbb{C})$.

There is no published statement of such an explicit conjecture by Wigner, Yanase or even Dyson, and Wigner and Yanase wrote that these other candidates have *undesirable* features, and appear to have been dismissive of their further study.

Nonetheless, Jost gave a partially alternate proof [123] of Theorem 10.9 that was written on a visit he made to the I.A.S. in Princeton during the fall of 1968. He thanks his friend Freeman Dyson for many discussions on the subject, and mentions the generalized cases proposed by Dyson. Jost does not ascribe any explicit conjecture to Dyson, but it would appear that five years after the work of Wigner and Yanase, Dyson felt that his proposed functionals were worthy of investigation. In any case, this conjecture became known as the Wigner-Yanase-Dyson conjecture, and was finally proved in 1973 by Elliott Lieb in a ground-breaking paper [141]. This fundamental result is the **Lieb concavity theorem**:

Theorem 10.10. *For all $K \in M_n(\mathbb{C})$ and all $0 \le s, t, s + t \le 1$,*

$$(X, Y) \mapsto \operatorname{Tr}[K^*Y^{1-t}KX^t]$$

is concave on $M_n^+(\mathbb{C}) \times M_n^+(\mathbb{C})$.

Before turning to the proof and consequences of Theorem 10.10, we discuss another conjectured generalization of the Wigner-Yanase theorem. Lieb conjectured [141, p. 282] that for each fixed $B \in M_n(\mathbb{C})$ and each $m \in \mathbb{N}$, the function $X \mapsto \operatorname{Tr}[(B^*X^{1/m}B)^m]$ is concave on $M_n^+(\mathbb{C})$. The same conjecture was made at the same time by Uhlmann [213], who explicity noted that for $m = 2$ this is true by the Wigner-Yanase theorem. To see the connection, note that for any $B \in M_n(\mathbb{C})$, BB^* is self-adjoint, and even positive. Hence we may take $H = BB^*$ and $\rho = X \in M_n^+(\mathbb{C})$, and then by cyclicity of the trace

$$\operatorname{Tr}[HX^{1/2}HX^{1/2}] = \operatorname{Tr}[(B^*X^{1/2}B)^2].$$

Therefore, the concavity of $X \mapsto \operatorname{Tr}[HX^{1/2}HX^{1/2}]$ for all self-adjoint H implies the concavity of $X \mapsto \operatorname{Tr}[(B^*X^{1/2}B)^2]$. Epstein [79] proved the following more general result:

Theorem 10.11 (Epstein's theorem). *For any $n \times \ell$ matrix B, and all $0 < p < 1$, the function $A \mapsto \operatorname{Tr}[(B^*A^pB)^{1/p}]$ is concave on $M_n^+(\mathbb{C})$.*

10.4. The Lieb concavity theorem and its consequences

In this section we prove Theorem 10.10 starting from Uhlmann's theorem, Theorem 10.5. Historically, Theorem 10.5 came later, but Uhlmann appears to have been among the first to understand that a good route to many convexity or concavity theorems for trace functionals is through a monotonicity theorem such as Theorem 10.5.

Proof of Theorem 10.10. Let $\Phi : M_n(\mathbb{C}) \to M_{2n}(\mathbb{C})$ be given by

$$(10.4.1) \qquad \Phi(A) = \begin{bmatrix} A & 0 \\ 0 & A \end{bmatrix}.$$

Then

$$(10.4.2) \qquad \Phi^\dagger\left(\begin{bmatrix} A & B \\ C & D \end{bmatrix} \right) = A + D$$

is the partial trace. Since Φ is completely positive and unital, it is a Schwarz map. Then for X_1, X_2 and Y_1, Y_2 in $M_n^+(\mathbb{C})$, and $0 < \lambda < 1$, define

$$(10.4.3) \qquad \widehat{X} := \begin{bmatrix} (1-\lambda)X_1 & 0 \\ 0 & \lambda X_2 \end{bmatrix} \quad \text{and} \quad \widehat{Y} := \begin{bmatrix} (1-\lambda)Y_1 & 0 \\ 0 & \lambda Y_2 \end{bmatrix}.$$

Then

$$\operatorname{Tr}[\Phi(K^*)\widehat{Y}^{1-t}\Phi(K)\widehat{X}^t] = (1-\lambda)\operatorname{Tr}[K^*Y_1^{1-t}KX_1^t] + \lambda \operatorname{Tr}[K^*Y_2^{1-t}KX_2^t]$$

and

$$\operatorname{Tr}[K^*\Phi^\dagger(\widehat{Y})^{1-t}K\Phi^\dagger(\widehat{X})^t] = \operatorname{Tr}[K^*((1-\lambda)Y_1 + \lambda Y_2)^{1-t}K((1-\lambda)X_1 + \lambda X_2)^t].$$

Specializing (10.2.1) to this particular choice of Φ yields

$$(1-\lambda)\operatorname{Tr}[K^*Y_1^{1-t}KX_1^t] + \lambda \operatorname{Tr}[K^*Y_2^{1-t}KX_2^t]$$
$$\leq \operatorname{Tr}[K^*((1-\lambda)Y_1 + \lambda Y_2)^{1-t}K((1-\lambda)X_1 + \lambda X_2)^t].$$

This is the concavity asserted in Theorem 10.10 for the case $s = 1 - t$.

As Araki [12] observed, the case in which $0 \leq s, t$ and $s + t < 1$ is a simple consequence of what has just been proved. He argued as follows: Let r be such that $r(1 - t) = s$. Then for positive invertible X_1, X_2, Y_1, Y_2,

$$K^*((Y_1 + Y_2)/2)^sK = K^*(((Y_1 + Y_2)/2)^r)^{1-t}K \geq K^*((Y_1^r + Y_2^r)/2)^{1-t}K,$$

since $Y \mapsto Y^r$ is operator monotone increasing and concave for $0 < r < 1$. Take the trace against $((X_1 + X_2)/2)^t$ to conclude from the $s = 1 - t$ case that the general case is valid. $\qquad\square$

There are by now many proofs of the Lieb concavity theorem (see, e.g., [22, 78, 170]), and each new perspective brings new insight to the subject. One of the best ways to show a function F is convex (or $-F$ is concave) is to display F as a Legendre transform. This approach was taken up by Pusz and Woronowicz [183, 184], and further developed by Donald [77] and Kosaki [132]. The Legendre transform approach is developed in the exercises.

We have already seen in part (4) of Theorem 7.61 that the relative entropy function $(\rho, \sigma) \mapsto D(\rho \| \sigma)$ is jointly convex. This was originally deduced by Lindblad [148] as a consequence of the Lieb concavity theorem. He argued as follows:

For $X, Y \in M_n^+(\mathbb{C})$, $D(X \| Y) := \mathrm{Tr}[X(\log X - \log Y)]$ satisfies

$$(10.4.4) \qquad D(X \| Y) = \frac{\mathrm{d}}{\mathrm{d}t} \mathrm{Tr}[Y^{1-t}X^t]\Big|_{t=1} = \lim_{t \uparrow 1} \frac{1}{1-t}\left(1 - \mathrm{Tr}[Y^{1-t}X^t]\right).$$

By the Lieb concavity theorem in the special case in which $K = \mathbb{1}$, the right side of (10.4.4) is jointly convex in X and Y.

There are many other consequences of the Lieb concavity theorem. In particular, Lieb proved his Theorem 10.13 and Theorem 10.15, stated and proved in the next section, starting from the Lieb concavity theorem. Here we deduce Theorem 10.13 from Epstein's theorem, but then the passage from there to Theorem 10.15 is exactly as in [141].

10.5. The Epstein concavity theorem and its consequences

Just as the Lieb concavity theorem may be conveniently deduced from Uhlmann's monotonicity theorem, Epstein's theorem may be conveniently deduced from another monotonicity theorem, first proved in [54]:

Theorem 10.12. *Let* $0 < p \leq 1$*, and let* $\Phi : M_m(\mathbb{C}) \to M_n(\mathbb{C})$ *be a unital Schwarz map. Then for any* $X \in M_n^{++}(\mathbb{C})$ *and any* $B \in M_m(\mathbb{C})$,

$$\mathrm{Tr}[(\Phi(B)^* X^p \Phi(B))^{1/p}] \leq \mathrm{Tr}[(B^* \Phi^\dagger(X)^p B)^{1/p}].$$

Theorem 10.12 turns out to follow from Uhlmann's theorem by a short Legendre transform argument. First we show how Theorem 10.12 implies Theorem 10.11.

Proof of Theorem 10.11. Exactly as in the proof of Theorem 10.10, we let Φ be given by (10.4.1) so that $\Phi^\dagger$ is given by (10.4.2). As before, for $X_1, X_2 \in M_n^+(\mathbb{C})$ and $0 < \lambda < 1$, define $\widehat{X} := \left[\begin{smallmatrix} (1-\lambda)X_1 & 0 \\ 0 & \lambda X_2 \end{smallmatrix} \right]$. Then

$$\mathrm{Tr}[(\Phi(B)^*\widehat{X}^p\Phi(B))^{1/p}] = (1-\lambda)\mathrm{Tr}[(B^*X_1^pB)^{1/p}] + \lambda\,\mathrm{Tr}[(B^*X_2^pB)^{1/p}]$$

and

$$\mathrm{Tr}[(B^*\Phi^\dagger(\widehat{X})^pB)^{1/p}] = \mathrm{Tr}[(B^*((1-\lambda)X_1 + \lambda X_2)^pB)^{1/p}].$$

By Theorem 10.12,

$$(1-\lambda)\mathrm{Tr}[(B^*X_1^pB)^{1/p}] + \lambda\,\mathrm{Tr}[(B^*X_2^pB)^{1/p}]$$
$$\leq \mathrm{Tr}[(B^*((1-\lambda)X_1 + \lambda X_2)^pB)^{1/p}],$$

and this is the concavity asserted in Theorem 10.11. $\qquad\square$

Proof of Theorem 10.12. When $p = 1$, the inequality follows directly from the assumption that Φ is Schwarz. Now let $0 < p < 1$ and put $r := 1/p > 1$, so that

$$(\mathrm{Tr}[(\Phi(B)^*X^p\Phi(B))^{1/p}])^p = \|\Phi(B)^*X^p\Phi(B)\|_r.$$

Recall that

$$(10.5.1) \qquad \|Z\|_r = \max\{\mathrm{Tr}[ZY] \ : \ Y \geq 0, \quad \mathrm{Tr}[Y^{r/(r-1)}] = 1\}.$$

Therefore,

$$\|\Phi(B)^*X^p\Phi(B)\|_r$$
$$= \max\{\mathrm{Tr}[\Phi(B)^*X^p\Phi(B)Y] \ : \ Y \geq 0, \quad \mathrm{Tr}[Y^{r/(r-1)}] = 1\},$$

and making the change of variables $Y^{1/(p-1)} \mapsto Y$,

$$\|\Phi(B)^*X^p\Phi(B)\|_r$$
$$= \max\{\mathrm{Tr}[\Phi(B)^*X^p\Phi(B)Y^{1-p}] \ : \ Y \geq 0, \quad \mathrm{Tr}[Y] = 1\}.$$

By Uhlmann's theorem, Theorem 10.5,

$$\mathrm{Tr}[\Phi(B)^*X^p\Phi(B)Y^{1-p}] \leq \mathrm{Tr}[B^*(\Phi^\dagger(X))^pB(\Phi^\dagger(Y))^{1-p}].$$

Therefore,

$$(10.5.2) \quad (\mathrm{Tr}[(\Phi(B)^*X^p\Phi(B))^{1/p}])^p$$
$$\leq \max\{\mathrm{Tr}[B^*(\Phi^\dagger(X))^pB(\Phi^\dagger(Y))^{1-p}] \ : \ Y \geq 0, \quad \mathrm{Tr}[Y] = 1\}.$$

Since $\Phi^\dagger$ is positive and trace preserving,

$$\{\Phi^\dagger(Y) \ : \ Y \geq 0, \quad \mathrm{Tr}[Y] = 1\} \subseteq \{Z \ : \ Z \geq 0, \quad \mathrm{Tr}[Z] = 1\}.$$

Taking the supremum in (10.5.2) over the larger set yields

$$(\mathrm{Tr}[(\Phi(B)^* X^p \Phi(B))^{1/p}])^p$$
$$\leq \max\{\mathrm{Tr}[B^* \Phi^\dagger(X)^p B Z^{1-p}] \ : \ Z \geq 0, \quad \mathrm{Tr}[Z] = 1\}.$$

Using (10.5.1) once more,

$$\max\{\mathrm{Tr}[B^* \Phi^\dagger(X)^p B Z^{1-p}] \ : \ Z \geq 0, \quad \mathrm{Tr}[Z] = 1\} = \|B^* \Phi^\dagger(X)^p B\|_r.$$

Now take the pth root, which is the rth power, of both sides. $\qquad\square$

10.5.1. Consequences of Epstein's theorem.

Another celebrated theorem of Lieb [141] can be deduced from Epstein's theorem.

Theorem 10.13 (Lieb's second concavity theorem). *For all self-adjoint $H \in M_n(\mathbb{C})$, the function*

$$(10.5.3) \qquad\qquad X \mapsto \mathrm{Tr}\left[e^{H+\log X}\right]$$

is concave on $M_n^{++}(\mathbb{C})$.

Remark 10.14. It is a simple consequence of the Gibbs variational principle that $X \mapsto \log(\mathrm{Tr}[e^{H+\log X}])$ is concave on $M_n^{++}(\mathbb{C})$. But the exponential of a concave function is not in general concave, so that Theorem 10.13 expresses a much deeper fact.

Proof of Theorem 10.13. By Theorem 10.11, for all $m \in \mathbb{N}$ and $B \in M_n^+$, $X \mapsto \mathrm{Tr}[(BX^{1/m}B)^m]$ is concave on $M_n^{++}(\mathbb{C})$. Apply this with $B = e^{H/2m}$. Then for each m, $X \mapsto \mathrm{Tr}[(e^{H/m} e^{\log X/m})^m]$ is concave. By the Lie product formula, so is $X \mapsto \mathrm{Tr}[e^{H+\log X}] = \lim_{m\to\infty} \mathrm{Tr}[(e^{H/m} e^{\log X/m})^m]$. $\qquad\square$

10.5.2. Lieb's triple matrix inequality.

Another of the results in [141] is an extension of the Golden-Thompson inequality to three matrices instead of only two. Early work [15, 213] aimed at proving the then-conjectured SSA had motivated the search for an upper bound on $\mathrm{Tr}[e^{H+K+L}]$, $H, K, L \in M_n^{\mathrm{s.a.}}(\mathbb{C})$. That reduces to $\mathrm{Tr}[e^H e^{K+L}]$ when K and L commute. The bound cannot simply be $\mathrm{Tr}[e^H e^K e^L]$ since this is not in general a real number, and there was not even a conjecture as to what form such a generalization might take before Lieb proved the following theorem.

Theorem 10.15 (Lieb's triple matrix inequality). *For all $H, K, L \in M_n^{\mathrm{s.a.}}(\mathbb{C})$,*

$$(10.5.4) \qquad\qquad \mathrm{Tr}[e^{H+K+L}] \leq \mathrm{Tr}[e^H T_{e^{-K}}(e^L)],$$

where

$$(10.5.5) \qquad T_A(B) = \int_0^\infty \frac{1}{s+A} B \frac{1}{s+A} \, ds = \frac{d}{dt} \log(A + tB)\Big|_{t=0}.$$

Proof of Theorem 10.15. The second equality in (10.5.5) follows from the integral representation of the logarithm. The function (10.5.3) is homogeneous of degree one as well as concave, so that Lemma 6.11 may be applied, with the result that

$$\frac{d}{dx}\operatorname{Tr}\left[\exp(H + K + \log(e^{-K} + xe^{L}))\right]\Big|_{x=0} \geq \operatorname{Tr}[e^{H+K+L}].$$

Since $\frac{d}{dx}\log(e^{-K} + xe^{L})\big|_{x=0} = T_{e^{-K}}(e^{L})$, with $T_{A}(B)$ defined as in (10.5.5), this proves (10.5.4). $\qquad\square$

As noted in [141], the second part of Lemma 6.11 allows one to reverse the argument that led from Theorem 10.13 to Theorem 10.15, and hence these two theorems are equivalent to one another.

There are other routes to Theorem 10.13 that relate it to the DPI. Tropp [210] gave a variational proof of Theorem 10.13, deducing the concavity from the joint convexity of the relative entropy. Alternatively, as shown in [43], a simple duality argument yields a general *monotonicity version* of Theorem 10.13, from which the concavity statement follows easily.

Theorem 10.16. *Let* $\Phi : M_n(\mathbb{C}) \to M_m(\mathbb{C})$ *be a unital positive map. For all self-adjoint* $H \in M_n(\mathbb{C})$ *and all* $Y \in M_m^{++}(\mathbb{C})$,

$$(10.5.6) \qquad \operatorname{Tr}\left[\exp(H + \log\Phi^{\dagger}(Y))\right] \geq \operatorname{Tr}\left[\exp(\Phi(H) + \log Y)\right].$$

Before giving the very simple proof, we explain how Theorem 10.16 implies Theorem 10.13. As before, let Φ be given by (10.4.1) so that Φ is unital and positive. Then $\Phi(H) = \begin{bmatrix} H & 0 \\ 0 & H \end{bmatrix}$ and $\Phi^{\dagger}\left(\begin{bmatrix} Y_1 & 0 \\ 0 & Y_2 \end{bmatrix}\right) = Y_1 + Y_2$. Inserting these in (10.5.6), $\operatorname{Tr}[e^{H+\log(Y_1+Y_2)}] \geq \operatorname{Tr}[e^{H+\log Y_1}] + \operatorname{Tr}[e^{H+\log Y_2}]$. Since $Y \mapsto \operatorname{Tr}[e^{H+\log Y}]$ is homogeneous of degree one, this is the same as concavity, and therefore Theorem 10.16 implies Theorem 10.13.

Proof of Theorem 10.16. By the Gibbs variational principle (7.3.9), restriction of the domain of maximization, and then the monotonicity of the relative entropy under positive unital maps, Corollary 6.45,

$$\log(\operatorname{Tr}[e^{H+\log\Phi^{\dagger}(Y)}]) = \sup\{\operatorname{Tr}[XH] - D(X\|\Phi^{\dagger}(Y)) : X \in \mathfrak{S}_n\}$$
$$\geq \sup\{\operatorname{Tr}[\Phi^{\dagger}(W)H] - D(\Phi^{\dagger}(W)\|\Phi^{\dagger}(Y)) : W \in \mathfrak{S}_n\}$$
$$\geq \sup\{\operatorname{Tr}[W\Phi(H)] - D(W\|Y) : W \in \mathfrak{S}_n\}$$
$$= \log\operatorname{Tr}\left[e^{\Phi(H)+\log Y}\right].$$

Exponentiating both sides of the inequality yields (10.5.6). $\qquad\square$

As one more application of Epstein's theorem, we prove the Lieb-Thirring trace inequality [145]. Special cases proved earlier in Corollary 6.67 were used to prove the Golden-Thompson inequality.

Theorem 10.17 (Lieb-Thirring trace inequality). *For all $A, B \in M_n^+(\mathbb{C})$ and all $t > 1$,*

$$\mathrm{Tr}\left[(B^{1/2}AB^{1/2})^t\right] \leq \mathrm{Tr}\left[B^{t/2}A^t B^{t/2}\right].$$

Proof. Define $C = A^t$ and $p = 1/t \leq 1$, so that $A = C^p$. Then

$$\mathrm{Tr}\left[(B^{1/2}AB^{1/2})^t\right] - \mathrm{Tr}\left[B^{t/2}A^t B^{t/2}\right]$$
$$= \mathrm{Tr}\left[(B^{1/2}C^p B^{1/2})^{1/p}\right] - \mathrm{Tr}\left[CB^{1/p}\right],$$

and by 10.11, the right hand side is a concave function of C.

Let $\mathcal{N}$ denote the von Neumann subalgebra of $M_n(\mathbb{C})$ consisting of all polynomials in B. Let $B = \sum_{\lambda \in \sigma(B)} \lambda P_\lambda$ be the spectral decomposition of B. The conditional expectation $\mathcal{E}_\mathcal{N}$ onto $\mathcal{N}$ is given by $\mathcal{E}_\mathcal{N}(X) = \sum_{\lambda \in \sigma(B)} P_\lambda X P_\lambda$, and this can be written in the form $\Phi(X) = \frac{1}{N} \sum_{j=1}^N U_j^* X U_j$, where each U_j is a unitary in $\mathcal{N}'$ and N is the cardinality of $\sigma(B)$.

Then since each U_j commutes with any function of B,

$$\mathrm{Tr}\left[\mathcal{E}_\mathcal{N}(C)B^{1/p}\right] = \frac{1}{N} \sum_{j=1}^N \mathrm{Tr}[U_j^* C U_j B^{1/p}] = \mathrm{Tr}\left[CB^{1/p}\right].$$

Similarly, by Epstein's theorem,

$$\mathrm{Tr}\left[(B^{1/2}(\mathcal{E}_\mathcal{N}(C))^p B^{1/2})^{1/p}\right] \geq \frac{1}{N} \sum_{j=1}^N \mathrm{Tr}\left[(B^{1/2}(U_j^* C U_j)^p B^{1/2})^{1/p}\right]$$
$$= \mathrm{Tr}\left[(B^{1/2}C^p B^{1/2})^{1/p}\right].$$

It follows that

$$\mathrm{Tr}\left[(B^{1/2}C^p B^{1/2})^{1/p}\right] - \mathrm{Tr}\left[CB^{1/p}\right]$$
$$\leq \mathrm{Tr}\left[(B^{1/2}(\mathcal{E}_\mathcal{N}(C))^p B^{1/2})^{1/p}\right] - \mathrm{Tr}\left[(\mathcal{E}_\mathcal{N}(C))B^{1/p}\right].$$

However, since $\mathcal{E}_\mathcal{N}(C)$ and B commute, the right hand side is zero. $\square$

10.6. The Lieb convexity theorem and its consequences

The second theorem in Lieb's fundamental 1973 paper [**141**] is the Lieb convexity theorem:

Theorem 10.18 (Lieb convexity theorem). *For all $0 \le s, t, \; 0 \le s + t \le 1$,*

$$(X, Y, K) \mapsto \mathrm{Tr}[K^* Y^{-s} K X^{-t}]$$

is jointly convex on $M_n^{++}(\mathbb{C}) \times M_n^{++}(\mathbb{C}) \times M_n(\mathbb{C})$.

The same argument that deduces the Lieb concavity theorem, Theorem 10.10 from Uhlmann's theorem, Theorem 10.5, may be used to deduce the Lieb concavity theorem from inequality (2) of Theorem 10.5, the dual to Uhlmann's inequality.

Proof of Theorem 10.18. Once again, let Φ be given by (10.4.1) so that $\Phi^\dagger$ is given by (10.4.2). Let $\widehat{X}$ and $\widehat{Y}$ again be given by (10.4.3). Define

$$\widehat{K} = \widehat{Y} := \begin{bmatrix} (1 - \lambda)K_1 & 0 \\ 0 & \lambda K_2 \end{bmatrix}.$$

Note that since $\Phi^\dagger(\mathbb{1}) = 2\mathbb{1} > 0$, $\Phi^\dagger(\widehat{X})$ and $\Phi^\dagger(\widehat{Y})$ are invertible. Hence we shall not need to consider generalized inverses. Finally, define

$$X_\lambda := (1 - \lambda)X_1 + \lambda X_2, \; Y_\lambda := (1 - \lambda)Y_1 + \lambda Y_2, \; \text{and} \; K_\lambda := (1 - \lambda)K_1 + \lambda K_2.$$

With the definitions,

$$\mathrm{Tr}[\widehat{K}^* \widehat{Y}^{t-1} \widehat{K} \widehat{X}^{-t}] = (1 - \lambda)\, \mathrm{Tr}[K_1^* Y_1^{t-1} K_1 X_1^{-t}] + \lambda\, \mathrm{Tr}[K_2^* Y_2^{t-1} K_2 X_2^{-t}]$$

and

$$\mathrm{Tr}[\Phi^\dagger(\widehat{K}^*)\Phi^\dagger(\widehat{Y})^{t-1}\Phi^\dagger(\widehat{K})\Phi^\dagger(\widehat{X})^{-t}] = \lambda\, \mathrm{Tr}[K_\lambda^* Y_\lambda^{t-1} K_\lambda X_\lambda^{-t}].$$

Therefore, by (2) of Theorem 10.5 applied with this choice of Φ, and to $\widehat{X}$, $\widehat{Y}$ and $\widehat{K}$ yields

$$\lambda\, \mathrm{Tr}[K_\lambda^* Y_\lambda^{t-1} K_\lambda X_\lambda^{-t}] \le (1 - \lambda)\, \mathrm{Tr}[K_1^* Y_1^{t-1} K_1 X_1^{-t}] + \lambda\, \mathrm{Tr}[K_2^* Y_2^{t-1} K_2 X_2^{-t}],$$

which is the convexity asserted in Theorem 10.18 for $s = 1 - t$. The same operator monotonicity argument used to prove the case $s + t < 1$ from the $s + t = 1$ case in Theorem 10.10 applies here as well. $\square$

Theorem 10.18 has an important consequence:

Corollary 10.19 (Ando's convexity theorem). *For all $K \in M_n(\mathbb{C})$ and all $1 \le q \le 2$ and $0 \le r \le 1$ with $q - r \ge 1$,*

$$(X, Y) \mapsto \mathrm{Tr}(K^* X^q K Y^{-r})$$

is convex on $M_n^{++}(\mathbb{C}) \times M_n^{++}(\mathbb{C})$.

Proof. Note that $\mathrm{Tr}(K^*X^qKY^{-r}) = \mathrm{Tr}((XK)^*X^{q-2}(XK)Y^{-r})$, and by the Lieb convexity theorem, the right side is jointly convex in X and Y whenever

$$(10.6.1) \qquad 0 \leq r, 2 - q \quad \text{and} \quad r - q + 2 \leq 1.$$

The second inequality in (10.6.1) says $q \geq 1 + r \geq 1$, and together with the first inequality, $1 \leq q \leq 2$. Then $0 \leq r \leq 1$. Finally, for $1 \leq q \leq 2$ and $0 \leq r \leq 1$, both inequalities in (10.6.1) are satisfied if and only if $q - r \geq 1$. $\qquad\square$

Theorem 10.19 was proved in [**8**] by a different argument, and the connection with the earlier Theorem 10.18 was not made.

The next theorem [**54**] stands in the same relation to (2) of Theorem 10.5 as does Theorem 10.12 to (1) of Theorem 10.5. As a consequence, we conclude some convexity analogues of Epstein's theorem, first proved in [**48**].

Theorem 10.20. *Let $0 < p < 1$, and let $\Phi : M_m(\mathbb{C}) \to M_n(\mathbb{C})$ be a unital Schwarz map. Then for any $A \in M_n^{++}(\mathbb{C})$ and any invertible $B \in M_m(\mathbb{C})$,*

$$(10.6.2) \qquad \mathrm{Tr}[(B^*A^{-p}B)^{1/(2-p)}] \geq \mathrm{Tr}[(\Phi^\dagger(B)^*\Phi^\dagger(A)^{-p}\Phi^\dagger(B))^{1/(2-p)}].$$

Proof. Define $r := 1/(2 - p)$ and note that $1/2 < r < 1$. Then

$$(10.6.3) \qquad (\mathrm{Tr}[(B^*A^{-p}B)^{1/(2-p)}])^{1/r} = \|B^*A^{-p}B\|_r.$$

Let $\mathfrak{S}_n^\circ$ denote the interior of the set of $n \times n$ density matrices. That is,

$$\mathfrak{S}_n^\circ = \{Y \in M_n^{++}(\mathbb{C}) : \quad \mathrm{Tr}[Y] = 1\}.$$

Then by (6.8.13),

$$
\begin{aligned}
\|B^*A^{-p}B\|_r &= \min\{\mathrm{Tr}[B^*A^{-p}BY] : \quad Y^{r/(r-1)} \in \mathfrak{S}_n^\circ\} \\
&= \min\{\mathrm{Tr}[B^*A^{-p}BY^{p-1}] : \quad Y \in \mathfrak{S}_n^\circ\} \\
&\geq \min\{\mathrm{Tr}[\Phi^\dagger(B)^*\Phi^\dagger(A)^{-p}\Phi^\dagger(B)\Phi^\dagger(Y)^{p-1}] : \quad Y \in \mathfrak{S}_n^\circ\} \\
&\geq \min\{\mathrm{Tr}[\Phi^\dagger(B)^*\Phi^\dagger(A)^{-p}\Phi^\dagger(B)Z^{p-1}] : \quad Z \in \mathfrak{S}_m^\circ\} \\
&= \min\{\mathrm{Tr}[\Phi^\dagger(B)^*\Phi^\dagger(A)^{-p}\Phi^\dagger(B)Z] : \quad Z^{r/(r-1)} \in \mathfrak{S}_m^\circ\} \\
&= (\mathrm{Tr}[(\Phi^\dagger(B)^*\Phi^\dagger(A)^{-p}\Phi^\dagger(B))^{1/(2-p)}])^{1/r}.
\end{aligned}
$$

The first inequality is (10.2.2), and the second results from taking the minimum over a larger set. Combine this with (10.6.3) and raise both sides to the rth power. $\qquad\square$

Theorem 10.21. *Let $0 \leq s < 1$ and let $1/(2 - s) \leq q < 1$. Then the map*

$$(10.6.4) \qquad (A, B) \mapsto \mathrm{Tr}[(B^*A^{-s}B)^q]$$

is jointly convex on $M_n^{++}(\mathbb{C}) \times M_n(\mathbb{C})$.

Proof. Take $n = 2m$. Let $A_1, A_2 \in M_n^{++}(\mathbb{C})$, and let $B_1, B_2 \in M_n(C)$ be invertible. Then define

$$A := \begin{bmatrix} A_1 & 0 \\ 0 & A_2 \end{bmatrix} \quad \text{and} \quad B := \begin{bmatrix} B_1 & 0 \\ 0 & B_2 \end{bmatrix}.$$

Again let Φ be given by (10.4.1) so that $\Phi^\dagger$ is given by (10.4.2). Let $p := 2 - 1/q$ so that $q = 1/(2 - p)$ and $0 < p < 1$. Define $0 < t < 1$ by $s = tp$. With these definitions,

$$\frac{1}{2} \operatorname{Tr}[(B^* A^{-s} B)^q] = \frac{1}{2} \left(\operatorname{Tr}[(B_1^* A_1^{-s} B_1)^q] + \operatorname{Tr}[(B_2^* A_2^{-s} B_2)^q] \right)$$

and

$$(10.6.5) \quad \operatorname{Tr}[(\Phi^\dagger(B)^* \Phi^\dagger(A)^{-s} \Phi^\dagger(B))^q]$$

$$= \operatorname{Tr}\left[\left(\left(\frac{B_1 + B_2}{2}\right)^* \left(\left(\frac{A_1 + A_2}{2}\right)^t\right)^{-p} \left(\frac{B_1 + B_2}{2}\right) \right)^q \right].$$

Since $X \mapsto X^t$ is concave on $M_n^+(\mathbb{C})$ and since $X \mapsto X^{-p}$ is monotone decreasing on $M_n^+(\mathbb{C})$, $\left(\left(\frac{A_1 + A_2}{2}\right)^t\right)^{-p} \leq \left(\frac{A_1^t + A_2^t}{2}\right)^{-p}$. Hence (10.6.5) yields

$$(10.6.6) \quad \operatorname{Tr}[(\Phi^\dagger(B)^* \Phi^\dagger(A)^{-s} \Phi^\dagger(B))^q]$$

$$\geq \operatorname{Tr}\left[\left(\left(\frac{B_1 + B_2}{2}\right)^* \left(\frac{A_1^t + A_2^t}{2}\right)^{-p} \left(\frac{B_1 + B_2}{2}\right) \right)^q \right].$$

Combing this with (10.6.6) and Theorem 10.20 yields the asserted convexity.
$\square$

As a corollary we recover a result first proved in [**48**]:

Corollary 10.22. *Let* $1 < p \leq 2$, *and let* $r \geq 1$. *Then for all* $B \in M_n(\mathbb{C})$, *the map*

$$(10.6.7) \qquad\qquad A \mapsto \operatorname{Tr}[(B^* A^p B)^{r/p}]$$

is convex on $M_n^+(\mathbb{C})$.

Proof. First suppose that $1 \leq r < p$. As in the proof of Ando's convexity theorem, replace B by AB in (10.6.4), so that $\operatorname{Tr}[(B^* A^{-s} B)^q]$ becomes $\operatorname{Tr}[(B^* A^{2-s} B)^q]$. Define $p := 2 - s$, and note that $1 < p \leq 2$. Then $1/(2 - s) \leq q < 1$ becomes $1/p \leq q < 1$. Finally, replacing q by r/p yields (10.6.7) in this case.

The case $r \geq p$ is elementary and does not depend on Theorem 10.21. By duality for the Schatten norms,

$$\|B^* A^p B\|_q = \sup\{\mathrm{Tr}[CB^* A^p B] \ : \ C^{q/(q-1)} \in \mathfrak{S}_n\},$$

and then since $A \mapsto A^p$ is operator convex, $A \mapsto \|B^* A^p B\|_q$ is convex on $M_n^+(\mathbb{C})$. Then $g(A) := \mathrm{Tr}[(B^* A^p B)^q] = (f(A))^q$ is convex since $q \geq 1$. $\qquad\square$

Theorem 10.23. *Let Φ be any completely positive map from $M_n(\mathbb{C})$ to $M_m(\mathbb{C})$. Then for all $1 \leq p \leq 2$, and for all $r \geq 1$,*

$$X \mapsto \mathrm{Tr}[(\Phi(X^p))^{r/p}]$$

is convex on $M_n^+(\mathbb{C})$.

Proof. By the Stinespring representation theorem, there is a representation π of $M_n(\mathbb{C})$ on $\mathbb{C}^k$ for some k, and an $k \times n$ matrix B such that $\Phi(X) = B^* \pi(X) B$. Then $\Phi(X^p) = B^* \pi(X^p) B = B^* (\pi(X))^p B$. Since $X \mapsto \pi(X)$ is linear, the claim now follows from Corollary 10.22. $\qquad\square$

Corollary 10.24. *Let $\mathcal{H}_1, \mathcal{H}_2$ be finite-dimensional Hilbert spaces. For $1 \leq p \leq 2$ and for $q \geq 1$, define the function $\Psi_{p,q}$ on $\mathcal{B}^+(\mathcal{H}_1 \otimes \mathcal{H}_2)$ by*

$$\Psi_{p,q}(X) := \mathrm{Tr}_{\mathcal{H}}[(\mathrm{Tr}_{\mathcal{K}}[X^p])^{q/p}]$$

for all $X \in \mathcal{B}^+(\mathcal{H} \otimes \mathcal{K})$. Then $\Psi_{p,q}$ is convex on $\mathcal{B}^+(\mathcal{H}_1 \otimes \mathcal{H}_2)$.

Proof. The map $X \mapsto \mathrm{Tr}_{\mathcal{K}}(X)$ is completely positive, and hence Theorem 10.23 applies. $\qquad\square$

Example 10.25. Suppose that $X \in \mathcal{B}(\mathcal{H}_1 \otimes \mathcal{H}_2)$ is a block diagonal matrix with entries $X_1, \ldots, X_k$ in $\mathcal{B}^+(\mathcal{H}_1)$; i.e.,

$$X = \begin{bmatrix} X_1 & & \\ & \ddots & \\ & & X_k \end{bmatrix}.$$

As a consequence of Corollary 10.24, for all $1 \leq p \leq 2$, and for all $q \geq 1$,

$$(X_1, \ldots, X_k) \mapsto \mathrm{Tr}\left[\left(\sum_{j=1}^k X_j^p\right)^{q/p}\right]$$

is jointly convex, as first proved in [**48**].

10.7. Convexity and concavity of $A \mapsto \mathrm{Tr}[(K^*A^pK)^s]$

For $p, s \in \mathbb{R}$, and $K \in M_n(\mathbb{C})$, define the function $\Upsilon_{K,p,s}$ on $M_n^{++}(\mathbb{C})$ by

$$\Upsilon_{K,p,s}(A) = \mathrm{Tr}[(K^*A^pK)^s].$$

For K invertible, $\mathrm{Tr}[((K^{-1})^*A^{-p}K^{-1})^{-s}] = \mathrm{Tr}[(K^*A^pK)^s]$, and hence we loose nothing by restricting our attention to $s > 0$. We have already encountered several results asserting concavity or convexity of $\Upsilon_{K,p,s}$ for particular values of p and s: Epstein's theorem says that for $0 < p \leq 1$, $\Upsilon_{K,p,1/p}$ is concave, while Theorem 10.23 says that for $1 \leq p \leq 2$, $s \geq 1/p$, $\Upsilon_{K,p,s}$ is convex. It is natural to seek the full range of $p \in \mathbb{R}$ and $s > 0$ for which $\Upsilon_{p,s}$ is convex or concave. We first consider necessary conditions. The following lemma summarizes results obtained in [25, 48, 102].

Lemma 10.26. *Let $s > 0$ and $p \neq 0$.*

(1) *If $\Upsilon_{K,p,s}(A)$ is concave on $M_n^{++}(\mathbb{C})$ for all n and all $K \in M_n(\mathbb{C})$, then $0 < p \leq 1$, and $s \leq 1/p$.*

(2) *If $\Upsilon_{K,p,s}(A)$ is convex on $M_n^{++}(\mathbb{C})$ for all n and all $K \in M_n(\mathbb{C})$, then either $-1 \leq p < 0$ and $s > 0$ or $1 \leq p \leq 2$ and $s \geq 1/p$.*

Proof. Concavity of $\Upsilon_{K,p,s}$ on $M_1^{++}(\mathbb{C})$ implies that $a \mapsto a^{ps}$ is concave on $(0, \infty)$ and therefore $0 < ps \leq 1$. Next, let

$$A = \begin{bmatrix} x & 0 \\ 0 & y \end{bmatrix} \quad \text{and} \quad K = \begin{bmatrix} 1 & 0 \\ 1 & \epsilon \end{bmatrix}$$

with numbers $x, y, \epsilon > 0$. Letting $\epsilon \to 0$, we deduce from concavity on $M_2^{++}(\mathbb{C})$ that $(x, y) \mapsto (x^p + y^p)^s$ is jointly concave. Hence for all $t > 0$, $f(x) = (tx^p + y^p)^s$ is concave. Then since $f(x) = y^{ps} + sy^{p(s-1)}x^p t + o(t)$. Hence x^p is a convex function of x, and therefore $p \leq 1$. Then since $ps \leq 1$, $s \leq 1/p$. This proves (1).

Convexity of $\Upsilon_{K,p,s}$ on $M_1^{++}(\mathbb{C})$ implies that $a \mapsto a^{ps}$ is convex on $(0, \infty)$ and therefore $ps \geq 1$ or $ps < 0$. We show that if $\Upsilon_{K,p,s}(A)$ is convex on $M_2^{++}(\mathbb{C})$ for all choices of $B \in M_2(\mathbb{C})$, then $A \mapsto A^p$ is convex on $M_2^{++}(\mathbb{C})$. This is evident for $s = 1$. For $s \neq 1$, let

$$A = \begin{bmatrix} X & 0 \\ 0 & Y \end{bmatrix} \quad \text{and} \quad K = \begin{bmatrix} 1 & 0 \\ 1 & \epsilon 1 \end{bmatrix},$$

where now $X, Y \in M_2^{++}(\mathbb{C})$. Letting $\epsilon \to 0$, we deduce from convexity on $M_4^{++}(\mathbb{C})$ that $(X, Y) \mapsto \mathrm{Tr}(X^p + Y^p)^s$ is jointly convex on $M_2^{++}(\mathbb{C})$. In particular, for any $t > 0$, $X \mapsto \mathrm{Tr}(tX^p + Y^p)^s$ is convex. Using

$$\mathrm{Tr}(tX^p + Y^p)^s = \mathrm{Tr}\, Y^{ps} + st\, \mathrm{Tr}\, Y^{p(s-1)}X^p + o(t),$$

we deduce that $A \mapsto \mathrm{Tr}\, B^{p(s-1)}A^p$ is convex on $M_2^{++}(\mathbb{C})$. Then since $s \neq 1$, $A \mapsto A^p$ is convex on $M_2^{++}(\mathbb{C})$, and hence $-1 \leq p < 0$ or $1 \leq p \leq 2$. By what was noted at the beginning, we must also have $ps \geq 1$ or $ps < 0$. $\qquad \square$

The following theorem shows that these necessary conditions are also sufficient. Part (1) is due to Epstein for $s = 1/p$ [79], Carlen and Lieb for $1 \leq s \leq 1/p$ [48], and Hiai for $0 < s < 1$ [103]. Part (2) is due to Carlen and Lieb [48]. Part (3) is due to Hiai [103].

Theorem 10.27. *For all $n \in \mathbb{N}$, and all $K \in M_n(\mathbb{C})$:*

    (1) *If $0 \leq p \leq 1$ and $0 < s \leq 1/p$, then $\Upsilon_{K,p,s}$ is concave on $M_n^{++}(\mathbb{C})$.*

    (2) *If $1 \leq p \leq 2$ and $s \geq 1/p$, then $\Upsilon_{K,p,s}$ is convex on $M_n^{++}(\mathbb{C})$.*

    (3) *If $-1 \leq p \leq 0$ and $s > 0$, then $\Upsilon_{K,p,s}$ is convex on $M_n^{++}(\mathbb{C})$.*

For no other values of p and s is $\Upsilon_{K,p,s}$ either concave or convex on $M_n^{++}(\mathbb{C})$ for all choices of $K \in M_n(\mathbb{C})$.

We shall use two sets variational formulas. The first is [48, Lemma 2.2]:

Lemma 10.28. *For all $Y \in M_n(\mathbb{C})$ and all $0 < s < 1$,*

$$(10.7.1) \quad \mathrm{Tr}\left[(Y^*Y)^s\right] = \inf\{\mathrm{Tr}\left[sY^*X^{(s-1)/s}Y + (1-s)X\right] \;:\; X \in M_n^{++}(\mathbb{C})\}.$$

Likewise, if the infimum is replaced by a supremum, the resulting formula is valid for $s > 1$.

Proof. By continuity, we may assume that $Y^*Y > 0$. Then for $0 < s < 1$, there is a minimizing X. Define $Z := X^{(s-1)/s}$, and then minimizing $\mathrm{Tr}\left[sY^*X^{(1-s)/s}Y + (1-s)X\right]$ over $X > 0$ is the same as minimizing $\mathrm{Tr}\left[sYY^*Z + (1-s)Z^{s/(s-1)}\right]$. Differentiating, we find that the minimizer Z satisfies $Z^{1/(s-1)} = YY^*$, and hence $Z = (YY^*)^{s-1}$. For this choice of Z, $\mathrm{Tr}\left[sYY^*Z + (1-s)Z^{s/(s-1)}\right] = \mathrm{Tr}[(Y^*Y)^s]$. The formula for $s > 1$ may be proved in the same way. $\qquad \square$

The second set of variational formulas is due to Zhang [237]. Recall that for all $r > 0$, $\|X\|_r^r := \mathrm{Tr}\left[|X|^r\right]$.

Theorem 10.29. *Let $\mathcal{H}$ be a finite-dimensional Hilbert space. For $X, Y \in \mathcal{B}(\mathcal{H})$, and $s, t, u > 0$ such that $\frac{1}{u} = \frac{1}{s} + \frac{1}{t}$,*

$$(10.7.2) \qquad \|XY\|_u^u = \inf_{|Z|>0} \left\{\frac{u}{s}\|XZ\|_s^s + \frac{u}{t}\|Z^{-1}Y\|_t^t\right\}.$$

Moroever, for Y invertible,

$$\|XY\|_s^s = \sup_{|Z|>0} \left\{\frac{s}{u}\|XZ\|_u^u - \frac{s}{t}\|Y^{-1}Z\|_t^t\right\}.$$

Proof. By the extended Hölder inequality, Theorem 11.28, to be proved in the next chapter, $\|XY\|_u = \|XZZ^{-1}Y\|_r \le \|XZ\|_s\|Z^{-1}Y\|_t$. Then by the arithmetic-geometric mean inequality,

$$(10.7.3) \qquad \|XY\|_u^u \le \|XZ\|_s^u\|Z^{-1}Y\|_t^u \le \frac{u}{s}\|XZ\|_s^s + \frac{u}{t}\|Z^{-1}Y\|_t^t.$$

This shows that the infimum in (10.7.2) is at least as large as the left side. To show that there is equality, suppose $X, Y \in \mathcal{B}(\mathcal{H})$ are invertible. Let $U|Y^*X^*|$ be the polar decomposition of Y^*X^* so that $XYU = |Y^*X^*|$. Define $Z := YU|Y^*X^*|^{-s/(s+t)}$. Then simple calculations show

$$XZ = |Y^*X^*|^{t/(s+t)} \quad \text{and} \quad Z^{-1}Y = |Y^*X^*|^{s/(s+t)}U^*.$$

Therefore, $\|XZ\|_s^s = \mathrm{Tr}[|Y^*X^*|^{st/(s+t)}] = \|Y^*X^*\|_u^u = \|XY\|_u^u$ and likewise, $\|Z^{-1}Y\|_t^t = \|XY\|_u^u$. This proves (10.7.2) when X and Y are invertible, and shows that in this case the infimum in (10.7.2) is achieved; that is, it is a minimum. Since invertible operators are dense in $\mathcal{B}(\mathcal{H})$, a simple approximation yields the general statement.

Exchanging Y and Z in (10.7.3) yields, for Y invertible,

$$\|XZ\|_u^u \le \frac{r}{p}\|XY\|_s^s + \frac{u}{t}\|Y^{-1}Z\|_t^t,$$

or equivalently $\|XY\|_s^s \ge \frac{s}{u}\|XZ\|_s^s - \frac{s}{t}\|Y^{-1}Z\|_t^t$. It remains to show that there is equality for an appropriate choice of Z. As before, let $U|Y^*X^*|$ be the polar decomposition of Y^*X^*. Define $Z := YU|Y^*X^*|^{s/t}$. Simple computations show that this choice yields equality. $\qquad \square$

Proof of Theorem 10.27. The validity of (2) is immediate from Corollary 10.22.

To prove (3), let $-1 \le p < 0$. We first consider $0 < s < 1$. Then by Lemma 10.28 with $Y^* := A^{p/2}K$, and then $B := X^{(s-1)/(s(1-p))}$,

$$\Upsilon_{K,p,s}(A) = \inf_{X>0}\left(s\,\mathrm{Tr}[K^*A^pKX^{(s-1)/s}] + (1-s)\,\mathrm{Tr}[X]\right)$$

$$= \inf_{B>0}\left(s\,\mathrm{Tr}[K^*A^pKB^{1-p}] + (1-s)\,\mathrm{Tr}[B^{s(1-p)/(s-1)}]\right).$$

By Ando's convexity theorem, $(A, B) \mapsto \mathrm{Tr}\,K^*A^pKB^{1-p}$ is jointly convex. Moreover, since $s(1-p)/(s-1) < 0$, $B \mapsto \mathrm{Tr}\,B^{\frac{s(1-p)}{s-1}}$ is convex. Since the partial infimum of a jointly convex function is convex by Lemma A.25, $\Upsilon_{p,s}$ is convex for $-1 \le p \le 0$ and $0 < s < 1$. The case $s = 1$ is trivial since $A \mapsto A^p$ is operator convex, and we next consider $s > 1$. Again by Lemma 10.28 with $Y^* := A^{p/2}K$,

$$\Upsilon_{K,p,s}(A) = \sup_{X>0}\left(s\,\mathrm{Tr}[K^*A^pKX^{(s-1)/s}] + (1-s)\,\mathrm{Tr}[X]\right).$$

Since $A \mapsto A^p$ is operator convex, this displays $\Upsilon_{K,p,s}(A)$ as the supremum of a family of convex functions in A. This proves (2).

To prove (1) following [237], we show that the general case of concavity for $0 < s < 1/p$ follows form the case $s = 1/p$, originally proved by Epstein. This argument uses the variational formula (10.7.2). Since the variable s already appears in (10.7.2), we shall prove the concavity of $\Upsilon_{K,p,v}$ for $0 < p < 1$ and $0 < v \leq 1/p$. Applying (10.7.2) with $X := A^{p/2}$, $Y := K$, $u = 2v$, $s = 2/p$ and $t = 2v/(1 - vp)$, we have

$$\Upsilon_{K,p,v}(A) = \inf_{|Z|>0} \{vp \operatorname{Tr}[(Z^*A^pZ)^{1/p}] + \frac{1-vp}{2} \operatorname{Tr}[(K^*|Z|^{-2}K)^{2v/(1-vp)}]\}.$$

This displays $\Upsilon_{p,v}(A)$ as the infimum of a family of concave functions, each of which is concave by Epstein's theorem. Thus $\Upsilon_{K,p,v}(A)$ is concave for $0 < p < 1$ and $0 < v \leq 1/p$. $\qquad\square$

Remark 10.30. The proof of part (3) using Lemma 10.28 follows [45]. The proof of part (1), from [237], shows the utility of Zhang's theorem 10.29: while (10.7.1) has only one variable on the left, (10.7.2) has two, and this is what permits the factors of $A^{p/2}$ and K to be moved into two different pieces on the right, which is essential in the reduction to Epstein's theorem.

10.8. Quantum Rényi relative entropy inequalities

Recall that for two probability densities ρ and σ on a finite set $\mathcal{X}$ and $\alpha \in (0, 1)$, the α Rényi relative entropy of ρ with respect to σ, $D_\alpha(\rho\|\sigma)$ is given by

$$(10.8.1) \qquad D_\alpha(\rho\|\sigma) = \frac{1}{\alpha - 1} \log\left(\sum_{x \in \mathcal{X}} \rho^\alpha(x)\sigma^{1-\alpha}(x)\right).$$

For $\alpha \in (1, \infty)$, if $\rho(x) = 0$ for all x such that $\sigma(x) = 0$, define $D_\alpha(\rho\|\sigma)$ using the same formula, but summing only over x such that $\sigma(x) > 0$. If for some x, $\rho(x) > 0$ and $\sigma(x) = 0$, we set $D_\alpha(\rho\|\sigma) = \infty$.

Because of noncommutativity, there are many possible quantum analogues of the formula (10.8.1), and several have found application in quantum information theory. These applications depend on the mathematical properties of these quantum Rényi relative entropies, which is our focus here. The goal is to illustrate the power of the results obtained earlier in this chapter as much as it is to investigate quantum Rényi relative entropy.

10.8.1. α-Petz–Rényi relative entropy. One natural choice, the α-**Petz–Rényi relative entropy** $D_\alpha(\rho\|\sigma)$, is given by

$$(10.8.2) \qquad D_\alpha(\rho\|\sigma) = \frac{1}{\alpha - 1} \log(\operatorname{Tr}[\rho^\alpha\sigma^{1-\alpha}])$$

for $\alpha \in (0, 1)$. Note that $D_\alpha(\rho\|\sigma)$ is finite if and only if $\operatorname{supp}(\rho)$ and $\operatorname{supp}(\sigma)$ are not mutually orthogonal.

Define $Q_\alpha(\rho\|\sigma) := \mathrm{Tr}[\rho^\alpha \sigma^{1-\alpha}]$. Let $\rho, \sigma, \tilde{\rho}, \tilde{\sigma} \in \mathfrak{S}_{\mathcal{H}}$. Then $Q_\alpha(\rho\|\sigma) - Q_\alpha(\tilde{\rho}\|\tilde{\sigma}) = (\rho^\alpha - \tilde{\rho}^\alpha)\sigma^{1-\alpha} + \tilde{\rho}^\alpha(\sigma^{1-\alpha} - \tilde{\sigma}^{1-\alpha})$, and by Hölder's inequality, $|Q_\alpha(\rho\|\sigma) - Q_\alpha(\tilde{\rho}\|\tilde{\sigma})| \leq \|\rho^\alpha - \tilde{\rho}^\alpha\|_{\frac{1}{\alpha}} + \|\sigma^{1-\alpha} - \tilde{\sigma}^{1-\alpha}\|_{\frac{1}{1-\alpha}}$. By Theorem 6.51,

$$|Q_\alpha(\rho\|\sigma) - Q_\alpha(\tilde{\rho}\|\tilde{\sigma})| \leq \|\rho - \tilde{\rho}\|_1^\alpha + \|\sigma - \tilde{\sigma}\|_1^{1-\alpha}.$$

Hence for $0 < \alpha < 1$, $(\rho, \sigma) \mapsto D_\alpha(\rho\|\sigma)$ is continuous at any (ρ, σ) for which $D_\alpha(\rho\|\sigma) < \infty$.

For $\alpha > 1$, when $\sigma > 0$, we may again use (10.8.2) to define $D_\alpha(\rho\|\sigma)$. (What really matters for $\alpha > 1$ is that $\mathrm{supp}(\rho) \subseteq \mathrm{supp}(\sigma)$; see Exercise 9.)

For $\sigma > 0$ and $\alpha > 1$, $|Q_\alpha(\rho, \sigma) - Q_\alpha(\tilde{\rho}, \sigma)| \leq \|\sigma^{-1}\|^{\alpha-1}\|\rho - \tilde{\rho}\|_\alpha \leq \|\sigma^{-1}\|^{\alpha-1}\|\rho - \tilde{\rho}\|_1$, so that $\rho \mapsto D_\alpha(\rho, \sigma)$ is continuous. We now prove some monotonicity and convexity properties of $D_\alpha(\rho, \sigma)$.

Because of the continuity properties just proved, we may freely assume that ρ and σ are both strictly positive. In this case, the relative modular operator, $\Delta_{\rho,\sigma}$, is strictly positive. Then $Q_\alpha(\rho, \sigma)$ can be written as a Petz quasi relative entropy (hence the name α-Petz–Rényi relative entropy):

$$Q_\alpha(\rho\|\sigma) = \langle \sigma^{1/2} \Delta_{\rho,\sigma}^\alpha \sigma^{1/2} \rangle_{\mathrm{HS}}.$$

Since $Q_\alpha(\rho\|\sigma) > 0$ for all $\alpha > 0$, we may define $g(\alpha) := \log Q_\alpha(\rho\|\sigma)$. By the Cauchy-Schwarz inequality,

$$|\langle \sigma^{1/2} \log(\Delta_{\rho,\sigma})\Delta_{\rho,\sigma}^\alpha \sigma^{1/2} \rangle_{\mathrm{HS}}|$$

$$\leq \langle \sigma^{1/2}(\log(\Delta_{\rho,\sigma}))^2 \Delta_{\rho,\sigma}^\alpha \sigma^{1/2} \rangle_{\mathrm{HS}}^{1/2} \langle \sigma^{1/2} \Delta_{\rho,\sigma}^\alpha \sigma^{1/2} \rangle_{\mathrm{HS}}^{1/2},$$

and this is equivalent to $g''(\alpha) \geq 0$, with equality if and only if $\log(\Delta_{\rho,\sigma})\sigma^{1/2}$ is proportional to $\Delta_{\rho,\sigma}^\alpha \sigma^{1/2}$. This reduces to $\log \rho - \log \sigma = c\rho^\alpha \sigma^{-\alpha}$, which is satisfied if and only if $c = 0$ and then $\rho = \sigma$. This proves that $\log Q_\alpha(\rho\|\sigma)$ is strictly convex as a function of α on $[0, \infty)$, unless $\rho = \sigma$.

The following theorem is proved in [**157**].

Theorem 10.31. *For fixed $\rho, \sigma \in \mathfrak{S}_{\mathcal{H}}$, $\alpha \mapsto D_\alpha(\rho\|\sigma)$ is monotone increasing in α.*

Proof. Evidently, $\log Q_\alpha(\rho\|\sigma) = 0$ for $\alpha = 0$ and $\alpha = 1$. Assume $\rho, \sigma > 0$. Then with $g(\alpha)$ defined as above,

$$\frac{d}{d\alpha} D_\alpha(\rho\|\sigma) = \frac{1}{(\alpha-1)^2}\Big((\alpha-1)g'(\alpha) - (g(\alpha) - g(1))\Big),$$

and this is nonnegative since g is convex. (See Theorem A.31). $\qquad\qquad\square$

Another simple calculation yields

$$\lim_{\alpha \to 1} D_\alpha(\rho\|\sigma) = D(\rho\|\sigma) = \mathrm{Tr}[\rho(\log \rho - \log \sigma)],$$

and hence we identify $D_1(\rho\|\sigma)$ with the Umegaki relative entropy $D(\rho\|\sigma)$. As an immediate consequence of this monotonicity and Pinsker's inequality,

$$D_\alpha(\rho\|\sigma) \geq \frac{1}{2}\|\rho - \sigma\|_1^2 \quad \text{for all} \quad \alpha \geq 1.$$

However, one can do better by considering convexity and monotonicity properties of $D_\alpha(\rho\|\sigma)$. We need the following simple lemma:

Lemma 10.32. *Let $\mathcal{H}$ and $\mathcal{K}$ be finite-dimensional Hilbert spaces. Let $\omega \in \mathfrak{S}_{\mathcal{K}}$. Let $\rho, \sigma \in \mathfrak{S}_{\mathcal{H}}$, Then for all $\alpha > 0$,*

$$D_\alpha(\rho \otimes \omega\|\sigma \otimes \omega) = D_\alpha(\rho\|\sigma).$$

Proof. For $\alpha > 1$, we suppose with $\operatorname{supp}(\rho) \subseteq \operatorname{supp}(\sigma)$ in case $\alpha > 1$. Then for all α,

$$\operatorname{Tr}[(\rho \otimes \omega)^\alpha(\sigma \otimes \omega)^{1-\alpha} = \operatorname{Tr}[\rho^\alpha \sigma^{1-\alpha} \otimes \omega] = \operatorname{Tr}[\rho^\alpha \sigma^{1-\alpha}]. \qquad \square$$

Theorem 10.33. *For $0 < \alpha \leq 2$, $(\rho, \sigma) \mapsto D_\alpha(\rho\|\sigma)$ is jointly convex, and for all quantum operations $\Phi^\dagger$,*

$$(10.8.3) \qquad D_\alpha(\Phi^\dagger\rho\|\Phi^\dagger\sigma) \leq D_\alpha(\rho\|\sigma).$$

Proof. For $0 \leq \alpha < 1$, $(\rho, \sigma) \mapsto Q_\alpha(\rho\|\sigma)$ is concave by the Lieb concavity theorem. Then since the logarithm is concave and monotone increasing, $(\rho, \sigma) \mapsto \log(Q_\alpha(\rho\|\sigma))$ is concave, and then since $\alpha - 1 < 0$, $(\rho, \sigma) \mapsto D_\alpha(\rho\|\sigma)$ is convex. The validity of (10.8.3), which is the DPI for the α-Petz–Rényi relative entropy, now follows by adapting Lindblad's original proof of the DPI, given in the proof of Theorem 7.62. This used two ingredients: the joint convexity of $D(\rho\|\sigma)$ and the identity $D(\rho \otimes \omega\|\sigma \otimes \omega) = D(\rho\|\sigma)$. Since we have these properties also for $D_\alpha(\rho\|\sigma)$ by the Lieb concavity theorem and Lemma 10.32, this proves (10.8.3) for $0 < \alpha < 1$.

Alternatively, for $0 < \alpha < 1$, by (10.2.1) of Uhlmann's theorem,

$$\operatorname{Tr}[(\Phi^\dagger(\rho))^\alpha(\Phi^\dagger(\sigma))^{1-\alpha}] \geq \operatorname{Tr}[\rho^\alpha \sigma^{1-\alpha}],$$

so that $Q_\alpha(\Phi^\dagger\rho\|\Phi^\dagger\sigma) \geq Q_\alpha(\rho\|\sigma)$. Then by the monotonicity of the logarithm, and the fact that $\alpha - 1 < 0$, we obtain (10.8.3).

For $1 < \alpha \leq 2$, we must argue differently since the logarithm of a convex function need not be convex. Still, by Ando's convexity theorem, $(\rho, \sigma) \mapsto Q_\alpha(\rho\|\sigma)$ is convex. The identity proved in Lemma 10.32 is equivalent to $Q_\alpha(\rho \otimes \omega\|\sigma \otimes \omega) = Q_\alpha(\rho\|\sigma)$. Hence exactly as in the proof of Theorem 7.62, we obtain $Q_\alpha(\Phi^\dagger\rho\|\Phi^\dagger\sigma) \leq Q_\alpha(\rho\|\sigma)$. Since the logarithm is monotone, and $\alpha - 1 > 0$, this yields (10.8.3). From here we deduce the joint convexity of ρ by the now familiar argument that was used, for example, to deduce the Lieb concavity theorem from Uhlmann's theorem. $\qquad \square$

As an application, we prove:

Theorem 10.34. *For all $\rho, \sigma \in \mathfrak{S}_{\mathcal{H}}$ and all $\alpha > 0$,*

$$D_\alpha(\rho\|\sigma) \geq \left(\min\left\{\tfrac{\alpha}{2}, \tfrac{1}{2}\right\}\right)\|\rho - \sigma\|_1^2.$$

Proof. We have already proved this for $\alpha = 1$, and we proceed in the same way, by a reduction to the classical case for probability densities on a 2 point set. $H := \rho - \sigma$ and let P be the spectral projection onto the positive part of H. Define $\Phi^\dagger$ to be the conditional expectation onto the span of $\{P, P^\perp\}$; that is, $\Phi^\dagger(X) = \mathrm{Tr}[XP]P + \mathrm{Tr}[XP^\perp]P^\perp$. Let $\hat\rho := \Phi^\dagger(\rho) = pP + (1-p)P^\perp$ and $\hat\sigma := \Phi^\dagger(\sigma) = qP + (1-q)P^\perp$. Then since $\Phi^\dagger$ is a quantum operation, $D_\alpha(\hat\rho\|\hat\sigma) \leq D_{\alpha,z}(\rho\|\sigma)$, and

$$D_\alpha(\hat\rho\|\hat\sigma) = \frac{1}{\alpha - 1}\log(p^\alpha q^{1-\alpha} + (1-p)^\alpha(1-q)^{1-\alpha}).$$

Then a calculation similar to the one made in the proof of Lemma 7.19 yields, for $0 < \alpha < 1$,

$$D_\alpha(\hat\rho\|\hat\sigma) \geq \frac{\alpha}{2}(2(p-q))^2 = \frac{\alpha}{2}\|\rho - \sigma\|_1^2.$$

For the classical calculation, see [91], which also contains a more refined result. Combining this inequality with Theorem 10.33 completes the proof for $0 < \alpha \leq 1$. Then the claim for $\alpha > 1$ follows from Theorem 10.31. $\square$

10.8.2. α-sandwiched and α-z Rényi relative entropies. There are other quantum analogues of the classical α-Rényi relative entropies that have been found to be useful in quantum information theory. Rewrite the classical formula (10.8.1) as

$$D_\alpha(\rho\|\sigma) = \frac{1}{\alpha - 1}\log\left(\sum_{x\in\mathcal{X}}\left(\frac{\rho(x)}{\sigma(x)}\right)^\alpha \sigma(x)\right).$$

For $\alpha \in (1, \infty)$ and $\sigma > 0$, a natural quantum analogue of $\sum_{x\in\mathcal{X}}\left(\frac{\rho(x)}{\sigma(x)}\right)^\alpha \sigma(x)$ is $\|\sigma^{-1/2}\rho\sigma^{-1/2}\|_{\alpha,\sigma}^\alpha$, where $\|\cdot\|_{\alpha,\sigma}$ denotes the $L^\alpha(\sigma)$ norm from Definition 6.34. Then from the definitions,

$$\|\sigma^{-1/2}\rho\sigma^{-1/2}\|_{\alpha,\sigma}^\alpha = \mathrm{Tr}\left[\left(\sigma^{\frac{1-\alpha}{2\alpha}}\rho\sigma^{\frac{1-\alpha}{2\alpha}}\right)^\alpha\right].$$

Definition 10.35 (Sandwiched Rényi relative entropy [161, 233]). Let $\mathcal{H}$ be a finite-dimensional Hilbert space. Let $\alpha \in (0,1) \cup (1,\infty)$. For $\rho \in \mathfrak{S}_{\mathcal{H}}$ and $\sigma \in \mathfrak{S}_{\mathcal{H}}^\circ$, the α-**sandwiched Rényi relative entropy of ρ with respect to σ,** $D_{\alpha,\alpha}(\rho\|\sigma)$, is given by

$$(10.8.4) \qquad D_{\alpha,\alpha}(\rho\|\sigma) = \frac{1}{\alpha - 1}\log\left(\mathrm{Tr}\left[\left(\sigma^{\frac{1-\alpha}{2\alpha}}\rho\sigma^{\frac{1-\alpha}{2\alpha}}\right)^\alpha\right]\right).$$

The Araki-Lieb-Thirring inequality is a generalization of the Lieb-Thirring inequality, Theorem 10.17, that will be proved in the next chapter; see Theorem 11.30. It says in particular that for all $A, B \in \mathcal{B}^+(\mathcal{H})$ and all $0 < s < t$,

$$\mathrm{Tr}[(A^{s/2}B^s A^{s/2})^{1/s}] \leq \mathrm{Tr}[(A^{t/2}B^t A^{t/2})^{1/t}].$$

Therefore, for fixed $\alpha > 0$, $f(z) := \mathrm{Tr}\left[\left(\sigma^{\frac{1-\alpha}{2z}} \rho^{\frac{\alpha}{z}} \sigma^{\frac{1-\alpha}{2z}}\right)^z\right]$ is a decreasing function of $z > 0$. Note that $D_{\alpha,\alpha}(\rho\|\sigma) = \frac{1}{\alpha-1}\log(f(\alpha))$ and $D_\alpha(\rho\|\sigma) = \frac{1}{\alpha-1}\log(f(1))$. It follows that for all α, $D_\alpha(\rho\|\sigma) \leq D_{\alpha,\alpha}(\rho\|\sigma)$.

As we have just seen, there is a natural interpolation between the α-Petz–Rényi relative entropies and the α-sandwiched Rényi entropies:

Definition 10.36 (α-z Rényi relative entropy, [**19**, **118**, **120**]). Let $\mathcal{H}$ be a finite-dimensional Hilbert space. Let $\alpha \in (0,1) \cup (1,\infty)$ and $z \in (0,\infty)$. For $\rho \in \mathfrak{S}_\mathcal{H}$ and $\sigma \in \mathfrak{S}_\mathcal{H}^\circ$, the α-z **Rényi relative entropy of ρ with respect to σ**, $D_{\alpha,z}(\rho\|\sigma)$, is given by

$$(10.8.5) \qquad D_{\alpha,z}(\rho\|\sigma) = \frac{1}{\alpha-1}\log\left(\mathrm{Tr}\left[\left(\sigma^{\frac{1-\alpha}{2z}} \rho^{\frac{\alpha}{z}} \sigma^{\frac{1-\alpha}{2z}}\right)^z\right]\right).$$

For $z = \alpha$, (10.8.5) reduces to (10.8.4), and for $z = 1$, (10.8.5) reduces to (10.8.2). Continuity and monotonicity in α and z have been studied in the full range of parameters. In addition to the papers cited above, see [**39**, **158**].

The rest of this section is devoted to the central issue regarding these relative entropies. The fundamental inequality concerning the Umegaki relative entropy is the DPI, and these quantum Rényi relative entropies really only deserve the name if they satisfy the analogue of the DPI. This leads to the question: For which values of α, z is it the case that

$$(10.8.6) \qquad D_{\alpha,z}(\Phi^\dagger(\rho)\|\Phi^\dagger(\sigma)) \leq D_{\alpha,z}(\rho\|\sigma)$$

whenever Φ is a CP unital map?

Many people worked on this problem which was finally solved completely by Zhang [**237**]. Hiai [**102**] had given necessary conditions and some positive conditions. Audenaert and Datta [**19**] conjectured the following theorem, various special cases of which were proved, in particular, the cases corresponding to the Petz and sandwiched Rényi relative entropy [**26**, **85**]. Finally, Zhang [**237**, Theorem 1.2] proved the conjecture in full. For more on the history, see his paper and [**44**, **45**] to which he refers for some of the history.

Theorem 10.37. *Let $\mathcal{H}, \mathcal{K}$ be finite-dimensional Hilbert spaces. Then the DPI (10.8.6) is satisfied for all quantum operations $\Phi^\dagger : \mathcal{B}(\mathcal{H}) \to \mathcal{B}(\mathcal{K})$ if and only if*

    (1) $0 < \alpha < 1$ *and* $z \geq \max\{\alpha, 1-\alpha\}$, *or*

    (2) $\alpha > 1$ *and* $\max\{\frac{\alpha}{2}, \alpha-1\} \leq z \leq \alpha$.

Henceforth, when we refer to α-z Rényi relative entropies, we assume that α and z satisfy one of the conditions in Theorem 10.37, though limiting cases, such as $\alpha \to 0$ and $\alpha \to \infty$, are of interest and are investigated in [**19**].

Proof of Theorem 10.37. Let

$$(10.8.7) \qquad Q_{\alpha,z}(\rho\|\sigma) := \mathrm{Tr}\left[\left(\sigma^{\frac{1-\alpha}{2z}}\rho^{\frac{\alpha}{z}}\sigma^{\frac{1-\alpha}{2z}}\right)^{z}\right].$$

We first determine conditions on z under which this function is jointly concave or convex in ρ and σ in the given ranges of α, and then apply Lindblad's argument to deduce the monotonicity from the joint convexity.

For (1), assume $0 < \alpha < 1$. Define $u := 2z$, $X := \rho^{\frac{\alpha}{2z}}$, $Y := \sigma^{\frac{1-\alpha}{2z}}$. Then $Q_{\alpha,z}(\rho\|\sigma) = \|XY\|_u^u$. Define $s := \frac{2z}{\alpha}$ and $t := \frac{2z}{1-\alpha}$. Note that $s, t, u > 0$ such that $\frac{1}{u} = \frac{1}{s} + \frac{1}{t}$. Then by (1) of Theorem 10.29,

$$Q_{\alpha,z}(\rho\|\sigma) = \inf_{|Z|>0}\left\{\frac{u}{s}\|XZ\|_s^s + \frac{u}{t}\|Z^{-1}Y\|_t^t\right\}$$

$$= \inf_{|Z|>0}\left\{\alpha\,\mathrm{Tr}[(Z^*\rho^{\frac{\alpha}{z}}Z)^{\frac{z}{\alpha}}] + (1-\alpha)\,\mathrm{Tr}\left[\left(\frac{1}{Z}\sigma^{\frac{1-\alpha}{z}}\frac{1}{Z^*}\right)^{\frac{z}{1-\alpha}}\right]\right\}$$

$$(10.8.8) \qquad = \inf_{|Z|>0}\left\{\alpha\Upsilon_{Z,\frac{\alpha}{z},\frac{z}{\alpha}}(\rho) + (1-\alpha)\Upsilon_{\frac{1}{Z^*},\frac{1-\alpha}{z},\frac{z}{1-\alpha}}(\sigma)\right\}.$$

By part (1) of Theorem 10.27, for $\frac{\alpha}{z}, \frac{1-\alpha}{z} \in (0,1)$, both $\Upsilon_{Z,\frac{\alpha}{z},\frac{z}{\alpha}}$ and $\Upsilon_{\frac{1}{Z^*},\frac{1-\alpha}{z},\frac{z}{1-\alpha}}$ are concave. Then (10.8.8) displays $Q_{\alpha,z}(\rho\|\sigma)$ as the infimum of a family of jointly convex functions of ρ and σ, and hence $Q_{\alpha,z}(\rho\|\sigma)$ is a jointly convex function of ρ and σ when $z \geq \max\{\alpha, 1-\alpha\}$.

For (2), assume $z > 1$. Define $s := 2z$. Then with X and Y as before, $Q_{\alpha,z}(\rho\|\sigma) = \|XY\|_s^s$. Define $t := \frac{2z}{\alpha-1}$ and $u := \frac{2z}{\alpha}$. Note that $s, t, u > 0$ such that $\frac{1}{u} = \frac{1}{s} + \frac{1}{t}$. Then by (2) of Theorem 10.29,

$$Q_{\alpha,z}(\rho\|\sigma) = \sup_{|Z|>0}\left\{\frac{s}{u}\|XZ\|_u^u - \frac{s}{t}\|Y^{-1}Z\|_t^t\right\}$$

$$= \sup_{|Z|>0}\left\{\alpha\,\mathrm{Tr}[(Z^*\rho^{\frac{\alpha}{z}}Z)^{u/2}] - (\alpha-1)\,\mathrm{Tr}[(Z^*\sigma^{\frac{\alpha-1}{z}}Z)^{t/2}]\right\}$$

$$(10.8.9) \qquad = \sup_{|Z|>0}\left\{\alpha\Upsilon_{Z,\frac{\alpha}{z},\frac{z}{\alpha}}(\rho) - (\alpha-1)\Upsilon_{Z,\frac{\alpha-1}{z},\frac{z}{\alpha-1}}(\sigma)\right\}.$$

By Theorem 10.27 $\rho \mapsto \Upsilon_{Z,\frac{\alpha}{z},\frac{z}{\alpha}}(\rho)$ is convex if and only if $1 \leq \frac{\alpha}{z} \leq 2$. Also by Theorem 10.27 $\sigma \mapsto \Upsilon_{\frac{\alpha-1}{z},\frac{t}{2}}(\sigma)$ is concave if and only if $\frac{\alpha-1}{z} \leq 1$. Therefore, (10.8.9) displays $Q_{\alpha,z}(\rho\|\sigma)$ as a jointly convex function of ρ and σ provided $\max\{\frac{\alpha}{2}, \alpha-1\} \leq z \leq \alpha$.

Now, applying Lindblad's argument exactly as in the proof of Theorem 10.33, which is a special case of this theorem, completes the proof of the monotonicity.

To see the necessity of these conditions, note that by a now familiar argument, the monotonicity also implies the concavity or convexity as the case may be, and hence it suffices to show that the conditions under which concavity or convexity were proved are necessary as well as sufficient. Since

$$Q_{\alpha,z}(\rho\|\sigma) = \Upsilon_{\sigma^{\frac{1-\alpha}{2z}},\frac{\alpha}{z},z}(\rho) = \Upsilon_{\rho^{\frac{\alpha}{2z}},\frac{1-\alpha}{z},z}(\sigma),$$

when $Q_{\alpha,z}(\rho\|\sigma)$ is jointly concave, and hence separately concave, $\Upsilon_{\sigma^{\frac{1-\alpha}{2z}},\frac{\alpha}{z},z}$ and $\Upsilon_{\rho^{\frac{\alpha}{2z}},\frac{1-\alpha}{z},z}$ must be concave. Then by Theorem 10.37 it is necessary that $\frac{\alpha}{z}, \frac{1-\alpha}{z} \in (0,1)$, and hence $0 < \alpha < 1$ and $z \geq \max\{\alpha, 1-\alpha\}$.

Likewise, when $Q_{\alpha,z}(\rho\|\sigma)$ is jointly convex, $\Upsilon_{\sigma^{\frac{1-\alpha}{2z}},\frac{\alpha}{z},z}$ and $\Upsilon_{\rho^{\frac{\alpha}{2z}},\frac{1-\alpha}{z},z}$ must be convex. For $\alpha, z > 0$, by Theorem 10.37 it is necessary that $\frac{\alpha}{z} \in (1,2]$ and $\frac{1-\alpha}{z} \in [-1,0)$. Hence for the joint convexity of $Q_{\alpha,z}(\rho\|\sigma)$, it is necessary that $\alpha > 1$ and $\max\{\frac{\alpha}{2}, \alpha-1\} \leq z \leq \alpha$. $\qquad\square$

Note that we did not use all of Theorem 10.27 in proving Theorem 10.37. In fact, we used only the convexity of $\Upsilon_{Z,p,1/p}$ for $1 < p \leq 2$, [48] and the concavity of $\Upsilon_{Z,p,1/p}$ for $0 < p < 1$, [79]. For further applications of this variational method, see [238].

It is a remarkable fact that the requirement that the map Φ in Theorem 10.37 be completely positive is superfluous; it is enough to require that Φ be positive and unital. This was known for some time for $D_{\alpha,\alpha}$, $\alpha > 1$ by the work of Beigi [26] and Müller-Hermes and Reeb [160], but the proof in the other cases is more recent. The following extends Theorem 10.37 to the case of positive maps.

Theorem 10.38. *Let $\mathcal{H}, \mathcal{K}$ be finite-dimensional Hilbert spaces. Then the DPI (10.8.6) is satisfied for all trace preserving positive maps $\Phi^\dagger : \mathcal{B}(\mathcal{H}) \to \mathcal{B}(\mathcal{K})$ if and only if*

(1) $0 < \alpha < 1$ *and* $z \geq \max\{\alpha, 1-\alpha\}$, *or*

(2) $\alpha > 1$ *and* $\max\{\frac{\alpha}{2}, \alpha-1\} \leq z \leq \alpha$.

We will prove two lemmas and a corollary, from which the theorem will follow. Note that the necessity of the conditions on α and z is already established. The following lemma is due to Kato [126].

Lemma 10.39. *Let $\mathcal{H}, \mathcal{K}$ be finite-dimensional Hilbert spaces. Let $\Phi : \mathcal{B}(\mathcal{K}) \to \mathcal{B}(\mathcal{H})$ be a positive unital map. Then for all $\alpha \in (0,1)$ and all $z > \max\{\alpha, 1-\alpha\}$, (10.8.6) is satisfied.*

Proof. It suffices to show that under the stated conditions on α and z,

$$(10.8.10) \qquad Q_{\alpha,z}(\Phi^\dagger(\rho)\|\Phi^\dagger(\sigma)) \geq Q_{\alpha,z}(\rho\|\sigma),$$

where $Q_{\alpha,z}$ is defined in (10.8.7).

Let $s := \frac{\alpha}{z}$ and $t := \frac{1-\alpha}{z}$ so that $s + t = \frac{1}{z}$. Let $r := \frac{2}{s+t}$, $p := \frac{2}{s}$ and $q := \frac{2}{t}$ so that $\frac{1}{r} = \frac{1}{p} + \frac{1}{q}$. Then $Q_{\alpha,z}(\Phi^\dagger(\rho)\|\Phi^\dagger(\sigma)) = \|\Phi^\dagger(\rho)^{s/2}\Phi^\dagger(\sigma)^{t/2}\|_r^r$.

By Theorem 10.29,

$$Q := \|\Phi^\dagger(\rho)^{\frac{s}{2}}\Phi^\dagger(\sigma)^{\frac{t}{2}}\|_r^r = \inf_{|Z|>0}\left\{\frac{s}{s+t}\|\Phi^\dagger(\rho)^{\frac{s}{2}}Z\|_p^p + \frac{t}{s+t}\|Z^{-1}\Phi^\dagger(\sigma)^{\frac{t}{2}}\|_q^q\right\}.$$

Using the polar factorization of Z^*, write $Z = |Z^*|U^*$, and define $A := |Z^*|^2 > 0$. Then $Z = A^{1/2}U^*$ and $Z^{-1} = UA^{-1/2}$. Substituting,

$$Q = \inf_{A>0}\left\{\frac{s}{s+t}\|\Phi^\dagger(\rho)^{\frac{s}{2}}A^{\frac{1}{2}}\|_p^p + \frac{t}{s+t}\|A^{-\frac{1}{2}}\Phi^\dagger(\sigma)^{\frac{t}{2}}\|_q^q\right\}$$

$$= \inf_{A>0}\left\{\alpha\|\Phi^\dagger(\rho)^{\frac{\alpha}{2z}}A^{\frac{1}{2}}\|_p^p + (1-\alpha)\|A^{-\frac{1}{2}}\Phi^\dagger(\sigma)^{\frac{1-\alpha}{2z}}\|_q^q\right\}$$

$$= \inf_{A>0}\left\{\alpha\left\|\Phi^\dagger(\rho)^{\frac{\alpha}{2z}}A\Phi^\dagger(\rho)^{\frac{\alpha}{2z}}\right\|_{\frac{z}{\alpha}}^{\frac{z}{\alpha}} + (1-\alpha)\left\|\Phi^\dagger(\sigma)^{\frac{1-\alpha}{2z}}A^{-1}\Phi^\dagger(\sigma)^{\frac{1-\alpha}{2z}}\right\|_{\frac{z}{1-\alpha}}^{\frac{z}{1-\alpha}}\right\}.$$

Since $\frac{z}{\alpha} \geq 1$, by Beigi's inequality, Theorem 6.43,

$$\left\|\Phi^\dagger(\rho)^{\frac{\alpha}{2z}}A\Phi^\dagger(\rho)^{\frac{\alpha}{2z}}\right\|_{\frac{z}{\alpha}}^{\frac{z}{\alpha}} \geq \left\|\rho^{\frac{\alpha}{2z}}\Phi(A)\rho^{\frac{\alpha}{2z}}\right\|_{\frac{z}{\alpha}}^{\frac{z}{\alpha}}.$$

Again by Theorem 6.43 and Choi's inequality $\Phi(A^{-1}) \geq \Phi(A)^{-1}$, Theorem 5.65,

$$\left\|\Phi^\dagger(\sigma)^{\frac{1-\alpha}{2z}}A^{-1}\Phi^\dagger(\sigma)^{\frac{1-\alpha}{2z}}\right\|_{\frac{z}{1-\alpha}}^{\frac{z}{1-\alpha}} \geq \left\|\sigma^{\frac{1-\alpha}{2z}}\Phi(A^{-1})\sigma^{\frac{1-\alpha}{2z}}\right\|_{\frac{z}{1-\alpha}}^{\frac{z}{1-\alpha}}$$

$$\geq \left\|\sigma^{\frac{1-\alpha}{2z}}\Phi(A)^{-1}\sigma^{\frac{1-\alpha}{2z}}\right\|_{\frac{z}{1-\alpha}}^{\frac{z}{1-\alpha}}.$$

Let $\epsilon > 0$, and choose $A > 0$ so that

$$Q + \epsilon \geq \alpha\left\|\Phi^\dagger(\rho)^{\frac{\alpha}{2z}}A\Phi^\dagger(\rho)^{\frac{\alpha}{2z}}\right\|_{\frac{z}{\alpha}}^{\frac{z}{\alpha}} + (1-\alpha)\left\|\Phi^\dagger(\sigma)^{\frac{1-\alpha}{2z}}A^{-1}\Phi^\dagger(\sigma)^{\frac{1-\alpha}{2z}}\right\|_{\frac{z}{1-\alpha}}^{\frac{z}{1-\alpha}}.$$

Then by what we have just seen using Theorem 6.43, defining $B := \Phi(A)$,

$$Q + \epsilon \geq \alpha\left\|\rho^{\frac{\alpha}{2z}}B\rho^{\frac{\alpha}{2z}}\right\|_{\frac{z}{\alpha}}^{\frac{z}{\alpha}} + (1-\alpha)\left\|\sigma^{\frac{1-\alpha}{2z}}B^{-1}\sigma^{\frac{1-\alpha}{2z}}\right\|_{\frac{z}{1-\alpha}}^{\frac{z}{1-\alpha}}.$$

By the variational formula deduced above, taking the infimum over $B > 0$ on the right side yields $Q_{\alpha,z}(\rho\|\sigma)$ and since $\epsilon > 0$ is arbitrary, this proves (10.8.10).

$$\square$$

The following corollary is due to Hiai and Jenčova [**104**].

Corollary 10.40. *Let $\mathcal{H}, \mathcal{K}$ be finite-dimensional Hilbert spaces. Let $\rho \in \mathfrak{S}_{\mathcal{H}}$. Let $\Phi : \mathcal{B}(\mathcal{K}) \to \mathcal{B}(\mathcal{H})$ be a positive unital map. Then for $\frac{1}{2} \leq p \leq 1$, and $B \in \mathcal{B}^+(\mathcal{H})$,*

$$\|\Phi^\dagger(\rho)^{\frac{1}{2p}} B \Phi^\dagger(\rho)^{\frac{1}{2p}}\|_p^p \leq \|\rho^{\frac{1}{2p}} \Phi(B) \rho^{\frac{1}{2p}}\|_p^p.$$

Remark 10.41. By homogeneity, we may replace ρ by $A \in \mathcal{B}^+(\mathcal{H})$, and then we obtain

$$\|\Phi^\dagger(A)^{\frac{1}{2p}} B \Phi^\dagger(A)^{\frac{1}{2p}}\|_p^p \leq \|A^{\frac{1}{2p}} \Phi(B) A^{\frac{1}{2p}}\|_p^p.$$

which is the reverse of Beigi's inequality (6.4.14), which he proved for $p \geq 1$. In other words, the corollary of Hiai and Jenčova says that Beigi's inequality extends, in reverse form, to $\frac{1}{2} \leq p < 1$. We will soon see that this is best possible.

Proof of Corollary 10.40. Start on the right. We may suppose $\rho > 0$,

$$(10.8.11) \qquad \|\rho^{\frac{1}{2p}} \Phi(B) \rho^{\frac{1}{2p}}\|_p^p = \|\rho^{\frac{1-p}{2p}} (\rho^{\frac{1}{2}} \Phi(B) \rho^{\frac{1}{2}}) \rho^{\frac{1-p}{2p}}\|_p^p.$$

To simplify the formulas that follow, define $\hat{\rho} = \Phi^\dagger(\rho)$, and then define

$$C := \hat{\rho}^{\frac{1}{2}} B \hat{\rho}^{\frac{1}{2}} \quad \text{and} \quad \mathcal{R}_\rho(X) = \rho^{\frac{1}{2}} \Phi(\hat{\rho}^{-\frac{1}{2}} X \hat{\rho}^{-\frac{1}{2}}) \rho^{\frac{1}{2}}.$$

Then $\mathcal{R}_\rho$ is the Petz recovery map introduced in Definition 8.8. In particular, it is a positive trace preserving map, and $\mathcal{R}_\rho(\hat{\rho}) = \rho$. (That is, $\mathcal{R}_\rho$ recovers ρ from $\hat{\rho}$.) Now we can rewrite (10.8.11) as

$$\|\rho^{\frac{1}{2p}} \Phi(B) \rho^{\frac{1}{2p}}\|_p^p = \|\rho^{\frac{1-p}{2p}} (\rho^{\frac{1}{2}} \Phi(B) \rho^{\frac{1}{2}}) \rho^{\frac{1-p}{2p}}\|_p^p$$

$$= \|\rho^{\frac{1-p}{2p}} (\mathcal{R}_\rho(C)) \rho^{\frac{1-p}{2p}}\|_p^p = \|\mathcal{R}_\rho(\hat{\rho})^{\frac{1-p}{2p}} (\mathcal{R}_\rho(C)) \mathcal{R}_\rho(\hat{\rho})^{\frac{1-p}{2p}}\|_p^p$$

$$= Q_{p,p}(\mathcal{R}_\rho(C) \| \mathcal{R}_\rho(\hat{\rho})),$$

where we have used the notation (10.8.7).

By (10.8.10), since $p \in [0,1]$, $Q_{p,p}(\mathcal{R}_\rho(C) \| \mathcal{R}_\rho(\hat{\rho})) \geq Q_{p,p}(C \| \hat{\rho})$, and then

$$Q_{p,p}(C \| \hat{\rho}) = \|\hat{\rho}^{\frac{1-p}{2p}} C \hat{\rho}^{\frac{1-p}{2p}}\|_p^p = \|\hat{\rho}^{\frac{1}{2p}} B \hat{\rho}^{\frac{1}{2p}}\|_p^p = \|\Phi^\dagger(\rho)^{\frac{1}{2p}} B \Phi^\dagger(\rho)^{\frac{1}{2p}}\|_p^p. \qquad \square$$

The next lemma is also due to Hiai and Jenčova [**104**].

Lemma 10.42. *Let $\mathcal{H}, \mathcal{K}$ be finite-dimensional Hilbert spaces. Let $\sigma, \rho \in \mathfrak{S}_{\mathcal{H}}^\circ$. Let $\Phi : \mathcal{B}(\mathcal{K}) \to \mathcal{B}(\mathcal{H})$ be a positive unital map. Then for all $\alpha > 1$ and $\max\{\frac{\alpha}{2}, \alpha - 1\} \leq z \leq \alpha$, (10.8.6) is satisfied.*

Proof. It suffices to show that under the stated conditions on α and z,

$$(10.8.12) \qquad Q_{\alpha,z}(\Phi^\dagger(\rho) \| \Phi^\dagger(\sigma)) \leq Q_{\alpha,z}(\rho \| \sigma),$$

where $Q_{\alpha,z}$ is defined in (10.8.7).

Define $s := 2z$. Then with $X := \rho^{\frac{\alpha}{2z}}$, $Y := \sigma^{\frac{1-\alpha}{2z}}$, $Q_{\alpha,z}(\rho\|\sigma) = \|XY\|_s^s$. Define $t := \frac{2z}{\alpha-1}$ and $u := \frac{2z}{\alpha}$. Note that $s, t, u > 0$ such that $\frac{1}{u} = \frac{1}{s} + \frac{1}{t}$. Then by (2) of Theorem 10.29 and a polar factorization, $Z = |Z^*|U^*$, define $A := |Z^*|^2 > 0$. Substituting,

$$Q_{\alpha,z}(\rho\|\sigma) = \sup_{|Z|>0}\left\{\frac{s}{u}\|XZ\|_u^u - \frac{s}{t}\|Y^{-1}Z\|_t^t\right\}$$

$$= \sup_{A>0}\left\{\alpha\operatorname{Tr}[(A^{\frac{1}{2}}\rho^{\frac{\alpha}{z}}A^{\frac{1}{2}})^{\frac{z}{\alpha}}] - (\alpha-1)\operatorname{Tr}[(A^{\frac{1}{2}}\sigma^{\frac{\alpha-1}{z}}A^{\frac{1}{2}})^{\frac{z}{\alpha-1}}]\right\}$$

$$(10.8.13) \qquad = \sup_{A>0}\left\{\alpha\operatorname{Tr}[(\rho^{\frac{1}{2p}}A\rho^{\frac{1}{2p}})^p] - (\alpha-1)\operatorname{Tr}[(\sigma^{\frac{1}{2q}}A\sigma^{\frac{1}{2q}})^q]\right\},$$

where $p := \frac{z}{\alpha}$ and $q := \frac{z}{\alpha-1}$. Note that $p \in [\frac{1}{2}, 1]$ and $q \geq 1$. Therefore, for any $B > 0$,

$$Q_{\alpha,z}(\rho\|\sigma) \geq \alpha\operatorname{Tr}[(\rho^{\frac{1}{2p}}\Phi(B)\rho^{\frac{1}{2p}})^p] - (\alpha-1)\operatorname{Tr}[(\sigma^{\frac{1}{2q}}\Phi(B)\sigma^{\frac{1}{2q}})^q]$$

$$\geq \alpha\operatorname{Tr}[(\Phi^\dagger(\rho)^{\frac{1}{2p}}B\Phi^\dagger(\rho)^{\frac{1}{2p}})^p] - (\alpha-1)\operatorname{Tr}[(\Phi^\dagger(\sigma)^{\frac{1}{2q}}B\Phi^\dagger(\sigma)^{\frac{1}{2q}})^q],$$

where the first inequality is from (10.8.13) and the second uses Corollary 10.40 and the fact that this inequality reverses for $p > 1$; see Remark 10.41. Now taking the supremum over $B > 0$ and using (10.8.13) once more, (10.8.12) is proved. $\qquad\square$

Proof of Theorem 10.38. This is immediate from Lemma 10.39 and Lemma 10.42. $\qquad\square$

Exercises

(1) With B_{XY} defined as in Lemma 10.7, show that for $X, Y \in M_n^+(\mathbb{C})$ and $K \in M_n(\mathbb{C})$ such that $\ker(K) \subseteq \ker(X)$ and $\ker(K^*) \subseteq \ker(Y)$,

$$B_{Y,X}^+(K) = \int_0^\infty \frac{1}{\lambda+Y}K\frac{1}{\lambda+X}\mathrm{d}\lambda.$$

Give an extension of (2) of Theorem 10.8 to the case in which X and Y are only assumed to be positive semidefinite.

(2) Define $g(x) := \frac{x-1}{\log(x)}$. Show that for this choice of g, the two inequalities (10.2.10) and (10.2.11) are the two inequalities of Theorem 10.8.

(3) Show that for $g = a + bx$, $a, b \geq 0$, $a + b > 0$, (10.2.10) and (10.2.11) are valid for all unital Schwarz maps Φ and all $X, Y > 0$, and all K in the relevant spaces.

(4) For $A \in M_n^{++}(\mathbb{C})$ and $B \in M_n^{\text{s.a.}}(\mathbb{C})$, let $T_A(B)$ be defined as in Theorem 10.15. Let $H, K, L \in M_n^{\text{s.a.}}(\mathbb{C})$. Show that if K and L commute $\operatorname{Tr}[e^H T_{e^{-K}}(e^L)]| = \operatorname{Tr}[e^H e^{K+L}] = \operatorname{Tr}[e^H e^K e^L]$.

(5) Complete the proof of Theorem 10.18, showing how the cases $s + t < 1$ follow from the cases $s + t = 1$.

(6) One might expect that there would be a monotonicity variant of Ando's convexity theorem stating that for all $0 \leq t \leq 1$, all unital CP maps $\Phi : M_n(\mathbb{C}) \to M_m(\mathbb{C})$, all $X, Y \in M_m^+(\mathbb{C})$ with $\Phi^\dagger(X) \in M_n^{++}(\mathbb{C})$, and all $K \in M_n(\mathbb{C})$,

$$\mathrm{Tr}[\Phi(K^*)Y^{1+t}\Phi(K)X^{-t}] \geq \mathrm{Tr}[K^*\Phi^\dagger(Y)^{1+t}K\Phi^\dagger(X)^{-t}]. \qquad (*)$$

   (a) Show that $(*)$ is not true in general.
   (b) Show that $(*)$ is valid under the additional assumption that K belongs to the multiplicative domain of Φ, and that this follows from (10.2.2) using the result of Exercise 21 in Chapter 5 by the same argument that was used in this chapter to derive Ando's convexity theorem from Lieb's convexity theorem.

(7) (Carlen and Lieb) Use Theorem 10.17 to show $\|X^{1/p}Y^{1/p}\|_p^p \leq \mathrm{Tr}[XY]$ for all $p \geq 2$ for all $X, Y \in M_n^+(\mathbb{C})$. Then use the identity

$$\mathrm{Tr}[X^t Y^s X^{1-t} Y^{1-s}] = \mathrm{Tr}[(X^{t+s-1}Y^{t+s-1})(Y^{1-t}X^{1-t})(Y^{1-s}X^{1-s})]$$

and Hölder's inequality to prove that for $0 \leq s, t \leq 1$, $s + t \leq \frac{3}{2}$,

$$|\mathrm{Tr}[X^t Y^s X^{1-t} Y^{1-s}]| \leq \mathrm{Tr}[XY],$$

which is an inequality of Ando, Hiai, and Okubo [11], whose proof used the majorization methods discussed in the next chapter. (The restriction that $s + t \leq \frac{3}{2}$ is essential.)

(8) Let $\Psi_{p,1}$ be the function on $\mathcal{B}^+(\mathcal{H}_1 \otimes \mathcal{H}_2)$ considered in Corollary 10.24 for $1 \leq p \leq 1$ and $q = 1$. Show that for the density matrix $\rho_{12} > 0$ on $\mathcal{H}_1 \otimes \mathcal{H}_2$,

$$\left.\frac{d}{dp}\Psi_{p,1}(\rho_{12})\right|_{p=1} = S(\rho_2) - S(\rho_{12}),$$

and use this to give a proof of the concavity of the conditional entropy.

(9) (Carlen and Lieb) Let A be a positive operator on the tensor product of three finite-dimensional Hilbert spaces $\mathcal{H}_1 \otimes \mathcal{H}_2 \otimes \mathcal{H}_3$.
   (a) Let the dimension of $\mathcal{H}_1$ be n. Let $\mathcal{N}$ denote the subalgebra $\mathbb{1}_1 \otimes \mathcal{B}(\mathcal{H}_2 \otimes \mathcal{H}_3)$ of $\mathcal{B}(\mathcal{H}_1 \otimes \mathcal{H}_2 \otimes \mathcal{H}_3)$. Then for an $A \in \mathcal{B}(\mathcal{H}_1 \otimes \mathcal{H}_2 \otimes \mathcal{H}_3)$, the conditional expectation $\mathcal{E}_\mathcal{N}(A)$ is given by

$$\mathcal{E}_\mathcal{N}(A) = \frac{1}{n}\mathbb{1}_1 \otimes \mathrm{Tr}_1[A].$$

   Now take $\mathcal{H} := \mathcal{H}_2 \otimes \mathcal{H}_3$ and $\mathcal{K} = \mathcal{H}_2$, and let $\Psi_{p,1}$ be the function on $\mathcal{B}^+(\mathcal{H} \otimes \mathcal{K}) = \mathcal{B}^+((\mathcal{H}_1 \otimes \mathcal{H}_2 \otimes \mathcal{H}_3)$ defined in Corollary 10.24. Show that

$$\mathrm{Tr}_3\left[\mathrm{Tr}_2(\mathrm{Tr}_1 A)^p\right]^{1/p} = \mathrm{Tr}_{13}\left[\mathrm{Tr}_2\left(\frac{1}{n}\mathbb{1}_1 \otimes \mathrm{Tr}_1 A\right)^p\right]^{1/p}$$

$$= \Psi_{p,1}(\mathcal{E}_\mathcal{N}(A)).$$

(b) By Theorem 4.40, there is a finite group $\mathcal{G}$ of unitaries W on $\mathcal{H}_1$ so that $\mathcal{E}_{\mathcal{N}}(A) = \frac{1}{|\mathcal{G}|} \sum_{W \in \mathcal{G}} (W^* \otimes \mathbb{1}_{23}) A (W \otimes \mathbb{1}_{23})$. Use this and the convexity of $\Psi_{p,1}$, proved in Corollary 10.24, to prove that

$$\mathrm{Tr}_3[(\mathrm{Tr}_2[(\mathrm{Tr}_1[A])^p])^{1/p}] \le \mathrm{Tr}_{13}[(\mathrm{Tr}_2[A^p])^{1/p}]. \tag{$*$}$$

(c) Note that for $p = 1$, both sides of $(*)$ reduce to $\mathrm{Tr}[A]$. Take $A = \rho_{123}$, a tripartite density matrix. Then

$$\frac{1}{p-1}\left(\mathrm{Tr}_3[(\mathrm{Tr}_2[\mathrm{Tr}_1[\rho_{123}]^p])^{1/p}] - 1\right)$$

$$\le \frac{1}{p-1}\left(\mathrm{Tr}_{13}[(\mathrm{Tr}_2[\rho_{123}^p])^{1/p}] - 1\right).$$

Show that taking the limit $p \to 1$ yields $S_{13} + S_{23} \ge S_{123} + S_3$, giving another proof of SSA in the quantum case.

(d) A classical analogue of the inequality $(*)$ may be easily deduced from Minkowski's inequality for sums. Do this to give a proof of classical SSA for a density matrix on the product of three finite sets that does not refer to conditional probability.

(10) Let $\mathcal{H}$ be a d-dimensional Hilbert space, and let $\tau = d^{-1}\mathbb{1}$. For $0 < \epsilon < 1$ and $\sigma \in \mathfrak{S}_{\mathcal{H}}$, define $\sigma_\epsilon := (1 - \epsilon)\sigma + \epsilon\tau$ so that $\sigma_\epsilon > 0$. Then for $\alpha > 1$, define $D_\alpha(\rho\|\sigma) := \lim_{\epsilon \downarrow 0} D_\alpha(\rho\|\sigma_\epsilon)$. Show that this limit exists, and if $\mathrm{supp}(\rho) \subseteq \mathrm{supp}(\sigma)$,

$$D_\alpha(\rho\|\sigma) = \frac{1}{\alpha - 1} \log\left(\mathrm{Tr}[\rho^\alpha(\sigma^+)^{\alpha-1}]\right),$$

where σ^+ is the generalized inverse of σ, and otherwise $D_\alpha(\rho\|\sigma) = \infty$.

(11) Use the argument in the proof of Theorem 10.10 to prove that for $p > 1$, $(A, B) \mapsto \mathrm{Tr}\left[\left(A^{\frac{1-p}{2p}} BA^{\frac{1-p}{2p}}\right)^p\right]$ is jointly convex on $\mathcal{B}^{++}(\mathcal{H}) \times \mathcal{B}^{++}(\mathcal{H})$. Then use the result of Exercise 12 in Chapter 5 to prove that for $p > 1$, $0 < r \le 1$, $(A, C) \mapsto \mathrm{Tr}\left[\left(A^{\frac{1+p}{2p}} C^{-r} A^{\frac{1+p}{2p}}\right)^p\right]$ is jointly convex on $\mathcal{B}^{++}(\mathcal{H}) \times \mathcal{B}^{++}(\mathcal{H})$. However, since this functional is not homogeneous of degree one, there is no corresponding monotonicity inequality.

(12) Use the method of Exercises 18–20 in Chapter 5 to compute the Legendre transform of $(X, Y) \mapsto -\mathrm{Tr}[K^* Y^r K X^{1-r}]$, which by the Lieb concavity theorem is convex.

(13) Prove (10.2.10) for $g(x) = a + bx$, $a, b \ge 0$.

(14) Show that the classical Fisher information metric (10.2.14) has the property (10.2.13).

(15) Let $\rho \in \mathfrak{S}_{\mathcal{H}}^\circ$ and let $K \in V$ be the traceless self-adjoint operators on $\mathcal{H}$. For all t sufficiently close to zero, $\rho + tK \in \mathfrak{S}_{\mathcal{H}}^\circ$. Define

$$\gamma_\rho(K, K) := -\frac{\mathrm{d}^2}{\mathrm{d}t^2} S(\rho + tK)\Big|_{t=0}.$$

Show that $\gamma_\rho(K, K)$ is a monotone metric.

(16) From the integral representation of $x^r, 0 \le r \le 1$, proved in Exercise 13 of Chapter 3, $x^{r+1} = \int_0^1 \left(\frac{x^2}{t+(1-t)x} \right) \rho_r(t)dt$ where $\rho_r(t)$ is a probability density specified in the exercise.

(17) Let $\rho + 0, \rho_1 \in \mathfrak{S}_{\mathcal{H}}^{\circ}$ where $\mathcal{H}$ is a finite-dimensional Hilbert space. Let γ be a monotone metric on $\mathfrak{S}_{\mathcal{H}}^{\circ}$, and suppose that $\rho(t), t \in [0, 1]$, is a minimal length path with $\rho(0) = \rho_0$ and $\rho(1) = \rho_1$. Show that if ρ_0 and ρ_1 commute, then for each t, $\rho(t)$ commutes with ρ_0 and ρ_1, and thus that on the *classical sector*, and quantum monotome metric reduces to a multiple of the classical Fisher information metric.

   (a) Show that $R_X f(L_Y R_X^{-1})(K) = Y^{1+r}KX^{-r}$ for all $X, Y \in M_n^{++}(\mathbb{C})$, $K \in M_n(\mathbb{C})$, and therefore that

$$\mathrm{Tr}[K^* Y^{1+r}KX^{-r}] = \int_0^1 \left\langle K, \frac{L_Y^2}{tR_X + (1-t)L_Y} K \right\rangle \rho_r(t)dt.$$

   (b) Use the result of part (a) and Kiefer's theorem to give a proof of Ando's convexity theorem.

(18) Fix $K \in M_n(\mathbb{C})$, and for $0 < r < 1$, define the function $G(X, Y)$ on $M_n(\mathbb{C}) \times M_n(\mathbb{C})$ by

$$G(X, Y) := \begin{cases} \mathrm{Tr}[K^* Y^{1+r}KX^{-r}] & (X, Y) \in M_n^{++}(\mathbb{C}) \times M_n^{++}(\mathbb{C}) \\ \infty & (X, Y) \notin M_n^{++}(\mathbb{C}) \times M_n^{++}(\mathbb{C}). \end{cases}$$

Then G is convex and proper by Ando's convexity theorem. Since G is homogeneous of degree one, Theorem 6.10 says that its Legendre transform G^* has the form

$$G^*(H, K) := \begin{cases} 0 & (H, K) \in \Omega \\ \infty & (H, K) \in \Omega \end{cases}$$

for some nonempty, closed convex set Ω in $M_n(\mathbb{C}) \times M_n(\mathbb{C})$.

   (a) Use the idea outlined in Exercises 18 to 20 in Chapter 6 to compute an explicit formula for G^*.

   (b) Compute an explicit formula for G^{**}.

(19) Fix $K \in M_n(\mathbb{C})$, and for $0 < r < 1$, define the function $G(X, Y)$ on $M_n(\mathbb{C}) \times M_n(\mathbb{C})$ by

$$G(X, Y) := \begin{cases} -\mathrm{Tr}[K^* Y^{1-r}KX^{r}] & (X, Y) \in M_n^{++}(\mathbb{C}) \times M_n^{++}(\mathbb{C}) \\ \infty & (X, Y) \notin M_n^{++}(\mathbb{C}) \times M_n^{++}(\mathbb{C}). \end{cases}$$

Then G is convex and proper by Lieb's concavity theorem. Since G is homogeneous of degree one, Theorem 6.10 says that its Legendre transform G^* has the form

$$G^*(H, K) := \begin{cases} 0 & (H, K) \in \Omega \\ \infty & (H, K) \in \Omega \end{cases}$$

for some nonempty, closed convex set Ω in $M_n(\mathbb{C}) \times M_n(\mathbb{C})$.

   (a) Use the idea outlined in Exercises 18 to 20 in Chapter 6 to compute an explicit formula for G^*.

   (b) Compute an explicit formula for G^{**}.

Chapter 11

Majorization methods

11.1. Majorization in $\mathbb{R}^n$

Majorization concerns the theory of a certain preorder on vectors in $\mathbb{R}^n$ that is very useful for proving matrix inequalities. The subject and its history are well explained in the text [152] by Marshal and Olkin. The two surveys of Ando [9,10] are another excellent source for the general theory. Here we briefly cover the parts that are most relevant to the topics treated in this text, including some very recent developments.

Definition 11.1. Let A be an $n \times n$ matrix. Then:

  (1) $\vec{\delta}(A) = (A_{1,1}, \ldots, A_{n,n})$ denotes the vector in $\mathbb{C}^n$ consisting of the **diagonal entries of** A.

  (2) $\vec{\lambda}(A) = (\lambda_1, \ldots, \lambda_n)$ denotes a vector in $\mathbb{C}^n$ whose entries are the **eigenvalues of** A, repeated according to their algebraic multiplicity, in some order.

  (3) $\vec{\sigma}(A) = (\sigma_1, \ldots, \sigma_n)$ where if the rank of A is r, the first r entries are the **singular values of** A arranged in decreasing order, and, if $r < n$, the final $n - r$ entries are zero.

A number of important theorems that involve the notion of majorization concern relations between these vectors for one or more matrices. First, some notation. Let $\mathbf{x} = (x_1, \ldots, x_n)$ be a vector in $\mathbb{R}^n$, and let $\mathbf{x}^*$ denote its decreasing rearrangement; that is, choose a permutation $\pi \in \mathcal{S}_n$ such that $x_{\pi(j)} \geq x_{\pi(j+1)}$ for all $j \in \{1, \ldots, n-1\}$, and define the jth component of $\mathbf{x}^*$, x_j^*, to be $x_{\pi(j)}$. For $f : \mathbb{R} \to \mathbb{R}$, define $f(\mathbf{x}) = (f(x_1), \ldots, f(x_n))$. For $\mathbf{x}, \mathbf{y} \in \mathbb{R}^n$, $\mathbf{x} \leq \mathbf{y}$ means that $x_j \leq y_j$ for each j.

Definition 11.2. For $\mathbf{x}, \mathbf{y} \in \mathbb{R}^n$, we say that $\mathbf{y}$ **weakly majorizes** $\mathbf{x}$, written $\mathbf{x} \prec_w \mathbf{y}$ in case

$$(11.1.1) \qquad \sum_{j=1}^{k} x_j^* \le \sum_{j=1}^{k} y_j^* \quad \text{for all} \quad k \le n-1 \quad \text{and} \quad \sum_{j=1}^{n} x_j \le \sum_{j=1}^{n} y_j.$$

We say that $\mathbf{y}$ **majorizes** $\mathbf{x}$, written $\mathbf{x} \prec \mathbf{y}$ in case (11.1.1) is satisfied with equality in the final inequality.

For $J \subseteq \{1, \ldots, n\}$, let $|J|$ denote the cardinality of J. Then an equivalent formulation of the inequality in (11.1.1) is for all $k \in \{1, \ldots, n-1\}$ and all J with $|J| = k$,

$$\sum_{j \in J} x_j \le \sum_{j=1}^{k} y_j^* = \max_{|I|=k} \sum_{i \in I} y_i.$$

11.2. Majorization and double stochasticity

A fundamental theorem of Alfred Horn [**114**] relates the vectors $\vec{\lambda}(A)$ and $\vec{\delta}(A)$ introduced at the beginning of this chapter.

Theorem 11.3 (Horn's theorem). *Let* $\mathbf{x}, \mathbf{y} \in \mathbb{R}^n$. *Then* $\mathbf{x} \prec \mathbf{y}$ *if and only if there exists a self-adjoint* $A \in M_n(\mathbb{C})$ *such that* $\mathbf{x} = \vec{\delta}(A)$ *and* $\mathbf{y} = \vec{\lambda}(A)$.

Moreover, let $Y \in M_n(\mathbb{R})$ *be the diagonal matrix such that* $Y_{j,j} = y_j$. *Then whenever* $\mathbf{x} \prec \mathbf{y}$, *there is an orthogonal matrix* R *such that* $\mathbf{x} = \vec{\delta}(R^*YR)$.

Theorem 11.3 is a consequence of several lemmas of independent interest. The first is due to Hardy, Littlewood, and Polya [**97**], and is part of a theorem of theirs proved below.

Lemma 11.4. *Let* $S \in M_n(\mathbb{R})$, *and let* $\mathbf{x}, \mathbf{y} \in \mathbb{R}^n$ *satisfy* $\mathbf{x} = S\mathbf{y}$. *If* S *is doubly stochastic, then* $\mathbf{x} \prec \mathbf{y}$. *If* $\mathbf{y} \ge 0$ *and* S *is doubly substochastic, then* $\mathbf{x} \prec_w \mathbf{y}$.

Proof. Suppose first that S is a permutation matrix. Then if $\mathbf{x} = S\mathbf{y}$, the entries of $\mathbf{x}$ are simply some permutation of the entries of $\mathbf{y}$, and hence for every $J \subset \{1, \ldots, n\}$ such that $|J| = k$,

$$(11.2.1) \qquad \sum_{j \in J} x_j = \sum_{j \in J} (\Pi \mathbf{y})_j \le \sum_{j=1}^{k} y_j^*.$$

Now suppose that S is doubly stochastic. By Birkhoff's theorem, S is a convex combination of permutation matrices, and then by (11.2.1), $\mathbf{x} = S\mathbf{y}$ implies that for $|J| = k$, $\sum_{j \in J} x_j \le \sum_{j=1}^{k} y_j^*$, and since $\sum_{j=1}^{n} S_{j,k} = 1$ for all k, there is equality when $k = n$. This proves the first statement.

Now suppose S is doubly substochastic. By Lemma A.21 there is a doubly stochastic matrix T such that $S_{j,k} \le T_{j,k}$ for all j, k. Define $\mathbf{z} = T\mathbf{y}$, and then by the first part $\mathbf{z} \prec \mathbf{y}$. Since $\mathbf{x} \le \mathbf{z}$, $\mathbf{x} \prec_w \mathbf{y}$, and this proves the second part. $\qquad\square$

A result of Schur, proves the sufficiency in Horn's theorem.

Lemma 11.5. *Let $A \in M_n(\mathbb{C})$ be self-adjoint. Then $\vec{\delta}(A) \prec \vec{\lambda}(A)$.*

Proof. By the spectral theorem, there is a unitary $U \in M_n(\mathbb{C})$ such that $A = UYU^*$ where Y is diagonal and $\vec{\delta}(Y) = \vec{\lambda}(A)$. The ith diagonal entry of A is then $\sum_{j=1}^{n} |U_{i,j}|^2 \lambda_j(A)$, and since U is unitary $S_{i,j} = |U_{i,j}|^2$ is doubly stochastic. $\qquad\square$

Lemma 11.6. *Let $\mathbf{x}, \mathbf{y} \in \mathbb{R}^n$ satisfy $\mathbf{x} \prec \mathbf{y}$. Then there is an orthogonal matrix $R \in M_n(\mathbb{R})$ such that for $i = 1, \ldots, n$,*

$$(11.2.2) \qquad x_i = \sum_{j=1}^{n} |R_{i,j}|^2 y_j.$$

Our proof of Lemma 11.6, which is the heart of the proof of Horn's theorem, follows Chan and Li [**58**].

Proof. If π and π' are any permutations on $\{1, \ldots, n\}$ and R is any orthogonal matrix in $M_n(\mathbb{C})$, then $\widetilde{R} \in M_n(\mathbb{C})$ defined by $\widetilde{R}_{i,j} := R_{\pi(i),\pi'(j)}$ is also orthogonal. Then because $\mathbf{x}^*$ and $\mathbf{y}^*$ are permutations of $\mathbf{x}$ and $\mathbf{y}$, respectively, we may assume that $\mathbf{x} = \mathbf{x}^*$ and $\mathbf{y} = \mathbf{y}^*$.

First consider the case $n = 2$. Define $Y \in M_2(\mathbb{R})$ by $Y = \begin{bmatrix} y_1 & 0 \\ 0 & y_2 \end{bmatrix}$. Under our hypotheses on $\mathbf{x}$ and $\mathbf{y}$, $y_1 \ge x_1 \ge x_2 \ge y_2$. If $y_1 = y_2$, then $\mathbf{x} = \mathbf{y}$ and we may take $U = \mathbb{1}$. Hence we may assume that $y_1 > y_2$. Then

$$R := (y_1 - y_2)^{-1/2} \begin{bmatrix} (x_1 - y_2)^{1/2} & -(y_1 - x_1)^{1/2} \\ (y_1 - x_1)^{1/2} & (x_1 - y_2)^{1/2} \end{bmatrix}$$

is orthogonal and

$$R^*YR = \begin{bmatrix} x_1 & ((y_1 - x_1)(x_1 - y_2))^{1/2} \\ ((y_1 - x_1)(x_1 - y_2))^{1/2} & x_2 \end{bmatrix}.$$

Thus, for $i = 1, 2$, $x_i = \sum_{j=1}^{2} |R_{i,j}|^2 y_j$, and this proves the lemma for $n = 2$.

Now for $n \geq 3$ make the inductive hypothesis that the lemma is proved with $n-1$ in place of n. Since $\sum_{j=1}^{n} x_j = \sum_{j=1}^{n} y_j$, we must have $x_{j+1} \geq y_{j+1}$ for some $1 \leq j \leq n-1$. Let k denote the first such index j, so that $y_k \geq x_k \geq x_{k+1} \geq y_{k+1}$. Define $\mathbf{a} = (y_k, y_{k+1})$ and $\mathbf{b} = (x_k, x'_{k+1})$ where $x'_{k+1} = y_k + y_{k+1} - x_k$. Then $\mathbf{b} \prec \mathbf{a}$.

By what has been proved for $n = 2$, there is a 2×2 orthogonal matrix $\widetilde{R}$ such that with $\widetilde{Y} := \begin{bmatrix} y_k & 0 \\ 0 & y_{k+1} \end{bmatrix}$,

$$(11.2.3) \qquad \vec{\delta}(\widetilde{R}^* \widetilde{Y} \widetilde{R}) = (x_k, x'_{k+1}).$$

Now define an $n \times n$ orthogonal matrix V by setting $V_{k,k} = \widetilde{R}_{1,1}$, $V_{k+1,k+1} = \widetilde{R}_{2,2}$, $V_{k,k+1} = \widetilde{R}_{1,2}$ and $V_{k+1,k} = \widetilde{R}_{2,1}$, with all other entries of U being the corresponding entries of the identity matrix $\mathbb{1}$. Let Y be the diagonal matrix with $\delta_j(Y) = y_j$. Then $V^* Y V$ is an $n \times n$ matrix whose off-diagonal entries, other than $(V^* Y V)_{k,k+1}$ and $(V^* Y V)_{k+1,k}$, are all zero, and moreover $(V^* Y V)_{k,k} = x_k$. Define $\mathbf{z} := \vec{\delta}(V^* Y V)$. We now claim

$$(11.2.4) \qquad \qquad \mathbf{x} \prec \mathbf{z}.$$

To see this, note that by construction, $z_j = y_j$ except possibly for $j = k$ and $j = k+1$, but in any case $z_k + z_{k+1} = y_k + y_{k-1}$. Moreover, since $\mathbf{y} = \mathbf{y}^*$, and since

$$y_k \geq \max\{x_k, x'_{k+1}\} \geq \min\{x_k, x'_{k+1}\} \geq y_{k+1},$$

which implies $z_{k-1} \geq \max\{z_k, z_{k+1}\} \geq \max\{z_k, z_{k+1}\} \geq z_{k+2}$, either $\mathbf{z}^* = \mathbf{z}$, or else $\mathbf{z}^*$ is obtained from $\mathbf{z}$ by swapping the the positions of z_k and z_{k+1}.

Thus, $z_j^* = y_j$, except possibly for $j = k$ and $j = k+1$, but in any case $z_k^* + z_{k+1}^* = y_k + y_{k-1}$. It follows immediately that $\sum_{j=1}^{\ell} z_j^* = \sum_{j=1}^{\ell} y_j$ for all $\ell \neq k$. Since $\mathbf{x} \prec \mathbf{y}$, $\sum_{j=1}^{\ell} y_j \geq \sum_{j=1}^{\ell} x_j$ for all ℓ, and this proves that at least when $\ell \neq k$,

$$(11.2.5) \qquad \qquad \sum_{j=1}^{\ell} z_j^* \geq \sum_{j=1}^{\ell} x_j.$$

We now show (11.2.5) is also true for $\ell = k$. Indeed, since $z_k^* = \max\{x_k, x'_{k+1}\} \geq x_k$,

$$\sum_{j=1}^{k} z_j^* = \sum_{j=1}^{k-1} y_j + z_k^* \geq \sum_{j=1}^{k-1} x_j + x_k = \sum_{j=1}^{k} x_j,$$

and this completes the proof of (11.2.4).

Now consider permutations of $\mathbf{x}$, $\mathbf{y}$, and $\mathbf{z}$ defined as follows. Let $\hat{\mathbf{x}}$ be given by $\hat{x}_1 = x_k$, $\hat{x}_2 = x_{k+1}$, and then with the remaining entires of $\mathbf{x}$ listed in decreasing order. Define $\hat{\mathbf{x}}$ and $\hat{\mathbf{z}}$ in the analogous manner, and note that $\hat{x}_1 = \hat{z}_1 = x_k$.

Now let W_1 be the $n \times n$ orthogonal matrix that is the identity with its upper-left 2×2 block replaced by the orthogonal matrix $\widetilde{R}$ figuring in (11.2.3). Then evidently $\hat{\mathbf{z}} = \vec{\delta}(W_1^* \widehat{Y} W_1)$, where $\widehat{Y}$ is the diagonal matrix with $\vec{\delta}(\widehat{Y}) = \hat{\mathbf{y}}$.

Next, since $\hat{x}_1 = \hat{z}_1 = x_k$, the vectors $\tilde{\mathbf{x}}$ and $\tilde{\mathbf{z}}$ in $\mathbb{R}^{n-1}$ obtained by deleting the first entries of $\hat{\mathbf{x}}$ and $\hat{\mathbf{z}}$, respectively, satisfy $\tilde{\mathbf{x}} \prec \tilde{\mathbf{z}}$. Then by the inductive hypothesis, there exists an orthogonal matrix $T \in M_{n-1}(\mathbb{R})$ such that for $i = 1, \dots, n-1$, $\tilde{x}_i = \sum_{j=1}^{n-1} |T_{i,j}|^2 \tilde{z}_j$ or equivalently such that $\tilde{\mathbf{x}} = \vec{\delta}(T^* \widetilde{Z}' T)$, where $\widetilde{Z}$ is the diagonal matrix in $M_{n-1}(\mathbb{R})$ with $\vec{\delta}(\widetilde{Z}) = \tilde{\mathbf{z}}$.

Define W_2 to be the $n \times n$ orthogonal matrix that is the identity with its lower right $(n-1) \times (n-1)$ block replaced by T. Then since the upper left entry of $W_1^* \widehat{Y} W_1$ is $\hat{x}_1 = x_k$, and since the lower right $(n-1) \times (n-1)$ block of this matrix diagonal with the vector diagonal entries being $\tilde{\mathbf{z}}$,

$$\hat{\mathbf{x}} = \vec{\delta}((W_1 W_2)^* \widehat{Y}(W_1 W_2)).$$

Therefore, with $R := W_1 W_2$, which is orthogonal, for all $i = 1, \dots, n$, $\hat{x}_i = \sum_{j=1}^n |R_{i,j}|^2 \hat{y}_j$. Finally by initial remarks in the proof, since $\hat{\mathbf{x}}$ and $\hat{\mathbf{y}}$ are permutations of $\mathbf{x}$ and $\mathbf{y}$, respectively, this implies (11.2.2) for another orthogonal matrix R. $\qquad\square$

The next theorem, an earlier result of Hardy, Littlewood, and Pólya [**97**], gives another characterization of majorization and also a characterization of weak majorization.

Theorem 11.7 (Hardy, Littlewood, and Pólya). *Let $\mathbf{x}, \mathbf{y} \in \mathbb{R}^n$. Then $\mathbf{x} \prec \mathbf{y}$ if and only if there is a doubly stochastic matrix S such that $\mathbf{x} = S\mathbf{y}$. If $\mathbf{x}, \mathbf{y} \geq 0$, $\mathbf{x} \prec_w \mathbf{y}$ if and only if there is a doubly substochastic matrix S such that $\mathbf{x} = S\mathbf{y}$, and there is a $\mathbf{z} \in \mathbb{R}^n$ such that*

$$\mathbf{x} \leq \mathbf{z} \prec \mathbf{y}.$$

Proof. Suppose that $\mathbf{x} \prec \mathbf{y}$. Then by Horn's theorem, there is an orthogonal matrix $R \in M_n(\mathbb{R})$ such that for $i = 1, \dots, n$, $x_i = \sum_{j=1}^n |R_{i,j}|^2 y_j$. The matrix $S \in M_n(\mathbb{R})$ defined by $S_{i,j} = |R_{i,j}|^2$ is doubly stochastic. Therefore, when $\mathbf{x} \prec \mathbf{y}$, there is a doubly stochastic matrix S such that $\mathbf{x} = S\mathbf{y}$. Next suppose that there is a doubly stochastic matrix S such that $\mathbf{x} = S\mathbf{y}$. Then by Lemma 11.4, $\mathbf{x} \prec \mathbf{y}$. This proves the first part.

For the rest, suppose $\mathbf{x}, \mathbf{y} \geq 0$. If there is a doubly substochastic matrix S such that $\mathbf{x} = S\mathbf{y}$, then by Lemma 11.4, $\mathbf{x} \prec_w \mathbf{y}$. If $\mathbf{x} \prec \mathbf{y}$, define $a := \sum_{j=1}^n y_j - \sum_{j=1}^n x_j$ and vectors $\hat{\mathbf{x}}$ and $\hat{\mathbf{y}}$ in $\mathbb{R}^{n+1}$ by $\hat{\mathbf{x}} = (x_1, \dots, x_n, a)$ and $\hat{\mathbf{y}} = (y_1, \dots, y_n, 0)$. Then $\hat{\mathbf{x}} \prec \hat{\mathbf{y}}$, and then by Horn's theorem, there is a doubly stochastic matrix $\hat{S} \in M_{n+1}(\mathbb{R})$ such that $\hat{\mathbf{x}} = \hat{S}\hat{\mathbf{y}}$. Then define S to be the upper left $n \times n$ block of $\hat{S}$, $\mathbf{x} = S\mathbf{y}$, and S to be doubly substochastic. Finally, by Lemma A.21, there exists a doubly stochastic matrix $\tilde{S}$ such that for each i, j, $S_{i,j} \leq \tilde{S}_{i,j}$. Define $\mathbf{z} = \tilde{S}\mathbf{y}$. $\qquad\qquad\square$

11.3. Majorization and convexity

The following theorem is due to Hardy, Littlewood, and Pólya [**96**].

Theorem 11.8. *Let* $f : \mathbb{R} \to \mathbb{R}$ *be a convex function. Then*

$$\mathbf{x} \prec \mathbf{y} \quad \Rightarrow \quad f(\mathbf{x}) \prec_w f(\mathbf{y}),$$

and moreover, if f is strictly convex, $f(\mathbf{x}) \prec f(\mathbf{y})$ if and only if $\mathbf{x}$ is a permutation of $\mathbf{y}$. Moreover, if f is an increasing convex function, then

$$\mathbf{x} \prec_w \mathbf{y} \quad \Rightarrow \quad f(\mathbf{x}) \prec_w f(\mathbf{y}),$$

and if f is strictly increasing and strictly convex, $f(\mathbf{x}) \prec f(\mathbf{y})$ if and only if $\mathbf{x}$ is a permutation of $\mathbf{y}$.

Proof. Suppose that $\mathbf{x} \prec \mathbf{y}$ and write $\mathbf{x}$ as a convex combination of permutations of $\mathbf{y}$: $\mathbf{x} = \sum_{k=1}^{n} t_k \Pi^{(k)} \mathbf{y}$. Define $S := \sum_{k=1}^{n} t_k \Pi^{(k)}$, and note that S is doubly stochastic. Hence for each $1 \leq i \leq n$, Jensen's inequality yields

$$(11.3.1) \qquad f(x_i) = f\left(\sum_{j=1}^{n} S_{i,j} y_j\right) \leq \sum_{j=1}^{n} S_{i,j} f(y_j);$$

that is, $f(\mathbf{x}) \leq Sf(\mathbf{y})$. By Theorem 11.7, $Sf(\mathbf{y}) \prec f(\mathbf{y})$, and therefore, $f(\mathbf{x}) \prec_w f(\mathbf{y})$.

Now suppose that f is strictly convex. Let $\mathcal{A}$ denote the set of distinct values in $\{y_1, \ldots, y_n\}$. By Jensen's inequality for strictly convex functions, if there is equality in (11.3.1), then y_j has the same value for all j such that $S_{i,j} > 0$, and in this case $x_i \in \mathcal{A}$. Since this is true for each i, every entry of $\mathbf{x}$ belongs to $\mathcal{A}$.

Next, for each $a \in \mathcal{A}$ define $Y_a := \{j \ : \ y_j = a\}$ and $X_a := \{i \ : \ x_i = a\}$. Let $|X_a|$ and $|Y_a|$ denote cardinality of X_a and Y_a, respectively. Then by what we have said above, for each $i \in A_a$, $S_{i,j} > 0$ only if $j \in Y_a$, and hence for such i, $\sum_{j \in Y_a} S_{i,j} = 1$. Therefore, $\sum_{i \in X_a, j \in Y_a} S_{i,j} = \sum_{i \in X_a} 1 = |X_a|$. Summing in the other order, $\sum_{i \in X_a, j \in Y_a} S_{i,j} \leq \sum_{j \in Y_a} \left(\sum_{i=1}^{n} S_{i,j}\right) = \sum_{j \in Y_a} 1 = |Y_a|$. Hence $|X_a| \leq |Y_a|$ for all a. Then since $\sum_{a \in \mathcal{A}} |X_a| = \sum_{a \in \mathcal{A}} |Y_a| = n$, $|X_a| = |Y_a|$ for all $a \in \mathcal{A}$, and thus $\mathbf{x}$ is a permutation of $\mathbf{y}$. This proves the first part.

Suppose that f is increasing and convex. If $\mathbf{x} \prec_w \mathbf{y}$, by Theorem 11.7 there exists $\mathbf{z}$ such that $\mathbf{x} \leq \mathbf{z} \prec \mathbf{y}$. By the first part, $f(\mathbf{z}) \prec_w f(\mathbf{y})$. Since $\mathbf{x} \leq \mathbf{z}$, $f(\mathbf{x}) \leq f(\mathbf{z})$ so that $f(\mathbf{x}) \prec_w f(\mathbf{z})$. If $f(\mathbf{x}) \prec f(\mathbf{y})$, we must have $f(\mathbf{x}) = f(\mathbf{z})$, and since f is strictly increasing, this means that $\mathbf{x} = \mathbf{z}$, and hence $\mathbf{x} \prec \mathbf{y}$. By the first part, $\mathbf{x}$ is a permutation of $\mathbf{y}$. $\qquad\qquad\square$

Definition 11.9 (Majorization for Hermitian matrices). Let A and B be $n \times n$ Hermitian matrices. Then B *majorizes* A, written $A \prec B$, in case $\vec{\lambda}(A) \prec \vec{\lambda}(B)$, and B *weakly majorizes* A, written $A \prec_w B$, in case $\vec{\lambda}(A) \prec_w \vec{\lambda}(B)$.

Theorem 11.10. *Let* $f : \mathbb{R} \to \mathbb{R}$ *be a convex function. Then for all* $n \times n$ *Hermitian matrices* A *and* B,

$$A \prec B \quad \Rightarrow \quad f(A) \prec_w f(B),$$

and moreover, if f *is strictly convex,* $f(A) \prec f(B)$ *if and only if there is a unitary matrix* U *such that* $A = U^*BU$.

If f *is an increasing convex function, then*

$$A \prec_w B \quad \Rightarrow \quad f(A) \prec_w f(B),$$

and if f *is strictly increasing and strictly convex,* $f(A) \prec f(B)$ *if and only if there is a unitary matrix* U *such that* $A = U^*BU$.

Proof. We need only apply Theorem 11.8, and recall that $\vec{\lambda}(A)$ is a permutation of $\vec{\lambda}(B)$ if and only if there is a unitary matrix U such that $A = U^*BU$. $\qquad\square$

The following corollary is the source of many trace inequalities.

Corollary 11.11. *Let* $A, B \in M_n(\mathbb{C})$ *be self-adjoint, and let* f *be a convex function on* $\mathbb{R}$. *If* $A \prec B$, *or if* $A \prec_w B$ *and* f *is increasing as well as convex,*

$$\mathrm{Tr}[f(A)] \leq \mathrm{Tr}[f(B)].$$

Proof. Whenever $f(A) \prec_w f(B)$, meaning $\vec{\lambda}(A) \prec_w \vec{\lambda}(B)$,

$$\sum_{j=1}^{n} \lambda_j(A) \leq \sum_{j=1}^{n} \lambda_j(B). \qquad\square$$

11.4. Ky Fan's formula and matrix majorization

To apply Theorem 11.10, we need conditions that imply $A \prec B$ or $A \prec_w B$ for $n \times n$ Hermitian matrices A and B. One important source of these is a variational formula due to Ky Fan [80, 81].

Definition 11.12. For each $k = 1, \ldots, n$, let $\mathcal{C}_k$ denote the set of $n \times n$ matrices C such that

$$0 \leq C \leq \mathbb{1} \qquad \text{and} \qquad \mathrm{Tr}[C] = k.$$

Let $\widehat{\mathcal{C}}_k$ denote the set of sequences $\mathbf{c} \in \mathbb{R}^n$ with $0 \leq c_j \leq 1$ for all j, and $\sum_{k=1}^{n} c_j = k$.

Theorem 11.13. *Let $A \in M_n^{\mathrm{s.a.}}(\mathbb{C})$, and let $(\lambda_1,\ldots,\lambda_n) := \vec{\lambda}^*(A)$, the decreasing rearrangement of the eigenvalue sequence of A. For each $k \in (1,\ldots,n)$, define $S_k(A) = \sum_{j=1}^{k} \lambda_j$. Then*

$$S_k(A) = \max\{\mathrm{Tr}[AC] \ : \ C \in \mathcal{C}_k\} = \max\{\mathrm{Tr}[AP] \ : \ P \in \mathcal{P}_k\},$$

where $\mathcal{P}_k$ is the set of all orthogonal projections onto k-dimensional subspaces of $\mathbb{C}^n$.

We first prove the following lemma:

Lemma 11.14. *The set $\mathcal{C}_k$ is compact and convex, and the set of extreme points of $\mathcal{C}_k$ is the set $\mathcal{P}_k$. The set $\widehat{\mathcal{C}}_k$ is compact and convex, and the set of its extreme points is the set of vectors $\{\mathbf{e}_J \ : \ J \subset \{1,\ldots,n\}, \ |J| = k\}$, where $(\mathbf{e}_J)_j = 1$ if $j \in J$ and $(\mathbf{e}_J)_j = 0$ if $j \notin J$.*

Proof. The convexity and compactness is evident in both cases. Suppose $\mathbf{c} \in \widehat{\mathcal{C}}_k$ and $c_j > 0$ for more than k values of j. Then there are at least two indices j_1, j_2 such that $0 < c_{j_1}, c_{j_2} < 1$. Let $a := \min\{c_{j_1}, c_{j_2}, 1 - c_{j_1}, 1 - c_{j_2}\} > 0$. Define two elements $\mathbf{c}_\pm$ of $\mathbb{R}^n$ by

$$(\mathbf{c}_\pm)_j = \begin{cases} c_{j_1} \pm a & j = j_1 \\ c_{j_2} \mp a & j = j_2 \\ c_j & j \neq j_1, j_2. \end{cases}$$

Then $\mathbf{c} = \frac{1}{2}(\mathbf{c}_= + \mathbf{c}_-)$ and $\mathbf{c}_+$ and $\mathbf{c}_-$ are distinct elements of $\widehat{\mathcal{C}}_k$. Therefore, $\mathbf{c}$ is not an extreme point of $\widehat{\mathcal{C}}_k$, and any extreme point must have exactly k nonzero entries, each of which must be 1.

Let $C \in \mathcal{C}_k$, and let $\{\phi_1, \ldots, \phi_n\}$ be an orthonormal basis of $\mathbb{C}^n$ consisting of eigenfunctions of C with $C\phi_j = c_j \phi_j$. Then $C = \sum_{k=1}^{n} c_k |\phi_k\rangle\langle\phi_k|$. The vector $\mathbf{c} = (c_1,\ldots,c_n)$ belongs to $\widehat{\mathcal{C}}_k$, and if $\mathbf{c}$ is not extreme in $\widehat{\mathcal{C}}_k$, C is not extreme in $\mathcal{C}_k$. Hence for C to be extreme, it is necessary that $\mathbf{c} = \mathbf{e}_J$ for some $J \subset \{1,\ldots,n\}$ with $|J| = k$. In this case C is an orthogonal projection onto a k-dimensional subspace. It is evident that any orthogonal projection onto a k-dimensional subspace is extreme in $\mathcal{C}_k$. Evidently, C is extreme in $\mathcal{C}_k$ only if $\mathbf{c}$ is extreme in $\widehat{\mathcal{C}}_k$, and this is the case only if $\mathbf{c}$ is such that $c_j \in \{0, 1\}$ for each j. $\qquad\square$

Proof of Theorem 11.13. Let $C \in \mathcal{C}_k$, and let $\{\phi_1, \ldots, \phi_n\}$ be an orthonormal basis of $\mathbb{C}^n$ consisting of eigenvectors of A such that $A\phi_j = \lambda_j$. Define $c_j = \langle \phi_j, C\phi_j \rangle$. Then

$$(11.4.1) \qquad\qquad \mathrm{Tr}[AC] = \sum_{j=1}^{n} \lambda_j c_j.$$

Since $\mathbf{c} = (c_1, \ldots, c_n) \in \widehat{\mathcal{C}}_k$, Lemma 11.14 says that $\mathbf{c}$ is a convex combination of vectors $\mathbf{e}_J$ with $|J| = k$, and then by (11.4.1), $\mathrm{Tr}[AC]$ is a convex combination of numbers of the form $\sum_{j \in J} \lambda_j$, and each of these is no larger than $S_k(A)$. Thus for all $C \in \mathcal{C}_k$, $\mathrm{Tr}[AC] \leq S_k(A)$. This upper bound is achieved by choosing $C = \sum_{j=1}^{k} |\phi_k\rangle\langle\phi_k|$, which also belongs to $\mathcal{P}_k$, proving the second equality. $\square$

Corollary 11.15. *Let* $A, B \in M_n^{\mathrm{s.a.}}(\mathbb{C})$. *Then*

$$A \leq B \quad \Rightarrow \quad \vec{\lambda}^*(A) \leq \vec{\lambda}^*(B) \quad \Rightarrow \quad A \prec_w B.$$

Furthermore, if P is any orthogonal projection,

$$\vec{\lambda}^*(PAP) \leq \vec{\lambda}^*(A),$$

and hence $PAP \prec_w A$.

Proof. For any $P \in \mathcal{P}_k$, $\mathrm{Tr}[AP] = \mathrm{Tr}[PAP] \leq \mathrm{Tr}[PBP] = \mathrm{Tr}[BP]$, and now $S_k(A) \leq S_k(B)$ follows from Theorem 11.13. Next observe that PAP and $A^{1/2}PA^{1/2}$ have the same eigenvalues with the same multiplicities, and $A^{1/2}PA^{1/2} \leq A$, so the second statement follows from the first. $\square$

Corollary 11.16. *For* $A, B \in M_n^{\mathrm{s.a.}}(\mathbb{C})$,

$$\vec{\lambda}^*(A + B) \prec \vec{\lambda}^*(A) + \vec{\lambda}^*(B).$$

Proof. Let $P \in \mathcal{P}_k$ be such that $S_k(A + B) = \mathrm{Tr}[P(A + B)]$. Then

$$S_k(A + B) = \mathrm{Tr}[PA] + \mathrm{Tr}[PB] \leq S_k(A) + S_k(B)$$

for each $1 \leq k \leq n$. $\square$

11.5. The Courant-Fischer theorem and matrix majorization

The Courant-Fischer theorem, recalled next, is another important source of matrix majorization criteria.

Theorem 11.17 (Courant-Fischer theorem). *Let* $A \in M_n(\mathbb{C})$ *be self-adjoint, and let $\vec{\lambda}^*(A)$ be the eigenvalue sequence of A arranged in decreasing order. Let W denote a subspace of $\mathbb{C}^n$. Then for each $1 \leq k \leq n$,*

$$(11.5.1) \qquad \lambda_k^*(A) = \max_{\{\dim(W)=k\}} \left(\min_{\{\mathbf{u} \in W,\, \|\mathbf{u}\|=1\}} \langle \mathbf{u}, A\mathbf{u} \rangle \right).$$

Proof. Let $\{\mathbf{u}_1, \ldots, \mathbf{u}_n\}$ be an orthonormal basis of $\mathbb{C}^n$ such that for each $j = 1, \ldots, n$, $A\mathbf{u}_j = \lambda_j^*(A)\mathbf{u}_j$. For each $1 \leq k \leq n$, let $\mathcal{U}_k := \mathrm{span}(\{\mathbf{u}_j \mid j \geq k\})$, and note that $\dim(\mathcal{U}_k) = n - k + 1$. Therefore, if $\dim(W) = k$, $\dim(W \cap \mathcal{U}_k) \geq 1$. Let $\mathbf{u}$ be a unit vector in $W \cap \mathcal{U}_k$. Then since $\mathbf{u} \in \mathcal{U}_k$, $\mathbf{u} = \sum_{j=k}^{n} a_j \mathbf{u}_j$ with

$\sum_{j=k}^{n} |a_j|^2 = 1$, and hence $\langle \mathbf{u}, A\mathbf{u} \rangle = \sum_{j=k}^{n} \lambda_j^*(A)|a_j|^2 \leq \lambda_k^*(A)$, showing that for all subspaces W with $\dim(W) = k$, $\min_{\{\mathbf{u} \in W, \, \|\mathbf{u}\|=1\}} \langle \mathbf{u}, A\mathbf{u} \rangle \leq \lambda_k^*(A)$. This proves that the left hand side of (11.5.1) is no larger than λ_k^*.

On the other hand, define $W_0 = \mathrm{span}(\{\mathbf{u}_1, \ldots, \mathbf{u}_k\}$. If $\mathbf{u}$ is a unit vector in W_0, then $\mathbf{u} = \sum_{j=1}^{k} b_j \mathbf{u}_j$ with $\sum_{j=1}^{k} |b_j|^2 = 1$, so that $\langle \mathbf{u}, A\mathbf{u} \rangle = \sum_{j=1}^{k} \lambda_j^* |b_j|^2 \geq \lambda_k^*(A)$, with equality if $\mathbf{u} = \eta_k$. Since $\dim(W_0) = k$, this proves that the left side of (11.5.1) is at least as large as $\lambda_k^*(A)$. $\square$

Remark 11.18. Note that $\lambda_k^*(A) = -\lambda_{n-k+1}^*(-A)$, and for this reason the following is equivalent to (11.5.1):

$$(11.5.2) \qquad \lambda_k^*(A) = \min_{\{\dim(W)=n-k+1\}} \left(\max_{\{\mathbf{u} \in W, \, \|\mathbf{u}\|=1\}} \langle \mathbf{u}, A\mathbf{u} \rangle \right).$$

For all $A \in M_n(\mathbb{C})$, there are two Hermitian matrices naturally associated to A, namely

$$\mathrm{Re}(A) = \frac{1}{2}(A + A^*) \qquad \text{and} \qquad |A| = (A^*A)^{1/2}.$$

Theorem 11.19. *For all $n \times n$ matrices A,*

$$\vec{\lambda}^*(\mathrm{Re}(A)) \leq \vec{\lambda}^*(|A|),$$

or, what is the same thing, there exists a unitary matrix U such that

$$\mathrm{Re}(A) \leq U^*|A|U.$$

Proof. Let $\{\mathbf{u}_1, \ldots, \mathbf{u}_n\}$ be an orthonormal basis of $\mathbb{C}^n$ consisting of eigenvectors of $|A|$ with $|A|\mathbf{u}_j = \lambda_j^*(|A|)\mathbf{u}_j$ and $\lambda_j^*(|A|) \geq \lambda_{j+1}^*(|A|)$ for all $0 \leq j \leq n-1$. Fix k with $1 \leq k \leq n$, and define $W := \mathrm{span}(\{\mathbf{u}_k, \ldots, \mathbf{u}_n\})$, and note that $\dim(W) = n - k + 1$. Then by (11.5.2), and the Cauchy-Schwarz inequality,

$$\lambda_k^*(\mathrm{Re}(A)) \leq \max_{\{\mathbf{u} \in W, \, \|\mathbf{u}\|=1\}} \langle \mathbf{u}, \tfrac{1}{2}(A + A^*)\mathbf{u} \rangle \leq \max_{\{\mathbf{u} \in W, \, \|\mathbf{u}\|=1\}} \|A\mathbf{u}\| = \lambda_k^*(|A|). \quad \square$$

Corollary 11.20. *For all $A, B \in M_n(\mathbb{C})$, there exist unitary matrices V and W such that*

$$|A + B| \leq V^*|A|V + W^*|B|W.$$

Proof. Let $A + B = U|A + B|$ be the polar decomposition of $A + B$. Then

$$|A + B| = U^*(A + B) = \mathrm{Re}(U^*A) + \mathrm{Re}(U^*B).$$

Since $|U^*A| = |A|$ and $|U^*B| = |B|$, it follows from Theorem 11.19 that there exist unitary matrices V and W such that $\mathrm{Re}(U^*A) \leq V^*|A|V$ and $\mathrm{Re}(U^*B) \leq W^*|B|W$. $\square$

The Courant-Fischer theorem has many other applications to eigenvalue inequalities that do not directly involve majorization. However, since these results are useful, we prove two of the more important of these here.

Theorem 11.21 (Weyl's inequality). *For all self-adjoint A and B in $M_n(\mathbb{C})$, and all $i, j > 0$ with $i + j \leq n + 1$, and all $1 \leq k \leq n$,*

$$(11.5.3) \qquad \lambda^*_{i+j-1}(A + B) \leq \lambda^*_j(A) + \lambda^*_i(B).$$

As a consequence, for all $k = 1, \ldots, n$,

$$(11.5.4) \qquad |\lambda^*_k(A + B) - \lambda^*_k(A)| \leq \|B\|.$$

Proof. As in the proof of Theorem 11.17, let $\{\mathbf{u}_1, \ldots, \mathbf{u}_n\}$ be an orthonormal basis of $\mathbb{C}^n$ such that for each $j = 1, \ldots, n$, $A\mathbf{u}_j = \lambda^*_j(A)\mathbf{u}_j$, and for each $1 \leq k \leq n$, let $\mathcal{U}_k := \mathrm{span}(\{\mathbf{u}_j \mid j \geq k\})$. Likewise, let $\{\mathbf{v}_1, \ldots, \mathbf{v}_n\}$ be an orthonormal basis of $\mathbb{C}^n$ such that for each $j = 1, \ldots, n$, $B\mathbf{v}_j = \lambda^*_j(B)\mathbf{v}_j$, and for each $1 \leq k \leq n$, let $\mathcal{V}_k := \mathrm{span}(\{\mathbf{v}_j \mid j \geq k\})$.

Let $i, j \in \mathbb{N}$, satisfying $n \geq i+j-1$. Then $\dim(U_j)+\dim(V_i) = 2n-j-i+2$, and hence $\dim((\mathcal{U}_j \cap \mathcal{V}_i)^{\perp}) \leq i+j-2$, and hence $\dim(\mathcal{U}_j \cap \mathcal{V}_i) \geq n-i-j+2$.

Now let $\mathcal{W}$ be any subspace with $\dim(\mathcal{W}) = i+j-1$. Since $\dim(\mathcal{U}_j \cap \mathcal{V}_i)+\dim(\mathcal{W}) \geq n+1$, there exists a nonzero unit vector $\mathbf{u} \in \mathcal{U}_j \cap \mathcal{V}_i \cap \mathcal{W}$. Let $\mathbf{u}$ be any such unit vector. Then since $\mathbf{u} \in \mathcal{U}_j \cap \mathcal{V}_i$,

$$\langle \mathbf{u}, A\mathbf{u} \rangle \leq \lambda^*_j(A) \quad \text{and} \quad \langle \mathbf{u}, B\mathbf{u} \rangle \leq \lambda^*_i(B),$$

as in the proof of Theorem 11.17. Therefore,

$$\langle \mathbf{u}, (A + B)\mathbf{u} \rangle = \langle \mathbf{u}, A\mathbf{u} \rangle + \langle \mathbf{u}, B\mathbf{u} \rangle \leq \lambda^*_j(A) + \lambda^*_i(B),$$

and hence for all subspaces $\mathcal{W}$ of dimension $i + j - 1$,

$$\min_{\{\mathbf{u} \in \mathcal{W}, \, \|\mathbf{u}\|=1\}} \langle \mathbf{u}, A\mathbf{u} \rangle \leq \lambda^*_j(A) + \lambda^*_i(B),$$

and then (11.5.3) is an immediate consequence of the Courant-Fischer theorem.

To prove (11.5.4), take $i = k$ and $j = 1$, and then from (11.5.3), $\lambda^*_k(A + B) - \lambda^*_k(A) \leq \lambda^*_1(B)$. Taking $i = n + 1 - k$ and $j = 1$, and replacing A and B by $-A$ and $-B$, respectively, yields $\lambda^*_{n+1-k}(-A - B) - \lambda^*_{n+1-k}(-A) \leq \lambda^*_1(-B)$. Then using the fact that $\lambda^*_{n-k}(-A) = -\lambda^*_k(A)$, and likewise for the other matrices, this gives $\lambda^*_k(A) - \lambda^*_k(A + B) \leq -\lambda^*_n(B)$. Since $\max\{\lambda^*_1(B), -\lambda^*_n(B)\} = \|B\|$, these inequalities prove (11.5.4). $\qquad\square$

11.6. Log-majorization

Definition 11.22. For $\mathbf{x}, \mathbf{y} \in \mathbb{R}^n$, $\mathbf{x}, \mathbf{y} \geq 0$, we say that $\mathbf{y}$ **weakly log-majorizes** $\mathbf{x}$, $\mathbf{x} \prec_{w\log} \mathbf{y}$, in case for all $k = 1, \ldots, n$,

$$(11.6.1) \qquad \prod_{j=1}^{k} \mathbf{x}_j^* \leq \prod_{j=1}^{k} \mathbf{y}_j^*.$$

We say that $\mathbf{y}$ **log-majorizes** $\mathbf{x}$, $\mathbf{x} \prec_{\log} \mathbf{y}$, in case, in addition, there is equality in (11.6.1) for $k = n$. Extend these notions to self-adjoint matrices exactly as in Definition 11.9.

For $\mathbf{x}, \mathbf{y}$ such that all entries are strictly positive, it is evident that

$$(11.6.2) \qquad \mathbf{x} \prec_{w\log} \mathbf{y} \quad \Longleftrightarrow \quad \log(\mathbf{x}) \prec_w \log(\mathbf{y})$$

and

$$\mathbf{x} \prec_{\log} \mathbf{y} \quad \Longleftrightarrow \quad \log(\mathbf{x}) \prec \log(\mathbf{y}).$$

Lemma 11.23. *Suppose $\mathbf{x}, \mathbf{y} \in \mathbb{R}^n$ and $\mathbf{x} \prec_{w\log} \mathbf{y}$. There exists $\mathbf{z} \in \mathbb{R}^n$, $\mathbf{x}^* \leq \mathbf{z}$, such that $\mathbf{z} \prec_{w\log} \mathbf{y}$, and finally such that $\mathbf{z}$ and $\mathbf{y}$ have the same number of nonzero entries.*

Proof. Let ℓ be the maximum index such that $x_\ell^* > 0$, and let k be the maximum index such that $y_k^* > 0$. Then $k \geq \ell$. If $k = \ell$, we may take $\mathbf{z} = \mathbf{x}^*$. Therefore suppose that $k > \ell$, Define $z_j := x_j^*$ for $j = 1, \ldots, \ell$, and define $z_j = y_j^*$ for $j > \ell$. $\qquad\qquad\qquad\qquad\qquad\qquad\qquad\qquad\qquad\qquad\qquad$ $\square$

Theorem 11.24. *Let $\mathbf{x} \prec_{w\log} \mathbf{y}$. Then $\mathbf{x} \prec_w \mathbf{y}$ and if $\mathbf{x} \prec \mathbf{y}$, then $\mathbf{x}$ is a permutation of $\mathbf{y}$. Moreover, let $f : \mathbb{R} \to \mathbb{R}$ increasing and such that $f(e^t)$ is a convex function of t. Then*

$$\mathbf{x} \prec_{w\log} \mathbf{y} \quad \Rightarrow \quad f(\mathbf{x}) \prec_w f(\mathbf{y}),$$

and if $f(e^t)$ is a strictly convex function of t, $f(\mathbf{x}) \prec f(\mathbf{y})$ only when $\mathbf{x}$ is a permutation of $\mathbf{y}$.

Proof. If any entries of $\mathbf{x}$ are zero, by Lemma 11.23 there exists $\mathbf{z}$ with $\mathbf{z} \prec_{w\log} \mathbf{y}$, $\mathbf{x}^* \leq \mathbf{z}$ and such that $\mathbf{z}$ and $\mathbf{y}$ have the same number of nonzero entries. Let $\mathbf{z}'$ and $\mathbf{y}'$ be the vectors obtained from them by discarding the zero entries. Then $\mathbf{z}', \mathbf{y}' > 0$ and

$$(11.6.3) \qquad \log(\mathbf{z}') \prec_w \log(\mathbf{y}').$$

Since the exponential function is strictly convex and strictly increasing, $\mathbf{z}' \prec_w \mathbf{y}'$ by Theorem 11.8 and $\mathbf{z}' \prec \mathbf{y}'$ if and only if $\mathbf{z}'$ is a permutation of $\mathbf{y}'$. Evidently, this implies that $\mathbf{z} \prec_w \mathbf{y}$, and that $\mathbf{z} \prec \mathbf{y}$ if and only if $\mathbf{z}$ is a permutation of $\mathbf{y}$. Then since $\mathbf{x}^* \le \mathbf{z}$, $\mathbf{x} \prec_w \mathbf{y}$ and $\mathbf{x} \prec \mathbf{y}$ requires that $\mathbf{x}^* = \mathbf{z}$, and then $\mathbf{z} \prec \mathbf{y}$, this implies that $\mathbf{x}$ is a permutation of $\mathbf{y}$. This proves the first part.

Next, let $f : \mathbb{R} \to \mathbb{R}$ be increasing and such that $f(e^t)$ is a convex function of t. Then by (11.6.3) and Theorem 11.8, for any increasing convex function g on $\mathbb{R}$, $g(\log \mathbf{z}') \prec_w g(\log \mathbf{y}')$, and if g is strictly convex and $g(\log \mathbf{z}') \prec g(\log \mathbf{y}')$, then $\log \mathbf{z}'$ is a permutation of $\log \mathbf{y}'$, or, what is the same thing, $\mathbf{z}$ is a permutation of $\mathbf{y}$.

For any function $f : [0, \infty) \to \mathbb{R}$, define $g : \mathbb{R} \to R$ by $g(\log s) = f(s)$ for all $s > 0$, or equivalently, $f(e^t) = g(t)$ for all $t \in \mathbb{R}$. Note that g is increasing if and only if f is increasing, and g is convex if and only if $f(e^t)$ is convex as a function of t.

Thus when f is increasing with $f(e^t)$ convex, $\mathbf{x} \prec_{w\log} \mathbf{y}$ implies that $g(\log \mathbf{x}) \prec_w g(\log \mathbf{y})$ which is equivalent to $f(\mathbf{x}) \prec_w f(\mathbf{y})$, and when $f(e^t)$ is a strictly convex function of t, $f(\mathbf{x}) \prec f(\mathbf{y})$ only when $\mathbf{x}$ is a permutation of $\mathbf{y}$. $\square$

Corollary 11.25. *Let $f : \mathbb{R} \to \mathbb{R}$ be increasing and such that $f(e^t)$ is a convex function of t. For positive semidefinite $A, B \in M_n(\mathbb{C})$,*

$$A \prec_{w\log} B \quad \Rightarrow \quad \mathrm{Tr}[f(A)] \le \mathrm{Tr}[f(B)],$$

and there is equality if and only if for some unitary $U \in M_n(\mathbb{C})$, $A = U^*BU$.

Proof. Suppose first that each entry of $\vec{\lambda}(A)$ and $\vec{\lambda}(B)$ is strictly positive. Then by (11.6.2), $\log(\vec{\lambda}(A)) \prec_w \log(\vec{\lambda}(B))$, and then by Theorem 11.8, with $g(t) = f(e^t)$, $g(\log(\vec{\lambda}(A))) \prec_w g(\log(\vec{\lambda}(B)))$, or what is the the same thing, $f(\vec{\lambda}(A)) \prec_w f(\vec{\lambda}(B))$. Moreover, again by Theorem 11.8, $f(\vec{\lambda}(A)) \prec f(\vec{\lambda}(B))$ if and only if $\vec{\lambda}(A)$ is a permutation of $\vec{\lambda}(B)$, and this is the case if and only if A and B are unitarily equivalent.

To relax the assumption that $\vec{\lambda}(A)$ and $\vec{\lambda}(B)$ are strictly positive, we may assume without loss of generality that $\vec{\lambda}(A) = \vec{\lambda}(A)^*$ and $\vec{\lambda}(B) = \vec{\lambda}(B)^*$. Let k be the least index such that $y_k^* = 0$, and if there is no such index, set $k = n+1$. Let ℓ be the least index such that $x_\ell^* = 0$, and if there is no such index, we are in the case $\vec{\lambda}(A), \vec{\lambda}(B) > 0$ treated above. So we may suppose that $k \le n$. If $k = \ell$, we may discard the final $n - k + 1$ common zero terms in the vectors $\vec{\lambda}(A)$ and $\vec{\lambda}(B)$, leaving p strictly positive vectors to which the result proved above applies. If $k < \ell$, we define $\mathbf{z}$ by $z_j = x_j$ for $j < k$, and $z_j = y_j$ for $k \le j < \ell$. Then $\mathbf{z} \prec_{w\log} \vec{\lambda}(B)$ and truncating both vectors after the entry $\ell - 1$ if $\ell < n + 1$, we conclude $f(\mathbf{z}) \prec_w f(\vec{\lambda}(B))$ and $f(\mathbf{z}) \prec f(\vec{\lambda}(B))$ only if $\mathbf{z}$ is a permutation of $\vec{\lambda}(B)$. But since f is strictly increasing, $\sum_{j=1}^{n} f(x_j) < \sum_{j=1}^{n} f(z_j) = \sum_{j=1}^{n} f(y_j)$. Hence $f(\vec{\lambda}(A)) \prec f(\vec{\lambda}(B))$ is impossible in this final case. $\qquad\square$

11.7. Weyl's method

Weyl's fundamental paper [**229**] introduced a powerful method for proving matrix inequalities.

Lemma 11.26 (Weyl's log-majorization lemma). *Let $\mathcal{H}$ be a Hilbert space of dimension d. Let A and B be operators on $\mathcal{H}$. If*

$$\left\| \bigwedge^{k} A \right\| \leq \left\| \bigwedge^{k} B \right\| \qquad \text{for all} \qquad 1 \leq k \leq d,$$

then

$$(11.7.1) \qquad\qquad \vec{\sigma}(A) \prec_{w\log} \vec{\sigma}(B),$$

and consequently, for all functions $f : [0, \infty) \to \mathbb{R}$ such that f is monotone nondecreasing and such that $f(e^t)$ is a convex function of t,

$$(11.7.2) \qquad\qquad \sum_{j=1}^{d} f(\sigma_j(A)) \leq \sum_{j=1}^{d} f(\sigma_j(B)),$$

and there is equality if and only if $\vec{\sigma}(A)$ is a permutation of $\vec{\sigma}(B)$. Consequently, if $A, B \geq 0$,

$$(11.7.3) \qquad\qquad \mathrm{Tr}[f(A)] \leq \mathrm{Tr}[f(B)]$$

and there is equality if and only if for some $U \in \mathcal{B}(\mathcal{H})$, $A = UBU^*$.

Proof. Without loss of generality we may suppose that the singular value vectors $\vec{\sigma}(A)$ and $\vec{\sigma}(B)$ be arranged in decreasing order; i.e., $\vec{\sigma}(A) = \vec{\sigma}^*(A)$ and $\vec{\sigma}(B) = \vec{\sigma}^*(B)$. Then $\|A\| \leq \|B\|$ is the same as $\sigma_1(A) \leq \sigma_1(B)$ and $\left\| \bigwedge^{k} A \right\| \leq \left\| \bigwedge^{k} B \right\|$ is the same as

$$\prod_{j=1}^{k} \sigma_j(A) \leq \prod_{j=1}^{k} \sigma_j(B).$$

This proves (11.7.1), and then (11.7.2) follows from Corollary 11.25. If $A, B \geq 0$, $\vec{\sigma}(A) = \vec{\lambda}(A)$ and $\vec{\sigma}(B) = \vec{\lambda}(B)$, and then (11.7.2) becomes (11.7.3), and the condition for equality becomes that A and B have the same spectrum with the same multiplicities, which is the same as $A = UBU^*$. $\qquad\square$

To apply Weyl's lemma, we need norm inequalities for antisymmetric tensor products. The following simple observations shall provide many examples. For $X \in \mathcal{B}(\mathcal{H})$, let

$$X = V\Sigma U$$

be a singular value decomposition of X with V and U unitary in $\mathcal{B}(\mathcal{H})$. We suppose that the diagonal matrix Σ has the corresponding vectors $\vec{\sigma}(X)$ arranged in nonincreasing order; that is $\vec{\sigma}(X) = \vec{\sigma}^*(X)$. Then for each $1 \leq k \leq d$,

$$(11.7.4) \qquad \bigwedge^k X = \left(\bigwedge^k V\right)\left(\bigwedge^k \Sigma\right)\left(\bigwedge^k U\right)^*.$$

Since $\bigwedge^k U$ and $\bigwedge^k V$ are unitary, it follows that (11.7.4) gives a singular value decomposition of $\bigwedge^k X$. As a consequence,

$$(11.7.5) \qquad \left\|\bigwedge^k X\right\| = \prod_{j=1}^{k} \sigma_j(X).$$

Next, recall that for each $1 \leq k \leq d$, $(\bigwedge^k X)^* = \bigwedge^k X^*$, and that for all $X, Y \in \mathcal{B}(\mathcal{H})$, $\bigwedge^k(XY) = (\bigwedge^k X)(\bigwedge^k Y)$. If $X \geq 0$, then of course $\vec{\lambda}(X) = \vec{\sigma}(X)$, and hence these considerations also apply to the eigenvalue vectors of positive matrices.

The following application of Weyl's method was made by A. Horn [113].

Theorem 11.27 (Horn's theorem on singular values of matrix products). *Let $A, B \in M_n(\mathbb{C})$. Then*

$$\vec{\sigma}(AB) \prec_{w\log} \vec{\sigma}(A)\vec{\sigma}(B),$$

where $\vec{\sigma}(A)\vec{\sigma}(B)$ is the vector in $\mathbb{R}^d$ whose jth entry is $\sigma_j^*(A)\sigma_j^*(B)$. Consequently, for all functions $f : [0, \infty) \to \mathbb{R}$ such that f is monotone nondecreasing and such that $f(e^t)$ is a convex function of t,

$$(11.7.6) \qquad \sum_{j=1}^{d} f(\sigma_j(AB)) \leq \sum_{j=1}^{d} f(\sigma_j^*(A)\sigma_j^*(B)).$$

Proof. Assume that $\vec{\sigma}(A) = \vec{\sigma}^*(A)$ and $\vec{\sigma}(B) = \vec{\sigma}^*(B)$. Apply the elementary inequality $\|XY\| \leq \|X\|\|Y\|$ to $\bigwedge^k(AB) = (\bigwedge^k A)(\bigwedge^k B)$ to conclude that $\|\bigwedge^k(AB)\| \leq \|\bigwedge^k A\|\|\bigwedge^k B\|$, and hence by (11.7.5),

$$\prod_{j=1}^{k} \sigma_j^*(AB) \leq \prod_{j=1}^{k} \sigma_j(A)\sigma_j(B). \qquad \square$$

As a consequence of Theorem 11.27, we obtain a very useful extension of Hölder's inequality, Theorem 6.62, that applies beyond the case of Schatten norms *per se*. Recall that as in Section 6.8.1, for all $0 < p < \infty$, and all $X \in M_n(\mathbb{C})$, $\|X\|_p := \|X\|_p = \left(\sum_{j=1}^{r} \sigma_j^p(X) \right)^{1/p} = (\mathrm{Tr}[|X|^p])^{1/p}$ so that for $p > 1$, $\|X\|_p$ is the Schatten p-norm of X. The next theorem differs from Theorem 6.62 in that the condition there that $1 \le p, q, r$ is relaxed to the condition $0 < p, q, r$. In this form it may be found in [**35**, Exercise IV.2.7].

Theorem 11.28. *For all $A, B \in M_n(\mathbb{C})$, all $0 \le p, q, r$ such that $1/p + 1/q = 1/r$,*

$$(11.7.7) \qquad \|AB\|_r \le \|A\|_p \|B\|_q.$$

Under the further condition that $p, q < \infty$, there is equality in (11.7.7), if and only if for some $c \ge 0$,

$$(11.7.8) \qquad |A|^p = c|B^*|^q.$$

Proof. Take $f(x) = x^r, r > 0$ so that f is monotone nondecreasing and $f(e^t) = e^{tr}$ is a convex function of t. Then (11.7.6) becomes

$$(11.7.9) \qquad \sum_{j=1}^{d}(\sigma_j(AB))^r \le \sum_{j=1}^{d}(\sigma_j^*(A))^r(\sigma_j^*(B))^r.$$

Define $s := p/r$ and $t := q/r$ and note that $1/s + 1/t = 1$, and apply Hölder's inequality in $\mathbb{R}^d$ with indices s and t, and then take the rth root to conclude that

$$(11.7.10) \qquad \left(\sum_{j=1}^{d}(\sigma_j(AB))^r \right)^{1/r} \le \left(\sum_{j=1}^{d}(\sigma_j(A))^{rs} \right)^{1/rs} \left(\sum_{j=1}^{d}(\sigma_j(B))^{rt} \right)^{1/rt}.$$

which can be written as (11.7.7).

To have equality in (11.7.7), it is necessary and sufficient to have equality in (11.7.9) and (11.7.10). There is equality (11.7.10) if and only if there is some $c \ge 0$ so that for all j, $(\sigma_j^*(B))^r = c(\sigma_j^*(A))^{r(s-1)}$, or equivalently for some other c,

$$(11.7.11) \qquad (\sigma_j^*(A))^p = c(\sigma_j^*(B))^q.$$

This is certainly the case if (11.7.8) is satisfied, and moreover in this case, there are unitaries U and V such that $AB = U|A||B^*|V$, and hence $\sigma_j^*(AB) = \sigma_j^*(|A||B^*|) = \sigma_j^*(A)\sigma_j^*(B)$. This much shows that (11.7.8) is a sufficient condition for equality in (11.7.7), and that equality in (11.7.11) is necessary. It is left as an exercise, to show that (11.7.8) is also a necessary condition for equality in (11.7.7). $\qquad\square$

Next, Weyl's method will be used to prove the **Araki-Lieb-Thirring inequality (ALT inequality)**, a significant extension [14] of the Lieb-Thirring inequality.

Lemma 11.29. *Let $A, B > 0$ in $M_n(\mathbb{C})$. Then for all $0 < s < t$,*

$$(11.7.12) \qquad \|(A^{s/2}B^sA^{s/2})^{1/s}\| \leq \|(A^{t/2}B^tA^{t/2})^{1/t}\|.$$

If there is equality in (11.7.12) for some $s < t$, then there is equality for all $s < t$, and there is a unit vector $\mathbf{u}$ such that for some $\lambda > 0$ and all $t > 0$, $A\mathbf{u} = \lambda\mathbf{u}$, $B\mathbf{u} = \lambda^{-1}\mathbf{u}$.

Proof. Let $0 < s < t$, and note that by the homogeneity in A of B, we may assume without loss of generality that $\|A^{t/2}B^tA^{t/2}\| = 1$. Since $\|A^{t/2}B^tA^{t/2}\| = 1$ if and only if $A^{t/2}B^tA^{t/2} \leq \mathbb{1}$, we have that $B^t \leq A^{-t}$. But then since $f(x) = x^{s/t}$ is operator monotone on $(0, \infty)$, $B^s \leq A^{-s}$, which is the same as $A^{s/2}B^sA^{s/2} \leq \mathbb{1}$. Thus $\|A^{s/2}B^sA^{s/2}\| \leq 1$, proving (11.7.12).

Now we treat the cases of equality following [86]. Suppose that for some $0 < s < t$, $\|A^{s/2}B^sA^{s/2}\| = \|A^{t/2}B^tA^{t/2}\| = 1$. Then for all $u \in [s, t]$, $\|A^{u/2}B^uA^{u/2}\| = 1$. For all $u, \alpha \in \mathbb{R}$ such that $u \pm \alpha \in [s, t]$, define $X_u := B^{u/2}A^{u/2}$ so that $X_u^*X_u = A^{u/2}B^uA^{u/2}$. Then

$$A^{\alpha/2}(X_u^*X_u)A^{-\alpha/2} = X_{u+\alpha}^*X_{u-\alpha}.$$

Since $X_u^*X_u$ and $X_{u+\alpha}^*X_{u-\alpha}$ are similar, they have the same spectrum. Since $\|X_u^*X_u\| = 1$, 1 is an eigenvalue of both $X_u^*X_u$ and $X_{u+\alpha}^*X_{u-\alpha}$. Define $\mathcal{V}_u$ and $\mathcal{W}_{u,\alpha}$ to be the eigenspaces of $X_u^*X_u$ and $X_{u+\alpha}^*X_{u-\alpha}^*$, respectively, with eigenvalue 1. Then $A^{\alpha/2}\mathcal{V}_u = \mathcal{W}_{u,\alpha}$. In particular,

$$(11.7.13) \qquad \dim(\mathcal{V}_u) = \dim(\mathcal{W}_{u,\alpha})$$

for all u, α such that $u \pm \alpha \in [s, t]$. Now fix such u, α and let $\mathbf{u}$ be a unit vector in $\mathcal{W}_{u,\alpha}$. Then by the Cauchy-Schwarz inequality,

$$1 = \langle \mathbf{u}, X_{u+\alpha}^*X_{u-\alpha}\mathbf{u}\rangle \leq \langle \mathbf{u}, X_{u+\alpha}^*X_{u+\alpha}\mathbf{u}\rangle^{1/2}\langle \mathbf{u}, X_{u-\alpha}^*X_{u-\alpha}\mathbf{u}\rangle^{1/2} \leq 1.$$

Every inequality in the line above is an equality, and in particular by the conditions for equality in the Schwarz inequality, $X_{u+\alpha}\mathbf{u} = X_{u-\alpha}\mathbf{u}$ and $\mathbf{u} \in \mathcal{V}_{u+\alpha} \cap \mathcal{V}_{u-\alpha}$. In particular,

$$(11.7.14) \qquad \mathcal{W}_{u,\alpha} \subset \mathcal{V}_{u+\alpha} \cap \mathcal{V}_{u-\alpha},$$

and hence $\dim(\mathcal{W}_{u,\alpha}) \leq \dim(\mathcal{V}_{u+\alpha}) \wedge \dim(\mathcal{V}_{u-\alpha})$.

From (11.7.13), $\dim(\mathcal{V}_u) \leq \dim(\mathcal{V}_{u+\alpha}) \wedge \dim(\mathcal{V}_{u-\alpha})$, and this means that for some fixed d, $\dim(\mathcal{V}_u) = d$ for all $u \in [s, t]$, and from (11.7.13) again, for all u, α such that $u \pm \alpha \in [s, t]$, $\dim(\mathcal{W}_{u,\alpha}) = d$. Then from (11.7.14), for all such α, u, $\mathcal{V}_{u+\alpha} = \mathcal{V}_{u-\alpha}$ since otherwise the intersection would have a dimension strictly smaller than d. Therefore, the subspaces $\mathcal{V}_u$ not only have the same dimension, they are the same d-dimensional subspace $\mathcal{V}$, and moreover, by

(11.7.14) once more, for each u, α such that $u \pm \alpha \in [s,t]$, $\mathcal{W}_{u,\alpha} = \mathcal{V}$ as well. Now the similarity relation $A^{\alpha/2}\mathcal{V}_u = \mathcal{W}_{u,\alpha}$ becomes $A^{\alpha/2}\mathcal{V} = \mathcal{V}$. Since $\mathcal{V}$ is invariant under $A^{\alpha/2}$ for some $\alpha > 0$, and hence invariant under A, there is an orthonormal basis of $\mathcal{V}$, $\{\mathbf{u}_1, \ldots, \mathbf{u}_d\}$, consisting of eigenvectors of A: $A\mathbf{u}_j = \lambda_j\mathbf{u}_j$. Then since for each $u \in [s,t]$, $A^{u/2}B^u A^{u/2}\mathbf{u}_j = \mathbf{u}_j$, $B\mathbf{u}_j = \lambda_j^{-1}\mathbf{u}_j$. $\square$

Theorem 11.30 (Araki-Lieb-Thirring inequality). *Let $\mathcal{H}$ be a Hilbert space of dimension d, and let $A, B \in \mathcal{B}^+(\mathcal{H})$. Then for all $0 < s < t$,*

$$(11.7.15) \qquad \vec{\lambda}((A^{s/2}B^s A^{s/2})^{1/s}) \leq \vec{\lambda}((A^{t/2}B^t A^{t/2})^{1/t}),$$

and for all functions $f : [0, \infty) \to \mathbb{R}$ such that f is monotone nondecreasing and such that $f(e^u)$ is a convex function of u,

$$(11.7.16) \qquad \mathrm{Tr}[f((A^{s/2}B^s A^{s/2})^{1/s})] \leq \mathrm{Tr}[f((A^{t/2}B^t A^{t/2})^{1/t})].$$

If there is equality in (11.7.16) with $f(e^u)$ strictly convex as a function of u, then A and B commute.

Proof. Apply Lemma 11.29 with A replaced by $\bigwedge^k A$ and B replaced by $\bigwedge^k B$, to deduce from Lemma 11.29 that for all $1 \leq k \leq d$,

$$\left\|\left(\left(\bigwedge^k A\right)^{s/2}\left(\bigwedge^k B\right)^s\left(\bigwedge^k A\right)^{s/2}\right)^{1/s}\right\| \leq \left\|\left(\left(\bigwedge^k A\right)^{t/2}\left(\bigwedge^k B\right)^t\left(\bigwedge^k A\right)^{t/2}\right)^{1/t}\right\|.$$

Then (11.7.15) and (11.7.16) follow from Weyl's log-majorization lemma, and there is equality in (11.7.16) if and only if $\vec{\lambda}((A^{s/2}B^s A^{s/2})^{1/s})$ is a permutation of $\vec{\lambda}((A^{t/2}B^r A^{t/2})^{1/t})$. Then in particular, $\|(A^{s/2}B^s A^{s/2})^{1/s}\| = \|(A^{t/2}B^r A^{t/2})^{1/t}\|$, and (by Lemma 11.29) A and B have a common eigenvector. That is, there exists a unit vector $\mathbf{u}$ such that $A\mathbf{u} = \lambda\mathbf{u}$ and $B\mathbf{u} = \mu\mathbf{u}$. If $d = 2$ this means that any nonzero vector orthogonal to $\mathbf{u}$ is also a common eigenvector of A and B, and hence A and B can be simultaneously diagonalized, and therefore commute. We now prove that when there is equality in (11.7.16), then A and B commute, proceeding by induction on d, not that this is proved for $d = 2$. Making the inductive assumption that this is proved for $d - 1$.

Introduce an orthonormal basis $\{\mathbf{u}_1, \ldots, \mathbf{u}_d\}$ of $\mathcal{H}$ in which $\mathbf{u}_1 = \mathbf{u}$, the common eigenvector of A and B. Then in this basis, A and B have the matrix representations

$$A = \begin{bmatrix} \lambda & 0 \\ 0 & \tilde{A} \end{bmatrix} \quad \text{and} \quad B = \begin{bmatrix} \mu & 0 \\ 0 & \tilde{B} \end{bmatrix},$$

where $\tilde{A}, \tilde{B} \geq 0$ in $M_{d-1}(\mathbb{C})$. Then evidently for all $r > 0$,

$$\mathrm{Tr}[f((A^{r/2}B^r A^{r/2})^{1/r})] = f(\lambda\mu) + \mathrm{Tr}[f((\tilde{A}^{r/2}\tilde{B}^r \tilde{A}^{r/2})^{1/r})].$$

By the inductive hypothesis, $\tilde{A}$ and $\tilde{B}$ commute and may therefore be simultaneously diagonalized. Thus, A and B commute. $\square$

Corollary 11.31 (Araki-Lieb-Thirring inequality for powers). *Let $\mathcal{H}$ be a d-dimensional Hilbert space and let $A, B \geq 0$ in $\mathcal{B}(\mathcal{H})$. Then,*

$$(11.7.17) \qquad \mathrm{Tr}[(A^{1/2}BA^{1/2})^p] \leq \mathrm{Tr}[A^{p/2}B^pA^{p/2}] \qquad \textit{for all} \quad p > 1$$

and

$$(11.7.18) \qquad \mathrm{Tr}[(A^{1/2}BA^{1/2})^p] \geq \mathrm{Tr}[A^{p/2}B^pA^{p/2}] \qquad \textit{for all} \quad 0 < p < 1.$$

Furthermore, in both cases, there is equality if and only if A and B commute.

Proof. Take $f(x) = x$, $s = 1/p$, and $t = 1$ in (11.7.16) with A and B replaced by A^p and B^p, respectively, to obtain (11.7.17). Take $f(x) = x^s$, $s = p$, and $t = 1$ in (11.7.16) to obtain (11.7.18). In both cases, $f(e^u)$ is strictly convex in u, and the assertion about cases of equality follows from Theorem 11.30. $\qquad\square$

11.8. Multivariate ALT inequalities

Majorization methods can be used to efficiently prove some multivariate extensions of the Araki-Lieb-Thirring inequality that were proved by Sutter, Berta, and Tomamichel [205]. The proof presented here is based on the majorization approach of Hiai, König, and Tomamichel, who prove some generalizations of the results in [205], but it best serves our purposes to use their methods to prove the original results from [205].

Theorem 11.32 (Multivariate ALT inequality). *Let $\mathcal{H}$ be a finite-dimensional Hilbert space and for an integer $n \geq 2$, let $A_j \in \mathcal{B}^+(\mathcal{H})$, $1 \leq j \leq n$. Let $q > 1$ and for $r \in (0, 1]$, let μ_r be the probability measure on $\mathbb{R}$ defined by*

$$\mathrm{d}\mu_r := \frac{\sin(\pi r)}{2r(\cosh(\pi t) + \cos(\pi r))}\mathrm{d}t.$$

For any real valued function f on $(0, \infty)$ such that f increasing and such that $f(e^t)$ is convex,

$$(11.8.1) \qquad \vec{\lambda}\left(f\left(\left|\prod_{j=1}^{n} A_j^r\right|^{1/r}\right)\right) \prec_w \int_{\mathbb{R}} \vec{\lambda}\left(f\left(\left|\prod_{j=1}^{n} A_j^{1+it}\right|\right)\right) \mathrm{d}\mu_r(t)$$

and in particular,

$$(11.8.2) \qquad \mathrm{Tr}\left[f\left(\left|\prod_{j=1}^{n} A_j^r\right|^{1/r}\right)\right] \leq \mathrm{Tr}\left[\int_{\mathbb{R}} f\left(\left|\prod_{j=1}^{n} A_j^{1+it}\right|\right) \mathrm{d}\mu_r(t)\right].$$

Before proving the theorem, we explain some of its consequences. First, note that

$$\left|\prod_{j=1}^{n} A_j^r\right|^{1/r} = (A_n^r A_{n-1}^r \cdots A_2^r A_1^{2r} A_2^r \cdots A_{n-1}^r A_n^r)^{1/2r}$$

and

$$\left|\prod_{j=1}^{n} A_j^{1+it}\right| = A_n^{-it}(A_n A_{n-1}^{1-it} \cdots A_2^{1-it} A_1^2 A_2^{1+it} \cdots A_{n-1}^{i+it} A_n)^{1/2} A_n^{it},$$

since A_n^{it} is unitary. Therefore, we can rewrite (11.8.2) as

$$(11.8.3) \quad \mathrm{Tr}\big[f\big((A_n^r A_{n-1}^r \cdots A_2^r A_1^{2r} A_2^r \cdots A_{n-1}^r A_n^r)^{1/2r}\big)\big]$$

$$\leq \int_{\mathbb{R}} \mathrm{Tr}\big[f\big((A_n A_{n-1}^{1-it} \cdots A_2^{1-it} A_1^2 A_2^{1+it} \cdots A_{n-1}^{i+it} A_n)^{1/2}\big)\big]\, \mathrm{d}\mu_r(t).$$

Observe that if $n = 2$, the integrand on the right side of (11.8.3) is independent of t, and thus for $n = 2$ the inequality (11.8.2) reduces to

$$(11.8.4) \qquad \mathrm{Tr}\big[f((A_2^r A_1^{2r} A_2^r)^{1/2r})\big] \leq \mathrm{Tr}\big[f((A_2 A_1^2 A_2)^{1/2})\big].$$

This is simply the ALT inequaity (11.7.16) in other variables. To recover (11.7.16) from (11.8.4), let $A_1 = B^{t/2}$, $A_2 = A^{t/2}$, $r = s/t$, and replace $f(x)$ by $f(x^{2/t})$ which is a function of the same type. Therefore, we are justified in referring to (11.8.2) as a multivariate generalization of the ALT inequaity.

We next consider the case $n = 3$, for which the right side of (11.8.3) is

$$\int_{\mathbb{R}} \mathrm{Tr}\big[f\big((A_3 A_2^{1-it} A_1^2 A_2^{1+it} A_3)^{1/2}\big)\big]\, \mathrm{d}\mu_r(t).$$

Taking $f(x) = x^2$, the limit $r \downarrow 0$ yields, and then applying Lemma B.10,

$$\int_{\mathbb{R}} \mathrm{Tr}\big[A_3 A_2^{1-it} A_1^2 A_2^{1+it} A_3\big]\, \mathrm{d}\mu_0(t)$$

$$= \mathrm{Tr}\left[A_3\left(\int_{\mathbb{R}} A_2^{1-it} A_1^2 A_2^{1+it}\, \mathrm{d}\mu_0(t)\right)A_3\right] = \mathrm{Tr}\left[A_3 D_{\log, A_2^{-2}}(A_1^2)A_3\right].$$

For $n = 3$ and $f(x) = x^2$, the left side of (11.8.3) is

$$\mathrm{Tr}\big[(A_3^r A_2^r A_1^{2r} A_2^r A_3^r)^{1/r}\big],$$

$m \in \mathbb{N}$. Define $2H := \log A_1$, $2K := \log A_2$ and $2L := \log A_3$, and take $r = 1/m$. By the Lie product formula, $\lim_{m\to\infty}(A_2^{1/m} A_1^{2/m} A_2^{1/m})^m = e^{K+L}$, and then

$$\lim_{m\to\infty} (A_3^{1/m} A_2^{1/m} A_1^{2/m} A_2^{1/m} A_3^{1/m})^m = e^{H+K+L}.$$

Altogether, in the new variables, $\mathrm{Tr}\left[e^{H+K+L}\right] \le \mathrm{Tr}\left[e^{L} D_{\log, e^{-K}}(e^{H})\right]$. Using the *standard* form for $D_{\log, e^{-K}}(e^{H})$, this becomes Lieb's triple matrix inequality,

$$\mathrm{Tr}\left[e^{H+K+L}\right] \le \mathrm{Tr}\left[e^{L}\left(\int_{0}^{\infty} \frac{1}{t+e^{-K}} e^{H} \frac{1}{t+e^{-K}} dt\right)\right].$$

Thus, a special case of Theorem 11.32 yields another proof of Theorem 10.15.

We now apply the same argument to higher order products. For $f(x) = x^2$, the left side of (11.8.3) is

$$\mathrm{Tr}\left[(A_n^r A_{n-1}^r \cdots A_2^r A_1^{2r} A_2^r \cdots A_{n-1}^r A_n^r)^{1/r}\right].$$

Writing $A_j = e^{H_j/2}$, this becomes

$$\mathrm{Tr}\left[\left(e^{(r/2)H_n} e^{(r/2)H_{n-1}} \cdots e^{(r/2)H_2} e^{r H_1} e^{(r/2)H_2} \cdots e^{(r/2)H_{n-1}} e^{(r/2)H_n}\right)^{1/r}\right].$$

By the extended Lie product formula (see Exercise 10, Chapter 3), the limit $r \downarrow 0$ of this is $\mathrm{Tr}\left[e^{\sum_{j=1}^{n} H_j}\right]$. For $n = 4$, the right side of (11.8.3) becomes, in the limit $r \downarrow 0$,

$$\int_{\mathbb{R}} \mathrm{Tr}\left[e^{H_1} e^{\frac{1+it}{2} H_2} e^{\frac{1+it}{2} H_3} e^{H_4} e^{\frac{1-it}{2} H_3} e^{\frac{1-it}{2} H_2}\right] d\mu_0(t).$$

This gives us the 4-matrix Golden-Thompson inequality:

Theorem 11.33. *For all self-adjoint H_1, H_2, H_3 and H_4 in $\mathcal{B}^{\mathrm{s.a.}}(\mathcal{H})$,*

$$\mathrm{Tr}\left[e^{H_1+H_2+H_3+H_4}\right] \le \int_{\mathbb{R}} \mathrm{Tr}\left[e^{H_1} e^{\frac{1+it}{2} H_2} e^{\frac{1+it}{2} H_3} e^{H_4} e^{\frac{1-it}{2} H_3} e^{\frac{1-it}{2} H_2}\right] d\mu_0(t).$$

Proof of Theorem 11.32. We now apply Weyl's method starting from the operator norm inequality, which is proved in Appendix B, Corollary B.8,

$$\log\left\|\left\|\prod_{j=1}^{n} A_j^r\right\|^{1/r}\right\| \le \int_{\mathbb{R}} \log\left\|\prod_{j=1}^{n} A_j^{1+it}\right\| d\mu_r(t).$$

Note that for each k, $\left|\prod_{j=1}^{n} \wedge^{k} A_j^r\right|^{1/r} = \wedge^{k}\left|\prod_{j=1}^{n} A_j^r\right|^{1/r}$ by the properties of the antisymmetric tensor product. Likewise, for each k, $\left|\prod_{j=1}^{n} \wedge^{k} A_j^{1+it}\right| = \wedge^{k}\left|\prod_{j=1}^{n} A_j^{1+it}\right|$.

Weyl's lemma yields

$$\vec{\lambda}\left(\log\left|\prod_{j=1}^{n} A_j^r\right|^{1/r}\right) \prec_{w} \int_{\mathbb{R}} \vec{\lambda}\left(\log\left|\prod_{j=1}^{n} A_j^{1+it}\right|\right) d\mu_r(t).$$

Now define $g(t) = f(e^t)$ so that $g(t)$ is convex and increasing. Then by Theorem 11.8,

$$g\left(\vec{\lambda}\left(\log\left|\prod_{j=1}^{n} A_j^r\right|^{1/r}\right)\right) \prec_w g\left(\int_{\mathbb{R}} \vec{\lambda}\left(\log\left|\prod_{j=1}^{n} A_j^{1+it}\right|\right) d\mu_r(t)\right).$$

By Jensen's inequality,

$$(11.8.5) \quad g\left(\int_{\mathbb{R}} \vec{\lambda}\left(\log\left|\prod_{j=1}^{n} A_j^{1+it}\right|\right) d\mu_r(t)\right) \leq \int_{\mathbb{R}} g\left(\vec{\lambda}\left(\log\left|\prod_{j=1}^{n} A_j^{1+it}\right|\right)\right) d\mu_r(t),$$

and then since $\mathbf{x} \prec_w \mathbf{y}$ and $\mathbf{y} \leq \mathbf{z}$ together imply that $\mathbf{x} \prec_w \mathbf{z}$, (11.8.5) implies

$$g\left(\vec{\lambda}\left(\log\left|\prod_{j=1}^{n} A_j^r\right|^{1/r}\right)\right) \prec_w \int_{\mathbb{R}} g\left(\vec{\lambda}\left(\log\left|\prod_{j=1}^{n} A_j^{1+it}\right|\right)\right) d\mu_r(t).$$

Now (11.8.1) follows from the fact that $g(\vec{\lambda}(X)) = \vec{\lambda}(g(X))$ for all $X \in \mathcal{B}^{\text{s.a.}}(\mathcal{H})$ and the identity $g(\log(s)) = f(s)$, and (11.8.1) is an immediate consequence of (11.8.1). $\qquad\square$

Exercises

(1) Prove that $\mathbf{x} \prec_w \mathbf{y} \iff \sum_{j=1}^{n}(x_j - t)_+ \leq \sum_{j=1}^{n}(y_j - t)_+$ for all $t \in \mathbb{R}$.

(2) Prove that $\mathbf{x} \prec \mathbf{y} \iff \sum_{j=1}^{n}|x_j - t| \leq \sum_{j=1}^{n}|y_j - t|$ for all $t \in \mathbb{R}$.

(3) Let $\mathbf{x}, \mathbf{y} \in \mathbb{R}^n$, $\mathbf{z} \in \mathbb{R}^m$. Define $\mathbf{x}', \mathbf{y}' \in \mathbb{R}^{n+m}$ by appending $\mathbf{z}$ to $\mathbf{x}, \mathbf{y}$: $\mathbf{x}' = (\mathbf{x}, \mathbf{z})$ and $\mathbf{y}' = (\mathbf{y}, \mathbf{z})$. Prove that $\mathbf{x} \prec_w \mathbf{y} \iff (\mathbf{x}, \mathbf{z}) \prec_w (\mathbf{y}, \mathbf{z})$ and $\mathbf{x} \prec \mathbf{y} \iff (\mathbf{x}, \mathbf{z}) \prec (\mathbf{y}, \mathbf{z})$.

(4) (Alberti and Uhlmann [4]) Let $A, B \in M_n^{\text{s.a.}}(\mathbb{C})$. Show that $A \prec B$ if and only if

$$A = \sum_{j=1}^{n} p_j U_j B U_j^*,$$

where $\{U_1, \dots, U_n\}$ is a set of unitaries, and $\{p_1, \dots, p_n\}$ are probabilities.

(5) Let $\Phi : M_n(\mathbb{C}) \to M_n(\mathbb{C})$ be positive, unital, and trace preserving. Show for all $A \in M_n^{\text{s.a.}}$ (such linear maps are called **doubly stochastic**). Show that $\Phi(X) \prec X$.

(6) For $X \in M_n^{++}(\mathbb{C})$, $K \in M_n(C)$, define $\Phi_X(K) := X^{1/2}D_{\log,X}(K)X^{1/2}$, so that by (3.2.5)

$$\Phi_X(K) = \int_0^\infty \left(\frac{X^{1/2}}{\lambda + X} K \frac{X^{1/2}}{\lambda + X} \right) d\lambda.$$

Show that Φ is doubly stochastic in the sense of Exercise 5. In Lemma B.10, a stronger result will be proved, namely that

$$\Phi_X(K) = \int_{\mathbb{R}} X^{(1-it)/2}KX^{(1-it)/2}d\mu_0(t)$$

for a certain probability measure μ_0 on $\mathbb{R}$.

(7) (Nielsen) Let $\mathcal{H}_A, \mathcal{H}_B$ be finite-dimensional Hilbert spaces. Let $|\psi\rangle\langle\psi|$ and $|\phi\rangle\langle\phi|$ be pure states on $\mathcal{H}_A \otimes \mathcal{H}_B$. Let $\rho_\psi := \mathrm{Tr}_B[|\psi\rangle\langle\psi|]$, and let $\rho_\phi := \mathrm{Tr}_B[|\phi\rangle\langle\phi|]$. A theorem of Nielsen [167] says that $|\psi\rangle\langle\psi|$ may be transformed into $|\phi\rangle\langle\phi|$ by means of an LOCC operation if and only if $\rho_\psi \prec \rho_\phi$. This exercise steps through a proof of this.

   (a) Let Ψ be an LOCC1 quantum operation on $\mathcal{H}_A \otimes \mathcal{H}_B$ so that for some finite set $\mathcal{X}$, some quantum operator Φ_A on $\mathcal{B}(\mathcal{H}_A)$ given by $\Phi_A(Y) = \sum_{x \in \mathcal{X}} A_x Y A_x^*$, and some set $\{\Phi_B^{(x)} : x \in \mathcal{X}\}$ of quantum operations on $\mathcal{B}(\mathcal{H}_B)$,

$$\Psi(X) := \sum_{x \in \mathcal{X}} \Psi_x(X) := \sum_{x \in \mathcal{X}} \mathbb{1}_A \otimes \Phi_B^{(x)}((A_x \otimes \mathbb{1}_B)X(A_x^* \otimes \mathbb{1}_B)).$$

Suppose that $|\phi\rangle\langle\phi| = \Psi(|\psi\rangle\langle\psi|)$. Show that for each x, there is a number $p_x \geq 0$ such that $\Psi_x(|\psi\rangle\langle\psi|) = p_x|\phi\rangle\langle\phi|$, and that $A_x\rho_\psi A_x^* = p_x\rho_\phi$. Show also that $\sum_{x \in \mathcal{X}} p_x = 1$.

   (b) Let $U_x|\rho_\psi^{1/2}A_x^*|$ be the polar decomposition of $\rho_\psi^{1/2}A_x^*$ so that $A_x\rho_\psi^{1/2} = (A_x\rho_\psi A_x^*)^{1/2}U_x^*$. Show that

$$\rho_\psi = \sum_{x \in \mathcal{X}} \rho_\psi^{1/2}A_x^*A_x\rho_\psi^{1/2} = \sum_{x \in \mathcal{X}} p_x U_x \rho_\phi U_x^*,$$

and then show that $\rho_\psi \prec \rho_\phi$.

   (c) Conversely, suppose $\rho_\psi \prec \rho_\phi$. Show that there is an LOCC operation Ψ that transforms $|\phi\rangle\langle\phi|$ into $|\phi\rangle\langle\phi|$. Do this first in the case where $\mathcal{H}_A$ and $\mathcal{H}_B$ are two dimensional. Then leverage this as in the proof of Lemma 11.6.

Chapter 12

Tomita-Takesaki theory and operator inequalities

12.1. The GNS construction and purification

Let $\mathcal{H}$ denote a Hilbert space with inner product $\langle \cdot, \cdot \rangle$, and let $\mathcal{M}$ be a von Neumann algebra acting on $\mathcal{H}$. Let $\mathcal{M}'$ denote the commutant of $\mathcal{M}$.

Definition 12.1. A vector $\Omega \in \mathcal{H}$ is a **cyclic vector** for $\mathcal{M}$ in case the subspace $\{A\Omega \ : \ A \in \mathcal{M}\}$ is norm dense in $\mathcal{H}$. A vector $\Omega \in \mathcal{H}$ is a **separating vector** for $\mathcal{M}$ in case the only $A \in \mathcal{M}$ such that $A\Omega = 0$ is $A = 0$.

Lemma 12.2. *A vector Ω is cyclic for $\mathcal{M}$ if and only if it is separating for $\mathcal{M}'$, and a vector Ω is separating for $\mathcal{M}$ if and only if it is cyclic for $\mathcal{M}'$. Consequently, if Ω is cyclic and separating for $\mathcal{M}$, it is also cyclic and separating for $\mathcal{M}'$.*

Proof. Since $\mathcal{M}'' = \mathcal{M}$, we need only prove that a vector Ω is cyclic for $\mathcal{M}$ if and only if it is separating for $\mathcal{M}'$.

Suppose that Ω is cyclic for $\mathcal{M}$. If $A'\Omega = 0$ for some $A' \in \mathcal{M}'$, then for all $B \in \mathcal{M}$, $0 = BA'\Omega = A'B\Omega$, and hence A' vanishes on a dense subset of $\mathcal{H}$, proving that $A' = 0$.

Next, suppose that Ω is separating for $\mathcal{M}'$. Let P be the orthogonal projection in $\mathcal{H}$ onto the closure of $V := \{A\Omega \ : \ A \in \mathcal{M}\}$. For all $B \in \mathcal{M}$ and $A\Omega \in V$, $BA\Omega \in V$ so that $\overline{V}$ is invariant under $\mathcal{M}$. Let $\Psi \in V^\perp$. Then for all $A, B \in \mathcal{M}$, $0 = \langle \Psi, B^*A\Omega \rangle = \langle B\Psi, A\Omega \rangle$, and therefore $B\Psi \in V^\perp$. Since $B \in \mathcal{M}$ is arbitrary, $V^\perp$ is invariant under $\mathcal{M}$.

Now for any $\Psi \in \mathcal{H}$ and any $B \in \mathcal{M}$,

$$PB\Psi = PB(P\Psi + (\mathbb{1} - P)\Psi) = PBP\Psi = BP\Psi.$$

Since Ψ is arbitrary, $PB = BP$ and hence $P \in \mathcal{M}'$, and also $\mathbb{1} - P \in \mathcal{M}'$. Then since $\Omega \in \mathcal{V}$, $(\mathbb{1} - P)\Omega = 0$. Since Ω is separating for $\mathcal{M}'$, $P = \mathbb{1}$ and hence Ω is cyclic for $\mathcal{M}$. $\square$

Given any unit vector $\Psi \in \mathcal{H}$, define a linear functional ω_Ψ on $\mathcal{M}$ by

$$\omega_\Psi(X) := \langle \Psi, X\Psi \rangle$$

for all $X \in \mathcal{M}$. Evidently, ω_Ψ is a state on $\mathcal{M}$. It may or may not be faithful.

Given any faithful state ω on a von Neumann algebra $\mathcal{M}$ of operators on some Hilbert space, the **Gelfand-Naimark-Segal (GNS) construction** yields a new Hilbert space $\mathcal{H}$, built out of $\mathcal{M}$ and ω, together with a $*$-isomorphism π of $\mathcal{M}$ into $\mathcal{B}(\mathcal{H})$. The image of $\mathcal{M}$ may of course be identified with $\mathcal{M}$, and then there is always a cyclic and separating vector $\Omega \in \mathcal{H}$ such that $\omega(X) = \langle \Omega, X\Omega \rangle = \omega_\Omega(X)$ for all $X \in \mathcal{M}$.

The GNS construction was introduced in a special case in Chapter 4. We now describe it in general. It allows us to regard any von Neumann algebra with a faithful state as a von Neumann algebra with a cyclic and separating vector.

Let ω be a faithful state on $\mathcal{M}$. Define an inner product $\langle \cdot, \cdot \rangle_\omega$ on $\mathcal{M}$ by

$$(12.1.1) \qquad\qquad \langle X, Y \rangle_\omega := \omega(X^*Y).$$

Since ω is faithful, $\langle X, X \rangle_\omega > 0$ for all $X \in \mathcal{M}$ except $X = 0$. Thus, $\langle X, Y \rangle_\omega$ is a nondegenerate inner product on $\mathcal{M}$. Define $\mathcal{H}$ to be the completion of $\mathcal{M}$ in this inner product. Of course, if $\mathcal{M}$ is finite dimensional, there is no need to take a completion; $\mathcal{M}$ is already complete in this inner product. Let $\| \cdot \|_\omega$ denote the corresponding Hilbert space norm.

For each $X \in \mathcal{M}$, define L_X to be the operator of left multiplication on $\mathcal{H}$; that is, $L_X(A) = XA$ for all $A \in \mathcal{M}$. Then

$$\|L_X(A)\|_\omega^2 = \omega(A^*X^*XA) \leq \|X\|^2 \|A\|_\omega^2,$$

so that $L_X \in \mathcal{B}(\mathcal{H})$, and $\|L_X\|_{\mathcal{B}(\mathcal{H})} = \|X\|$. Since each L_X is bounded, it has a continuous extension from $\mathcal{M}$ to its completion $\mathcal{H}$, but again, in finite dimensions there is no extension to make. We now claim that the map $X \mapsto L_X$ from $\mathcal{M}$ into $\mathcal{B}(\mathcal{H})$ is a representation. It is clear that it is a homomorphism since for all $X, Y \in \mathcal{M}$ and all $z, w \in \mathbb{C}$,

$$L_{zX+wY} = zL_X + wL_Y \qquad \text{and} \quad L_{XY} = L_X L_Y.$$

Moreover, since $\|L_X\|_{\mathcal{B}(\mathcal{H})} = \|X\|$, this map is injective. To show that it is a representation, it remains to show that for all $X \in \mathcal{M}$, $(L_X)^* = L_{X^*}$: Let $A, B \in \mathcal{M}$. Then

$$\langle A, L_X B \rangle_\omega = \omega(A^* X B) = \omega((X^* A)^* B) = \langle L_{X^*} A, B \rangle_\omega,$$

so that indeed $(L_X)^* = L_{X^*}$. Define Ω in $\mathcal{H}$ to be $\mathbb{1}$, considered as a vector in $\mathcal{M}$. Then for all $X \in \mathcal{M}$,

$$\langle \Omega, L_X \Omega \rangle_\omega = \langle \mathbb{1}, X\mathbb{1} \rangle_\omega = \omega(X),$$

which proves (12.1.1).

The vector Ω may be regarded as a kind of purification of the state ω, and later we will relate this sort of purification to the purification of states determined by density matrices that has been discussed earlier. It turns out that the purification of one faithful state ω using the GNS construction yields a canonical purification of all other states.

The following lemma is due to Sakai.

Lemma 12.3. *Let $\mathcal{M}$ be a von Neumann algebra on $\mathcal{H}$ with a cyclic and separating vector Ω. Let φ be a positive linear functional on $\mathcal{M}$ such that for some $\lambda < \infty$,*

$$\varphi(X) \leq \lambda \langle \Omega, X\Omega \rangle$$

is valid for all $X \in \mathcal{M}^+$. Then there is an operator $T \in \mathcal{M}'$ such that $0 \leq T \leq \lambda$ and such that for all $X \in \mathcal{M}$,

$$\varphi(X) = \langle \Omega, TX\Omega \rangle.$$

In particular, defining $\Psi = T^{1/2}\Omega$, $\varphi(X) = \langle \Psi, X\Psi \rangle$ for all $X \in \mathcal{M}$, so that

$$\varphi = \omega_\Psi.$$

Proof. Define a sesquilinear form f_φ on $\mathcal{H} \times \mathcal{H}$ by

$$f_\varphi(X\Omega, Y\Omega) = \varphi(X^* Y)$$

for all $X, Y \in \mathcal{M}$. Then since $|\varphi(Y^* X)| \leq \varphi(Y^* Y)^{1/2} \varphi(X^* X)^{1/2}$,

$$|f_\varphi(X\Omega, Y\Omega)| \leq \lambda \|X\Omega\| \|Y\Omega\|.$$

By the Riesz representation theorem, there is an operator T on $\mathcal{H}$ such that $\|T\| \leq \lambda$ and $f_\varphi(X\Omega, Y\Omega) = \langle X\Omega, TY\Omega \rangle$ for all $X, Y \in \mathcal{M}$. Therefore, $|\langle X\Omega, TY\Omega \rangle| \leq \lambda \|X\Omega\| \|Y\Omega\|$ and since Ω is cyclic for $\mathcal{M}$, $\|T\| \leq \lambda$ and since φ is positive, so is T.

Finally, for all $X, Y, Z \in \mathcal{M}$,

$$\langle X\Omega, TZY\Omega \rangle = \varphi(X^* ZY) = \varphi((Z^* X)^* Y)$$
$$= \langle Z^* X\Omega, TY\Omega \rangle = \langle X\Omega, ZTY\Omega \rangle.$$

It follows that T commutes with every $Z \in \mathcal{M}$; that is $T \in \mathcal{M}'$. $\qquad\square$

Let φ be a positive linear funcrtional on $\mathcal{M}$ that has the form $\varphi = \omega_\Psi$ for some unit vector Ψ in $\mathcal{H}$. Let U be any unitary in $\mathcal{M}'$. Then for all $X \in \mathcal{M}$,

$$\langle U\Psi, XU\Psi \rangle = \langle \Psi, U^*XU\Psi \rangle = \langle \Psi, XU^*U\Psi \rangle = \langle \Psi, X\Psi \rangle = \varphi(X).$$

That is, $\varphi = \omega_{U\Psi}$ for all unitaries $U \in \mathcal{M}'$.

Lemma 12.4. *Let φ be a positive linear functional on $\mathcal{M}$ that has the form $\varphi = \omega_\Psi$ for some cyclic and separating vector Ψ in $\mathcal{H}$. Then $\varphi = \omega_{\Psi'}$ if and only if $\Psi' = U\Psi$ for some unitary $U \in \mathcal{M}'$.*

Proof. We have already seen that if $\Psi' = U\Psi$ for some unitary $U \in \mathcal{M}'$, then $\omega_{\Psi'} = \omega_\Psi$, even without assuming that Ψ is cyclic and separating. Therefore, suppose that Ψ is cyclic and separating and that $\omega_{\Psi'} = \omega_\Psi$. Since Ψ is cyclic and separating also for $\mathcal{M}'$, there is some $A \in \mathcal{M}'$ such that $\|A\Psi - \Psi'\|$ is arbitrarily small. Of course, in finite dimensions, we do not need to approximate, and there is an $A \in \mathcal{M}'$ such that $\Psi' = A\Psi$. Let us first assume that this is the case. Then for all $X, Y \in \mathcal{M}$,

$$\langle X\Psi, Y\Psi \rangle = \langle \Psi, X^*Y\Psi \rangle = \langle \Psi', X^*Y\Psi' \rangle$$
$$= \langle A\Psi, X^*YA\Psi \rangle = \langle X\Psi, A^*AY\Psi \rangle.$$

Since Ψ is cyclic for $\mathcal{M}$, $A^*A = \mathbb{1}$, and hence A is unitary. The extension to the case in which $\mathcal{M}$ is only dense in $\mathcal{H}$ is left as an exercise for those who are familiar with the necessary facts about topologies on $\mathcal{B}(\mathcal{H})$. $\qquad\square$

All of the results stated so far hold for general von Neumann algebras and, with the exception of Lemma 12.4, complete proofs applicable in this case have been given, and the extension of the proof of Lemma 12.4 to the general case is not hard.

In the remaining sections of this chapter, *all theorems are true as stated in the general case of a von Neumann algebra with a cyclic and separating vector.* We will give proofs that apply in the finite-dimensional case, taking advantage of the technical simplifications this affords (as seen already in the proof of Lemma 12.4), but we will be careful to point out when we are using a finite-dimensional argument, and what can be done to avoid doing so.

The finite-dimensional theory is of interest in and of itself. It provides a way to deal with trace inequalities that avoids mentioning traces explicitly. This is necessary in the context of Type III von Neumann algebras where nothing has a trace. But as Petz showed, the framework of the Tomita-Takesaki theory provides a powerful tool for proving new trace inequalities in finite dimensions. If only for this reason, this book would not be complete without a finite-dimensional treatment of the Tomita-Takesaki theory. However, this theory provides the gateway to extend the results discussed so far in this book into a much more general setting, necessary in many branches of physics, and not only does it yield *trace free* formulations of important trace inequalities, the proofs in the tracial cases can often be readily adapted.

The point of departure from the finite-dimensional theory discussed so far is provided by Theorem 5.30. This says that every state ω on a von Neumann algebra $\mathcal{M}$ of operators on a finite-dimensional Hilbert space $\mathcal{H}$ has the form

$$(12.1.2) \qquad \omega(X) = \mathrm{Tr}[\rho_\omega X]$$

for a density matrix ρ on $\mathcal{M}$. Theorem 5.30 is not true for general von Neumann algebras. There are von Neumann algebras in which no nonzero element has a trace. Nonetheless, a number of theorems pertaining to states that one can readily prove using the representation (12.1.2) when it is available are true even when it is not available. The Tomita-Takesaki theory is built out of such theorems.

We next prove a lemma showing that in finite dimensions, *every* state φ satisfies the condition in Sakai's lemma, Lemma 12.3, with respect to every faithful state ω. Thus, applying the GNS construction using ω not only purifies ω, it simultaneously purifies every other state, faithful or not. This is a finite-dimensional result, but it illustrates the utility of Theorem 5.30 in dealing with states in terms of density matrices—when this is an option.

Lemma 12.5. *Let $\mathcal{M}$ be a von Neumann algebra on a finite-dimensional Hilbert space $\mathcal{H}$. Let ω be a faithful state on $\mathcal{M}$, and let φ be any state on $\mathcal{M}$. Then there exists $0 < \lambda < \infty$ such that for all $X \in \mathcal{M}^+$,*

$$\varphi(X) \leq \lambda\omega(X).$$

Proof. Since ω is faithful, $\mathrm{supp}(\rho_\varphi) \subseteq \mathrm{supp}(\rho_\omega)$, and hence with $A := \rho_\varphi^{1/2}(\rho_\omega^{-1})^{1/2}$,

$$\varphi(X) = \mathrm{Tr}[\rho_\varphi^{1/2} X \rho_\varphi^{1/2}] = \mathrm{Tr}[A\rho_\omega^{1/2} X \rho_\omega^{1/2} A^*]$$
$$\leq \|A\|_\infty \|A^*\|_\infty \|\rho_\omega^{1/2} X \rho_\omega^{1/2}\|_1 = \|\rho_\varphi \rho_\omega^{-1}\|\omega(X).$$

$$\square$$

We close this section with a fundamental example.

Example 12.6. Take $\mathcal{M} = M_n(\mathbb{C})$, and let ω denote the trace on $\mathcal{M}$; $\omega(A) = \mathrm{Tr}[A]$ for all $A \in \mathcal{M}$. Evidently, ω is a faithful positive linear functional on $\mathcal{M}$. The Hilbert space $\mathcal{H}$ produced by the GNS construction is then $M_n(\mathbb{C})$ equipped with the Hilbert-Schmidt inner product.

Recall from Section 4.8 that the GNS representation π of $M_n(\mathbb{C})$ on $\mathcal{H}$ is the map $\pi : A \mapsto L_A$, where L_A denotes left multiplication by A. Take Ω to any invertible element of $M_n(\mathbb{C})$. Then Ω is cyclic for $\mathcal{M}$ since for any $X \in \mathcal{H}$, $X = L_{X\Omega^{-1}}\Omega$ and $L_{X\Omega^{-1}} \in \mathcal{M}$. Likewise, $L_A\Omega = A\Omega$, and since Ω is invertible, if $L_A\Omega = 0$, then $A = 0$, showing that Ω is separating for $\mathcal{M}$, which we identify with $\mathcal{M}_L$, its image under the GNS representation. Of course, we can take $\Omega = \mathbb{1}$, but we are not limited to this choice.

Now define $\mathcal{M}_R$ to be the subalgebra of $\mathcal{B}(\mathcal{H})$ consisting of all operators of the form R_B, $B \in M_n(\mathbb{C})$, where R_B denotes right multiplication by B.

It is evident that $\mathcal{M}_R \subseteq \mathcal{M}'_L$ and $\mathcal{M}_L \subseteq \mathcal{M}'_R$, and in fact by Theorem 4.53, $\mathcal{M}_R = \mathcal{M}'_L = \mathcal{M}'$. That is, $\mathcal{M}' = \mathcal{M}_R$. Moreover, Ω is also a cyclic and separating vector for $\mathcal{M}_R = \mathcal{M}'$ by Lemma 12.2.

In this example, there is a compelling symmetry between $\mathcal{M}$ and $\mathcal{M}'$. It turns out that this is always the case for any von Neumann algebra with a cyclic and separating vector, and this is the main theme of the Tomita-Takesaki theory. To bring out the symmetry in this example more clearly and to open the way towards generalization, define the map $J : \mathcal{H} \to \mathcal{H}$ by

$$J(X) = X^* \qquad \text{for all} \quad X \in M_n(\mathbb{C}) = \mathcal{H}.$$

Evidently, J is an involution and an isometry. It is also *antilinear*. That is, for all $\Phi, \Psi \in \mathcal{H}$ and $\alpha, \beta \in \mathbb{C}$,

$$J(\alpha\Phi + \beta\Psi) = \overline{\alpha}J(\Phi) + \overline{\beta}J\Psi.$$

With Ω as above, and $A, B \in \mathcal{M}_n(\mathbb{C})$, let $\Psi := B\Omega$. Then

$$JL_AJ\Psi = JAJB\Omega = JL_A\Omega^*B^* = JA\Omega^*B^* = B\Omega A^* = R_{A^*}B\Omega = R_{A^*}\Psi.$$

Since Ψ can be any vector in $\mathcal{H}$, this shows $JL_AJ = R_{A^*}$, and therefore, since L_A is any element of $\mathcal{M}$, and $\mathcal{M}' = \mathcal{M}_R$, $J\mathcal{M}J = \mathcal{M}'$. Since J is an involution, $J\mathcal{M}'J = \mathcal{M}$. That is,

$$(12.1.3) \qquad\qquad J\mathcal{M}J = \mathcal{M}' \qquad \text{and} \qquad J\mathcal{M}'J = \mathcal{M}.$$

12.2. The Tomita-Takesaki theory

One of the main results of the Tomita-Takesaki theory is that for any von Neumann algebra $\mathcal{M}$ on a Hilbert space $\mathcal{H}$ with a cyclic and separating vector Ω, there is an antilinear operator J on $\mathcal{H}$ that is an involution and an isometry such that (12.1.3) is valid. This is only one part of the story, but already it explains that we shall be working with antilinear operators, and before going further, we need to record some simple facts about antilinear operators.

The following device of van Daele [**218**] is useful in this regard: Define a Hilbert space $\overline{\mathcal{H}}$ by taking $\overline{\mathcal{H}}$ to equal $\mathcal{H}$ as a set, and to have the same addition rule. Define the scalar multiplication

$$\mathbb{C} \times \overline{\mathcal{H}} \ni (z, \Psi) \mapsto \overline{z}\Psi,$$

and define the inner product on $\overline{\mathcal{H}}$ to be

$$(12.2.1) \qquad \langle \Psi, \Phi \rangle_{\overline{\mathcal{H}}} := \overline{\langle \Psi, \Phi \rangle} = \langle \Phi, \Psi \rangle.$$

Now let X be any map from $\mathcal{H}$ to $\mathcal{H}$. Since $\mathcal{H}$ and $\overline{\mathcal{H}}$ agree as sets, X may also be regarded as a map from $\mathcal{H}$ to $\overline{\mathcal{H}}$ or as a map from $\overline{\mathcal{H}}$ to $\mathcal{H}$. It is obvious from the definitions that X is antilinear from $\mathcal{H}$ to $\mathcal{H}$ if and only if it is linear both as a map from $\mathcal{H}$ to $\overline{\mathcal{H}}$ and as a map from $\overline{\mathcal{H}}$ to $\mathcal{H}$.

Definition 12.7. The **adjoint** of an antilinear map X on $\mathcal{H}$, denoted X^*, is the usual adjoint of X considered as a linear map from $\mathcal{H}$ to $\overline{\mathcal{H}}$.

Lemma 12.8. *Let X be an antilinear map on $\mathcal{H}$. Then X^* is the unique antilinear map on $\mathcal{H}$ such that for all $\Phi, \Psi \in \mathcal{H}$*

$$(12.2.2) \qquad \langle X\Psi, \Phi \rangle = \langle X^*\Phi, \Psi \rangle.$$

Proof. We may regard Φ as an element of $\overline{\mathcal{H}}$ and X as a transformation from $\mathcal{H}$ to $\overline{\mathcal{H}}$. Then

$$\Psi \mapsto \langle \Phi, X\Psi \rangle_{\overline{\mathcal{H}}} = \langle X\Psi, \Phi \rangle$$

is a bounded linear functional on $\mathcal{H}$. Hence, there is a unique vector $X^*\Phi \in \mathcal{H}$ such that (12.2.2) is valid for all $\Psi \in \mathcal{H}$. If $z \in \mathbb{C}$,

$$\langle z\Phi, X\Psi \rangle_{\overline{\mathcal{H}}} = \overline{z}\langle \Phi, X\Psi \rangle_{\overline{\mathcal{H}}} = \overline{z}\langle X\Psi, \Phi \rangle = \overline{z}\langle X^*\Phi, \Psi \rangle = \langle zX^*\Phi, \Psi \rangle,$$

and therefore $\Phi \mapsto X^*\Phi$ is linear in Φ as an operator from $\overline{\mathcal{H}}$ to $\mathcal{H}$ since the additivity is trivial. Thus, X^* is antilinear on $\mathcal{H}$. $\qquad \square$

The standing assumption going forward is that $\mathcal{M}$ is a von Neumann algebra of operators on a Hilbert space $\mathcal{H}$, and that Ω is a cyclic and separating vector for $\mathcal{M}$ and hence also $\mathcal{M}'$.

Definition 12.9 (The operators S and F). Define an antilinear operator $S : \mathcal{H} \to \mathcal{H}$ by $S(A\Omega) = A^*\Omega$ for $A \in \mathcal{M}$. Likewise, define an antilinear operator $F : \mathcal{H} \to \mathcal{H}$ by $F(A'\Omega) = (A')^*\Omega$ for $A' \in \mathcal{M}'$.

Remark 12.10. In the finite-dimensional setting, *every* $\Psi \in \mathcal{H}$ is of the form $A\Omega$ for some $A \in \mathcal{M}$. Then evidently $S^2 = F^2 = \mathbb{1}$. In particular, $\ker(S) = \ker(F) = 0$. Therefore, S^*S and F^*F are strictly positive operators on $\mathcal{H}$, and hence the polar decomposition of S is given by

$$S = (S(S^*S)^{-1/2})(S^*S)^{1/2},$$

and likewise for F.

In the infinite-dimensional setting, $\{A\Omega \ : \ A \in \mathcal{M}\}$ may be only dense in $\mathcal{H}$, and S and F may be unbounded densely defined operators. However they are always *closed*, and then again the polar decomposition is again well defined and unique. (The closure is an easy consequence of Lemma 12.11.) There is an alternative approach to the Tomita-Takesaki theory using only bounded operators that is due to Rieffel and van Daele [**185**]. However, here we follow the original development of the theory, which remains what one is most likely to encounter in fully general treatments.

Lemma 12.11. *The antilinear operators S and F are adjoint to one another. That is*

$$S^* = F \quad and \quad F^* = S.$$

Proof. Let $\Psi, \Phi \in \mathcal{H}$. Then there exist unique $A \in \mathcal{M}$ and $A' \in \mathcal{M}'$ such that $\Psi = A\Omega$ and $\Phi = A'\Omega$. Regard Φ as an element of $\overline{\mathcal{H}}$, and compute

$$\begin{aligned}
\langle \Phi, S\Psi \rangle_{\overline{\mathcal{H}}} &= \langle S(A\Omega), A'\Omega \rangle = \langle A^*\Omega, A'\Omega \rangle \\
&= \langle \Omega, AA'\Omega \rangle = \langle \Omega, A'A\Omega \rangle = \langle (A')^*\Omega, A\Omega \rangle \\
&= \langle F(A'\Omega), A\Omega \rangle = \langle F\Phi, \Psi \rangle.
\end{aligned}$$

That is $\langle S\Psi, \Phi \rangle = \langle F\Phi, \Psi \rangle$. By Lemma 12.8, $F = S^*$, and then $S = F^*$ follows automatically. $\qquad\square$

Definition 12.12 (The modular operator and modular conjugation). Define

$$S =: J\Delta^{1/2}$$

to be the polar decomposition of S, so that

$$\Delta = S^*S.$$

The operator Δ is called the **modular operator** corresponding to Ω, and the operator J is called the **modular conjugation** corresponding to Ω.

By Lemma 12.11, $\Delta = FS$, and by definition $S\Omega = F\Omega = \Omega$. Therefore,

$$\Delta\Omega = \Omega \quad and \quad J\Omega = \Omega.$$

Theorem 12.13 (Tomita-Takesaki theorem). *The following are valid:*

(1) *For all $t \in \mathbb{R}$,*

$$(12.2.3) \qquad e^{it\Delta}\mathcal{M}e^{-it\Delta} = \mathcal{M}.$$

(2)

$$(12.2.4) \qquad J\mathcal{M}J = \mathcal{M}'.$$

Lemma 12.14. *J is an involutive self-adjoint, antilinear isometry on $\mathcal{H}$ and Δ is invertible and*

$$(12.2.5) \qquad J\Delta^{1/2}J = \Delta^{-1/2}.$$

Proof. Note that $S^2 = \mathbb{1}$, so that $J\Delta^{1/2}J\Delta^{1/2} = \mathbb{1}$ which implies (12.2.5). Since $JJ^* = \mathbb{1}$, this yields

$$J(JJ^*)\Delta^{1/2}J = J^2(J^*\Delta^{1/2}J) = \Delta^{-1/2}.$$

Since J^2 is unitary and $J^*\Delta^{1/2}J$ is positive, $J^2(J^*\Delta^{1/2}J) = \Delta^{-1/2}$ gives a polar decomposition of $\Delta^{-1/2}$. By the uniqueness of the polar decomposition,

$$J^2 = \mathbb{1} \quad \text{and} \quad J^*\Delta^{1/2}J = \Delta^{-1/2}.$$

Combining this last identity with (12.2.5) yields $J = J^*$. $\qquad\square$

Lemma 12.15. *The following identities are valid:*

(1) $S = J\Delta^{1/2} = \Delta^{-1/2}J$ *and* $F = J\Delta^{-1/2} = \Delta^{1/2}J$.

(2) *For all $z \in \mathbb{C}$,*

$$(12.2.6) \qquad J\Delta^{\bar{z}}J = \Delta^{-z}.$$

Proof. By definition, $S = J\Delta^{1/2}$. By (12.2.5), $J\Delta^{1/2} = J\Delta^{1/2}J^2 = \Delta^{-1/2}J$. By Lemma 12.11, taking adjoints yields the claims for F.

It remains to prove (2). By (12.2.5) and the spectral theorem, for any continuous function f on $(0, \infty)$, $Jf(\Delta^{1/2})J = f(\Delta^{-1/2})$. Taking $f(\lambda) = \log(\lambda)$, $J\log(\Delta)J = -\log\Delta$. Let $z \in \mathbb{C}$. Since J is antilinear,

$$J\bar{z}\log(\Delta)J = zJ\log(\Delta)J = -z\log\Delta.$$

Exponentiating yields (12.2.6). $\qquad\square$

Lemma 12.16. $S\mathcal{M}S \subset \mathcal{M}'$ *and* $F\mathcal{M}'F \subset \mathcal{M}$.

Proof. Let $A \in \mathcal{M}$, and let $\Psi \in \mathcal{H}$. Let X be the unique element of $\mathcal{M}$ such that $\psi = X\Omega$. Then

$$SAS\Psi = SASX\Omega = S(AX^*\Omega) = XA^*\Omega.$$

Hence for any $B \in \mathcal{M}$,

$$(SAS)B\Psi = (SAS)BX\Omega = S(AX^*B^*\Omega) = BXA^*\Omega = B(SAS)\Psi,$$

where the last step is the calcualation just above. Since A and B are arbitrary elements of $\mathcal{M}$, it follows that $SAS \in \mathcal{M}'$ for all $A \in \mathcal{M}$, and the same reasoning shows that $F\mathcal{M}'F \subset \mathcal{M}$. $\qquad\square$

Lemma 12.17. *Let ρ_ω be the density matrix in $\mathcal{M}$ provided by Theorem 5.30 such that for all $X \in \mathcal{M}$,*

$$(12.2.7) \qquad\qquad \langle \Omega, X\Omega \rangle = \mathrm{Tr}[X\rho_\Omega].$$

Then for all $A \in \mathcal{M}$,

$$(12.2.8) \qquad\qquad \Delta A\Omega = S^*SA\Omega = \rho_\Omega A \rho_\Omega^{-1}\Omega.$$

Proof. Let $\Psi, \Phi \in \mathcal{H}$ and let $\Psi = A\Omega$ and $\Phi = B\Omega$. By Lemma 12.8 and the definition of S, $\langle \Phi, S^*S\Psi \rangle = \langle S\Psi, S\Phi \rangle = \langle A^*\Omega, B^*\Omega \rangle$. Therefore, using (12.2.7), and then cyclicity of the trace twice and then (12.2.7) once more,

$$\langle \Phi, S^*S\Psi \rangle = \mathrm{Tr}[\rho_\Omega AB^*] = \mathrm{Tr}[B^*\rho_\Omega A]$$
$$= \mathrm{Tr}[B^*\rho_\Omega A\rho_\Omega^{-1}\rho_\Omega] = \mathrm{Tr}[\rho_\Omega B^*(\rho_\Omega A\rho_\Omega^{-1})] = \langle \Phi, (\rho_\Omega A\rho_\Omega^{-1})\Omega \rangle.$$

That is, for all $\Phi \in \mathcal{H}$ and all $A \in \mathcal{M}$, $\langle\langle \Phi, \Delta A\Omega \rangle = \langle \Phi, (\rho_\Omega A\rho_\Omega^{-1})\Omega \rangle$. This proves (12.2.8). $\qquad\square$

Lemma 12.18. *For all $t \in \mathbb{R}$,*

$$(12.2.9) \qquad \Delta^t\mathcal{M}\Delta^{-t} = \mathcal{M} \qquad and \qquad \Delta^t\mathcal{M}'\Delta^{-t} = \mathcal{M}'.$$

Proof. We first show that $\Delta^t\mathcal{M}\Delta^{-t} \subseteq \mathcal{M}$. Let $A \in \mathcal{M}$. We will show that

$$(12.2.10) \qquad\qquad \Delta^t A\Delta^{-t} = \rho_\Omega^t A\rho_\Omega^{-t},$$

and the right hand side is in $\mathcal{M}$ since $\rho_\Omega \in \mathcal{M}$.

Since Δ is a positive operator, we have from (12.2.8) that for all $t \in \mathbb{R}$,

$$(12.2.11) \qquad\qquad \Delta^t A\Omega = \rho_\Omega^t A\rho_\Omega^{-t}\Omega.$$

Let $\psi \in \mathcal{H}$ be given, and let $X \in \mathcal{M}$ be such that $\psi = X\Omega$. Then for $A \in \mathcal{M}$, since $\rho_\Omega^t \in \mathcal{M}$ for all $t \in \mathbb{R}$, using (12.2.11) twice,

$$\Delta^t A\Delta^{-t}\psi = \Delta^t A\Delta^{-t}X\Omega = \Delta^t A\rho_\Omega^{-t}X\rho_\Omega^t\Omega = \rho_\Omega^t A\rho_\Omega^{-t}X\Omega = \rho_\Omega^t A\rho_\Omega^{-t}\psi.$$

Since $\rho_\Omega^t \in \mathcal{M}$ for all $t \in \mathbb{R}$, this proves (12.2.10), and hence $\Delta^t\mathcal{M}\Delta^{-t} \subseteq \mathcal{M}$. To see that equality holds, let $A \in \mathcal{M}$ and define $B := \Delta^{-t}A\Delta^t$ which belongs to $\mathcal{M}$ by what we have just proved, But then $A = \Delta^t B\Delta^{-t}$, so every $A \in \mathcal{M}$ belongs to $\Delta^t\mathcal{M}\Delta^{-t}$.

Finally, let $t \in \mathbb{R}$ and let $A' \in \mathcal{M}'$. Then for every $A \in \mathcal{M}$ by what we have just proved, $\Delta^t A \Delta^{-t}$ commutes with A' so that

$$(\Delta^t A \Delta^{-t})A' = A'(\Delta^t A \Delta^{-t}).$$

Multiplying on the left by Δ^{-t} and on the right by Δ^t, this becomes $A(\Delta^{-t}A'\Delta^t) = (\Delta^{-t}A'\Delta^t)A$, and hence $\Delta^{-t}A'\Delta^t$ commutes with every $A \in \mathcal{M}$. This shows that for all t, $\Delta^t \mathcal{M}' \Delta^{-t} \subseteq \mathcal{M}'$, and then arguing as above, we see that in fact $\Delta^t \mathcal{M}' \Delta^{-t} = \mathcal{M}'$. $\qquad\square$

Proof of Theroem 12.13. By the left and right polar decompositions of S and F as given in part (1) of Lemma 12.15, for all $A \in \mathcal{M}$ and all $A' \in \mathcal{M}'$,

$$SAS = J(\Delta^{1/2}A\Delta^{-1/2})J \qquad \text{and} \qquad FA'F = J(\Delta^{-1/2}A'\Delta^{1/2})J.$$

Thus by Lemma 12.16, for all $A \in \mathcal{M}$ and all $A' \in \mathcal{M}'$,

$$(12.2.12) \qquad J(\Delta^{1/2}A\Delta^{-1/2})J \in \mathcal{M}' \qquad \text{and} \qquad J(\Delta^{-1/2}A'\Delta^{1/2})J \in \mathcal{M}.$$

Now for all $B \in \mathcal{M}$ and all $B' \in \mathcal{M}'$, define $A := \Delta^{-1/2}B\Delta^{1/2}$ and $A' := \Delta^{1/2}B'\Delta^{-1/2}$. By Lemma 12.18, $A \in \mathcal{M}$ and $A' \in \mathcal{M}'$, and with these choices, (12.2.12) becomes

$$JBJ \in \mathcal{M}' \qquad \text{and} \qquad JB'J \in \mathcal{M},$$

and this shows $J\mathcal{M}J \subseteq \mathcal{M}'$ and $J\mathcal{M}'J \subseteq \mathcal{M}$. To see that equality holds, let $B' \in \mathcal{M}'$. Then $JB'J \in \mathcal{M}$, and then since J is an involution, $B' = J(JB'J)J$ so that $B' \in J\mathcal{M}J$. This completes the proof of (12.2.4).

Since Δ is a positive operator, Δ^z is analytic on all of $\mathbb{C}$, and hence (12.2.3) follows from (12.2.9). In fact, in this finite-dimensional setting we have $e^{z\Delta}\mathcal{M}e^{-z\Delta} = \mathcal{M}$ for all $Z \in \mathbb{C}$. In infinite dimensions when Δ is unbounded, it is necessary to take z to be pure imaginary; that is $z = it, t \in \mathbb{R}$, since then Δ^{it} is unitary. $\qquad\square$

12.3. The relative modular operator

Let $\mathcal{M}$ be a von Neumann algebra of operators on a Hilbert space $\mathcal{H}$, and let Ω_1 and Ω_2 be two cyclic and separating vectors in $\mathcal{H}$. With more than one cyclic and separating vector in play, we need a more elaborate notation than in the previous section.

Define antilinear maps S_{Ω_2,Ω_1} and F_{Ω_2,Ω_1} on $\mathcal{H}$ by

$$(12.3.1) \qquad S_{\Omega_2,\Omega_1}A\Omega_1 = A^*\Omega_2 \qquad \text{and} \qquad F_{\Omega_2,\Omega_1}A'\Omega_1 = A'^*\Omega_2$$

for all $A \in \mathcal{M}$ and all $A' \in \mathcal{M}'$.

Remark 12.19. If Ω_1 and Ω_2 are not unit vectors, define $\widehat{\Omega}_j = \|\Omega_j\|^{-1}\Omega_j$, $j = 1, 2$. Then

$$S_{\Omega_2,\Omega_1} = \|\Omega_1\|\|\Omega_2\|S_{\widehat{\Omega}_2,\widehat{\Omega}_1},$$

and hence there would be no loss of generality in assuming that Ω_1 and Ω_2 were unit vectors, as is sometimes done, but it is convenient to avoid imposing this requirement.

Lemma 12.20. *The antilinear operators S_{Ω_2,Ω_1} and F_{Ω_2,Ω_1} are adjoint to one another. That is*

$$S^*_{\Omega_2,\Omega_1} = F_{\Omega_2,\Omega_1} \quad and \quad F^*_{\Omega_2,\Omega_1} = S_{\Omega_2,\Omega_1}.$$

Proof. For any $\Psi \in \mathcal{H}$, let $A \in \mathcal{M}$ be such that $\Psi = A\Omega_1$. For any $\Phi \in \mathcal{H}$, regarded as an element of $\overline{\mathcal{H}}$, let $A' \in \mathcal{M}'$ be such that $\Phi = A'\Omega_1$. Regarding S_{Ω_2,Ω_1} as a linear map from $\mathcal{H}$ to $\overline{\mathcal{H}}$, the definitions yield

$$\langle \phi, S_{\Omega_2,\Omega_1}\psi \rangle_{\overline{\mathcal{H}}} = \langle A'\Omega_1, S_{\Omega_2,\Omega_1}A\Omega_1 \rangle_{\overline{\mathcal{H}}}$$
$$= \langle S_{\Omega_2,\Omega_1}A\Omega_1, A'\Omega_1 \rangle = \langle A^*\Omega_2, A'\Omega_1 \rangle.$$

Then since A and A' commute,

$$\langle A^*\Omega_2, A'\Omega_1 \rangle = \langle \Omega_2, AA'\Omega_1 \rangle = \langle \Omega_2, A'A\Omega_1 \rangle$$
$$= \langle A'^*\Omega_2, A\Omega_1 \rangle = \langle F_{\Omega_2,\Omega_1}\Phi, \Psi \rangle.$$

This shows that the adjoint of the linear map $S_{\Omega_2,\Omega_1} : \mathcal{H} \to \overline{\mathcal{H}}$ is the linear map $F_{\Omega_2,\Omega_1} : \overline{\mathcal{H}} \to \mathcal{H}$. $\quad\square$

Definition 12.21 (Relative modular operator). Let $\mathcal{M}$ be a von Neumann algebra of operators on a Hilbert space $\mathcal{H}$, and let Ω_1 and Ω_2 be two cyclic and separating vectors in $\mathcal{H}$. Let S_{Ω_2,Ω_1} be defined by (12.3.1). Then the **relative modular operator** $\Delta_{\Omega_2,\Omega_1}$ on $\mathcal{H}$ is defined by

$$\Delta_{\Omega_2,\Omega_1} = S^*_{\Omega_2,\Omega_1}S_{\Omega_2,\Omega_1}.$$

Lemma 12.22. *Let $\mathcal{M}$ be a von Neumann algebra of operators on a Hilbert space $\mathcal{H}$, and let Ω_1 and Ω_2 be two cyclic and separating vectors in $\mathcal{H}$, and assume that both are unit vectors. Let ω_1 and ω_2 be the corresponding states on $\mathcal{M}$.*

Then for all $A \in \mathcal{M}$, with ρ_{ω_1} and ρ_{ω_2} defined as in Theorem 5.30,

$$(12.3.2) \qquad \Delta_{\Omega_2,\Omega_1}A\Omega_1 = S^*_{\Omega_2,\Omega_1}S_{\Omega_2,\Omega_1}A\Omega_1 = \rho_{\omega_2}A\rho_{\omega_1}^{-1}\Omega_1.$$

In particular, $\Delta_{\Omega_2,\Omega_1}$ depends on Ω_1 only through the state ω_1.

Proof. Let $\Psi, \Phi \in \mathcal{H}$ and let $\Psi = A\Omega_1$ and $\Phi = B\Omega_1$. Then by (12.2.1) and the definition of S_{Ω_2,Ω_1},

$$\langle \Phi, S^*_{\Omega_2,\Omega_1}S_{\Omega_2,\Omega_1}\Psi \rangle = \langle S_{\Omega_2,\Omega_1}\Phi, S_{\Omega_2,\Omega_1}\Psi \rangle_{\overline{\mathcal{H}}}$$
$$= \langle S_{\Omega_2,\Omega_1}\Psi, S_{\Omega_2,\Omega_1}\Phi \rangle = \langle \Omega_2, AB^*\Omega_2 \rangle.$$

Therefore, using (12.2.7), and then cyclicity of the trace twice and then (12.2.7) once more,

$$\langle \Phi, S^*_{\Omega_2,\Omega_1} S_{\Omega_2,\Omega_1} \Psi \rangle = \mathrm{Tr}[\rho_{\Omega_2} AB^*] = \mathrm{Tr}[B^* \rho_{\Omega_2} A]$$
$$= \mathrm{Tr}[B^* \rho_{\Omega_2} A \rho^{-1}_{\Omega_1} \rho_{\Omega_1}] = \mathrm{Tr}[\rho_{\Omega_1} B^* (\rho_{\Omega_2} A \rho^{-1}_{\Omega_1})]$$
$$= \langle B\Omega_1, (\rho_{\Omega_2} A \rho^{-1}_{\Omega_1})\Omega_1 \rangle = \langle \Phi, (\rho_{\Omega_2} A \rho^{-1}_{\Omega_1})\Omega_1 \rangle.$$

That is, for all $\Phi \in \mathcal{H}$ and all $A \in \mathcal{M}$, $\langle \Phi, \Delta_{\Omega_2,\Omega_1} A\Omega_1 \rangle = \langle \Phi, (\rho_{\Omega_2} A \rho^{-1}_{\Omega_1})\Omega_1 \rangle$. This proves (12.3.2). $\qquad \square$

Lemma 12.23. *For all $t \in \mathbb{R}$,*

$$\Delta^t_{\Omega_2,\Omega_1} \mathcal{M} \Delta^{-t}_{\Omega_2,\Omega_1} = \mathcal{M} \qquad and \qquad \Delta^t_{\Omega_2,\Omega_1} \mathcal{M}' \Delta^{-t}_{\Omega_2,\Omega_1} = \mathcal{M}'.$$

Proof. Let $A \in \mathcal{M}$. Since $\Delta_{\Omega_2,\Omega_1}$ is a positive operator, by (12.3.2) for all $t \in \mathbb{R}$,

$$(12.3.3) \qquad \Delta^t_{\Omega_2,\Omega_1} A\Omega_1 = \rho^t_{\Omega_2} A \rho^{-t}_{\Omega_1} \Omega_1.$$

Let $\Psi \in \mathcal{H}$ be given, and let $X \in \mathcal{M}$ be such that $\Psi = X\Omega_1$. Then for $A \in \mathcal{M}$, using (12.3.3) twice,

$$\Delta^t_{\Omega_2,\Omega_1} A\Delta^{-t}_{\Omega_2,\Omega_1} \Psi = \Delta^t_{\Omega_2,\Omega_1} A\Delta^{-t}_{\Omega_2,\Omega_1} X\Omega_1$$
$$= \Delta^t_{\Omega_2,\Omega_1} A \rho^{-t}_{\Omega_2} X \rho^t_{\Omega_1} \Omega_1 = \rho^t_{\Omega_2} A \rho^{-t}_{\Omega_2} X\Omega_1 = \rho^t_{\Omega_2} A \rho^{-t}_{\Omega_2} \Psi.$$

That is, for all $A \in \mathcal{M}$, $\Delta^t_{\Omega_2,\Omega_1} A\Delta^{-t}_{\Omega_2,\Omega_1} = \rho^t_{\Omega_2} A \rho^{-t}_{\Omega_2} \in \mathcal{M}$. Therefore, $\Delta^t_{\Omega_2,\Omega_1} \mathcal{M} \Delta^{-t}_{2.1} \subseteq \mathcal{M}$, but since the map is invertible, $\Delta^t_{\Omega_2,\Omega_1} \mathcal{M} \Delta^{-t}_{2.1} = \mathcal{M}$. This proves the first part of (12.2.9) and the second part follows in the same way. $\qquad \square$

12.4. Trace inequalities without traces

The following definition due to Petz generalizes a construction of Araki. As we shall see, it allows one to write many trace functionals on $M_n(\mathbb{C})$ in a form that makes sense in any von Neumann algebra, even in von Neumann algebras in which nothing has a trace.

First, some context. Let $\mathcal{M}$ be a von Neumann algebra. Every von Neumann algebra is the dual of a Banach space $\mathcal{M}_*$, called its *predual*. In our finite-dimensional setting, $\mathcal{M}_*$ may be identified with $\mathcal{M}^*$, the dual of $\mathcal{M}$, but the distinction matters more generally. In any case, for $\omega \in \mathcal{M}_*$, we write $\omega(X) = X(\omega)$, identifying ω with an element of $\mathcal{M}^*$, and we say $\omega \in \mathcal{M}^+_*$ if it is positive as an element of $\mathcal{M}^*$.

Let $\omega \in \mathcal{M}^+_*$ be faithful. Let Ω be the cyclic and separating vector corresponding to ω so that $\omega = \omega_\Omega$. Let $\psi \in \mathcal{M}^+_*$, and let Ψ be the unique vector representative of ψ in the positive cone determined by Ω. Let $\Delta_{\Psi,\Omega}$ be the relative modular operator which we recall depends on Ψ only through ψ.

Definition 12.24 (Petz quasi relative entropies). Let f be a function on $[0, \infty)$ with values in $\mathbb{R}$. Let $K \in \mathcal{M}$. Let $\omega, \psi \in \mathcal{M}_*^+$ with ω faithful. Then with Ψ, Ω and $\Delta_{\Psi, \Omega}$ determined by ψ and ω as described above, the **Petz quasi relative entropy** determined by ω, ψ, K, and f is the quantity

$$S_f^K(\psi, \omega) := \langle K\Omega, f(\Delta_{\Psi, \Omega}) K\Omega \rangle,$$

where Ω is the cyclic and separating vector for the GNS representation of $\mathcal{M}$ determined by ω. For $K = \mathbb{1}$, we simply write $S_f(\psi, \omega)$ in place of $S_f^{\mathbb{1}}(\psi, \omega)$. That is

$$S_f(\psi, \omega) := \langle \Omega, f(\Delta_{\Psi, \Omega}) \Omega \rangle.$$

Example 12.25. Let $\mathcal{M} = M_n(\mathbb{C})$, and let $X, Y \in M_n^+(\mathbb{C})$ be such that for all $A \in \mathcal{M}$,

$$\psi(A) = \mathrm{Tr}[AX] \quad \text{and} \quad \omega(A) = \mathrm{Tr}[AY].$$

Since ω is faithful, $Y > 0$. Then $\Psi = X^{1/2}$ and $\Omega = Y^{1/2}$, and for all $K \in \mathcal{M}$,

$$\Delta_{\Psi, \Omega} = L_X R_{Y^{-1}}.$$

Consider for example $f(x) = x^p, 0 < p < 1$. Then

$$f(\Delta_{\Psi, \Omega}) = L_{X^r} R_{Y^{-r}}.$$

Then $f(\Delta_{\Psi, \Omega})(K\Omega) = X^r K Y^{1/2-r}$ so that

$$S_f^K(\psi, \omega) = \mathrm{Tr}[(KY^{1/2})^* X^r K Y^{1/2-r}] = \mathrm{Tr}[K^* X^r K Y^{1-r}].$$

This is the functional appearing in the Lieb concavity theorem, and of course Uhlmann's theorem.

Let $\mathcal{M}$ and $\mathcal{N}$ be von Neumann algebras, and let α be a linear map from $\mathcal{M}$ to $\mathcal{N}$. so that $\alpha(X^*X) \geq \alpha(X)^*\alpha(X)$ for all $X \in \mathcal{M}$. Any such map is positive. For any linear map $\alpha : \mathcal{M} \to \mathcal{N}$, let $\alpha^\dagger$ denote the dual map from $\mathcal{N}_* \to \mathcal{M}_*$ defined by

$$\alpha^\dagger(\varphi)(X) = \varphi(\alpha(X))$$

for all $X \in \mathcal{M}$ and all $\varphi \in \mathcal{N}_*$. If $\omega_{\mathcal{N}} \in \mathcal{N}_*^+$ is faithful, and if α is positive and such that $\alpha(X^*X) = $ only for $X = 0$, then $\omega_{\mathcal{M}} := \alpha^\dagger(\omega_{\mathcal{N}}) \in \mathcal{M}_*^+$ is faithful. We then perform the GNS construction using these two faithful positive linear functionals, constructing the Hilbert spaces $\mathcal{H}$ and $\mathcal{K}$ on which $\mathcal{M}$ and $\mathcal{N}$ act. By construction,

$$\alpha^\dagger(\omega_{\mathcal{N}}) = \omega_{\mathcal{M}}.$$

Let $\Omega_{\mathcal{M}}$ and $\Omega_{\mathcal{N}}$ be the corresponding cyclic and separating vectors.

Lemma 12.26 (Petz's first lemma). *Define a map $V_\alpha : \mathcal{H} \to \mathcal{K}$ by*

$$V_\alpha(X\Omega_{\mathcal{M}}) = \alpha(X)\Omega_{\mathcal{N}}.$$

If α is a Schwarz map, then V_α is a contraction.

Proof. For any $X \in \mathcal{M}$,

$$\|V_\alpha(X\Omega_\mathcal{M})\|_\mathcal{K}^2 = \langle \Omega_\mathcal{N}, \alpha(X)^*\alpha(X)\Omega_\mathcal{N}\rangle \le \langle \Omega_\mathcal{N}, \alpha(X^*X)\Omega_\mathcal{N}\rangle$$
$$= \alpha^\dagger(\omega_{\Omega_\mathcal{N}})(X^*X) = \omega_{\Omega_\mathcal{M}}(X^*X) = \|X\Omega_\mathcal{M}\|_\mathcal{H}^2. \qquad \square$$

Continuing with the same notation, consider any $\psi \in \mathcal{N}_*^+$ so that $\alpha^\dagger(\psi) \in \mathcal{M}_*^+$. Let $\Psi_\mathcal{N}$ and $\Psi_\mathcal{M}$ be vector representatives of these positive linear functionals on $\mathcal{N}$ and $\mathcal{M}$.

Then we have the two relative modular operators $\Delta_{\Psi_\mathcal{N},\Omega_\mathcal{N}}$ and $\Delta_{\Psi_\mathcal{M},\Omega_\mathcal{M}}$ Recall that by the definition of the relative modular operator, for all X,

$$(12.4.1) \qquad \langle \Omega_\mathcal{M}, X^*\Delta_{\Psi_M,\Omega_M}X\Omega_\mathcal{M}\rangle = \langle X^*\Psi_\mathcal{M}, X^*\Psi_\mathcal{M}\rangle,$$

and likewise with $\mathcal{N}$ in place of $\mathcal{M}$.

Lemma 12.27 (Petz's second lemma). *Let α and V_α be specified as in Lemma 12.26. Then*

$$V_\alpha^*\Delta_{\Psi_\mathcal{N},\Omega_\mathcal{N}}V \le \Delta_{\Psi_\mathcal{M},\Omega_\mathcal{M}}.$$

Proof. For any $X \in \mathcal{M}$,

$$\langle X\Omega_\mathcal{M}, (V_\alpha^*\Delta_{\Psi_\mathcal{N},\Omega_\mathcal{N}}V)X\Omega_\mathcal{M}\rangle = \langle \alpha(X)\Omega_\mathcal{N}, \Delta_{\Psi_\mathcal{N},\Omega_\mathcal{N}}(\alpha(X)\Omega_\mathcal{N})\rangle$$
$$= \langle \alpha(X^*)\Psi_\mathcal{N}, (\alpha(X^*)\Psi_\mathcal{N})\rangle$$
$$\le \langle \Psi_\mathcal{N}, \alpha(XX^*)\Psi_\mathcal{N}\rangle$$
$$= \langle \Psi_\mathcal{M}, XX^*\Psi_\mathcal{M})\rangle$$
$$= \langle X\Omega_\mathcal{M}, \Delta_{\Psi_\mathcal{M},\Omega_\mathcal{M}}X\Omega_\mathcal{M}\rangle.$$

The first equality is from the definition if V_α, the second and final are from (12.4.1) or its analogue with $\mathcal{N}$ in place of $\mathcal{M}$. The rest is the Schwarz inequality and the definition of $\Psi_\mathcal{M}$. Since $X \in \mathcal{M}$ is arbitrary, this proves the lemma. $\qquad \square$

Combining the two lemmas yields the following theorem:

Theorem 12.28 (Petz's theorem). *Let $\mathcal{M}$ and $\mathcal{N}$ be von Neumann algebras, and let α be a Schwarz map from $\mathcal{M}$ to $\mathcal{N}$ such that $\alpha(X^*X) = 0$ only for $X = 0$. Let $\omega \in \mathcal{N}_*^+$ be faithful on $\mathbb{N}$, and then $\alpha^\dagger(\omega)$ is faithful on $\mathcal{M}$. Let $\psi \in \mathcal{N}_*^!$, so that $\alpha^\dagger(\psi) \in \mathcal{M}_*^+$. Let f be operator monotone with $f(0) = 0$. Then for all $K \in \mathcal{M}$,*

$$(12.4.2) \qquad S_f^K(\alpha^\dagger(\psi), \alpha^\dagger(\omega)) \ge S_f^{\alpha(K)}(\psi, \omega).$$

Proof. Starting from the definition of $S_f^K(\alpha^\dagger(\psi), \alpha^\dagger(\omega))$,

$$
\begin{aligned}
S_f^K(\alpha^\dagger(\psi), \alpha^\dagger(\omega)) &= \langle K\Omega_{\mathcal{M}}, f(\Delta_{\Psi_{\mathcal{M}}, \Omega_{\mathcal{M}}}) K\Omega_{\mathcal{M}} \rangle \\
&\geq \langle K\Omega_{\mathcal{M}}, V_\alpha^* f(\Delta_{\Psi_{\mathcal{N}}, \Omega_{\mathcal{N}}}) V_\alpha K\Omega_{\mathcal{M}} \rangle \\
&= \langle V_\alpha K\Omega_{\mathcal{M}}, f(\Delta_{\Psi_{\mathcal{N}}, \Omega_{\mathcal{N}}}) V_\alpha K\Omega_{\mathcal{M}} \rangle \\
&= \langle \alpha(K)\Omega_{\mathcal{N}}, f(\Delta_{\Psi_{\mathcal{N}}, \Omega_{\mathcal{N}}}) \alpha(K)\Omega_{\mathcal{N}} \rangle = S_f^{\alpha(K)}(\psi, \omega).
\end{aligned}
$$

The first and fourth equalities are from the definition of the Petz quasi relative entropy. The only inequality is from Petz's second lemma. The second equality is trivial, and the third is from Petz's first lemma. $\qquad\square$

Applying Petz's theorem in the context of Example 12.25 yields another proof of Theorem 10.5, Uhlmann's monotonicity theorem. One can also give a direct proof of the DPI for Schwarz maps in this way, and indeed, this is the approach that was taken in Section 8.2, which, as we have seen, is well adapted to the study of cases of equality. Hopefully the present brief discussion of these ideas provides an understanding of where the constructions used in Section 8.2, such as the relative modular operator, come from, and also show that methods developed in this book in the matrix setting can be applied in a general von Neumann algebra setting. After all, the only two inequalities used in the proof of Petz's theorem were the Schwarz inequality and the operator Jensen inequality which readily extend to the infinite-dimensional setting.

Exercises

(1) Let $\mathcal{H}$ be $M_n(\mathbb{C})$ equipped with the Hilbert-Schmidt inner product. Let ρ be a strictly positive density matrix in $M_n(\mathbb{C})$. Let $\mathcal{M}$ denote the image of $M_n(\mathbb{C})$ under the GNS construction. That is, $\mathcal{M}$ is the von Neumann subalgebra of $\mathcal{B}(\mathcal{H})$ consisting of the operators L_X, $X \in M_n(\mathbb{C})$.

   (a) Define $\Omega \in \mathcal{H}$ by $\Omega := \rho^{1/2}$. Show that Ω is cyclic and separating for $\mathcal{M}$.

   (b) Let S be the operator specified in Definition 12.9. Show that for all $X \in M_n(\mathbb{C})$,

$$
S(X\Omega) = S(X\rho^{1/2}) = (\rho^{1/2}(X\rho^{1/2})\rho^{-1/2})^*.
$$

   (c) The identity derived in (b) expresses S as the composition of two maps: the linear map $Y \mapsto \rho^{1/2} Y \rho^{-1/2}$ (where $Y = \rho^{1/2} X$) and the antilinear map $Z \mapsto Z^*$. Show that the first map is positive, and then use the uniqueness of the polar decomposition to show that the first map is $\Delta^{1/2}$ and the second is J.

   (d) Prove that in this case, the modular operator Δ and modular conjugation J are given by

$$
\Delta(X) = \rho X \rho^{-1} \quad \text{and} \quad J(X) = X^*.
$$

(2) Continue with the setting specified in Exercise 1, with $\Omega = \rho^{1/2}$ for invertible $\rho \in M_n(\mathbb{C})$. Show that for $f(x) = x^p$, $0 \le p \le 1$, (12.4.2) becomes

$$\mathrm{Tr}[K^*(\alpha^\dagger(X))^p K(\alpha^\dagger(Y))^{1-p}] \ge \mathrm{Tr}[\alpha(K^*)X^p\alpha(K)Y^{1-p}],$$

which, apart from the change in notation, is Uhlmann's theorem, Theorem 10.5.

Convex geometry

A.1. Convex sets

Let $\mathcal{H}$ be a real Hilbert space with norm $\|\mathbf{x}\|$ and inner product $\langle \mathbf{x}, \mathbf{y} \rangle$ for $\mathbf{x}, \mathbf{y} \in \mathcal{H}$. In many of applications in this book, $\mathcal{H}$ will be the space $M_n(\mathbb{C})$, considered as a real Hilbert space equipped with the inner product $\Re(\mathrm{Tr}[X^*Y])$ for $X, Y \in M_n(\mathbb{C})$, or some subspace of this real Hilbert space. In this case, $\mathcal{H}$ will be finite dimensional. However, most of the theorems discussed in this appendix are true in infinite dimensions, and often the simplest proofs apply also in this case. We shall be clear about when the assumption that $\mathcal{H}$ is finite-dimensional matters.

Definition A.1 (Convex set). A nonempty set $C \subset \mathcal{H}$ is **convex** in case for all $\mathbf{x}, \mathbf{y} \in C$ and all $0 < \lambda < 1$, $(1 - \lambda)\mathbf{x} + \lambda\mathbf{y} \in C$. It is convenient to define the empty set to be convex as well.

The following fundamental lemma is a consequence of the parallelogram identity for Hilbert space norms; namely, for all $\mathbf{x}, \mathbf{y} \in \mathcal{H}$,

$$\text{(A.1.1)} \qquad \left\| \frac{\mathbf{x} + \mathbf{y}}{2} \right\|^2 + \left\| \frac{\mathbf{x} - \mathbf{y}}{2} \right\|^2 = \frac{\|\mathbf{x}\|^2 + \|\mathbf{y}\|^2}{2}.$$

Lemma A.2 (Projection lemma). *Let C be a nonempty closed convex subset of $\mathcal{H}$. Then there exists a unique $\mathbf{c}_0 \in C$ such that*

$$\|\mathbf{c}_0\| \le \|\mathbf{c}\| \quad \text{for all} \quad \mathbf{c} \in C.$$

Proof. Let $d := \inf\{\|\mathbf{c}\| \; : \; \mathbf{c} \in C\}$. If $d = 0$, then $0 \in C$ since C is closed, and this is the unique element of minimal norm. Hence we may suppose that $d > 0$. Let $\{\mathbf{c}_n\}_{n \in \mathbb{N}}$ be a sequence in C such that $\lim_{n \to \infty} \|\mathbf{c}_n\| = d$. By the parallelogram identity

$$\left\|\frac{\mathbf{c}_m + \mathbf{c}_n}{2}\right\|^2 + \left\|\frac{\mathbf{c}_m - \mathbf{c}_n}{2}\right\|^2 = \frac{\|\mathbf{c}_m\|^2 + \|\mathbf{c}_n\|^2}{2}.$$

By the convexity of C, and the definition of d, $\left\|\frac{\mathbf{c}_m + \mathbf{c}_n}{2}\right\|^2 \geq d^2$, and so

$$\left\|\frac{\mathbf{c}_m - \mathbf{c}_n}{2}\right\|^2 \leq \frac{\left(\|\mathbf{c}_m\|^2 - d^2\right) + \left(\|\mathbf{c}_n\|^2 - d^2\right)}{2}.$$

By construction, the right side tends to zero, and hence $\{\mathbf{c}_n\}_{n\in\mathbb{N}}$ is a Cauchy sequence. Then, by the completeness that is a defining property of Hilbert spaces, $\{\mathbf{c}_n\}_{n\in\mathbb{N}}$ is a convergent sequence. Let $\mathbf{c}_0$ denote the limit. By the continuity of the norm, $\|\mathbf{c}_0\| = \lim_{n\to\infty}\|\mathbf{c}_n\| = d$. Finally, if $\mathbf{c}_1$ is any other vector in C with $\|\mathbf{c}_1\| = d$, $(\mathbf{c}_0 + \mathbf{c}_1)/2 \in C$, so that $\|(\mathbf{c}_0 + \mathbf{c}_1)/2\| \geq d$. Then by (A.1.1), $\|(\mathbf{c}_0 - \mathbf{c}_1)/2\| = 0$. This proves the uniqueness. $\qquad\square$

For nonempty closed and convex $K \subset \mathcal{H}$ and $\mathbf{x} \in \mathcal{H}$, define the distance from $\mathbf{x}$ to K, $d(\mathbf{x}, K)$, by

$$d(\mathbf{x}, K) = \inf\{\|\mathbf{y} - \mathbf{x}\| \,:\, \mathbf{y} \in K\}.$$

Define $C = K - \mathbf{x} = \{\mathbf{k} - \mathbf{x} \,:\, \mathbf{k} \in K\}$. Since C is merely a translate of C, it is also closed and convex, and since $\mathbf{x} \notin K$, $0 \notin C$. Let $\mathbf{c}_\mathbf{x}$ be the element of C of minimal norm provided by the projection lemma. Evidently $d(0, C) = \|\mathbf{c}_\mathbf{x}\|$. Define $\mathbf{v}_\mathbf{x} = \mathbf{c}_\mathbf{x} + \mathbf{x} \in K$, noting that $\mathbf{v}_\mathbf{x}$ is the element of K that is closest to $\mathbf{x}$. Then $d(\mathbf{x}, K) = \|\mathbf{x} - \mathbf{v}_\mathbf{x}\|$. Now consider any two points $\mathbf{x}, \mathbf{y} \notin K$. By the optimality property of $\mathbf{v}_\mathbf{x}$,

$$d(\mathbf{x}, K) = \|\mathbf{x} - \mathbf{v}_\mathbf{x}\| \leq \|\mathbf{x} - \mathbf{v}_\mathbf{y}\| = \|(\mathbf{x} - \mathbf{y}) + (\mathbf{y} - \mathbf{v}_\mathbf{y})\|$$
$$\leq \|\mathbf{x} - \mathbf{y}\| + \|\mathbf{y} - \mathbf{v}_\mathbf{y}\| = \|\mathbf{x} - \mathbf{y}\| + d(\mathbf{y}, K).$$

By symmetry in $\mathbf{x}$ and $\mathbf{y}$, we then have

$$|d(\mathbf{x}, K) - d(\mathbf{y}, K)| \leq \|\mathbf{x} - \mathbf{y}\|.$$

That is, for a nonempty closed convex set K, the function $d(\mathbf{x}, K)$ is Lipschitz continuous on $\mathcal{H}$ with Lipschitz constant 1.

Theorem A.3. *Let K be a nonempty closed convex subset of $\mathcal{H}$. Then for all $\mathbf{x}_0 \in \mathcal{H}$, $\mathbf{x}_0 \notin K$, there exists a unit vector $\mathbf{u} \in \mathcal{H}$ and $d > 0$ such that*

$$\langle \mathbf{x}, \mathbf{u} \rangle \geq \langle \mathbf{x}_0, \mathbf{u} \rangle + d(\mathbf{x}_0, K) \quad \text{for all} \quad \mathbf{x} \in K.$$

Proof. As above, let $C = K - \mathbf{x}_0 = \{\mathbf{x} - \mathbf{x}_0 \,:\, \mathbf{x} \in K\}$. Since C is merely a translate of C, it is also closed and convex, and since $\mathbf{x}_0 \notin K$, $0 \notin C$. By Lemma A.2, there exists $\mathbf{c}_0 \in C$ of minimal norm, and since $0 \notin C$, $\|\mathbf{c}_0\| \neq 0$. Let $\mathbf{c} \in C$, and $0 < \lambda < 1$. Then since C is convex, $(1 - \lambda)\mathbf{c}_0 + \lambda\mathbf{c} \in C$, and hence $\|(1-\lambda)\mathbf{c}_0 + \lambda\mathbf{c}\|^2 \geq \|\mathbf{c}_0\|^2$. Consequently, $\langle \mathbf{c} - \mathbf{c}_0, \mathbf{c}_0 \rangle + \dfrac{\lambda}{2}\|\mathbf{c}\|^2 \geq 0$, and since this is true for all $0 < \lambda < 1$,

$$(A.1.2) \qquad\qquad \langle \mathbf{c} - \mathbf{c}_0, \mathbf{c}_0 \rangle \geq 0.$$

Now writing $\mathbf{c} = \mathbf{x} - \mathbf{x}_0$, $\mathbf{x} \in K$, (A.1.2) becomes $\langle \mathbf{x}, \mathbf{c}_0 \rangle \geq \langle \mathbf{x}_0, \mathbf{c}_0 \rangle + \|\mathbf{c}_0\|^2$, valid for all $\mathbf{x} \in K$. We take $\mathbf{u} = \|\mathbf{c}_0\|^{-1}\mathbf{c}_0$ and $d = \|\mathbf{c}_0\|$. $\qquad\square$

A **closed half-space** H in $\mathcal{H}$ is a subset of $\mathcal{H}$ of the form

$$H = \{\mathbf{x} \;:\; \langle \mathbf{x}, \mathbf{v} \rangle \geq \alpha\}$$

for some nonzero $\mathbf{v} \in \mathcal{H}$ and some $\alpha \in \mathbb{R}$. For any closed convex set K and any $\mathbf{x}_0 \notin K$, Theorem A.3 provides a half-space H such that $K \subset H$ and $\mathbf{x}_0 \notin H$. (Simply take $\alpha = \langle \mathbf{x}_0, \mathbf{v} \rangle + d(\mathbf{x}_0, K)/2$.) It follows that K is the intersection of the closed half-spaces containing it.

A variant of Theorem A.3 will be important in what follows, and we now prepare for this. For $r > 0$ and $\mathbf{x} \in \mathcal{H}$, $B_r(\mathbf{x})$ denotes the open ball of radius r about $\mathbf{x}$; i.e., $B_r(\mathbf{x}) = \{\mathbf{y} \;:\; \|\mathbf{y} - \mathbf{x}\| < r\}$. Let K be a convex set with nonempty interior. For $\epsilon > 0$, define

$$K_\epsilon := \{\mathbf{x} \in K \;:\; \mathbf{x} + B_\epsilon(\mathbf{x}) \in K\}.$$

While for some $\epsilon > 0$, K_ϵ may be empty, since the interior of K is not empty, for all ϵ sufficiently small, K_ϵ is not empty.

Lemma A.4. *Let K be a convex set with nonempty interior. Then for all $\epsilon > 0$, K_ϵ is an open convex subset of K, and the interior of K, $\mathrm{Int}(K)$ is given by*

$$\mathrm{Int}(K) = \bigcup_{\epsilon > 0} K_\epsilon,$$

which is also convex.

Proof. Fix $\epsilon > 0$ sufficiently small that $K_\epsilon \neq \emptyset$. Let $\mathbf{x}, \mathbf{y} \in K_\epsilon$. Let $\mathbf{u}$ be any unit vector in $\mathcal{H}$, and let $0 < r < \epsilon$. Then $\mathbf{x} + r\mathbf{u}$ and $\mathbf{y} + r\mathbf{u}$ both belong to K. Hence for any $0 < \lambda < 1$,

$$(1-\lambda)(\mathbf{x} + r\mathbf{u}) + \lambda(\mathbf{y} + r\mathbf{u}) = ((1-\lambda)\mathbf{x} + \lambda\mathbf{y}) + r\mathbf{u} \in K,$$

and this shows that $(1-\lambda)\mathbf{x} + \lambda\mathbf{y} + B_\epsilon((1-\lambda)\mathbf{x} + \lambda\mathbf{y}) \in K$. Hence K_ϵ is convex. Clearly, for each $\epsilon > 0$, $K_\epsilon \subset \mathrm{Int}(K)$, and each $\mathbf{x} \in \mathrm{Int}(K)$ belongs to K_ϵ for some $\epsilon > 0$. $\qquad\square$

In the next theorem, we shall assume that $\mathcal{H}$ is finite dimensional, as it will be in all of our applications. When $\mathcal{H}$ is finite dimensional, the elementary Bolzano-Wierstrass theorem says that any bounded infinite set in $\mathcal{H}$ contains a convergent sequence. By using the Hahn-Banach theorem instead, one can prove the same result without restriction on the dimension.

Theorem A.5. *Suppose that $\mathcal{H}$ is finite dimensional. Let K be a convex set with a nonempty interior. Let $\mathbf{x}_0$ be any boundary point of K. Then there exists a unit vector $\mathbf{u}$ such that*

$$(A.1.3) \qquad \langle \mathbf{x}, \mathbf{u} \rangle \geq \langle \mathbf{x}_0, \mathbf{u} \rangle \quad \text{for all} \quad \mathbf{x} \in K.$$

Proof. Let $\epsilon > 0$ be sufficiently small that $K_\epsilon \neq \emptyset$. Then its closure, $\overline{K_\epsilon}$ is a closed convex set contained in $\mathrm{Int}(K)$, and then since $\mathbf{x}_0$ is a boundary point of K, $\mathbf{x}_0 \notin \overline{K_\epsilon}$. We apply Theorem A.3 to K_ϵ and $\mathbf{x}_0$ to conclude that there exists a unit vector $\mathbf{u}_\epsilon \in \mathcal{H}$ such that

$$(A.1.4) \qquad \langle \mathbf{x}, \mathbf{u}_\epsilon \rangle \geq \langle \mathbf{x}_0, \mathbf{u}_\epsilon \rangle + d(\mathbf{x}_0, \overline{K_\epsilon}) \quad \text{for all} \quad \mathbf{x} \in \overline{K_\epsilon}.$$

By the Bolzano-Weirerstrass theorem, there exists a sequence $\{\epsilon_n\}_{n \in \mathbb{N}}$ monotonically decreasing to 0, with ϵ_1 sufficiently small that $K_{\epsilon_1} \neq \emptyset$, and along which $\mathbf{u}_{\epsilon_n}$ converges to a unit vector $\mathbf{u}$, and then (A.1.3) follows from (A.1.4) in the limit $n \to \infty$. $\qquad \square$

Let $C \subset \mathcal{H}$ be convex and contain more than one point. For any $\mathbf{x}_0 \in C$, define

$$\mathcal{V}(C) = \mathrm{span}(C - \mathbf{x}_0) \quad \text{and} \quad \mathcal{S}(C) := \mathcal{V}(C) + \mathbf{x}_0.$$

It is left as an exercise to show that $\mathcal{V}(C)$ and $\mathcal{S}(C)$ do not depend on the choice of $\mathbf{x}_0 \in C$. The dimension of C, $\dim(C)$, is then defined to be the dimension of the subspace $\mathcal{V}(C)$. It is also left as an exercise to show that $\mathcal{S}(C)$ is the minimal affine subspace of $\mathcal{H}$ containing C. $\mathcal{S}(C)$ is given the topology it inherits from $\mathcal{H}$.

Lemma A.6. *Let C be convex subset of $\mathcal{H}$ with more than one point and $\dim(C) < \infty$. Then as a subset of $\mathcal{S}(C)$, C has a nonempty interior.*

Proof. Let $d = \dim(C)$, and let $\mathbf{x}_0 \in C$, and choose $\{\mathbf{x}_1, \ldots, \mathbf{x}_d\} \subset C$ so that $\{\mathbf{x}_1 - \mathbf{x}_0, \ldots, \mathbf{x}_d - \mathbf{x}_0\}$ is a linearly independent set. Then the simplex with vertices $\mathbf{x}_0, \mathbf{x}_1, \ldots, \mathbf{x}_d$ has a nonempty interior and is contained in C. $\qquad \square$

Definition A.7 (Relative interior). Let C be a nonempty convex set in $\mathcal{H}$. The **relative interior** of C is the interior of C considered as a subset of $\mathcal{S}(C)$.

Example A.8. We shall consider many convex functions of the set $\mathfrak{S}_n$ of $n \times n$ density matrices; i.e., the set of positive semidefinite matrices ρ such that $\mathrm{Tr}[\rho] = 1$. It is left as an exercise to show that $\mathcal{V}(\mathfrak{S}_n)$ is the subspace of $M_n(\mathbb{C})$ consisting of self-adjoint matrices with zero trace, and that $\mathcal{S}(\mathfrak{S}_n)$ is the affine subspace of $M_n(\mathbb{C})$ consisting of self-adjoint matrices with unit trace. The dimension of $\mathfrak{S}_n$ is $n^2 - 1$ while the dimension of the ambient space $M_n(\mathbb{C})$, considered as a real vector space, is $2n^2$.

A.2. Extreme points

Definition A.9. Let $K \subset \mathcal{H}$ be convex. A point $\mathbf{z} \in \mathcal{H}$ is **extreme** if for all $\mathbf{x}, \mathbf{y} \in K$ and $0 < \lambda < 1$,

$$(A.2.1) \qquad \mathbf{z} = (1 - \lambda)\mathbf{x} + \lambda\mathbf{y} \quad \Rightarrow \quad \mathbf{x} = \mathbf{y} = \mathbf{z}.$$

The set of extreme points of K is denoted by $E(K)$.

Remark A.10. $E(K)$ is also the set of points $\mathbf{z} \in K$ such that whenver $\mathbf{v}, \mathbf{w} \in K$

$$(A.2.2) \qquad \mathbf{z} = \frac{1}{2}(\mathbf{v} + \mathbf{w}) \quad \Rightarrow \quad \mathbf{v} = \mathbf{w} = \mathbf{z}.$$

To see this, suppose $\mathbf{z}$ is such that for all $\mathbf{v}, \mathbf{w} \in K, (A.2.2)$ is satisfied. Suppose that for some $\mathbf{x}, \mathbf{y} \in K, \mathbf{z} = (1 - \lambda)\mathbf{x} + \lambda\mathbf{y}$ with $0 < \lambda < 1$. If $\lambda = 1/2$, Consider the case $1/2 < \lambda < 1$. Define $\mathbf{w} = \mathbf{y}$ and $\mathbf{v} = (2 - 2\lambda)\mathbf{x} + (2\lambda - 1)\mathbf{x} \in K$. Then $\mathbf{z} = \frac{1}{2}(\mathbf{v} + \mathbf{w})$, and then by (A.2.2), $\mathbf{v} = \mathbf{w} = \mathbf{z}$. But then $\mathbf{y} = \mathbf{w} = \mathbf{z}$, and hence $\mathbf{x} = \mathbf{z}$, so that (A.2.1) is satisfied. The case $0 < \lambda < 1/2$ is similar.

Definition A.11. Let M be an arbitrary subset of $\mathcal{H}$. The **convex hull** of M, $C(M)$, is the set of a finite convex combinations of points of M. That is, $\mathbf{x} \in C(M)$ if and only if for some $m \in \mathbb{N}$,

$$\mathbf{x} = \sum_{j=1}^{m} \lambda_j \mathbf{x}_j,$$

where $\{\mathbf{x}_1, \ldots, \mathbf{x}_m\} \subset M, \{\lambda_1, \ldots, \lambda_m\} \subset (0, 1)$, and $\sum_{j=1}^{m} \lambda_j = 1$. The **closed convex hull** of $M, \overline{C(M)}$, is the closure of $C(M)$.

It is evident that for any set $M, C(M)$ is convex, and hence $\overline{C(M)}$ is also convex. The following theorem is fundamental. It was proved in three dimensions by Minkowski [154] and in finite dimensions by Steinitz [201].

Theorem A.12 (Minkowski-Steinitz theorem). *Let K be a nonempty compact convex set in $\mathcal{H}$. Then $K = \overline{C(E(K))}$.*

Remark A.13. The set $E(K)$ may be finite or infinite, even in finite dimensions. For example, consider the case in which K is the closed unit ball of $\mathcal{H}$. Then $E(K)$ consists of all unit vectors in $\mathcal{H}$. However, if $E(K)$ is finite, then $C(E(K))$ is automatically closed, and in this case K is the convex hull of its extreme points.

The proof of Theorem A.12 given here; as in [201], will use induction on dimension, and hence it is strictly finite dimensional. The infinite-dimensional version of the theorem is known as the Krein-Milamn theorem. In proving it, the heart of the matter is to show that the set $E(K) \neq \emptyset$. Then the following lemma applies in any dimension.

Lemma A.14. *Suppose that for all K nonempty, compact, and convex, $E(K) \neq \emptyset$. Then for all K nonempty, compact, and convex, $K = \overline{C(E(K))}$.*

Proof. Let $\dim(K) = d > 1$, and make the inductive hypothesis that the theorem is proved for dimensions less than d, dimension 1 being trivial. Since $E(K) \subseteq K$ and K is convex, $C(E(K)) \subseteq K$. Then since K is closed, $\overline{C(E(K))} \subseteq K$. Supppose $\overline{C(E(K))} \neq K$. Then there exists $\mathbf{x}_0 \in K$ such that $\mathbf{x}_0 \notin \overline{C(E(K))}$. By Theorem A.3, there exists a unit vector $\mathbf{u}$ such that

$$(\text{A.2.3}) \qquad \langle \mathbf{x}, \mathbf{u} \rangle \geq \langle \mathbf{x}_0, \mathbf{u} \rangle + d(\mathbf{x}_0, \overline{C(E(K))}) \quad \text{for all} \quad \mathbf{x} \in \overline{C(E(K))}.$$

Define

$$\alpha := \inf\{\langle \mathbf{x}, \mathbf{u} \rangle \ : \ \mathbf{x} \in K\} \quad \text{and} \quad D := \{\mathbf{x} \in K \ : \ \langle \mathbf{x}, \mathbf{u} \rangle = \alpha\}.$$

Since $\mathbf{x}_0 \in K$, $\alpha \leq \langle \mathbf{x}_0, \mathbf{u} \rangle$, and then by (A.2.3),

$$(\text{A.2.4}) \qquad D \cap \overline{C(E(K))} = \emptyset.$$

Since K is nonempty, compact, and convex, D has these properties as well. By the inductive hypothesis, D has an extreme point $\mathbf{z}$. We claim that $\mathbf{z}$ is also extreme in K: Suppose that there exist $\mathbf{x}, \mathbf{y} \in K$ and $0 < \lambda < 1$ such that $\mathbf{z} = (1 - \lambda)\mathbf{x} + \lambda\mathbf{y}$. Then

$$\alpha = \langle \mathbf{z}, \mathbf{u} \rangle = (1 - \lambda)\langle \mathbf{x}, \mathbf{u} \rangle + \lambda\langle \mathbf{y}, \mathbf{u} \rangle$$

but $\langle \mathbf{x}, \mathbf{u} \rangle, \langle \mathbf{y}, \mathbf{u} \rangle \geq \alpha$. Consequently, $\langle \mathbf{x}, \mathbf{u} \rangle = \langle \mathbf{x}, \mathbf{u} \rangle = \alpha$ and $\mathbf{x}, \mathbf{y} \in D$, and then $\mathbf{x} = \mathbf{y} = \mathbf{z}$. Hence $\mathbf{z} \in E(K)$, which contradicts (A.2.4), and shows that there cannot exist $\mathbf{x}_0 \in K, \mathbf{x}_0 \notin \overline{C(E(K))}$. $\qquad \square$

Proof of Theorem A.12. By Lemma A.14, it remains to show that every nonempty compact, convex set K in a finite-dimensional Hilbert space $\mathcal{H}$ contains at least one extreme point. In one dimension this is obvious; K is a closed interval and the two endpoints are extreme.

We now suppose that K is a closed convex set in a d-dimensional Hilbert space $\mathcal{H}$, and that the claim is valid in dimension $d-1$. Let $\mathbf{u}$ be any unit vector in $\mathcal{H}$, and define $\alpha = \min\{\langle \mathbf{x}, \mathbf{u} \rangle \ : \ \mathbf{x} \in K\}$, and define $D =: \{\mathbf{x} \in K : \langle \mathbf{x}, \mathbf{u} \rangle = \alpha\}$. Then D is a nonempty compact convex set that is contained in a translate of a $(d-1)$-dimensional subspace of $\mathcal{H}$, namely the subspace of vectors orthogonal to $\mathbf{u}$. Then D contains extreme points, and by the argument used in the final part of the proof of Lemma A.14, any extreme point of D is an extreme point of K. $\qquad \square$

By definition, if S is a subset of a finite-dimensional real Hilbert space, and $\mathbf{x} \in C(S)$, then $\mathbf{x}$ can be written as a convex combination of the elements in some finite set $\{\mathbf{x}_1, \ldots, \mathbf{x}_M\} \subseteq S$. A theorem of Carathéodory gives a bound on M in terms of d: $\mathbf{x}$ can be written as a convex combination of $d + 1$ or fewer elements of S.

Theorem A.15 (Carathéodory's theorem). *Let $\mathcal{H}$ be a real Hilbert space of finite dimension d. Let $S \subset \mathcal{H}$, and let $\mathbf{x} \in C(S)$. Then there exists $\{\mathbf{x}_1, \ldots, \mathbf{x}_{d+1}\} \subset S$ and nonnegative numbers $\{\lambda_1, \ldots, \lambda_{d+1}\}$ with $\sum_{j=1}^{d+1} \lambda_j = 1$ such that $\mathbf{x} = \sum_{j=1}^{d+1} \lambda_j \mathbf{x}_j$.*

Proof. Let $\mathbf{x} \in C(S)$, and let $\mathbf{x} = \sum_{j=1}^{M} \lambda_j \mathbf{x}_j$ be a representation of $\mathbf{x}$ as a convex combination of M elements of S. Suppose that M is minimal for $\mathbf{x}$. We must show that $M \leq d + 1$. Since M is minimal, $\lambda_j > 0$ for each j.

Suppose on the contrary that $M > d + 1$. For $j = 1, \ldots, M - 1$, define $\mathbf{v}_j := \mathbf{x}_j - \mathbf{x}_M$. Then

$$(A.2.5) \qquad \mathbf{x} - \mathbf{x}_M = \sum_{j=1}^{M-1} \lambda_j \mathbf{v}_j.$$

A subset of at most d of the vectors in $\{\mathbf{v}_1, \ldots, \mathbf{v}_{M-1}\}$ can be linearly independent, and $M - 1 > d$ by assumption. Relabeling as needed, we may assume that for each $k > d$, $\mathbf{v}_k$ is in the span of $\{\mathbf{v}_1, \ldots, \mathbf{v}_d\}$. Then for real coefficients $t_{k,j}$, with $d + 1 \leq k \leq M - 1$ and $1 \leq j \leq d$, $\mathbf{v}_k = \sum_{j=1}^{d} t_{k,j} \mathbf{v}_j$. For $0 \leq \theta \leq 1$, and $1 \leq j \leq M - 1$, define

$$\mu_j(\theta) := \begin{cases} \lambda_j + \theta \sum_{k=d+1}^{M-1} t_{k,j} \lambda_k & 1 \leq j \leq d \\ (1 - \theta)\lambda_j & d + 1 \leq j \leq M - 1 \end{cases}$$

and

$$\mu_M(\theta) := 1 - \sum_{j=1}^{M-1} \mu_j(\theta).$$

Then (A.2.5) can be written as $\mathbf{x} = \sum_{j=1}^{M} \mu_j(\theta) \mathbf{v}_j$. Note that for all θ sufficiently small,

$$(A.2.6) \qquad \mu_j(\theta) > 0 \quad \text{for} \quad j = 1, \ldots, M,$$

and that for all θ, $\sum_{j=1}^{M} \mu_j(\theta) = 1$. Define θ_0 to be the least value of $\theta \in [0, 1]$ such that one of the conditions in (A.2.6) is violated, or, if there is no such θ, set $\theta_0 = 1$. Then $\mathbf{x} = \sum_{j=1}^{M} \mu_j(\theta_0) \mathbf{x}_j$ expresses $\mathbf{x}$ as a convex combination of the $\{\mathbf{x}_1, \ldots, \mathbf{x}_M\}$, and at least one of the coefficients is zero, contradicting the minimality of M. Hence the minimal value of M cannot be greater than $d + 1$. $\square$

In more than two dimensions, the set of extreme points of a compact convex set need not be closed, though when it is the case, Carathéodory's theorem has a useful corollary:

Corollary A.16. *Let $\mathcal{H}$ be a real Hilbert space, and let K be a nonempty, compact convex set in $\mathcal{H}$ of finite dimension d. Suppose that $E(K)$ is closed. Then every $\mathbf{x} \in K$ is a convex combination of at most $d + 1$ extreme points of K.*

Proof. Let $\mathbf{x} \in K$. By Theorem A.12, there is a sequence $\{\mathbf{x}_n\}_{n \in \mathbb{N}}$ contained in $C(E(K))$ such that $\lim_{n \to \infty} \mathbf{x}_n = \mathbf{x}$. By Carathéodory's theorem, each $\mathbf{x}_n$ has an expression as a convex combination of $d + 1$, possibly repeated, elements of $E(K)$: $\mathbf{x}_n = \sum_{j=1}^{d+1} \lambda_{n,j} \mathbf{e}_{n,j}$. By the Heine-Borel theorem, we may pass to a subsequence in n along which for each j, $\lambda_j := \lim_{n \to \infty} \lambda_{n,j}$ and $\mathbf{e}_j := \lim_{n \to \infty} \mathbf{e}_{n,j}$ both exist. Evidently, for all j, $\lambda_j \geq 0$, and since d is finite, $\sum_{j=1}^{d+1} \lambda_j = 1$. Since $E(K)$ is closed by hypothesis, $\mathbf{e}_j \in E(K)$ for each j. Therefore $\mathbf{x} = \sum_{j-1}^{d+1} \lambda_j \mathbf{e}_j$ expresses $\mathbf{x}$ as a convex combination of at most $d + 1$ elements of $E(K)$. $\square$

A.2.1. Birkhoff's theorem. We shall encounter many compact convex sets of matrices, and it will be useful to determine their extreme points. A theorem of Birkhoff [**38**] specifying the extreme points of the set of doubly stochastic matrices provides a fundamental example.

Definition A.17 (Doubly substochastic and doubly stochastic matrices). A matrix $S \in M_n(\mathbb{R})$ is **doubly substochastic** if each entry $S_{i,j}$ is nonnegative and for each $1 \leq i, j \leq n$,

$$(\text{A.2.7}) \qquad \sum_{k=1}^{n} S_{i,k} \leq 1 \quad \text{and} \quad \sum_{k=1}^{n} S_{k,j} \leq 1,$$

and S is **doubly stochastic** in case equality holds in (A.2.7) for each $1 \leq i, j \leq n$. An $n \times n$ **permutation matrix** is an $n \times n$ doubly stochastic matrix Π each of whose entries is either 0 or 1, and evidently in each row and column, exactly one entry is 1.

Let $\mathcal{S}_n$ denote the group of invertible maps from $\{1, \dots, n\}$ onto itself. This group is often called the symmetric group on n letters, and its elements are often called permutations. The map sending $\pi \in \mathcal{S}_n$ to the matrix $\Pi := S(\pi) \in M_n(\mathbb{R})$ defined by

$$S(\pi)_{i,j} = \begin{cases} 1 & j = \pi(i) \\ 0 & j \neq \pi(i) \end{cases}$$

is a bijection with the set of $n \times n$ permutation matrices.

One way that doubly stochastic matrices will frequently enter consideration here is the following:

Example A.18. Let $U \in M_n(\mathbb{C})$ be unitary. Then the matrix S defined by $S_{i,j} := |U_{i,j}|^2$ is doubly stochastic.

Theorem A.19 (Birkhoff's theorem). *Every doubly stochastic $n \times n$ matrix is a convex combination of permutation matrices.*

Dantzig [67] gave a short proof that uses a key lemma of his from the theory of linear programming. Consider a nonempty set $P \subset \mathbb{R}^n$ defined to be the set of vectors $\mathbf{x} \in \mathbb{R}^n$ satisfying the $m < n$ equations

$$(A.2.8) \qquad \mathbf{a}_j \cdot \mathbf{x} = b_j, \quad j = 1, \ldots, m,$$

and

$$(A.2.9) \qquad x_k \geq 0, \quad k = 1, \ldots, m,$$

where $\{\mathbf{a}_1, \ldots, \mathbf{a}_m\}$ are given nonzero vectors in $\mathbb{R}^n$ and $(b_1, \ldots, b_m) \in \mathbb{R}^m$. We suppose further that P is bounded, and therefore compact. This is ensured, for example, if each entry of each $\mathbf{a}_j$ is nonnegative. Then by Theorem A.12 and Remark A.13, P is the convex hull of its extreme points.

Lemma A.20 (Dantzig's lemma). *Let P be the solution set of* (A.2.8) *and* (A.2.9). *If $\mathbf{x} \in E(P)$, $x_k > 0$ for at most m values of k.*

Proof. Suppose that $\mathbf{x} \in P$, and for some set $K \subset \{1, \ldots, n\}$ of cardinality $m+1$, $x_k > 0$ for all $k \in K$. Let A be the $m \times n$ matrix whose ith row is $\mathbf{a}_i$. Let $\mathbf{c}_k$ denote the kth column of A. Since $\{\mathbf{c}_k \; : \; k \in K\}$ is linearly dependent. there are numbers, $\{y_k \; : \; k \in K\}$, not all zero, such that $\sum_{k \in K} y_k \mathbf{c}_k = 0$. Define a vector $\mathbf{y} \in \mathbb{R}^n$ be setting its kth component equal to y_k if $k \in K$ and equal to 0 otherwise. Then $A\mathbf{y} = 0$, so that for all $t \in \mathbb{R}$ and all $j = 1, \ldots, m$, $\mathbf{a}_j \cdot (\mathbf{x} + t\mathbf{y}) = b_j$. But for some $\epsilon > 0$, all components of $\mathbf{x} + t\mathbf{y}$ are nonnegative for all $|t| < \epsilon$, since $x_k > 0$ whenever $y_k \neq 0$. Therefore, $\mathbf{x}$ is not extreme. $\square$

Proof of Birkhoff's theorem. Regard elements of $M_n(\mathbb{R})$ as elements of $\mathbb{R}^{n^2}$ in the obvious way. For $X, Y \in M_n(\mathbb{R})$, define $X \cdot Y = \sum_{i,j=1}^n X_{i,j} Y_{i,j}$. For $j = 1, \ldots, n$, let $A_{(j)} \in M_n(\mathbb{R})$ be such that each entry in the jth column is 1, and all others are 0. For $j = n + 1, \ldots, 2n$, let $A_{(j)} \in M_n(\mathbb{R})$ be such that each entry in the jth row is 1, and all others are 0. Then the condition for an $n \times n$ matrix S to be doubly stochastic can be written

$$(A.2.10) \qquad A_{(j)} \cdot S = 1, \quad j = 1, \ldots, 2n,$$

and

$$S_{i,j} \geq 0, \quad i, j \in \{1, \ldots, n\}.$$

However, since $\sum_{j=1}^{n} A_{(j)} = \sum_{j=n+1}^{2n} A_{(j)}$, (A.2.10) is satisfied if $A_j(j) \cdot X = 1, j = 1, \ldots, 2n-1$. That is, the last equation in (A.2.10) is redundant, and may be deleted.

By Dantzig's theorem, an extreme point S of the set of doubly stochastic matrices has at most $2n - 1$ nonzero entries. In particular, there is at least one row with only a single nonzero entry, which must be 1, and then the column containing this entry has only one nonzero entry. Hence there are permutation matrices Π_1 and Π_2 such that $(\Pi_1 X \Pi_2)_{1,1} = 1$. The lower right $(n-1) \times (n-1)$ block of $\Pi_1 X \Pi_2$ is doubly stochastic, and now an easy induction on n completes the proof. $\qquad\square$

The condition for $R \in M_n(\mathbb{R})$ to be doubly substochastic can be expressed using the notation from the previous proof as

$$A_{(j)} \cdot R \leq 1, \quad j = 1, \ldots, 2n, \quad \text{and} \quad R_{i,j} \geq 0, \quad i, j \in \{1, \ldots, n\}.$$

The following lemma was proved by von Neumann [224]. A particularly simple proof was given by Dantzig shortly afterwards [67]; his proof is given here.

Lemma A.21. *$R \in M_n(\mathbb{R})$ is doubly substochastic if and only if it has nonnegative entries and there is another doubly substochastic matrix R' such that $R + R'$ is doubly stochastic. In particular, R is doubly substochastic if and only if there is a doubly stochastic matrix S with $R_{j,k} \leq S_{j,k}$ for all j, k.*

Proof. Let R be doubly substochastic. For $j \in \{1, \ldots, n\}$, define $r_j = \sum_{k=1}^{n} R_{j,k}$ and $c_k = \sum_{j=1}^{n} R_{j,k}$. Note that $\sum_{j=1}^{n} r_j = \sum_{k=1}^{n} c_k =: s$. Suppose that $s < n$, and define the $n \times n$ matrix R' by $R'_{j,k} := \frac{(1-r_j)(1-c_k)}{n-s}$ which is nonnegative since R is doubly substochastic. Then evidently $R + R'$ is doubly stochastic. $\qquad\square$

A.3. Convex functions and their Legendre transforms

This section collects some parts of convex geometry centered on the Legendre transform that will be of frequent use. Much of the material is well known and can be found in Rockafellar [188], or more briefly, Fenchel [83]. However, short proofs of statements tailored to our needs are presented here.

Definition A.22 (Convex function). A function F from $\mathcal{H}$ to $(-\infty, \infty]$ is **convex** in case for all $\mathbf{x}_0, \mathbf{x}_1 \in \mathcal{X}$ and all $\lambda \in [0, 1]$,

$$(A.3.1) \qquad F((1-\lambda)\mathbf{x}_0 + \lambda \mathbf{x}_1) \leq (1-\lambda)F(\mathbf{x}_0) + \lambda F(\mathbf{x}_1).$$

The function F is **strictly convex** in case equality holds in in (A.3.1) only when $\lambda = 0, \lambda = 1$ or $x_0 = x_1$.

A convex function F from $\mathcal{H}$ to $(-\infty, \infty]$ is **proper** in case it is not identically $+\infty$. The **domain** of a proper convex function F on $\mathcal{H}$ is the set

$$D(F) := \{x \in \mathcal{H} \ : \ F(x) < \infty\}.$$

It is clear that $D(F)$ is convex when F is convex. Conversely, given a convex set D and a function $F : D \to \mathbb{R}$ such that (A.3.1) holds for all $\mathbf{x}_0, \mathbf{x}_1 \in \mathcal{X}$ and all $\lambda \in [0,1]$, we may extend the definition of F to all of $\mathcal{H}$ by $F(\mathbf{x}) = \infty$ for $\mathbf{x} \notin D$. For this reason, we always consider convex functions as defined on all of $\mathcal{H}$.

Lemma A.23 (Jensen's inequality). *Let F be convex on $\mathcal{H}$. Then for all $m \geq 2$, and $\{\lambda_1, \ldots, \lambda_m\}$ with $\lambda_j \geq 0$ for all j and $\sum_{j=1}^{m} \lambda_j = 1$, and all $\{\mathbf{x}_1, \ldots, \mathbf{x}_m\} \subset \mathcal{H}$,*

$$(A.3.2) \qquad F\left(\sum_{j=1}^{m} \lambda_j \mathbf{x}_j\right) \leq \sum_{j=1}^{m} \lambda_j F(\mathbf{x}_j).$$

Proof. Changing the indices if needed, we may assume that $\lambda_m < 1$. For $j = 1, \ldots, m-1$, define $\mu_j := \lambda_j / (1 - \lambda_m)$. Then with $\mathbf{y} := \sum_{j=1}^{m-1} \mu_j \mathbf{x}_j$,

$$(A.3.3) \quad F\left(\sum_{j=1}^{m} \lambda_j \mathbf{x}_j\right) = F((1 - \lambda_m)\mathbf{y} + \lambda_m \mathbf{x}_m) \leq (1 - \lambda_m)F(\mathbf{y}) + \lambda_m F(\mathbf{x}_m),$$

using (A.3.1). By definition, (A.3.2) is valid for $m = 2$. Making the inductive assumption (A.3.2) is valid with m replaced by $m - 1$,

$$F(\mathbf{y}) = F\left(\sum_{j=1}^{m-1} \mu_j \mathbf{x}_j\right) \leq \sum_{j=1}^{m-1} \mu_j F(\mathbf{x}_j),$$

and combining this with (A.3.3) yields (A.3.2). $\qquad \square$

The geometric implications of convexity are often clarified by considering another convex set associated to a proper convex function F, namely its epigraph:

Definition A.24 (Epigraph). Let F be a proper convex function on $\mathcal{H}$. The **epigraph** of F is the set

$$\mathrm{Epi}(F) = \{(x, t) \in \mathcal{H} \oplus \mathbb{R} \ : \ t \geq F(x)\}.$$

It is left as an exercise to show that $\mathrm{Epi}(F)$ is closed in $\mathcal{H} \oplus \mathbb{R}$ if and only if F is lower semicontinuous. For this reason, lower semicontinuous convex functions a referred to as **closed convex functions**.

The following lemma is frequently useful.

Lemma A.25. *Let $\mathcal{K}$ and $\mathcal{L}$ be subspaces of $\mathcal{H}$ so that $\mathcal{H} = \mathcal{K} \oplus \mathcal{L}$ so that $\mathcal{H}$ can be identified, as a set, with $\mathcal{K} \times \mathcal{L}$. Let F be proper convex on $\mathcal{H}$. Define a function G on $\mathcal{K}$ by*

$$G(\mathbf{x}) := \inf\{F(\mathbf{x}, \mathbf{y}) \ : \ \mathbf{y} \in \mathcal{L}\}.$$

Then G is convex on $\mathcal{K}$.

Proof. Suppose $\mathbf{x}_1, \mathbf{x}_2 \in \mathcal{K}$ and $0 < \lambda < 1$. Let $\epsilon > 0$. Then there exist $\mathbf{y}_1, \mathbf{y}_2 \in \mathcal{L}$ such that $F(\mathbf{x}_j, \mathbf{y}_j) \leq G(\mathbf{x}_j) + \epsilon, j = 1, 2$. Since F is convex,

$$F((1 - \lambda)(\mathbf{x}_1, \mathbf{y}_1) + \lambda(\mathbf{x}_2, \mathbf{y}_2))$$
$$\leq (1 - \lambda)F(\mathbf{x}_1, \mathbf{y}_1) + \lambda F(\mathbf{x}_2, \mathbf{y}_2) \leq G(\mathbf{x}_1) + G(\mathbf{x}_2) + 2\epsilon.$$

Since ϵ is arbitrary, this proves $G((1 - \lambda)\mathbf{x}_1 + \lambda\mathbf{x}_2) \leq (1 - \lambda)G(\mathbf{x}_1) + \lambda G(\mathbf{x}_2)$. $\square$

It is evident that G is not identically ∞, but it may take on the value $-\infty$, and for this reason, G may not be a proper convex function on $\mathcal{K}$.

Definition A.26 (Legendre transform). The **Legendre transform** of a function $F : \mathcal{H} \to (-\infty, \infty]$ is the function F^* on $\mathcal{H}$ defined by

$$(A.3.4) \qquad F^*(\mathbf{y}) = \sup_{\mathbf{x} \in \mathcal{X}}\{\langle \mathbf{x}, \mathbf{y} \rangle - F(\mathbf{x})\}.$$

The function F^*, being a supremum of a family of continuous linear functions, is lower semicontinuous and convex, even if F has neither of these properties. We now come to the central theorem of convex geometry.

Theorem A.27 (Fenchel-Moreau theorem). *Let $F : \mathcal{H} \to (-\infty, \infty]$ be a proper function. Then its Legendre transform $F^* : \mathcal{H} \to (-\infty, \infty]$ is a proper, lower semicontinuous convex function. The Legendre transform of F^*, namely $F^{**}(\mathbf{x}) = \sup_{\mathbf{y} \in \mathcal{H}}\{\langle \mathbf{x}, \mathbf{y} \rangle - F^*(\mathbf{y})\}$, always satisfies*

$$F^{**}(\mathbf{x}) \leq F(\mathbf{x}) \quad \text{for all} \quad \mathbf{x} \in \mathcal{H},$$

and $F^{**} = F$ if and only if F is convex and lower semicontinuous. In particular, the Legendre transform is an involution on the set of proper lower semicontinuous convex functions on $\mathcal{H}$.

Before proving Theorem A.27, we prove a fundamental lemma:

Lemma A.28. *Let $F : \mathcal{H} \to (-\infty, \infty]$ be a proper, lower-semicontinuous convex function. Suppose that $\mathbf{x}_0 \in \mathcal{H}$ and $s \in \mathbb{R}$ are such that $(\mathbf{x}_0, s) \notin \mathrm{Epi}(F)$. Then either:*

    (i) *there exists $\mathbf{y}_0 \in \mathcal{H}$ such that*

$$(A.3.5) \qquad F^*(\mathbf{y}_0) < \langle \mathbf{x}_0, \mathbf{y}_0 \rangle - s$$

      or else

    (ii) *there exists $\mathbf{y}_0 \in \mathcal{H}$ such that*

$$(A.3.6) \qquad \sup\{\langle \mathbf{x}, \mathbf{y}_0 \rangle \ : \ \mathbf{x} \in D(F)\} < \langle \mathbf{x}_0, \mathbf{y}_0 \rangle,$$

      and in particular $F(\mathbf{x}_0) = \infty$.

Proof. By the hypotheses on F, $\mathrm{Epi}(F)$ is nonempty, closed, and convex. By Theorem A.3, there exists $\mathbf{y}_0 \in \mathcal{H}$ and $a, \alpha \in \mathbb{R}$ such that

$$(A.3.7) \quad \langle \mathbf{x}_0, \mathbf{y}_0 \rangle - sa > \alpha \quad \text{and} \quad \langle \mathbf{x}, \mathbf{y}_0 \rangle - ta \le \alpha \quad \text{for all } (\mathbf{x}, t) \in \mathrm{Epi}(F).$$

Since t can be arbitrarily large for each $\mathbf{x} \in D(F)$, $a \ge 0$. Suppose first that $a > 0$. Then we may divide through by a, and hence in this case we may assume without loss of generality that $a = 1$. Take $t = F(\mathbf{x})$ in (A.3.5) to obtain

$$\langle \mathbf{x}_0, \mathbf{y}_0 \rangle - s > \alpha \quad \text{and} \quad \langle \mathbf{x}, \mathbf{y}_0 \rangle - F(\mathbf{x}) \le \alpha \quad \text{for all } (\mathbf{x}, t) \in \mathrm{Epi}(F).$$

By the definition (A.3.4), the second part yields $F^*(\mathbf{y}_0) \le \alpha$, and this together with the first part yields (A.3.5).

Next, suppose that (A.3.7) holds with $a = 0$. Then from (A.3.7),

$$\langle \mathbf{x}_0, \mathbf{y}_0 \rangle > \alpha \quad \text{and} \quad \langle \mathbf{x}, \mathbf{y}_0 \rangle \le \alpha \quad \text{for all } \mathbf{x} \in D(F),$$

and this implies (A.3.6), which says, in particular, that $\mathbf{x}_0 \notin D(F)$, and hence $F(\mathbf{x}_0) = \infty$. $\qquad\qquad \square$

Proof of Theorem A.27. Since F is proper, there exists some $\mathbf{x}_0 \in D(F)$, and then applying Lemma A.28 to $(\mathbf{x}_0, s)$ with $s < F(\mathbf{x}_0)$, case (i) must apply since $F(\mathbf{x}_0) < \infty$. Then by (A.3.5), there exists $\mathbf{y}_0$ with $F^*(\mathbf{y}_0) < \infty$, and hence F^* is proper. We have already seen that it is lower semicontinuous and convex.

Let F^{**} denote the Legendre transform of F^*. For all $\mathbf{x} \in \mathcal{H}$, $F^{**}(\mathbf{x}) \le F(\mathbf{x})$: observe that for all $\mathbf{y} \in \mathcal{H}$, the definition (A.3.4) yields $F^*(\mathbf{y}) \ge \langle \mathbf{x}, \mathbf{y} \rangle - F(\mathbf{x})$ and then

$$F^{**}(\mathbf{x}) = \sup_{\mathbf{y} \in \mathcal{H}}\{\langle \mathbf{x}, \mathbf{y} \rangle - F^*(\mathbf{y})\} \le \sup_{\mathbf{y} \in \mathcal{H}}\{\langle \mathbf{x}, \mathbf{y} \rangle - (\langle \mathbf{x}, \mathbf{y} \rangle - F(\mathbf{x}))\} = F(\mathbf{x}).$$

It remains to prove that $F^{**}(\mathbf{x}) \geq F(\mathbf{x})$. Suppose on the contrary that for some $\mathbf{x}_0 \in \mathcal{H}$, $F^{**}(\mathbf{x}_0) < F(\mathbf{x}_0)$. Then $(\mathbf{x}_0, F^{**}(\mathbf{x}_0)) \notin \mathrm{Epi}(F)$, and we may apply Lemma A.28 with $s = F^{**}(\mathbf{x}_0)$. Suppose that case (i) applies. Then by (A.3.5), there exists $\mathbf{y}_0 \in \mathcal{H}$ such that $\langle \mathbf{x}_0, \mathbf{y}_0 \rangle - F(\mathbf{x}_0) > F^*(\mathbf{y}_0)$. Hence $F^{**}(\mathbf{x}_0) \geq \langle \mathbf{x}_0, \mathbf{y}_0 \rangle - F^*(\mathbf{y}_0) > \langle \mathbf{x}_0, \mathbf{y}_0 \rangle - (\langle \mathbf{x}_0, \mathbf{y}_0 \rangle - F(\mathbf{x}_0)) = F(\mathbf{x}_0)$, which is impossible. Suppose next that case (ii) applies. Since F^* is proper, there exists $\mathbf{y}_1 \in \mathcal{H}$ such that $F^*(\mathbf{y}_1) < \infty$. Then for any $u > 0$,

$$F^*(u\mathbf{y}_0 + \mathbf{y}_1) = \sup_{\mathbf{x} \in D(F)} \{u\langle \mathbf{x}, \mathbf{y}_0 \rangle + \langle \mathbf{x}, \mathbf{y}_1 \rangle - F(\mathbf{x})\}$$

$$\leq u \sup_{\mathbf{x} \in D(F)} \{\langle \mathbf{x}, \mathbf{y}_0 \rangle\} + F^*(\mathbf{y}_1).$$

Then

$$F^{**}(\mathbf{x}_0) \geq \langle \mathbf{x}_0, u\mathbf{y}_0 + \mathbf{y}_1 \rangle - F^*(u\mathbf{y}_0 + \mathbf{y}_1)$$

$$= u\left(\langle \mathbf{x}_0, \mathbf{y}_0 \rangle - \sup_{\mathbf{x} \in D(F)} \{\langle \mathbf{x}, \mathbf{y}_0 \rangle\}\right) + \langle \mathbf{x}_0, \mathbf{y}_1 \rangle - F^*(\mathbf{y}_1).$$

Since $\langle \mathbf{x}_0, \mathbf{y}_1 \rangle - \sup_{\mathbf{x} \in D(F)}\{\langle \mathbf{x}, \mathbf{y}_0 \rangle\} > 0$ by (A.3.6), and u can be taken arbitrarily large, this implies $F^{**}(\mathbf{x}_0) = \infty$. Thus, $F^{**}(\mathbf{x}_0) < F(\mathbf{x}_0)$ is impossible in this case as well. This completes the proof that $F^{**} = F$. $\qquad\square$

A consequence of Theorem A.27 is that for all $F : \mathcal{H} \to (-\infty, \infty]$,

$$(\text{A.3.8}) \qquad \langle \mathbf{x}, \mathbf{y} \rangle \leq F(\mathbf{x}) + F^*(\mathbf{y}) \quad \text{for all} \quad \mathbf{x} \in \mathcal{H} \quad \text{and} \quad \mathbf{y} \in \mathcal{H}.$$

The cases of equality will be of interest. Recall that the subgradient ∂F of a proper function F on a Hilbert space $\mathcal{H}$ has been defined in Definition 6.3.

Theorem A.29 (Fenchel-Young inequality). *Let F be a proper lower semicontinuous convex function on $\mathcal{H}$. Then* (A.3.8) *is valid, and moreover:*

(1) *Whenever there is equality in* (A.3.8), $\mathbf{y} \in \partial F(\mathbf{x})$ *and* $\mathbf{x} \in \partial F^*(\mathbf{y})$.

(2) *Whenever* $\mathbf{y} \in \partial F(\mathbf{x})$, *or* $\mathbf{x} \in \partial(F^*)$, *there is equality in* (A.3.8), *and consequently,*

$$\mathbf{y} \in \partial F(\mathbf{x}) \quad \Longleftrightarrow \quad \mathbf{x} \in \partial F^*(\mathbf{y}).$$

Proof. We need only explain the cases of equality. If there is equality in (A.3.8), then for all $\mathbf{y}_1 \in \mathcal{H}$,

$$\langle \mathbf{x}, \mathbf{y}_1 \rangle - \langle \mathbf{x}, \mathbf{y} \rangle \leq (F(\mathbf{x}) + F^*(\mathbf{y}_1)) - (F(\mathbf{x}) + F^*(\mathbf{y})) = F^*(\mathbf{y}_1) - F^*(\mathbf{y}),$$

which is the same as $F^*(\mathbf{y}_1) \geq F^*(\mathbf{y}) + \langle \mathbf{x}, \mathbf{y}_1 - \mathbf{y} \rangle$, which means that $\mathbf{y} \in \partial F(\mathbf{x})$. An analogous argument shows that $\mathbf{x} \in \partial F^*(\mathbf{y})$ when there is equality in (A.3.8).

Conversely, if $\mathbf{y} \in \partial F(\mathbf{x})$, then for all $\mathbf{x}_1 \in \mathcal{H}$, $F(\mathbf{x}_1) \geq F(\mathbf{x}) + \langle \mathbf{x}_1 - \mathbf{x}, \mathbf{y} \rangle$, and rearranging terms, we obtain $\langle \mathbf{x}, \mathbf{y} \rangle \geq F(\mathbf{x}) + \langle \mathbf{x}_1, \mathbf{y} \rangle - F(\mathbf{x}_1)$. Taking the supremum over $\mathbf{x}_1$, we obtain $\langle \mathbf{x}, \mathbf{y} \rangle \geq F(\mathbf{x}) + F^*(\mathbf{y})$, and equality must hold in (A.3.8). An analogous argument shows that argument when $\mathbf{x} \in \partial F^*(\mathbf{y})$, then $\langle \mathbf{x}, \mathbf{y} \rangle \leq F^{**}(\mathbf{x}) + F^*(\mathbf{y})$, which is the same as (A.3.8) since $F^{**} = F$. $\qquad\square$

For $\mathbf{x}_0 \in D(F)$, $y \in \partial F(\mathbf{x}_0)$ if and only if the hyperplane

$$\{(\mathbf{x}, t) \ : \ t = F(\mathbf{x}_0) + \langle \mathbf{x} - \mathbf{x}_0, \mathbf{y} \rangle\}$$

lies below the graph of $\mathbf{x} \mapsto F(\mathbf{x})$, and intersects it at $(\mathbf{x}_0, F(\mathbf{x}_0))$. It is easy to see that if F is convex and differentiable at $\mathbf{x}_0$, with derivative $\nabla F(\mathbf{x}) \in \mathcal{H}$, then $\partial F(\mathbf{x}_0) = \{\nabla F(\mathbf{x}_0)\}$. If F is not differentiable at $\mathbf{x}_0 \in D(F)$, then $\partial F(\mathbf{x}_0)$ may consist of more than one point.

Lemma A.30. *Let F be a proper closed convex function on $\mathcal{H}$. Then for each $\mathbf{x}_0 \in \mathrm{Int}(D(F))$, $\partial F(\mathbf{x}_0)$ is a nonempty compact convex set.*

Proof. Let $\mathbf{x}_0 \in \mathrm{Int}(D(F))$. Let $\{\mathbf{v}_1, \ldots, \mathbf{v}_d\}$ be a basis for $\mathcal{H}$. Since $\mathbf{x}_0 \in \mathrm{Int}(D(F))$, there is an $\epsilon > 0$ such that $\mathbf{x}_0 + \sum_{j=1}^{d} t_j \mathbf{v}_j \in \mathrm{Int}(D(F))$ whenever $|t_j| \leq \epsilon$ for all j. F, being convex, is bounded above on the set $\{\mathbf{x}_0 + \sum_{j=1}^{d} t_j \mathbf{v}_j \ : \ |t_j| \leq \epsilon, 1 \leq j \leq d\}$ by the maximum of its values at the 2^d vertices of this set. Hence there exist $r > 0$ and $M < \infty$ such that

$$(A.3.9) \qquad F(\mathbf{x}) \leq F(\mathbf{x}_0) + M \text{ for all } \|\mathbf{x} - \mathbf{x}_0\|_{\mathcal{H}} < r.$$

Let $\{s_n\}$, $n \in \mathbb{N}$, be a strictly increasing sequence of real numbers with $\lim_{n\to\infty} s_n = F(\mathbf{x}_0)$. Then for all n, $(\mathbf{x}_0, s_n) \notin \mathrm{Epi}(F)$, and Lemma A.28 applies. Since $F(\mathbf{x}_0) < \infty$, case (i) applies and there exists $\mathbf{y}_n \in \mathcal{H}$ such that $\langle \mathbf{x}_0, \mathbf{y}_n \rangle - s_n > F^*(\mathbf{y}_n) \geq \langle \mathbf{x}, \mathbf{y}_n \rangle - F(\mathbf{x})$ for all $\mathbf{x} \in \mathcal{H}$. That is,

$$(A.3.10) \qquad F(\mathbf{x}) \geq \langle \mathbf{x} - \mathbf{x}_0, \mathbf{y}_n \rangle + s_n \quad \text{for all} \quad \mathbf{x} \in \mathcal{H}.$$

Combining (A.3.9) and (A.3.10), we have that for all n sufficently large

$$\langle \mathbf{x} - \mathbf{x}_0, \mathbf{y}_n \rangle \leq M + 1 \quad \text{for all} \quad \|\mathbf{x} - \mathbf{x}_0\| < r.$$

Since $\|\mathbf{y}\|_{\mathcal{H}} = \sup\{\langle \mathbf{x}, \mathbf{y} \rangle \ : \ \|\mathbf{x}\|_{\mathcal{H}} \leq 1\}$, it follows that for all n sufficiently large, $\|\mathbf{y}_n\|_{\mathcal{H}} \leq (M + 1)r$. Passing to a subsequence, we may suppose that $\lim_{n\to\infty} \mathbf{y}_n = \mathbf{y}$. Then we obtain from (A.3.10) that

$$F(\mathbf{x}) \geq \langle \mathbf{x} - \mathbf{x}_0, \mathbf{y} \rangle + F(\mathbf{x}_0) \quad \text{for all} \quad \mathbf{x} \in \mathcal{H},$$

and hence $\mathbf{y} \in \partial F(\mathbf{x}_0)$. Moreover, for any $\mathbf{y} \in \partial F(\mathbf{x}_0)$, we have $\langle \mathbf{x} - \mathbf{x}_0, \mathbf{y} \rangle \leq M$ for all $\mathbf{x}$ with $\|\mathbf{x} - \mathbf{x}_0\|_{\mathcal{H}} < r$, and hence $\|y\|_{\mathcal{H}} \leq M/r$. This shows that $\partial F(\mathbf{x}_0)$ is a nonempty bounded set, and it is clear from the definition that it is convex and closed. $\qquad\square$

One can say more for convex function f on the real line.

Theorem A.31 (Increasing slopes of secant lines). *Let f be a convex function on $\mathbb{R}$. Let $x_1, x_2, x_3 \in D(f)$ with $x_1 < x_2 < x_3$. Then*

$$\frac{F(x_2) - F(x_1)}{x_2 - x_1} \leq \frac{F(x_3) - F(x_1)}{x_3 - x_1} \quad and \quad \frac{F(x_2) - F(x_1)}{x_2 - x_1} \leq \frac{F(x_3) - F(x_1)}{x_3 - x_1}.$$

Proof. Since $x_2 \in (x_1, x_3)$, there exists $\lambda := \frac{x_2 - x_1}{x_3 - x_1} \in (0, 1)$ is such that $x_2 - x_1 = \lambda(x_3 - x_1)$. Then from the inequality defining convexity, $F(x_2) \leq (1 - \lambda)F(x_1) + \lambda F(x_3)$, we deduce

$$(A.3.11) \qquad F(x_2) \leq \frac{x_3 - x_2}{x_3 - x_1}F(x_1) + \frac{x_2 - x_1}{x_3 - x_1}F(x_3).$$

Subtracting $F(x_1)$ from both sides of (A.3.11) and dividing by $x_2 - x_1$ yields the inequality on the left. Subtracting $F(x_3)$ from both of (A.3.11) sides and dividing by $x_2 - x_3$ yields the inequality on the right. $\qquad\square$

It follows that for any proper convex function f on $\mathbb{R}$ and all $x \in D(f)$, the right and left derivatives

$$f'_+(x) = \lim_{h\downarrow 0} \frac{f(x + h) - f(x)}{h} \quad and \quad f'_-(x) = \lim_{h\downarrow 0} \frac{f(x) - f(x) - h}{h}$$

both exist. It is then left as an exercise to show that for $x \in \text{Int}(D(F))$, $\partial F(x) = [f'_-(x), f'_+(x)]$, although if x belongs to the boundary of $D(F)$, $\partial F(x)$ need not be compact.

We close this appendix with a final theorem about convex functions f on $\mathbb{R}$ and their Legendre transforms f^*.

Theorem A.32. *Let f be convex on $\mathbb{R}$, and let (a, b) be the interior of the set on which f is finite. Let f be twice continuously differentiable and strictly convex on (a, b). Define $c := \lim_{t\downarrow a} f'(t)$ and $d := \lim_{t\uparrow b} f'(t)$. Then (c, d) is the interior of the set on which f^* is finite, and on (c, d), f^* is twice continuously differentiable and strictly convex. Moreover, the functions f' and $(f^*)'$ are inverse to one another; that is, for all $t \in (a, b)$ and all $u \in (c, d)$, $(f^*)'(f'(t)) = t$ and $f'((f^*)'(u)) = u$, and hence $(f^*)''(f'(t)) = \frac{1}{f''(t)}$ and $f''((f^*)'(u)) = \frac{1}{(f^*)''(u)}$.*

Proof. By definition, $f^*(u) = \sup_{t \in \mathbb{R}}\{tu - f(t)\}$. The function $t \mapsto tu - f(t)$ is strictly concave, and hence it is maximized at any critical point. Differentiating yields $u = f'(t)$. For $u \in (c, d)$, there is a unique solution of this equation since f' increases strictly and continuously with values in (c, d). If $d > \infty$ and $u > d$, $t \mapsto tu - f(t)$ increases without bound as t increases, and hence $f^*(u) = \infty$. A similar argument yiedls $f^*(u) = \infty$ in case $c > -\infty$ and $u < c$.

Thus, for any $t \in (a, b)$, $f^*(f'(t)) = tf'(t) - f(t)$. Since f and f' are continuously differentiable and f' is strictly increasing, f^* is differentiable, and differentiating yields $(f^*)'(f'(t))f''(t) = tf''(t)$. Since $f''(t) > 0$, $(f^*)'(f'(t)) = t$. This shows that $(f^*)'$ is the inverse function of f', and then of course f' is the inverse function of $(f^*)'$. Since f' is continuously differentiable, the inverse function theorem says that $(f^*)'$ is differentiable, and differentiating $(f^*)'(f'(t)) = t$ yields $(f^*)''(f'(t)) = \frac{1}{f''(t)}$ which shows that $(f^*)''$ is continuous and strictly positive. Hence f^* is also strictly convex. $\qquad\square$

Complex interpolation

B.1. The three lines theorem and Hirschman's lemma

A sharpened version of Hadmard's *three lines theorem* [**94**] due to Hirschman [**110**] is a powerful source of inequalities for operators. However, the simple original version by Hadamard suffices for many purposes, and we begin with this:

Theorem B.1 (Hadmard's three lines theorem). *Let $f(z)$ be an analytic function of $z = x + iy$ on the strip*

$$S := \{(x, y) \ : \ 0 < x < 1, \ y \in \mathbb{R}\}$$

that is continuous and bounded on $\overline{S}$, the closure of S. Define the function $M(x)$ on $[0, 1]$ by

$$M(x) := \sup_{y \in \mathbb{R}} |f(x + iy)|.$$

Then $\log M(x)$ is convex. In particular, for $0 < \theta < 1$

$$(\text{B.1.1}) \qquad\qquad |f(\theta)| \le M(0)^{1-\theta} M(1)^{\theta}.$$

Proof. We may suppose that $M(0), M(1) < \infty$. Fix $\epsilon > 0$, and define $F(z) := f(z)e^{\epsilon(z^2-1)}M(0)^{z-1}M(1)^{-z}$. Then F is analytic in S,

$$\sup_{y \in \mathbb{R}} |F(iy)| \le 1 \quad \text{and} \quad \sup_{y \in \mathbb{R}} |F(1 + iy)| \le 1.$$

Since $F(z)$ goes to zero as $|z|$ goes to infinity in $\overline{S}$, there exists $z_0 \in \overline{S}$ such that $|F(z_0)| \ge |F(z)|$ for all $z \in \overline{S}$. We now claim that z_0 must belong to the boundary of S, or else F is identically zero.

To see this, suppose that for some $z_0 \in S$, $F(z_0)$, $|F(z_0)| \geq |F(z)|$ for all $z \in \overline{S}$. We must show that F is identically zero. Recall that by construction, $F(z)$ goes to zero as $|z|$ goes to infinity in S, and so it suffices to show that F is constant on a neighborhood of z_0.

Let R be the distance from z_0 to the boundary of S. For all $r < R$, let $\gamma_{z_0,r}$ denote the circle of radius r centered at z_0. By the Cauchy integral formula, $F(z_0) = \frac{1}{2\pi} \oint_{\gamma_{z_0,r}} \frac{F(w)}{w-z} dw$, and hence

$$|F(z_0)| \leq \frac{1}{2\pi r} \oint_0^{2\pi} |F(z_0 + re^{i\phi})| d\phi \leq |F(z_0)|.$$

Equality requires: (1) the phase of $F(z_0 + re^{i\phi})$ is independent of ϕ; and (2) $|F(z_0 + re^{i\phi})| = |F(z_0)|$ for all ϕ. That is, $F(w)$ is constant on $\gamma_{z_0,r}$, and the modulus of this constant value is $|F(z_0)|$.

Since $0 < r < R$ was arbitrary, it follows that $|F(z)| = |F(z_0)|$ for all points $z \in D_{r,z_0}$, the open disk of radius R about z_0, and the phase is constant along all circles in D_{R,z_0} centered at z_0. Then, since for any $z_1 \in D_{R,z_0}$, the modulus of F is maximal at z_1, we may repeat the argument to conclude that the phase of F is constant on all circles in D_{R,z_0} no matter what the center is. Therefore, the phase as well as the modulus of F are constant in D_{R,z_0}. Since F is analytic in S, this means that F is constant on D_{R,z_0}, and hence everywhere in $\overline{S}$.

By construction $|F(z)|$ is bounded by 1 on the boundary of S. By the definition of F $|f(\theta)e^{\epsilon(\theta^2-1)}M(0)^{\theta-1}M(1)^{-\theta}| = F(\theta) \leq 1$, and then since $\epsilon > 0$ is arbitrary, (B.1.1) is proved. $\qquad\square$

B.2. Hirschman's refined three lines theorem

We have given a proof of Theorem B.1 using the elementary Cauchy integral formula to express $f(z)$ as an average of the values of $f(w)$ on any circle in $\overline{S}$ centered on z, and we used the fact that any average of a set of values is bounded above by the maximum of these values.

Hirschman observed that one can do better by using the theory of **harmonic functions**; that is, of solutions to Laplace's equation,

$$\left(\frac{\partial^2}{\partial x^2} + \frac{\partial^2}{\partial y^2} \right) u(x, y) = 0,$$

to represent $\log|f(z)|$ for $z \in S$ as an average over the values of $\log|f(w)|$ for w in the boundary of S.

This is possible because the real and imaginary parts of an analytic function are harmonic. Hence, if $f(z)$ is analytic in S, since $\log f(z) = \log|f(z)| + i\arg f(z)$, $u(x,y) = \log|f(x+iy)|$ is harmonic in S away from any zeros of f, and as (x,y) approaches any zero, $u(x,y)$ approaches $-\infty$. Such points are not interesting if we seek an upper bound on $u(x,y)$.

Moreover, the value of u at (x,y) is an **average** of the boundary values of u, with the average being taken with respect to so-called **harmonic measure**, or, what is the same thing, with respect to the **Poisson kernel** for S. To develop this idea, we require some information about the solutions of Laplace's equation on S in terms of data prescribed on the boundary of S.

We require a formula for the harmonic measure on the boundary of S.

Definition B.2. For $(x,y) \in S$, define the probability measure $d\mu_{x,y}$ on $\mathbb{R}$ by

$$(B.2.1) \qquad d\mu_{(x,y)}(t) = \frac{1}{2(1-x)} P(\pi x, \pi t - \pi y)dt,$$

where

$$(B.2.2) \qquad P(x,y) := \frac{\sin x}{\cosh y - \cos x}.$$

Simple calculations show that

$$P(x,y) := \frac{\sin x}{\cosh y - \cos x} = 2\Im\left((e^{-i(x+iy)} + 1)^{-1}\right).$$

Hence P is harmonic on $S_\pi := \{(x,y) \;:\; 0 < x < \pi\}$. Evidently, $P \geq 0$ on S_π and

$$\frac{\partial}{\partial y} \cot^{-1}\left(\frac{\cos x - e^y}{\sin x}\right) = \frac{1}{2}P(x,y).$$

Therefore,

$$\frac{1}{2(1-x)} \int_{\mathbb{R}} P(\pi x, \pi y)dy = 1.$$

This shows that $d\mu_{(x,y)}(t)$ defined in (B.2.2) is indeed a probability measure, and yields a formal demonstration that for any bounded continuous function ϕ on $\mathbb{R}$,

$$h(x,y) := (1-x) \int_{\mathbb{R}} \phi(t)d\mu_{(x,y)}(t)$$

is harmonic in S. Moreover, it is easy to see that as $x \downarrow 0$, $d\mu_{(x,y)}(t)$ is more and more concentrated near $t = y$ so that

$$\lim_{t \downarrow 0} h(x,y) = \phi(y).$$

It is not hard to make all of this rigorous, and in fact to prove somewhat more. The following lemma is due to Widder [**226**], which can be consulted for all the details.

Lemma B.3. *Let ϕ_0 and ϕ_1 be continuous real valued functions on $\mathbb{R}$ such that $\int_{\mathbb{R}}(|\phi_0(y)| + |\phi_1(y)|)e^{-|y|}\mathrm{d}y < \infty$. Then*

$$H(x,y) := (1-x)\int_{\mathbb{R}} \phi_0(t)\mathrm{d}\mu_{(x,y)}(t) + x \int_{\mathbb{R}} \phi_1(t)\mathrm{d}\mu_{(1-x,y)}(t)$$

is well defined and harmonic on S. Moreover, for all $y_0 \in \mathbb{R}$,
(B.2.3)
$$\lim_{(x,y)\to(0,y_0)} H(x,y) = \phi_0(y_0) \qquad and \qquad \lim_{(x,y)\to(\pi,y_0)} H(x,y) = \phi_1(y_0).$$

Moreover, $H(x,y)$ is the unique harmonic function in S that is continuous on $\overline{S}$ and satisfies (B.2.3).

Remark B.4. The probability measure $\mathrm{d}\mu_{(x,y)}(t)$ has the following probabilistic interpretation in terms of Brownian motion: A particle undergoing Brownian motion starting at $(x,y) \in S$ will eventually hit the boundary of S, for the first time, at some point $(0,t)$ on the left, or some point $(1,t)$ on the right.

The probability that the particle first hits the boundary on the left is $1 - x$, while the probability that it first hits the boundary on the right is x.

The conditional probability distribution of the particle's arrival point, given that the particle first hits the boundary on the left at $(0,t)$, is $\mathrm{d}\mu_{(x,y)}(t)$. By symmetry, the conditional probability distribution of the particle's arrival point, given that the particle first hits the boundary on the right at $(1,t)$, is $\mathrm{d}\mu_{(1-x,y)}(t)$. A concise but full account of all of this can be found in [**153**].

We may now state Hirschman's lemma:

Lemma B.5 (Hirschman's lemma). *Let $f(z)$ be analytic in the strip $\{(x + iy) : -1 \le x \le 1, \, y \in \mathbb{R}\}$ such that for some $C < \infty$ and some $0 < a < \pi$,*

$$\max_{0 \le x \le 1}\{|f(x + it)|\} \le Ce^{a|t|}$$

for all $|t|$ sufficiently large.

Let ϕ_1 and ϕ_2 be two continuous functions on $\mathbb{R}$ with values in $[0, \infty)$ such that

$$\int_{\mathbb{R}} (\phi_1(y) + \phi_2(y))e^{|y|}\mathrm{d}y < \infty$$

and such that for all y

$$\log|f(iy)| \le \phi_1(y) \qquad and \qquad \log|f(1 + iy)| \le \phi_2(y).$$

Then for all $0 < r < 1$,

(B.2.4)
$$\log|f(r)| \le \int_{\mathbb{R}} \phi_1(y)\mathrm{d}\mu_{(r,0)}(y) + \int_{\mathbb{R}} \phi_2(y)\mathrm{d}\mu_{(1-r,0)}(y).$$

Hirschman's proof makes use of the theory of harmonic majorization, which is discussed in Chapter 3 of [**166**]. The key idea is known as the **principle of harmonic majorization**.

Theorem B.6 (Principle of harmonic majorization). *Let D be an open subset of $\mathbb{C}$ with compact closure $\overline{D}$. Let $f(z)$ be analytic on D and continuous on $\overline{D}$. Suppose that $h(x,y)$ is harmonic on D and continuous on $\overline{D}$, with both sets also considered as subsets of $\mathbb{R}^2$ under the identification of $z = x + iy$ with (x,y). If for all $(x_0, y_0) \in \overline{D}\backslash D$,*

$$(B.2.5) \qquad \log|f(x_0, y_0)| \leq h(x_0, y_0).$$

Then for all $z = x + iy \in D$,

$$\log|f(x + iy)| \leq h(x, y).$$

Proof. Let $L := \inf_{(x,y)\in\overline{D}}\{|h(x,y)|\}$. Define U to be the open set

$$U := \{z \in D \ : \ |f(z)| > e^L\}.$$

If for some $z = (x + iy) \in D$, it were the case that $\log|f(x + iy)| > h(x,y)$, then $|f(x + iy)| > e^L > 0$, and hence $(x,y) \in U$. Therefore, if $U = \emptyset$, there is nothing to prove, and we proceed under the assumption that $U \neq \emptyset$.

Since $\log f(z) = \log|f(z)| + i \arg f(z)$, $u(x,y) = \log|f(x+iy)|$ is harmonic in D away from any zeros of f and, in particular, is harmonic in U. Therefore, the function $g(x,y)$ defined by $g(x,y) := \log|f(x + iy)| - h(x,y)$ is harmonic in U. The set $\overline{U} \subset \overline{D}$ is compact, and if $(x_0, y_0) \in \overline{U}\backslash U$, then either $\log|f(x_0 + iy_0)| = L$ or $(x_0, y_0) \in \overline{D}\backslash D$. By the definition of L and (B.2.5), in either case, $g(x_0, y_0) \leq 0$. By the maximum principle for harmonic functions, $g(x,y) \leq 0$ everywhere in U. $\qquad\square$

Proof of Lemma B.5. For $T > 0$, let D_T denote the open rectangle consisting of (x,y) such that $0 < x < 1$ and $-T < y < T$. Define

$$H_T(x,y) = \int_{-T}^{T} \phi_1(t)\mathrm{d}\mu_{(x,y)}(t) + \int_{-T}^{T} \phi_2(t)\mathrm{d}\mu_{(1-x,y)}(t),$$

which by Lemma B.3 is harmonic on the strip $\{(x + iy) \ : \ -1 \leq x \leq 1, \ y \in \mathbb{R}\}$, and hence in D_T.

For $b > 0$, the function h_b defined by

$$h_b(x, y) := \cos\left(b\left(x - \tfrac{1}{2}\right)\right)\cosh(by)$$

is harmonic on $\mathbb{R}^2$, and for $0 < b < a$, $\cos\left(b\left(x - \tfrac{1}{2}\right)\right)$ has a strictly positive lower bound on $[-1, 1]$. Then for any $\epsilon > 0$ and all T sufficiently large, depending on ϵ and $0 < b < a$,

$$\log|f(x + iy)| - H_T(x, y) - h_b(x, y)$$

is negative on the boundary of D_T, and then by the principle of harmonic majorization, on all of $\overline{D_T}$. Then by Lebesgue's monotone convergence theorem, taking the limit $T \uparrow \infty$, and then the limit $\epsilon \downarrow 0$,

$$\log|f(x + iy)| \le \int_{\mathbb{R}} \phi_0(t)\,\mathrm{d}\mu_{(r,y)}(t) + \int_{\mathbb{R}} \phi_1(t)\,\mathrm{d}\mu_{(1-r,y)}(t).$$

Then taking $y = 0$ yields (B.2.4). $\qquad\qquad\qquad\qquad\qquad\qquad\qquad\qquad\square$

B.3. Applications of Hirschman's lemma

The following application may be found in Lemma 3.2 of [124], though as noted there, similar statements making the same application of Hirschman's lemma are well known.

Theorem B.7. *Let $X(z)$ be a continuous function on closed strip $\overline{S}$ with values in $\mathcal{B}(\mathcal{H})$, and let $\mathcal{H}$ be a finite-dimensional Hilbert space. Suppose that $X(z)$ is analytic in S and the $\|X(z)\|$ is bounded on $\overline{S}$. Let $1 \le p_0, p_1 \le \infty$, and for $0 < \theta < 1$, define*

$$\frac{1}{p_\theta} = (1 - \theta)\frac{1}{p_0} + \theta\frac{1}{p_1}.$$

Then with $\|\cdot\|_p$, $1 \le p \le \infty$, denoting the Schatten p norm,

$$(B.3.1) \qquad \log\|X(\theta)\|_{p_\theta} \le (1 - \theta)\int_{\mathbb{R}} \log\|X(it)\|_{p_0}\,\mathrm{d}\mu_{(\theta,0)}(t)$$

$$+ \theta\int_{\mathbb{R}} \log\|X(1 + it)\|_{p_0}\,\mathrm{d}\mu_{(1-\theta,0)}(t).$$

where $\mathrm{d}\mu_{(x,y)}(t)$ is defined in (B.2.1).

Proof. Let p_θ' be the dual index to p_θ so that there exists $Y \in \mathcal{B}(\mathcal{H})$ such that $\|Y\|_{p_\theta'} = 1$ and

$$\|X(\theta)\|_{p_\theta} = \mathrm{Tr}[YX(\theta)].$$

Write the SVD of Y in the form $Y = UD^{1/p_\theta'}V$ so that D is a density matrix. Now define

$$Y(z) := UD^{(1-z)/p_0' + z/p_1'}V \qquad \text{and} \qquad f(z) = \mathrm{Tr}[Y(z)X(z)].$$

Note that for all t, $\|Y(it)\|_{p_0'} = \|D^{1/p_0'}\|_{p_0'} = 1$ and $\|Y(1+it)\|_{p_1'} = \|D^{1/p_1'}\|_{p_1'} = 1$. Therefore, by Hölder's inequality,

$$|f(it)| \leq \|X(it)\|_{p_0} \quad \text{and} \quad |f(1+it)| \leq \|X(1+it)\|_{p_1}.$$

Since $\log|f(\theta)| = \log\|G(\theta)\|_{p_\theta}$, (B.3.1) now follows from Hirschman's lemma.

$\square$

The following inequality is a special case of an inequality proved in [**205**], which was isolated in [**105**] as the key to proving a generalization of the full family of inequalities proved in [**205**]. Hirschman's lemma is used in both [**205**] and [**105**].

Corollary B.8. *Let* $\{A_1, \ldots, A_m\} \subset \mathcal{B}^{++}(\mathcal{H})$. *Then for all* $0 < \theta < 1$,

$$\text{(B.3.2)} \qquad \log\left\|\left|\prod_{j=1}^{m} A_j^\theta\right|^{1/\theta}\right\| \leq \int_{\mathbb{R}} \log\left\|\prod_{j=1}^{m} A_j^{1+it}\right\| \, d\mu_{(1-\theta,t)}(t).$$

Proof. Define $F(z) = \prod_{j=1}^{m} A_j^z$. Take $p_0 = p_1 = \infty$, so that all norms involved are the operator norm. For all $t \in \mathbb{R}$, $F(it)$ is unitary and hence $\|F(it)\| = 1$, so that $\log\|F(it)\| = 0$. Also,

$$F(\theta) = \prod_{j=1}^{m} A_j^\theta \quad \text{so that} \quad \log\|F(\theta)\| = \log\left\|\prod_{j=1}^{m} A_j^\theta\right\| = \theta \log\left\|\left|\prod_{j=1}^{m} A_j^\theta\right|^{1/\theta}\right\|.$$

Now (B.3.2) follows directly from (B.3.1). $\square$

Theorem B.7 has been used, together with Weyl's log majorization method, to prove Theorem 11.33, a "4 matrix" version of the Golden-Thompson inequality. This in turn has be used to prove a strengthened form of SSA, Theorem 8.20, involving the *twirled Petz recovery map*

$$\widetilde{R}_\sigma(X) := \int_{\mathbb{R}} \sigma_{12}^{\frac{1+it}{2}} \sigma_1^{-\frac{1+it}{2}} X \sigma_1^{-\frac{1-it}{2}} \sigma_{12}^{\frac{1-it}{2}} \, d\mu_0(t),$$

where

$$\text{(B.3.3)} \qquad\qquad d\mu_0 := \lim_{x \uparrow 1} d\mu_{(x,0)}$$

so that

$$d\mu_0(t) := \frac{\pi}{2} \frac{1}{\cosh(\pi t) + 1} dt.$$

B.4. An alternate form for the derivative of the logarithm

There is an integral representation for the logarithmic mean in terms of the measure $d\mu_0(t)$, as defined in (B.3.3), that leads to an alternate integral formula for the derivative of the logarithm due to Sutter, Berta, and Tomamichael [205].

Lemma B.9. *The characteristic function (Fourier transform) of μ_0, the function $\hat{\mu}_0(k) := \int_{\mathbb{R}} e^{-ikt} d\mu_0(t)$ is given by*

$$\text{(B.4.1)} \qquad\qquad \hat{\mu}_0(k) = \frac{\pi\omega}{\sinh(\pi\omega)}.$$

In particular, for $a, b > 0$,

$$\text{(B.4.2)} \qquad\qquad \hat{\mu}_0\left(\tfrac{1}{2\pi}\log(b/a)\right) = \frac{\sqrt{ab}}{\Lambda(a,b)},$$

where $\Lambda(a,b)$ denotes the logarithmic mean of a and b.

Proof. Making the change of variable $s = \pi t$, and defining $\omega := k/\pi$,

$$\hat{\mu}_0(k) = \int_R \phi(s)ds \quad \text{where for } z \in \mathbb{C}, \quad \phi(z) := \frac{e^{(1-i\omega)z}}{(e^z + 1)^2}.$$

Note that $\phi(z)$ has poles at $z = i\pi(1 + 2k)$, $k \in \mathbb{Z}$, and that $\phi(z + i2\pi) = \phi(z)e^{2\pi\omega}$. It is therefore easy to evaluate the integral using the residue formula and the contour C_R counterclockwise around the rectangle with vertices at $-R$, R, $R + i2\pi$ and $-R + i2\pi$. By what was noted above,

$$\text{(B.4.3)} \qquad\qquad \oint_{C_R} \phi(z)dz = (1 - e^{2\pi\omega}) \int_{\mathbb{R}} \phi(s)ds + \mathcal{O}(e^{-R}).$$

Then since

$$\oint_{C_R} \phi(z)dz = -\oint_{C_R} e^{i\omega z} \frac{d}{dz}\frac{1}{e^z + 1}dz = i\omega \int_{C_R} e^{i\omega z} \frac{1}{e^z + 1}dz$$

and for the integral on the right, C_R encloses a single pole at $z = i\pi$, the residue formula yields $\oint_{C_R} \phi(z)dz = 2\pi\omega e^{-\pi\omega}$, which together with (B.4.3) yields (B.4.1), and then (B.4.2) follows directly. $\qquad\qquad \square$

Lemma B.10. *For $X \in \mathcal{B}^{++}(\mathcal{H})$, the derivative of the logarithm function at X, $D_{\log,X}$, is given by*

$$D_{\log,X}(K) = \int_{\mathbb{R}} X^{(1-it)/2}KX^{(1-it)/2}d\mu_0(t)$$

for all $K \in \mathcal{B}^{\text{s.a.}}(\mathcal{H})$, where $d\mu_0$ is the probability measure on $\mathbb{R}$ specified in (B.3.3).

Proof. By (3.2.6) and Definition 3.14, it suffices to prove that for $a, b > 0$,

$$(B.4.4) \qquad \int_R a^{(1-it)/2} b^{(1+it)/2} \, d\mu_0(t) = \Lambda^{-1}(a, b).$$

The left side of (B.4.4) equals

$$\sqrt{ab} \int_{\mathbb{R}} e^{i(\frac{\pi t}{2\pi}(\log(b) - \log(a)))} \frac{1}{\cosh(\pi t) + 1} \pi \, dt,$$

and hence (B.4.4) is a consequence of Lemma B.9. $\qquad\qquad\square$

Remark B.11. For $x > 0$ and $h \in \mathbb{R}$, $\frac{d}{dt} \log(x + th)\big|_{t=0} = \frac{h}{x}$. Therefore, by the spectral theorem, for $X \in \mathcal{B}^{++}(\mathcal{H})$ and $K \in \mathcal{B}^{\text{s.a.}}(\mathcal{H})$ such that X and K commute,

$$\frac{d}{dt} \log(X + tK)\Big|_{t=0} = \frac{1}{X^{1/2}} K \frac{1}{X^{1/2}}.$$

Although this formula is not true in general when X and K do not commute, there is a connection between the two sides that is valid in general: Define $L := \log(X) \in \mathcal{B}^{\text{s.a.}}(\mathcal{H})$ and $U(t) = e^{-itL/2}$. Then by Lemma B.10,

$$\frac{d}{dt} \log(X + tK)\Big|_{t=0} = \int_{\mathbb{R}} \left[U^*(t) \left(\frac{1}{X^{1/2}} K \frac{1}{X^{1/2}} \right) U(t) \right] d\mu_0(t).$$

Bibliography

[1] L. Accardi, *Non commutative Markov chains*, Proc. School of Math. Phys. Camerino (1974).

[2] L. Accardi and C. Cecchini, *Conditional expectations in von Neumann algebras and a theorem of Takesaki*, J. Functional Analysis **45** (1982), no. 2, 245–273, DOI 10.1016/0022-1236(82)90022-2. MR647075

[3] A. Albert, *Conditions for positive and nonnegative definiteness in terms of pseudoinverses*, SIAM J. Appl. Math. **17** (1969), 434–440, DOI 10.1137/0117041. MR245582

[4] P. M. Alberti and A. Uhlmann, *Stochasticity and partial order: Doubly stochastic maps and unitary mixing*, Mathematische Monographien [Mathematical Monographs], vol. 18, VEB Deutscher Verlag der Wissenschaften, Berlin, 1981. MR649773

[5] R. Alicki and M. Fannes, *Continuity of quantum conditional information*, J. Phys. A **37** (2004), no. 5, L55–L57, DOI 10.1088/0305-4470/37/5/L01. MR2043448

[6] W. N. Anderson Jr., *Shorted operators*, SIAM J. Appl. Math. **20** (1971), 520–525, DOI 10.1137/0120053. MR287970

[7] W. N. Anderson Jr. and R. J. Duffin, *Series and parallel addition of matrices*, J. Math. Anal. Appl. **26** (1969), 576–594, DOI 10.1016/0022-247X(69)90200-5. MR242573

[8] T. Ando, *Concavity of certain maps on positive definite matrices and applications to Hadamard products*, Linear Algebra Appl. **26** (1979), 203–241, DOI 10.1016/0024-3795(79)90179-4. MR535686

[9] T. Ando, *Majorization, doubly stochastic matrices, and comparison of eigenvalues*, Linear Algebra Appl. **118** (1989), 163–248, DOI 10.1016/0024-3795(89)90580-6. MR995373

[10] T. Ando, *Majorizations and inequalities in matrix theory*, Linear Algebra Appl. **199** (1994), 17–67, DOI 10.1016/0024-3795(94)90341-7. MR1274407

[11] T. Ando, F. Hiai, and K. Okubo, *Trace inequalities for multiple products of two matrices*, Math. Inequal. Appl. **3** (2000), no. 3, 307–318, DOI 10.7153/mia-03-32. MR1768492

[12] H. Araki, *Inequalities in von Neumann algebras*, in *Les rencontres physiciens-mathématiciens de Strasbourg* RCP25, **22** (1975), 1–25.

[13] H. Araki, *Relative entropy of states of von Neumann algebras*, Publ. Res. Inst. Math. Sci. **11** (1975/76), no. 3, 809–833, DOI 10.2977/prims/1195191148. MR425631

[14] H. Araki, *On an inequality of Lieb and Thirring*, Lett. Math. Phys. **19** (1990), no. 2, 167–170, DOI 10.1007/BF01045887. MR1039525

[15] H. Araki and E. H. Lieb, *Entropy inequalities*, Comm. Math. Phys. **18** (1970), 160–170. MR266563

[16] H. Araki and M. M. Yanase, *Measurement of quantum mechanical operators*, Phys. Rev. (2) **120** (1960), 622–626. MR122377

[17] K. M. R. Audenaert, *A sharp continuity estimate for the von Neumann entropy*, J. Phys. A **40** (2007), no. 28, 8127–8136, DOI 10.1088/1751-8113/40/28/S18. MR2344161

[18] K. M. R. Audenaert, *A note on the $p \to q$ norms of 2-positive maps*, Linear Algebra Appl. **430** (2009), no. 4, 1436–1440, DOI 10.1016/j.laa.2008.09.040. MR2489405

[19] K. M. R. Audenaert and N. Datta, $\alpha - z$-*Rényi relative entropies*, J. Math. Phys. **56** (2015), no. 2, 022202, 16, DOI 10.1063/1.4906367. MR3390876

[20] K. M. R. Audenaert, J. Calsamiglia, R. Munoz-Tapia, E. Bagan, Ll. Masanes, A. Acin, and F. Verstraete, *Discriminating states: the quantum Chernoff bound*, Phys. Rev. Lett. **98** (2007), 160501.

[21] K. M. R. Audenaert, M. Nussbaum, A. Szkoła, and F. Verstraete, *Asymptotic error rates in quantum hypothesis testing*, Comm. Math. Phys. **279** (2008), no. 1, 251–283, DOI 10.1007/s00220-008-0417-5. MR2377635

[22] J. S. Aujla, *A simple proof of Lieb concavity theorem*, J. Math. Phys. **52** (2011), no. 4, 043505, 3, DOI 10.1063/1.3573594. MR2964191

[23] K. Ball, E. A. Carlen, and E. H. Lieb, *Sharp uniform convexity and smoothness inequalities for trace norms*, Invent. Math. **115** (1994), no. 3, 463–482, DOI 10.1007/BF01231769. MR1262940

[24] H. Barnum, C. Caves, C. A. Fuchs, R. Jozsa, and B. Schumacher, *Noncommuting mixed states cannot be broadcast*. Phys. Rev. Letters. **76** (1996), 2818–2821.

[25] T. N. Bekjan, *On joint convexity of trace functions*, Linear Algebra Appl. **390** (2004), 321–327, DOI 10.1016/j.laa.2004.05.002. MR2083663

[26] S. Beigi, *Sandwiched Rényi divergence satisfies data processing inequality*, J. Math. Phys. **54** (2013), no. 12, 122202, 11, DOI 10.1063/1.4838855. MR3867202

[27] V. P. Belavkin and P. Staszewski, C^*-*algebraic generalization of relative entropy and entropy*, Ann. Inst. H. Poincaré Sect. A (N.S.) **37** (1982), no. 1, 51–58. MR667882

[28] E. Beltrami, *Sulle funzioni bilineari*. Giornalle di Matematiche ad Uso degli Studenti delle Uninversita, **11** (1873), 98–106.

[29] J. Bendat and S. Sherman, *Monotone and convex operator functions*, Trans. Amer. Math. Soc. **79** (1955), 58–71, DOI 10.2307/1992836. MR82655

[30] C. H. Bennett, G. Brassard, C. Crépeau, R. Jozsa, A. Peres, and W. K. Wootters, *Teleporting an unknown quantum state via dual classical and Einstein-Podolsky-Rosen channels*, Phys. Rev. Lett. **70** (1993), no. 13, 1895–1899, DOI 10.1103/PhysRevLett.70.1895. MR1208247

[31] C. H. Bennett, H. J. Bernstein, S. Popescu, and B. Schumacher, *Concentrating partial entanglement by local operations*, Phys. Rev. A, **53** (1996), 2046–2052.

[32] C. H. Bennett, D. P. DiVincenzo, J. A. Smolin, and W. K. Wootters, *Mixed-state entanglement and quantum error correction*, Phys. Rev. A (3) **54** (1996), no. 5, 3824–3851, DOI 10.1103/PhysRevA.54.3824. MR1418618

[33] M. Berta, O. Fawzi, and M. Tomamichel, *On variational expressions for quantum relative entropies*, Lett. Math. Phys. **107** (2017), no. 12, 2239–2265, DOI 10.1007/s11005-017-0990-7. MR3719640

[34] M. Berta and M. Tomamichel, *Entanglement monogamy via multivariate trace inequalities*, Comm. Math. Phys. **405** (2024), no. 2, Paper No. 29, 30, DOI 10.1007/s00220-023-04920-5. MR4699595

[35] R. Bhatia, *Matrix analysis*, Graduate Texts in Mathematics, vol. 169, Springer-Verlag, New York, 1997, DOI 10.1007/978-1-4612-0653-8. MR1477662

[36] R. Bhatia, *Positive definite matrices*, Princeton Series in Applied Mathematics, Princeton University Press, Princeton, NJ, 2007. MR2284176

[37] R. Bhatia and J. Holbrook, *Riemannian geometry and matrix geometric means*, Linear Algebra Appl. **413** (2006), no. 2-3, 594–618, DOI 10.1016/j.laa.2005.08.025. MR2198952

[38] G. Birkhoff, *Three observations on linear algebra* (Spanish), Univ. Nac. Tucumán. Revista A. **5** (1946), 147–151. MR20547

[39] A. Bluhm, Á. Capel, P. Gondolf, and A. Pérez-Hernńdez, *General continuity bounds for quantum relative entropies*, 2023 IEEE International Symposium on Information Theory (ISIT), 2023.

[40] N. N. Bogoliubov, *On a variational principle in the many-body problem*, Soviet Physics Dokl. **119(3)** (1958), 292–294 (244–246 Dokl. Akad. Nauk SSSR). MR101796

[41] F. G. S. L. Brandão, M. Christandl, and J. Yard, *Faithful squashed entanglement*, Comm. Math. Phys. **306** (2011), no. 3, 805–830, DOI 10.1007/s00220-011-1302-1. MR2825510

[42] D. Bures, *An extension of Kakutani's theorem on infinite product measures to the tensor product of semifinite W^*-algebras*, Trans. Amer. Math. Soc. **135** (1969), 199–212, DOI 10.2307/1995012. MR236719

[43] E. A. Carlen, *A monotonicity version of a concavity theorem of Lieb*, Arch. Math. (Basel) **119** (2022), no. 5, 525–529, DOI 10.1007/s00013-022-01774-6. MR4496982

[44] E. A. Carlen, R. L. Frank, and E. H. Lieb, *Some operator and trace function convexity theorems*, Linear Algebra Appl. **490** (2016), 174–185, DOI 10.1016/j.laa.2015.11.006. MR3429039

[45] E. A. Carlen, R. L. Frank, and E. H. Lieb, *Inequalities for quantum divergences and the Audenaert-Datta conjecture*, J. Phys. A **51** (2018), no. 48, 483001, 23, DOI 10.1088/1751-8121/aae8a3. MR3878510

[46] E. A. Carlen and E. H. Lieb, *Optimal hypercontractivity for Fermi fields and related noncommutative integration inequalities*, Comm. Math. Phys. **155** (1993), no. 1, 27–46. MR1228524

[47] E. A. Carlen and E. H. Lieb, *A Minkowski type trace inequality and strong subadditivity of quantum entropy*, Differential operators and spectral theory, Amer. Math. Soc. Transl. Ser. 2, vol. 189, Amer. Math. Soc., Providence, RI, 1999, pp. 59–68, DOI 10.1090/trans2/189/05. MR1730503

[48] E. A. Carlen and E. H. Lieb, *A Minkowski type trace inequality and strong subadditivity of quantum entropy. II. Convexity and concavity*, Lett. Math. Phys. **83** (2008), no. 2, 107–126, DOI 10.1007/s11005-008-0223-1. MR2379699

[49] E. A. Carlen and E. H. Lieb, *Bounds for entanglement via an extension of strong subadditivity of entropy*, Lett. Math. Phys. **101** (2012), no. 1, 1–11, DOI 10.1007/s11005-012-0565-6. MR2935474

[50] E. A. Carlen and M. P. Loss, *On an inequality of Lin, Kim, and Hsieh and strong subadditivity*, Lett. Math. Phys. **114** (2024), no. 5, Paper No. 109, 7, DOI 10.1007/s11005-024-01857-1. MR4795830

[51] E. A. Carlen and J. Maas, *Gradient flow and entropy inequalities for quantum Markov semigroups with detailed balance*, J. Funct. Anal. **273** (2017), no. 5, 1810–1869, DOI 10.1016/j.jfa.2017.05.003. MR3666729

[52] E. A. Carlen and A. Müller-Hermes, *A tracial Schwarz inequality and a theorem of Hiai and Petz*, arXiv:2203.03433, 2022.

[53] E. A. Carlen and A. Vershynina, *Recovery map stability for the data processing inequality*, J. Phys. A **53** (2020), no. 3, 035204, 17, DOI 10.1088/1751-8121/ab5ab7. MR4054724

[54] E. A. Carlen and H. Zhang, *Monotonicity versions of Epstein's concavity theorem and related inequalities*, Linear Algebra Appl. **654** (2022), 289–310, DOI 10.1016/j.laa.2022.09.001. MR4483360

[55] M. Caspers, J. Parcet, M. Perrin, and É. Ricard, *Noncommutative de Leeuw theorems*, Forum Math. Sigma **3** (2015), Paper No. e21, 59, DOI 10.1017/fms.2015.23. MR3482270

[56] C. M. Caves, C. A. Fuchs, and R. Schack, *Unknown quantum states: the quantum de Finetti representation*, J. Math. Phys. **43** (2002), no. 9, 4537–4559, DOI 10.1063/1.1494475. Quantum information theory. MR1924454

[57] N. N. Čencov, *Statistical decision rules and optimal inference*, Translations of Mathematical Monographs, vol. 53, American Mathematical Society, Providence, RI, 1982. Translation from the Russian edited by Lev J. Leifman. MR645898

[58] N. N. Chan and K. H. Li, *Diagonal elements and eigenvalues of a real symmetric matrix*, J. Math. Anal. Appl. **91** (1983), no. 2, 562–566, DOI 10.1016/0022-247X(83)90171-3. MR690890

[59] V. Chayes, *Reverse Hölder, Minkowski, and Hanner inequalities for matrices*, arXiv:2103.09915, 2021.

[60] M. D. Choi, *Positive linear maps on C^*-algebras*, Canadian J. Math. **24** (1972), 520–529, DOI 10.4153/CJM-1972-044-5. MR315460

[61] M. D. Choi, *A Schwarz inequality for positive linear maps on C^*-algebras*, Illinois J. Math. **18** (1974), 565–574. MR355615

[62] M. D. Choi, *Completely positive linear maps on complex matrices*, Linear Algebra Appl. **10** (1975), 285–290, DOI 10.1016/0024-3795(75)90075-0. MR376726

[63] M. D. Choi, *Some assorted inequalities for positive linear maps on C^*-algebras*, J. Operator Theory **4** (1980), no. 2, 271–285. MR595415

[64] M. Christandl and A. Winter, *"Squashed entanglement": an additive entanglement measure*, J. Math. Phys. **45** (2004), no. 3, 829–840, DOI 10.1063/1.1643788. MR2036165

[65] I. Csiszár, *Information-type measures of difference of probability distributions and indirect observations*, Studia Sci. Math. Hungar. **2** (1967), 299–318. MR219345

[66] B. Dacorogna and P. Maréchal, *The role of perspective functions in convexity, polyconvexity, rank-one convexity and separate convexity*, J. Convex Anal. **15** (2008), no. 2, 271–284. MR2422989

[67] G. Dantzig, *Comments on J. von Neumann's "The problem of optimal assignment in a two person game"* Rand Corporation Report P-435, July 21, 1952.

[68] E. B. Davies, *Quantum theory of open systems*, Academic Press [Harcourt Brace Jovanovich, Publishers], London-New York, 1976. MR489429

[69] C. Davis, *A Schwarz inequality for convex operator functions*, Proc. Amer. Math. Soc. **8** (1957), 42–44, DOI 10.2307/2032808. MR84120

[70] C. Davis, *Various averaging operations onto subalgebras*, Illinois J. Math. **3** (1959), 538–553. MR112042

[71] C. Davis, *Notions generalizing convexity for functions defined on spaces of matrices*, Proc. Sympos. Pure Math., Vol. VII, Amer. Math. Soc., Providence, RI, 1963, pp. 187–201. MR155837

[72] M. Delbrürk and G. Moleìere, *Statistische quantenmechanik und thermodynamik*, Abh. Preuss. Akad. Wissenschaften **1** (1936), 1–42.

[73] D. Dieks, *Communication by EPR devices.* Phys. Letters A, **92** (1982), 271–272.

[74] D. Dieks, *Overlap and distinguishability of quantum states*, Phys. Lett. A **126** (1988), no. 5-6, 303–306, DOI 10.1016/0375-9601(88)90840-7. MR924511

[75] J. Dixmier, *Formes linéaires sur un anneau d'opérateurs* (French), Bull. Soc. Math. France **81** (1953), 9–39. MR59485

[76] A. C. Doherty, P. A. Parrilo, and F. M. Spedalieri, **TITLE??**, Phys. Rev. Lett., **88** (2002), 187904.

[77] M. J. Donald, *On the relative entropy*, Comm. Math. Phys. **105** (1986), no. 1, 13–34. MR847125

[78] E. G. Effros, *A matrix convexity approach to some celebrated quantum inequalities*, Proc. Natl. Acad. Sci. USA **106** (2009), no. 4, 1006–1008, DOI 10.1073/pnas.0807965106. MR2475796

[79] H. Epstein, *Remarks on two theorems of E. Lieb*, Comm. Math. Phys. **31** (1973), 317–325. MR343073

[80] K. Fan, *On a theorem of Weyl concerning eigenvalues of linear transformations. I, II*, Proc. Nat. Acad. Sci. U.S.A. **35, 36** (1949, 1950), 652–655; 31–35, DOI 10.1073/pnas.35.11.652. MR34519

[81] K. Fan, *Maximum properties and inequalities for the eigenvalues of completely continuous operators*, Proc. Nat. Acad. Sci. U.S.A. **37** (1951), 760–766, DOI 10.1073/pnas.37.11.760. MR45952

[82] O. Fawzi and R. Renner, *Quantum conditional mutual information and approximate Markov chains*, Comm. Math. Phys. **340** (2015), no. 2, 575–611, DOI 10.1007/s00220-015-2466-x. MR3397027

[83] W. Fenchel, *Convex cones, sets and functions*, Lecture Notes from a Princeton University course, Spring Term, 1951.

[84] R. A. Fisher, *Theory of statistical estimation*, Proc. Cambridge Philos. Soc. **22** (1925), 700–725.

[85] R. L. Frank and E. H. Lieb, *Monotonicity of a relative Rényi entropy*, J. Math. Phys. **54** (2013), no. 12, 122201, 5, DOI 10.1063/1.4838835. MR3156109

[86] S. Friedland and W. So, *On the product of matrix exponentials*, Linear Algebra Appl. **196** (1994), 193–205, DOI 10.1016/0024-3795(94)90324-7. MR1273984

[87] C. A. Fuchs and C. M. Caves, *Mathematical techniques for quantum communication theory*, Open Systems and Information Dynamics, **3** (1995), 345–356.

[88] C. A. Fuchs and J. van de Graaf, *Cryptographic distinguishability measures for quantum-mechanical states*, IEEE Trans. Inform. Theory **45** (1999), no. 4, 1216–1227, DOI 10.1109/18.761271. MR1686254

[89] J. Fujii and E. Kamei, *Relative operator entropy in noncommutative information theory*, Math. Japon. **34** (1989), no. 3, 341–348. MR1003918

[90] I. Gelfand and M. Neumark, *On the imbedding of normed rings into the ring of operators in Hilbert space* (English, with Russian summary), Rec. Math. [Mat. Sbornik] N.S. **12(54)** (1943), 197–213. MR9426

[91] G. L. Gilardoni, *On Pinsker's and Vajda's type inequalities for Csiszár's f-divergences*, IEEE Trans. Inform. Theory **56** (2010), no. 11, 5377–5386, DOI 10.1109/TIT.2010.2068710. MR2808583

[92] S. Golden, *Lower bounds for the Helmholtz function*, Phys. Rev. (2) **137** (1965), B1127–B1128. MR189691

[93] V. Gorini, A. Kossakowski, and E. C. G. Sudarshan, *Completely positive dynamical semi-groups of N-level systems*, J. Math. Phys. **17** (1976), 821–825.

[94] J. Hadamard, *Sur les fonctions entières*, Bull. Soc. Math. Fr., **24** (1896), 186–187.

[95] F. Hansen, *The fast track to Löwner's theorem*, Linear Algebra Appl. **438** (2013), no. 11, 4557–4571, DOI 10.1016/j.laa.2013.01.022. MR3034551

[96] G. H. Hardy, J. E. Littlewood and G. Polya, *Some simple inequalities satisfied by convex functions*, Messenger Math. **58** (1929), 145–152.

[97] G. H. Hardy, J. E. Littlewood, and G. Pólya, *Inequalities*, Cambridge, at the University Press, 1952. 2d ed. MR46395

[98] M. B. Hastings. *Superadditivity of communication capacity using entangled inputs*, Nature Physics **5** (2009), 255.

[99] P. Hayden, R. Jozsa, D. Petz, and A. Winter, *Structure of states which satisfy strong subadditivity of quantum entropy with equality*, Comm. Math. Phys. **246** (2004), no. 2, 359–374, DOI 10.1007/s00220-004-1049-z. MR2048562

[100] P. M. Hayden, M. Horodecki, and B. M. Terhal, *The asymptotic entanglement cost of preparing a quantum state: Quantum information and computation*, J. Phys. A **34** (2001), no. 35, 6891–6898, DOI 10.1088/0305-4470/34/35/314. MR1863123

[101] N. Herbert, *FLASH—A superluminal communicator based upon a new kind of quantum measurement*. Found. of Phys., **12** (1982), 1171–79.

[102] F. Hiai, *Concavity of certain matrix trace and norm functions*, Linear Algebra Appl. **439** (2013), no. 5, 1568–1589, DOI 10.1016/j.laa.2013.04.020. MR3067823

[103] F. Hiai, *Concavity of certain matrix trace and norm functions. II*, Linear Algebra Appl. **496** (2016), 193–220, DOI 10.1016/j.laa.2015.12.032. MR3464069

[104] F. Hiai and A. Jenčová, *α-z-Rényi divergences in von Neumann algebras: data processing inequality, reversibility, and monotonicity properties in α, z*, Comm. Math. Phys. **405** (2024), no. 11, Paper No. 271, 43, DOI 10.1007/s00220-024-05124-1. MR4814085

[105] F. Hiai, R. König, and M. Tomamichel, *Generalized log-majorization and multivariate trace inequalities*, Ann. Henri Poincaré **18** (2017), no. 7, 2499–2521, DOI 10.1007/s00023-017-0569-y. MR3665221

[106] F. Hiai, M. Mosonyi, D. Petz, and C. Bény, *Quantum f-divergences and error correction*, Rev. Math. Phys. **23** (2011), no. 7, 691–747, DOI 10.1142/S0129055X11004412. MR2826462

[107] F. Hiai and D. Petz, *The proper formula for relative entropy and its asymptotics in quantum probability*, Comm. Math. Phys. **143** (1991), no. 1, 99–114. MR1139426

[108] F. Hiai and D. Petz, *From quasi-entropy to various quantum information quantities*, Publ. Res. Inst. Math. Sci. **48** (2012), no. 3, 525–542, DOI 10.2977/PRIMS/79. MR2973391

[109] S. Hill and W. Wootters, *Entanglement of a pair of quantum bits*, Phys. Rev. Lett. **78** (1997), 5022–5025.

[110] I. I. Hirschman Jr., *A convexity theorem for certain groups of transformations*, J. Analyse Math. **2** (1953), 209–218, DOI 10.1007/BF02825637. MR57936

[111] M. Horodecki, J. Oppenheim, and A. Winter, *Quantum state merging and negative information*, Comm. Math. Phys. **269** (2007), no. 1, 107–136, DOI 10.1007/s00220-006-0118-x. MR2274464

[112] R. Horodecki, P. Horodecki, M. Horodecki, and K. Horodecki, *Quantum entanglement*, Rev. Modern Phys. **81** (2009), no. 2, 865–942, DOI 10.1103/RevModPhys.81.865. MR2515619

[113] A. Horn, *On the singular values of a product of completely continuous operators*, Proc. Nat. Acad. Sci. U.S.A. **36** (1950), 374–375, DOI 10.1073/pnas.36.7.374. MR45316

[114] A. Horn, *Doubly stochastic matrices and the diagonal of a rotation matrix*, Amer. J. Math. **76** (1954), 620–630, DOI 10.2307/2372705. MR63336

[115] R. A. Horn and C. R. Johnson, *Matrix analysis*, Cambridge University Press, Cambridge, 1985, DOI 10.1017/CBO9780511810817. MR832183

[116] R. A. Horn and C. R. Johnson, *Matrix analysis*, Cambridge University Press, Cambridge, 1985, DOI 10.1017/CBO9780511810817. MR832183

[117] L. P. Hughston, R. Jozsa, and W. K. Wootters, *A complete classification of quantum ensembles having a given density matrix*, Phys. Lett. A **183** (1993), no. 1, 14–18, DOI 10.1016/0375-9601(93)90880-9. MR1248347

[118] V. Jaksic, Y. Ogata, Y. Pautrat, and C. Pillet. *Entropic fluctuations in quantum statistical mechanics. An introduction*, in *Quantum Theory from Small to Large Scales: Lecture Notes of the Les Houches Summer School: Volume 95, August 2010*. Oxford University Press, 2012.

[119] I. D. Ivanović, *How to differentiate between nonorthogonal states*, Phys. Lett. A **123** (1987), no. 6, 257–259, DOI 10.1016/0375-9601(87)90222-2. MR905418

[120] A. Jenčová, *Rényi relative entropies and noncommutative L_p-spaces*, Ann. Henri Poincaré **19** (2018), no. 8, 2513–2542, DOI 10.1007/s00023-018-0683-5. MR3830221

[121] C. Jordan, *Mémoire sur les formes bilinéaires* Jour. Math. Pures et Appl. **19** (1874) 35–54.

[122] C. Jordan, *Sur le réduction les formes bilinéaires* C.R.A.S. Paris **78** (1874) 614–617, 54.

[123] R. Jost, *Ueber eine Ungleichung voin E. P. Wigner und M. M. Yanase*, pp. 13-19 in *Quanta: Essays in Theoretical Physics Dedicated to Gregor Wentzel*, edited by P. G. O. Freund, C. I. Goebel and Y. Nambu, University of Chicago Press, Chicago, 1970.

[124] M. Junge, R. Renner, D. Sutter, M. M. Wilde, and A. Winter, *Universal recovery maps and approximate sufficiency of quantum relative entropy*, Ann. Henri Poincaré **19** (2018), no. 10, 2955–2978, DOI 10.1007/s00023-018-0716-0. MR3851777

[125] R. V. Kadison, *A generalized Schwarz inequality and algebraic invariants for operator algebras*, Ann. of Math. (2) **56** (1952), 494–503, DOI 10.2307/1969657. MR51442

[126] S. Kato, *On α-z-Rényi divergence in the von Neumann algebra setting*, J. Math. Phys. **65** (2024), no. 4, Paper No. 042202, 16, DOI 10.1063/5.0186552. MR4727519

[127] J. Kiefer, *Optimum experimental designs*, J. Roy. Statist. Soc. Ser. B **21** (1959), 272–319. MR113263

[128] I. Kim and M. B. Ruskai, *Bounds on the concavity of quantum entropy*, J. Math. Phys. **55** (2014), no. 9, 092201, 5, DOI 10.1063/1.4895757. MR3390783

[129] A. Yu. Kitaev, A. H. Shen, and M. N. Vyalyi, *Classical and quantum computation*, Graduate Studies in Mathematics, vol. 47, American Mathematical Society, Providence, RI, 2002. Translated from the 1999 Russian original by Lester J. Senechal, DOI 10.1090/gsm/047. MR1907291

[130] O. Klein, *Zur Quantenmechanischen Begründung des zweiten Hauptsatzes der Wärmelehre*, Z. Physik **72** (1931), 767–775.

[131] M. Koashi and A. Winter, *Monogamy of quantum entanglement and other correlations*, Phys. Rev. A (3) **69** (2004), no. 2, 022309, 6, DOI 10.1103/PhysRevA.69.022309. MR2041765

[132] H. Kosaki, *Relative entropy of states: a variational expression*, J. Operator Theory **16** (1986), no. 2, 335–348. MR860352

[133] K. Kraus, *General state changes in quantum theory*, Ann. Physics **64** (1971), 311–335, DOI 10.1016/0003-4916(71)90108-4. MR292434

[134] F. Kubo and T. Ando, *Means of positive linear operators*, Math. Ann. **246** (1979/80), no. 3, 205–224, DOI 10.1007/BF01371042. MR563399

[135] S. Kullback *Lower bound for discrimination information in terms of variation*, IEEE Transactions on Information Theory, **13** 126-127 (1967)

[136] S. Kullback and R. A. Leibler, *On information and sufficiency*, Ann. Math. Statistics **22** (1951), 79–86, DOI 10.1214/aoms/1177729694. MR39968

[137] E. C. Lance, *Ergodic theorems for convex sets and operator algebras*, Invent. Math. **37** (1976), no. 3, 201–214, DOI 10.1007/BF01390319. MR428060

[138] O. Lanford and D. Robinson, *Mean entropy of states in quantum statistical mechanics*, J. Math. Phys. **9** (1968), 1120.

[139] A. Lesniewski and M. B. Ruskai, *Monotone Riemannian metrics and relative entropy on noncommutative probability spaces*, J. Math. Phys. **40** (1999), no. 11, 5702–5724, DOI 10.1063/1.533053. MR1722334

[140] K. Li and A. Winter, *Squashed entanglement, k-extendibility, quantum Markov chains, and recovery maps*, Found. Phys. **48** (2018), no. 8, 910–924, DOI 10.1007/s10701-018-0143-6. MR3846918

[141] E. H. Lieb, *Convex trace functions and the Wigner-Yanase-Dyson conjecture*, Advances in Math. **11** (1973), 267–288, DOI 10.1016/0001-8708(73)90011-X. MR332080

[142] E. H. Lieb, *Some convexity and subadditivity properties of entropy*, Bull. Amer. Math. Soc. **81** (1975), 1–13, DOI 10.1090/S0002-9904-1975-13621-4. MR356797

[143] E. H. Lieb and M. B. Ruskai, *Proof of the strong subadditivity of quantum-mechanical entropy*, J. Mathematical Phys. **14** (1973), 1938–1941, DOI 10.1063/1.1666274. With an appendix by B. Simon. MR345558

[144] E. H. Lieb and M. B. Ruskai, *Some operator inequalities of the Schwarz type*, Advances in Math. **12** (1974), 269–273, DOI 10.1016/S0001-8708(74)80004-6. MR336406

[145] E. H. Lieb and W. Thirring, *Inequalities for the moments of the eigenvalues of the Schrödinger Hamiltonian and their relation to Sobolev inequalities*, in *Studies in Mathematical Physics*, E.H. Lieb, B. Simon, A. Wightman eds., Princeton University Press, 269-303 (1976).

[146] T.-C. Lin, I. H. Kim, and M.-H. Hsieh, *A new operator extension of strong subadditivity of quantum entropy*, Lett. Math. Phys. **113** (2023), no. 3, Paper No. 68, 9, DOI 10.1007/s11005-023-01688-6. MR4604968

[147] G. Lindblad, *Entropy, information and quantum measurements*, Comm. Math. Phys. **33** (1973), 305–322. MR337240

[148] G. Lindblad, *Expectations and entropy inequalities for finite quantum systems*, Comm. Math. Phys. **39** (1974), 111–119. MR363351

[149] G. Lindblad, *Completely positive maps and entropy inequalities*, Comm. Math. Phys. **40** (1975), 147–151. MR376050

[150] G. Lindblad, *A general no-cloning theorem*, Lett. Math. Phys. **47** (1999), no. 2, 189–196, DOI 10.1023/A:1007581027660. MR1682306

[151] K. Löwner, *Über monotone Matrixfunktionen* (German), Math. Z. **38** (1934), no. 1, 177–216, DOI 10.1007/BF01170633. MR1545446

[152] A. W. Marshall and I. Olkin, *Inequalities: theory of majorization and its applications*, Mathematics in Science and Engineering, vol. 143, Academic Press, Inc. [Harcourt Brace Jovanovich, Publishers], New York-London, 1979. MR552278

[153] H. P. McKean Jr., *Stochastic integrals*, Probability and Mathematical Statistics, No. 5, Academic Press, New York-London, 1969. MR247684

[154] H. Minkowski, *Theorie der Konvexen Korper, Insbesondere Begründung ihres Oberflächenbegriffs*, Gessamelte Abhandlungen II, 1911, Leipzig.

[155] E. H. Moore *On the reciprocal of the general algebraic matrix*, Bull. Amer. Math. Soc. **26** (1920) 394–395.

[156] E. A. Morozova and N. N. Chentsov, *Markov invariant geometry on state manifolds* (Russian), Current problems in mathematics. Newest results, Vol. 36 (Russian), Itogi Nauki i Tekhniki, Akad. Nauk SSSR, Vsesoyuz. Inst. Nauchn. i Tekhn. Inform., Moscow, 1989, pp. 69–102, 187. Translated in J. Soviet Math. **56** (1991), no. 5, 2648–2669. MR1057197

[157] M. Mosonyi and F. Hiai, *On the quantum Rényi relative entropies and related capacity formulas*, IEEE Trans. Inform. Theory **57** (2011), no. 4, 2474–2487, DOI 10.1109/TIT.2011.2110050. MR2809103

[158] M. Mosonyi and F. Hiai, *Some continuity properties of quantum Rényi divergences*, IEEE Trans. Inform. Theory **70** (2024), no. 4, 2674–2700, DOI 10.1109/tit.2023.3324758. MR4730060

[159] F. J. Murray and J. Von Neumann, *On rings of operators*, Ann. of Math. (2) **37** (1936), no. 1, 116–229, DOI 10.2307/1968693. MR1503275

[160] A. Müller-Hermes and D. Reeb, *Monotonicity of the quantum relative entropy under positive maps*, Ann. Henri Poincaré **18** (2017), no. 5, 1777–1788, DOI 10.1007/s00023-017-0550-9. MR3635967

[161] M. Müller-Lennert, F. Dupuis, O. Szehr, S. Fehr, and M. Tomamichel, *On quantum Rényi entropies: a new generalization and some properties*, J. Math. Phys. **54** (2013), no. 12, 122203, 20, DOI 10.1063/1.4838856. MR3156110

[162] M. A. Neumark, *On a representation of additive operator set functions*, C. R. (Doklady) Acad. Sci. USSR (N.S.) **41** (1943), 359–361. MR10789

[163] M. Navascués, M Owari and M B Plenio, *Power of symmetric extensions for entanglement detection*, Phys. Rev. A **80** (2009), 052306.

[164] E. Nelson, *Topics in dynamics. I: Flows*, Mathematical Notes, Princeton University Press, Princeton, NJ; University of Tokyo Press, Tokyo, 1969. MR282379

[165] E. Nelson, *Notes on non-commutative integration*, J. Functional Analysis **15** (1974), 103–116, DOI 10.1016/0022-1236(74)90014-7. MR355628

[166] R. Nevanlinna, *Analytic functions*, Die Grundlehren der mathematischen Wissenschaften, Band 162, Springer-Verlag, New York-Berlin, 1970. Translated from the second German edition by Phillip Emig. MR279280

[167] M. A. Nielsen, *Conditions for a class of entanglement transformations*, Phys. Rev. Lett. **83** (1999), 436.

[168] M. A. Nielsen, *Continuity bounds for entanglement*, Phys. Rev. A (3) **61** (2000), no. 6, 064301, 4, DOI 10.1103/PhysRevA.61.064301. MR1767484

[169] M. A. Nielsen and I. L. Chuang, *Quantum computation and quantum information*, Cambridge University Press, Cambridge, 2000. MR1796805

[170] I. Nikoufar, A. Ebadian, and M. Eshaghi Gordji, *The simplest proof of Lieb concavity theorem*, Adv. Math. **248** (2013), 531–533, DOI 10.1016/j.aim.2013.07.019. MR3107520

[171] Y. Ogata, *A generalization of Powers-Størmer inequality*, Lett. Math. Phys. **97** (2011), no. 3, 339–346, DOI 10.1007/s11005-011-0504-y. MR2826816

[172] T. Osborne and F. Verstraete, *General Mmnogamy inequality for bipartite qubit entanglement*, Phys. Rev. Lett. **96** (2006), 220503.

[173] V. Paulsen, *Completely bounded maps and operator algebras*, Cambridge Studies in Advanced Mathematics, vol. 78, Cambridge University Press, Cambridge, 2002. MR1976867

[174] R. Penrose, *A generalized inverse for matrices*, Proc. Cambridge Philos. Soc. **51** (1955), 406–413. MR69793

[175] A. Peres, *How to differentiate between nonorthogonal states*, Phys. Lett. A **128** (1988), no. 1-2, 19, DOI 10.1016/0375-9601(88)91034-1. MR935913

[176] D. Petz, *Quasi-entropies for states of a von Neumann algebra*, Publ. Res. Inst. Math. Sci. **21** (1985), no. 4, 787–800, DOI 10.2977/prims/1195178929. MR817164

[177] D. Petz, *Quasi-entropies for finite quantum systems*, Rep. Math. Phys. **23** (1986), no. 1, 57–65, DOI 10.1016/0034-4877(86)90067-4. MR868631

[178] D. Petz, *Monotone metrics on matrix spaces*, Linear Algebra Appl. **244** (1996), 81–96, DOI 10.1016/0024-3795(94)00211-8. MR1403277

[179] M. Piani, *Relative entropy of entanglement and restricted measurements*, Phys. Rev. Lett. **103** (2009), no. 16, 160504, 4, DOI 10.1103/PhysRevLett.103.160504. MR2558601

[180] M. S. Pinsker, *Information and information stability of random variables and processes*, Holden-Day, Inc., San Francisco, Calif.-London-Amsterdam, 1964. Translated and edited by Amiel Feinstein. MR213190

[181] R. T. Powers and E. Størmer, *Free states of the canonical anticommutation relations*, Comm. Math. Phys. **16** (1970), 1–33. MR269230

[182] E. Progovečki, *Information-theoretical aspects of quantum measurement*, Int. Jour. Theor. Phys., **16** (1977), 321-331.

[183] W. Pusz and S. L. Woronowicz, *Functional calculus for sesquilinear forms and the purification map*, Rep. Mathematical Phys. **8** (1975), no. 2, 159–170, DOI 10.1016/0034-4877(75)90061-0. MR420302

[184] W. Pusz and S. L. Woronowicz, *Form convex functions and the WYDL and other inequalities*, Lett. Math. Phys. **2** (1977/78), no. 6, 505–512, DOI 10.1007/BF00398504. MR513118

[185] M. A. Rieffel and A. van Daele, *A bounded operator approach to Tomita-Takesaki theory*, Pacific J. Math. **69** (1977), no. 1, 187–221. MR438147

[186] M. Riesz, *Sur les maxima des formes bilinéaires et sur les fonctionnelles linéaires* (French), Acta Math. **49** (1927), no. 3-4, 465–497, DOI 10.1007/BF02564121. MR1555250

[187] D. W. Robinson and D. Ruelle, *Mean entropy of states in classical statistical mechanics*, Comm. Math. Phys. **5** (1967), 288–300. MR225553

[188] R. T. Rockafellar, *Convex analysis*, Princeton Mathematical Series, No. 28, Princeton University Press, Princeton, NJ, 1970. MR274683

[189] B. Russo and H. A. Dye, *A note on unitary operators in C*-algebras*, Duke Math. J. **33** (1966), 413–416. MR193530

[190] M. B. Ruskai, *Inequalities for quantum entropy: a review with conditions for equality: Quantum information theory*, J. Math. Phys. **43** (2002), no. 9, 4358–4375, DOI 10.1063/1.1497701. MR1924445

[191] E. Schmidt, *Zur Theorie der linearen und nichtlinearen Integralgleichungen, I. Teil: Entwicklung willkürlicher Funktionen nach Systemen vorgeschriebener* (German), Math. Ann. **63** (1907), no. 4, 433–476, DOI 10.1007/BF01449770. MR1511415

[192] P. W. Shor, *Equivalence of additivity questions in quantum information theory*, Comm. Math. Phys. **246** (2004), no. 3, 453–472, DOI 10.1007/s00220-003-0981-7. MR2053939

[193] E. Schrödinger, *Discussion of probability relations between separated systems*, Proc. Cambridge Phil. Soc. **31** (1935) 555–563.

[194] J. Schwinger, *Unitary operator bases*, Proc. Nat. Acad. Sci. U.S.A. **46** (1960), 570–579, DOI 10.1073/pnas.46.4.570. MR115648

[195] I. E. Segal, *Irreducible representations of operator algebras*, Bull. Amer. Math. Soc. **53** (1947), 73–88, DOI 10.1090/S0002-9904-1947-08742-5. MR20217

[196] I. Segal, *Algebraic integration theory*, Bull. Amer. Math. Soc. **71** (1965), 419–489, DOI 10.1090/S0002-9904-1965-11284-8. MR178384

[197] C. E. Shannon, *A mathematical theory of communication*, Bell System Tech. J. **27** (1948), 379–423, 623–656, DOI 10.1002/j.1538-7305.1948.tb01338.x. MR26286

[198] B. Simon, *Trace ideals and their applications*, 2nd ed., Mathematical Surveys and Monographs, vol. 120, American Mathematical Society, Providence, RI, 2005, DOI 10.1090/surv/120. MR2154153

[199] B. Simon, *Loewner's theorem on monotone matrix functions*, Grundlehren der mathematischen Wissenschaften [Fundamental Principles of Mathematical Sciences], vol. 354, Springer, Cham, 2019, DOI 10.1007/978-3-030-22422-6. MR3969971

[200] E. M. Stein, *Interpolation of linear operators*, Trans. Amer. Math. Soc. **83** (1956), 482–492, DOI 10.2307/1992885. MR82586

[201] E. Steinitz, *Bedingt konvergente Reihen und konvexe Systeme, VI, VII* (German), J. Reine Angew. Math. **146** (1916), 1–52, DOI 10.1515/crll.1916.146.1. MR1580921

[202] G. W. Stewart, *On the early history of the singular value decomposition*, SIAM Rev. **35** (1993), no. 4, 551–566, DOI 10.1137/1035134. MR1247916

[203] W. F. Stinespring, *Positive functions on C^*-algebras*, Proc. Amer. Math. Soc. **6** (1955), 211–216, DOI 10.2307/2032342. MR69403

[204] E. Størmer, *Positive linear maps of operator algebras*, Springer Monographs in Mathematics, Springer, Heidelberg, 2013, DOI 10.1007/978-3-642-34369-8. MR3012443

[205] D. Sutter, M. Berta, and M. Tomamichel, *Multivariate trace inequalities*, Comm. Math. Phys. **352** (2017), no. 1, 37–58, DOI 10.1007/s00220-016-2778-5. MR3623253

[206] W. Thirring, *A course in mathematical physics. Vol. 4: Quantum mechanics of large systems*, Springer-Verlag, New York-Vienna, 1983. Translated from the German by Evans M. Harrell. MR681697

[207] C. J. Thompson, *Inequality with applications in statistical mechanics*, J. Mathematical Phys. **6** (1965), 1812–1813, DOI 10.1063/1.1704727. MR189688

[208] G. O. Thorin, *Convexity theorems generalizing those of M. Riesz and Hadamard with some applications*, Comm. Sém. Math. Univ. Lund [Medd. Lunds Univ. Mat. Sem.] **9** (1948), 1–58. MR25529

[209] J. Tomiyama, *On the geometry of positive maps in matrix algebras. II*, Linear Algebra Appl. **69** (1985), 169–177, DOI 10.1016/0024-3795(85)90074-6. MR798371

[210] J. A. Tropp, *From joint convexity of quantum relative entropy to a concavity theorem of Lieb*, Proc. Amer. Math. Soc. **140** (2012), no. 5, 1757–1760, DOI 10.1090/S0002-9939-2011-11141-9. MR2869160

[211] H. F. Trotter, *On the product of semi-groups of operators*, Proc. Amer. Math. Soc. **10** (1959), 545–551, DOI 10.2307/2033649. MR108732

[212] R. Tucci, *Entanglement of distillation and conditional mutual information*, eprint quant-ph/0202144, 2002.

[213] A. Uhlmann, *Endlich-dimensionale Dichtematrizen. II*, Wiss. Z. Karl-Marx-Univ. Leipzig Math.-Natur. Reihe **22** (1973), 139–177. MR400997

[214] A. Uhlmann, *The "transition probability" in the state space of a *-algebra*, Rep. Mathematical Phys. **9** (1976), no. 2, 273–279, DOI 10.1016/0034-4877(76)90060-4. MR423089

[215] A. Uhlmann, *Relative entropy and the Wigner-Yanase-Dyson-Lieb concavity in an interpolation theory*, Comm. Math. Phys. **54** (1977), no. 1, 21–32. MR479224

[216] A. Uhlmann, *Optimizing entropy relative to a channel or a subalgebra*, in *Proc. XXI Int. Coll. on Group Theoretical Methods in Physics*, H. D. Doebner, P. Nattermann and W. Scherer, eds., Vol. I, World Scientific, Singapore, 1997, 343–348.

[217] H. Umegaki, *Conditional expectation in an operator algebra. IV. Entropy and information*, Kōdai Math. Sem. Rep. **14** (1962), 59–85. MR142006

[218] A. van Daele, *A new approach to the Tomita-Takesaki theory of generalized Hilbert algebras*, J. Functional Analysis **15** (1974), 378–393, DOI 10.1016/0022-1236(74)90029-9. MR346539

[219] V. Vedral, M. B. Plenio, M. A. Rippin, and P. L. Knight, *Quantifying entanglement*, Phys. Rev. Lett. **78** (1997), no. 12, 2275–2279, DOI 10.1103/PhysRevLett.78.2275. MR1438269

[220] V. Vedral and M. B. Plenio, *Entanglement measures and purification procedures*, Phys. Rev. A **57** (1998), 1619.

[221] G. Vidal, *Entanglement monotones*, J. Modern Opt. **47** (2000), no. 2-3, 355–376, DOI 10.1080/095003400148268. Physics of quantum information. MR1756860

[222] K. G. H. Vollbrecht and R. F. Werner, *Entanglement measures under symmetry*, Phys. Rev. A **64** (2001), 062307.

[223] J. von Neumann, *Zur Algebra der Funktionaloperationen und Theorie der normalen Operatoren* (German), Math. Ann. **102** (1930), no. 1, 370–427, DOI 10.1007/BF01782352. MR1512583

[224] J. von Neumann, *A certain zero-sum two-person game equivalent to the optimal assignment problem*, Contributions to the theory of games, vol. 2, Ann. of Math. Stud., no. 28, Princeton Univ. Press, Princeton, NJ, 1953, pp. 5–12. MR54920

[225] J. Watrous, *Notes on super-operator norms induced by Schatten norms*, Quantum Inf. Comput. **5** (2005), no. 1, 58–68. MR2123899

[226] D. V. Widder, *Functions harmonic in a strip*, Proc. Amer. Math. Soc. **12** (1961), 67–72, DOI 10.2307/2034126. MR132838

[227] A. Wehrl, *General properties of entropy*, Rev. Modern Phys. **50** (1978), no. 2, 221–260, DOI 10.1103/RevModPhys.50.221. MR496300

[228] H. Weyl, *The theory of groups and quantum mechanics*, Dover Publications, Inc., New York, 1950. Translated from the second (revised) German edition by H. P. Robertson; Reprint of the 1931 English translation. MR3363447

[229] H. Weyl, *Inequalities between the two kinds of eigenvalues of a linear transformation*, Proc. Nat. Acad. Sci. U.S.A. **35** (1949), 408–411, DOI 10.1073/pnas.35.7.408. MR30693

[230] E. P. Wigner and M. M. Yanase, *On the positive semidefinite nature of certain matrix expressions*, Proc. Nat. Acad. Sci. **49** (1963), 910–918.

[231] E. P. Wigner and M. M. Yanase, *Information contents of distributions* Canad. J. Math., **16** (1964), 397–406.

[232] M. M. Wilde, *Quantum information theory*, 2nd ed., Cambridge University Press, Cambridge, 2017, DOI 10.1017/9781316809976. MR3645110

[233] M. M. Wilde, A. Winter, and D. Yang, *Strong converse for the classical capacity of entanglement-breaking and Hadamard channels via a sandwiched Rényi relative entropy*, Comm. Math. Phys. **331** (2014), no. 2, 593–622, DOI 10.1007/s00220-014-2122-x. MR3238525

[234] A. Winter, *Tight uniform continuity bounds for quantum entropies: conditional entropy, relative entropy distance and energy constraints*, Comm. Math. Phys. **347** (2016), no. 1, 291–313, DOI 10.1007/s00220-016-2609-8. MR3543185

[235] W. Wotters and W. Zurek, *A single quantum cannot be cloned* Nature **299** (1982), 802–803.

[236] W. Wootters, *Entanglement of formation of an arbitrary state of two qubits*, Phys. Rev. Lett. **80** (1998), 2245

[237] H. Zhang, *From Wigner-Yanase-Dyson conjecture to Carlen-Frank-Lieb conjecture*, Adv. Math. **365** (2020), 107053, 18, DOI 10.1016/j.aim.2020.107053. MR4064777

[238] H. Zhang, *Some convexity and monotonicity results of trace functionals*, Ann. Henri Poincaré **25** (2024), no. 4, 2087–2106, DOI 10.1007/s00023-023-01345-7. MR4721700

SELECTED PUBLISHED TITLES IN THIS SERIES

For a complete list of titles in this series, visit the
AMS Bookstore at **www.ams.org/bookstore/gsmseries/**.